Artificial Intelligence-Driven Methods for Remote Sensing Target and Object Detection

Artificial Intelligence-Driven Methods for Remote Sensing Target and Object Detection

Guest Editors

Yanni Dong
Xiaochen Yang
Qian Du

MDPI

Basel • Beijing • Wuhan • Barcelona • Belgrade • Novi Sad • Cluj • Manchester

Guest Editors

Yanni Dong
School of Resource and Environment Sciences
Wuhan University
Wuhan
China

Xiaochen Yang
School of Mathematics and Statistics
University of Glasgow
Glasgow
UK

Qian Du
Department of Electrical and Computer Engineering
Mississippi State University
Starkville
USA

Editorial Office
MDPI AG
Grosspeteranlage 5
4052 Basel, Switzerland

This is a reprint of the Special Issue, published open access by the journal *Remote Sensing* (ISSN 2072-4292), freely accessible at: https://www.mdpi.com/journal/remotesensing/special_issues/Driven_Methods.

For citation purposes, cite each article independently as indicated on the article page online and as indicated below:

Lastname, A.A.; Lastname, B.B. Article Title. *Journal Name* **Year**, *Volume Number*, Page Range.

ISBN 978-3-7258-3231-6 (Hbk)
ISBN 978-3-7258-3232-3 (PDF)
https://doi.org/10.3390/books978-3-7258-3232-3

Contents

About the Editors

Yanni Dong

Yanni Dong is a professor and Ph.D. supervisor at the School of Resources and Environmental Science, Wuhan University, and a recipient of funding from the National Excellent Young Scientist Fund. Her research primarily focuses on the intelligent interpretation of remote sensing imagery and geospatial intelligence applications. She has published more than 60 SCI-indexed papers in prestigious journals such as *IEEE Transactions on Image Processing* (TIP), *IEEE Transactions on Cybernetics* (TCybern), and *IEEE Transactions on Geoscience and Remote Sensing* (TGRS). She serves as an associate editor for the SCI journals *IEEE TGRS* and *IEEE GRSL*, a senior program committee member for international image analysis conferences such as ICCV, AAAI, and IJCAI, and a senior member of IEEE. She is also the Vice Chair of the IEEE GRSS Wuhan Chapter and the Vice President of the Excellent Doctoral Club of the China Society of Image and Graphics.

Xiaochen Yang

Xiaochen Yang is a lecturer at the School of Mathematics and Statistics of the University of Glasgow. She received her Ph.D. in statistical science from University College London in 2020. Her research focuses on statistical machine learning and image analysis, with a strong interest in medical applications. Her research has been published in top conferences and journals such as *Neural Information Processing Systems* (NeurIPS), the International Conference on Learning Representations (ICLR), and *Transactions on Pattern Analysis and Machine Intelligence* (TPAMI), and she is also on the Editorial Boards of *IEEE Transactions on Circuits and Systems for Video Technology* (TCSVT) and *Neurocomputing*.

Qian Du

Qian Du received her Ph.D. degree in electrical engineering from the University of Maryland, Baltimore, MD, USA, in 2000. She is currently a Bobby Shackouls Professor at the Department of Electrical and Computer Engineering, Mississippi State University, Starkville, MS, USA. Her research interests include hyperspectral image analysis and applications, pattern classification, data compression, and neural networks. She is a Fellow of the SPIE-International Society for Optics and Photonics. From 2016 to 2020, she was the Editor-in-Chief for the *IEEE Journal of Selected Topics in Applied Earth Observations and Remote Sensing* (JSTARS). She is currently the Editor-in-Chief of the *Journal of Applied Remote Sensing*.

remote sensing

MDPI

Technical Note

Multi-Prior Twin Least-Square Network for Anomaly Detection of Hyperspectral Imagery

Jiaping Zhong [1], Yunsong Li [1,*], Weiying Xie [1], Jie Lei [1] and Xiuping Jia [2]

[1] State Key Laboratory of Integrated Services Networks, Xidian University, Xi'an 710071, China; jpzhong@stu.xidian.edu.cn (J.Z.); wyxie@xidian.edu.cn (W.X.); jielei@mail.xidian.edu.cn (J.L.)

[2] School of Engineering and Information Technology, The University of New South Wales, Canberra, ACT 2600, Australia; x.jia@adfa.edu.au

* Correspondence: ysli@mail.xidian.edu.cn; Tel.: +86-029-8820-3116

Abstract: Anomaly detection of hyperspectral imagery (HSI) identifies the very few samples that do not conform to an intricate background without priors. Despite the extensive success of hyperspectral interpretation techniques based on generative adversarial networks (GANs), applying trained GAN models to hyperspectral anomaly detection remains promising but challenging. Previous generative models can accurately learn the complex background distribution of HSI and typically convert the high-dimensional data back to the latent space to extract features to detect anomalies. However, both background modeling and feature-extraction methods can be improved to become ideal in terms of the modeling power and reconstruction consistency capability. In this work, we present a multi-prior-based network (MPN) to incorporate the well-trained GANs as effective priors to a general anomaly-detection task. In particular, we introduce multi-scale covariance maps (MCMs) of precise second-order statistics to construct multi-scale priors. The MCM strategy implicitly bridges the spectral- and spatial-specific information and fully represents multi-scale, enhanced information. Thus, we reliably and adaptively estimate the HSI label to alleviate the problem of insufficient priors. Moreover, the twin least-square loss is imposed to improve the generative ability and training stability in feature and image domains, as well as to overcome the gradient vanishing problem. Last but not least, the network, enforced with a new anomaly rejection loss, establishes a pure and discriminative background estimation.

Keywords: anomaly detection; hyperspectral imagery; adversarial learning; multi-scale covariance map; least-square; unsupervised learning

Citation: Zhong, J.; Li, Y.; Xie, W.; Lei, J.; Jia, X. Multi-Prior Twin Least-Square Network for Anomaly Detection of Hyperspectral Imagery. *Remote Sens.* **2022**, *14*, 2859. https://doi.org/10.3390/rs14122859

Academic Editors: Qian Du, Yanni Dong and Xiaochen Yang

Received: 7 May 2022
Accepted: 12 June 2022
Published: 15 June 2022

1. Introduction

Due to hundreds of continuous bands, hyperspectral imagery (HSI) obtains high spectral resolution and abundant spectral information, which has been widely used in anomaly detection, target detection, and classification [1,2]. Generally, anomaly detection refers to identifying instances that do not conform to the expected normal distribution, which plays an essential role in the computer version area [3,4]. Among various interpretation techniques for earth observation and deep space exploration, hyperspectral anomaly detection aims to locate anomaly instances without any prior information in different spatial and spectral features, i.e., spectral signatures discrimination or probability difference from diverse backgrounds.

Traditional statistics- or representation-based detectors measure the observational data under the normal condition from assumed probabilistic distribution or representations, lacking the ability to characterize complex hyperspectral data [5,6]. The assumption of Gaussian distribution may not be consistent with the actual scene, and cannot completely cover the real-world data distribution in a complicated imaging environment. To address these issues, fruitful research based on deep learning has appeared, which has powerful and unique advantages in modeling complex hyperspectral data and digging out the underlying distribution [7–10]. To be specific, there are two main categories of employing generative models in anomaly detection, including AE- and GAN-embedded architectures to reconstruct in both latent and reconstructed domains according to minimizing the error between the original and reconstructed spectra [11]. However, though GAN performs well in anomaly-detection tasks according to the literature, the objective of GAN is supposed to capture more separable latent features between background and anomalies rather than minimizing the reconstruction error at the pixel level [12–14]. The gradient vanishing problem is partly caused by the hypothesis of the discriminator acting as a classifier with the sigmoid cross-entropy loss function in regular GANs, which is not conducive to the generation of background and discrimination of anomalies.

For hyperspectral anomaly detection, existing works based on deep learning focus mainly on feature extraction or dimension reduction [15–19]. With small input samples, limited by difficulties in annotation and collection of label training, supervised or semi-supervised methods aim to make a tradeoff between sample numbers and detection performance [20–22]. However, they still require hard-to-attain pure background training samples for representation [23–26]. Unsupervised learning has become a new trend, but detection accuracy is unsatisfactory with insufficient prior knowledge [27–30]. Moreover, complex characteristics of HSIs in real scenarios are not yet fully considered or plugged into the network. In contrast, the proposed MPN aims to address these issues while maintaining good performance. The contributions can be summarized as follows:

- To change the preconceptions of anomaly detection with insufficient samples, we propose a multi-prior strategy to reliably and adaptively generate prior dictionaries. Specifically, we calculate a series of multi-scale covariance matrices rather than traditional one-order statistics, taking advantage of second-order statistics to naturally model the distribution with integrated spectral and spatial information.
- The twin least-square loss in both the feature and image domains and differential expansion loss are jointly introduced into the architecture to fit the characteristics of high-dimensional and complex HSI data, which can overcome the gradient vanishing and training stability problem.
- To lease the generation ability of the model and reduce the false-alarm rate by an order of magnitude, we design a weakly supervised training pattern to enlarge the distribution diversity between background regions and anomalies, aiming to distinguish between background and anomalies in reconstruction. Experimental results illustrate that the AUC score of $\left(P_f, \tau\right)$ in MPN is one order of magnitude lower than other compared methods.

We divide the remainder of this paper into four sections. The hyperspectral anomaly detection and generative adversarial networks are reviewed in Section 2. The MPN methodology is concretely described in Section 3. In Section 4, we represent experiments and discuss the results. In Section 5, we draw our conclusions.

2. Related Work

2.1. Hyperspectral Anomaly Detection

Recently, plenty of anomaly-detection methods have been developed. The statistics-based methods and sparsity-based methods are two representative categories [31,32]. There are mainly two categories for traditional hyperspectral anomaly detection approaches, the first of which comprises RX-based algorithms [33]. The RX detector is based on the assumption that each spectral channel obeys Gaussian distribution. The non-RX-based methods include collaborative representations, sparse representations, and low-rank representations. The ADLR method obtains abundance vectors by spectral decomposition. It constructs a dictionary based on the mean value clustering of abundance vectors [34]. The low-rank and sparse constraints are imposed in the PAB_DC model in consideration of the homogeneity of background and the sparsity of anomalies to construct the background and potential anomaly dictionaries [35]. The emerging typical algorithm for background removal is AED, which removes the background by attribute filtering and difference operation [36]. Additionally, unlike single distribution modeling in traditional LSDM, the LSDM–MoG method combines the mixed-noise models and low-rank background to characterize complex distributions more accurately [37].

2.2. Generative Adversarial Networks (GANs)

By improvement of generation quality and training stability, generative adversarial networks (GANs) perform advanced image generation ability [38–41]. The capability of high-quality image generation accounts for the application of GANs to many image processing tasks [42,43]. For hyperspectral anomaly detection, existing works based on deep learning focus mainly on feature extraction or dimension reduction. With small input samples, limited by difficulties in annotation and collection of label training, supervised or semi-supervised methods aim to make a tradeoff between sample numbers and detection performance. However, they still require hard-to-attain pure background training samples for representation. Unsupervised learning has become a new trend, but detection accuracy is unsatisfactory with insufficient prior knowledge. Moreover, complex characteristics of HSIs in real scenarios are not yet fully considered or plugged into the network. In contrast, the proposed MPN aims to address these issues while maintaining a good performance.

3. Proposed Method

In this work, we elaborate on MPN for hyperspectral anomaly detection. The whole process is shown in Figure 1. We denote the HSI as $\mathbf{H} \in \mathbb{R}^{h \times w \times d}$. The number of the spectral bands is denoted by d. The spatial size of the data is described by h and w. For convenience, as the input of the network, the 3D cube $\mathbf{H}$ is converted into a 2D matrix $\mathbf{H} = \{\mathbf{h}_i\}_{i=1}^{n} \in \mathbb{R}^{d \times n}$. Each column of $\mathbf{H}$ is a spectral pixel vector in the HSI. $n = h \times w$ is the number of the pixels. The HSI data matrix can be decomposed into two parts—background and anomaly—which are denoted as $\mathbf{Y} = [\mathbf{y}_1, \mathbf{y}_2, \ldots, \mathbf{y}_{n_B}]$, $\mathbf{X} = [\mathbf{x}_1, \mathbf{x}_2, \ldots, \mathbf{x}_{n_A}]$, respectively, where $n_A + n_B = n$. $\mathbf{y}_i$ represents the ith vector in $\mathbf{Y}$, $i = 1, 2, \cdots, n_A$. $\mathbf{x}_i$ represents the ith vector in $\mathbf{X}$, $i = 1, 2, \cdots, n_B$. Furthermore, the reconstructed background spectral is denoted as $\hat{\mathbf{Y}} = \sigma(\mathbf{W}\mathbf{W}^T\mathbf{Y} + \mathbf{B})$, where $\sigma(\cdot)$ represents the activation function. $\hat{\mathbf{Y}}$ is the output of the network, and $\mathbf{W}$ is the weight of the encoder. $\mathbf{B}$ is the bias of the whole network. The pixel can be represented by the combination of other surrounding background pixels since there is a correlation between background pixels. The linear decomposition model of HSI is formulated as

$$\mathbf{H} = \mathbf{B}_0\mathbf{W} + \mathbf{A}_0 + \mathbf{N}, \tag{1}$$

where $\mathbf{B}_0\mathbf{W}$ is the background component; $\mathbf{B}_0 = [b_1, b_2, \ldots, b_{n_B}]$ are the background dictionaries; $\mathbf{W} = [w_1, w_2, \ldots, w_n]$ are the corresponding representation coefficients; and $\mathbf{A}_0 = [a_1, a_2, \ldots, a_{n_a}]$ is anomaly component. For simplicity, we consider $\mathbf{N}$ as zero.

Different from the previous single-distribution-based anomaly-detection method RX, we calculate the covariance between the two vectors by estimating the probabilities of the testing pixel belong to the anomaly class by calculating the Mahalanobis distance with neighboring pixels. Inspired by the distance-based method and the multi-scale approach [44,45], we consider that, through background dictionaries modeling, the constructed covariance maps can reuse the MCMs and represent the spatial and spectral information of HSI as the prior for GAN training. Thus, the decomposition can be regarded as a nonlinear problem, which has the following expression:

$$\mathbf{H} = f_{MCM}(\mathbf{B}, \mathbf{A}) \tag{2}$$

where $f_{MCM}(\cdot)$ represents the learning process of MCM algorithm. $\mathbf{A}$ and $\mathbf{B}$ denote the background and anomaly dictionaries, respectively.

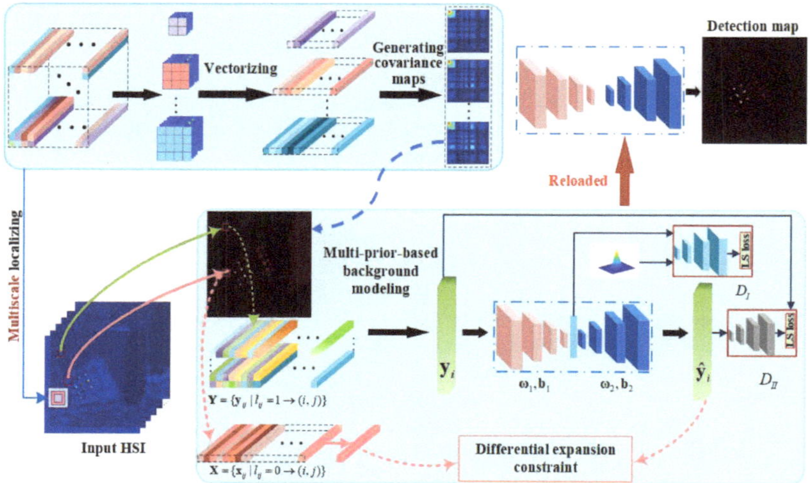

Figure 1. Flowchart of the detection algorithm based on MPN, including covariance maps generation, multi-prior-based background modeling, and differential expansion constraint-imposed network.

3.1. Network Architecture

Instead of single-scale Gaussian distribution with a generative adversarial network method, we present a deep network, i.e., MPN, to directly model the complex HSI and its dictionary, which characterizes the essential attribute of HSI implicitly. In this way, both the multi-prior dictionary scheme and the jointly separable loss can contribute to detecting the anomalies from the complex HSI data in a weakly supervised pattern. Thus, according to the defined dictionaries, the anomaly-detection task can be formulated as

$$\begin{aligned} L(\mathbf{Y},\mathbf{X}) &= L_{TLS}(\mathbf{Y}) + L_{au}(\mathbf{Y},\hat{\mathbf{Y}}) + L_{en}(\hat{\mathbf{Y}},\mathbf{X}) \\ &= L_{LS_1}(\mathbf{Z}) + L_{LS_2}(\mathbf{Y},\hat{\mathbf{Y}}) + \|\mathbf{Y}-\hat{\mathbf{Y}}\| - \alpha\|\hat{\mathbf{Y}}-\mathbf{X}\|, \end{aligned} \tag{3}$$

$$s.t. \begin{cases} \mathbf{Z} = Enc(\mathbf{Y}; \arg(G)) \\ \hat{\mathbf{Y}} = Dec(\mathbf{Z}; \arg(D)) \\ \alpha \sim N(0, I) \end{cases},$$

where $\hat{\mathbf{Y}}$ is decoder output, and $\mathbf{Z}$ is the latent feature as encoder output. The twin least-square loss L_{TLS} added for the two discriminators are denoted by L_{LS_1} and L_{LS_2}, respectively, which make up the whole least-square-based loss L_{TLS}. L_{au} represents the spectral reconstruction loss, which is aimed at mapping the input to the output space. L_{en} denotes the separability loss between background and anomaly dictionary which aims to enlarge the difference, respectively. $\|\mathbf{Y}-\hat{\mathbf{Y}}\|$ denotes the reconstruction error of basic generative adversarial network. *Enc* and *Dec* represent the encoder and decoder, respectively.

3.2. Multi-Prior for Background Construction

We generate pseudo priors for GAN training through multi-scale covariance map construction. This strategy was inspired by the estimation of the Mahalanobis distance between the test pixels and the constructed pixels on one scale. Thus, we can obtain sufficient prior information and take advantage of both the spatial and spectral information of HSI. The progress of multi-prior construction strategy is shown in Figure 2.

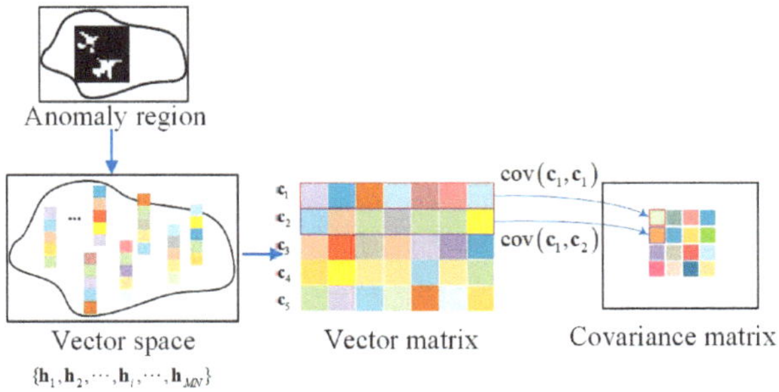

Figure 2. The progress of multi-prior construction strategy.

3.2.1. Multi-Scale Localizing

For each central pixel, we try to realize multi-scale localizing first based on the Euclidean distance with a classical classifier, i.e., K nearest neighbors (KNN), to obtain the local pixel cubes at different scales. Then, we generate a series of gradually increasing cubes of different scales. For each of the cubes, we transfer it to a vector. After that, the covariance matrix is calculated between the vectors. Furthermore, we take into account the situation of M scales $T_k, k = 1, \cdots, M$. Thus, the extraction of M covariance maps are attained to represent the center pixel $\mathbf{x}_k$, denoted by $\mathbf{C}_k$, respectively. Additionally, for the central pixel $\mathbf{h}_i$, taking the scale $R \times R$ as an example, the covariance map of $\mathbf{h}_i$ on the fixed one scale is extracted as

$$\mathbf{C}_k = \frac{1}{R^2 - 1} \sum_{i=1}^{R^2} (\mathbf{h}_i - \mu)(\mathbf{h}_i - \mu)^T \in \mathbb{R}^{L \times L}, \tag{4}$$

where μ represents the mean of input HSI vectors $\{\mathbf{h}_i | i = 1, 2, \cdots, R\}$. Furthermore, $\{\mathbf{h}_i | i = 2, \cdots, R^2\}$ represents the corresponding adjacent pixels in a window of $R \times R$ pixels. In addition, we extend to M scales of R_k , i.e., $k = 1, \cdots, M$. We denote the covariance maps of other scales as $\mathbf{C}_k, k = 1, \cdots, M$, which make up the covariance pool to construct the background.

3.2.2. Generating Covariance Maps

Aiming at the original pixel, we construct a series of increasingly larger cubes in consideration of the increasing window size of the spatial neighborhood around each one. Furthermore, for each of these cubes, after vectorizing, we calculate the CM of the two vectors to represent the information contained in the central pixel and enhanced by considering gradually expanding neighborhoods. Such CMs obtained from increasingly larger spatial proportions are denoted by MCM, which can be used to train the detector more effectively.

3.3. Two-Branch Cascaded Architecture with Least-Square Losses

Though GAN performs well in anomaly-detection tasks according to the literature [45], the real objective of GAN is supposed to capture more separable latent features between background and anomalies instead of minimizing the pixel-wise reconstruction error. Additionally, conventional GAN style architectures obtain blurry reconstructions because of multiple modes in the actual normal distribution falsifying reconstruction errors. Moreover, the unstable training process and limitation of the capacity of GAN-style architecture remain to be improved.

Therefore, we propose a twin least-square GAN under the semi-supervised pattern with background vectors reconstructed to address the above issues. Compared with conventional GANs, an auxiliary discriminator, anomaly rejection loss, and least-square losses are adopted to the framework. The anomaly rejection loss aims to gather the latent background dictionaries and make them more concentrated. Thus, the differentiation between the background and anomalies can be enlarged. Additionally, the auxiliary image discriminator plays a role in achieving a better balance in the training process.

3.3.1. Stability Branch

When the generator is updated, the non-saturation loss and min–max loss will hardly lead to gradients to false samples since the fake samples still differ from the real data [40]. To solve the problem, we consider the least-squares loss to penalize the samples which are far from the decision boundary, even if they are on the right side. We introduce least-squares loss to move false vectors to the decision boundary. The least-square loss added network can reconstruct vectors that are similar to the real data. Furthermore, the vanishing gradient is alleviated while the GANs learning is more stable. Thus, we introduce least-square losses to the architecture and define them in both the two discriminators with enhanced background anomaly discrimination from the input space to the reconstructed one. Consequently, the background spectral vectors are adequately and stably estimated with better performance. The objective function of the least-squares-loss-based network can be expressed as follows.

$$\min_{G} V_{LS}(G) = \frac{1}{2}E_{\mathbf{x}\sim p_{data}(\mathbf{x})}[(D(\mathbf{x}) - n)^2] + \frac{1}{2}E_{\mathbf{z}\sim p_{\mathbf{z}}(\mathbf{z})}[(D(G(\mathbf{z})) - m)^2] \tag{5}$$

$$\min_{D} V_{LS}(D) = \frac{1}{2}E_{\mathbf{x}\sim p_{data}(\mathbf{x})}[(D(\mathbf{x}) - l)^2] + \frac{1}{2}E_{\mathbf{z}\sim p_{\mathbf{z}}(\mathbf{z})}[(D(G(\mathbf{z})) - l)^2] \tag{6}$$

where m and n are the labels for fake and real data, respectively. Furthermore, l denotes the value that G wants D to believe for fake data. As shown in Figure 1, the MPN architecture contains one generator and two discriminators. The generator network includes the encoder En and decoder De. Firstly, the mapping relationship between input background dictionaries and the network output can be regarded as the mapping function. The training objective of $\mathcal{L}(\mathbf{Y}, D_R, D_F, G)$ is to capture the distribution of background dictionaries in both the reconstructed and deep-spectrum vector space. Motivated by this, $\mathcal{L}(\mathbf{Y}, D_R, D_F, G)$ intends to produce a minimum anomaly score for background, but a higher score for the anomaly. To achieve this goal, in addition to the basic reconstruction loss in the equation, we also add spatial adversarial loss, potential spatial adversarial loss, and anomaly rejection loss, which all contribute to the whole learning objective.

Considering the discriminator as a classifier, the sigmoid cross-entropy loss function is adopted in regular basic GANs. When updating the generator, the loss function will cause vanishing gradients because the samples on the correct side of the decision boundary are still far from the real data. To remedy this problem, we introduce the least-square loss function into the architecture. As shown in Figure 1, the MPN network consists of two discriminators and a generator. The twin least-square loss can be expressed as

$$L_{LS_1} = \frac{1}{2}E_{\mathbf{y}\sim p_{data}(\mathbf{y})}[(D_F(\mathbf{y}) - 1)^2] + \frac{1}{2}E_{\mathbf{z}\sim p_{\mathbf{z}}(\mathbf{z})}[(D_F(G(\mathbf{z})) + 1)^2], \tag{7}$$

$$L_{LS_2} = \frac{1}{2}E_{\mathbf{y}\sim p_{data}(\mathbf{y})}[(D_R(\mathbf{y}) - 1)^2] + \frac{1}{2}E_{\hat{\mathbf{y}}\sim p_{\hat{\mathbf{y}}}(\hat{\mathbf{y}})}[(D_R(G(\hat{\mathbf{y}})) + 1)^2], \tag{8}$$

From the binary classification point of view, we introduce the twin least-square loss to move false samples to the decision boundary of being anomaly or background. Thus, we perform adversarial training on the two least-square-loss imposed discriminators D_R and D_F against the generator. Consequently, the problem of gradient vanishing can be overcome and the consistency with the distribution of the reconstructed $\hat{\mathbf{y}}$ and the input data $\mathbf{y}$ can be ensured.

3.3.2. Separability Branch

As mentioned earlier, we impose the spectral reconstruction loss using mean squared error (MSE) so that the deviation between the reconstructed and the input can be optimized:

$$\begin{aligned} L_{au}(\mathbf{Y},\hat{\mathbf{Y}}) &= \left\|\mathbf{Y}-\hat{\mathbf{Y}}\right\| \\ &= \left\|\mathbf{Y}-\sigma\left(\mathbf{W}\mathbf{W}^T\mathbf{Y}+\mathbf{B}\right)\right\|, \end{aligned} \tag{9}$$

To ensure that the learning samples come entirely from the background, we introduce the following distance function based on L_{au}. The second item of this is expected to be as large as possible:

$$\begin{aligned} L_{an}(\mathbf{Y},\hat{\mathbf{Y}}) &= L_{au}(\mathbf{Y},\hat{\mathbf{Y}}) + L_{en}(\hat{\mathbf{Y}},\mathbf{X}) \\ &= \left\|\mathbf{Y}-\hat{\mathbf{Y}}\right\| - \alpha\left\|\hat{\mathbf{Y}}-\mathbf{X}\right\| \\ &= \left\|\mathbf{Y}-\sigma\left(\mathbf{W}\mathbf{W}^T\mathbf{Y}+\mathbf{B}\right)\right\| \\ &\quad - \alpha\left\|\sigma\left(\mathbf{W}\mathbf{W}^T\mathbf{Y}+\mathbf{B}\right)-\mathbf{X}\right\|, \end{aligned} \tag{10}$$

Let $\mathbf{x}_i$ and $\mathbf{y}_i$ represent the ith component in $\mathbf{X}$ and $\mathbf{Y}$, respectively. $\bar{\mathbf{x}}$ denotes the mean of all the $\mathbf{x}_i$ in $\mathbf{X}$. When the distance between the reconstructed spectrum vector $\hat{\mathbf{y}}_i$ and the average spectrum vector $\bar{\mathbf{x}}$ are small, $\mathbf{y}_i$ is suspected to be an anomaly vector. Then, the suppression coefficient α of the function aims to reduce it rapidly for adjustment. When the distance between $\bar{\mathbf{x}}$ and $\mathbf{y}_i$ is large, from a statistical point of view, $\mathbf{y}_i$ is the background dictionary to be estimated. Then, the value of the suppression coefficient α is approximately 1.

3.4. Solving the Cascaded Model

The twin least-square loss, spectral reconstruction loss, and differential expansion loss are jointly learned with the weighting coefficient of 1 by carrying out alternative updates of each component as follows. Subsequently, we obtain the detection maps by Gaussian-statistics-based discrimination.

- Minimize L_{LS_1} by updating parameters of D_F.
- Minimize L_{LS_2} by updating parameters of D_R.
- Minimize L_{an} by updating parameters of En and the decoder De.

where the first discriminator D_F with the least-square loss aims to reduce gradient vanishing in the latent space. The second D_R with the least-square loss tries to reconstruct images to approximate the input in the reconstructed image space. The model is trained using the Adam optimization algorithm by carrying out alternative updates of each component. Then we re-adopted the multi-prior algorithm on reconstruction results as follows.

$$\mathbf{J} = (\hat{\mathbf{y}}_i - \mu)^T \mathbf{C} (\hat{\mathbf{y}}_i - \mu). \tag{11}$$

where μ denotes the mean vector and $\mathbf{C}$ represents the multi-prior of the reconstructed $\hat{\mathbf{H}}$, respectively.

4. Experimental Results and Discussion

We design and execute experiments to verify the effectiveness of the proposed model. We qualitatively and quantitatively compared and analyzed the detection results of the MPN method with six state-of-the-art algorithms.

4.1. Datasets Description

The tested HSIs captured at different scenes are described for performance evaluation of the MPN and compared methods. The pseudo-color images and ground truth of the datasets are shown in Figure 3. There are part of the datasets provided online to access (http://xudongkang.weebly.com/data-sets.html (accessed on 6 May 2022)).

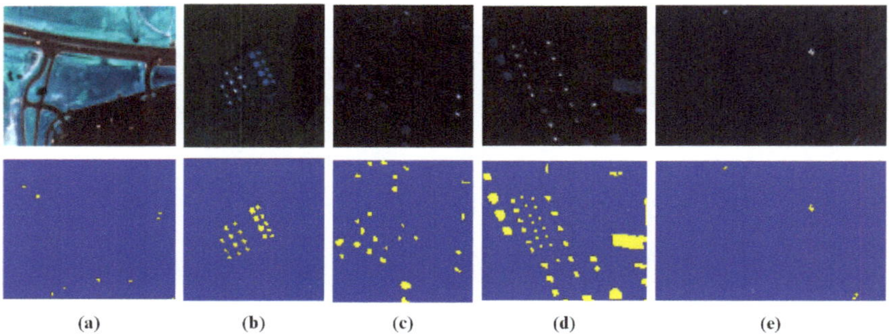

Figure 3. Pseudo-color images and ground truth for (**a**) HYDICE, (**b**) ABU-urban1, (**c**) ABU-urban2, (**d**) EI Segundo, and (**e**) Grand Island.

4.1.1. HYDICE

Captured by the hyperspectral digital image acquisition experiment (HYDICE) airborne sensors above the city, the first dataset describes urban scenes in the United States. There are 162 spectral channels in the original image with a wavelength from 400 to 2500 nanometers (nm). From the entire original image, a sub-image with a space size of 80×100 is cropped. In the spectral band, 162 bands are reserved. The total number of anomalous pixels is 19. The anomaly pixels represent cars and roofs.

4.1.2. Airport–Beach–Urban (ABU) Database

The airplane, city, and most beach scenes of the second dataset are recorded by the airborne visible/infrared imaging spectrometer (AVIRIS) sensors. One of the datasets in the beach scene is captured by the Reflective Optics System Imaging Spectrometer (ROSIS-03) sensor. The AVIRIS instrument contains 224 different detectors, each with a spectral bandwidth of approximately 10 nm, allowing it to cover the entire range between 380 and 2500 nm (http://aviris.jpl.nasa.gov/ (accessed on 6 May 2022)). The sample image with a size of 100×100 is manually extracted from the large image. The datasets are available online and can be downloaded from the AVIRIS website.

4.1.3. Grand Island

The Grand Island dataset was also acquired by the AVIRIS sensor. It was captured at the location of GrandIsland on the Gulf Coast. In spatial domain, it contains 300×480 pixels. While in spectral domain, it contains 224 spectral channels with a wavelength range of 366–2496 nm.

4.1.4. EI Segundo

The EI Segundo dataset was captured by AVIRIS sensors. The structure of the refinery is regarded as an anomaly to be detected. It has 250×300 pixels in the spatial domain. In the spectral domain, the band number is 224, ranging from 370 to 250 nm.

4.2. Evaluation Metrics

Let P_d, P_f, and τ represent the true positive rate, false positive rate, and threshold, respectively. We introduce the receiver operating characteristic curve (ROC) together with the area under it to quantitative evaluation. The area under the ROC curve of $\left(P_d, P_f\right)$ demonstrates the detection accuracy. The closer this value is to 1, the better the detection capability. In contrast, the closer the area value under the $\left(P_f, \tau\right)$ is to 0, the better the false alarm rate.

4.3. Experiment Setup

Our model is built on the basic GAN model, which is composed of encoder, decoder, and discriminators. We set the training iteration to 5000. The learning rate is chosen to be 0.0001 to balance the performance and iteration speed. There are three layers for each part. We set the number of the extracted feature to 20 after ranging the value from 10 to 100 for all the datasets. With a proper number of 20, the proposed model can represent an extracted feature well for the input spectral and promote the detection performance. We optimize all parameters by the Adam optimization algorithm with backpropagation. Varying from 30 to 50, the size of windows make the detection results differ from each other. It achieves the optimal accuracy when the size is set to between 15 and 25. Similarly, we set the value of the scales of multi-scale covariance maps as 5 to make a balance between the performance and the computation costs for the tested datasets. For the parameter settings of compared methods, the RX method is parameter-free by design. Other algorithms require parameters tuning to achieve upper bounds on the performance. For the Grand Island and Segundo datasets, the main parameters s is set to 10, and r is set to 0.6. There are two main parameters in the ADLR method. We set the c between 20 and 30. The *bw* varies from 0.2 to 0.55. Furthermore, *lambda* is set between 0.01 and 0.1. In the PAB_DC method, we select the parameters according to [35]. The LSDM–MoG algorithm uses a combination of mixed-noise models and low-rank backgrounds to more accurately characterize complex distributions. The parameter setting is consistent with [37]. In AAE, the three layers of each sub-network obtain 500, 500, and 20 units, respectively, [38]. In SAFL, the batch size is set as the number of spatial pixels of HSI. The training epoch is set to 10,000. The learning rates are set to 0.001 and 0.0001, respectively, for the encoder, decoder, and discriminator [39].

4.4. Ablation Study

To better understand the effect of each component on output detection result in MPN, we analyzed the performances under the following four scenes of training models: (1) MPN without multi-prior module; (2) MPN without twin least-square loss module; (3) MPN without differential expansion loss; (4) MPN.

From Tables 1 and 2, it can be concluded that the MPN achieves a higher AUC score of $\left(P_d, P_f\right)$ and a lower AUC score of $\left(P_f, \tau\right)$ than other models. The AUC score of $\left(P_d, P_f\right)$ on average compared with other configurations are improved by about 0.276%, 0.266%, and 0.156%, respectively. Furthermore, the AUC score of $\left(P_f, \tau\right)$ on average are optimized by about 10.425%, 6.950%, and 1.351%, respectively. The results indicate the effectiveness and necessity of multi-prior, twin least-square loss, and differential expansion loss, which contribute effectively to improve the detection performance.

Table 1. Evaluation AUC scores of $\left(P_d, P_f\right)$ for different scenes on different datasets.

Configuration	HYDICE	Urban-1	Urban-2	Grand Island	EI Segundo	Average
$Scene_1$	0.99716	0.99839	0.99234	0.99940	0.98943	0.99534
$Scene_2$	0.99804	0.99805	0.99305	0.99990	0.98816	0.99544
$Scene_3$	0.99790	0.99778	0.99302	**0.99991**	0.99372	0.99654
MPN	**0.99945**	**0.99816**	**0.99312**	**0.99991**	**0.99982**	**0.99809**

The best result is shown in bold.

Table 2. Evaluation AUC scores of $\left(P_f, \tau\right)$ for different scenes on different datasets.

Configuration	HYDICE	Urban-1	Urban-2	Grand Island	EI Segundo	Average
$Scene_1$	0.02059	0.00193	0.00090	0.00311	0.00207	0.00572
$Scene_2$	0.02169	**0.00124**	0.00159	0.00071	0.00246	0.00554
$Scene_3$	0.01965	0.00146	**0.00088**	0.00257	**0.00178**	0.00525
MPN	**0.01540**	0.002	0.00145	**0.00069**	0.00218	**0.00518**

The best result is shown in bold.

4.5. Discussion

4.5.1. Baseline Methods

We compare the performance with five commonly used methods according to their open-source codes and literature, including RX, PAB_DC, AAE, ADLR, SAFL, and LSDM–MoG. To further validate the effectiveness of the multi-scale covariance maps (MCMs) and least-square loss, the experiments were conducted on several public and widely used real-world HSI datasets, i.e., HYDICE, ABU datasets, Grand Island, EI Segundo datasets.

4.5.2. Quantitative Comparison

The quantified AUC scores of $\left(P_d, P_f\right)$, and $\left(P_f, \tau\right)$ are listed in Tables 3 and 4, respectively. Relative to the AUC score of $\left(P_d, P_f\right)$, MPN exceeds the well-performed methods ADLR and RX for about 0.493% and 1.063% on average, respectively. More notably, its AUC score of $\left(P_f, \tau\right)$ is mostly an order of magnitude less than that of others. Specifically, the average AUC score of $\left(P_d, P_f\right)$ of the proposed method is 0.99809. In comparison, the result of the second best method ADLR of 0.99321, with lower AUC scores of $\left(P_f, \tau\right)$ of 0.00518 compared with 0.05096. It can be concluded that the MPN method shows promising background suppression ability for most of the scene and better detection accuracy for small and embedded anomalies. Generally speaking, the comparison methods may achieve good results for some specific datasets, while MPN can achieve promising detection results in both indexes for all the datasets, which can also be reflected on the ROC curves of $\left(P_d, P_f\right)$ in Figure 4, respectively. To further verify the background suppression and target detection ability, we also perform the separation analysis through a boxplot. The detection results of all the detectors on different datasets are shown in Figure 5. The distance between the upper boundary of the yellow box and the lower boundary of the gray box indicates the separability of the background anomaly, in which the MPN obtains better discrimination results on most datasets among the compared methods.

Table 3. Evaluation AUC scores of $\left(P_d, P_f\right)$ obtained by MPN and six compared methods.

Dataset	MPN	ADLR	LSDM–MoG	RX	AAE	PAB_DC	SAFL
HYDICE	**0.99945**	0.99471	0.95643	0.97637	0.92185	<u>0.99760</u>	0.99621
Urban-1	**0.99816**	0.98774	0.99321	<u>0.99463</u>	0.96806	0.98327	0.99800
Urban-2	**0.99312**	<u>0.99284</u>	0.97504	0.98874	<u>0.99284</u>	0.54635	0.97188
Grand Island	<u>0.99991</u>	**0.99993**	0.99989	0.99990	0.99940	0.98768	0.99804
EI Segundo	**0.99982**	0.99145	0.96798	0.97826	<u>0.99260</u>	0.75911	0.99231
Average	**0.99809**	<u>0.99321</u>	0.97851	0.98757	0.97495	0.85480	0.99141

The best result is shown in bold, and the second best result is underlined.

Table 4. Evaluation AUC scores of $\left(P_f, \tau\right)$ obtained by MPN and six compared methods.

Dataset	MPN	ADLR	LSDM–MoG	RX	AAE	PAB_DC	SAFL
HYDICE	0.01540	<u>0.00316</u>	0.18799	0.03798	0.00962	0.10525	**0.00310**
Urban-1	**0.00200**	0.08011	0.04555	<u>0.01351</u>	0.04229	0.16974	0.00216
Urban-2	**0.00145**	0.07451	0.01338	0.01140	0.06192	0.32350	<u>0.01045</u>
Grand Island	**0.00069**	<u>0.00079</u>	0.02848	0.00738	0.04818	0.06799	0.01126
EI Segundo	**0.00218**	0.09622	0.14127	<u>0.00952</u>	0.04355	0.10506	0.01864
Average	**0.00518**	0.05096	0.08333	0.01596	0.0328	0.15431	<u>0.00912</u>

The best result is shown in bold, and the second best result is underlined.

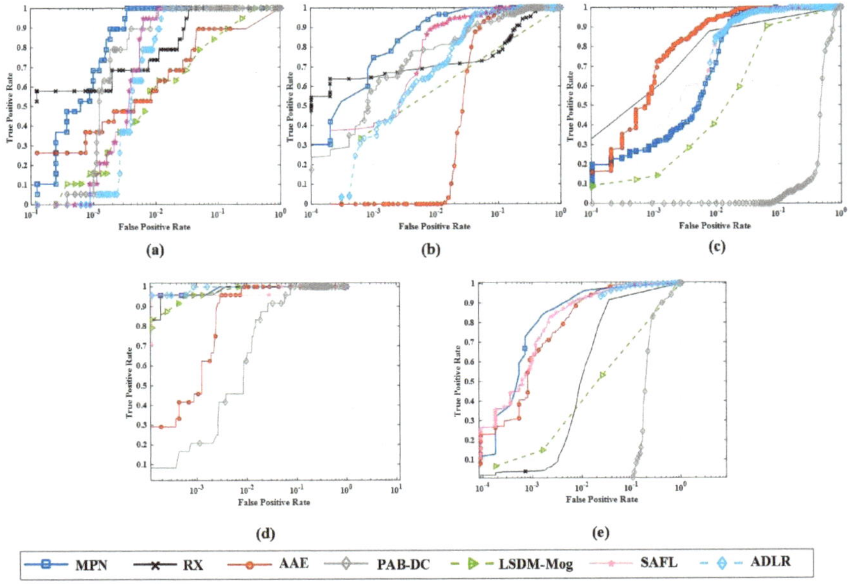

Figure 4. The ROC curves comparison of different methods for (**a**) HYDICE, (**b**) ABU-urban1, (**c**) ABU-urban2, (**d**) Grand Island, (**e**) EI Segundo models . When the curve is higher, the detection performance is better.

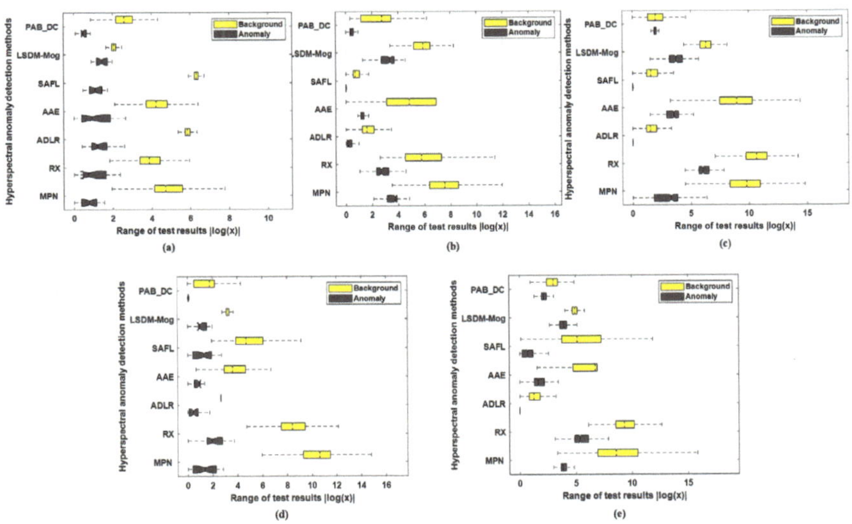

Figure 5. The separability analysis of different methods for (**a**) HYDICE, (**b**) ABU-urban1, (**c**) ABU-urban2, (**d**) Grand Island, (**e**) EI Segundo. The larger the distance, the more obvious discrimination.

4.5.3. Qualitative Comparison

For the ABU-urban1 dataset, the MPN performs well in detection, especially in some small-sized anomalies with higher intensity. Meanwhile, the false-alarm rate is restricted to a lower level in urban scenes. The RX, AAE, and ADLR methods can exactly locate the anomalies. However, the spatial shape preservation is not good enough. As for LSDM–MoG and PAB_DC methods, the anomalies are detected although the edges information around anomalies is not well obtained. For the ABU-urban2 dataset, the ADLR and the MPN can detect anomalies well and lower false-alarm rate. The AAE method presents a good ability in detecting all anomalies, but the intensity is relatively high. In RX and ADLR methods, the background suppression ability performs well but some pixel-level anomalies are relatively blurred. For the LSDM–MoG method, some of the anomalies are lost and cannot be figured out visually. The RX, LSDM–MoG, and PAB_DC methods perform well on the Grand Island dataset in detection accuracy, but they suffer from the interference of the complex background. The ADLR method can avoid false alarms, but some edge details are not obtained. Both AAE and the MPN method can protect anomalies with high detection accuracy and low false-alarm rates. In Figure 6, it can be observed that the PAB_DC method for HYDICE dataset keeps targets relatively clearly although some part of anomalies is missed. In ADLR and LSDM–MoG methods, most anomalies are visually mixed with the background. RX can detect almost all anomalies with high intensity. However, the suppression ability of background interference is not effective. Both PAB_DC and the MPN method can suppress the background well and obtain high detection accuracy. For the Segundo dataset, MPN can well detect the anomalies in the scene among the compared methods with the best results. The RX, AAE, and ADLR methods have favorable identification of the location of the anomalies but lost some of the shape information. In conclusion, the MPN represents promising performance with the detection maps closest to the ground truths.

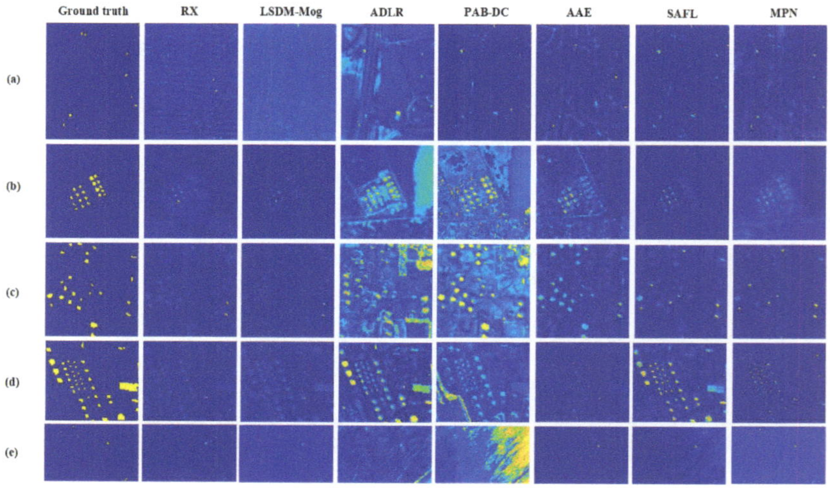

Figure 6. Detection results of MPN and the compared methods for (**a**) HYDICE, (**b**) ABU-urban1, (**c**) ABU-urban2, (**d**) EI Segundo, and (**e**) Grand Island.

5. Conclusions

In this work, a novel multi-prior strategy and joint separable loss scheme embedded network is proposed for anomaly detection of HSI. Some points of our work deserve consideration. For the first time, we propose the multi-prior strategy to construct multi-scale prior information and reuse the covariance map at multiple scales to adaptively create reliable and stable priors. Therefore, we can solve the lack of priors with pseudo-labeling and fully take advantage of spectral and spatial information. To overcome the problem of gradient vanishing and improve the generative ability during training, we introduce twin least-square loss to the network in both feature and image domains. Finally, the differential expansion loss added MPN establishes a pure and discriminative background estimation, separating the background and the anomalous spectral vectors to a greater extent. Through experiments, we have proved that the MPN exhibits superior performance in background reconstruction and outperforms the state-of-the-art methods.

Author Contributions: J.Z., Y.L. and W.X. provided the conceptualization of the work, designed the methodology, and conducted the experiments; Y.L., W.X., J.L. and X.J. gave advice on optimization and reviewed the manuscript; writing, J.Z. All authors have read and agreed to the published version of the manuscript.

Funding: This work was supported in part by the National Natural Science Foundation of China under Grant 62121001, Grant 62071360, and Grant 61801359.

Data Availability Statement: Publicly available datasets were analyzed in this study. The ABU dataset can be found here: [http://xudongkang.weebly.com/data-sets.html accessed on 6 May 2022]. The Grand Island and EI Segundo datasets are obtained from Xudong Kang. We would like to express our appreciation to Xudong Kang for providing these datasets.

Conflicts of Interest: The authors declare no conflict of interest.

Abbreviations

The following abbreviations are used in this manuscript:

GANs	generative adversarial networks
HSI	hyperspectral imagery
MCMs	multi-scale covariance maps
KNN	K nearest neighbors
MSE	mean squared error
ROC	the receiver operating characteristic curve

References

1. Camps-Valls, G.; Tuia, D.; Bruzzone, L.; Benediktsson, J. Advances in hyperspectral image classifification: Earth monitoring with statistical learning methods. *IEEE Signal Process. Mag.* **2013**, *31*, 45–54. [CrossRef]
2. Okwuashi, O.; Christopher, E. Deep support vector machine for hyperspectral image classification. *Pattern Recognit.* **2020**, *103*, 107298. [CrossRef]
3. Yan, X.; Zhang, H.; Xu, X.; Hu, X.; Heng, P. Learning semantic context from normal samples for unsupervised anomaly detection. In Proceedings of the AAAI Conference on Artificial Intelligence (AAAI), Virtual, 8 February 2021; p. 425.
4. Abati, D.; Porrello, A.; Calderara, S.; Rita, C. Latent space autoregression for novelty detection. In Proceedings of the IEEE Conference on Computer Vision and Pattern Recognition (CVPR), Long Beach, CA, USA, 15–20 June 2019; pp. 481–490.
5. Ma, M.; Mei, S.; Wan, S.; Hou, J.; Wang, Z.; Feng, D. Video summarization via block sparse dictionary selection. *Neorocomputing* **2020**, *378*, 197–209. [CrossRef]
6. Luo, W.; Liu, W.; Lian, D.; Tang, J.; Gao, S. Video anomaly detection with sparse coding inspired deep neural networks. *IEEE Trans. Pattern Anal. Mach. Intell.* **2019**, *129*, 1070–1084. [CrossRef]
7. Stanislaw, J.; Maciej, S.; Stanislav, F.; Devansh, A.; Jacek, T.; Kyunghyun, C.; Krzysztof, G. The break-even point on optimization trajectories of deep neural networks. In Proceedings of the International Conference on Learning Representations (ICLR), Addis Ababa, Ethiopia, 26–30 April 2020.
8. Chalapathy, R.; Chawla, S. Deep learning for anomaly detection: A survey. *arXiv* **2019**, arXiv:1901.03407.
9. Eyal, G.; Aryeh, K.; Sivan, S.; Ofer, B.; Oded, S. Temporal anomaly detection: Calibrating the surprise. In Proceedings of the AAAI Conference on Artificial Intelligence (AAAI). Honolulu, HI, USA, 27 January–1 February 2019; pp. 689–692.
10. Wen, T.; Keyes, R. Time series anomaly detection using convolutional neural networks and transfer learning. In Proceedings of the International Joint Conference on Artifificial Intelligence (IJCAI), Macao, China, 10–16 August 2019.
11. Ansari, A.; Scarlett, J.; Soh, H. A characteristic function approach to deep implicit generative modeling. In Proceedings of the IEEE Conference on Computer Vision and Pattern Recognition (CVPR), Seattle, WA, USA, 13–19 June 2020; pp. 7478–7487.
12. Villa, A.; Chanussot, J.; Benediktsson, J.; Jutten, C.; Dambreville, R. Unsupervised methods for the classification of hyperspectral images with low spatial resolution. *Pattern Recognit.* **2013**, *46*, 1556–1568. [CrossRef]
13. Fowler, J.; Du, Q. Anomaly detection and reconstruction from random projections. *IEEE Trans. Image Process.* **2011**, *21*, 184–195. [CrossRef]
14. Gong, Z.; Zhong, P.; Hu, W. Statistical loss and analysis for deep learning in hyperspectral image classification. *IEEE Trans. Neural Netw. Learn. Syst.* **2020**, *32*, 322–333. [CrossRef]
15. Zhang, L.; Cheng, B. A stacked autoencoders-based adaptive subspace model for hyperspectral anomaly detection. *Infrared Phys. Technol.* **2019**, *96*, 52–60. [CrossRef]
16. Erfani, S.; Rajasegarar, S.; Karunasekera, S.; Leckie, S.; Leckie, C. High dimensional and large-scale anomaly detection using a linear one-class SVM with deep learning. *Pattern Recognit.* **2016**, *58*, 121–134. [CrossRef]
17. Wang, L.; Xiong, Z.; Shi, G.; Wu, F.; Zeng, W. Adaptive nonlocal sparse representation for dual-camera compressive hyperspectral imaging. *IEEE Trans. Pattern Anal. Mach. Intell.* **2016**, *39*, 2104–2111. [CrossRef] [PubMed]

18. Velasco-Forero, S.; Angulo, J. Classification of hyperspectral images by tensor modeling and additive morphological decomposition. *Pattern Recognit.* **2013**, *46*, 566–577. [CrossRef]
19. Malpica, J.; Rejas, J.; Alonsoa, M. A projection pursuit algorithm for anomaly detection in hyperspectral imagery. *Pattern Recognit.* **2008**, *41*, 3313–3327. [CrossRef]
20. Wang, Z.; Li, Y.; Guo, Y.; Fang, L.; Wang, S. Data-uncertainty guided multiphase learning for semi-supervised object detection. In Proceedings of the IEEE Conference on Computer Vision and Pattern Recognition (CVPR), Nashville, TN, USA, 20–25 June 2021.
21. Taghipour, A.; Ghassemian, H. Unsupervised hyperspectral target detection using spectral residual of deep autoencoder networks. In Proceedings of the International Conference on Pattern Recognition and Image Analysis (IPAS), Tehran, Iran, 6–7 March 2019; pp. 52–57.
22. Liu, Y.; Li, Z.; Zhou, C.; Jiang, Y.; Sun, J.; Wang, M.; He, X. Generative adversarial active learning for unsupervised outlier detection. *IEEE Trans. Knowl. Data Eng.* **2019**, *32*, 1517–1528. [CrossRef]
23. Yang, X.; Dong, M.; Wang, Z.; Gao, L.; Zhang, L.; Xue, J. Data-augmented matched subspace detector for hyperspectral subpixel target detection. *Pattern Recognit.* **2020**, *106*, 107464. [CrossRef]
24. Ergen, T.; Kozat, S. Unsupervised anomaly detection with LSTM neural networks. *IEEE Trans. Neural Netw. Learn. Syst.* **2020**, *31*, 3127–3141. [CrossRef]
25. Wu, H.; Prasad, S. Semi-supervised deep learning using pseudo labels for hyperspectral image classifification. *IEEE Trans. Image Process.* **2018**, *27*, 1259–1270. [CrossRef]
26. Antonio, P.; Atli, B.; Jason, B.; Lorenzo, B.; Gustavo, C.; Jocelyn, C.; Mathieu, F.; Paolo, G.; Anthony, G. Recent advances in techniques for hyperspectral image processing. *Remote Sens. Environ.* **2009**, *113*, S110–S122.
27. Yang, X.; Deng, C.; Zheng, F.; Yan, J.; Liu, W. Deep spectral clustering using dual autoencoder network. In Proceedings of the IEEE Conference on Computer Vision and Pattern Recognition (CVPR), Long Beach, CA, USA, 15–20 June 2019; pp. 4066–4075.
28. Dincalp, U.; G zel, M.; Sevine, O.; Bostanci, E.; Askerzade, I. Anomaly based distributed denial of service attack detection and prevention with machine learning. In Proceedings of the 2018 2nd International Symposium on Multidisciplinary Studies and Innovative Technologies (ISMSIT), Ankara, Turkey, 19–21 October 2018; pp. 1–4.
29. Wang, R.; Guo, H.; Davis, L.; Dai, Q. Covariance discriminative learning: A natural and efficient approach to image set classification. In Proceedings of the IEEE Conference on Computer Vision and Pattern Recognition (CVPR), Providence, RI, USA, 16–21 June 2012; pp. 2496–2503.
30. He, N.; Paoletti, M.; Haut, J.; Fang, L.; Li, S.; Plaza, A.; Plaza, J. Feature extraction with multiscale covariance maps for hyperspectral image Classification. *IEEE Trans. Geosci. Remote Sens.* **2019**, *57*, 755–769. [CrossRef]
31. Li, A.; Miao, Z.; Cen, Y.; Zhang, X.-P.; Zhang, L.; Chen, S. Abnormal event detection in surveillance videos based on low-rank and compact coefficient dictionary learning. *Pattern Recognit.* **2020**, *108*, 107355. [CrossRef]
32. Fang, L.; He, N.; Li, S.; Plaza, A.; Plaza, J. A new spatial-spectral feature extraction method for hyperspectral images using local covariance matrix representation. *IEEE Trans. Geosci. Remote Sens.* **2018**, *56*, 3534–3546. [CrossRef]
33. Reed, I.; Yu, X. Adaptive multiple band CFAR detection of an optical pattern with unknown spectral distribution. *IEEE Trans. Acoust. Speech Signal Process.* **1990**, *38*, 1760–1770. [CrossRef]
34. Qu, Y.; Wang, W.; Guo, R.; Ayhan, B.; Kwan, C.; Vance, S.; Qi, H. Hyperspectral anomaly detection through spectral unmixing and dictionary-based low-rank decomposition. *IEEE Trans. Geosci. Remote Sens.* **2018**, *56*, 4391-4405. [CrossRef]
35. Huyan, N.; Zhang X.; Zhou, H.; Jiao, L. Hyperspectral anomaly detection via background and potential anomaly dictionaries construction. *IEEE Trans. Geosci. Remote Sens.* **2019**, *57*, 2263–2276. [CrossRef]
36. Kang, X.; Zhang, X.; Li, S.; Li, K.; Li, J.; Benediktsson, J. Hyperspectral anomaly detection with attribute and edge-preserving filters. *IEEE Trans. Geosci. Remote Sens.* **2017**, *55*, 5600–5611. [CrossRef]
37. Li, L.; Li, W.; Du, Q.; Tao, R. Low-rank and sparse decomposition with mixture of gaussian for hyperspectral anomaly detection. *IEEE Trans. Cybern.* **2020**, *51*, 4363–4372. [CrossRef]
38. Xie, W.; Lei, J; Liu, B.; Li, Y.; Jia, X. Spectral constraint adversarial autoencoders approach to feature representation in hyperspectral anomaly detection. *Neural Netw.* **2019**, *119*, 222–234. [CrossRef]
39. Xie, W.; Liu, B.; Li, Y.; Lei, J.; Chang, C.-I.; He, G. Spectral adversarial feature learning for anomaly detection in hyperspectral imagery. *IEEE Trans. Geosci. Remote Sens.* **2019**, *58*, 2352–2365. [CrossRef]
40. Mao, X.; Li, Q.; Xie, H.; Lau, R. YK; Wang, Z.; Paul Smolley, S. Least squares generative adversarial networks. In Proceedings of the IEEE International Conference on Computer Vision (ICCV), Venice, Italy, 22–29 October 2017; pp. 2794–2802.
41. Goodfellow, I.; Pouget-Abadie, J.; Mirza, M.; Xu, B.; Warde-Farley, D.; Ozair, S.; Courville, A.; Bengio, Y. Generative adversarial nets. In Proceedings of the Advances in Neural Information Processing Systems (NIPS), Montreal, QC, Canada, 8–13 December 2014; p. 27.
42. Karras, T.; Laine, S.; Aila, T. A style-based generator architecture for generative adversarial networks. In Proceedings of the IEEE/CVF Conference on Computer Vision and Pattern Recognition (CVPR), Long Beach, CA, USA, 15–20 June 2019; pp. 4401–4410.
43. Brock, A.; Donahue, J.; Simonyan, K. Large scale GAN training for high fidelity natural image synthesis. *arXiv* **2018**, arXiv:1809.11096.

44. Carreira, J.; Caseiro, R.; Batista, J.; Sminchisescu, C. Free-form region description with second-order pooling. *IEEE Trans. Pattern Anal. Mach. Intell.* **2015**, *37*, 1177–1189. [CrossRef]
45. Chen, D.; Yue, L.; Chang, X.; Xu, M.; Jia, T. NM-GAN: Noisemodulated generative adversarial network for video anomaly detection. *Pattern Recognit.* **2021**, *116*, 107969. [CrossRef]

remote sensing

MDPI

Article

A Study on the Dynamic Effects and Ecological Stress of Eco-Environment in the Headwaters of the Yangtze River Based on Improved DeepLab V3+ Network

Chunsheng Wang [1], Rui Zhang [2,*] and Lili Chang [1]

1 Institute of Geophysics & Geomatics, China University of Geosciences, Wuhan 430079, China; atccll@cug.edu.cn (C.W.); lilichang@cug.edu.cn (L.C.)
2 Land Satellite Remote Sensing Application Center, Ministry of Natural Resources of the People's Republic of China, Beijing 100048, China
* Correspondence: zhangrui@radi.ac.cn

Abstract: The headwaters of the Yangtze River are a complicated system composed of different eco-environment elements. The abnormal moisture and energy exchanges between the atmosphere and earth systems caused by global climate change are predicted to produce drastic changes in these eco-environment elements. In order to study the dynamic effect and ecological stress in the eco-environment, we adapted the Double Attention Mechanism (DAM) to improve the performance of the DeepLab V3+ network in large-scale semantic segmentation. We proposed Elements Fragmentation (*EF*) and Elements Information Content (*EIC*) to quantitatively analyze the spatial distribution characteristics and spatial relationships of eco-environment elements. In this paper, the following conclusions were drawn: (1) we established sample sets based on "Sentinel-2" remote sensing images using the interpretation signs of eco-environment elements; (2) the *mAP*, *mIoU*, and *Kappa* of the improved DeepLab V3+ method were 0.639, 0.778, and 0.825, respectively, which demonstrates a good ability to distinguish the eco-environment elements; (3) between 2015 and 2021, *EF* gradually increased from 0.2234 to 0.2394, and *EIC* increased from 23.80 to 25.32, which shows that the eco-environment is oriented to complex, heterogeneous, and discontinuous processes; (4) the headwaters of the Yangtze River are a community of life, and thus we should build a multifunctional ecological management system with which to implement well-organized and efficient scientific ecological rehabilitation projects.

Keywords: eco-environment elements; the headwaters of the Yangtze River; improved DeepLab V3+ network; dynamic effects; ecological stress

Citation: Wang, C.; Zhang, R.; Chang, L. A Study on the Dynamic Effects and Ecological Stress of Eco-Environment in the Headwaters of the Yangtze River Based on Improved DeepLab V3+ Network. *Remote Sens.* **2022**, *14*, 2225. https://doi.org/10.3390/rs14092225

Academic Editor: Gang Chen

Received: 15 March 2022
Accepted: 4 May 2022
Published: 6 May 2022

1. Introduction

The headwaters of the Yangtze River are a complicated system composed of different eco-environment elements, including glaciers, lakes, grasslands, and bare land [1]. Over the past half-century, the glaciers in the headwaters of the Yangtze River have been shrinking [2,3]. Surface water is an important component of the cryosphere and the unique underlying surface of the plateau, and its abnormal circulation is directly threatening the ecological security of the headwaters of the Yangtze River and the basin as a whole [4–6]. Therefore, it is of great geoscientific and ecological significance to study the ecological stress and dynamic effects of eco-environment elements in the headwaters of the Yangtze River.

There is a series of studies on the relationship between glaciers and climate change, the influence of melting glaciers on runoff, and the control of grassland development [7,8]. With the growth of human needs and ecological requirements, the demand for normalization, intelligence, and elaboration in monitoring the eco-environment is increasing [9,10]. The introduction of remote sensing has expanded the horizons of eco-environment researches and improved our understanding of the region [11,12]. The research methods utilized to study the eco-environment in the headwaters of the Yangtze River mainly include the basin hydrological model, the climate model, and the landscape ecology model [13–15]. The basin hydrological model focuses on the combined influences of changes in various environmental factors associated with climatic and underlying surface properties [16,17]. Han et al., evaluated the evolution of the headwaters of the Yangtze River under climate change based on remote sensing data of the snow water equivalent and the snow cover area for the first time [18]. Guo et al., designed a series of descending experiments based on a grid-based hydrological model to quantify the combined influences of multiple environmental factors on runoff changes [19]. The climate model assesses the impacts of climate change on glacier melt, the degradation of soils, and active layer thickness [20–22]. Ahmed et al., analyzed climate changes based on trends in air temperature variables, the diurnal temperature range, and elevation-dependent warming on annual and seasonal scales in the headwaters of Yangtze River [23]. Sang et al., analyzed the regional water resources and ecological security using typical climate variables [24]. A modified Mann–Kendall trend test, Pettitt test, wavelet analysis, and multivariate correlation analysis were deployed to assess the effects of global climate change [25]. The landscape ecology model mainly studies the overall spatial structure, interaction, coordination function, and dynamic change in many different ecosystems based on the principle of landscape scale, landscape dynamics, and landscape connectivity [26–29]. Wang et al., proposed a quantitative method and a digital model for the headwaters of Yangtze River that aimed to demonstrate that evolving landscape patterns inevitably cause an evolution in ecosystem functionality [30]. In summary, scholars have established a hierarchical ecosystem framework for the headwaters of Yangtze River, and have developed an indices system for the assessment of the ecosystem integrity of the headwaters. This provides conditions for the transformation and development of qualitative and quantitative as well as static and dynamic eco-environment analyses from process to model [31]. The development of landscape ecology provides a theoretical basis for long-term and high-precision studies [32], which may provide the foundation for studying dynamic effects and ecological stress.

The spatial distribution pattern of elements reflects the spatial variation characteristics of the eco-environment [33]. In order to better study the impact of global climate change on the eco-environment, we need to first accurately extract the components of the eco-environment [34,35]. Traditional methods for extracting the eco-environment elements based on remote sensing images mainly include visual interpretation, single-band threshold segmentation, supervised classification, and unsupervised classification [36]. With the rapid development of computer vision, deep learning algorithms have shown great applicability in the automatic extraction of remote sensing image information. These methods can be divided into the following classes: (a) Feature encoder-based methods. Feature encoder-based networks are mainly represented by VGGNet and ResNet [37], which explore the relationship between the depth and performance of the Convolutional Neural Network (CNN); (b) Regional proposal-based methods. The remote sensing classification prediction of the Region-based Convolutional Neural Network (R-CNN) is carried out according to the detection results of the color-space and similarity matrix [38]; (c) Recurrent neural network-based methods. Recurrent neural networks (RNN), mainly represented by ReSeg and MDRNNs, are composed of Long Short-Term Memory (LSTM) blocks, which can also complete remote sensing semantic segmentation and data annotation tasks [39]; (d) Up-sampling-based methods. Missing information can be estimated by up-sampling, as the accurate segmentation boundary can be obtained to a certain extent [40]. The most common algorithms are Fully Convolutional Networks (FCNs), UNet, and so on; (e) Feature resolution improvement-based methods. Feature resolution improvement-based networks are mainly represented by the DeepLab series. DeepLab networks usually integrate shallow features with deep features to improve semantic segmentation efficiency. DeepLab V3+ combines Deep Convolutional Neural Networks (DCNNs) and DenseCRFs and exhibits impressive segmentation performance. It can recover the reduced resolution in the DCNN and obtain more contextual information [41]. The emergence of the DeepLab V3+ method addresses the problems of shadow and threshold in traditional algorithms, and it has a high application value in element classification.

In order to explore the eco-environment situation in the study area, we conducted in-depth research focused on dynamic effects and ecological stress [42]. The main contributions are as follows: (1) The DeepLab V3+ network improved by DAM, which has an excellent segmentation ability, is introduced and a "Sentinel-2" remote sensing image dataset is established. (2) The headwaters of the Yangtze River comprise a community of life; in this study, we attempt to analyze the dynamic effects in this region from a systematic, holistic, and multiscale perspective. (3) We propose Elements Fragmentation (*EF*) and Elements Information Content (*EIC*) to quantitatively analyze the spatial distribution characteristics and spatial relationships of eco-environment elements. (4) This study aims to build a multifunctional ecological management system with which to implement well-organized and efficient scientific ecological rehabilitation projects.

2. Material and Methods

2.1. Study Aera and Data Source

2.1.1. Study Aera

The headwaters of the Yangtze River are located in the hinterland of the Qinghai-Tibet Plateau. Figure 1a,b show the location of the study area on the Qinghai-Tibet Plateau and in the Yangtze River Basin. The geographical coordinates of the study area are longitude 32°30′~35°35′E, latitude 90°43′~96°45′N, and the basin covers an area of approximately 138.2×10^3 km^2 [43]. As the geomorphological and drainage map of the headwaters of the Yangtze River shows (Figure 1c), there are various landforms in the basin, which include mountains, valleys, basins, lakes, and marshes. The Qinghai-Tibet Plateau climate system is a typical continental plateau climate, which is characterized by alternating wet-hot and dry-cold seasons, and the precipitation varies greatly with the altitude [44]. Glaciers are concentrated in the headwaters of the Yangtze River, which are mainly distributed in the Keladandong, Gaqiadirugang, and Gangqin snow mountain group. The largest areas covered by glaciers are found in the Dangqu Basin, followed by the Tuotuo River Basin; the smallest is located in the Chumar River Basin [45,46]. Furthermore, an interchange of materials and energy is constantly occurring between the glaciers, lakes, grasslands, and bare land [47].

2.1.2. Data Source

In this paper, the experimental dataset was produced from "Sentinel-2" high-resolution multispectral images, which were mainly used for monitoring the land environment, terrestrial vegetation, land cover, and watershed environment. This images were especially advantageous in the target segmentation in long-time series. The "Sentinel-2" images in this article were downloaded from the Copernicus Open Access Hub [48]. In order to comprehensively study the dynamic effects and the ecological stress of the eco-environment, the "Sentinel-2" images selected were mainly taken in July when the glacier line was the lowest. When the images were poor quality, such as at times of high cloud cover, they would be replaced by images taken in June and August [49,50].

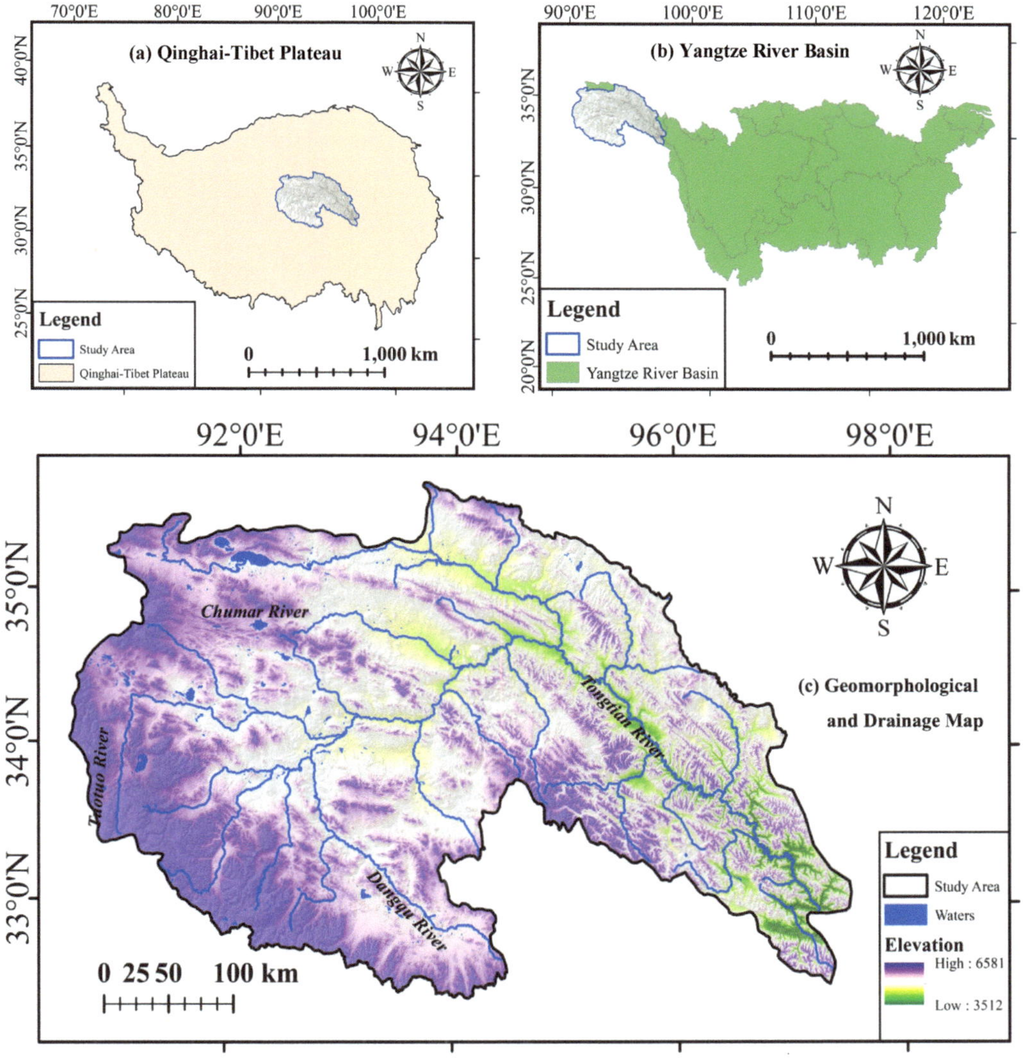

Figure 1. The location on the headwaters of the Yangtze River. (**a**) The location of the study area on the Qinghai-Tibet Plateau; (**b**) the location of the study area in the Yangtze River Basin; (**c**) the geomorphological and drainage map of the headwaters of the Yangtze River.

2.2. Model

2.2.1. DeepLab V3+ Network

The semantic segmentation of remote sensing images is a full-pixel classification task, which includes categorical information and location information [51]. The traditional object-oriented segmentation algorithm uses low-level semantic information, such as the color, texture, and shape of the image being segmented [52], which is lacking intermediate semantics, such as pixel contrast and orientation. DeepLab V3+ introduced an encoder–decoder, which is commonly used to integrate multiscale information [53,54]. The decoder combines the up-sampling results of the depth map with shallow features [55]. Ultimately, the layer features are used to optimize location information that cannot be retrospectively sampled (Figure 2); the decoder up-samples the depth features and combines them with the shallow features [56,57]. The DeepLab V3+ network uses up-sampling to reconstruct the feature image, the results of which can include the classification and location information of glaciers, lakes, grasslands, and bare land [58,59].

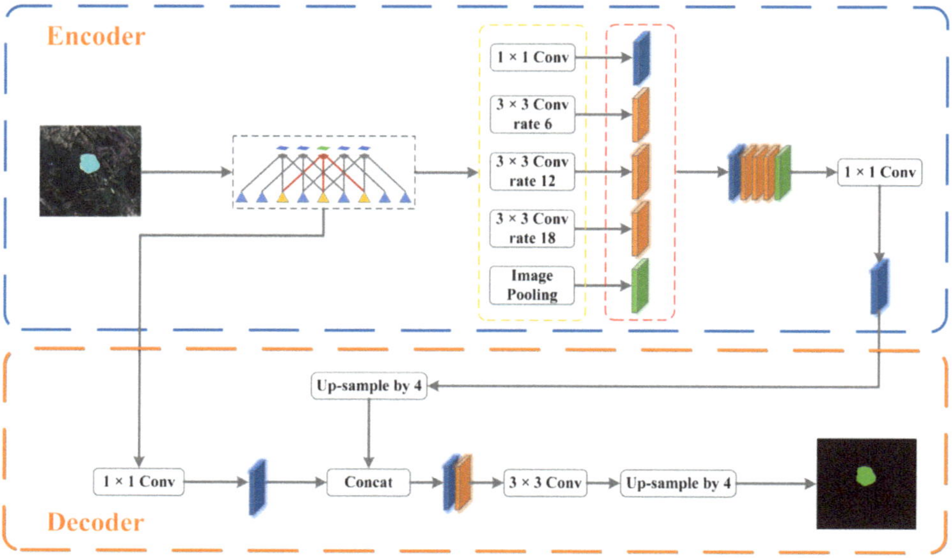

Figure 2. DeepLab V3+ network generalization diagram.

2.2.2. Improved DeepLab V3+ Model

DeepLab networks usually integrate shallow features with deep features, thus obtaining the high-level semantics between pixels and entity categories [60]. The encoder mainly extracts the depth feature information of the image, which is the key to achieving high-precision semantic segmentation. The encoder is mainly composed of DCNN and Atrous Spatial Pyramid Pooling (ASPP). We added the Double Attention Mechanism (DAM) to the ASPP module, so as to reduce accuracy loss (Figure 3). The specific operation is as follows: First, the global dependency relationship between features is captured in the spatial dimension and the channel dimension of features, respectively, to capture the context feature information and enhance the expression ability of features [61]; Then, the output of the two attention modules is added to further improve feature representation; Finally, a more accurate segmentation result is obtained.

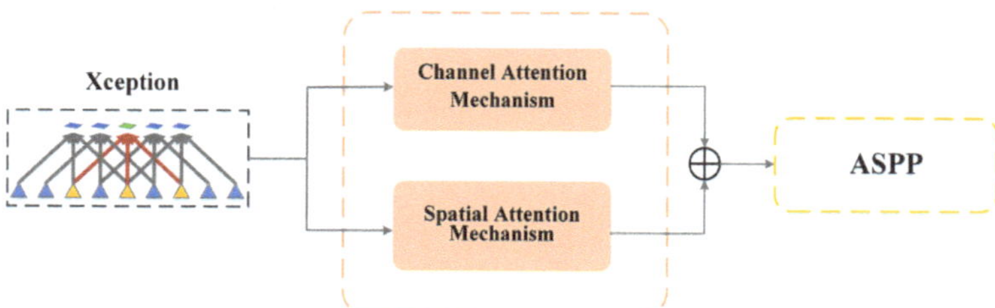

Figure 3. The DAM module structure diagram of the improved DeepLab V3+ network.

The Channel Attention Mechanism (CAM) module selectively emphasizes the interconnected channel map by integrating the relevant characteristics in all channel maps. Therefore, this paper explicitly models the interdependence between channels by adding the CAM module. As Figure 4 shows, the original input feature map is multiplied by elements to obtain the channel attention weighted map.

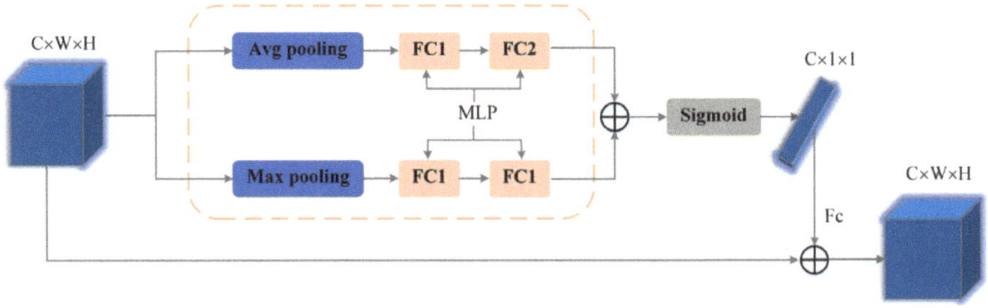

Figure 4. Channel attention mechanism module.

The Spatial Attention Mechanism (SAM) module selectively aggregates the features of each location through the weighted sum of features of all locations. Regardless of distance, similar features are interrelated (Figure 5). As a consequence, in order to establish a richer context between local features, the SAM module is introduced.

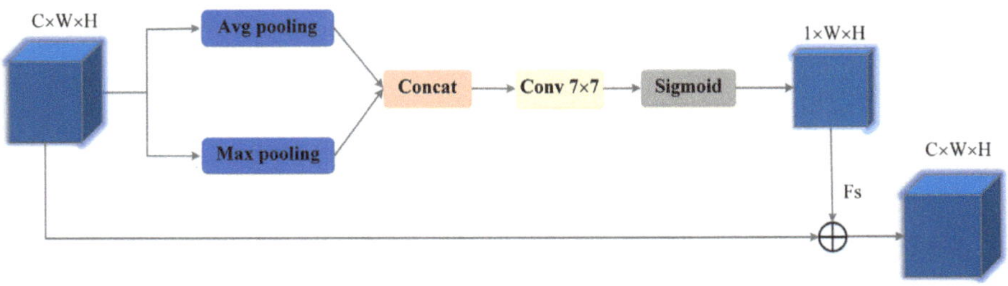

Figure 5. Spatial attention mechanism module.

2.2.3. Accuracy Assessment

Pixel accuracy is an evaluation index with which to calculate the proportion of the true value matched by the predicted value [62]. The mean Pixel Accuracy (*mPA*) represents the proportion of the number of classification pixels for each class to the average [63]. The mean Intersection over Union (*mIoU*) uses the mean value to represent the ratio of the intersection and union of the predicted set to the true set of the correct classification for each class [64]. The *Kappa* coefficient represents the coincidence degree between the classified image and the reference image, and it is an objective evaluation standard to test their consistency [65]. Both the *mPA* and *mIoU* are between 0 and 1, and the *Kappa* also usually falls between 0 and 1. Therefore, the closer the value is to 1, the better the semantic segmentation performance will be. The above three indicators are widely used in the evaluation of image semantic segmentation accuracy because of their simplicity and representation. All calculation formulae are shown below:

$$mPA = \frac{1}{N}\frac{\sum_{i=1}^{N} n_{ii}}{\sum_{i=1}^{N}\sum_{j=1}^{N} n_{ij}} \quad (1)$$

$$mIoU = \frac{1}{N}\sum_{i=1}^{N}\frac{n_{ii}}{\sum_{j=1}^{N} n_{ij} + \sum_{j=1}^{N} n_{ji} - n_{ii}} \quad (2)$$

$$Kappa = \frac{p_o - p_e}{1 - p_e} \quad (3)$$

where, N represents the total number of semantic segmentation classes, and n_{ij} is the pixel of class i divided into class j. The p_o is the number of the correctly classified samples divided by the total number of samples, and p_e is the number of the misclassified samples divided by the total number of samples.

2.2.4. Elements Spatial Analysis

On the basis of the accurate extraction of eco-environment elements, a quantitative analysis of the spatial distribution characteristics and spatial relations can be conducted. Landscape connectivity is mainly controlled by the spatial distribution characteristics and spatial relationships of eco-environment elements, which provided the foundation from which to study the dynamic effects and ecological stress in the study area. Among them, Landscape Fragmentation (*LF*) and Shannon's Diversity Index (*SHDI*) were used for the quantitative description [66]. The calculations of *SHDI* and *LF* were performed according to Formulas (4) and (5). *LF* refers to the process through which the landscape, from a single, homogeneous, and continuous whole, tends to become a complex, heterogeneous, and discontinuous patched mosaic [67]. *LF* reflects the complexity of the landscape spatial structure and the degree of interference caused by natural or human factors. *SHDI* is a measurement index based on information theory, the serious fragmentation of the landscape, and the lower mobility of the patches. This provides the conditions for the transformation and development of eco-environment analyses, from qualitative to quantitative, from static to dynamic, and from process to model. We propose the *EF* and the *EIC* based on the *LF* and *SHDI* to calculate eco-environment element connectivity [68–70]; the calculations were performed using Formulas (6) and (7).

$$LF = \frac{N_i}{A_i} \tag{4}$$

where, N_i is the number of patches of landscape i, and A_i is the total area of landscape i.

$$SHDI = \sum_{i=1}^{S} p_i \ln(p_i) \tag{5}$$

where, S is the classes of landscape, and p_i is the proportion of i landscape to the total landscape.

$$EF = \frac{1}{nm} \sum_{i=1}^{n} \sum_{j=1}^{m} \frac{N_{ij}}{A_{ij}} \tag{6}$$

where, n is the number of eco-environment elements, m is the number of classification scales, and N_{ij} and A_{ij} are the numbers and the areas of elements belonging to the m scale in class i, respectively.

$$EIC = \frac{\sum_{i=1}^{n} \sum_{j=1}^{m} (q_{ij} \ln(q_{ij}))}{\sum_{i=1}^{n} (q_i \ln(q_i))} \tag{7}$$

where, n is the number of eco-environment elements, m is the number of classification scales, q_{ij} is the proportion of the j scale to the total number of class i, and q_i is the proportion of i element to the total elements.

3. Results

3.1. Experiment Data and Parameter

3.1.1. Identification Index

After radiation calibration and atmospheric correction processing, the radiation errors caused by atmospheric scattering and absorption in the "Sentinel-2" raw images were eliminated [71]. The samples obtained based on "Sentinel-2" images were selected to be exported into TIF datasets. We converted all samples from TIF format to IMG format, which was used as the input for the training [72]. According to the distribution, the identification indexes of eco-environment elements were established according to the surface reflection and spatial texture characteristics, which were used to establish a label system that the network was able to recognize [73]. The interpretation marks established in the eco-environment elements were verified in related articles. This laid the foundation for building sample sets (Figure 6). The interpretation signs are summarized as follows:

(1) Glaciers have a relatively simple textural structure, with a bright tone in both true-color and false-color images. Glaciers have a high contrast with the surrounding environment. In regards to morphological structure, glaciers often appear with an arc-shaped boundary;
(2) Water has a certain boundary in the remote sensing image. According to the mineral content, the depth of the water, and the imaging time, water is predominantly cyan or bluish green both in true-color and false-color images;
(3) Grassland and bare land, as the largest portion of the study area, are relatively similar in terms of spatial distribution. In true-color images, grassland mainly has a green color and bare land mainly a tan and flesh pink color. In false-color images, grassland mainly has a red color, and bare land mainly has a tan and flesh pink color similar to in true-color images.

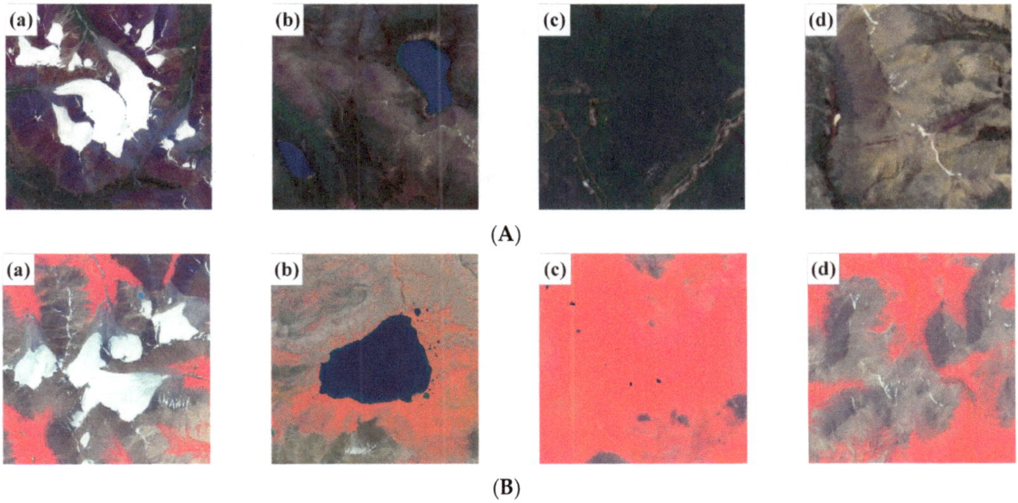

Figure 6. The interpretation signs of eco-environment elements. (**a**) Glacier; (**b**) water; (**c**) grassland; (**d**) bare land. (**A**) is the true-color image and (**B**) is the false-color image.

3.1.2. Sample Sets

There was a total of 5717 pictures annotated with eco-environment elements. All pictures with resolution of 10 m were collected from the Source of Three Rivers (Table 1, in which the number and size are non-dimensional). Sample sets were divided into mutual exclusion training sets and test sets, and the validation sets were included in the training sets. We trained the model on the training sets, and verified the accuracy of the extractive effect using the validation sets. We implemented model selection using the validation sets. Once the best parameters were identified, final training could be performed on the test sets. Finally, the model was used to train the remote sensing images in the study area, and the classification of eco-environment elements was obtained. The eco-environment elements in the training sets and test sets demonstrated spatial differences and feature similarity, so the generalization performance of the model was well reflected. The ratio of training sets to test sets was 7:3. The dataset ratio R of training sets to validation sets is discussed in Section 4.3. Eco-environmental element monitoring research was conducted in the study area.

Table 1. The parameter of sample sets.

Name	Number	Resolution (m)	Size
sample sets	5717	10	521 × 521

3.2. Parameters Setting

3.2.1. Parameters Index

In the process of pixel segmentation of the eco-environment elements, the parameter selection of the dataset ratio R of the training sets to validation sets, the learning rate Y_t, and *batch_size* had the most impact on performance [74]. In this paper, the control variable method was used to comprehensively select the three parameters. We selected the momentum optimizer, which uses the objective function to optimize training parameters. The appropriate learning rate causes the optimizer to optimize the model parameters more effectively, and the poly learning rate was selected for optimization [75]. The training objective function includes the log-likelihood function and $L2$ regularization term [76–78]. The results calculated by Softmax represent the difference between the true label and the predicted label, and the regularization term can prevent overfitting to a certain extent [79]. The formulae are as follows:

$$Y_t = X\left(1 - \frac{t}{T}\right)^p \tag{8}$$

$$Loss = -\frac{1}{K}\sum_{n=1}^{N}\sum_{i=1}^{K} y_{true}^{ni} \log y_{pred}^{ni} \tag{9}$$

where, Y_t is the current learning rate, X is the initial learning rate, t is the current training times, and T is the total number of trainings. K is the total number of pixels with labels, N is the number of classes, y_{true}^{ni} is the probability of pixel i corresponding to class N, and y_{pred}^{ni} is the probability of pixel i being output by the network corresponding to class N.

3.2.2. Parameters Selection

As a result of the influence of computer GUP memory, the *batch_size* was fixed at 10 during the experiment. When Y_t= 0.9 and the proportion of samples in the training sets and validation sets was 6:4, the training accuracy and validation accuracy were the highest (Table 2), so the parameter R was taken as 6:4. Using the same method, we set *batch_size* = 10, R = 6:4, and Y_t as 0.25, 0.9, 0.1, and 2, respectively, for training. In the initial stage of training, the Loss values of all curves were large. With the increase in iterations, the Loss value continued to oscillate and decline slowly. When the model was iterated 60,000 times, the Loss values of all curves were basically stable and remained so until the end of the iteration. When Y_t= 0.9, the *Loss* value finally stabilized at 0.21, so the parameter Y_t was taken as 0.9 (Figure 7). In order to avoid the introduction of a large amount of noise into the parameters, the remaining parameters that had little influence on the model were used in the training with their initial values. The final parameter values are shown in Table 3.

Table 2. Determination of the dataset ratio R.

R	Y_t	*Batch_Size*	Training Accuracy	Validation Accuracy
6:4	0.9	10	0.913	0.901
7:3	0.9	10	0.823	0.837
8:2	0.9	10	0.879	0.864

Table 3. Values of various parameters in the DeepLab V3+ networks.

R	Y_t	*Batch_Size*	Regularization Term	*Eval_Scales*	Iterations
6:4	0.9	10	0.0001	[0.5:0.25:1.75]	120,000

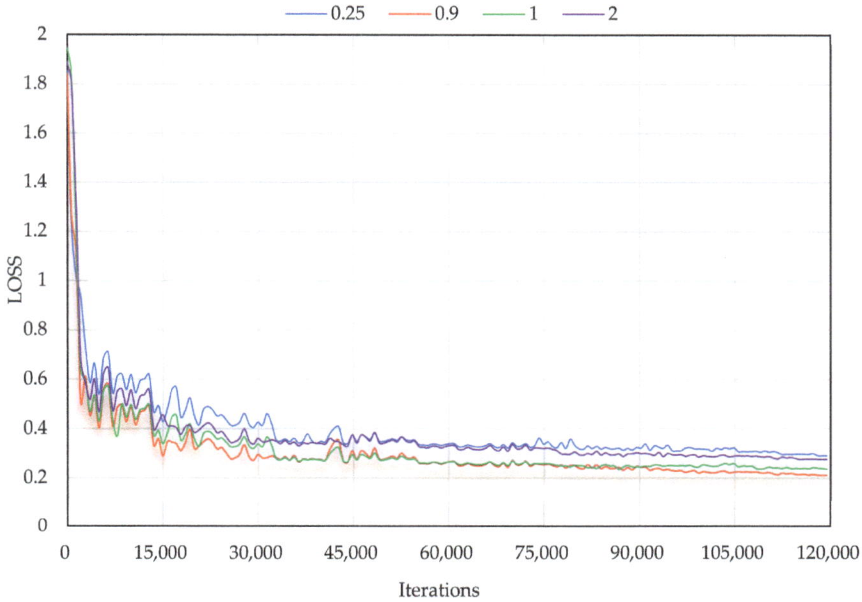

Figure 7. Evaluation of *Loss* curve with the Y_t changes.

3.3. Results Analysis

In order to verify the performance of the DeepLab V3+ network and demonstrate its application value, we validated DeepLab V3+ and Support Vector Machine (SVM), UNet, and DeepLab V3 on the sample sets. The automatic identification results of different models are shown in Figure 8. Generally speaking, the classification based on the SVM and UNet methods demonstrated better extraction integrity. The eco-environmental elements were prone to misclassification. As compared with the other four methods, the extraction results of the eco-environment elements based on the DeepLab V3+ network exhibited better performance in terms of comprehensive positioning accuracy and extraction boundary integrity. The comparison results are summarized in the following:

(1) SVM extracted the pixels that conformed to the glaciers, lakes, grasslands, and bare land to a certain degree. However, there was obvious misclassification in the extraction results of different classes. For this method, the *mPA* and *Kappa* of the SVM segmentation results were the worst, with values of 0.463 and 0.641, respectively. The segmentation results of SVM are greatly affected by other surface reflectivity features;

(2) The extraction results of UNet were greatly affected by background interference and spectral features, and some frozen lakes were mistakenly classified as glaciers. As a result of the high altitude, some lakes were still frozen during this period, but they had various different shape characteristics as compared to glaciers. As the Table 4 shows, the lowest index of *mIoU* was recorded for UNet, indicating that this method could not semantically segment the eco-environment elements well. As a result, the extraction of grassland was more fragmented and the accuracy was lower;

(3) DeepLab V3 had a good ability to identify the eco-environment elements. However, it needed to train many times to achieve better results for complex eco-environment elements. DeepLab V3 was able to accurately classify the eco-environment elements in the spatial position through high-cost training;

(4) The performance of each index for DeepLab V3+ was superior to those of DeepLab V3, with the *mAP*, *mIoU*, and *Kappa* of the former being 0.639, 0.778, and 0.825, respectively. The extraction results based on DeepLab V3+ had a complete structure and obvious

edge features, and it did not produce missing or wrong divisions for small areas of grassland. The DeepLab V3+ method demonstrated a good ability to distinguish the eco-environment elements in the headwaters of the Yangtze River.

Table 4. The segmentation evaluation results of eco-environment elements.

	SVM	UNet	DeepLab V3	DeepLab V3+
mPA	0.478	0.463	0.597	0.639
mIoU	0.493	0.517	0.739	0.778
Kappa	0.674	0.641	0.802	0.825

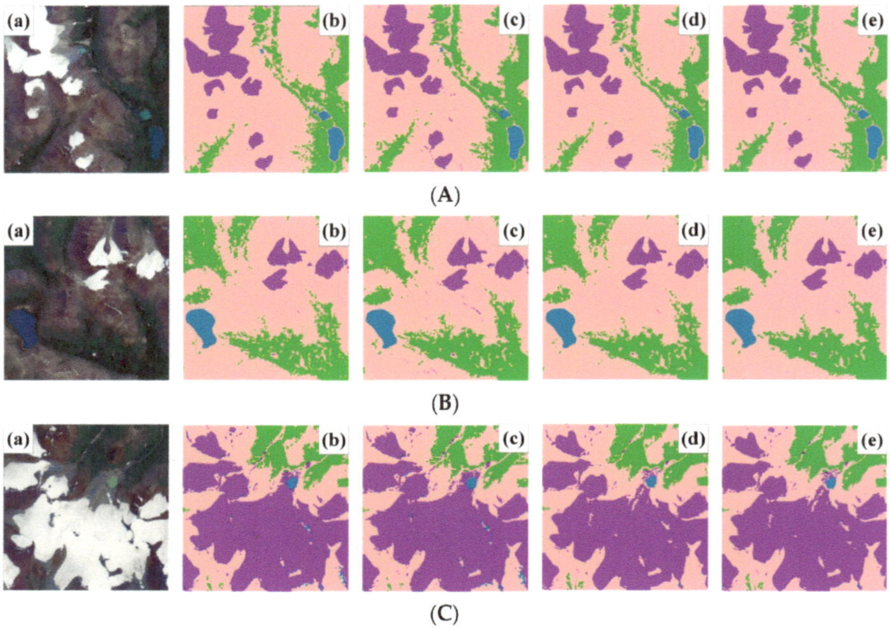

Figure 8. The extraction results of eco-environment elements. (**a**) The raw image; (**b**) SVM; (**c**) UNet; (**d**) DeepLab V3; (**e**) DeepLab V3+. Where (**A–C**) represent the three typical regions of the headwaters of the Yangtze River.

4. Discussion

4.1. Elements Changes

Using the DeepLab V3+ network, each pixel was accurately extracted. The eco-environment elements extraction experimental results were pieced together from the original geographic information, and each pixel had its own function because of the spatial relationship. The changes in these elements in the study area under the different space-time states and scales of coverage were used to produce comparative statistics, as illustrated in Table 5. In this paper, every two years was set as an interval (Table 6). Taking Yushu City as an example, we found that the glaciers, lakes, grasslands, and bare land underwent constant development and changes. In the headwaters of the Yangtze River, the glaciers, lakes, grasslands, and bare land are interdependent and interlaced.

Table 5. The scale classification results of each eco-environment elements.

Scale	Large	Medium	Small	Smaller
Area (km^2)	>100	10–100	1–10	<1

Table 6. Time series representation.

Period	2015–2017	2017–2019	2019–2021
Representation	I	II	III

4.1.1. Glaciers Changes

As shown in Figure 9, the glaciers are mainly distributed to the southwest of Golmud City. There are other glaciers scattered in Southern Zaduo County, Southern Zhiduo County, Sortheastern Yushu City, Northern Qumalai County, and Western Chengduo County (Figure 10). By studying the distribution of glaciers, the overall trend of glacier change could be obtained. In general, the glaciers in the study area were decreasing and breaking up during the study period.

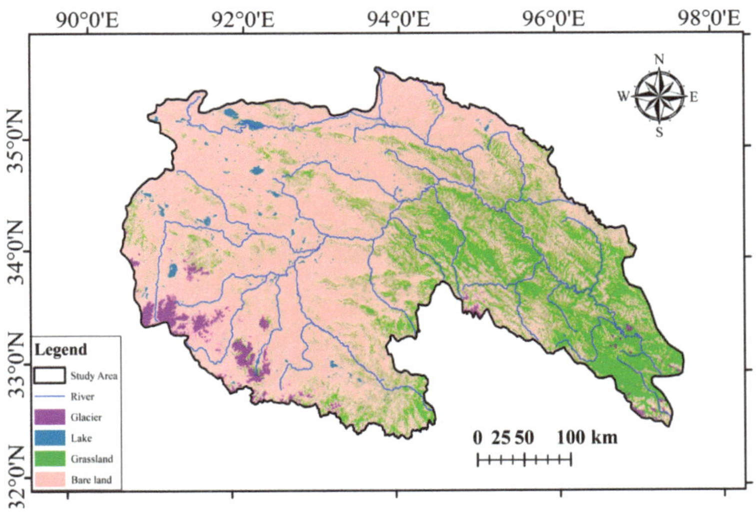

Figure 9. The spatial distribution of each eco-environment element in the initial stage.

(1) The total area of glaciers in each region is shrinking. The total area of glaciers in the initial stage (2015) was 2818.98 km^2, and the areas of glaciers in Golmud City, Yushu City, Chengduo County, Qumalai County, Zaduo County, and Zhiduo County were 2249.58, 170.99, 39.08, 39.22, 200.84, and 119.27, respectively. However, in 2021, the areas were 2201.33, 167.36, 38.26, 38.37, 196.97, and 116.85, respectively.

(2) The decrease in the glacier area was accompanied by an increase in the snowline altitude. As shown in Figure 11, the elevation distribution and change rate of the snowline altitude in each region was not the same (the changing rate was non-dimensional), with the snowline altitude equaling 5234, 4322, 4164, 5144, 5218, and 4982 m in 2021.

(3) The glacier changes at different scales in the headwaters of the Yangtze River generally exhibited different change rates, as shown in Figure 12. The number of large-scale glaciers in the study area remained unchanged, and the downward trend for medium-scale glaciers was slight as compared with the dramatic change seen in smaller-scale glaciers, as shown in Figure 13 (the unit of area is square kilometers (km^2), while the unit of elevation is meters (m)).

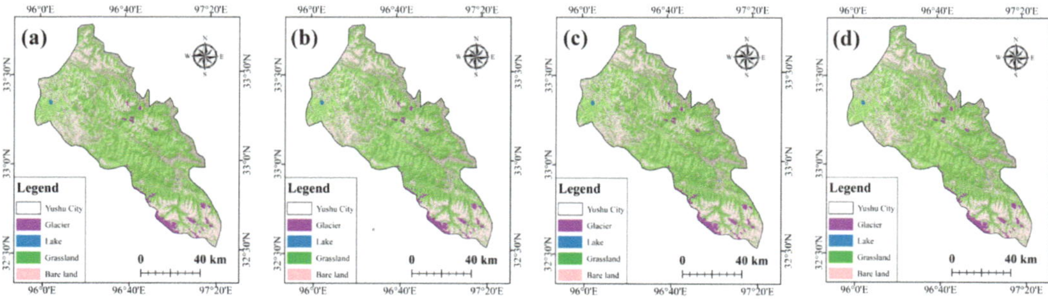

Figure 10. The changes in the eco-environment elements in Yushu City. (**a**) 2015; (**b**) 2017; (**c**) 2019; (**d**) 2021.

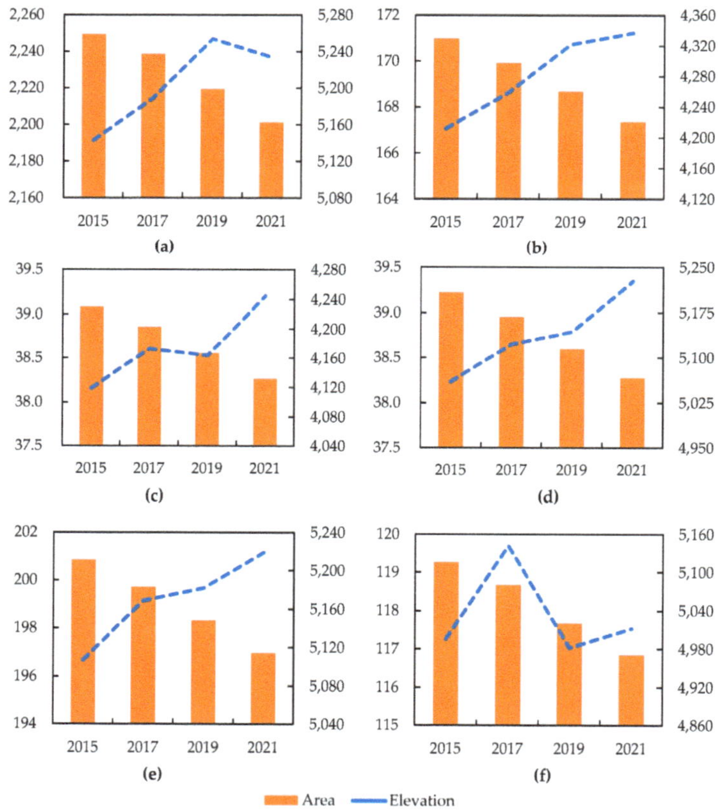

Figure 11. The glacier area and elevation distribution. (**a**) Golmud City; (**b**) Yushu City; (**c**) Chengduo County; (**d**) Qumalai County; (**e**) Zaduo County; (**f**) Zhiduo County.

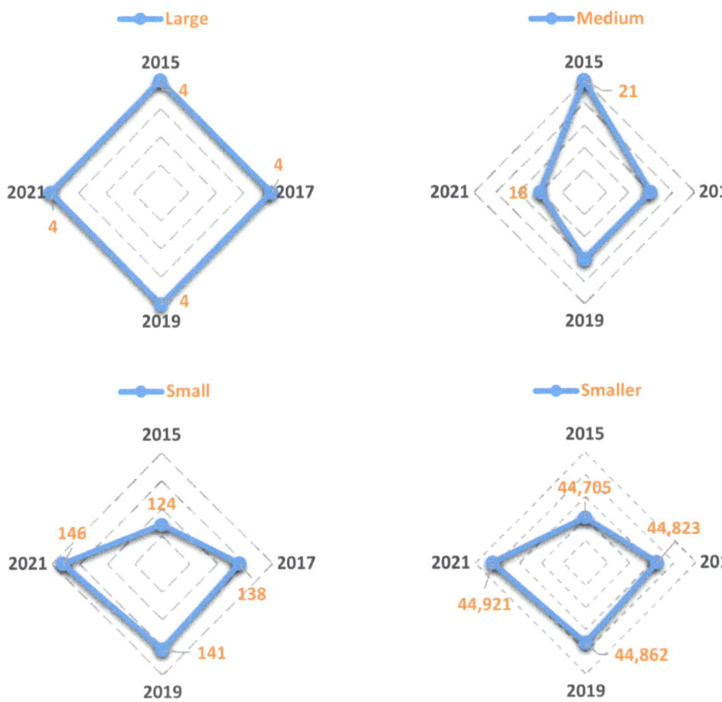

Figure 12. The changes in the number of glaciers at different scales.

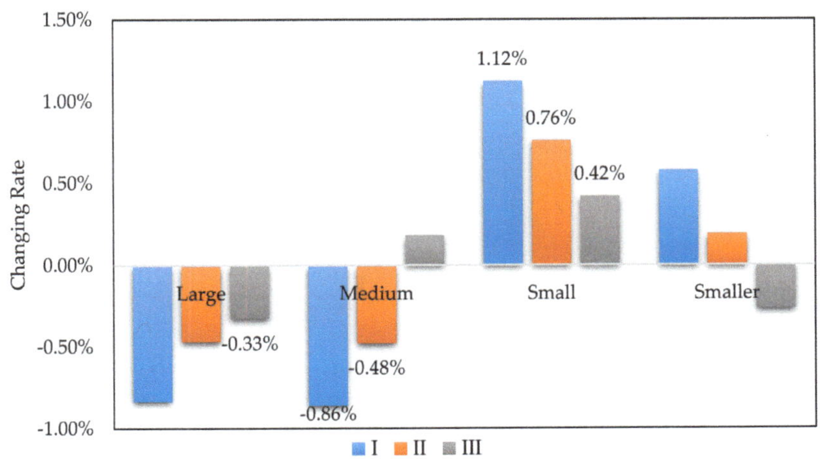

Figure 13. The glacier change statistics at different scales.

4.1.2. Lakes Variation

Lakes are mainly located in the central or western regions of the study area, and there are relatively few lakes in Eastern Zhiduo County, Eastern Qumalai County, and most parts of Yushu City. Geographically, most lakes in the headwaters of the Yangtze River are located in uninhabited areas with high altitudes and are less affected by human activities. In the three periods, the changing trend for scales was different from that of area. Generally, increasing temperature and precipitation were conducive to the expansion of the area of glacier-replenished lakes.

As shown in Figures 14 and 15, there were 4072 smaller-scale lakes, 89 small lakes, 15 medium lakes, and 2 large lakes in 2015. In the initial stage, the total number of large lakes was two, representing 21.28% of all lakes, and the proportions of medium-scale, small-scale, and smaller-scale lakes were 34.98%, 20.87%, and 22.87%, respectively. By 2021, the number of large-scale, medium-scale, small-scale, and smaller-scale lakes was 2, 17, 93, and 3927, accounting for 23.64%, 37.27%, 20.67%, and 18.42% of the total area. As a result of the influence of various factors, the total number of lakes continued to increase, but the quantity of large-scale lakes remained unchanged, and the quantity and area of small lakes was reduced to different degrees.

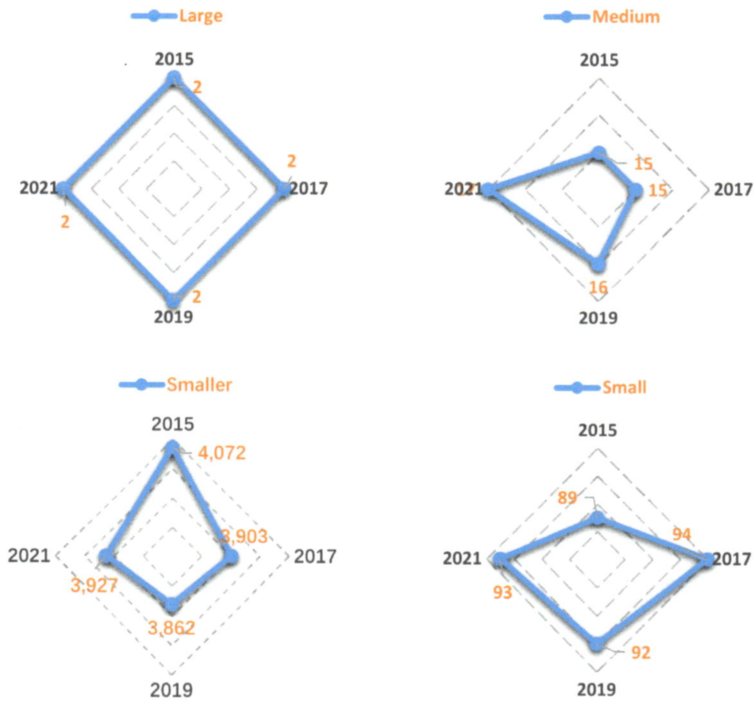

Figure 14. Changes in the number of lakes at different scales.

4.1.3. Grasslands and Bare Land

The grasslands represent the main area covered with vegetation in the headwaters of the Yangtze River and the coverage is an important index to measure the surface vegetation coverage in the study area. As a result of the typical continental plateau climate, the sparsely covered and uncovered vegetation areas, which we called bare land, were extremely fragile and sensitive, with complicated changes in vegetation cover.

(1) Grasslands are widely distributed in the Eastern Zhiduo County, Eastern Qumalai County, Western Chengduo County, Southeastern Zaduo County, and most parts of Yushu City. In this region, bare land mainly exists in the alpine desert areas, such as most parts of Golmud City, Northern Qumalai County, and Northern Zaduo County. According to the statistics in Figure 16 (the unit of area is square kilometers), the area of grassland in the study area was 48,796 km^2 in 2015, and 47,448, 48,127, and 48,817 km^2 in 2017, 2019, and 2021, respectively, exhibiting a slight increase overall.

(2) According to the statistics in Table 7 and Figure 17, small-scale and smaller-scale grassland areas increased and the larger-scale areas decreased slightly. However, the bare land area did not continue to increase as originally envisaged. A comprehensive comparison showed that the grassland coverage in the study area was directly or indirectly affected by unique climate factors.

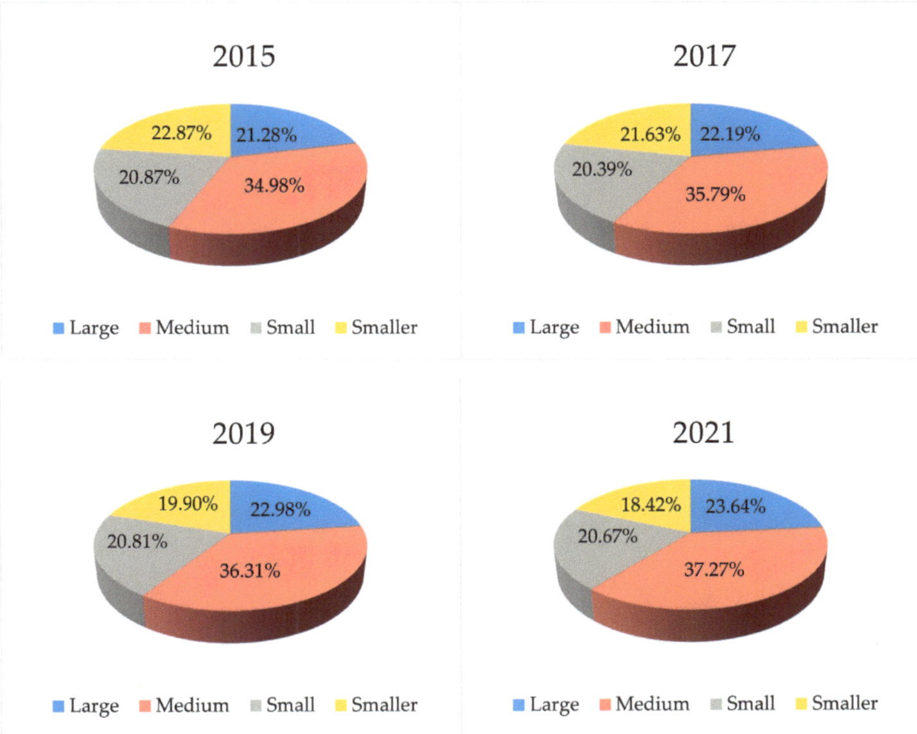

Figure 15. Changes in proportion of lake areas at different scales.

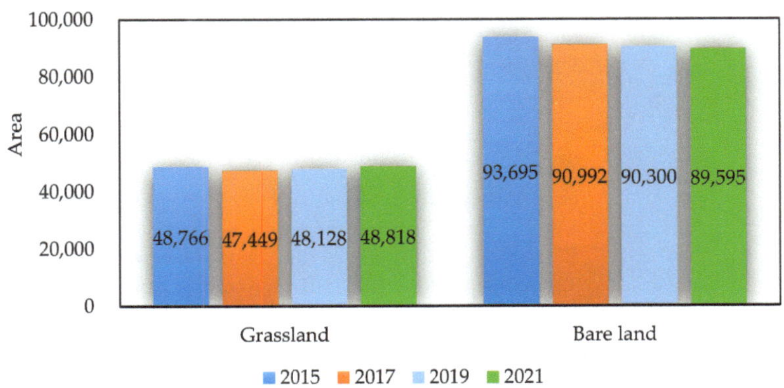

Figure 16. The changes to grassland and bare land areas.

Table 7. The changes to bare land areas at different scales.

	Large	Medium	Small	Smaller
2015	95.13%	1.59%	0.98%	2.30%
2017	95.34%	1.61%	0.97%	2.08%
2019	95.29%	1.69%	1.01%	2.01%
2021	95.41%	1.64%	0.96%	1.99%

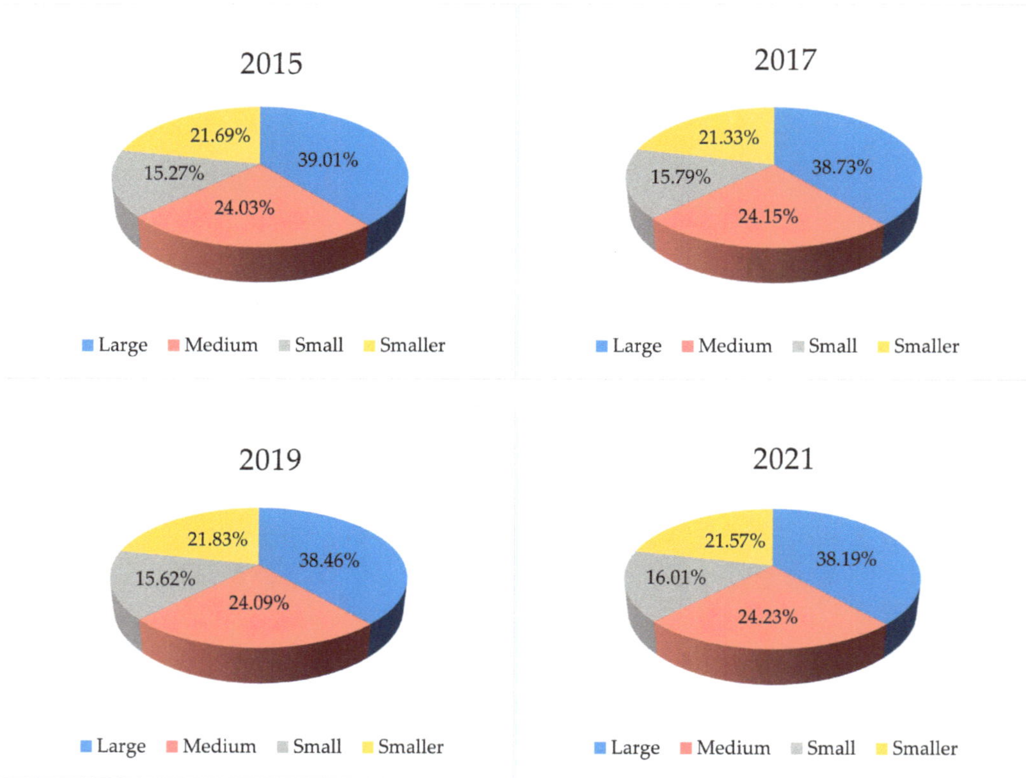

Figure 17. The changes to grassland areas at different scales.

4.2. Dynamic Effect

Global climate change affects the moisture and energy exchange between atmosphere and earth systems and finally influences the distribution of eco-environment elements [80,81]. As an individual whole, the study area is frequently involved in complex material circulation and energy flow with the outside world. The headwaters of the Yangtze River are a community of life. The system scientifically guides mankind to explore the endogenous relationship between human beings and the natural environment in a highly condensed and vividly illustrative way. Therefore, this paper attempts to analyze the dynamic effects of this region from a systematic, holistic, and multiscale perspective.

4.2.1. Systematic

The community of life in the headwaters of the Yangtze River has a distinct gradation. The indicators of different dimensions can be divided into the indicator layer, the criterion layer, and the target layer from low to high, and there is a clear logical relationship between different levels, i.e., the dynamic systematic effects [82]. As shown in Figure 18, the individual eco-environment element is the criterion layer of the system, and it reflects a certain aspect of the development of the community of life in the headwaters of the Yangtze River, which is represented by the target layer [83]. Ultimately, the comprehensive eco-environment elements reflect the overall situation of the region. The indicator layer mainly includes temperature, extreme climate, and precipitation efficiency. There are differences in the location, function, and other attributes of different elements in the eco-environment, and there are also differences in the basic characteristics, such as the structure and function of different types of eco-environment.

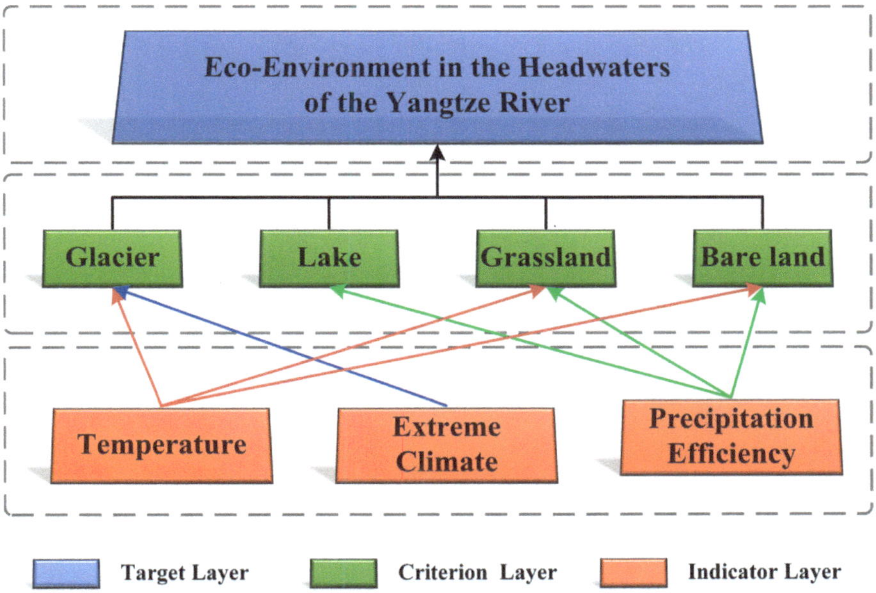

Figure 18. The systematic indicators of the headwaters of the Yangtze River.

4.2.2. Holistic

There are relationships, interplay, and restrictions in eco-environment elements. Once the eco-environment is damaged to a certain extent, it will affect the normal operation of the whole eco-environment. From 2015 to 2021, as Table 8 shows (the unit of area is km^2), the areas of glacier and bare land reduced from 2818.99 to 2759.04 and from 93,695.39 to 89,595.21, respectively. Moreover, the areas of lake and grassland increased from 1424.37 to 1532.69 and from 46,766.02 to 48,817.83, respectively. The overall eco-environment elements are constantly transforming. Taking the glaciers as an example, changes in these areas can cause variation in the energy exchange between the atmosphere and earth systems and the snowmelt runoff. The continuous melting of glaciers leads to an increase in runoff, which is conducive to the expansion of glacial-replenished lakes and grasslands. The increase in lakes has the most direct effect on the development of the surrounding grassland ecosystem. Furthermore, the changes in grasslands and bare land can affect water retention and groundwater storage in soil. As shown in Figure 19, the holistic of dynamic effects in the headwaters of the Yangtze River are ultimately reflected in the interaction and mutual development of the overall eco-environment elements.

Table 8. The area changes in the different eco-environment elements.

	Glacier	Lake	Grassland	Bare Land
2015	2818.99	1424.37	46,766.02	93,695.39
2017	2804.91	1458.70	47,448.81	90,992.35
2019	2781.35	1495.61	48,127.65	90,300.16
2021	2759.04	1532.69	48,817.83	89,595.21

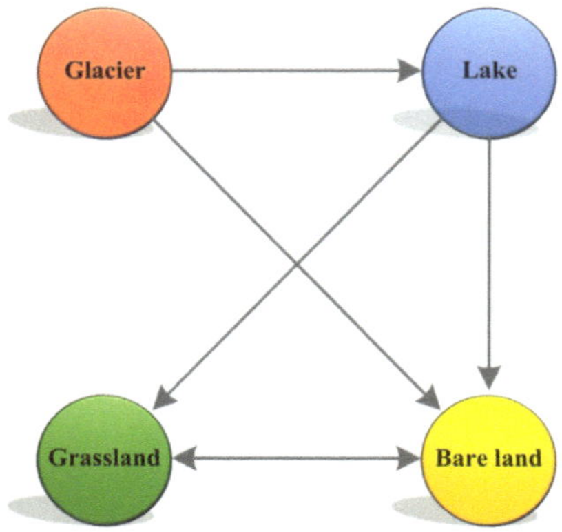

Figure 19. The holistic changes in the headwaters of the Yangtze River.

4.2.3. Multiscale

There are large differences in the structural complexity and functional diversity of ecosystems at different scales. The change rates of various eco-environment elements in the headwaters of the Yangtze River at different scales in the three time periods are shown in Figure 20. The area change rate of each scale element in the three periods exhibits a similar trend. With the increase in the change rate of the smaller-scale glaciers, the smaller-scale lakes, the grasslands, and bare land generally increased. Thus, it can be concluded that the changes in areas of glaciers, lakes, and grasslands were consistent in the smaller-scale. This also appears in small-scale, medium-scale, and large-scale eco-environment element changes. There are differences in the position, function, and other attributes of the eco-environment elements on different scales in the ecosystem, and the results of their interaction also differ.

4.3. Ecological Stress

The ecological stress in the headwaters of the Yangtze River represents a natural disturbance to the structure and function of the ecosystem caused by human action, and this direct or indirect interference often exceeds the scope of ecological rehabilitation [84]. In this paper, the ecological landscape was introduced in order to scientifically define ecological stress [85,86]. Ecologically, the landscape is a spatially heterogeneous region that repeatedly appears in a similar form, and it is a natural complex with classification significance [87]. Therefore, the different eco-environment elements are also the different landscapes that have different continuities in terms of spatial structural features.

With the intensification of global warming and the increasing intensity of the Indian summer monsoon, the climate in the headwaters of the Yangtze River exhibits a warm and humid trend. Therefore, the eco-environment in this area has experienced drastic changes. Specifically, rising temperatures lead to accelerating glacial melting, a reduction in permafrost, fluctuations in the number and areas of small lakes, and serious desertification of the land. As Figure 21 shows, as time goes on, although the trends of the two differ, the *EF* and *EIC* keep increasing, which indicates that, during to the combined effect of nature and humans, the eco-environment elements are oriented towards complex, heterogeneous, and discontinuous processes (*EF* and *EIC* are non-dimensional). Eco-environment element fragmentation is serious, and the community of life has a relatively low level of connectivity between the eco-environment elements.

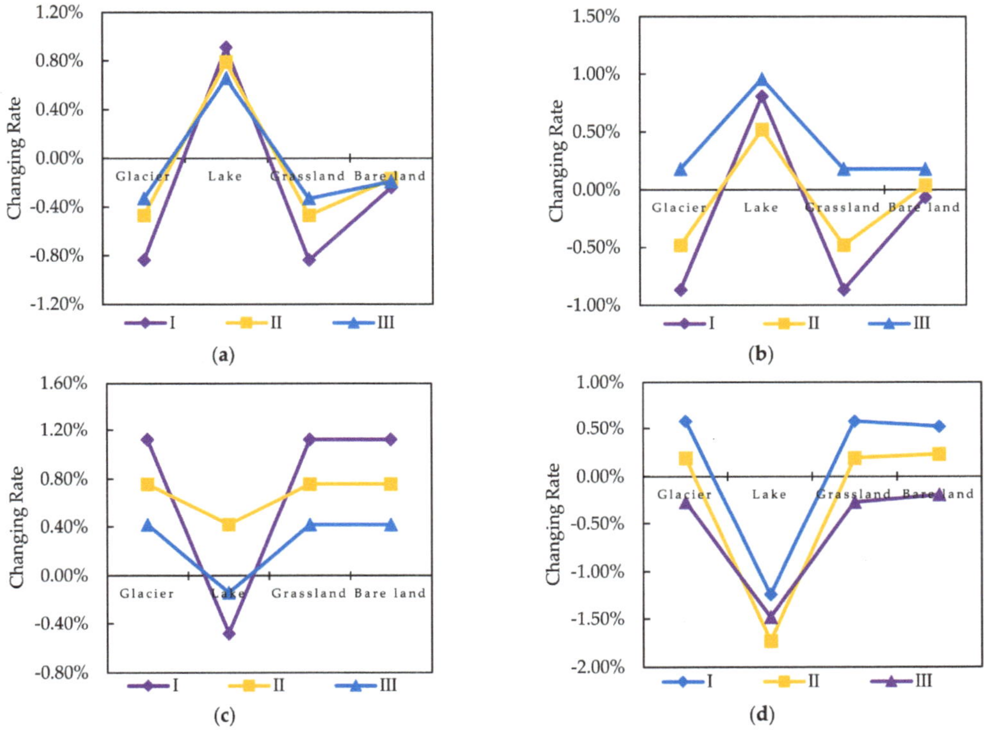

Figure 20. The multiscale change rate of the different eco-environment elements. (**a**) Large scale; (**b**) medium scale; (**c**) small scale; (**d**) smaller scale.

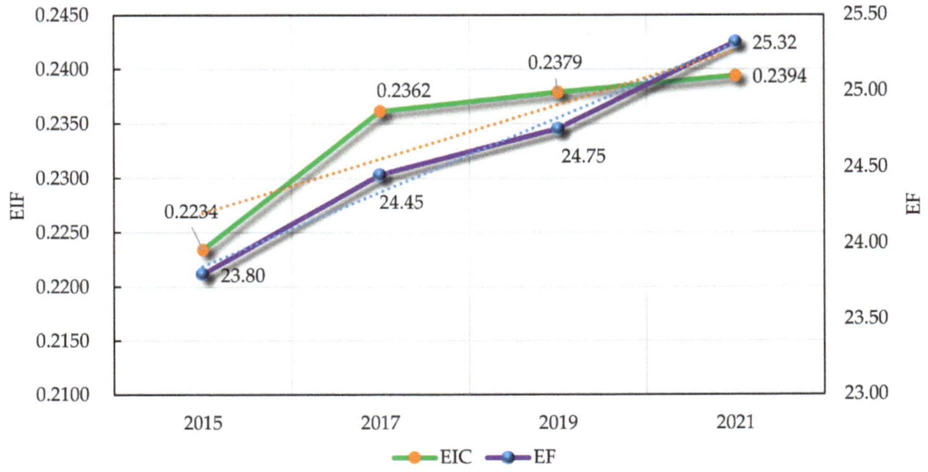

Figure 21. The changes in the *EF* and *EIC* in the headwaters of the Yangtze River.

5. Conclusions

The headwaters of the Yangtze River are a complicated system composed of different eco-environment elements, and there is constant multilevel and multiscale material circulation and energy exchange in such elements. On the basis of a series of studies on the dynamic effects of ecological stress on eco-environment elements, we came to the following main conclusions:

(1) In the process of eco-environment element identification, the improved DeepLab V3+ network was used to efficiently identify glaciers, lakes, grasslands, and bare land elements on the dataset established by "Sentinel-2" remote sensing images in the headwaters of the Yangtze River. The *mAP*, *mIoU*, and *Kappa* of the improved DeepLab V3+ method were 0.639, 0.778, and 0.825, respectively, which demonstrate a good ability to distinguish eco-environment elements;
(2) We propose using the *EF* and *EIC* to calculate the connectivity between eco-environment elements against the background of change and transformation. Between 2015 and 2021, *EF* gradually increased from 0.2234 to 0.2394, and *EIC* increased from 23.80 to 25.32, which indicates that the study area has a relatively low level of eco-environment elements connectivity. The eco-environment is oriented towards complex, heterogeneous, and discontinuous processes;
(3) As a community of life, the study area is frequently involved in a complex material circulation and energy flow with the outside world. The eco-environment elements in the headwaters of the Yangtze River are a systematic, holistic, and multiscale whole within a constantly transforming system, and each of them is universally connected.

Author Contributions: Conceptualization, C.W.; Formal analysis, L.C.; Methodology, C.W.; Project administration, R.Z.; Software, L.C.; Supervision, R.Z.; Validation, C.W. and L.C.; Writing—original draft, C.W. All authors have read and agreed to the published version of the manuscript.

Funding: This research received no external funding.

Data Availability Statement: Due to the nature of this research, participants of this study did not agree for their data to be shared publicly, so supporting data is not available.

Conflicts of Interest: The authors declare no conflict of interest.

References

1. Yue, W.Z.; Wang, T.Y. Rethinking on the Basic Issues of Territorial and Spatial Use Control in China. *China Land Sci.* **2019**, *33*, 8–15.
2. Gao, H.; Feng, Z.; Zhang, T.; Wang, Y.; He, X.; Li, H.; Pan, X.; Ren, Z.; Chen, X.; Zhang, W.; et al. Assessing glacier retreat and its impact on water resources in a headwater of Yangtze River based on CMIP6 projections. *Sci. Total Environ.* **2021**, *765*, 142774. [CrossRef] [PubMed]
3. Das, S.; Fime, A.A.; Siddique, N.; Hashem, M.M.A. Estimation of Road Boundary for Intelligent Vehicles Based on DeepLabV3+ Architecture. *IEEE Access* **2021**, *9*, 121060–121075. [CrossRef]
4. Chen, B.; Zhang, X.; Tao, J.; Wu, J.; Wang, J.; Shi, P.; Zhang, Y.J.; Yu, C. The impact of climate change and anthropogenic activities on alpine grassland over the Qinghai-Tibet Plateau. *Agric. For. Meteorol.* **2014**, *189–190*, 11–18. [CrossRef]
5. Cheng, G.D.; Wu, T.H. Responses of permafrost to climate change and their environmental significance, Qinghai-Tibet Plateau. *J. Geophys. Res. Earth Surf.* **2007**, *112*, F02S03. [CrossRef]
6. Liu, J.; Chen, J.; Xu, J.; Lin, Y.; Yuan, Z.; Zhou, M. Attribution of runoff variation in the headwaters of the Yangtze River based on the Budyko hypothesis. *Int. J. Environ. Res. Public Health* **2019**, *16*, 2506. [CrossRef]
7. Wang, J.; Schweizer, D.; Liu, Q. Three-dimensional landslide evolution model at the Yangtze River. *Eng. Geol.* **2021**, *292*, 106275. [CrossRef]
8. Du, S.; Du, S.; Liu, B.; Zhang, X. Incorporating DeepLabv3+ and object-based image analysis for semantic segmentation of very high resolution remote sensing images. *Int. J. Digit. Earth* **2021**, *14*, 357–378. [CrossRef]
9. Guo, B.; Luo, W.; Wang, D.L.; Jiang, L. Spatial and temporal change patterns of freeze-thaw erosion in the three-river source region under the stress of climate warming. *J. Mt. Sci.* **2017**, *14*, 1086–1099. [CrossRef]
10. Makarevich, K.G. Hydrological aspects of the glacier regime in the north tien shan in the anomalously arid period of 1974–1978. *Int. Assoc. Hydrol. Sci. Publ.* **1982**, *138*, 43–50.
11. Mellit, A.; Kalogirou, S. Artificial intelligence and internet of things to improve efficacy of diagnosis and remote sensing of solar photovoltaic systems: Challenges, recommendations and future directions. *Renew. Sust. Energy Rev.* **2021**, *143*, 110889. [CrossRef]

12. Haefner, N.; Wincent, J.; Parida, V. Artificial intelligence and innovation management: A review, framework, and research agenda. *Technol. Forecast Soc.* **2021**, *162*, 120392. [CrossRef]
13. Lary, D.J.; Alavi, A.H.; Gandomi, A.H. Machine learning in geosciences and remote sensing. *Geosci. Front.* **2016**, *7*, 3–10. [CrossRef]
14. Bradley, B.A.; Mustard, J.F. Characterizing the landscape dynamics of an invasive plant and risk of invasion using remote sensing. *Ecol. Appl.* **2006**, *16*, 1132–1147. [CrossRef]
15. Jiang, C.; Zhang, L.B. Climate Change and Its Impact on the Eco-Environment of the Three-Rivers Headwater Region on the Tibetan Plateau, China. *Int. J. Environ. Res. Public Health* **2015**, *12*, 12057–12081. [CrossRef]
16. Menz, M.H.M.; Dixon, K.W.; Hobbs, R.J. Hurdles and opportunities for landscape-scale restoration. *Science* **2013**, *339*, 526–527. [CrossRef] [PubMed]
17. O'Neill, R.V.; Hunsaker, C.T.; Jones, K.B.; Riitters, K.H.; Wickham, J.D.; Schwartz, P.M.; Goodman, I.A.; Jackson, B.L.; Baillargeon, W.S. Monitoring environmental quality at the landscape scale: Using landscape indicators to assess biotic diversity, watershed integrity, and landscape stability. *BioScience* **1997**, *47*, 513–519. [CrossRef]
18. Han, P.; Long, D.; Han, Z.; Du, M.; Dai, L.; Hao, X. Improved understanding of snowmelt runoff from the headwaters of China's Yangtze River using remotely sensed snow products and hydrological modeling. *Remote Sens. Environ.* **2019**, *224*, 44–59. [CrossRef]
19. Guo, M.; Li, J.; Wang, Y.; Bai, P.; Wang, J. Distinguishing the relative contribution of environmental factors to runoff change in the headwaters of the Yangtze River. *Water* **2019**, *11*, 1432. [CrossRef]
20. He, C.; Zhao, Y.; Huang, Q.; Zhang, Q.; Zhang, D. Alternative future analysis for assessing the potential impact of climate change on urban landscape dynamics. *Sci. Total Environ.* **2015**, *532*, 48–60. [CrossRef]
21. Baguette, M.; Blanchet, S.; Legrand, D.; Stevens, V.M.; Turlure, C. Individual dispersal, landscape connectivity and ecological networks. *Biol. Rev.* **2013**, *88*, 310–326. [CrossRef] [PubMed]
22. Mitchell, M.G.E.; Bennett, E.M.; Gonzalez, A. Linking landscape connectivity and ecosystem service provision: Current knowledge and research gaps. *Ecosystems* **2013**, *16*, 894–908. [CrossRef]
23. Ahmed, N.; Wang, G.X.; Oluwafemi, A.; Munir, S.; Hu, Z.Y.; Shakoor, A.; Imran, M.A. Temperature trends and elevation dependent warming during 1965–2014 in headwaters of Yangtze River, Qinghai Tibetan Plateau. *J. Mt. Sci.* **2020**, *17*, 556–571. [CrossRef]
24. Sang, Y.F.; Wang, Z.; Liu, C.; Gong, T. Temporal–Spatial Climate Variability in the Headwater Drainage Basins of the Yangtze River and Yellow River, China. *J. Clim.* **2013**, *26*, 5061–5071. [CrossRef]
25. Ahmed, N.; Wang, G.; Booij, M.J.; Oluwafemi, A.; Hashmi, M.Z.-U.; Ali, S.; Munir, S. Climatic variability and periodicity for upstream sub-basins of the Yangtze River, China. *Water* **2020**, *12*, 842. [CrossRef]
26. Coulon, A.; Cosson, J.F.; Angibault, J.M.; Cargnelutti, B.; Galan, M.; Morellet, N.; Petit, E.; Aulagnier, S.; Hewison, A.J.M. Landscape connectivity influences gene flow in a roe deer population inhabiting a fragmented landscape: An individual-based approach. *Mol. Ecol.* **2004**, *13*, 2841–2850. [CrossRef]
27. Su, H.; Peng, Y.; Xu, C.; Feng, A.; Liu, T. Using improved DeepLabv3+ network integrated with normalized difference water index to extract water bodies in Sentinel-2A urban remote sensing images. *J. Appl. Remote Sens.* **2021**, *15*, 018504. [CrossRef]
28. Landi, D.; Michele, G.; Marco, M. Analyzing the environmental sustainability of glass bottles reuse in an Italian wine consortium. *Procedia CIRP* **2019**, *80*, 399–404. [CrossRef]
29. Silvestri, L.; Forcina, A.; Di Bona, G.; Silvestri, C. Circular economy strategy of reusing olive mill wastewater in the ceramic industry: How the plant location can benefit environmental and economic performance. *J. Clean. Prod.* **2021**, *326*, 129388. [CrossRef]
30. Wang, G.X.; Wang, Y.B.; Kubota, J. Land-Cover Changes and Its Impacts on Ecological Variables in the Headwaters Area of the Yangtze River, China. *Environ. Monit. Assess.* **2006**, *120*, 361–385. [CrossRef]
31. Yao, Z.; Liu, Z.; Huang, H.; Liu, G.; Wu, S. Statistical estimation of the impacts of glaciers and climate change on river runoff in the headwaters of the Yangtze River. *Quat. Int.* **2014**, *336*, 89–97. [CrossRef]
32. Lovell, S.T.; Johnston, D.M. Designing landscapes for performance based on emerging principles in landscape ecology. *Ecol. Soc.* **2009**, *14*, 44. [CrossRef]
33. Liu, D.; Cao, C.; Dubovyk, O.; Tian, R.; Chen, W.; Zhuang, Q.; Zhao, Y.; Menz, G. Using fuzzy analytic hierarchy process for spatio-temporal analysis of eco-environmental vulnerability change during 1990–2010 in Sanjiangyuan region, China. *Ecol. Indic.* **2017**, *73*, 612–625. [CrossRef]
34. Garajeh, M.K.; Malakyar, F.; Weng, Q.H.; Feizizadeh, B.; Blaschke, T.; Lakes, T. An automated deep learning convolutional neural network algorithm applied for soil salinity distribution mapping in Lake Urmia, Iran. *Sci. Total Environ.* **2021**, *778*, 146253. [CrossRef]
35. Kussul, N.; Lavreniuk, M.; Skakun, S.; Shelestov, A. Deep Learning Classification of Land Cover and Crop Types Using Remote Sensing Data. *IEEE Geosci. Remote Sens. Lett.* **2017**, *14*, 778–782. [CrossRef]
36. Maggiori, E.; Tarabalka, Y.; Charpiat, G.; Alliez, P. Convolutional Neural Networks for Large-Scale Remote-Sensing Image Classification. *IEEE Trans. Geosci. Remote Sens.* **2017**, *55*, 645–657. [CrossRef]
37. Benediktsson, J.A.; Pesaresi, M.; Arnason, K. Classification and feature extraction for remote sensing images from urban areas based on morphological transformations. *IEEE Trans. Geosci. Remote Sens.* **2003**, *41*, 1940–1949. [CrossRef]

38. Fauvel, M.; Benediktsson, J.A.; Chanussot, J.; Sveinsson, J.R. Spectral and Spatial Classification of Hyperspectral Data Using SVMs and Morphological Profiles. *IEEE Trans. Geosci. Remote Sens.* **2008**, *46*, 3804–3814. [CrossRef]
39. Cao, Y.; Zhang, W.; Bai, X.; Chen, K. Detection of excavated areas in high-resolution remote sensing imagery using combined hierarchical spatial pyramid pooling and VGGNet. *Remote Sens. Lett.* **2021**, *12*, 1269–1280. [CrossRef]
40. Xie, J.; He, N.; Fang, L.; Plaza, A. Scale-free convolutional neural network for remote sensing scene classification. *IEEE Trans. Geosci. Remote Sens.* **2019**, *57*, 6916–6928. [CrossRef]
41. Ocer, N.E.; Kaplan, G.; Erdem, F.; Kucuk Matci, D.; Avdan, U. Tree extraction from multi-scale UAV images using Mask R-CNN with FPN. *Remote Sens. Lett.* **2020**, *11*, 847–856. [CrossRef]
42. Wang, C.; Chang, L.; Zhao, L.; Liu, R. Automatic Identification and Dynamic Monitoring of Open-Pit Mines Based on Improved Mask R-CNN and Transfer Learning. *Remote Sens.* **2020**, *12*, 3474. [CrossRef]
43. Francis, L.M.; Sreenath, N. Live detection of text in the natural environment using convolutional neural network. *Future Gen. Comput. Syst.* **2019**, *98*, 444–455. [CrossRef]
44. Mao, T.X.; Wang, G.X.; Zhang, T. Impacts of Climatic Change on Hydrological Regime in the Three-River Headwaters Region, China, 1960–2009. *Water Resour. Manag.* **2016**, *30*, 115–131. [CrossRef]
45. Jiang, L.G.; Yao, Z.J.; Liu, Z.F.; Wang, R.; Wu, S. Hydrochemistry and its controlling factors of rivers in the source region of the Yangtze River on the Tibetan Plateau. *J. Geochem. Explor.* **2015**, *155*, 76–83. [CrossRef]
46. Yang, J.P.; Ding, Y.J.; Chen, R.S. Causes of glacier change in the source regions of the Yangtze and Yellow rivers on the Tibetan Plateau. *J. Glaciol.* **2003**, *49*, 539–546.
47. Wu, W.H.; Yang, J.D.; Xu, S.J.; Yin, H. Geochemistry of the headwaters of the Yangtze River, Tongtian He and Jinsha Jiang: Silicate weathering and CO_2 consumption. *Appl. Geochem.* **2008**, *23*, 3712–3727. [CrossRef]
48. Gascon, F.; Bouzinac, C.; Thépaut, O.; Jung, M.; Francesconi, B.; Louis, J.; Lonjou, V.; Lafrance, B.; Massera, S.; Gaudel-Vacaresse, A.; et al. Copernicus Sentinel-2A Calibration and Products Validation Status. *Remote Sens.* **2017**, *9*, 584. [CrossRef]
49. Naegeli, K.; Damm, A.; Huss, M.; Wulf, H.; Schaepman, M.; Hoelzle, M. Cross-Comparison of Albedo Products for Glacier Surfaces Derived from Airborne and Satellite (Sentinel-2 and Landsat 8) Optical Data. *Remote Sens.* **2017**, *9*, 110. [CrossRef]
50. Williamson, A.G.; Banwell, A.F.; Willis, I.C.; Arnold, N.S. Dual-satellite (Sentinel-2 and Landsat 8) remote sensing of supraglacial lakes in Greenland. *Cryosphere* **2018**, *12*, 3045–3065. [CrossRef]
51. Diakogiannis, F.I.; Waldner, F.; Caccetta, P.; Wu, C. ResUNet-a: A deep learning framework for semantic segmentation of remotely sensed data. *ISPRS J. Photogramm. Remote Sens.* **2020**, *162*, 94–114. [CrossRef]
52. Frohn, R.C.; Autrey, B.C.; Lane, C.R.; Reif, M. Segmentation and object-oriented classification of wetlands in a karst Florida landscape using multi-season Landsat-7 ETM+ imagery. *Int. J. Remote Sens.* **2021**, *32*, 1471–1489. [CrossRef]
53. Zhong, Y.F.; Zhu, Q.Q.; Zhang, L.P. Scene Classification Based on the Multifeature Fusion Probabilistic Topic Model for High Spatial Resolution Remote Sensing Imagery. *IEEE Trans. Geosci. Remote Sens.* **2015**, *53*, 6207–6222. [CrossRef]
54. Chen, L.C.; Zhu, Y.K.; Papandreou, G. Encoder-Decoder with Atrous Separable Convolution for Semantic Image Segmentation. In Proceedings of the 2018 European Conference on Computer Vision (ECCV), Munich, Germany, 8–14 September 2018; pp. 801–818.
55. Wu, D.; Yin, X.; Jiang, B.; Jiang, M.; Li, Z.; Song, H. Detection of the respiratory rate of standing cows by combining the Deeplab V3+ semantic segmentation model with the phase-based video magnification algorithm. *Biosyst. Eng.* **2020**, *192*, 72–89. [CrossRef]
56. Peng, H.; Xue, C.; Shao, Y.; Chen, K.; Xiong, J.; Xie, Z.; Zhang, L. Semantic segmentation of litchi branches using DeepLabV3+ model. *IEEE Access* **2020**, *8*, 164546–164555. [CrossRef]
57. He, H.; Yang, D.; Wang, S.; Wang, S.; Li, Y. Road Extraction by Using Atrous Spatial Pyramid Pooling Integrated Encoder-Decoder Network and Structural Similarity Loss. *Remote Sens.* **2019**, *11*, 1015. [CrossRef]
58. Krizhevsky, A.; Sutskever, I.; Hinton, G.E. ImageNet Classification with Deep Convolutional Neural Networks. *Commun. ACM* **2017**, *60*, 84–90. [CrossRef]
59. Barmpoutis, P.; Stathaki, T.; Dimitropoulos, K.; Grammalidis, N. Early Fire Detection Based on Aerial 360-Degree Sensors, Deep Convolution Neural Networks and Exploitation of Fire Dynamic Textures. *Remote Sens.* **2020**, *12*, 3177. [CrossRef]
60. Chen, L.C.; Papandreou, G.; Kokkinos, I.; Murphy, K.; Yuille, A.L. DeepLab: Semantic Image Segmentation with Deep Convolutional Nets, Atrous Convolution, and Fully Connected CRFs. *IEEE Trans. Pattern Anal. Mach. Intell.* **2018**, *40*, 834–848. [CrossRef]
61. Zhan, Z.Q.; Zhang, X.M.; Liu, Y.; Sun, X.; Pang, C.; Zhao, C.B. Vegetation Land Use/Land Cover Extraction from High-Resolution Satellite Images Based on Adaptive Context Inference. *IEEE Access* **2020**, *8*, 21036–21051. [CrossRef]
62. Romera, E.; Alvarez, J.M.; Bergasa, L.M.; Arroyo, R. ERFNet: Efficient Residual Factorized ConvNet for Real-Time Semantic Segmentation. *IEEE Trans. Intell. Transp. Syst.* **2018**, *19*, 263–272. [CrossRef]
63. Volpi, M.; Tuia, D. Dense Semantic Labeling of Subdecimeter Resolution Images with Convolutional Neural Networks. *IEEE Trans. Geosci. Remote Sens.* **2017**, *55*, 881–893. [CrossRef]
64. Ghorbanzadeh, O.; Blaschke, T.; Gholamnia, K.; Meena, S.R.; Tiede, D.; Aryal, J. Evaluation of Different Machine Learning Methods and Deep-Learning Convolutional Neural Networks for Landslide Detection. *Remote Sens.* **2019**, *11*, 196. [CrossRef]
65. Devalla, S.K.; Chin, K.S.; Mari, J.M.; Tun, T.A.; Strouthidis, N.G.; Aung, T.; Thiéry, A.H.; Girard, M.J.A. A Deep Learning Approach to Digitally Stain Optical Coherence Tomography Images of the Optic Nerve Head. *Investig. Ophthalmol. Vis. Sci.* **2018**, *59*, 63–74. [CrossRef] [PubMed]

66. Gorelick, R. Combining richness and abundance into a single diversity index using matrix analogues of Shannon's and Simpson's indices. *Ecography* **2006**, *29*, 525–530. [CrossRef]
67. Chao, A.; Shen, T.J. Nonparametric estimation of Shannon's index of diversity when there are unseen species in sample. *Environ. Ecol. Stat.* **2003**, *10*, 429–443. [CrossRef]
68. Fischer, J.; Lindenmayer, D.B. Landscape modification and habitat fragmentation: A synthesis. *Glob. Ecol. Biogeogr.* **2007**, *16*, 265–280. [CrossRef]
69. Leinster, T.; Cobbold, C.A. Measuring diversity: The importance of species similarity. *Ecology* **2012**, *93*, 477–489. [CrossRef]
70. Tscharntke, T.; Klein, A.M.; Kruess, A.; Steffan-Dewenter, I.; Thies, C. Landscape perspectives on agricultural intensification and biodiversity—ecosystem service management. *Ecol. Lett.* **2005**, *8*, 857–874. [CrossRef]
71. Huang, L.; Wu, X.; Peng, Q.; Yu, X. Depth Semantic Segmentation of Tobacco Planting Areas from Unmanned Aerial Vehicle Remote Sensing Images in Plateau Mountains. *J. Spectrosc.* **2021**, *2021*, 6687799. [CrossRef]
72. Wang, C.; Du, P.; Wu, H.; Li, J.; Zhao, C.; Zhu, H. A cucumber leaf disease severity classification method based on the fusion of DeepLabV3+ and U-Net. *Comput. Electron. Agric.* **2021**, *189*, 106373. [CrossRef]
73. Czajkowska, J.; Badura, P.; Korzekwa, S.; Płatkowska-Szczerek, A. Automated segmentation of epidermis in high-frequency ultrasound of pathological skin using a cascade of DeepLab v3+ networks and fuzzy connectedness. *Comput. Med. Imaging Graph.* **2022**, *95*, 102023. [CrossRef] [PubMed]
74. Akcay, O.; Kinaci, A.C.; Avsar, E.O.; Aydar, U. Semantic Segmentation of High-Resolution Airborne Images with Dual-Stream DeepLabV3+. *ISPRS Int. J. Geo-Inf.* **2022**, *11*, 23. [CrossRef]
75. Wang, W.J.; Su, C. Convolutional Neural Network-Based Pavement Crack Segmentation Using Pyramid Attention Network. *IEEE Access* **2020**, *8*, 206548–206558. [CrossRef]
76. Liu, F.Y.; Shen, C.H.; Lin, G.S.; Reid, I. Learning Depth from Single Monocular Images Using Deep Convolutional Neural Fields. *IEEE Trans. Pattern Anal. Mach. Intell.* **2016**, *38*, 2024–2039. [CrossRef] [PubMed]
77. Manukian, H.; Traversa, F.L.; Di Ventra, M. Accelerating deep learning with memcomputing. *Neural Netw.* **2019**, *110*, 1–7. [CrossRef]
78. Yeo, K.; Melnyk, I. Deep learning algorithm for data-driven simulation of noisy dynamical system. *J. Comput. Phys.* **2019**, *376*, 1212–1231. [CrossRef]
79. Tomczak, J.M. Learning Informative Features from Restricted Boltzmann Machines. *Neural Process. Lett.* **2016**, *44*, 735–750. [CrossRef]
80. Qin, B.Q.; Huang, Q. Evaluation of the climatic change impacts on the inland lake—A case study of Lake Qinghai, China. *Clim. Change* **1998**, *39*, 695–714. [CrossRef]
81. Xu, W.X.; Liu, X.D. Response of vegetation in the Qinghai-Tibet Plateau to global warming. *Chin. Geogr. Sci.* **2007**, *17*, 151–159. [CrossRef]
82. Ma, Z.; Zhou, L.; Yu, W.; Yang, Y.; Teng, H.; Shi, Z. Improving TMPA 3B43 V7 Data Sets Using Land-Surface Characteristics and Ground Observations on the Qinghai-Tibet Plateau. *IEEE Geosci. Remote Sens. Lett.* **2018**, *15*, 178–182. [CrossRef]
83. Sun, T.; Li, H.; Wu, K.; Chen, F.; Zhu, Z.; Hu, Z. Data-Driven Predictive Modelling of Mineral Prospectivity Using Machine Learning and Deep Learning Methods: A Case Study from Southern Jiangxi Province, China. *Minerals* **2020**, *10*, 102. [CrossRef]
84. Walter, J.; Jentsch, A.; Beierkuhnlein, C.; Kreyling, J. Ecological stress memory and cross stress tolerance in plants in the face of climate extremes. *Environ. Exp. Bot.* **2013**, *94*, 3–8. [CrossRef]
85. Chesson, P.; Huntly, N. The roles of harsh and fluctuating conditions in the dynamics of ecological communities. *Am. Nat.* **1997**, *150*, 519–553. [CrossRef]
86. Palmer, M.A.; Ambrose, R.F.; Poff, N.L. Ecological theory and community restoration ecology. *Restor. Ecol.* **1997**, *5*, 291–300. [CrossRef]
87. Sheriff, M.J.; Bell, A.; Boonstra, R.; Dantzer, B.; Lavergne, S.G.; McGhee, K.E.; MacLeod, K.J.; Winandy, L.; Zimmer, C.; Love, O.P. Integrating ecological and evolutionary context in the study of maternal stress. *Integr. Comp. Biol.* **2017**, *57*, 437–449. [CrossRef] [PubMed]

remote sensing

MDPI

Article

A Context Feature Enhancement Network for Building Extraction from High-Resolution Remote Sensing Imagery

Jinzhi Chen [1], Dejun Zhang [1,*], Yiqi Wu [1], Yilin Chen [2] and Xiaohu Yan [3]

[1] School of Computer Science, China University of Geosciences, Wuhan 430074, China; chenjinzhi@cug.edu.cn (J.C.); wuyq@cug.edu.cn (Y.W.)
[2] Hubei Key Laboratory of Intelligent Robot (Wuhan Institute of Technology), Wuhan 430205, China; yilinchen@wit.edu.cn
[3] School of Artificial Intelligence, Shenzhen Polytechnic, Shenzhen 518055, China; yanxiaohu@szpt.edu.cn
* Correspondence: zhangdejun@cug.edu.cn

Abstract: The complexity and diversity of buildings make it challenging to extract low-level and high-level features with strong feature representation by using deep neural networks in building extraction tasks. Meanwhile, deep neural network-based methods have many network parameters, which take up a lot of memory and time in training and testing. We propose a novel fully convolutional neural network called the Context Feature Enhancement Network (CFENet) to address these issues. CFENet comprises three modules: the spatial fusion module, the focus enhancement module, and the feature decoder module. First, the spatial fusion module aggregates the spatial information of low-level features to obtain buildings' outline and edge information. Secondly, the focus enhancement module fully aggregates the semantic information of high-level features to filter the information of building-related attribute categories. Finally, the feature decoder module decodes the output of the above two modules to segment the buildings more accurately. In a series of experiments on the WHU Building Dataset and the Massachusetts Building Dataset, our CFENet balances efficiency and accuracy compared to the other four methods we compared, and achieves optimality on all five evaluation metrics: PA, PC, F1, IoU, and FWIoU. This indicates that CFENet can effectively enhance and fuse buildings' low-level and high-level features, improving building extraction accuracy.

Keywords: CFENet; fully convolutional neural network; remote sensing images; building extraction

Citation: Chen, J.; Zhang, D.; Wu, Y.; Chen, Y.; Yan, X. A Context Feature Enhancement Network for Building Extraction from High-Resolution Remote Sensing Imagery. *Remote Sens.* **2022**, *14*, 2276. https://doi.org/10.3390/rs14092276

Academic Editor: Mohammad Awrangjeb

Received: 1 April 2022
Accepted: 6 May 2022
Published: 9 May 2022

1. Introduction

With the advancement of earth observation technology, the quality and quantity of high-resolution remote sensing data are constantly improving. The generation of high-resolution remote sensing images has made more convenient and detailed data sources available for its applications. Building extraction from high-resolution remote sensing images is a conversion process from data to information. In urban remote sensing, building extraction can be applied to urban and rural planning [1,2]. In surveying and mapping engineering, building extraction is widely used as a means of acquiring data [3,4]. At the same time, it is also an important part of the fields of precision agriculture and environmental monitoring [5–7]. However, the buildings in high-resolution remote sensing images have variable shapes, rich details, and inconsistent texture colors. The complexity of imaging conditions and spatial environment poses severe challenges to the automatic extraction of buildings [8,9].

The automated extraction of buildings from high-resolution images has been a challenging topic in the field of remote sensing image processing [10]. The ground object environment in high-resolution remote sensing images is far more complex than the target environment in general images. The proportion of targets to be segmented in remote sensing images is far less than that in general images. For the segmentation task of remote sensing images, the main difficulty lies in the changeable environment of the objects and the small and dense characteristics of the objects to be segmented. At present, building extraction methods can be broadly divided into two categories: the traditional building extraction method and the learning-based building extraction method.

The traditional building extraction methods mainly use hand-designed features such as texture features, geometric features, and spatial information features of buildings in remote sensing images to solve the problem of building extraction. Li et al. [11] utilized integrating saliency cues to extract buildings from remote sensing images in two steps. First, classical features (color, shadow, shape, etc.) were used to identify the initial candidate buildings, and then a significance estimation algorithm was used to filter out the final buildings. Chen et al. [12] first segmented the images to generate candidate buildings by an algorithm combining threshold watershed transformation and hierarchical merging, and then used their proposed edge regularity index and shadow line index to capture the characteristics of buildings, and finally they employed three classifiers to recognize the buildings in the images. Inglada et al. [13] used a machine learning method to perform pixel-level classification of high-resolution remote sensing images with different geometric features to distinguish buildings. Ding et al. [14] presented a building extraction model based on an image segmentation algorithm and geometric feature constraints combined with the Morphological Building Index (MBI). Katartzis et al. [15] used a semantic segmentation algorithm combined with the Markov model and edge corner detection to extract buildings. Awrangjeb et al. [16] proposed a threshold-based automatic evaluation system to segment buildings according to the correspondence between buildings.

The main drawback of the traditional building extraction method is that they heavily rely on the design and implementation of hand-crafted features and require some complex strategies to overcome noise and outlier variations, which increase computational costs and have poor generalization capabilities. Moreover, high-resolution remote sensing images have complex ground object information. It is difficult to accurately describe building information based only on manual-designed features, which results in low building extraction accuracy.

As a widely used deep learning architecture [17–19], combining remote sensing technology and deep learning has become a prevalent method in building extraction from high-resolution remote sensing images. The mainstream CNNs are VGGNet [20], ResNet [21], InceptionNet [22], DenseNet [23], AlexNet [24], etc. They have also become the commonly used baseline networks for many newly proposed segmentation models. Jonathan Long et al. [25] proposed the Fully Convolutional Network (FCN) for image pixel-level classification to overcome the limitations of CNN in fine image segmentation. The core difference between FCN and CNN is that FCN replaces the fully connected layer at the end of the CNN-based classification network with the convolution layer. After FCN was proposed, many researchers also proposed some improved image semantic segmentation models based on FCN, such as SegNet [26], DeconvNet [27], U-Net [28], DeepLab [29], PSPNet [30], HRNet [31], etc. Guo et al. [32] added an attention layer to U-Net and used multiple losses to extract complete buildings. Chen et al. [33] proposed a new neural network by combining DeepLabv3+Net with a densely connected convolutional neural network and residual structure. Shao et al. [34] introduced atrous convolution in U-Net to extract buildings and proposed a new residual refinement module to refine the extraction results further. Xu et al. [35] applied the guided filter to a deep neural network to improve the extraction accuracy of diverse objects in urban areas. Zhu et al. [36] proposed a multiple attending path network (MAP-NET) to extract building footprints by learning multi-scale features through multiple parallel paths, followed by applying an attention module and a pyramidal space pool module to obtain global dependencies. Liao et al. [37] combined building contour information and multi-scale semantic information to enhance the robustness of building extraction. Yang et al. [38] presented a novel encoder-decoder structure combining attention mechanism and DenseNets, which fuses features of buildings from different levels through a spatial attention module. Wen et al. [39] proposed an improved method based on the Mask Region Convolutional Neural Network (Mask R-CNN) to detect the boundaries of buildings in complex backgrounds. Zheng et al. [40] added a self-created symmetric information collection module in Deeplabv3+ to reduce noise in images for accurate segmentation of side-scan sonar remote sensing images. Yu et al. [41] proposed a method based on Transformer and YOLOv5 for real-time object detection on side-scan sonar remote sensing images.

Although these FCN-based methods have made considerable progress in the extraction accuracy of buildings, there are still some problems. First, high-resolution remote sensing images contain complex ground object information; different buildings have different shapes, inconsistent colors, and textures, making FCN-based networks fail to extract strong feature representations of buildings. Secondly, FCN-based networks often contain a large number of parameters, which causes the network model to occupy a large memory space during training and testing.

To sum up, the traditional building extraction methods rely excessively on manual-designed features, which are deficient in robustness and accuracy. The learning-based building extraction methods cannot effectively enhance and fuse low-level and high-level features from neural networks and often contain a large number of network structure parameters. In this paper, to better address the above issues, we design a Context Feature Enhancement Network (CFENet) that can learn more contextual information to achieve accurate localization segmentation of buildings. The contributions of this paper can be summarized as follows:

- An end-to-end context feature enhancement network, namely CFENet, is proposed to address the challenges of complexity and diversity of buildings encountered in building extraction from remote sensing images.
- CFENet achieves more accurate building extraction results on the WHU Building Dataset [42] and the Massachusetts Building Dataset [43] by explicitly establishing rich contextual relationships on low-level and high-level features.
- CFENet balances efficiency and accuracy by employing dilated convolution in the spatial fusion module and asymmetric convolution in the focus enhancement module.

The following sections are arranged as follows. Our CFENet is introduced in detail in Section 2. The experiments results of CFENet and the other four methods are analyzed in detail in Section 3, including comparative experiments and ablation experiments. Section 4 discusses the efficiency of different methods and the heatmaps verifying the effectiveness of our proposed modules. Finally, a summary of the entire text is presented in Section 5.

2. The Proposed Context Feature Enhancement Network

2.1. Model Description

Automated extraction of buildings from high-resolution remote sensing images poses serious challenges. The observation mentioned above motivates us to introduce an efficient end-to-end Context Feature Enhancement Network (CFENet), to learn more contextual information to achieve accurate localization segmentation of buildings. The pipeline of CFENet is shown in Figure 1.

CFENet adopts ResNet-101 as the backbone, and the input image size is 512×512 pixels. After performing feature extraction through ResNet-101, four different size feature maps are obtained, denoted as Feature 1, Feature 2, Feature 3, and Feature 4. Feature 1 and Feature 2 are low-level features, while Feature 3 and Feature 4 are high-level features. The low-level features contain larger spatial resolution and more detailed information (color, outline, texture, etc.) by employing fewer convolution layers and downsampling operations. However, the low-level features are not rich in semantic information and contain much noisy information. On the contrary, the high-level features have more abstract and rich semantic information (attributes, categories, etc.) by employing more convolution layers and downsampling operations. However, the spatial resolution of the high-level features is small, and spatial information loss is serious.

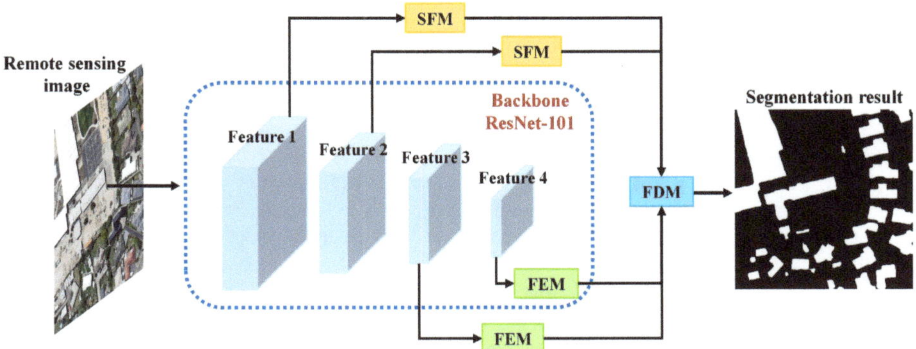

(**a**) An overall structure of CFENet.

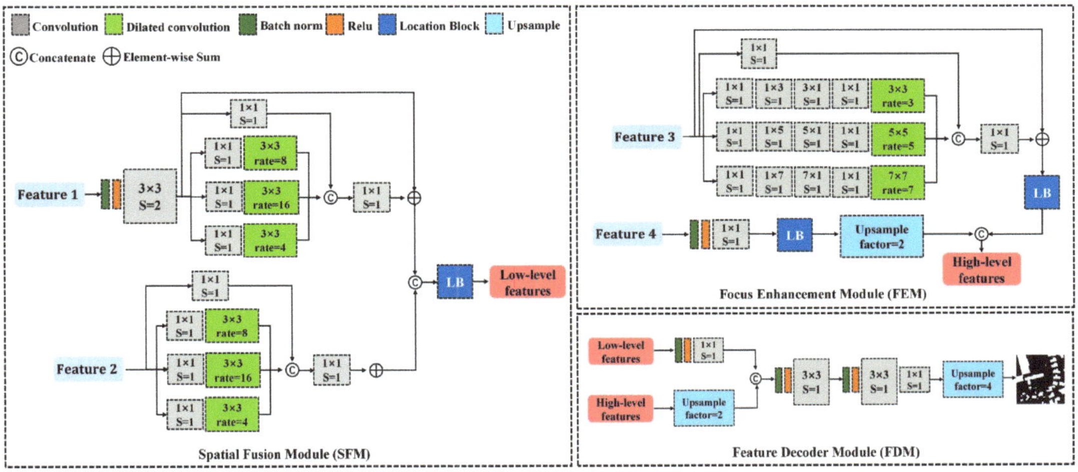

(**b**) Structure diagram of SFM, FEM, and FDM

Figure 1. Our network Context Feature Enhancement Network (CFENet) adopts a ResNet-101 as the backbone. It consists of three novel modules: Spatial Fusion Module (SFM), Focus Enhancement Module (FEM), and Feature Decoder Module (FDM). The input image size of our network is 512 × 512 pixels, and the outputs of the four stages of ResNet-101 are used as Feature 1, Feature 2, Feature 3, and Feature 4, respectively. Then Feature 1 and Feature 2 are fed into the SFM, and Feature 3 and Feature 4 are fed into the FEM. Finally, we input the FDM to obtain the final segmentation result. (**a**) An overall structure of CFENet. (**b**) Structure diagram of SFM, FEM, and FDM.

Our CFENet aggregates the spatial information of low-level features through the spatial fusion module to obtain buildings' outline and edge information. The focus enhancement module fully integrates the semantic information of the last two high-level features extracted from ResNet-101 to obtain the attribute category information of the building. After the feature encoding of Feature 1 and Feature 2 by the spatial fusion module and the feature encoding of Feature 3 and Feature 4 by the focus enhancement module, CFENet applies the feature decoder module to the outputs of the two modules and performs final decoding to segment accurate buildings.

2.2. Spatial Fusion Module

The proposed dilated convolution [44] significantly reduces the drawbacks associated with CNNs. Aiming at the problem of image resolution reduction and information loss caused by downsampling in the task of image semantic segmentation, dilated convolution proposes a new convolution idea. This idea expands the receptive field and reduces the number of network parameters. In this module, we superimpose dilated convolutions with different dilated rates to simulate different receptive fields according to the characteristics of dilated convolution and enhance the representational ability of features by concatenating the features of different receptive fields.

Because the low-level features are applied with fewer convolution layers and down-sampling operations, they usually contain greater spatial resolution and richer spatial information (color, outline, texture, etc.). To better extract the spatial information of buildings from low-level features, we increase the receptive field by using dilated convolutions with different dilation rates. The details of the spatial fusion module are illustrated in Figure 1b.

In our work, $\{F_i | i = \{1,2,3,4\}\}$ are the features of different stages extracted by the backbone network. F_1 and F_2 are respectively fed into the two branches, and then fed into the Location Block after concatenation to obtain enhanced low-level features f_{low}:

$$f_{low} = \Phi(C\{\delta_\theta(F_1), \varphi_\theta(F_2)\}) \quad (1)$$

where δ_θ represents the first branch of the spatial fusion module, φ_θ represents the second branch of the spatial fusion module. C represents concatenation operation, and Φ represents the Location Block.

The design of the spatial fusion module is based on multi-branch dilated convolution. The input of this module is Feature 1 (number of channels: 256, feature map size: 128 × 128) and Feature 2 (number of channels: 512, feature map size: 64 × 64). First, by using a 3 × 3 convolution layer with a step size of 2 to keep Feature 1 the same size as Feature 2, and then through a parallel branch, which has four branches $\{branch_k, k = 1,2,3,4\}$. A 1 × 1 convolution layer is used in each branch to reduce the number of channels in the feature map. When $k > 1$, a 3 × 3 dilated convolution layer is added after the 1 × 1 convolution layer in each branch, and the dilation ratio is set to 2^k. Subsequently, the output features of the four branches are aggregated through a concatenation operation, and then the dimensionality is reduced by a 1 × 1 convolution layer. Finally, the dimensionality reduction features are added element-by-element with the features before entering the branch. Like Feature 1, Feature 2 also needs to go through the same parallel branch. The difference is that Feature 2 does not need to perform any operations before entering the parallel branch. Finally, Feature 1 and Feature 2 are concatenated together, and then the concatenated features are fed into the Location Block for feature enhancement. The Location Block is described in detail in Section 2.3.

2.3. Location Block

In this paper, we propose a unit named Location Block (LB) based on the idea of the Dual Attention Network [45] combined with the channels Squeeze-Extraction [46] operation, which improves the performance of the network by enabling it to execute feature recalibration adaptively. The main function of Location Block is to enhance the essential features and weaken the non-essential features, so as to make the extracted features more powerfully representational. In addition, a parallel module enables the network to learn the mapping relationship between interdependent channels and the correlation degree between each pixel in the feature map. The architecture of the Location Block is shown in Figure 2.

The Location Block first performs a global pooling operation on the input features, followed by the first 1 × 1 convolution layer to squeeze the C channels into C/r channels to reduce the amount of computation, and the second 1 × 1 convolution layer is restored back to C channels, where r refers to the compression ratio, which is set to 16 in this paper. The next step is the reweight operation. The weight output by the sigmoid function is considered to the feature channels' importance after feature reselection. The weight is then multiplied channel-by-channel to the previous features, thus completing the recalibration of the original features in the channel dimension. We perform the above operations to model the feature channel-wise dependencies as completely as possible.

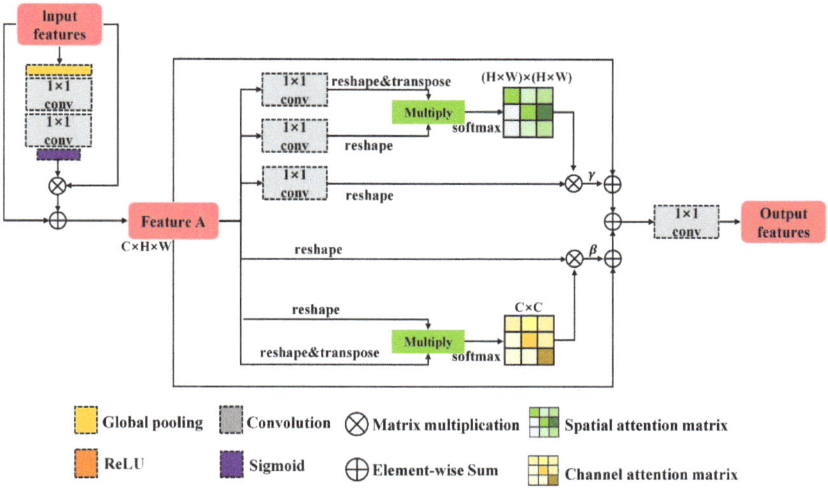

Figure 2. The architecture of Location Block.

After completing the above steps, we design two parallel branches to capture rich long-range contextual relationships in spatial and channel dimensions, in order to obtain better intra-class consistent feature representation. We first introduce the first branch. $A \in \mathbb{R}^{C\times H\times W}$ is the feature map of the input parallel module. First, feature A is converted into new features B, C, and D by the 1×1 convolution layer, where $\{B, C, D\} \in \mathbb{R}^{C\times H\times W}$. Then both feature maps B and C are reshaped to $\mathbb{R}^{C\times N}$, where $N = H \times W$. The transposed B is then multiplied by C and a softmax activation function is applied to obtain a spatial attention map $S \in \mathbb{R}^{N\times N}$:

$$s_{ji} = \frac{exp\left(B_i^T \cdot C_j\right)}{\sum_{i=1}^{N} exp\left(B_i^T \cdot C_j\right)} \tag{2}$$

where the impact of the jth pixel on the ith pixel denotes the value of s_{ji}. The more similar the feature representations of the two positions are, the greater their correlation; that is, the greater the value of s_{ji}.

At the same time, feature D is reshaped to $\mathbb{R}^{C\times N}$, multiplied by the transpose of S, and then reshaped back to $\mathbb{R}^{C\times H\times W}$. Finally, it is multiplied by a scale parameter γ and then employs element-wise summation with A to obtain the feature matrix map $E \in \mathbb{R}^{C\times H\times W}$:

$$E = \gamma \sum_{i=1}^{N} \left(S^T \cdot D_i\right) + A_j \tag{3}$$

where γ is a gradual learning scale parameter with an initial value of 0. The weighted summation of the features at all locations and the original features yields the feature matrix map E for each location. Therefore, the above branch can cluster rich contextual information on local features. It helps to facilitate semantic consistency and intra-class compactness of features.

For high-resolution remote sensing images, fully mining the interdependencies between channels can improve the representation ability of specific semantic information of features. Therefore, this branch aims to model the interdependencies between different channels by using a module capable of aggregating channel attention. Unlike the previous branch, no convolution layer is applied to the feature map A, but the associated operation is performed directly on the feature map A. Similarly, feature A is reshaped to $\mathbb{R}^{C\times N}$. Then the reshaped A is multiplied by the transpose of A, and a softmax activation function is applied to get the channel attention map $X \in \mathbb{R}^{C\times C}$:

$$x_{ji} = \frac{exp\left(A_i^T \cdot A_j\right)}{\sum_{i=1}^{C} exp\left(A_i^T \cdot A_j\right)} \tag{4}$$

where the impact of the jth pixel on the ith pixel denotes the value of x_{ji}. The transpose of X is multiplied by A and the result is reshaped to $\mathbb{R}^{C\times H\times W}$. After multiplying by a scale parameter β, it is summed with A element-by-element to obtain the feature matrix map E that fuses the channel information. β is a gradual learning scale parameter with an initial value of 0.

$$E = \beta \sum_{i=1}^{C} \left(X^T \cdot A_i\right) + A_j. \tag{5}$$

After the input features are processed separately by the two branches, element-wise summation is used to fuse the two feature maps, and a 1 × 1 convolution layer is used for dimensionality reduction to finally complete the enhancement of the input features.

2.4. Focus Enhancement Module

Asymmetric convolution is a means of compressing and accelerating networks. Previous works [22,47,48] have proved that 1 × n convolution plus n × 1 convolution can be used to replace n × n convolution to greatly reduce the amount of network parameters while ensuring accuracy. 1 × n convolution and n × 1 convolution can also explore more valuable information in the horizontal and vertical directions of features. In this module, asymmetric convolutions are used to reduce computational complexity and enable features to learn more valuable information in both horizontal and vertical directions.

In the focus enhancement module, F_3 and F_4 are respectively fed into the two branches of the focus enhancement module, and then fed into the Location Block respectively, and then concatenated to get the enhanced high-level feature f_{high}:

$$f_{high} = C\{\Phi(\psi_\theta(F_3)), \Phi(\xi_\theta(F_4))\} \tag{6}$$

where ψ_θ represents the first branch of the focus enhancement module, ξ_θ represents the second branch of the focus enhancement module, C represents concatenation operation, and Φ represents the Location Block.

The structure diagram of the focus enhancement module is shown in Figure 1b. The input of this module is Feature 3 (number of channels: 1024, feature map size: 32 × 32) and Feature 4 (number of channels: 2048, feature map size: 16 × 16) extracted by ResNet-101. We replace the original n x n convolution layer with a 1 × n convolution layer plus an n × 1 convolution layer to reduce the number of parameters and achieve deeper nonlinearity. Finally, we use the shortcut operation adopted by ResNet [21]. Feature 3 is first fed into a parallel branch. In each branch of the parallel branch, a 1 × 1 convolution layer is used first, followed by a 1 × n convolution layer, an n × 1 convolution layer, a 1 × 1 convolution layer, and an n × n dilated convolution layer, respectively. The dilation rate of dilated convolution in each branch is set to 3, 5, and 7, respectively. These hyperparameters are adjusted in the experiments according to the validation set.

After concatenating the outputs of the four branches, the number of feature channels is reduced to 48 by using a 1 × 1 convolution layer. Then the shortcut design from ResNet is applied, and the ReLU function is applied to obtain a new feature. The obtained new features are then input into the Location Block for feature enhancement. Meanwhile, Feature 4 is first applied with a convolution layer (followed by a BN layer and a ReLU activation function layer) for dimension reduction. Then the output is fed to the Location Block to obtain the enhanced feature. Finally, a 2-fold upsampling operation is performed to obtain the same size as Feature 3, and a concatenated operation is performed with the enhanced Feature 3 to obtain the final enhanced high-level feature.

We design multi-branch convolution layers with different kernels to simulate multi-scale receptive fields. The design principle of the focus enhancement module lies in that the high-level features have rich semantic information and larger receptive fields. Therefore, we only enhance Feature 3 and do not expand the receptive field for Feature 4. Meanwhile, we design the parallel branches with low dilation rates to prevent the gridding effect.

2.5. Feature Decoder Module

The structure of the feature decoder module is illustrated in Figure 1b. This module reduces the number of channels of low-level fusion features to 48, whereas the number of channels of high-level fusion features remains at 96. The goal is to keep the high-level features dominant while the low-level features play an auxiliary role. The reason is that the high-level features have abundant semantic information, which can better provide the category attribute information of buildings. Then the two features are concatenated together and passed sequentially through two 3 × 3 convolution layers (followed by a BN layer and a ReLU activation function layer). Ultimately, a 1 × 1 convolution layer is employed to reduce the number of feature map channels to 1, and it is upsampled four times to the same size as the input image and then fed into the sigmoid function to obtain the final building extraction result.

2.6. Loss Function

Cross-Entropy Loss is a frequently applied loss function in two-dimensional semantic segmentation tasks. The learning-based remote sensing building extraction task aims to train a binary classifier. The positive samples are pixels containing the building, whereas the negative samples are pixels containing the background. Because the building extraction task belongs to the two-class semantic segmentation task, we also employ Binary Cross-Entropy Loss (BCELoss) [49] in the training process. The calculation formula of BCELoss is defined as:

$$L_{bceloss} = -\frac{1}{N_{pixels}} \sum_{i=1}^{N_{pixels}} (y_i log p_i + (1 - y_i) log(1 - p_i)) \tag{7}$$

where N_{pixels} denotes the total number of pixels in the remote sensing image, y_i denotes the expected value of the ith pixel, and p_i denotes the predicted value of the ith pixel.

3. Experiments and Results

3.1. Evaluation Metrics

For our experiments, we adopted pixel accuracy (*PA*), precision (*PC*), *F*1 score (*F*1), intersection over union (*IoU*), and frequency weighted intersection over union (*FWIoU*) as evaluation metrics to evaluate the effectiveness of CFENet. The calculation formulas of these five evaluation metrics are defined as follows:

$$PA = \frac{TP + TN}{TP + FP + TN + FN} \tag{8}$$

$$PC = \frac{TP}{TP + FP} \tag{9}$$

$$F1 = 2 \times \frac{PC \times Recall}{PC + Recall} \tag{10}$$

$$IoU = \frac{TP}{TP + FN + FP} \tag{11}$$

$$FWIoU = \frac{TP + TN}{TP + FP + TN + FN} \times \frac{TP}{TP + FP + FN}. \tag{12}$$

For the remote sensing building extraction task, the positive sample is the building and the negative sample is the background. Where *TP* refers to the number of cases in which the true case is positive and the predicted case is also positive, and *TN* refers to the number of cases in which the true case is negative and the predicted case is also negative. *FP* refers to the number of cases in which the true case is the negative case and the predicted case is the positive case. *FN* refers to the number of cases in which the true case is the positive case and the predicted case is the negative case. The larger the value of these five evaluation indicators, the better the experimental effect.

3.2. Dataset and Implementation Details

WHU Building Dataset [42]: The dataset adopted for our experiments is the aerial image dataset of the WHU Building Dataset. The dataset covers a 450 km^2 area in Christchurch, New Zealand, which includes residential, industrial, cultural, and rural areas, and contains 187,000 buildings with different textures, shapes, and colors. This area is evenly divided into 8189 images of 512 × 512 pixels in tif format. The dataset for our experiment is divided randomly, and the training set, validation set, and test set contain 4736, 1036, and 2416 images, respectively. The original images and ground-truth values in the WHU Building Dataset are shown in Figure 3.

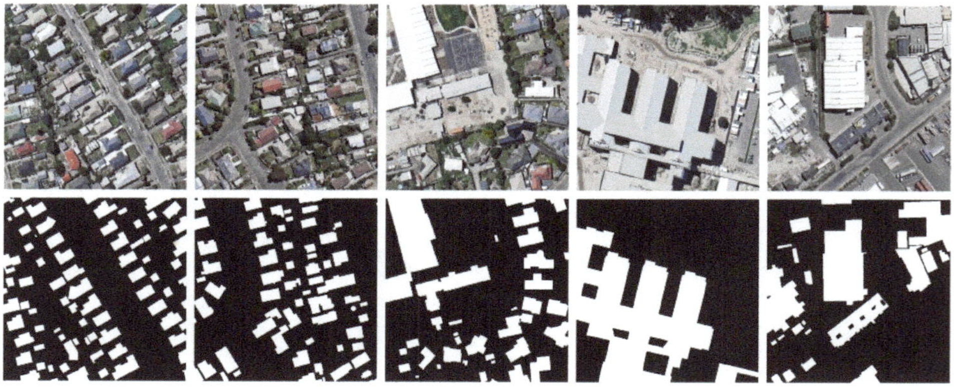

Figure 3. Some examples of the WHU Building Dataset. The first row is the original image and the second row is the ground truth.

Massachusetts Building Dataset [43]: The dataset contains 151 high-resolution aerial images of downtown and suburban Boston. The size of each image in the dataset is 1500 × 1500 pixels covering an area of 2250 × 2250 m^2. The dataset has 137 images in the training set, 10 images in the test set, and 4 images in the validation set. It is worth noting that the ground resolution of the images in this dataset is 1 m, which increases the difficulty of building extraction. When using the original images in this dataset, each image is evenly divided into several patches with a size of 256 × 256 pixels. The original images and ground-truth values in the Massachusetts Building Dataset are shown in Figure 4.

Implementation details: Our proposed method is implemented based on pytorch 1.7.0 and cuda 11.0, and the corresponding code is available from our community site (https://github.com/djzgroup/CFENet, accessed on 30 March 2022). An Adam optimizer is applied to train our network with a learning rate initialized to 0.0001, and then decayed to 0.1 of the current learning rate every 50 epochs. Our network is trained on the GPU for a number of 200 epochs.

(a) (b)

Figure 4. An example of the Massachusetts Building Dataset. (**a**) Original image; (**b**) Ground truth.

3.3. Comparative Experiments

3.3.1. Results on WHU Building Dataset

CFENet is compared with four methods, U-Net [28], Deeplabv3+ [50], PSPNet [30], and HRNet [31]. All these methods have good results in the semantic segmentation task. U-Net started as a semantic segmentation network applied to medical images, and due to its elegant network structure is widely used in semantic segmentation tasks of images in various fields. Deeplabv3+ is based on Deeplabv3, adding a simple and efficient decoder to refine the segmentation results. PSPNet reduces error segmentation by introducing more contextual and multi-scale information. HRNet is used for the representation of high-resolution images, which preserves the high-resolution representation by fusing features from different branches. HRNet has achieved great success in tasks such as semantic segmentation, image-level classification, and object detection. The performance of different methods on the WHU Building Dataset is shown in Table 1. Our method outperforms other methods on all five evaluation metrics. In particular, our method improves 0.0158 on PA, 0.1004 on PC, 0.0827 on F1, 0.0266 on FWIoU and even more on IoU by 0.1303 over U-Net which is currently used mainly for building extraction tasks. The experimental results of CFENet also show some improvement over HRNet which excels in semantic segmentation, especially the PC improvement of 0.0261.

We qualitatively assess the building extraction results of different methods on the WHU Building Dataset. In the visualization, green areas represent building pixels that are correctly predicted, red areas represent pixels that are actually background but incorrectly predicted as buildings, and blue areas represent pixels that are actually buildings but incorrectly predicted as background. As shown in Figure 5, our method shows better performance in dealing with small-sized and complex-shaped buildings, whereas the other four methods show more or less prediction bias. The buildings in the first row of Figure 6 are large buildings with rich textures and colors. Our method can extract buildings with richer details. The other four methods have the problem that the details of the extraction results are not prominent when extracting such buildings. The cistern shown in Figure 7a has a ground texture that is highly similar to the building footprint. Figure 7b,c are some containers where the surface texture is highly confused with the building. Our method can effectively identify accurate building footprints when dealing with such ground facilities that are easily confused with buildings, while the other four methods suffer from misidentification. Therefore, our method effectively solves the problems of inaccurate extraction results and insufficient detail of extracted buildings when extracting complex buildings.

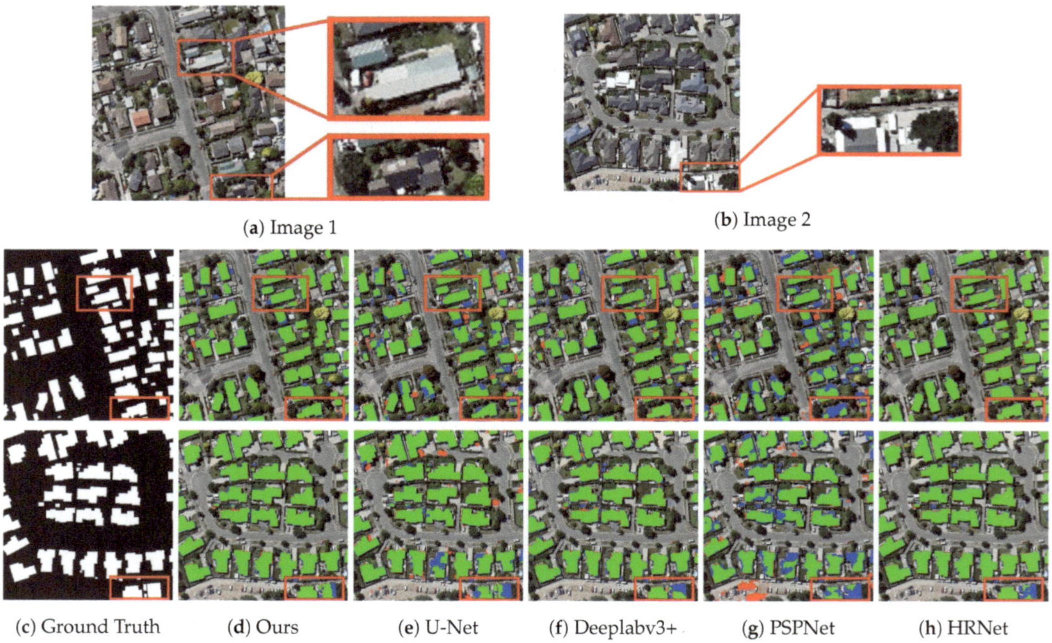

Figure 5. Results of different methods on buildings of small size and complex shapes. (**a**,**b**) are the original images; (**c**) ground truth; (**d**) the results of CFENet; (**e**–**h**) show the results of U-Net, Deeplabv3+, PSPNet and HRNet, respectively. Green, red, and blue pixels on the map represent predictions of true positives, false positives, and false negatives, respectively. The experimental results in the red box show the areas where the contrast effect is more obvious.

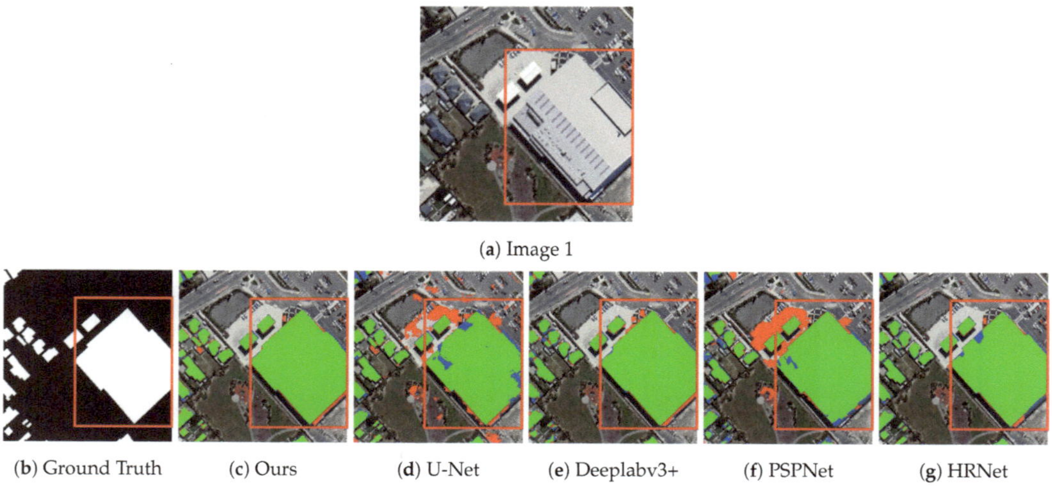

Figure 6. Results of different methods on large buildings with complex textures and colors. (**a**) Original image; (**b**) ground truth; (**c**) the results of our CFENet; (**d**–**g**) show the results of U-Net, Deeplabv3+, PSPNet, and HRNet, respectively. Green, red, and blue pixels on the map represent predictions of true positives, false positives, and false negatives, respectively. The experimental results in the red box show the areas where the contrast effect is more obvious.

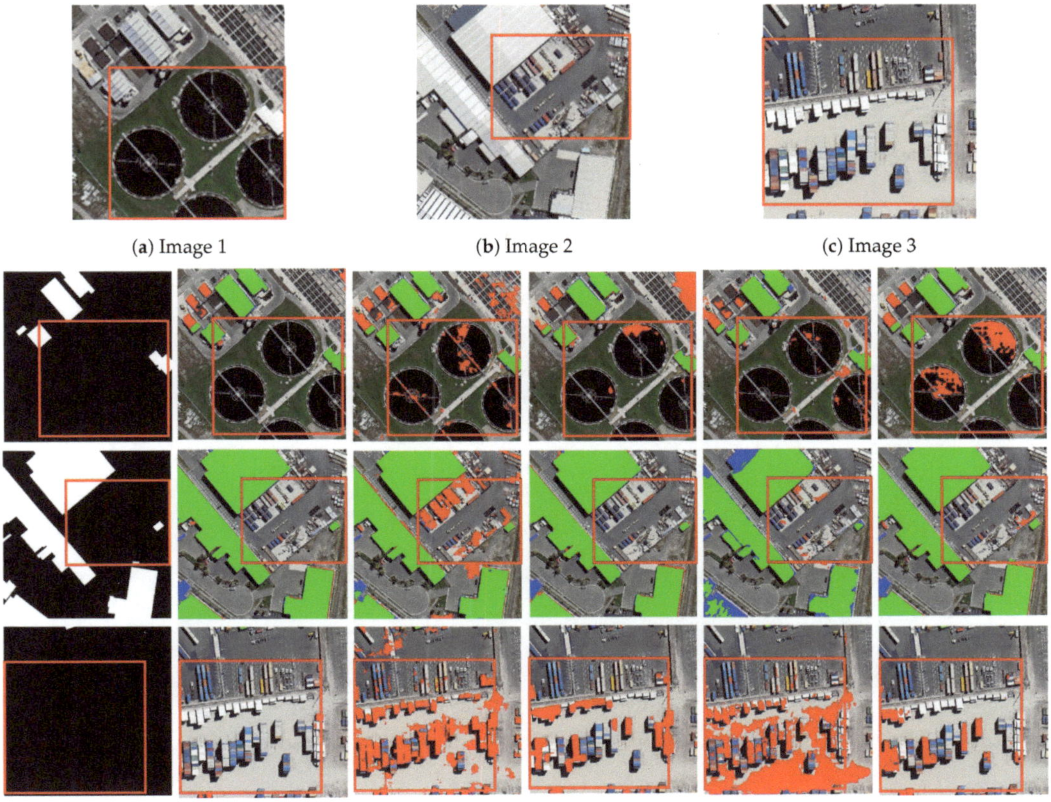

(**a**) Image 1 (**b**) Image 2 (**c**) Image 3

(**d**) Ground Truth (**e**) Ours (**f**) U-Net (**g**) Deeplabv3+ (**h**) PSPNet (**i**) HRNet

Figure 7. Results of different methods on easily confused buildings. (**a–c**) Original input images; (**d**) ground truth; (**e–i**) the results of our method and the other four methods, respectively. Green, red, and blue pixels on the map represent predictions of true positives, false positives, and false negatives, respectively. The experimental results in the red box show the areas where the contrast effect is more obvious.

Table 1. Performance of different methods on the WHU Building Dataset.

Method	PA	PC	F1	IoU	FWIoU
U-Net [28]	0.9713	0.8366	0.8435	0.7419	0.9485
Deeplabv3+ [50]	0.9816	0.9001	0.9010	0.8296	0.9651
PSPNet [30]	0.9586	0.8215	0.7734	0.6434	0.9247
HRNet [31]	0.9864	0.9109	0.9201	0.8647	0.9742
CFENet (Ours)	0.9871	0.9370	0.9262	0.8722	0.9751

3.3.2. Results on Massachusetts Building Dataset

We conduct experiments on the Massachusetts Buildings Dataset to further validate the effectiveness of CFENet. The experimental results of different methods on the Massachusetts Building Dataset are shown in Table 2. Our method also shows impressive performance on all five evaluation metrics. Among them, on IoU, CFENet improves the performance to 0.7486. On PA, our method achieves 0.9626, significantly outperforming the other methods we compared.

Table 2. Performance of different methods on the Massachusetts Building Dataset.

Method	PA	PC	F1	IoU	FWIoU
U-Net	0.9517	0.8635	0.7701	0.6626	0.9099
Deeplabv3+	0.9163	0.7153	0.6506	0.5266	0.8559
PSPNet	0.9116	0.7375	0.6084	0.4704	0.8455
HRNet	0.9581	0.8292	0.7910	0.6968	0.9225
CFENet (Ours)	0.9626	0.8277	0.8304	0.7486	0.9317

The qualitative assessment results on the Massachusetts Building Dataset are shown in Figure 8. The visualization result shows that CFENet has high extraction accuracy, and most of the buildings are predicted as true positives (green). Row 1, 2, and 4 are large, irregularly shaped buildings that other methods misjudge (red and blue) in predicting such buildings. Row 3 is of complex textured buildings, and Row 5 of buildings is small and dense. CFENet is basically accurate in predicting such buildings, whereas other methods predict some false positives (red) and false negatives (blue).

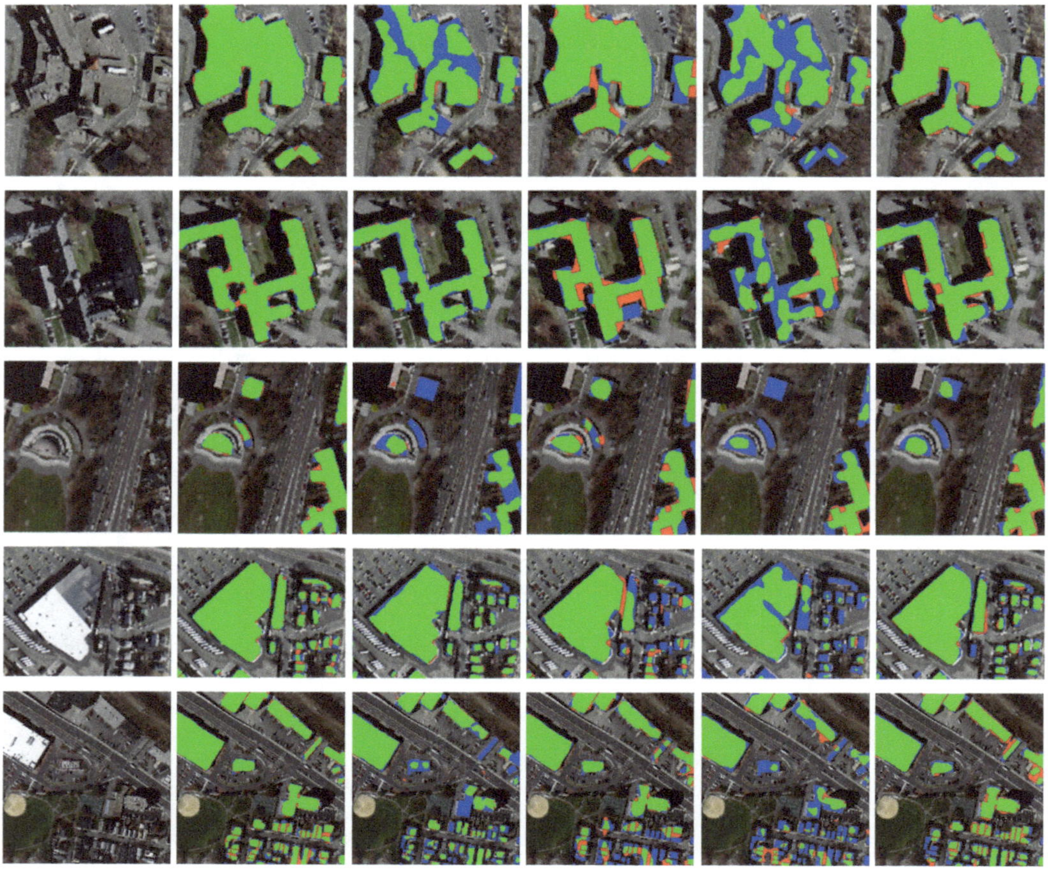

(**a**) Original images (**b**) Ours (**c**) U-Net (**d**) Deeplabv3+ (**e**) PSPNet (**f**) HRNet

Figure 8. Extraction results of different methods for different types of buildings on the Massachusetts Building Dataset. (**a**) Original input image; (**b–f**) the results of our method and the other four methods, respectively. Green, red, and blue pixels on the map represent predictions of true positives, false positives, and false negatives, respectively.

3.4. Ablation Experiments

As shown in Table 3, our proposed three modules significantly improve the performance. Compared with the baseline FCN (ResNet-101), we improve PA/PC/F1/IoU/FWIoU by 0.0168/0.1376/0.05/0.0839/0.0275 by using only the spatial fusion module. The effect of using both the spatial fusion module and the focus enhance module improved over baseline on all five evaluation metrics. When three modules are integrated together, the PA/PC/F1/IoU/FWIoU of CFENet is further improved to 0.9871/0.9370/0.9262/0.8722/0.9751. The experimental results are significantly improved when the three modules we designed are gradually added, proving that our proposed modules are effective. Therefore, our proposed three modules are of great help for building extraction from high-resolution remote sensing images.

Table 3. Ablation study on WHU Building Dataset.

Method	BaseNet	Component			PA	PC	F1	IoU	FWIoU
		SFM	FEM	FDM					
FCN	ResNet-101				0.9638	0.7578	0.8387	0.7309	0.9365
CFENet	ResNet-101	✓			0.9806	0.8954	0.8887	0.8148	0.9640
CFENet	ResNet-101	✓	✓		0.9864	0.9182	0.9161	0.8608	0.9744
CFENet	ResNet-101	✓	✓	✓	0.9871	0.9370	0.9262	0.8722	0.9751

The effect of each module is shown in Figure 9. After adding the spatial fusion module to the base network, the boundary of the building is clearer, indicating that the spatial fusion module aggregates the spatial information of low-level features to a great extent. After adding the focus enhancement module on this basis, the details of the building are more prominent, indicating that the focus enhancement module integrates the rich semantic information in the high-level features. The completeness and accuracy of the building are further improved when the three modules are integrated together.

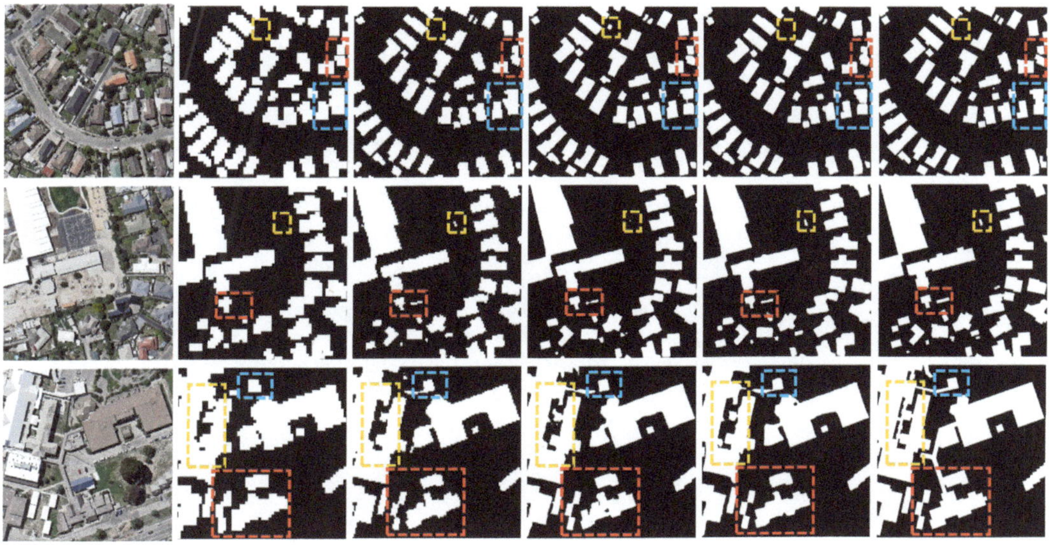

(a) (b) (c) (d) (e) (f)

Figure 9. Visualization results of ablation experiments. (**a**) Original image; (**b**) the output of BaseNet (ResNet-101); (**c**) the output of BaseNet + SFM; (**d**) the output of BaseNet + SFM + FEM; (**e**) the output of CFENet (BaseNet + SFM + FEM + FDM); (**f**) ground truth. The experimental results in the box show areas where the contrast effect is more obvious.

4. Discussion

4.1. Comparison of the Efficiency of Different Methods

In addition to the five evaluation metrics mentioned in Section 3.1, we analyze the efficiency of different methods from several aspects under the same batch size. The column "Model Size" represents the total model size for each method. The larger the model, the more memory it takes up during training and testing, and the smaller the model, the faster the model runs. As shown in the column "Model Size" in Table 4, the model size of CFENet is only larger than that of PSPNet. Figure 10 shows the model size and the experimental results of different methods on the WHU Building Dataset. Although the model size of CFENet is larger than that of PSPNet, our experimental results are the best among the five methods, and the experimental results of PSPNet are the worst. This indicates that other methods require more memory during training and testing. The reason why CFENet has fewer total parameters is that we use a lot of dilated convolution layers in our network, and use 1 × n convolution layers plus n × 1 convolution layers instead of n × n convolution layers, and use 1 × 1 convolution layers for dimensionality reduction when appropriate. The advantage is that the model size can be reduced, and the performance can also be improved. Therefore, CFENet has a relatively small model to ensure good results.

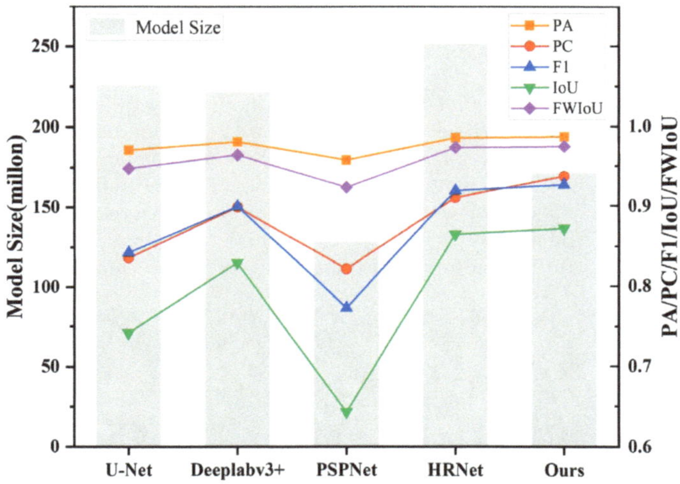

Figure 10. Comparison of experimental results and model sizes of different methods. Line graphs with different colors represent the results of different evaluation metrics. The histogram represents the model sizes of different methods.

Table 4. Comparison of the efficiency of different methods. Different methods are trained on the WHU Building Dataset for 200 epochs, and the batch size is set to 8. The column "Model Size" represents the total model size of each method. The column "Time" represents the time required to train an epoch.

Method	Batch Size	Model Size (million)	Time (min/epoch)
U-Net (ResNet-101)	8	226	20.050
Deeplabv3+ (ResNet-101)	8	222	10.417
PSPNet (ResNet-101)	8	128	20.383
HRNet	8	252	20.450
CFENet (ResNet-101)	8	171	13.250

The column "Time" of Table 4 represents the time required to train an epoch. CFENet takes 13.250 min to train an epoch, which is only slower than Deeplabv3+, and much lower than U-Net, Deeplabv3+, and PSPNet. It is worth mentioning that although the model of PSPNet is smaller than CFENet, PSPNet takes 20.383 min to train an epoch, which is much slower than our method.

4.2. Heatmaps for Validating the Effectiveness of CFENet's Modules

To further validate the effectiveness of our proposed modules, we perform a heatmap analysis of the features on the test images. The heatmap visualization of the features is shown in Figure 11. For the low-level features Feature 1 and Feature 2 extracted by ResNet-101, their spatial information is more abundant and leads to information redundancy because they are applied with fewer convolution layers. Therefore, many background pixels in the heatmap appear as highlighted cases. These highlighted background pixels are significantly improved after being processed by the spatial fusion module. The high-level features Feature 3 and Feature 4 extracted by ResNet-101 have more abstract feature information and rich semantic information (categories, attributes, etc.) due to the large number of convolutions. It can be observed from Figure 11 that the features processed by the focus enhancement module are significantly more similar to the ground truth, which proves the effectiveness of the focus enhancement module. The enhanced low-level and high-level features are fed into the feature decoder module for fusion and decoding. The output of the feature decoder module is very close to the ground truth.

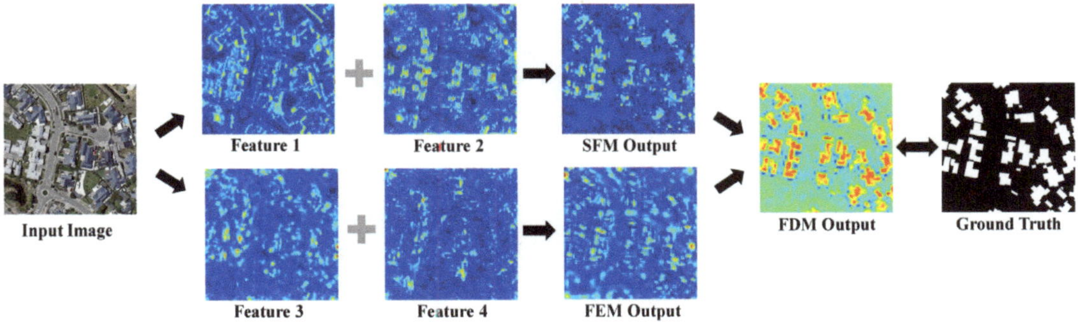

Figure 11. Heatmaps of CFENet's feature maps. SFM Output represents the feature maps output by the spatial fusion module. FEM Output represents the feature maps output by the focus enhancement module. FDM Output represents the feature maps output by the feature decoder module.

5. Conclusions

This paper proposes an end-to-end neural network (CFENet) that can effectively enhance low-level and high-level features and fuse them to obtain strong feature representations for building extraction tasks. Specifically, we introduce the spatial fusion and focus enhancement modules to efficiently enhance the low-level features and high-level features extracted by ResNet-101, respectively. Then the feature decoder module further fuses and enhances the features generated by the spatial fusion module and the focus enhancement module to obtain the final probability map. CFENet achieves superior performance over other methods we have compared on the WHU Building Dataset and the Massachusetts Building Dataset, and significantly improves the extraction accuracy while controlling the computational cost. Moreover, ablation experiments show that our proposed three modules can improve the accuracy of building extraction from remote sensing images.

In future work, our method could be further explored and improved in the following aspects. The generalization ability of each module of CFENet could be further improved. The structure of the loss function could be modified to improve the ability to discriminate difficult samples. The remote sensing images could be preprocessed before being input into our network to further improve the performance of CFENet. Meanwhile, our method would be explored in more application areas, such as side scan sonar remote sensing images, etc.

Author Contributions: J.C. conceived of and designed the algorithm and the experiments; J.C. and D.Z. analyzed the data; J.C. wrote the manuscript; D.Z. supervised the research; Y.W., Y.C. and X.Y. provided suggestions for the proposed method and its evaluation and assisted in the preparation of the manuscript. All authors have read and agreed to the published version of the manuscript.

Funding: This work is supported by the National Science Foundation of China (grant No. 62102268 and 61802355) and Hubei Key Laboratory of Intelligent Robot (HBIR 202105), the Research Foundation of Shenzhen Polytechnic under Grants 6022312044K and 6021271004K.

Data Availability Statement: Our experiments employ open datasets in [42,43]. The datasets can be downloaded from http://gpcv.whu.edu.cn/data/building_dataset.html (accessed on 30 March 2022) and https://www.cs.toronto.edu/~vmnih/data/ (accessed on 30 March 2022).

Conflicts of Interest: The authors declare no conflict of interest.

Abbreviations

The following abbreviations are used in this manuscript:

CFENet	Context Feature Enhancement Network
CNN	Convolutional Neural Network
FCN	Fully Convolutional Network
SFM	Spatial Fusion Module
FEM	Focus Enhancement Module
FDM	Feature Decoder Module
LB	Location Block
BN	Batch Normalization
Adam	Adaptive Moment Estimation
TP	True Positive
TN	True Negative
FP	False Positive
FN	False Negative
GPU	Graphics Processing Unit
PA	Pixel Accuracy
PC	Precision
F1	F1 score
IoU	Intersection over Union
FWIoU	Frequency Weighted Intersection over Union
MBI	Morphological Building Index
Mask R-CNN	Mask Region Convolutional Neural Network

References

1. Ma, L.; Liu, Y.; Zhang, X.; Ye, Y.; Yin, G.; Johnson, B.A. Deep learning in remote sensing applications: A meta-analysis and review. *ISPRS J. Photogramm. Remote Sens.* **2019**, *152*, 166–177. [CrossRef]
2. Georganos, S.; Grippa, T.; Vanhuysse, S.; Lennert, M.; Shimoni, M.; Kalogirou, S.; Wolff, E. Less is more: Optimizing classification performance through feature selection in a very-high-resolution remote sensing object-based urban application. *GISci. Remote Sens.* **2018**, *55*, 221–242. [CrossRef]
3. Yang, H.L.; Yuan, J.; Lunga, D.; Laverdiere, M.; Rose, A.; Bhaduri, B. Building Extraction at Scale Using Convolutional Neural Network: Mapping of the United States. *IEEE J. Sel. Top. Appl. Earth Obs. Remote Sens.* **2018**, *11*, 2600–2614.
4. Kuras, A.; Brell, M.; Rizzi, J.; Burud, I. Hyperspectral and Lidar Data Applied to the Urban Land Cover Machine Learning and Neural-Network-Based Classification: A Review. *Remote Sens.* **2021**, *13*, 3393. [CrossRef]
5. Munawar, H.S.; Ullah, F.; Qayyum, S.; Heravi, A. Application of deep learning on uav-based aerial images for flood detection. *Smart Cities* **2021**, *4*, 1220–1242. [CrossRef]
6. Wang, Y.; Cui, L.; Zhang, C.; Chen, W.; Xu, Y.; Zhang, Q. A Two-Stage Seismic Damage Assessment Method for Small, Dense, and Imbalanced Buildings in Remote Sensing Images. *Remote Sens.* **2022**, *14*, 1012. [CrossRef]
7. Peng, B.; Ren, D.; Zheng, C.; Lu, A. TRDet: Two-Stage Rotated Detection of Rural Buildings in Remote Sensing Images. *Remote Sens.* **2022**, *14*, 522. [CrossRef]

8. Mahabir, R.; Croitoru, A.; Crooks, A.T.; Agouris, P.; Stefanidis, A. A critical review of high and very high-resolution remote sensing approaches for detecting and mapping slums: Trends, challenges and emerging opportunities. *Urban Sci.* **2018**, *2*, 8. [CrossRef]
9. Ball, J.E.; Anderson, D.T.; Chan Sr, C.S. Comprehensive survey of deep learning in remote sensing: Theories, tools, and challenges for the community. *J. Appl. Remote Sens.* **2017**, *11*, 042609. [CrossRef]
10. Luo, L.; Li, P.; Yan, X. Deep Learning-Based Building Extraction from Remote Sensing Images: A Comprehensive Review. *Energies* **2021**, *14*, 7982. [CrossRef]
11. Li, E.; Xu, S.; Meng, W.; Zhang, X. Building extraction from remotely sensed images by integrating saliency cue. *IEEE J. Sel. Top. Appl. Earth Obs. Remote Sens.* **2016**, *10*, 906–919. [CrossRef]
12. Chen, R.; Li, X.; Li, J. Object-Based Features for House Detection from RGB High-Resolution Images. *Remote Sens.* **2018**, *10*, 451. [CrossRef]
13. Inglada, J. Automatic recognition of man-made objects in high resolution optical remote sensing images by SVM classification of geometric image features. *ISPRS J. Photogramm. Remote Sens.* **2007**, *62*, 236–248. [CrossRef]
14. Ding, Z.; Wang, X.; Li, Y.; Zhang, S. Study on building extraction from high-resolution images using Mbi. *Int. Arch. Photogramm. Remote Sens. Spat. Inf. Sci* **2018**, *42*. [CrossRef]
15. Katartzis, A.; Sahli, H.; Nyssen, E.; Cornelis, J. Detection of buildings from a single airborne image using a Markov random field model. In Proceedings of the IGARSS 2001, Scanning the Present and Resolving the Future, IEEE 2001 International Geoscience and Remote Sensing Symposium (Cat. No. 01CH37217), Sydney, NSW, Australia, 9–13 July 2001; Volume 6, pp. 2832–2834.
16. Awrangjeb, M.; Fraser, C.S. An automatic and threshold-free performance evaluation system for building extraction techniques from airborne LIDAR data. *IEEE J. Sel. Top. Appl. Earth Obs. Remote Sens.* **2014**, *7*, 4184–4198. [CrossRef]
17. Tu, Z.; Li, H.; Zhang, D.; Dauwels, J.; Li, B.; Yuan, J. Action-stage emphasized spatiotemporal VLAD for video action recognition. *IEEE Trans. Image Process.* **2019**, *28*, 2799–2812. [CrossRef]
18. Zhang, D.; He, L.; Tu, Z.; Han, F.; Zhang, S.; Yang, B. Learning motion representation for real-time spatio-temporal action localization. *Pattern Recognit.* **2020**, *103*, 107312. [CrossRef]
19. Zhang, D.; He, F.; Tu, Z.; Zou, L.; Chen, Y. Pointwise geometric and semantic learning network on 3D point clouds. *Integr.-Comput.-Aided Eng.* **2020**, *27*, 57–75. [CrossRef]
20. Simonyan, K.; Zisserman, A. Very deep convolutional networks for large-scale image recognition. *arXiv* **2014**, arXiv:1409.1556.
21. He, K.; Zhang, X.; Ren, S.; Sun, J. Deep residual learning for image recognition. In Proceedings of the IEEE Conference on Computer Vision and Pattern Recognition, Las Vegas, NV, USA, 27–30 June 2016; pp. 770–778.
22. Szegedy, C.; Vanhoucke, V.; Ioffe, S.; Shlens, J.; Wojna, Z. Rethinking the inception architecture for computer vision. In Proceedings of the IEEE Conference on Computer Vision and Pattern Recognition, Las Vegas, NV, USA, 27–30 June 2016; pp. 2818–2826.
23. Huang, G.; Liu, Z.; Van Der Maaten, L.; Weinberger, K.Q. Densely connected convolutional networks. In Proceedings of the IEEE Conference on Computer Vision and Pattern Recognition, Honolulu, HI, USA, 21–26 July 2017; pp. 4700–4708.
24. Krizhevsky, A.; Sutskever, I.; Hinton, G.E. Imagenet classification with deep convolutional neural networks. *Adv. Neural Inf. Process. Syst.* **2012**, *25*, 1097–1105. [CrossRef]
25. Long, J.; Shelhamer, E.; Darrell, T. Fully convolutional networks for semantic segmentation. In Proceedings of the IEEE Conference on Computer Vision and Pattern Recognition, Boston, MA, USA, 7–12 June 2015; pp. 3431–3440.
26. Badrinarayanan, V.; Kendall, A.; SegNet, R.C. A deep convolutional encoder-decoder architecture for image segmentation. *arXiv* **2015**, arXiv:1511.00561.
27. Zeiler, M.D.; Fergus, R. Visualizing and understanding convolutional networks. In Proceedings of the European Conference on Computer Vision, Zurich, Switzerland, 6–12 September 2014; pp. 818–833.
28. Ronneberger, O.; Fischer, P.; Brox, T. U-net: Convolutional networks for biomedical image segmentation. In Proceedings of the International Conference on Medical Image Computing and Computer-Assisted Intervention, Munich, Germany, 5–9 October 2015; pp. 234–241.
29. Chen, L.C.; Papandreou, G.; Kokkinos, I.; Murphy, K.; Yuille, A.L. Deeplab: Semantic image segmentation with deep convolutional nets, atrous convolution, and fully connected crfs. *IEEE Trans. Pattern Anal. Mach. Intell.* **2017**, *40*, 834–848. [CrossRef] [PubMed]
30. Zhao, H.; Shi, J.; Qi, X.; Wang, X.; Jia, J. Pyramid scene parsing network. In Proceedings of the IEEE Conference on Computer Vision and Pattern Recognition, Honolulu, HI, USA, 21–26 July 2017; pp. 2881–2890.
31. Sun, K.; Xiao, B.; Liu, D.; Wang, J. Deep high-resolution representation learning for human pose estimation. In Proceedings of the IEEE/CVF Conference on Computer Vision and Pattern Recognition, Long Beach, CA, USA, 15–20 June 2019; pp. 5693–5703.
32. Guo, M.; Liu, H.; Xu, Y.; Huang, Y. Building extraction based on U-Net with an attention block and multiple losses. *Remote Sens.* **2020**, *12*, 1400. [CrossRef]
33. Chen, M.; Wu, J.; Liu, L.; Zhao, W.; Tian, F.; Shen, Q.; Zhao, B.; Du, R. DR-Net: An improved network for building extraction from high resolution remote sensing image. *Remote Sens.* **2021**, *13*, 294. [CrossRef]
34. Shao, Z.; Tang, P.; Wang, Z.; Saleem, N.; Yam, S.; Sommai, C. BRRNet: A fully convolutional neural network for automatic building extraction from high-resolution remote sensing images. *Remote Sens.* **2020**, *12*, 1050. [CrossRef]
35. Xu, Y.; Wu, L.; Xie, Z.; Chen, Z. Building extraction in very high resolution remote sensing imagery using deep learning and guided filters. *Remote Sens.* **2018**, *10*, 144. [CrossRef]

36. Zhu, Q.; Liao, C.; Hu, H.; Mei, X.; Li, H. MAP-Net: Multiple Attending Path Neural Network for Building Footprint Extraction from Remote Sensed Imagery. *IEEE Trans. Geosci. Remote Sens.* **2021**, *59*, 6169–6181. [CrossRef]
37. Liao, C.; Hu, H.; Li, H.; Ge, X.; Chen, M.; Li, C.; Zhu, Q. Joint Learning of Contour and Structure for Boundary-Preserved Building Extraction. *Remote Sens.* **2021**, *13*, 1049. [CrossRef]
38. Yang, H.; Wu, P.; Yao, X.; Wu, Y.; Wang, B.; Xu, Y. Building extraction in very high resolution imagery by dense-attention networks. *Remote Sens.* **2018**, *10*, 1768. [CrossRef]
39. Wen, Q.; Jiang, K.; Wang, W.; Liu, Q.; Guo, Q.; Li, L.; Wang, P. Automatic building extraction from Google Earth images under complex backgrounds based on deep instance segmentation network. *Sensors* **2019**, *19*, 333. [CrossRef]
40. Zheng, G.; Zhang, H.; Li, Y.; Zhao, J. A Universal Automatic Bottom Tracking Method of Side Scan Sonar Data Based on Semantic Segmentation. *Remote Sens.* **2021**, *13*, 1945. [CrossRef]
41. Yu, Y.; Zhao, J.; Gong, Q.; Huang, C.; Zheng, G.; Ma, J. Real-Time Underwater Maritime Object Detection in Side-Scan Sonar Images Based on Transformer-YOLOv5. *Remote Sens.* **2021**, *13*, 3555. [CrossRef]
42. Ji, S.; Wei, S.; Lu, M. Fully convolutional networks for multisource building extraction from an open aerial and satellite imagery data set. *IEEE Trans. Geosci. Remote Sens.* **2018**, *57*, 574–586. [CrossRef]
43. Mnih, V. Machine Learning for Aerial Image Labeling. Ph.D. Thesis, University of Toronto, Toronto, ON, Canada, 2013.
44. Yu, F.; Koltun, V. Multi-scale context aggregation by dilated convolutions. *arXiv* **2015**, arXiv:1511.07122.
45. Fu, J.; Liu, J.; Tian, H.; Li, Y.; Bao, Y.; Fang, Z.; Lu, H. Dual attention network for scene segmentation. In Proceedings of the IEEE/CVF Conference on Computer Vision and Pattern Recognition, Long Beach, CA, USA, 15–20 June 2019; pp. 3146–3154.
46. Hu, J.; Shen, L.; Sun, G. Squeeze-and-excitation networks. In Proceedings of the IEEE Conference on Computer Vision and Pattern Recognition, Salt Lake City, UT, USA, 18–23 June 2018; pp. 7132–7141.
47. Denton, E.L.; Zaremba, W.; Bruna, J.; LeCun, Y.; Fergus, R. Exploiting linear structure within convolutional networks for efficient evaluation. *Adv. Neural Inf. Process. Syst.* **2014**, *27*, 1269–1277.
48. Jaderberg, M.; Vedaldi, A.; Zisserman, A. Speeding up convolutional neural networks with low rank expansions. *arXiv* **2014**, arXiv:1405.3866.
49. De Boer, P.T.; Kroese, D.P.; Mannor, S.; Rubinstein, R.Y. A tutorial on the cross-entropy method. *Ann. Oper. Res.* **2005**, *134*, 19–67. [CrossRef]
50. Chen, L.C.; Zhu, Y.; Papandreou, G.; Schroff, F.; Adam, H. Encoder-decoder with atrous separable convolution for semantic image segmentation. In Proceedings of the European Conference on Computer Vision (ECCV), Munich, Germany, 8–14 September 2018; pp. 801–818.

remote sensing

MDPI

Article

Spatial–Spectral Cross-Correlation Embedded Dual-Transfer Network for Object Tracking Using Hyperspectral Videos

Jie Lei [1], Pan Liu [1], Weiying Xie [1,*], Long Gao [1], Yunsong Li [1] and Qian Du [2]

[1] State Key Laboratory of Integrated Service Networks, School of Telecommunications Engineering, Xidian University, Xi'an 710071, China; jielei@mail.xidian.edu.cn (J.L.); liupan@stu.xidian.edu.cn (P.L.); lgao@xidian.edu.cn (L.G.); ysli@mail.xidian.edu.cn (Y.L.)

[2] The Department of Electrical and Computer Engineering, Mississippi State University, Starkville, MS 39762, USA; du@ece.msstate.edu

* Correspondence: wyxie@xidian.edu.cn; Tel.: +86-029-8820-3116

Abstract: Hyperspectral (HS) videos can describe objects at the material level due to their rich spectral bands, which are more conducive to object tracking compared with color videos. However, the existing HS object trackers cannot make good use of deep-learning models to mine their semantic information due to limited annotation data samples. Moreover, the high-dimensional characteristics of HS videos makes the training of a deep-learning model challenging. To address the above problems, this paper proposes a spatial–spectral cross-correlation embedded dual-transfer network (SSDT-Net). Specifically, first, we propose to use transfer learning to transfer the knowledge of traditional color videos to the HS tracking task and develop a dual-transfer strategy to gauge the similarity between the source and target domain. In addition, a spectral weighted fusion method is introduced to obtain the inputs of the Siamese network, and we propose a spatial–spectral cross-correlation module to better embed the spatial and material information between the two branches of the Siamese network for classification and regression. The experimental results demonstrate that, compared to the state of the art, the proposed SSDT-Net tracker offers more satisfactory performance based on a similar speed to the traditional color trackers.

Keywords: hyperspectral (HS) object tracking; dual-transfer learning; spectral weighting; spatial-spectral cross-correlation

Citation: Lei, J.; Liu, P.; Xie, W.; Gao, L.; Li, Y.; Du, Q. Spatial–Spectral Cross-Correlation Embedded Dual-Transfer Network for Object Tracking Using Hyperspectral Videos. *Remote Sens.* **2022**, *14*, 3512. https://doi.org/10.3390/rs14153512

Academic Editor: Anup Basu

Received: 16 May 2022
Accepted: 13 July 2022
Published: 22 July 2022

Publisher's Note: MDPI stays neutral with regard to jurisdictional claims in published maps and institutional affiliations.

1. Introduction

Great strides have been made regarding object tracking in traditional color videos due in part to the availability of large-scale datasets with rich annotations [1]. However, this has limitations in many practical scenarios where one may encounter objects with the same color or shape as the background [2]. In other words, tracking in traditional color videos fails to track camouflaged or hidden objects. Fortunately, hyperspectral (HS) data can distinguish similar appearances of different materials ascribed to the abundant spectra, and this has been widely and successfully applied to many practical applications [3,4]. In particular, object tracking in HS videos can estimate the bounding box of an interesting object that significantly differs spatially or spectrally from plentiful background categories.

A series of efforts dedicated to achieving tracking in HS videos have been explored [5–12]. For example, Qian et al. [5] extracted a group of patches around the object as a convolution kernel to extract the features of each band but ignored the correlation between the bands. A material-based HS tracker (MHT) was proposed by Xiong et al. [6], which embeds spectral–spatial information of HS images (HSIs) into the histogram of multi-dimensional oriented gradients and then combines it with the global material abundance features. The above methods take advantage of handcrafted features, and in contrast, HS tracking methods based on deep feature extraction have also emerged.

Uzkent et al. [7] transformed HSIs into three-channel data and performed deep feature extraction through the VGGNet [8]. However, spectral information will inevitably be lost in the three-channel data after conversion. As an alternative, multiple bands can be selected for feature extraction; however, how to choose these bands is a crucial problem. Based on the contribution of each band in HSI, a band attention mechanism network (BAE-NET) similar to an autoencoder was proposed by Li et al. [9], which learns nonlinear spectral relationships and generates band weights according to the importance of each band.

Under the guidance of these weights, an HSI is grouped into several three-channel data. These converted data are used by several VITAL [10] trackers to generate weak trackers, which are then ensemble learned to obtain the object position. BAE-Net only considers spectral information but ignores the valuable temporal and spatial information, which leads to unstable tracking. Temporal and spatial information can be applied as a supplement to spectral information to constitute stronger object representation features, and thus Li et al. [11] proposed a spectral–spatial–temporal attention neural network (SST-Net) for HS object tracking, which can combine temporal, spectral, and spatial attention to better select appropriate bands for deep ensemble tracking.

Although the above methods have gained hopeful results, there are still many challenges that impede the advance of deep-learning models for HS object tracking. First, learning a deep convolutional neural network (CNN) usually requires estimating millions of parameters, which requires a mass of annotation data. This bottleneck currently limits the performance of CNNs on HS object tracking. Moreover, HS data have a lower spatial resolution due to the imaging mechanism and high-dimensional nature caused by massive spectral bands. The rich spectral information provides details of material composition, and spatial information can be utilized as a supplement to spectral information to improve the ability to identify objects. However, the high-dimensional nature of HS data makes training a deep CNN model consume a large amount of resources.

To overcome the limitation of fewer annotation data while taking full advantage of HS data, we propose a novel HS tracker based on transfer learning, i.e., spatial–spectral cross-correlation embedded dual-transfer network (SSDT-Net), for the first time in the field of HS tracking. Specifically, we first analyze the applicability of transfer learning on HS videos—that is, to explore whether the trained CNN can be extended to HS videos [4,13].

Then, in order to effectively apply transfer learning to HS videos, we propose a new dual-transfer strategy to address the problem of object tracking in HS videos. In addition, to fully and effectively make use of the rich spectral information brought by the high-dimensional characteristics of HS videos, we introduce a spectral weighted fusion method based on the structure tensor (ST), which considers the contribution of each band.

In particular, inspired by the successful application of Siamese CNNs on traditional color videos [14–19], we chose Siamese CNNs as feature extractors for HS videos. Finally, considering the difference in the spatial structure recognition ability of the spatial dimension and the material recognition ability of the spectral dimension in HS videos, we propose a novel cross-correlation module to effectively embed the spatial–spectral information of the two branches for classification and regression.

To summarize, compared with the existing HS object tracking algorithms, the main contributions of this work are the following.

- For the first time, transfer learning technology was successfully applied in the field of HS object tracking. In particular, a new dual-transfer strategy is proposed—that is, the transfer learning method is adaptively selected according to the prior knowledge of the sample category, which not only solves the thorny problem of a lack of labeled data when learning deep models in the HS field but also verifies the applicability of transfer learning technology in the field of HS object tracking to a certain extent. This provides a flexible direction for future HS tracking research.
- For fully utilizing the spatial structure identification ability of the spatial dimension and the material identification ability of the spectral dimension in the HSIs, a novel

spatial–spectral cross-correlation module is designed to better embed the spatial information and material information between the two branches of the Siamese network.

- Considering the high-dimensional characteristic of HS videos, we introduce an effective spectral weighted fusion method based on ST to gain the inputs of CNNs that consider the contribution rate to the significant information of each selected band to make our network more efficient.
- The experimental results demonstrate that, compared to the state of the art, the proposed SSDT-Net tracker offers more satisfactory performance based on a similar speed to the traditional color trackers. We provide an efficient and enlightening scheme for HS object tracking.

The rest of the paper is structured as follows. The basic theories related to transfer learning and trackers based on Siamese network structure are reviewed in Section 2. Section 3 details the proposed SSDT-Net. Section 4 displays our experimental results and comprehensive analyses of the method. Sections 5 and 6 discuss and conclude our method, respectively.

2. Related Work

In this section, we briefly review two key aspects relevant to our work: hyperspectral video technology and the Siamese network-based trackers.

2.1. Hyperspectral Video Technology

Due to the limitations of sensors, traditional hyperspectral (HS) data are collected by sensors on satellites and are mostly used for remote sensing. However, recent advances in sensors have made it possible to collect HS sequences at close range at video rates. In paper [6], Xiong et al. published the first HS dataset captured by an HS camera in a close computer vision setting. They used a snapshot-spliced HS camera to collect the video. The HS camera can capture up to 180 HS images per second with each image having a spatial dimension of 512 × 256 pixels and a spectral dimension of 16 wavelengths ranging from 470 nm to 620 nm. During the collection of this dataset, the HS camera captures video at 25 frames per second (FPS).

To ensure high-quality hyperspectral images (HSIs), HS cameras also require preprocessing steps. This step includes spectral correction and image registration. Spectral correction mainly includes two steps: dark correction and spectral correction. The purpose of dark calibration is to eliminate the influence of noise generated by camera sensors by subtracting the dark frame from the captured image, where the dark frame is captured when the lens of an HS camera is covered by a lid. Spectral correction is designed to suppress the contribution of unnecessary second-order responses, such as the response to the wavelength leaking into the filter. The specific operation is to make the obtained spectrum more consistent with the expected spectrum by carrying out the sensor-specific spectral correction matrix for the obtained reflectivity.

At the same time, to more distinctly verify the advantages of material characteristics in target tracking and ensure a fair comparison with color-based tracking methods, Xiong et al. also obtained the RGB video corresponding to the HS video at the same frame rate from a close viewpoint and registered the HS sequence and color video so that they described almost the same scene. However, even then, there may be subtle differences between the two videos. Therefore, the authors of this dataset used the CIE color-matching function (CMFs) to convert HS videos into false color videos to ensure the complete consistency of data scenes. The CMFs represents the weight of each wavelength used to generate red, green, and blue channels in an HSI. Given an HSI $H \in \mathbb{R}^{W\times H\times B}$ with $W \times H$ pixels and B bands, we can convert the HSI to a CIE XYZ image using the following formula

$$I = \sum_{b=1}^{B} H(:,b)W(b,:) \tag{1}$$

where $W \in \mathbb{R}^{B\times3}$ are the CMFs. The color transformation method in [20] is used to make the color intensity of the transformed image close to the corresponding RGB image collected.

Finally, in this paper, all experiments were performed based on this dataset, which contains videos of different modalities, namely HS, color, and false color. This enables our proposed HS tracker to be compared not only with other advanced HS trackers but also with other advanced color video trackers, which can more comprehensively demonstrate the advantages of our proposed algorithm.

2.2. Siamese Network-Based Trackers

The Siamese network is composed of two branches, which implicitly encode the original patches into another space and then fuse the two sets of information through a specific information embedding method. It is usually applied in contrastive tasks to compare the two branch features in the implicitly embedded space. Recently, the Siamese network has received extensive attention from the visual tracking community due to its well-balanced accuracy and speed [14,15,18].

Inspired by correlation-based approaches, recent mainstream trackers based on Siamese networks define visual tracking as a cross-correlation problem. They are anticipated to fully use the advantages of deep networks in end-to-end learning. SiamFC [14] first includes a fully connected convolution Siamese network to train the tracker and introduced the correlation layer as the fusion tensors, which greatly improved the fusion accuracy. Inspired by its success, many researchers have continued to study and propose novel models [17,21,22].

CFNet [17] introduces a correlation filter layer based on the SiamFC framework and conducts online tracking to improve the accuracy. DSiam [23] uses a dynamic Siamese network learning method through two online transformations of Siamese network branches, which obtains better tracking accuracy under the premise of balanced speed. SAsiam [21] constructs a double Siamese network consisting of semantic branches and appearance branches, which are trained separately for keeping the heterogeneity of features and integrated during testing for increasing the tracking accuracy.

To settle the problem of scale change, these Siamese networks adopt multi-scale search; however, this is time-consuming. To avoid the time-consuming steps in the stage of extracting multi-scale feature maps, the SiamRPN [16] tracker, inspired by the target detection region proposal network [24], utilizes the output of the Siamese network to extract region proposal through joint learning of the classification branch and regression branch. However, it is challenging to deal with disturbances that look similar to the object.

According to the structure of SiamRPN, DaSiamRPN [19] improves the resolution of the tracker through data enhancement, thus, obtaining more robust tracking results. Although these methods improve on the original SiamFC in many ways, the performance stall is mainly due to the weak backbone network (AlexNet) they use. SiamRPN++ [22] replaces AlexNet with ResNet [25]. During the training process, the position of the training object in the search area is moved randomly to eliminate the center deviation. This modification improves the tracking accuracy.

However, the performance evaluation of these existing Siamese network-based trackers is based on traditional color videos. To unleash the potential of the Siamese network for the processing of HS video, in this paper, we chose the Siamese network architecture as the backbone of our solution to solve the HS video tracking problem.

3. Proposed Method

We detail our proposed SSDT-Net for HS object tracking in this section. As shown in Figures 1 and 2, our algorithm consists of five parts, dual-transfer strategy, spectral weighted fusion, feature extraction network, spatial–spectral cross-correlation module, and prediction head. The dual transfer strategy is a training method acting on the tracker. This strategy divides the target domain into two categories by measuring the similarity between the source domain and the target domain and adopts different training methods, while the inference method is unified.

The spectral weighted fusion is used as the data preprocessing module of the network. It calculates the significant information contribution rate of each band of HSI through ST and uses the contribution rate to perform adaptive weighted fusion on the original high-dimensional HSI data to make our network more efficient. The feature extraction network extracts the features of the search area and template and then uses the spatial–spectral cross-correlation module to obtain the similarity score matrix of the features. Finally, the prediction head performs binary classification and bounding box regression on the similarity matrix to generate the final tracking results. We present each component of SSDT-Net in detail in this section with some illustrations and discussions.

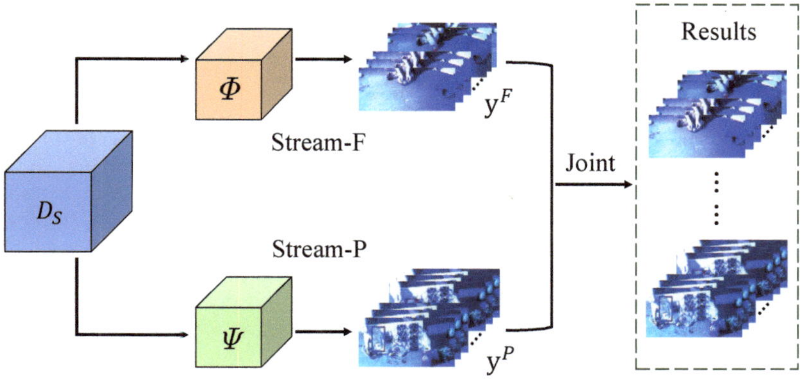

Figure 1. Our proposed dual-transfer module. D_S represents the source domain of transfer learning. In stream-F, we utilize the feature-based transfer learning method. In stream-P, we utilize the parameter-based transfer learning method.

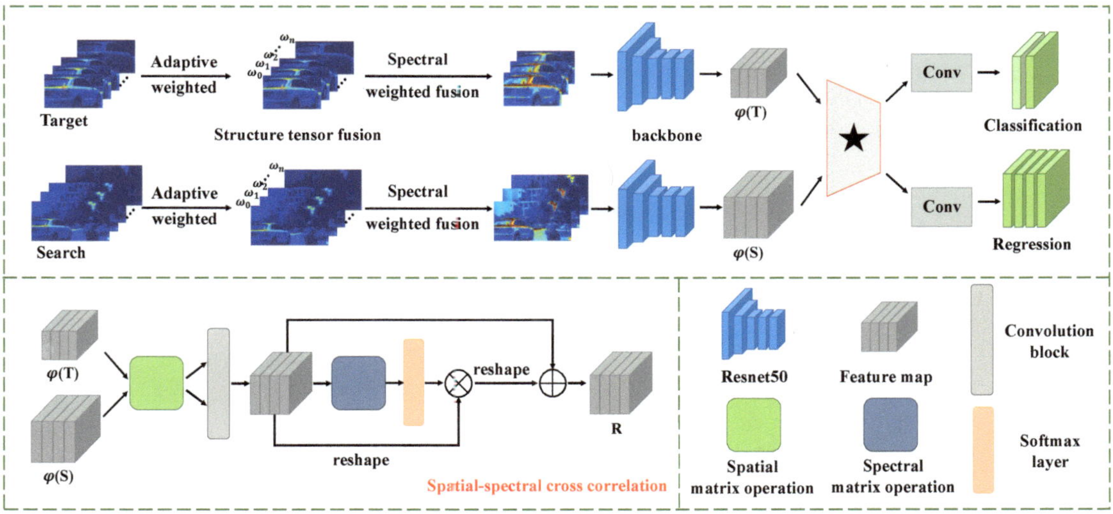

Figure 2. Illustration of the proposed SSDT-Net. Given the search region and target template, the network first reduces the dimensionality of HSIs through a spectral adaptive weighted fusion module, then uses a Siamese convolutional network for multi-channel response mapping extraction, and finally utilizes our proposed spatial–spectral cross-correlation module (denoted by ★) to embed the information between the two branches of the Siamese network to obtain response maps for regression and classification.

3.1. Dual Transfer

As is already known, traditional color videos contain millions of labeled natural images with thousands of categories. We use traditional color videos as the source domain for this transfer learning task. We first use our proposed spatial–spectral cross-correlation embedded network model to fully train on the source domain to obtain a knowledge weight, and then we fine-tune it on the HS dataset by freezing different numbers of network layers based on this knowledge weight. By analyzing the experimental results, we found that, for some categories, such as vehicles, animals, and people, the tracking accuracy increased as the number of frozen layers increased, while for other categories, such as paper, hard disk, circuit boards, and toys, as the number of frozen layers increased, the tracking accuracy became increasingly lower.

Then, we analyzed the reasons for this phenomenon in combination with the source domain and came to the conclusion that it is because the source domain we use is traditional RGB videos, and most of these datasets shoot scenes, such as vehicles, animals, and people. Therefore, in terms of the knowledge weight, we realize from the source domain training for these scenarios that the knowledge it learns is already very robust when used to test similar scenarios. On the contrary, if the feature extraction network is trained with very few such scenarios, it will lead to overfitting of the network, thereby, reducing the tracking accuracy. However, for scenarios where there is little or no presence in the source domain—that is, if the number of frozen layers is too many, a network on the dataset will not converge.

Based on the above analysis and inspired by [26], we propose a dual-transfer strategy that transfers traditional color video knowledge to the HS tracking task in different ways by measuring the similarity between the source domain and target domain. The specific operation is to use two different transfer learning methods according to the known prior information of each video category. As shown in Figure 1, the dual-transfer framework consists of two transfer learning streams.

Before formally introducing the strategy, we first introduce the notations used in this subsection. Let the source domain of transfer learning, i.e., traditional color videos, be represented by D_S, and the target domain of transfer learning, i.e., HS videos, be represented by D_T. The samples in the target domain that are similar to the source domain are defined as D_{TF}, and the samples in the target domain that are not similar to the source domain are defined as D_{TP}. The given test set of the target task is denoted as X. Similarly, the test set X can be divided into X_F and X_P. Y represents the test results, consisting of Y_F and Y_P. The following formula is established

$$
\begin{aligned}
D_S &= \{\{x_S^1, y_S^1\}, \cdots, \{x_S^M, y_S^M\}\} \\
D_T &= \{\{x_T^1, y_T^1\}, \cdots, \{x_T^N, y_T^N\}\} \\
D_{TF} &= \{\{x_T^1, y_T^1\}, \cdots, \{x_T^i, y_T^i\}\} \\
D_{TP} &= \{\{x_T^{i+1}, y_T^{i+1}\}, \cdots, \{x_T^N, y_T^N\}\} \\
X &= \{\{x^1\}, \cdots, \{x^K\}\} \\
X_F &= \{\{x^1\}, \cdots, \{x^j\}\} \\
X_P &= \{\{x^{j+1}\}, \cdots, \{x^K\}\} \\
Y &= \{\{y^1\}, \cdots, \{y^K\}\} \\
Y_F &= \{\{y^1\}, \cdots, \{y^j\}\} \\
Y_P &= \{\{y^{j+1}\}, \cdots, \{y^K\}\}
\end{aligned}
\tag{2}
$$

where x is the data instance and y is the corresponding label or test results. The subscript represents the domain index and the superscript is the sample index. With these notations, the objective of HS tracking is to find a predict function f such that

$$Y = f(X, D_T, D_S) \tag{3}$$

For D_{TF}, we utilize the proposed spatial–spectral cross-correlation embedded network, which is first trained on D_S and then used as knowledge to predict the outcome of the target task—that is, it is equivalent to transferring the knowledge of the source domain to obtain the features of the target domain. We call this a feature-based transfer learning method, which corresponds to stream-F in Figure 1. Denoting the classification and regression hypothesis as $\Phi(D_T, D_S)$ and considering Equation (3), the results of stream-F can be represented as

$$Y_F = f_F(X_F; \Phi(D_{TF}, D_S)) \quad (4)$$

On the other hand, for D_{TP}, the knowledge parameters of the source domain are transferred to initialize our network, and D_{TP} are trained based on knowledge weight parameters. We call this a parameter-based transfer learning method, which corresponds to stream-P in Figure 2. We denote the classification and regression in stream-P as $\Psi(D_T, D_S)$. Then, Equation (3) becomes

$$Y_P = f_P(X_P; \Psi(D_{TP}, D_S)) \quad (5)$$

Combining Equations (4) and (5), the dual-transfer strategy is formulated as

$$Y = f_F(X_F; \Phi(D_{TF}, D_S)) + f_P(X_P; \Psi(D_{TP}, D_S)) \quad (6)$$

to integrate the two parts of the results together.

This method is different from some previous transfer learning methods. In this new scheme, we take full advantage of the prior knowledge of sample categories so that the existing knowledge is capable of solving different but related problems to a greater extent. We show that the transfer learning method is effective for the problem of object tracking in HS videos and that the proposed dual-transfer strategy is more effective than single transfer learning.

3.2. Spectral Weighted Fusion

To reduce the spectral dimensionality, we apply a spectral adaptive weighted fusion method. A band after spectral adaptive fusion is obtained by:

$$\hat{B}_n = \sum_{l=1}^{L} \omega_l B_l, \quad n = 1, 2, \cdots, N \quad (7)$$

where B_l refers to the lth band and $[\omega_1, \omega_2, \cdots \omega_L]^T$is the weight vector of each band. In particular, ST is exploited to obtain the weight vector adaptively, and the structure information of the image is represented with effect. According to Equation (7), the ST of the lth band at the ith pixel in an HSI can be expressed as:

$$S_i^l = \begin{bmatrix} \left(Y_i^l\right)_x^2 & \left(Y_i^l\right)_x \left(Y_i^l\right)_y \\ \left(Y_i^l\right)_x \left(Y_i^l\right)_y & \left(Y_i^l\right)_y^2 \end{bmatrix} \quad (8)$$

where S_i^l is the ST of the lth band at the ith pixel. $\left(Y_i^l\right)_x = \left(\partial H^l / \partial x\right)$ and $\left(Y_i^l\right)_y = \left(\partial H^l / \partial y\right)$ stand for the gradient of the lth band at ith pixel along with the $x-$ and $y-$ directions. This ST S_i^l can be decomposed as:

$$S_i^l = \begin{bmatrix} \eta_{1,i}^l & \eta_{2,i}^l \end{bmatrix} \begin{bmatrix} \lambda_{1,i}^l & 0 \\ 0 & \lambda_{2,i}^l \end{bmatrix} \begin{bmatrix} \eta_{1,i}^l \\ \eta_{2,i}^l \end{bmatrix} \quad (9)$$

where $\eta_{1,i}^{l}$ and $\eta_{2,i}^{l}$ are the corresponding eigenvectors of the lth band at ith pixel and $\lambda_{1,i}^{l}, \lambda_{2,i}^{l}$ are the nonnegative eigenvalues of the lth band at the ith pixel. Assuming that $\lambda_{1,i}^{l}$ is the larger eigenvalue of the lth band at the ith pixel, i.e., $\lambda_{1,i}^{l} \geqslant \lambda_{2,i}^{l}$, then the larger eigenvalue, $\lambda_{1,i}^{l}$ indicates the edge strength of the lth band at the ith pixel, and the corresponding eigenvector, $\eta_{1,i}^{l}$, illustrates the direction in which the maximum gray value fluctuates.

If $\lambda_{1,i}^{l} > \lambda_{1,i}^{l+1}$, the edge strength of the lth band at the ith pixel is larger—that is, the lth band contains more structural information [27]. Therefore, its weight should be larger at the ith pixel. Similarly, the $(l+1)$th band should have a larger proportion of weight at the ith pixel in the case of $\lambda_{1,i}^{l} < \lambda_{1,i}^{l+1}$. When the eigenvalues of all bands are equal at the ith pixel, the weights of all bands are the same at this ith pixel, i.e., $1/l$.

Based on the analysis above, the weight vector for each band at the ith pixel can be gained by

$$\omega_{d,i} = \frac{\lambda_{1,i}^{l}}{\sum_{d=1}^{l} \lambda_{1,i}^{l}}, \quad d = 1, 2, \cdots l \tag{10}$$

where $\lambda_{1,i}^{l}$ expresses the weight of the lth band at the ith pixel. In this way, the weight vector of each band can be obtained by Equation (10), and the appropriate bands can be adaptively weighted by Equation (7). Figure 3 shows the visualization results of the spectral weighted fusion of the l bands. It is clear that the fused bands can better represent the corners of the target. Finally, the sample band set consists of $\hat{B}_1, \hat{B}_2, \cdots, \hat{B}_n$. Considering the balance between computational cost and tracking performance, we set $n = 3$.

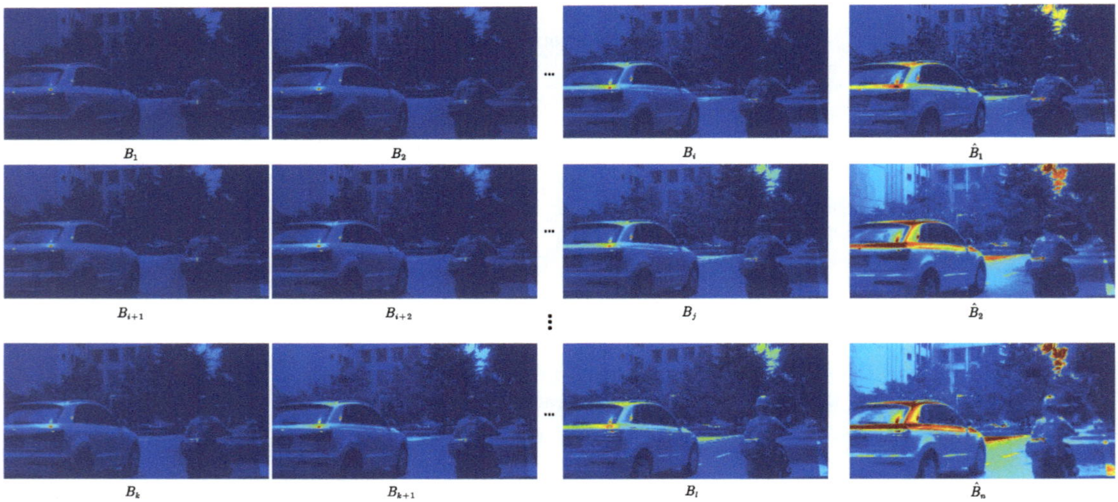

Figure 3. Visualization results of the spectral weighted fusion of the l bands.

3.3. Common Feature Extraction

Here, we make use of the CNN to structure the Siamese network structure to extract the common features. The Siamese network structure is comprised of two branches: one branch that learns the feature expression of the target T, and the other one that learns the feature presentation of the search area S. The two branches share the same CNN structure as their backbone networks, producing two feature maps $\varphi(T)$ and $\varphi(S)$. Low-level features, such as corners, shapes, edges, and colors, are indispensable visual attributes in location recognition, while high-level features are the key to semantic attribute recognition.

Therefore, many methods utilize the fusion of high-level and low-level features to better characterize the feature information of an image [22,28]. Here, we also consider the aggregate of multi-layer deep features to extract the common features. We utilize the modified ResNet-50 in [22] as our backbone model. To accomplish better recognition and inference discrimination, we associate the features extracted from the last three residual blocks of the backbone, which are expressed as $\Gamma_3(S), \Gamma_4(S)$, and $\Gamma_5(S)$. Specifically, the channel-wise concatenation is expressed as:

$$\varphi(S) = Cat(\Gamma_3(S), \Gamma_4(S), \Gamma_5(S)) \tag{11}$$

Then, to take full advantage of the abundant spatial–spectral information of two common feature maps, $\varphi(T)$ and $\varphi(S)$, we propose a spatial–spectral cross-correlation module to embed the information of these two branches as discussed in the next subsection.

3.4. Spatial–Spectral Cross-Correlation

The cross-correlation module is the essential operation of embedding the information of two branches. As we know, in some deep learning trackers for traditional color videos, such as SiamFC, SiamRPN, and SiamRPN++, some cross-correlation modules are proposed. SiamFC gains a single-channel response map for object localization by taking advantage of a cross-correlation layer. SiamRPN scales the channels by adding a large convolution layer so that the cross-correlation module can embed higher levels of information (UP-XCorr). In SiamRPN++, the depth-wise cross-correlation (DW-XCorr) is proposed, which performs the cross-correlation operation channel by channel to achieve efficient information association.

Figure 2 shows a spatial–spectral cross-correlation (SS-XCorr) module. The module first performs spatial matrix operations on the output features $\varphi(T)$ and $\varphi(S)$ of the feature extraction network. Then, we subject them to a grouped convolution operation to spatially correlate each channel to generate a response map $R_1^{W\times H\times B}$. The response map $R_1^{W\times H\times B}$ has the same number of channels as the common feature maps $\varphi(T)$ and $\varphi(S)$. The specific operation process is shown in Figure 4.

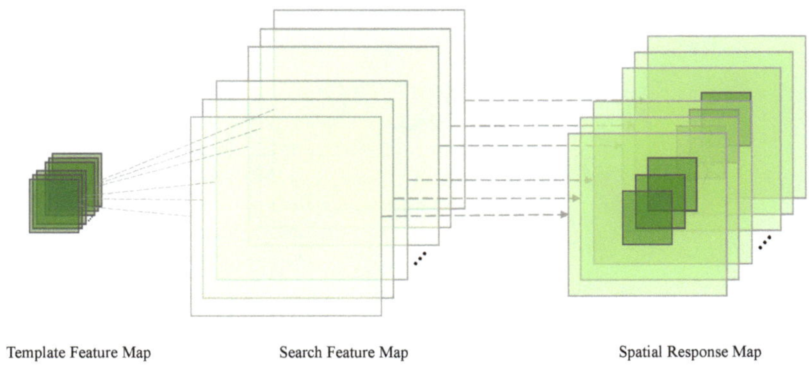

Figure 4. Spatial cross-correlation operation.

Here, we use the template feature map as the convolution kernel of the convolution, the search feature map as the input of the convolution, and use grouped convolution to perform channel-by-channel spatial cross-correlation operations on the template and search feature maps to obtain the spatial response map. For HS data, the channel dimension can be equivalent to the spectral dimension. Taking advantage of the hyperspectral correlation between adjacent bands, we can highlight the channels with more important information in the spatial response map. Therefore, we perform spectral matrix operations on the response map $R_1^{W\times H\times B}$, and we use a softmax layer to gain the spectral attention map $X^{B\times B}$.

Moreover, we perform a matrix multiplication between the transformation of $R_1^{W\times H\times B}$ and $X^{B\times B}$ and reshape their result to $R_1^{\prime W\times H\times B}$. Then, we perform an element-wise sum operation between R_1 and R_1' to gain the final response map $R^{W\times H\times B}$. The generated response map R models the long-range semantic correlation between spectral–spatial feature maps, and it includes massive information for classification and regression.

Figure 5a,b are the visualization results of the similarity matrix using the DW-XCorr module and the SS-XCorr module, respectively. It is evident that our proposed SS-XCorr module enhances the discrimination of distractors in the target tracking task.

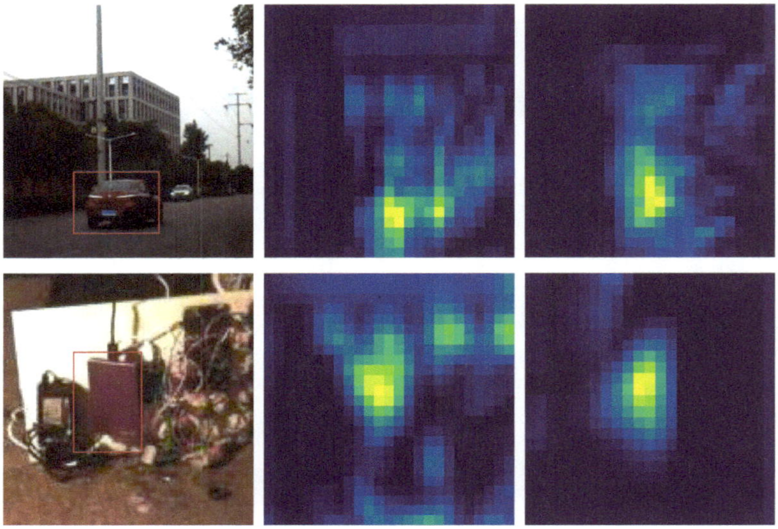

(a)With DW-XCorr (b)With SS-XCorr

Figure 5. Visualization results of the similarity matrix using the (**a**) DW-XCorr module and the (**b**) SS-XCorr module.

3.5. Bounding Box Prediction

Every position in the response map R can be mapped back to the input search area S. The bounding box prediction task is divided into two subtasks: one is the classification of target and background, and the other is the regression branch for the positioning of the target bounding box. For the response map $R^{W\times H\times L}$, the classification branch outputs a feature map $D^{cls}_{W\times H\times 2}$ for classification, and the regression branch produces a regression feature map $D^{reg}_{W\times H\times 4}$. As shown in Figure 2, each point in $D^{cls}_{W\times H\times 2}$ involves a $2D$ vector, which represents positive and negative scores of each anchor at the corresponding position on the input search region. Similarly, each point in $D^{reg}_{W\times H\times 4}$ involves a $4D$ vector, which expresses the distances between the anchor and the relevant ground truth.

Inspired by [24], we also utilize the cross-entropy loss for classification and the smooth L_1 loss with normalized coordinates for regression. Let A_x, A_y, A_w, and A_h represent the center point and shape of the anchor boxes, respectively, and G_x, G_y, G_w, and G_h represent those of the ground truth boxes, respectively. The normalized distance is:

$$\begin{aligned} d_x &= (G_x - A_x)/A_w, d_y = (G_y - A_y)/A_h \\ d_w &= \ln(G_w/A_w), d_h = \ln(G_h/A_h) \end{aligned} \tag{12}$$

The regression loss is:

$$L_{reg} = \sum_{i\in\{x,y,w,h\}} smooth_{L_1}(d_i,\sigma) \tag{13}$$

where

$$smooth_{L_1}(x,\sigma) = \begin{cases} 0.5\sigma^2 x^2, & |x| < \frac{1}{\sigma^2} \\ |x| - \frac{1}{2\sigma^2}, & |x| \geqslant \frac{1}{\sigma^2} \end{cases} \tag{14}$$

is a smooth L_1 loss.

The overall loss function is

$$Loss = L_{cls} + \lambda L_{reg} \quad (15)$$

where L_{cls} is the cross-entropy loss for classification, and the weighting parameter λ is empirically set to be 1.2 for stable tracking performance.

4. Experiments

In this section, we describe our experimental setup, compare the performance of our proposed SSDT-Net tracker against several typical color trackers and HS trackers, and present an ablation study to emphasize the importance of different components of our model.

4.1. Experimental Setup

Data Description: The color video data sets commonly used by traditional color video trackers act as the source domain for transfer learning, including COCO [29], YouTube-BoundingBoxes Dataset [30], ImageNet VID, and ImageNet DET [31]. COCO contains 80 categories and 200,000 images, the ImageNet DET and ImageNet VID datasets contain 200 object categories, and the YouTube-BoundingBoxes Dataset contains 10.5 million manually labeled frames on 23 object categories. The dataset provided by the HS object tracking competition is used as the object domain and target task for transfer learning.

The dataset is acquired by a XIMEA SNm4x4 VIS Camera and consisted of 40 training videos and 35 testing videos, each containing HS, RGB videos, and false-color in the same scene. The HS datasets were shot at 25 frames per second (FPS). Each frame was initially captured from a 2D image with 16 bands arranged in a mosaic pattern. Each frame was then converted to 3D, with the first two dimensions representing the position of each pixel and the third dimension representing the band number. The 16 bands cover the range from 470 to 620 nm, and each band is 512 $\times$ 256 pixels in size. RGB video was also obtained at the same frame rate at viewpoints very close to the HS video. The dataset also provides false-color videos converted from HS videos.

Evaluation Metrics: We employ the four most commonly used object tracking evaluation indexes, including success plot, precision plot, area under the curve (AUC), and the average distance precision scores at a threshold of 20 pixels (DP@20P), to compare the performance of the proposed method with its comparison methods quantitatively. However, it should be noted that most of the representative object tracking algorithm codes based on HS videos have not yet become open source, and thus we only use AUC as the evaluation index for quantitative comparison.

The success plot indicates the percentages of successful frames whose predicted bounding box and ground-truth overlap ratio is greater than a certain threshold within 0 and 1. A precision plot records the percentage of video frames whose distance between the center point of the estimated object position and the center point of the ground truth is less than a given threshold. Different thresholds result in different percentages; therefore, a curve can be obtained. All the results are presented through a one-pass evaluation (OPE), where the tracker runs through the entire test sequence and is initialized from the ground truth position in the initial frame.

Implementation Details: Our proposed approach was implemented in Python with PyTorch and trained on four RTX 2080Ti cards. The backbone network was pre-trained on ImageNet for image labeling, which has a good initialization effect on other tasks [32,33]. In stream-F, we train our network on the source domain and then use it as a knowledge model to predict the outcome of the object task. In stream-P, we initialize our network with the knowledge model of the source domain and then train it on the target domain to estimate the knowledge weight, which is utilized to predict the result of the target task. During this process, we set the batch size to 128, utilize Stochastic Gradient Descent (SGD) with an initial learning rate of 0.005, and perform a total of 20 epochs.

For the first five epochs, the parameters of the Siamese network are frozen while training the regression and classification branches. For the last 15 epochs, the entire network is trained end-to-end with an exponential drop in learning rate from 0.01 to 0.0005.

4.2. Ablation Study

An ablation study was conducted to demonstrate the validity of the proposed components by evaluating several variants of the SSDT-Net tracker on HS videos. We utilize the results of the network training from scratch as the baseline.

Dual Transfer: Figure 6 shows the success plots and precision plots of the baseline, feature-based transfer learning, parameter-based transfer learning, and dual-transfer learning strategy. The three transfer learning methods achieved significant improvement compared with the baseline. Here, feature-based transfer represents the result of overfitting of similar scenes in the source and target domains, while parameter-based transfer represents the result of network divergence in different scenes. As shown in Table 1, the value of AUC obtained reached 0.4848 and 0.5233 for the baseline when utilizing the DW-XCorr module and SS-XCorr module to embed information. When employing feature-based transfer learning, the performance of each group was significantly improved by 5.66% and 3.75%, respectively.

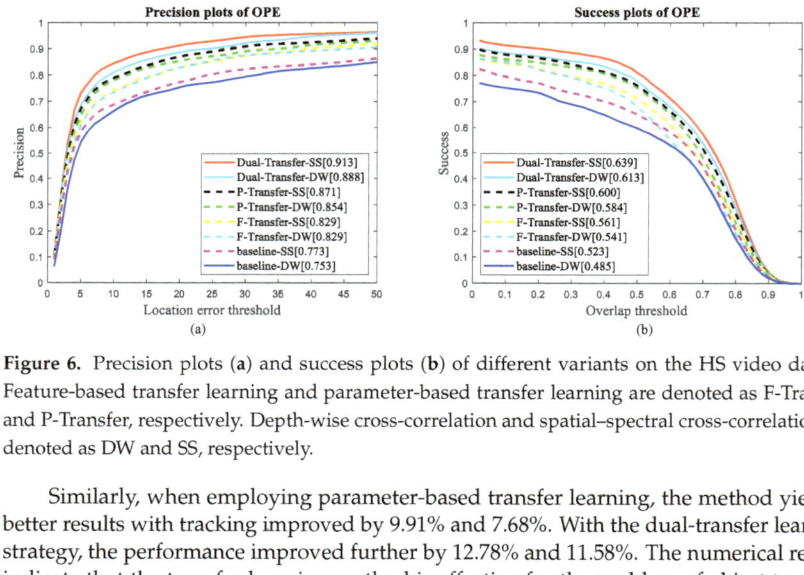

Figure 6. Precision plots (**a**) and success plots (**b**) of different variants on the HS video dataset. Feature-based transfer learning and parameter-based transfer learning are denoted as F-Transfer and P-Transfer, respectively. Depth-wise cross-correlation and spatial–spectral cross-correlation are denoted as DW and SS, respectively.

Similarly, when employing parameter-based transfer learning, the method yielded better results with tracking improved by 9.91% and 7.68%. With the dual-transfer learning strategy, the performance improved further by 12.78% and 11.58%. The numerical results indicate that the transfer learning method is effective for the problem of object tracking in HS videos, and this also confirms the correctness of our theoretical reasoning. Feature-based transfer learning may lead to the overfitting of similar scenes in the source and target domains, and parameter-based transfer learning may cause the divergence of networks in different scenes. The proposed dual-transfer strategy is more effective than single-transfer learning.

Spatial–Spectral Cross-Correlation (SS-XCorr): Figure 6 and Table 1 report the results of the network using DW-XCorr and SS-XCorr. Intuitively, it is expected that the SS-XCorr can be helpful because, for HS data, it is crucial to make good use of spatial–spectral joint information for the processing of its related tasks. This is indeed apparent in Table 1 where an increase in AUC is clear in the four states of the network with SS-XCorr. The value of the AUC obtained reached 0.4848, 0.5414, 0.5839, and 0.6126 for the baseline and three transfer learning strategies when utilizing the DW-XCorr module to embed information, respectively.

When utilizing the SS-XCorr module to embed information, the method yielded better results, with tracking results up to 3.85%, 1.94%, 1.62%, and 2.65%, respectively. Therefore, combining the spatial structure identification ability of the spatial dimension and the material identification ability of the spectral dimension in the HSIs is effective for more accurate object tracking.

Table 1. Ablation study of our proposed tracker on HS videos.

Transfer Strategy	Corr	AUC	DP@20P
Baseline	DW	0.4848	0.7533
	SS	0.5233	0.7727
Feature-based Transfer	DW	0.5414	0.8289
	SS	0.5608	0.8289
Parameter-based Transfer	DW	0.5839	0.8540
	SS	0.6001	0.8710
Dual-transfer	DW	0.6126	0.8875
	SS	**0.6391**	**0.9132**

DW and SS represent depth-wise cross-correlation and spatial–spectral cross-correlation, respectively. Red marks the top value.

4.3. Quantitative Comparison with the State-Of-The-Art Color Trackers

The proposed SSDT-Net tracker was compared with six state-of-the-art color trackers, including the BACF [34], fDSST [35], MCCT [36], SRDCF [37], SiamRPN++, and SiamCAR [38] methods. fDSST is a scale-adaptive tracking method that learns two distinguishing correlation filters for scale estimation and object localization. SRDCF and BACF aim to regularize the correlation filter on the basis of the spatial distribution to reduce the boundary effect on training samples prompted by periodic assumptions. MCCT brings together numerous experts to track objects through diverse levels of features.

SiamRPN++ is a modified method based on SiamRPN, which uses a deeper Siamese network to attach a sub-network for region proposal (RPN) extraction and performs visual tracking through joint training of a regression branch and a classification branch. By splitting the visual tracking problem into two sub-problems of pixel object bounding box regression and pixel classification, SiamCAR deploys a new fully convolutional Siamese network to settle the end-to-end visual tracking problem in a pixel-by-pixel manner.

The SSDT-Net tracker is tested on HS data. Since the contrast trackers selected were exploited for RGB data, they all run on both RGB data and false-color data converted from HS data. Figures 7 and 8 show the success plots and precision plots of the SSDT-Net tracker on HS videos and all color trackers on false-color videos generated from HS videos and RGB videos, respectively. The results indicate that the performance of the deep feature-based trackers in color video trackers is better than most of the handcrafted feature-based trackers. For trackers based on hand-crafted features, MCCT brings together multiple experts to track objects through different levels of features, resulting in better tracking performance.

In addition, as shown in Table 2 compared with the above color video trackers, the SSDT-Net tracker we proposed ranked the highest over a range of thresholds by achieving an AUC of 0.6391 and a DP@20P of 0.9132. When using traditional color videos, the highest performance color video tracker was SiamRPN, with the values of AUC and DP@20P reaching 0.5852 and 0.8371, respectively. In comparison, the AUC and DP@20P of the SSDT-Net tracker improved by 5.39% and 7.61%.

Analogously, when employing false-color videos, SiamCAR achieved the highest performance, with the values of AUC and DP@20 reaching 0.5441 and 0.7936. In comparison, the SSDT-Net tracker yielded better results, with tracking improving by 9.50% and 11.96%. Numerical results confirm that the SSDT-Net tracker can make good use of the knowledge transferred from the color videos as well as local spatial information and the detailed constituent material distribution information of the HS data to represent the image content, which helps the tracker to distinguish the object from the background.

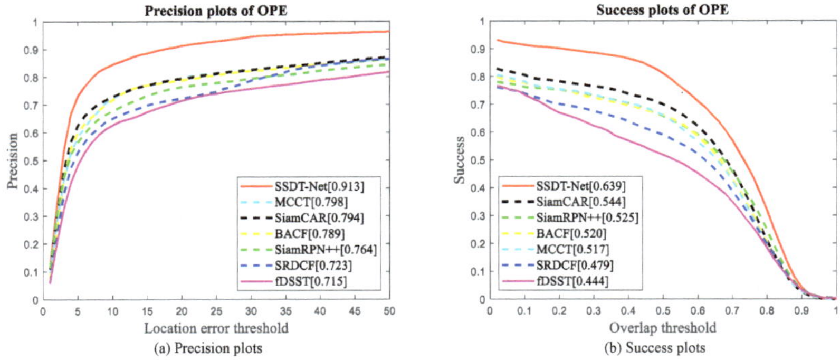

Figure 7. Comparisons with color video trackers on corresponding false-color videos. Our approach fulfills the best accuracy with an AUC of 63.9%.

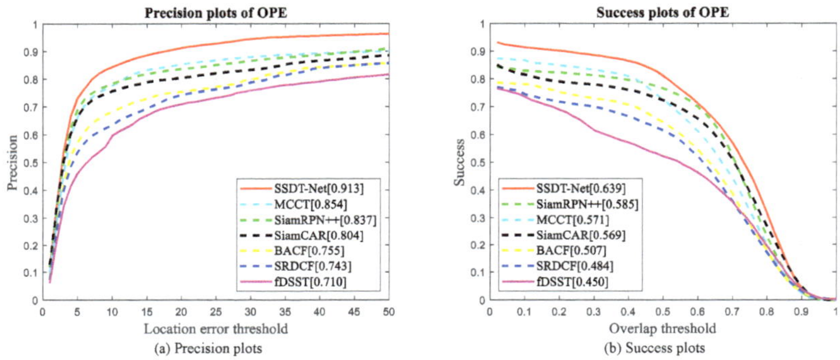

Figure 8. Comparisons with color video trackers on RGB videos. Our approach exceeds all the other trackers.

Table 2. Performance comparison with color video trackers in terms of the AUC and DP@20P.

Tracker	HS/False-Color Videos		Color Videos	
	AUC	**DP@20P**	**AUC**	**DP@20P**
BACF	0.5198	0.7889	0.5072	0.7549
fDSST	0.4442	0.7149	0.4499	0.7099
MCCT	0.5170	0.7984	0.5709	0.8542
SRDCF	0.4793	0.7229	0.4837	0.7433
SiamRPN++	0.5253	0.7643	0.5852	0.8371
SiamCAR	0.5441	0.7936	0.5688	0.8044
SSDT-Net	0.6391	0.9132	-	-

The top two values are marked by red and blue.

4.4. Quantitative Comparison with the State-Of-The-Art Hyperspectral Trackers

The SSDT-Net tracker was compared with three recent HS trackers, SST-Net, BAE-Net, and MHT. MHT is a tracker based on handcrafted features that embed the spectral–spatial information of the HS image into a multi-dimensional gradient histogram and combines it with the global material abundance feature to describe the target. Both BAE-Net and SST-Net take advantage of deep features. BAE-Net converts HSI based on spectral information into massive false-color images. These converted images generate several weak trackers through several VITAL trackers for subsequent ensemble learning to acquire the object positions. SST-Net combines spatial, spectral, and temporal attention based on BAE-Net to better select valuable HS bands for deep ensemble tracking.

Table 3 demonstrates the comparative results of these HS trackers. Although the BAE-Net tracker is based on data-driven deep features, the MHT tracker based on handcrafted features achieves higher tracking accuracy compared to the BAE-Net tracker because of its limited annotation data used for training and only utilizing the spectral information of HS data.

Table 3. Performance comparison with hyperspectral trackers of AUC.

Video	SSDT-Net	MHT	BAE-Net	SST-Net
Hyperspectral	0.6391	0.6080	0.6060	0.6230

The top two values are marked by red and blue.

Moreover, on the basis of BAE-Net, SST-Net increases the use of spatial-temporal information when converting HSI, which can describe the object more effectively and produce better tracking performance. For the SSDT-Net tracker, the prior knowledge of a large number of color tracking videos is fully utilized, and spectral–spatial information is considered simultaneously to extract object features, yielding the highest tracking accuracy, with an AUC value of 0.6391. The AUC value was further improved by 1.61%, compared with SST-Net, which is second only to the SSDT-Net tracker.

4.5. Running Time Comparison

Table 4 reports the tracking speed of six state-of-the-art color trackers and two state-of-the-art HS trackers on CPUs and GPUs concerning frames per second (FPS), respectively. BACF, fDSST, MCCT, SRDCF, and MHT were implemented on Windows machines equipped with Intel(R) Core(TM) i5-10500 CPU @ 3.10GHz and 32GB of RAM. SSDT-Net, SiamRPN++, and SiamCAR are performed in Python with PyTorch and trained on four RTX 2080Ti cards. fDSST attained the best speed on both videos on CPUs. SSDT-Net achieved similar speed with deep color trackers. The experimental results prove that our HS tracker can acquire better tracking performance based on a similar speed with state-of-the-art traditional color trackers.

Table 4. Running time comparison (FPS).

Videos	SSDT-Net	MHT	BACF	fDSST
RGB	-	-	60.6	120.6
Hyperspectral/False-Color	35.7	6.09	66.36	120.1
Videos	**MCCT**	**SRDCF**	**SiamRPN++**	**SiamCAR**
RGB	7.5	24.3	35.2	52.3
Hyperspectral/False-Color	7.3	25.8	35.5	52.5

4.6. Demonstrations of Visual Tracking Results

Figure 9 visually compares the tracking results of the SSDT-Net tracker with several color trackers on sample HS or false-color videos. It can be seen that the SSDT-Net tracker also provides the best visualization results in complex scenes, such as background clutter, target appearance similar to background, and target appearance changing rapidly. The results demonstrate the high robustness of the spatial structure identification ability and material identification ability brought by the rich spatial–spectral information of HS data, as well as the effectiveness of transfer learning applied to HS object tracking.

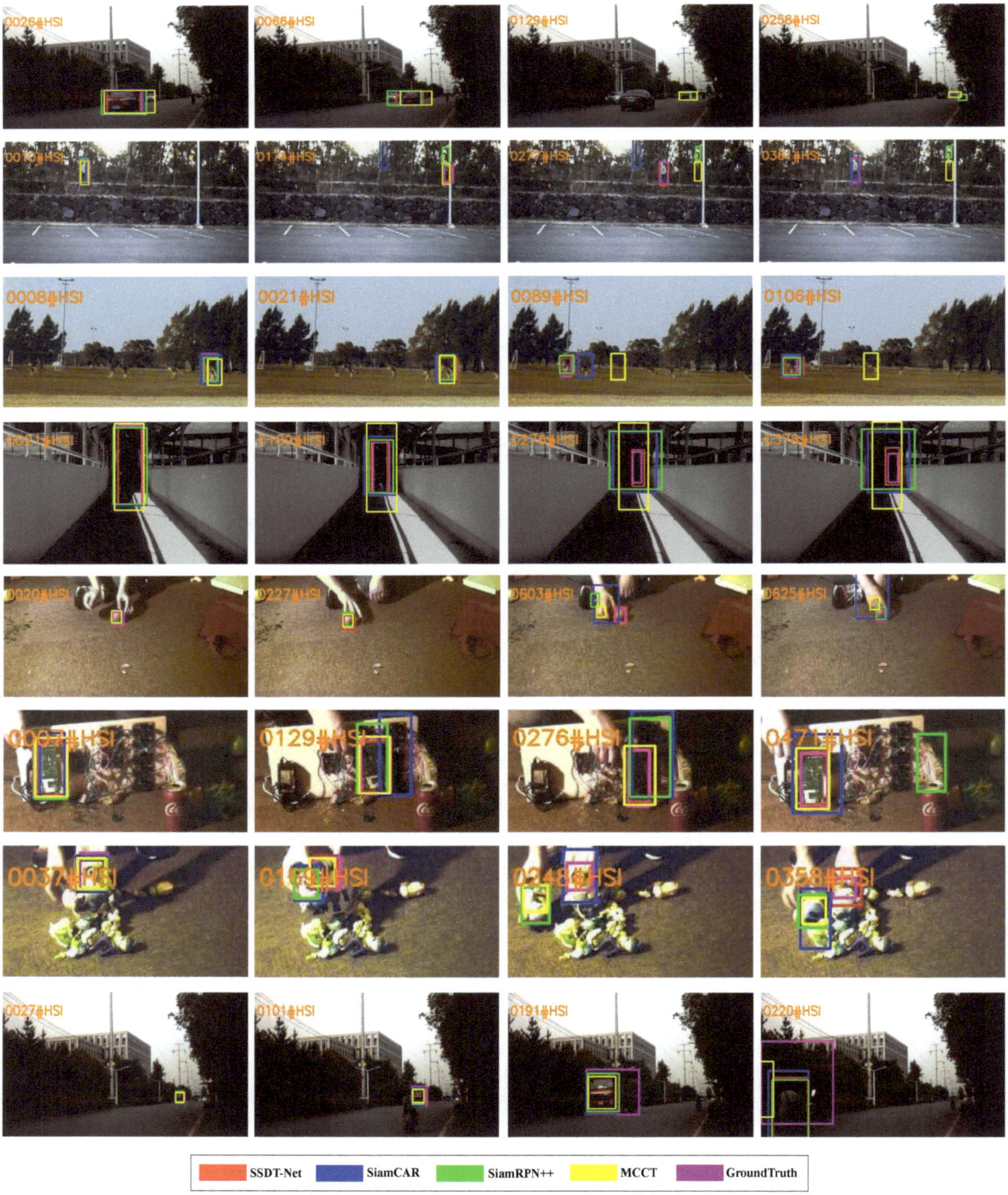

Figure 9. Visualization results of our proposed hyperspectral tracker and several color trackers (scenes: car3, forest2, kangaroo, student, ball, board, toy1, and truck).

5. Discussion

Although the effectiveness of the proposed SSDT-Net was verified by comprehensive and convincing experiments and analysis, there are still some interesting details that are further discussed below as future work:

1. As shown in Section 3.2, we directly use the spectral weighted fusion module based on the ST to fuse the data at the input end of the network to reduce the dimension and perform the fusion operation at the input of the network; although the tracker can obtain excellent speed, it will cause a great deal of loss of information, and the material identification characteristics of HS data will limit the performance improvement of the tracker. Therefore, in the following research, we can consider the fusion of feature maps after feature extraction on the network so as to make better use of the spectral information of HS data.
2. The experimental results show that the spectral cross-correlation module plays a role, improves the subjective and objective effects, and proves the effectiveness of spectral information for target recognition. Therefore, we can consider further investigation of spectral cross-correlation operations.
3. Regarding the cross-correlation module of the Siamese network, the correlation operation itself is a local linear matching process, and it is easy to lose semantic information and fall into a local optimum, which may become a bottleneck in designing high-precision tracking algorithms. Therefore, inspired by Transformer, we consider using the attention mechanism to perform cross-correlation operation, which requires further study.

6. Conclusions

In this paper, we proposed a novel HS tracker based on transfer learning, i.e., SSDT-Net, for the first time in the field of HS object tracking. The key points of this method are to transfer the knowledge of traditional color videos to the HS tracking task through transfer learning and to distinguish the target and background by combining the spatial structure information and material information of the HS data. Therefore, the superiority of the SSDT-Net tracker is due to several aspects.

First, the proposed dual-transfer strategy based on transfer learning makes the best use of color video information to solve the HS video tracking problem. Then, we take advantage of our proposed spatial–spectral cross-correlation module to better embed spatial and material information between the two branches of the Siamese network.

Furthermore, to make our network more efficient under the premise of ensuring high performance, a novel spectral weighted fusion method based on ST was adopted to gain the inputs of the network. Experiments performed on HS videos demonstrated that the proposed SSDT-Net tracker exhibited excellent performance in terms of quantitative evaluation and visual comparison based on its similar speed to state-of-the-art traditional color trackers.

Author Contributions: J.L. and P.L. provided conceptualization; J.L. and P.L. designed the methodology; P.L. performed the experiments and analyzed the result data; W.X. and L.G. investigated related work; W.X., Y.L. and Q.D. provided suggestions on paper revision; and P.L. and W.X. wrote the paper. All authors have read and agreed to the published version of the manuscript.

Funding: This research was funded by This work was supported in part by the National Natural Science Foundation of China under Grant 62071360, Grant 61571345, Grant 91538101, Grant 61501346, Grant 61502367 and Grant 61701360, in part by the Fundamental Research Funds for the Central Universities JB210103.

Data Availability Statement: Our experiments employ open datasets in [6].

Conflicts of Interest: The authors declare no conflict of interest.

Abbreviations

The following abbreviations are used in this manuscript:

HS	Hyperspectral
CNN	Convolutional neural network
ST	Structure tensor
DW-XCorr	Depth-wise cross-correlation
SS-XCorr	Spatial–spectral cross-correlation
AUC	Area under the curve
DP@20P	Distance precision scores at a threshold of 20 pixels
SGD	Stochastic gradient descent

References

1. Marvasti-Zadeh, S.M.; Cheng, L.; Ghanei-Yakhdan, H.; Kasaei, S. Deep Learning for Visual Tracking: A Comprehensive Survey. *IEEE Trans. Intell. Transp. Syst.* **2021**, *23*, 1–26. [CrossRef]
2. Gao, J.; Zhang, T.; Yang, X.; Xu, C. Deep Relative Tracking. *IEEE Trans. Image Process.* **2017**, *26*, 1845–1858. [CrossRef]
3. Liang, J.; Zhou, J.; Tong, L.; Bai, X.; Wang, B. Material based salient object detection from hyperspectral images. *Pattern Recognit.* **2018**, *76*, 476–490. [CrossRef]
4. Okwuashi, O.; Ndehedehe, C.E. Deep support vector machine for hyperspectral image classification. *Pattern Recognit.* **2020**, *103*, 107298. [CrossRef]
5. Qian, K.; Zhou, J.; Xiong, F.; Zhou, H.; Du, J. Object tracking in hyperspectral videos with convolutional features and kernelized correlation filter. In *International Conference on Smart Multimedia*; Springer: Berlin/Heidelberg, Germany, 2018; pp. 308–319.
6. Xiong, F.; Zhou, J.; Qian, Y. Material based object tracking in hyperspectral videos. *IEEE Trans. Image Process.* **2020**, *29*, 3719–3733. [CrossRef] [PubMed]
7. Uzkent, B.; Rangnekar, A.; Hoffman, M.J. Tracking in aerial hyperspectral videos using deep kernelized correlation filters. *IEEE Trans. Geosci. Remote Sens.* **2018**, *57*, 449–461. [CrossRef]
8. Simonyan, K.; Zisserman, A. Very Deep Convolutional Networks for Large-Scale Image Recognition. In Proceedings of the 3rd International Conference on Learning Representations, ICLR 2015, San Diego, CA, USA, 7–9 May 2015.
9. Li, Z.; Xiong, F.; Zhou, J.; Wang, J.; Lu, J.; Qian, Y. BAE-Net: A Band Attention Aware Ensemble Network for Hyperspectral Object Tracking. In Proceedings of the 2020 IEEE International Conference on Image Processing (ICIP), Abu Dhabi, United Arab Emirates, 25–28 October 2020; pp. 2106–2110.
10. Song, Y.; Ma, C.; Wu, X.; Gong, L.; Bao, L.; Zuo, W.; Shen, C.; Lau, R.W.; Yang, M.H. Vital: Visual tracking via adversarial learning. In Proceedings of the 2018 IEEE/CVF Conference on Computer Vision and Pattern Recognition, Salt Lake City, UT, USA, 18–23 June 2018; pp. 8990–8999.
11. Li, Z.; Ye, X.; Xiong, F.; Lu, J.; Zhou, J.; Qian, Y. Spectral-Spatial-Temporal Attention Network for Hyperspectral Tracking. Available online: http://www.ieee-whispers.com/wp-content/uploads/2021/03/WHISPERS_2021_paper_55.pdf (accessed on 26 March 2021).
12. Liu, Z.; Wang, X.; Shu, M.; Li, G.; Sun, C.; Liu, Z.; Zhong, Y. An Anchor-Free Siamese Target Tracking Network for Hyperspectral Video. Available online: http://www.ieee-whispers.com/wp-content/uploads/2021/03/WHISPERS_2021_paper_52.pdf (accessed on 26 March 2021).
13. Pan, S.J.; Yang, Q. A survey on transfer learning. *IEEE Trans. Geosci. Remote Sens.* **2009**, *22*, 1345–1359. [CrossRef]
14. Bertinetto, L.; Valmadre, J.; Henriques, J.F.; Vedaldi, A.; Torr, P.H. Fully-convolutional siamese networks for object tracking. In *European Conference on Computer Vision*; Springer: Berlin/Heidelberg, Germany, 2016; pp. 850–865.
15. Held, D.; Thrun, S.; Savarese, S. Learning to track at 100 fps with deep regression networks. In *European Conference on Computer Vision*; Springer: Berlin/Heidelberg, Germany, 2016; pp. 749–765.
16. Li, B.; Yan, J.; Wu, W.; Zhu, Z.; Hu, X. High performance visual tracking with siamese region proposal network. In Proceedings of the 2018 IEEE/CVF Conference on Computer Vision and Pattern Recognition, Salt Lake City, UT, USA, 18–23 June 2018; pp. 8971–8980.
17. Valmadre, J.; Bertinetto, L.; Henriques, J.; Vedaldi, A.; Torr, P.H. End-to-end representation learning for correlation filter based tracking. In Proceedings of the 2017 IEEE Conference on Computer Vision and Pattern Recognition (CVPR), Honolulu, HI, USA, 21–26 July 2017; pp. 2805–2813.
18. Wang, Q.; Gao, J.; Xing, J.; Zhang, M.; Hu, W. Dcfnet: Discriminant correlation filters network for visual tracking. *arXiv* **2017**, arXiv:1704.04057.
19. Zhu, Z.; Wang, Q.; Li, B.; Wu, W.; Yan, J.; Hu, W. Distractor-aware siamese networks for visual object tracking. In *European Conference on Computer Vision*; Springer: Berlin/Heidelberg, Germany, 2018; pp. 101–117.
20. Pitie, F.; Kokaram, A. The linear monge-kantorovitch linear colour mapping for example-based colour transfer. In Proceedings of the 4th European Conference on Visual Media Production, London, UK, 27–28 November 2007; pp. 1–9.
21. He, A.; Luo, C.; Tian, X.; Zeng, W. A twofold siamese network for real-time object tracking. In Proceedings of the 2018 IEEE/CVF Conference on Computer Vision and Pattern Recognition, Salt Lake City, UT, USA, 18–23 June 2018; pp. 4834–4843.

22. Li, B.; Wu, W.; Wang, Q.; Zhang, F.; Xing, J.; Yan, J. Siamrpn++: Evolution of siamese visual tracking with very deep networks. In Proceedings of the 2019 IEEE/CVF Conference on Computer Vision and Pattern Recognition (CVPR), Long Beach, CA, USA, 15–20 June 2019; pp. 4282–4291.
23. Guo, Q.; Feng, W.; Zhou, C.; Huang, R.; Wan, L.; Wang, S. Learning dynamic siamese network for visual object tracking. In Proceedings of the 2017 IEEE International Conference on Computer Vision (ICCV), Venice, Italy, 22–29 October 2017; pp. 1763–1771.
24. Ren, S.; He, K.; Girshick, R.; Sun, J. Faster R-CNN: Towards Real-Time Object Detection with Region Proposal Networks. *IEEE Trans. Pattern Anal. Mach. Intell.* **2017**, *39*, 1137–1149. [CrossRef] [PubMed]
25. He, K.; Zhang, X.; Ren, S.; Sun, J. Deep residual learning for image recognition. In Proceedings of the 2016 IEEE Conference on Computer Vision and Pattern Recognition (CVPR), Las Vegas, NV, USA, 27–30 June 2016; pp. 770–778.
26. Han, Y.; Huang, G.; Song, S.; Yang, L.; Wang, H.; Wang, Y. Dynamic neural networks: A survey. *IEEE Trans. Pattern Anal. Mach. Intell.* **2021**. [CrossRef] [PubMed]
27. Xie, W.; Jiang, T.; Li, Y.; Jia, X.; Lei, J. Structure tensor and guided filtering-based algorithm for hyperspectral anomaly detection. *IEEE Trans. Geosci. Remote Sens.* **2019**, *57*, 4218–4230. [CrossRef]
28. Ma, C.; Huang, J.B.; Yang, X.; Yang, M.H. Robust Visual Tracking via Hierarchical Convolutional Features. *IEEE Trans. Pattern Anal. Mach. Intell.* **2017**, 25, 670–674. [CrossRef] [PubMed]
29. Lin, T.Y.; Maire, M.; Belongie, S.; Hays, J.; Perona, P.; Ramanan, D.; Dollár, P.; Zitnick, C.L. Microsoft coco: Common objects in context. In *European Conference on Computer Vision*; Springer: Berlin/Heidelberg, Germany, 2014; pp. 740–755.
30. Real, E.; Shlens, J.; Mazzocchi, S.; Pan, X.; Vanhoucke, V. Youtube-boundingboxes: A large high-precision human-annotated data set for object detection in video. In Proceedings of the 2017 IEEE Conference on Computer Vision and Pattern Recognition (CVPR), Honolulu, HI, USA, 21–26 July 2017; pp. 5296–5305.
31. Russakovsky, O.; Deng, J.; Su, H.; Krause, J.; Satheesh, S.; Ma, S.; Huang, Z.; Karpathy, A.; Khosla, A.; Bernstein, M.; et al. Imagenet large scale visual recognition challenge. *Int. J. Comput. Vis.* **2015**, *115*, 211–252. [CrossRef]
32. He, K.; Gkioxari, G.; Dollár, P.; Girshick, R. Mask r-cnn. In Proceedings of the 2017 IEEE International Conference on Computer Vision (ICCV), Venice, Italy, 22–29 October 2017; pp. 2961–2969.
33. Long, J.; Shelhamer, E.; Darrell, T. Fully convolutional networks for semantic segmentation. In Proceedings of the 2015 IEEE Conference on Computer Vision and Pattern Recognition (CVPR), Boston, MA, USA, 7–12 June 2015; pp. 3431–3440.
34. Kiani Galoogahi, H.; Fagg, A.; Lucey, S. Learning background-aware correlation filters for visual tracking. In Proceedings of the 2017 IEEE International Conference on Computer Vision (ICCV), Venice, Italy, 22–29 October 2017; pp. 1135–1143.
35. Danelljan, M.; Häger, G.; Khan, F.S.; Felsberg, M. Discriminative scale space tracking. *IEEE Trans. Pattern Anal. Mach. Intell.* **2016**, *39*, 1561–1575. [CrossRef] [PubMed]
36. Wang, N.; Zhou, W.; Tian, Q.; Hong, R.; Wang, M.; Li, H. Multi-cue correlation filters for robust visual tracking. In Proceedings of the 2018 IEEE/CVF Conference on Computer Vision and Pattern Recognition, Salt Lake City, UT, USA, 18–23 June 2018; pp. 4844–4853.
37. Danelljan, M.; Hager, G.; Shahbaz Khan, F.; Felsberg, M. Learning spatially regularized correlation filters for visual tracking. In Proceedings of the 2015 IEEE International Conference on Computer Vision (ICCV), Santiago, Chile, 7–13 December 2015; pp. 4310–4318.
38. Guo, D.; Wang, J.; Cui, Y.; Wang, Z.; Chen, S. SiamCAR: Siamese fully convolutional classification and regression for visual tracking. In Proceedings of the 2020 IEEE/CVF Conference on Computer Vision and Pattern Recognition (CVPR), Seattle, WA, USA, 13–19 June 2020; pp. 6269–6277.

remote sensing

MDPI

Article

A Multiscale and Multitask Deep Learning Framework for Automatic Building Extraction

Jichong Yin [1], Fang Wu [1,*], Yue Qiu [1], Anping Li [2], Chengyi Liu [1] and Xianyong Gong [1]

[1] Institute of Geospatial Information, PLA Strategic Support Force Information Engineering University, Zhengzhou 450001, China
[2] 78098 Troops, Chengdu 610000, China
* Correspondence: wufang_630@126.com

Abstract: Detecting buildings, segmenting building footprints, and extracting building edges from high-resolution remote sensing images are vital in applications such as urban planning, change detection, smart cities, and map-making and updating. The tasks of building detection, footprint segmentation, and edge extraction affect each other to a certain extent. However, most previous works have focused on one of these three tasks and have lacked a multitask learning framework that can simultaneously solve the tasks of building detection, footprint segmentation and edge extraction, making it difficult to obtain smooth and complete buildings. This study proposes a novel multiscale and multitask deep learning framework to consider the dependencies among building detection, footprint segmentation, and edge extraction while completing all three tasks. In addition, a multitask feature fusion module is introduced into the deep learning framework to increase the robustness of feature extraction. A multitask loss function is also introduced to balance the training losses among the various tasks to obtain the best training results. Finally, the proposed method is applied to open-source building datasets and large-scale high-resolution remote sensing images and compared with other advanced building extraction methods. To verify the effectiveness of multitask learning, the performance of multitask learning and single-task training is compared in ablation experiments. The experimental results show that the proposed method has certain advantages over other methods and that multitask learning can effectively improve single-task performance.

Keywords: building extraction; multilevel and multitask; deep learning framework; high-resolution remote sensing image

Citation: Yin, J.; Wu, F.; Qiu, Y.; Li, A.; Liu, C.; Gong, X. A Multiscale and Multitask Deep Learning Framework for Automatic Building Extraction. *Remote Sens.* **2022**, *14*, 4744. https://doi.org/10.3390/rs14194744

Academic Editors: Qian Du, Yanni Dong and Xiaochen Yang

Received: 20 August 2022
Accepted: 19 September 2022
Published: 22 September 2022

1. Introduction

Buildings, as an integral part of human life, are among the most important elements on a map. Accordingly, building extraction is extremely important in urban planning, land use analysis, and map-making. With the rapid development of earth observation technology, the available spatial resolution of remote sensing imagery has increased year by year. By capturing rich and detailed information on ground objects, high-resolution remote sensing images enable fine extraction of ground objects (e.g., buildings and roads), providing important data support for the automatic extraction of large-scale buildings. However, because of the complexity of high-resolution remote sensing images and the lag in the development of extraction techniques [1], the automatic and precise extraction of buildings in high-resolution remote sensing images has remained a central challenge in remote sensing applications and cartography.

Most early building extraction methods including the mathematical morphology-based methods [2–5] and methods based on shape, color, and texture features [6–9], relied on manually extracted features for judgment. However, due to their limited capabilities for image expression, manually designed features are usually applicable only to specific regions and provide minimal support for model generalization. In recent years, deep learning technology has been widely applied in various fields related to computer vision and image processing, such as image classification [10], object detection [11], and image segmentation [12,13]. Considering the similarity between building extraction from remote sensing images and computer vision tasks, some building extraction methods based on deep learning technology [14–16] have been applied in remote sensing, enhancing the intelligence of building extraction methods to a new level. Compared with traditional building extraction methods that rely on manually designed features, deep learning methods have the advantage of powerful feature representation capabilities, enabling them to address more complex tasks.

In the computer vision field, building extraction tasks are commonly divided into three categories: building detection tasks, building footprint segmentation tasks, and building edge extraction tasks.

Building detection involves recognizing each building instance in a remote sensing image, applying object recognition techniques to obtain the location of each building instance as a rectangular bounding box, and determining the quantity of buildings. In recent years, as deep learning technology has rapidly advanced, a series of outstanding object detection algorithms have become available. These algorithms can be approximately divided into two categories. One category includes the two-stage object detection algorithms represented by the region-based convolutional neural network (RCNN) [17], Fast R-CNN [18] and Faster R-CNN [19] methods, whose main concept is to generate regional proposal boxes first and then input them into a convolutional neural network (CNN) for further classification. The other category includes single-stage object detection algorithms represented by the single-shot multibox detector (SSD) [20] and You Only Look Once (YOLO) [21] series of models, which constitute an end-to-end object detection framework that can directly output the category of each detected object. The aforementioned object detection methods have been applied for building detection in remote sensing images. Based on Faster R-CNN, Ding et al. [22] used deformable convolution to improve the adaptability to arbitrarily shaped collapsed buildings and proposed a new method of estimating the intersected proportion of objects (IPO) to describe the degrees to which bounding boxes intersect, thus offering better detection precision and recall. Bai et al. [23] proposed a Faster R-CNN method based on DRNet and ROI Align and utilized texture information to solve region mismatch problems. Building detection method can approximately recognize building locations but cannot achieve pixel-level segmentation, which is more accurate.

Building footprint segmentation refers to the pixel-level segmentation of remote sensing images, in which each pixel in an image is assigned either a building or nonbuilding label. In most building footprint segmentation methods, the fully convolutional network (FCN) architecture [12] or one of its variants is used as the basic architecture, and various measures are implemented to improve the multiscale learning capability of the model. Xie et al. [24] proposed MFCNN, a symmetric CNN with ResNet [25] as the feature extractor, which contains many complex designs, such as dilated convolution units and pyramid feature fusion. MAP-Net, proposed by Zhu et al. [26], has an HRNet-like [27] architecture with multiple feature encoding branches and a channel attention mechanism. Ma et al. [28] proposed the global and multiscale encoder–decoder network (GMEDN), which consists of a U-Net-like [29] network and a nonlocal modeling unit. These methods have greatly enhanced the accuracy of footprint segmentation for differently sized buildings in remote sensing images. However, the aforementioned methods focus only on distinguishing between building and nonbuilding pixel values and rarely closely observe building edge information, often resulting in blurred contours and failure to obtain regular boundaries in the segmentation results.

Building edge extraction refers to marking and extracting the outer boundaries of building instances in remote sensing images. Building edge extraction methods prioritize building boundary information and attempt to reach beyond pixel-level footprint segmentation to directly obtain regular and accurate building contour lines. Recently, some researchers have introduced building boundary information into deep learning networks to improve their building extraction accuracy. Lu et al. [30] adopted a deep learning network to extract building edge probability maps from remote sensing images and applied postprocessing to the edge probability maps based on geometrical morphological analysis to achieve refined building edge extraction. Wu et al. [31] proposed a novel deep FCN architecture, named the boundary regulated network (BRNet) architecture, which utilizes local and global information to simultaneously predict segments and contours to achieve better building segmentation and more accurate contour extraction. Jiwani et al. [32] improved the DeepLabV3+ [33] model by introducing a feature pyramid network (FPN) [34] module to achieve cross-scale feature extraction and designing a special weighted boundary diagram to penalize incorrect predictions of building boundaries. Li et al. [35] combined a graph-based conditional random field model with a segmentation network to preserve clear boundaries and fine-grained segmentation. However, due to the structural diversity of buildings and their complex environments, accurately locating and recognizing building edges remain significant challenges.

As a basis for interpreting remote sensing images, building detection provides a foundation for coping with higher-level interpretation tasks, such as building footprint segmentation and building edge extraction. Building detection determines the general locations for building footprint segmentation and edge extraction, while building footprint segmentation and edge extraction allow building shape features to be enhanced for building detection. Simultaneously, building footprint segmentation provides closed shape information for edge extraction, while building edge extraction yields exact boundary information for footprint segmentation. To an extent, building detection, building footprint segmentation, and building edge extraction have a symbiotic relationship of mutual dependence and information complementarity. Nevertheless, although many deep learning- based methods have been used to extract buildings and achieved good performance, the existing methods have been developed for certain tasks, e.g., footprint segmentation of buildings. Hence multitask learning frameworks are required to simultaneously perform multiple tasks, e.g., the detection, footprint segmentation, and edge extraction of the buildings.

Multitask learning can improve the performance on each task by learning better feature representations from the shared information for multiple related tasks. The classic instance segmentation framework Mask R-CNN [13] is based on the object detection framework Faster R-CNN with the addition of a branch for object mask prediction. It first locates the objects in an image and then segments the target objects in the positioning boxes, effectively combining the semantic segmentation and object detection tasks to facilitate their mutual performance enhancement. Considering the symbiotic relationship between road detection and centerline extraction, Lu et al. [36] proposed the MSMT-RE framework for performing these two tasks simultaneously, and this framework has delivered excellent road detection results. MultiNet [37] consists of a shared encoder and three independent decoders for simultaneously completing the three scene perception tasks: scene classification, object detection, and drivable area segmentation. Wu et al. [38] proposed the panoptic driving perception network YOLOP to simultaneously perform the tasks of traffic object detection, drivable area segmentation, and lane detection, significantly improving performance on each single task. Bischke et al. [39] adopted a multitask learning framework to combine the learning of boundaries and semantic information to improve the semantic segmentation of building boundaries. As seen above, multitask learning methods have been widely applied in segmentation tasks, but there is a lack of multitask learning frameworks that can solve the tasks of building detection, footprint segmentation and edge extraction simultaneously.

To address the abovementioned problems, this study proposes a multiscale and multitask deep learning framework called MultiBuildNet for simultaneously performing the building detection, footprint segmentation, and edge extraction tasks. This framework is also integrated with a multiscale feature fusion network to combine features from different scales, aiming to improve the robustness of feature extraction against complex backgrounds. In addition, to minimize the loss function across the three tasks, this study introduces a multitask loss function that fully considers any deviations between the predicted values and true values in all three tasks to obtain the best training effect.

The main contributions of this study are as follows:

(1) An effective multiscale and multitask deep learning framework is designed that can simultaneously handle the three key tasks in building extraction: building detection, footprint segmentation, and edge extraction.

(2) A multitask loss function is introduced that can address the imbalances between positive and negative samples and among sample categories in the three tasks.

The remainder of this study is organized as follows. The proposed MultiBuildNet framework is introduced in detail in Section 2. Then, Section 3 describes experiments conducted with the proposed method on open-source building datasets and large-scale high-resolution remote sensing images and presents comparisons with other advanced building extraction methods. Ablation experiments conducted to compare the performances of multitask learning and single-task training in order to verify the effectiveness of multitask learning are also reported. Section 4 discusses the performance improvements achieved by the MultiBuildNet framework compared with other deep learning methods as well as its limitations and prospects for future work. Finally, conclusions are presented in Section 5.

2. Materials and Methods

The MultiBuildNet framework essentially possesses an encoder–decoder architecture. The encoder part consists of a backbone network and a neck network, while the decoder primarily includes specific networks for the three building extraction tasks. As illustrated in Figure 1, the MultiBuildNet framework consists of three main parts:

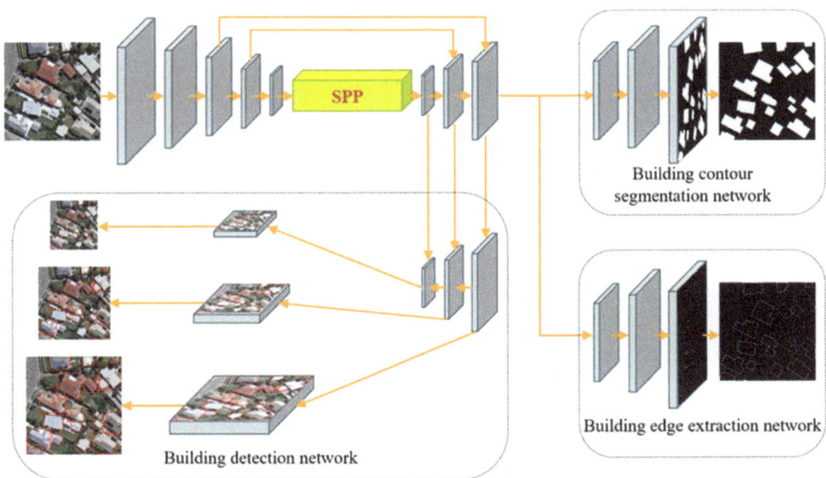

Figure 1. Flowchart of automatic building extraction based on the MultiBuildNet framework for very high-resolution remote sensing images.

(1) A multiscale feature extraction and fusion network. A spatial pyramid pooling (SPP) [40] module and a FPN module are integrated into the encoder portion of the MultiBuildNet framework. The framework can generate and fuse features at different scales and different semantic levels by performing multiscale feature extraction in the spatial context to improve the robustness of feature extraction.

(2) A building extraction module based on multitask learning. The MultiBuildNet framework includes a multitask learning module based on YOLACT [41], which can simultaneously train networks for building detection, footprint segmentation, and edge extraction.

(3) A multitask loss function. A multitask loss function is introduced into the Multi-BuildNet framework to avoid situations in which the multiple tasks are not sufficiently collaborative in relation to the network weights and to balance the joint training process for all tasks.

2.1. Multiscale Feature Extraction and Fusion Network

Convolutional layers of different sizes can generally extract feature information at different semantic levels. Therefore, this study introduces a multiscale feature extraction and fusion network, which incorporates a SPP module and a FPN module in the encoder portion while conducting multiscale encoding in the spatial context to yield rich spatial information that can be used to fuse features at different scales and different semantic levels, thus enhancing the robustness of feature extraction in complex environments.

Figure 2 shows the multiscale feature extraction and fusion network. YOLACT serves as the backbone network. The neck network fuses the features generated by the backbone network and is formed by combining a SPP module and a FPN module. In addition, multiple parallel convolutional layers of different sizes are combined in place of the resampling operation to achieve information extraction at different semantic levels and feature extraction at different scales. The convolutional layer for each level integrates convolution kernels of three different sizes, 1×1, 3×3, and 5×5, each of which generates a corresponding number of feature maps. The cascaded multiscale features are fed to the next convolution operation and subjected to a cubic convolution operation on the contracting path.

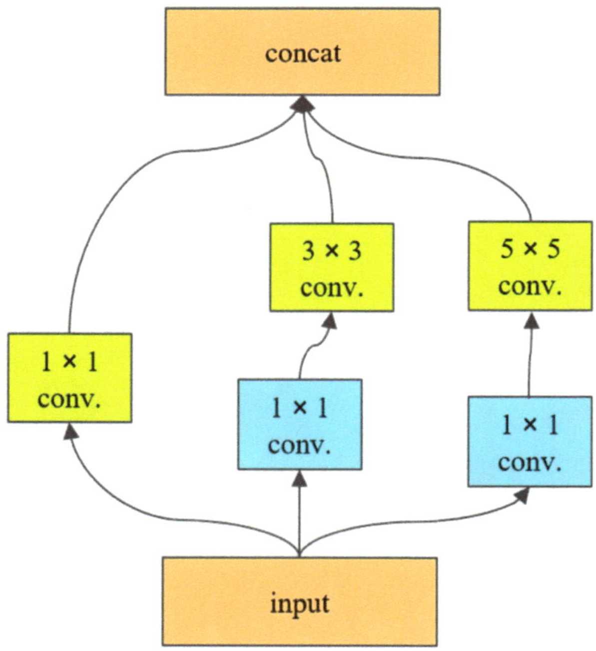

Figure 2. Multiscale feature extraction and fusion network.

2.2. Building Extraction Module Based on Multitask Learning

As shown in Figure 3, the building extraction module based on multitask learning consists of three networks, namely, a building detection network, a footprint segmentation network, and an edge extraction network, all of which run in parallel during training.

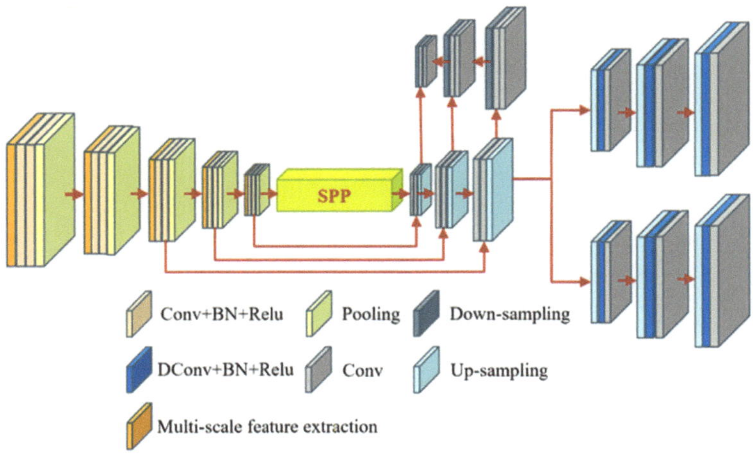

Figure 3. Building extraction module based on multitask learning.

The building extraction module takes red–green–blue (RGB) images as input. In the encoder part, there are four repeated groups of convolutional layers, where each group consists of two convolutional (Conv) layers, each followed by a batch normalization (BN) layer and a rectified linear unit (ReLU) activation function. Every group is followed by a max-pooling (Pooling) layer for downsampling. The decoder part primarily consists of the specific networks for each of the three building extraction tasks.

The building detection network adopts a multiscale detection scheme based on anchor boxes. The encoder part transfers semantic features from top to bottom through the FPN, while the decoder part uses a path aggregation network (PAN) [42] to transfer localization features from bottom to top. The combination of the FPN and PAN in the building detection network achieves an improved feature fusion effect and allows building detection to be performed using multiscale fusion feature maps in the PAN. Each grid point of the multiscale feature maps is used to generate three prior anchor boxes with different aspect ratios. The building detection network head predicts the position offset and the longitudinal and transverse scaling for each prior anchor box and the corresponding probabilities of building instances and predicted confidence values.

The building footprint segmentation network and building edge extraction network have the same network structure. The final prediction map is obtained by upsampling the underlying feature map from the FPN three times to restore the output feature map to the input image size. Every upsampling layer is followed by a deconvolutional (DConv) layer, a BN layer and a ReLU activation. Every group is followed by a convolutional layer. In addition, instead of the deconvolution operation, the bilinear interpolation method is adopted in the upsampling layers to reduce computational costs. Thus, the building contour segmentation network not only delivers high-performance segmentation results but also ensures an efficient inference speed.

2.3. Multitask Loss Function

Multitask learning is actually a problem of multiobjective optimization, and one of the complex challenges in solving such a problem is the optimization process itself, i.e., designing an appropriate loss function. In the extreme case in which the loss value for one task is considerable, but those for the other tasks are very small, multitask learning approximately degenerates into single-task learning and no longer has the advantage of multitask information sharing. Therefore, to avoid a situation in which one or more tasks dominate the network weights during the training process, an appropriate loss function is necessary to balance the loss values of each task. To this end, this study introduces a multitask loss function L_{ext}, which consists of three components: a building detection loss function L_{det}, a building footprint segmentation loss function L_{seg}, and a building edge extraction loss function L_{edg}. The symbols used for the loss functions in this paper and their meanings are shown in Table 1. The multitask loss function is a weighted sum of all three parts, as shown in Equation (1):

$$L_{ext} = \alpha_1 L_{det} + \alpha_2 L_{seg} + \alpha_3 L_{edg} \tag{1}$$

Table 1. The symbols used for the loss functions in this paper and their meanings.

Symbol	Meaning
L_{ext}	Multitask loss function
L_{det}	Building detection loss function
L_{seg}	Building footprint segmentation loss function
L_{edg}	Building edge extraction loss function
L_{cla}	Classification loss function
L_{obj}	Objective loss function
L_{box}	Bounding box loss function
L_{fl}	Focal loss function
L_{CIoU}	Complete intersection over union loss function
L_{ce}	Cross-entropy loss function

The building detection loss function is a weighted sum of the classification loss L_{cla}, the objective loss L_{obj}, and the bounding box loss L_{box}, as shown in Equation (2):

$$L_{det} = \beta_1 L_{cla} + \beta_2 L_{obj} + \beta_3 L_{box} \tag{2}$$

The focal loss L_{fl} is adopted for the classification loss and objective loss, which is used to control the weights of positive and negative samples and of difficult and easy samples. Two penalty factors are introduced to reduce the weights of easy samples so that the model can focus more on difficult samples during the training process, thus solving the issue of unbalanced sample categories. The classification loss is used to penalize problems in the binary classification of building/nonbuilding samples, while the objective loss is used to penalize confidence in building instance predictions. The complete intersection over union loss function L_{CIoU} is adopted for the bounding box loss. It is used to penalize the predicted frame confidence and takes into account three important geometric factors, namely, the overlap area between the predicted and real frames, the center point distance, and the aspect ratio, as shown in Figure 4. In this figure, the solid yellow rectangle represents the real frame for building detection, the solid green rectangle represents the predicted frame for building detection, the dotted black rectangle indicates the minimum closed area that can cover both the predicted and real frames, c is the diagonal length of the minimum closed area covering both the predicted and real frames, and d is the distance between the center points of the predicted and real frames. The focal loss and the complete intersection over union loss are calculated as shown in Equations (3) and (4), respectively:

$$L_{fl} = -(\alpha(1-\hat{p})^{\gamma} p \log(\hat{p}) + (1-\alpha)(1-p)\hat{p}^{\gamma} \log(1-\hat{p})) \tag{3}$$

$$\begin{cases} d = \rho^2(b, b^{gt}) \\ \beta = \frac{v}{1-IoU+v} \\ v = \frac{4}{\pi^2}\left(\arctan\frac{w^{gt}}{h^{gt}} - \arctan\frac{w}{h}\right)^2 \\ L_{CIoU} = 1 - IoU + \frac{d}{c^2} + \beta v \end{cases} \tag{4}$$

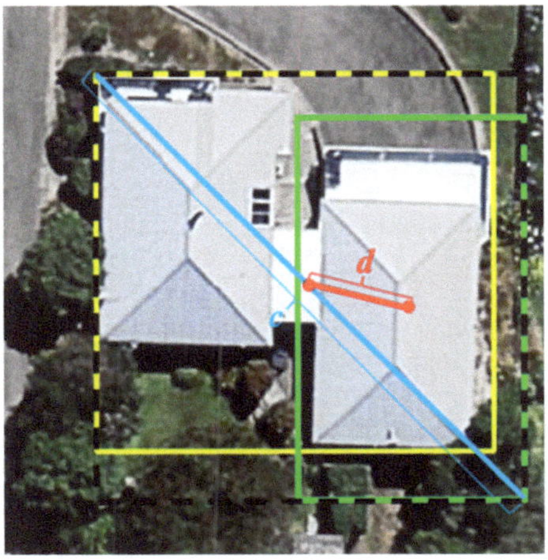

Figure 4. Calculation of geometric consistencies between predicted and real frames of building.

For the building footprint segmentation loss function, the cross-entropy loss L_{ce} is adopted to minimize the classification errors between the predicted pixels and building instances.

$$L_{ce} = -(p\log(\hat{p}) + (1-p)\log(1-\hat{p})) \tag{5}$$

On the basis of the cross-entropy loss, the building edge extraction loss function additionally includes the complete intersections over union loss to improve the prediction performance for building edges in sparse areas, as shown in Equation (6):

$$L_{edg} = L_{ce} + L_{CIoU} \tag{6}$$

where $\hat{p}$ is the probability for a model-predicted sample and p is the sample label. For positive samples, p is 1; otherwise, p is 0. w, h, w^{gt}, and h^{gt}, respectively, represent the widths and heights of the predicted and real frames; IoU is the intersection over union of the building pixels, i.e., the ratio between the intersection and union of the predicted and real building pixels; v is the similarity between the aspect ratios of the predicted and real building frames; α is a penalty factor for negative samples, with a range of [0, 1], which is used to control the ratio of positive to negative samples; γ is the modulation coefficient, with a range of [0, +∞], which is used to distinguish the complexity of samples and enables the model to focus more on difficult samples by reducing the weights of easy samples during training; β is a balance factor for the overlapping areas of the predicted and real building frame; b and b^{gt} represent the center points of the predicted frame and real frame, respectively; d is the distance between the center points of the predicted and the real frames, respectively; and c is the diagonal length of the smallest closed area that can cover both the predicted and the real frames.

3. Experiments and Results

3.1. Experimental Setup

In this study, three groups of experiments were conducted to verify the effectiveness of the proposed MultiBuildNet framework: experiments based on open-source building datasets, experiments based on large-scale high-resolution remote sensing images, and ablation experiments. The hardware used for the experiments has the following specifications: an Intel(R) Xeon(R) Gold 6230 CPU @ 2.10 GHz, 128 GB of RAM, and an NVIDIA Corporation (https://www.nvidia.cn/) TU102GL [Quadro RTX 6000/8000] graphics card, with Ubuntu 20.04 as the operating system and Python 3.7 and PyTorch (https://pytorch.org/) 1.7.1 as the programming environments.

3.2. Accuracy Evaluation Indicators

To thoroughly and quantitatively evaluate the performance of the MultiBuildNet framework, Recall and AP50 are adopted in this study as evaluation indicators for building detection accuracy. In addition, six indicators, specifically Recall, Precision, F1-score, IoU (intersection over union), mIoU (mean IoU), and Kappa, are used to evaluate the precision of building footprint segmentation and edge extraction. Recall refers to the proportion of correctly predicted building pixels among all real building pixels, Precision refers to the proportion of correctly predicted building pixels among all predicted building pixels, the F1-score is a combined metric considering both Recall and Precision, IoU refers to the ratio between the intersection and union of the predicted and real building pixels, mIoU is the mean of the IoU mean values for buildings and background, Kappa reflects the accuracy for both buildings and background, and AP50 represents the building detection accuracy when detected buildings with an IoU greater than 50% are regarded as correct. The above indicators are calculated using the following equations [43–45]:

$$Recall = \frac{TP}{TP + FN} \tag{7}$$

$$Precision = \frac{TP}{TP + FP} \tag{8}$$

$$F1 - score = \frac{2TP}{2TP + FN + FP} \tag{9}$$

$$IoU = \frac{TP}{TP + FP + FN} \tag{10}$$

$$mIoU = \frac{1}{2}\left(\frac{TP}{TP + FP + FN} + \frac{TN}{TN + FN + FP}\right) \tag{11}$$

$$\begin{cases} P_0 = \frac{TP+FP}{TP+TN+FP+FN} \\ P_e = \frac{(TP+FN)(TP+FP)+(FN+TN)(FP+TN)}{(TP+TN+FN+FP)^2} \\ Kappa = \frac{P_0-P_e}{1-P_e} \end{cases} \tag{12}$$

where TP represents the number of correctly predicted building pixels, TN represents the number of correctly predicted background pixels, FP represents the number of background pixels incorrectly predicted to be building pixels, and FN represents the number of building pixels incorrectly predicted to be background pixels.

3.3. Experiments Based on Open-Source Building Datasets

3.3.1. Dataset

In this experiment, we used the WHU building dataset (WHU) [14], the Massachusetts buildings dataset (Massachusetts) [46] and the remote sensing imagery for building extraction (RSIBE) dataset [47] to evaluate the performance of the proposed MultiBuildNet framework. The original images in these datasets are orthoimages derived from very high-resolution aerial imagery. The details of each dataset are listed in Table 2.

Table 2. Details of each dataset.

	Resolution (m)	Pixels	Area (km^2)	Tiles
WHU	0.3	512 × 512	450	8189
Massachusetts	1.0	1500 × 1500	240	151
RSIBE	0.075	1024 × 1024	525	103,439

Among the above datasets, the RSIBE dataset natively supports multitask learning. We reconstructed the detection labels and edge labels from the segmentation labels provided by the WHU and Massachusetts datasets for multitask building learning. Figure 5 shows the dataset images and labels used in our experiment.

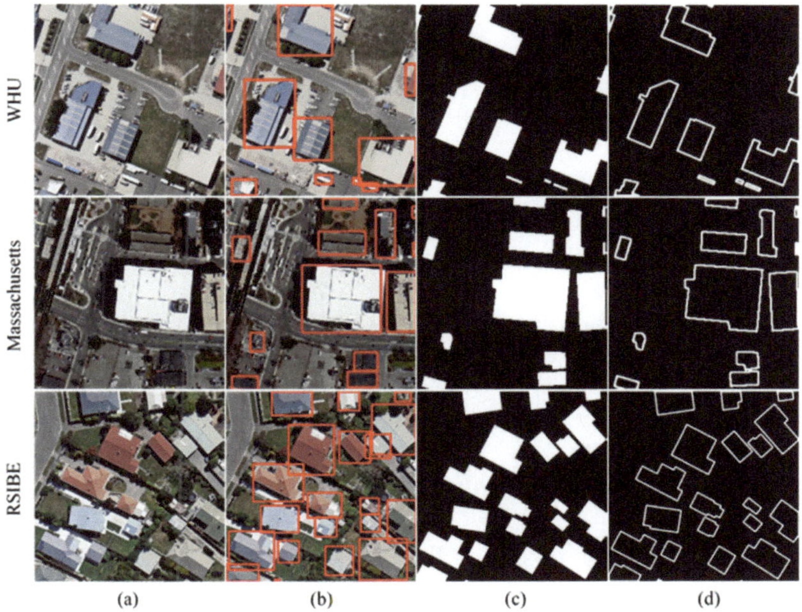

Figure 5. Examples of the experimental datasets. (**a**) Images. (**b**) Building detection labels. (**c**) Building footprint segmentation labels. (**d**) Building edge extraction labels.

3.3.2. Results of Building Detection

In this experiment, the MultiBuildNet framework was used to perform building detection on the three open-source building datasets and was compared with four classic object detection methods: YOLOv4, Faster R-CNN, Mask R-CNN, and YOLACT. The visualized results are shown in Figures 6–8. Among the tested methods, YOLOv4 and Faster RCNN can only detect buildings, while Mask RCNN, YOLACT and MultiBuildNet can simultaneously detect buildings and segment building footprints. Figures 6–8 indicate that the MultiBuildNet framework has an advantage over the other models when detecting small buildings, adjacent buildings, buildings with unique shapes, and building groups with large dimensional differences. Other models often fail to detect small buildings shaded by trees, while the MultiBuildNet framework can successfully adapt to building detection tasks in different scenarios. Tables 3–5 compare the quantitative results for the building detection accuracy of the tested methods. According to Tables 3–5, all models achieve satisfactory building detection results on the three open-source datasets, but the method proposed in this study is superior in terms of both Recall and AP50.

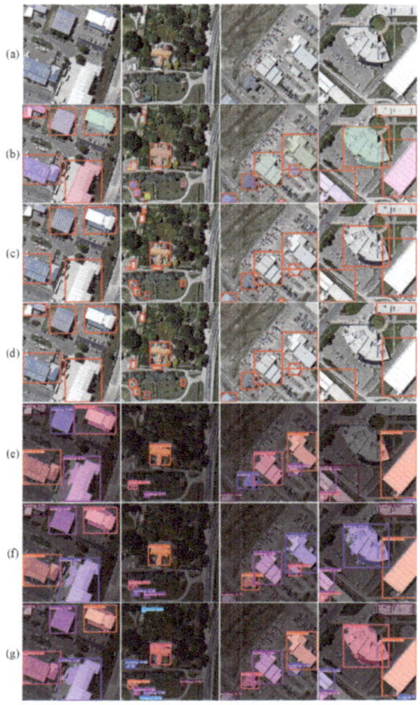

Figure 6. Results of building detection on the WHU dataset. (**a**) Images. (**b**) Labels. (**c**) YOLOv4. (**d**) Faster R-CNN. (**e**) YOLACT. (**f**) Mask R-CNN. (**g**) MultiBuildNet.

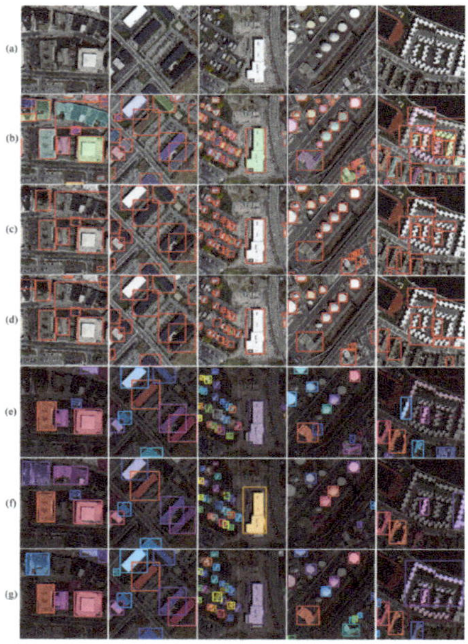

Figure 7. Results of building detection on the Massachusetts dataset. (**a**) Images. (**b**) Labels. (**c**) YOLOv4. (**d**) Faster R-CNN. (**e**) YOLACT. (**f**) Mask R-CNN. (**g**) MultiBuildNet.

Figure 8. Results of building detection on the RSIBE dataset. (**a**) Images. (**b**) Labels. (**c**) YOLOv4. (**d**) Faster R-CNN. (**e**) YOLACT. (**f**) Mask R-CNN. (**g**) MultiBuildNet.

Table 3. Accuracy of building detection on the WHU dataset.

Network	YOLOv4	Faster R-CNN	YOLACT	Mask R-CNN	MultiBuildNet
Recall (%)	77.4	75.3	77.2	75.9	**78.5**
AP50 (%)	79.5	77.6	79.6	78.4	**80.2**

Table 4. Accuracy of building detection on the Massachusetts dataset.

Network	YOLOv4	Faster RCNN	YOLACT	Mask RCNN	MultiBuildNet
Recall (%)	**66.8**	65.4	63.6	57.2	65.1
AP50 (%)	62.5	61.9	62.3	55.5	**63.4**

Table 5. Accuracy of building detection on the RSIBE dataset.

Network	YOLOv4	Faster RCNN	YOLACT	Mask RCNN	MultiBuildNet
Recall (%)	82.4	81.3	79.8	73.5	**83.7**
AP50 (%)	84.4	80.6	84.1	82.8	**85.1**

3.3.3. Results of Building Footprint Segmentation

In this experiment, the MultiBuildNet framework was utilized to conduct building footprint segmentation on the three open-source building datasets and was compared with four classic deep learning methods: U-Net, PSPNet [48], HRNet and DeepLabV3+. The visualized results are shown in Figures 9–11. Figures 9–11 indicate that these deep learning methods achieve satisfactory results on the three open-source datasets, and they all can fundamentally and accurately recognize buildings in remote sensing images. However, due to the range limitations of the receptive fields range obtained by their convolution kernels, U-Net, PSPNet, HRNet, and DeepLabV3+ are prone to producing holes in the extracted buildings, and fragmentation can easily occur in areas with dense buildings. In contrast, the MultiBuildNet framework adopts a multiscale feature fusion strategy. It can extract complete large buildings and obtain segmentation results with regular contours, and it does not often overlook small buildings shaded by trees. Even for circular buildings that are difficult to extract, MultiBuildNet can deliver better results than the other models.

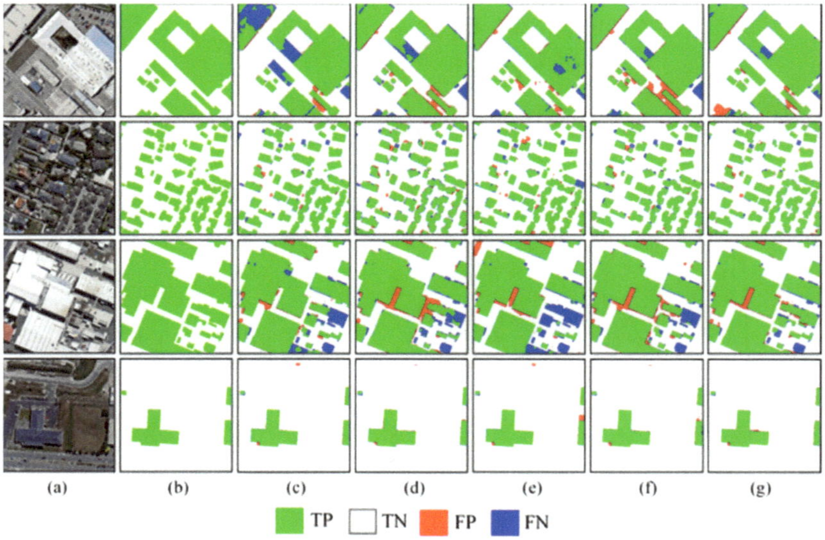

Figure 9. Results of building footprint segmentation on the WHU dataset. (**a**) Images. (**b**) Labels. (**c**) U-Net. (**d**) PSPNet. (**e**) HRNet. (**f**) DeepLabV3+. (**g**) MultiBuildNet.

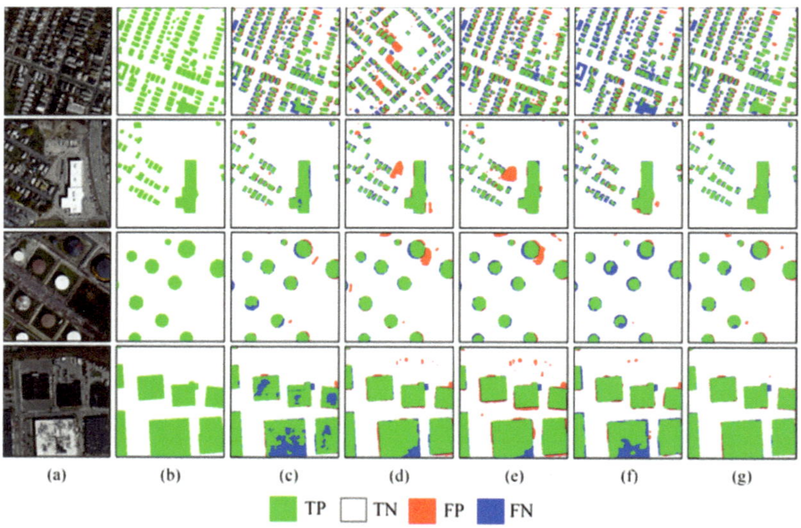

Figure 10. Results of building footprint segmentation on the Massachusetts dataset. (**a**) Images. (**b**) Labels. (**c**) U-Net. (**d**) PSPNet. (**e**) HRNet. (**f**) DeepLabV3+. (**g**) MultiBuildNet.

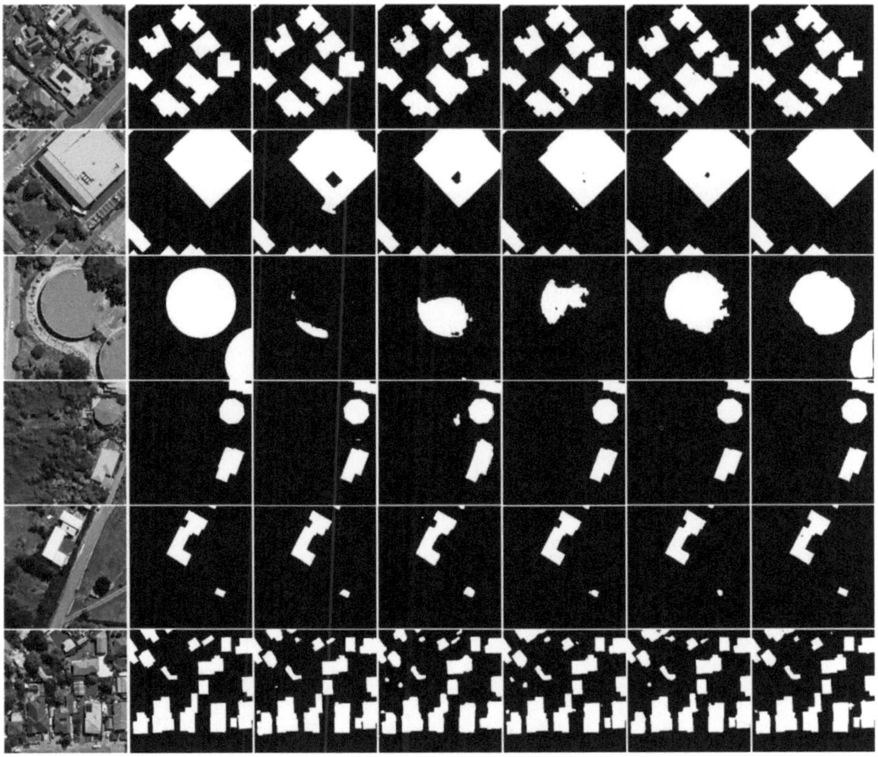

(a) (b) (c) (d) (e) (f) (g)

Figure 11. Results of building footprint segmentation on the RSIBE dataset. (**a**) Images. (**b**) Labels. (**c**) U-Net. (**d**) PSPNet. (**e**) DeepLabV3+. (**f**) HRNet. (**g**) MultiBuildNet.

Tables 6–8 compare the results for the building footprint segmentation accuracy of the above methods. As indicated in the tables, the MultiBuildNet framework is superior to the other deep learning models on the RSIBE dataset in terms of all indicators, with an IoU of 93.54%, a Precision of 95.23%, and a Recall of 96.34%, presenting significant performance advantages.

Table 6. Accuracy of building footprint segmentation on the WHU dataset.

Methods	Recall (%)	Precision (%)	F1-Score (%)	IoU (%)	mIoU (%)	Kappa (%)
U-Net	91.47	90.54	91.76	83.25	90.32	90.94
PSPNet	93.15	91.47	92.3	86.29	91.55	91.58
DeepLabV3+	92.85	91.91	92.38	85.84	91.96	91.41
HRNet	93.83	92.15	92.98	86.88	92.55	92.08
MultiBuildNet	**94.06**	**93.5**	**93.78**	**88.29**	**93.36**	**92.99**

Table 7. Accuracy of building footprint segmentation on the Massachusetts dataset.

Methods	Recall (%)	Precision (%)	F1-Score (%)	IoU (%)	mIoU (%)	Kappa (%)
U-Net	82.57	73.54	77.79	63.65	79.35	75.26
PSPNet	78.08	75.35	76.69	62.19	76.14	71.47
DeepLabV3+	81.47	**80.5**	80.99	68.05	79.88	76.67
HRNet	77.61	77.04	77.32	63.03	78.83	74.57
MultiBuildNet	**85.82**	80.08	**82.85**	**70.72**	**81.74**	**79.09**

Table 8. Accuracy of building footprint segmentation on the RSIBE dataset.

Methods	Recall (%)	Precision (%)	F1-Score (%)	IoU (%)	mIoU (%)	Kappa (%)
U-Net	94.96	93.86	94.16	91.48	93.61	93.53
PSPNet	95.13	94.03	94.33	91.80	93.80	93.73
DeepLabV3+	94.13	93.84	94.14	91.44	93.58	93.50
HRNet	94.78	93.25	94.39	91.92	93.88	93.81
MultiBuildNet	**96.34**	**95.23**	**96.15**	**93.54**	**95.06**	**95.88**

3.3.4. Results of Building Edge Extraction

In contrast to deep learning networks, such as U-Net, PSPNet, HRNet, and DeepLabV3+, the MultiBuildNet framework proposed in this study can refine the footprint segmentation results to additionally achieve building edge information extraction. Figures 12–14 show the visual results of building edge extraction on the three open-source building datasets obtained using the MultiBuildNet framework. As shown in Figures 12–14, the MultiBuildNet framework can improve unclear building edges to some extent and obtain complete boundaries for building instances of various shapes. However, it still has shortcomings. For example, it still has difficulty extracting the edges of small buildings. Moreover, although the right-angle features of buildings can be effectively maintained, the problem of how to construct the edges of irregular buildings still requires attention. Figure 15 shows the accuracy evaluation of building edge extraction on three open-source datasets using the MultiBuildNet framework. The IoU of buildings trained and tested in the RSIBE dataset is 19.86% higher than that of the Massachusetts dataset, because the RSIBE dataset has high-resolution images and high-accuracy labeling. Meanwhile, the Massachusetts dataset directly uses OSM data as labels, which leads to its relatively poor labeling quality.

Figure 12. Results of building edge extraction on the WHU dataset. (**a**) Images. (**b**) Labels. (**c**) Results.

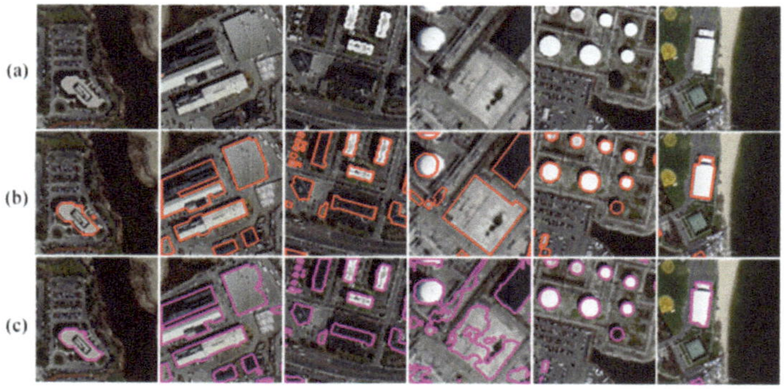

Figure 13. Results of building edge extraction on the Massachusetts dataset. (**a**) Images. (**b**) Labels. (**c**) Results.

Figure 14. Results of building edge extraction on the RSIBE dataset. (**a**) Images. (**b**) Labels. (**c**) Results.

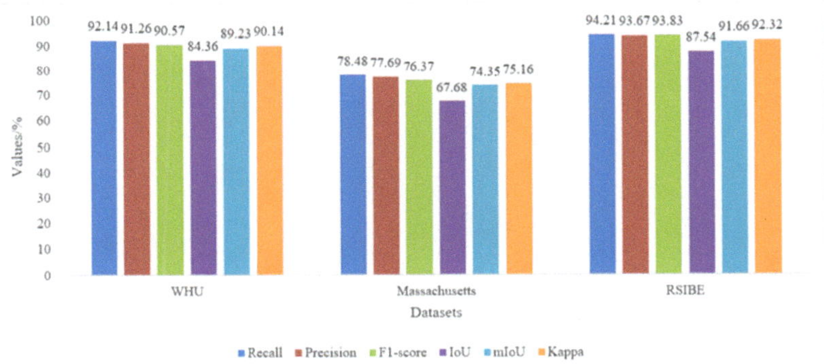

Figure 15. Accuracy of building edge extraction on the experimental datasets.

3.4. Experiments Based on Large-Scale High-Resolution Remote Sensing Images

Due to limited GPU memory, it is often necessary to crop remote sensing images into smaller images when deep learning methods are adopted for automatic building extraction. Hence, deep learning models perform well when inferring the contents of smaller cropped images but poorly when conducting direct inference for larger remote sensing images. In addition, when smaller images are spliced to obtain the building extraction results for a larger area, inaccurate segmentation of objects whose edges have been cut by the cropping process will commonly occur. To verify the MultiBuildNet framework's performance in building footprint segmentation on large-scale high-resolution remote sensing images, three large-scale aerial images of Christchurch in New Zealand were selected for experimentation, each with a resolution of 0.075 m. Residential, commercial, and industrial areas all appear in the three selected images, each of which has an area of 10,000 × 10,000 pixels. Figure 16 shows the visualized results of building footprint segmentation obtained from these large-scale high-resolution remote sensing images with U-Net, PSPNet, and MultiBuildNet. The building labels and the segmentation results of U-Net, PSPNet, and MultiBuildNet are shown from top to bottom on the original image. As seen from Figure 16, U-Net performs well when extracting relatively regular buildings in residential areas but yields a large number of missed extractions and incomplete segmentations in commercial and industrial areas. In comparison, although PSPNet can achieve relatively complete building extraction, its extraction effect is poor. It cannot maintain the inherent features of buildings, resulting in tortuous edges and unclear right-angle features in the extracted buildings. Meanwhile, MultiBuildNet demonstrates strong feature extraction capabilities in all three areas, with complete building edge extraction and few missed or incorrect extractions. The results indicate its significant advantages gained through multitask learning and multifeature fusion. Figure 17 shows the quantitative accuracy comparison of three deep learning methods when extracting building footprints from large-scale remote sensing images. Compared with U-Net and PSPNet, MultiBuildNet has obvious advantages in extracting buildings from large-scale remote sensing images.

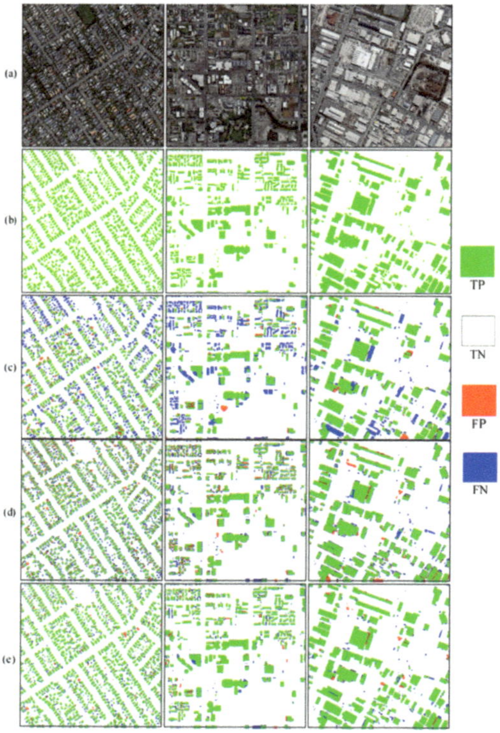

Figure 16. Results of building footprint segmentation from large-scale remote sensing images. (**a**) Images. (**b**) Labels. (**c**) U-Net. (**d**) PSPNet. (**e**) MultiBuildNet.

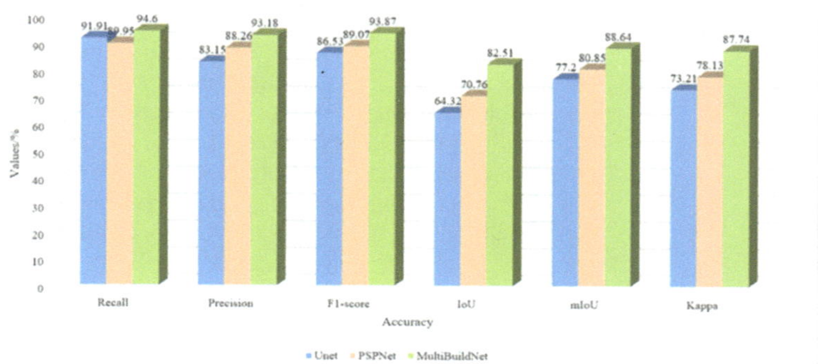

Figure 17. Accuracy of building footprint segmentation from large-scale remote sensing images.

3.5. Ablation Experiment

Ablation experiments were designed to compare the performances of the multitask and single-task learning strategies to verify the effectiveness of multitask learning and clarify the significance and role of each network in the multitask learning framework. In these experiments, we tested the building detection (Det), footprint segmentation (Seg) and edge extraction (Edg) modules of the MultiBuildNet framework and various combinations thereof on the RSIBE dataset. Table 9 compares the performance of the two strategies for a single specific task. The models using the multitask learning strategy perform better in terms of all indicators than the models adopting the single-task learning strategy. Similar tasks share some relevant features due to their underlying commonality. In the multitask learning strategy, the tasks of building footprint segmentation and edge extraction are highly similar; therefore, the combination of these two tasks is especially helpful in improving the network performance. The building detection network can use the global information of building objects to better identify buildings and to reduce false and missed detections of buildings. The footprint segmentation network can use the local information of surrounding pixels to better segment buildings and capture more details to achieve more complete segmentation. The building edge extraction network can make use of the global information of polygons to generate better contours and obtain smoother building boundaries.

Table 9. Comparison between the multitask learning strategy and single-task learning strategy.

Strategy	Recall (%)	AP50 (%)	Precision (%)	IoU (%)	mIoU (%)
Det	79.8	84.1	-	-	-
Seg	-	-	93.56	90.23	93.04
Edg	-	-	92.03	89.53	92.34
Det+Seg	80.16	**85.34**	94.28	93.31	95.38
Det+Edg	81.63	84.91	93.64	92.67	94.53
Seg+Edg	79.35	-	94.83	**93.76**	**95.68**
Det+Seg+Edg	**83.7**	85.1	**95.23**	93.54	95.06

4. Discussion

4.1. Regarding the Proposed MultiBuildNet Framework

The multitask learning framework proposed in this paper can accomplish building detection, building footprint segmentation and building edge extraction simultaneously. Such an architecture can obtain smooth and complete building extraction results. Compared with the classic building extraction methods, the proposed MultiBuildNet framework can capture more details and obtain more consistent results. The ablation experiments show that the mutual dependence among building detection, footprint segmentation and edge extraction is helpful for feature fusion and extraction. Because of its multitask learning capabilities, the MultiBuildNet framework can utilize both local information from surrounding pixels to segment buildings and global information from polygons to generate building outlines, thus achieving superior performance.

4.2. Limitations and Future Work

The MultiBuildNet framework is currently suitable for the pixel-level semantic segmentation of buildings, but it cannot generate regular building polygons. When the framework was tested on large-scale images for which the data distribution was different from that of the training set, its ability to extract buildings decreased because of the framework's limited transfer learning ability. In our future work, a more general deep learning framework will be studied for application to data from different sources. In addition to network learning, additional postprocessing could help to vectorize some irregular building edges and improve the network prediction results. For example, Zhao et al. [49] used building boundary regularization in the postprocessing stage to produce better regularized polygons and achieved good performance. Our team has previously carried out extensive research work based on building simplification [50]. Therefore, in our future research, we will consider integrating building simplification techniques into the network learning process to achieve end-to-end mapping.

5. Conclusions

This study proposes the MultiBuildNet framework to address the common problems associated with the extraction of buildings from high-resolution remote sensing images, such as incorrect extraction, missed extraction, insufficient integrity, and inaccurate boundaries. The proposed method can generate and fuse features at different scales and different semantic levels by means of its multiscale feature extraction and fusion network and learn rich semantic information in the spatial context, thereby improving the robustness of feature extraction against complex backgrounds. In addition, a multitask learning strategy is adopted to simultaneously perform the tasks of building detection, footprint segmentation, and edge extraction. Through the utilization of shared information among multiple tasks, this strategy allows better feature representations to be learned, thereby improving the performance on each task. Experiments on open-source building datasets and large-scale high-resolution remote sensing images indicate that the multiscale and multitask learning frameworks proposed in this study demonstrate considerable performance superiority over other deep learning methods and can better adapt to building extraction tasks in different scenarios. However, the method is currently applicable only for the raster-level semantic segmentation and instance segmentation of buildings and still produces pixel-based building patches. Methods to extend the building extraction process from the raster level to the vector level and to extract building vector polygons directly from remote sensing images merit further study.

Author Contributions: Conceptualization, J.Y.; methodology, J.Y. and F.W.; software, J.Y.; validation, J.Y. and Y.Q.; formal analysis, A.L.; investigation, C.L.; resources, X.G.; data curation, J.Y.; writing—original draft preparation, J.Y.; writing—review and editing, J.Y.; visualization, J.Y.; supervision, F.W.; project administration, F.W.; funding acquisition, F.W. All authors have read and agreed to the published version of the manuscript.

Funding: This research was funded by the Natural Science Foundation for Distinguished Young Scholars of Henan Province under grant number 212300410014.

Data Availability Statement: The data presented in this study are available on request from the corresponding author.

Acknowledgments: The authors would like to thank the anonymous reviewers for their constructive comments which improved considerably the quality of the final manuscript, and also express their gratitude to Springer Nature Author Services (https://authorservices.springernature.cn/, accessed on 15 September 2022) for the expert linguistic services provided.

Conflicts of Interest: The authors declare no conflict of interest.

Abbreviations

BN	Batch normalization
CIoU	Complete intersection over union
CNN	Convolutional neural network
Conv	Convolutional
CPU	Central processing unit
DConv	Deconvolutional
IPO	Intersected proportion of objects
FCN	Fully convolutional network
FN	False negative
FP	False positive
FPN	Feature pyramid network
mIoU	Mean intersection over union

ROI	Region of interest
GPU	Graphics processing unit
IoU	Intersection over union
PAN	Path aggregation network
RCNN	Regions with CNN features
ReLU	Rectified linear unit
SPP	Spatial pyramid pooling
SSD	Single shot multiBox detector
TP	True positive
TN	True negative

References

1. Grekousis, G.; Mountrakis, G.; Kavouras, M. An overview of 21 global and 43 regional land-cover mapping products. *Int. J. Remote Sens.* **2015**, *36*, 5309–5335. [CrossRef]
2. Gavankar, N.L.; Ghosh, S.K. Automatic building footprint extraction from high-resolution satellite image using mathematical morphology. *Eur. J. Remote Sens.* **2018**, *51*, 182–193. [CrossRef]
3. Huang, X.; Zhang, L. Morphological Building/Shadow Index for Building Extraction From High-Resolution Imagery Over Urban Areas. *IEEE J. Sel. Top. Appl. Earth Obs. Remote Sens.* **2012**, *5*, 161–172. [CrossRef]
4. Vu, T.T.; Yamazaki, F.; Matsuoka, M. Multi-scale solution for building extraction from LiDAR and image data. *Int. J. Appl. Earth Obs. Geoinf.* **2009**, *11*, 281–289. [CrossRef]
5. Bi, Q.; Qin, K.; Zhang, H.; Zhang, Y.; Li, Z.; Xu, K. A Multi-Scale Filtering Building Index for Building Extraction in Very High-Resolution Satellite Imagery. *Remote Sens.* **2019**, *11*, 482. [CrossRef]
6. Jabari, S.; Zhang, Y.; Suliman, A. Stereo-based building detection in very high resolution satellite imagery using IHS color system. In Proceedings of the 2014 IEEE Geoscience and Remote Sensing Symposium, Quebec City, QC, Canada, 13–18 July 2014.
7. Sirmacek, B.; Unsalan, C. Urban-Area and Building Detection Using SIFT Keypoints and Graph Theory. *IEEE Trans. Geosci. Remote Sens.* **2009**, *47*, 1156–1167. [CrossRef]
8. Tournaire, O.; Brédif, M.; Boldo, D.; Durupt, M. An efficient stochastic approach for building footprint extraction from digital elevation models. *ISPRS J. Photogramm. Remote Sens.* **2010**, *65*, 317–327. [CrossRef]
9. Argyridis, A.; Argialas, D.P. Building change detection through multi-scale GEOBIA approach by integrating deep belief networks with fuzzy ontologies. *Int. J. Image Data Fusion* **2016**, *7*, 148–171. [CrossRef]
10. Krizhevsky, A.; Sutskever, I.; Hinton, G.E. ImageNet classification with deep convolutional neural networks. *Commun. ACM* **2017**, *60*, 84–90. [CrossRef]
11. Gavrilescu, R.; Zet, C.; Foșalău, C.; Skoczylas, M.; Cotovanu, D. Faster R-CNN: An Approach to Real-Time Object Detection. In Proceedings of the 2018 International Conference and Exposition on Electrical And Power Engineering (EPE), Iasi, Romania, 18–19 October 2018.
12. Shelhamer, E.; Long, J.; Darrell, T. Fully Convolutional Networks for Semantic Segmentation. *IEEE Trans. Pattern Anal. Mach. Intell.* **2017**, *39*, 640–651. [CrossRef]
13. He, K.; Gkioxari, G.; Dollár, P.; Girshick, R. Mask R-CNN. *IEEE Trans. Pattern Anal. Mach. Intell.* **2020**, *42*, 386–397. [CrossRef] [PubMed]
14. Ji, S.; Wei, S.; Lu, M. Fully Convolutional Networks for Multisource Building Extraction From an Open Aerial and Satellite Imagery Data Set. *IEEE Trans. Geosci. Remote Sens.* **2019**, *57*, 574–586. [CrossRef]
15. Shi, Y.; Li, Q.; Zhu, X.X. Building segmentation through a gated graph convolutional neural network with deep structured feature embedding. *ISPRS J. Photogramm. Remote Sens.* **2020**, *159*, 184–197. [CrossRef] [PubMed]
16. Yuan, W.; Xu, W. MSST-Net: A Multi-Scale Adaptive Network for Building Extraction from Remote Sensing Images Based on Swin Transformer. *Remote Sens.* **2021**, *13*, 4743. [CrossRef]
17. Girshick, R.; Donahue, J.; Darrell, T.; Malik, J. Rich Feature Hierarchies for Accurate Object Detection and Semantic Segmentation. In Proceedings of the 2014 IEEE Conference on Computer Vision and Pattern Recognition, Columbus, OH, USA, 23–28 June 2014.
18. Girshick, R. Fast R-CNN. In Proceedings of the 2015 IEEE International Conference on Computer Vision (ICCV), Santiago, Chile, 7–13 December 2015.
19. Ren, S.; He, K.; Girshick, R.; Sun, J. Faster R-CNN: Towards Real-Time Object Detection with Region Proposal Networks. *IEEE Trans. Pattern Anal. Mach. Intell.* **2017**, *39*, 1137–1149. [CrossRef]
20. Liu, W.; Anguelov, D.; Erhan, D.; Szegedy, C.; Reed, S.; Fu, C.-Y.; Berg, A.C. SSD: Single Shot MultiBox Detector. In Proceedings of the Computer Vision–European Conference on Computer Vision 2016, Amsterdam, The Netherlands, 8–16 October 2016.
21. Bochkovskiy, A.; Wang, C.-Y.; Liao, H.-Y.M. YOLOv4: Optimal Speed and Accuracy of Object Detection. *arXiv* **2020**, arXiv:2004.10934.
22. Ding, J.; Zhang, J.; Zhan, Z.; Tang, X.; Wang, X. A Precision Efficient Method for Collapsed Building Detection in Post-Earthquake UAV Images Based on the Improved NMS Algorithm and Faster R-CNN. *Remote Sens.* **2022**, *14*, 663.
23. Bai, T.; Pang, Y.; Wang, J.; Han, K.; Luo, J.; Wang, H.; Lin, J.; Wu, J.; Zhang, H. An Optimized Faster R-CNN Method Based on DRNet and RoI Align for Building Detection in Remote Sensing Images. *Remote Sens.* **2020**, *12*, 762. [CrossRef]

24. Xie, Y.; Zhu, J.; Cao, Y.; Feng, D.; Hu, M.; Li, W.; Zhang, Y.; Fu, L. Refined Extraction Of Building Outlines From High-Resolution Remote Sensing Imagery Based on a Multifeature Convolutional Neural Network and Morphological Filtering. *IEEE J. Sel. Top. Appl. Earth Obs. Remote Sens.* **2020**, *13*, 1842–1855.
25. He, K.; Zhang, X.; Ren, S.; Sun, J. Deep Residual Learning for Image Recognition. In Proceedings of the 2016 IEEE Conference on Computer Vision and Pattern Recognition (CVPR), Las Vegas, NV, USA, 26 June–1 July 2016.
26. Zhu, Q.; Liao, C.; Hu, H.; Mei, X.; Li, H. MAP-Net: Multiple Attending Path Neural Network for Building Footprint Extraction From Remote Sensed Imagery. *IEEE Trans. Geosci. Remote Sens.* **2021**, *59*, 6169–6181. [CrossRef]
27. Sun, K.; Xiao, B.; Liu, D.; Wang, J. Deep High-Resolution Representation Learning for Human Pose Estimation. In Proceedings of the 2019 IEEE/CVF Conference on Computer Vision and Pattern Recognition (CVPR), Long Beach, CA, USA, 15–20 June 2019.
28. Ma, J.; Wu, L.; Tang, X.; Liu, F.; Zhang, X.; Jiao, L. Building Extraction of Aerial Images by a Global and Multi-Scale Encoder-Decoder Network. *Remote Sens.* **2020**, *12*, 2350. [CrossRef]
29. Ronneberger, O.; Fischer, P.; Brox, T. U-Net: Convolutional Networks for Biomedical Image Segmentation. In Proceedings of the Medical Image Computing and Computer-Assisted Intervention–MICCAI 2015, Munich, Germany, 5–9 October 2015.
30. Lu, T.; Ming, D.; Lin, X.; Hong, Z.; Bai, X.; Fang, J. Detecting Building Edges from High Spatial Resolution Remote Sensing Imagery Using Richer Convolution Features Network. *Remote Sens.* **2018**, *10*, 1496. [CrossRef]
31. Wu, G.; Guo, Z.; Shi, X.; Chen, Q.; Xu, Y.; Shibasaki, R.; Shao, X. A Boundary Regulated Network for Accurate Roof Segmentation and Outline Extraction. *Remote Sens.* **2018**, *10*, 1195. [CrossRef]
32. Jiwani, A.; Ganguly, S.; Ding, C.; Zhou, N.; Chan, D.M. A Semantic Segmentation Network for Urban-Scale Building Footprint Extraction Using RGB Satellite Imagery. *arXiv* **2021**, arXiv:2104.01263.
33. Chen, L.-C.; Zhu, Y.; Papandreou, G.; Schroff, F.; Adam, H. Encoder-Decoder with Atrous Separable Convolution for Semantic Image Segmentation. In Proceedings of the Computer Vision–European Conference on Computer Vision 2018, Munich, Germany, 8–14 September 2018.
34. Lin, T.-Y.; Dollár, P.; Girshick, R.; He, K.; Hariharan, B.; Belongie, S. Feature Pyramid Networks for Object Detection. In Proceedings of the 2017 IEEE Conference on Computer Vision and Pattern Recognition (CVPR), Honolulu, HI, USA, 21–26 July 2017.
35. Li, Q.; Shi, Y.; Huang, X.; Zhu, X.X. Building Footprint Generation by Integrating Convolution Neural Network With Feature Pairwise Conditional Random Field (FPCRF). *IEEE Trans. Geosci. Remote Sens.* **2020**, *58*, 7502–7519. [CrossRef]
36. Lu, X.; Zhong, Y.; Zheng, Z.; Liu, Y.; Zhao, J.; Ma, A.; Yang, J. Multi-Scale and Multi-Task Deep Learning Framework for Automatic Road Extraction. *IEEE Trans. Geosci. Remote Sens.* **2019**, *57*, 9362–9377. [CrossRef]
37. Teichmann, M.; Weber, M.; Zöllner, M.; Cipolla, R.; Urtasun, R. MultiNet: Real-time Joint Semantic Reasoning for Autonomous Driving. In Proceedings of the 2018 IEEE Intelligent Vehicles Symposium (IV), Changshu, China, 26–30 June 2018.
38. Wu, D.; Liao, M.; Zhang, W.; Wang, X.; Bai, X.; Cheng, W.; Liu, W. YOLOP: You Only Look Once for Panoptic Driving Perception. *arXiv* **2021**, arXiv:2108.11250.
39. Bischke, B.; Helber, P.; Folz, J.; Borth, D.; Dengel, A. Multi-Task Learning for Segmentation of Building Footprints with Deep Neural Networks. In Proceedings of the 2019 IEEE International Conference on Image Processing (ICIP), Taipei, Taiwan, 22–25 September 2019.
40. He, K.; Zhang, X.; Ren, S.; Sun, J. Spatial Pyramid Pooling in Deep Convolutional Networks for Visual Recognition. *IEEE Trans. Pattern Anal. Mach. Intell.* **2015**, *37*, 1904–1916. [CrossRef]
41. Bolya, D.; Zhou, C.; Xiao, F.; Lee, Y.J. YOLACT: Real-Time Instance Segmentation. In Proceedings of the 2019 IEEE/CVF International Conference on Computer Vision (ICCV), Seoul, Korea, 27 October–2 November 2019.
42. Liu, S.; Qi, L.; Qin, H.; Shi, J.; Jia, J. Path Aggregation Network for Instance Segmentation. In Proceedings of the 2018 IEEE Conference on Computer Vision and Pattern Recognition (CVPR), Salt Lake City, UT, USA, 18–21 June 2018.
43. Garcia-Garcia, A.; Orts-Escolano, S.; Oprea, S.; Villena-Martinez, V.; Garcia-Rodriguez, J. A Review on Deep Learning Techniques Applied to Semantic Segmentation. *arXiv* **2017**, arXiv:1704.06857.
44. Chen, J.; Zhang, D.; Wu, Y.; Chen, Y.; Yan, X. A Context Feature Enhancement Network for Building Extraction from High-Resolution Remote Sensing Imagery. *Remote Sens.* **2022**, *14*, 2276. [CrossRef]
45. Qiu, Y.; Wu, F.; Yin, J.; Liu, C.; Gong, X.; Wang, A. MSL-Net: An Efficient Network for Building Extraction from Aerial Imagery. *Remote Sens.* **2022**, *14*, 3914. [CrossRef]
46. Mnih, V. Machine Learning for Aerial Image Labeling. Ph.D. Thesis, University of Toronto, Toronto, ON, Canada, 2013.
47. YICHO-YUE. GitHub Repository. Available online: https://github.com/Yicho-Yue/RSIBE (accessed on 11 July 2022).
48. Zhao, H.; Shi, J.; Qi, X.; Wang, X.; Jia, J. Pyramid Scene Parsing Network. In Proceedings of the 2017 IEEE Conference on Computer Vision and Pattern Recognition (CVPR), Honolulu, HI, USA, 21–26 July 2017.
49. Zhao, K.; Kang, J.; Jung, J.; Sohn, G. Building Extraction from Satellite Images Using Mask R-CNN with Building Boundary Regularization. In Proceedings of the 2018 IEEE/CVF Conference on Computer Vision and Pattern Recognition Workshops (CVPRW), Salt Lake City, UT, USA, 18–22 June 2018.
50. Zhai, R.; Li, A.; Yin, J.; Du, J.; Qiu, Y. A Progressive Simplification Method for Buildings Based on Structural Subdivision. *ISPRS Int. J. Geo-Inf.* **2022**, *11*, 393. [CrossRef]

remote sensing

MDPI

Article

Oriented Ship Detection Based on Intersecting Circle and Deformable RoI in Remote Sensing Images

Jun Zhang [1,2], Ruofei Huang [1,2], Yan Li [1,2] and Bin Pan [3,*]

1 School of Artificial Intelligence, Hebei University of Technology, Tianjin 300401, China
2 Hebei Data Driven Industrial Intelligent Engineering Research Center, Tianjin 300401, China
3 School of Statistics and Data Science, Nankai University, Tianjin 300071, China
* Correspondence: panbin@nankai.edu.cn

Abstract: Ship detection is an important topic in the task of understanding remote sensing images. One of the challenges for ship detection is the large length–width ratio of ships, which may weaken the feature extraction ability. Simultaneously, ships inclining in any direction is also a challenge for ship detection in remote sensing images. In this paper, a novel Oriented Ship detection method is proposed based on an intersecting Circle and Deformable region of interest (OSCD-Net), which aims at describing the characteristics of a large length–width ratio and arbitrary direction. OSCD-Net is composed of two modules: an intersecting circle rotated detection head (ICR-head) and a deformable region of interest (DRoI). The ICR-head detects a horizontal bounding box and an intersecting circle to obtain an oriented bounding box. DRoI performs three RoIAlign with different pooled sizes for each feature candidate region. In addition, the DRoI module uses transformation and deformation operations to pay attention to ship feature information and align feature shapes. OSCD-Net shows promising performance on public remote sensing image datasets.

Keywords: deep learning; object detection; oriented bounding box; ship detection; remote sensing image

Citation: Zhang, J.; Huang, R.; Li, Y.; Pan, B. Oriented Ship Detection Based on Intersecting Circle and Deformable RoI in Remote Sensing Images. *Remote Sens.* **2022**, *14*, 4749. https://doi.org/10.3390/rs14194749

Academic Editors: Qian Du, Yanni Dong and Xiaochen Yang

Received: 25 August 2022
Accepted: 19 September 2022
Published: 22 September 2022

1. Introduction

Remote sensing technology, as one of the core earth observation technologies, plays an important role in the progress of human science and technology. At present, remote sensing technology involves many fields, such as land and marine resource surveys, urban change and vegetation surveys, port detection and so on. Benefiting from the development of remote sensing technology, there are more and more remote sensing images of ship, which makes ship detection develop rapidly. However, because the direction of the ship is usually inclined in any direction and the length–width ratio of the ship is overlarge, this task is quite challenging.

Object detection has always been an important task in the field of remote sensing technology. In recent years, with the development of deep learning, many researchers [1–6] have made great contributions to object detection in remote sensing images. However, ships manifest some special characteristics (complex background, different sizes, inclination at any angle, obvious geometry, etc.) in remote sensing images compared with other objects, which makes detection more difficult. Therefore, according to the characteristics of ships, many meaningful methods have been proposed. The factors that make the background of ship images complex and diverse include ports, ocean, land, weather conditions and day and night conditions. Aiming at the complex background of ships in remote sensing images, many researchers use feature fusion [7–9], attention mechanism [10–12] and other methods [13–15] in popular object detection [16–20] methods to enhance the features of ships and make the detection effect better. The types of ships or the height of aerial photos make the sizes of ships different in remote sensing images, and the unique geometry of ships will also increase the difficulty of the task. Therefore, many researchers have conducted significant research on the size of ships. For example, Xie et al. [21] and Kan et al. [22] conducted a profound study on small ship detection; Zhu et al. [23] and Sun et al. [24] used the anchor-free method to alleviate the problems caused by ship scale; Chen et al. [25] used a multiscale adaptive recalibration network to detect multiscale and arbitrarily oriented ships. Affected by the shooting angle or geographic environment, many ships tilt at any angle, and using the horizontal box will lead to too much redundant information. More seriously, if the targets are densely arranged, the intersection over union (IoU) of adjacent targets may be too large, which may lead to missed detection (Figure 1a). Therefore, the oriented bounding box is more suitable for rotated object detection (Figure 1b). In order to solve the problem of object rotation, researchers have performed a lot of meaningful research on the problem of rotated object detection [26–30]. How to represent orientation information is a difficulty in rotated object detection. In view of this difficulty, many researchers have made many contributions. For example, some researchers obtained object rotation information by detecting the rotation angle θ of an object relative to the horizontal axis of the image [23,27,31]; Xu et al. [32] use eight-parameter representation to obtain rotation information of the object; Liu et al. [33] use trilateral representation to detect rotating rectangles; Xue et al. [34,35] also regard angle information as classification information. Therefore, ship detection is suitable for rotated object detection. Many researchers have performed some interesting research in the field of rotated ship detection. For example, Shao et al. [36] proposed a rotated balanced feature-aligned network; Cui et al. and Yu et al. [37,38] used the anchor-free method to detect rotated ships; Zhu et al. [39] proposed a rotated RetinaNet with considerable accuracy.

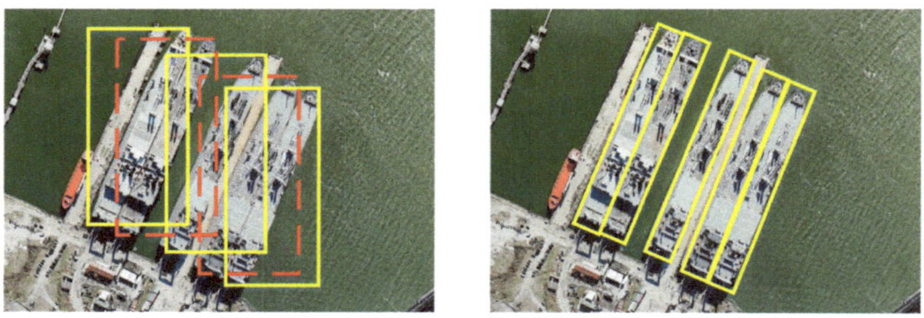

(a) (b)

Figure 1. (**a**) The situation of missed detection due to too dense targets. The yellow box indicates the detected targets, and the red box indicates the missed targets. (**b**) The result of using the oriented bounding box.

However, the ship in the remote sensing image also has obvious geometry, and the object rotation is still a problem worthy of discussion. For the method [37,40] of using rotation angle θ to detect directional ships, there is a boundary discontinuity problem [34], and the characteristics may be the same for some specific angles, which makes the learning unstable. The eight-parameter method [32,41] has the problem of too many parameters. In addition, in order to alleviate the resistance in water, the ship design is mostly a slender shape, which makes the length–width ratio overlarge, resulting in difficult feature extraction.

In this paper, we propose the OSCD-Net to solve the problems of ship inclination in any direction and an overlarge length–width ratio. OSCD-Net contains two modules: intersecting circle rotated detection head (ICR-head) and deformable region of interest (DRoI). In ICR-head, we intersect an oriented bounding box by detecting a horizontal rectangle, a circle and incline direction prediction. In addition, ICR-head also uses the form of double-heads [42] to reduce the coupling between classification and regression tasks. At the same time, we are inspired by the RoI poolings of different pooled sizes method of R2CNN [26] in the field of text detection. This method can effectively solve the problem of a too large length–width ratio of text by turning the feature pool into different proportions of region of interest poolings. This method may also be applicable to ship detection. However, the ship length–width ratio is far less than that of words. Therefore, according to the characteristics of ships, we designed a DRoI module. DRoI performs three RoIAlign [43] with different pooled sizes for each feature candidate region, and a self offset matrix is used to enhance the feature information of ship. Then, the deformation operation is carried out to obtain three features of the same size and concatenate them.

In summary, our main contributions are as follows:

- We develop an intersecting circle rotated detection head for detecting rotated ships in remote sensing images. This detection head is also suitable for detecting other rotated targets.
- We designed a DRoI module, which can obtain features of different pooled sizes. DRoI is used to solve the problem that the ship length–width ratio is overlarge.
- A new end-to-end algorithm framework for oriented ship detection in remote sensing images.

2. Methods

This section describes OSCD-Net for ship detection, which contains the overall architecture of OSCD-Net, ICR-Head, deformable RoI module, multitask loss, and inference stage.

2.1. Overall Architecture of OSCD-Net

The OSCD-Net's main framework is depicted in Figure 2. As the baseline of our network, we use the two-stage detector Faster R-CNN [17]. The reason why the two-stage network is selected is that the two-stage network's module is disassembled more carefully, and the space for change is large according to the characteristics of the object compared with other methods. OSCD-Net recodes the rotated ship labels and then extracts the feature map. We use the better performing ResNet-101 [44] as the backbone network. The residual structure of ResNet can solve the problem of gradient disappearance or gradient explosion caused by network deepening. For feature enhancement and adaptation to multi-scale ships, FPN [45] is also used in our network. The feature maps can be obtained by ResNet and FPN. The feature map is then forwarded to the region proposal network (RPN) network to acquire the RoI. The DRoI module then pools the feature map and RoI into three deformable features and uses transformation and deformation operations to transform and deform the features before fusing them. Finally, ICR-head is used for classification and regression and obtains a rotated rectangle by detecting a circle, a horizontal rectangle and incline direction prediction.

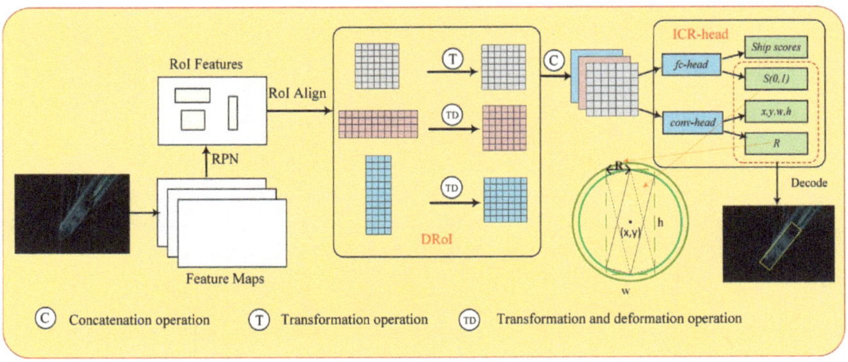

Figure 2. Overview of the our proposed OSCD-Net. Feature extraction and RPN are located on the left. The DRoI module is an intermediate grid rectangle used for enhancement to form the final detection feature. ICR-head on the right is used to detect the horizontal rectangle and a circle, which is decoded into a rotating rectangular box.

2.2. ICR-Head

ICR-head is a rotated detection head, which obtains a rotated rectangle by detecting a circle, a horizontal rectangle, and incline direction prediction. In addition, in order to alleviate the coupling between classification and location, we use double-head structures in ICR-head. Classification and incline direction prediction is performed utilizing fc-head. Conv-head is used to perform regression tasks. Here, we mainly describe three parts: obtaining a rotated rectangle through a horizontal rectangle and an intersecting circle (RHC), the encoding and decoding of a rotated rectangle and the use of double-heads in ICR-head.

2.2.1. Obtaining a Rotated Rectangle

If the rotated box label of a regular inclined object (rectangle-like) is known, the horizontal box label, which is the circumscribed horizontal rectangle of the spinning box, may be simply acquired and is unique. If we know a regular rectangle, on the other hand, we try to find its inscribed rectangle (a rectangle whose four vertices are on its edges). As long as we use the rectangle's center point as the the center of a circle, we can draw a circle intersecting the rectangle's four edges, link these intersections and obtain the normal rectangle's inscribed rectangle. Because the diameter of a circle to any point on the circumference is a right triangle. However, the circle intersects with the regular rectangle, and the inscribed rectangle is not one, but four (Figure 3a). If a circle is drawn from the top edge of a regular rectangle to its first intersection, only two rectangles are produced (Figure 3b). Therefore, when using a binary classification, we can determine a unique inclined rectangle, which is the target box we want. The final representation parameters are (x, y, w, h, R, S), where (x, y) is the center point of the horizontal rectangle and also the center of a circle. w and h are the width and length of the horizontal rectangle, respectively. R is the distance from the vertex of the upper corner of the rectangle to the intersection of the circle and the uppermost edge of the rectangle, which can determine the only circle. S is the inclined direction prediction of the inclined rectangle, which can determine the only rotated box.

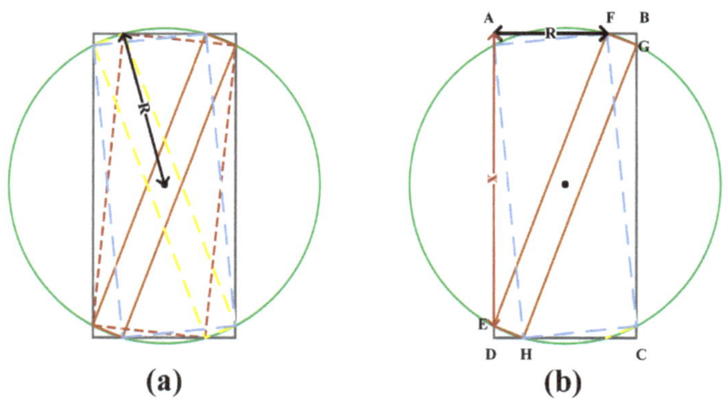

Figure 3. (**a**) The circle marked with the radius of the circle. The circle and the horizontal rectangle will intersect into four inclined rectangles. (**b**) The vertex of the horizontal rectangle and the intersection with the circle, and explains that there are only two inclined rectangles using AF to draw the circle.

2.2.2. Encode and Decode

At present, the annotation form of most labels for rotated object detection are (x, y, w, h, θ) or $(x_1, y_1, x_2, y_2, x_3, y_3, x_4, y_4)$, no matter which annotation is used, they can be converted to each other. Where θ is the angle of the ship relative to the horizontal axis of the image, and (x_i, y_i) $(i = 1, 2, 3, 4)$ is vertex coordinates. Here, we recode $(x_1, y_1, x_2, y_2, x_3, y_3, x_4, y_4)$ to obtain (x, y, w, h, R, S):

$$w = x_2 - x_4 \quad , \quad h = y_3 - y_1 \tag{1}$$

$$x = \frac{x_2 + x_4}{2} \quad , \quad y = \frac{y_1 + y_3}{2} \tag{2}$$

$$R = x_1 - x_4 \quad , \quad S = \begin{cases} 0 & y_4 - y_1 > \frac{h}{2} \\ 1 & \text{otherwise} \end{cases} \tag{3}$$

Decode is what we know (x, y, w, h, R, S) to solve $(x_1, y_1, x_2, y_2, x_3, y_3, x_4, y_4)$, where $r = AF$, let $X = AE$ and X is unknown, as shown in Figure 3b.

According to $\triangle AEF \sim \triangle BFG$, $FB/AE = BG/AF$, with formula:

$$\frac{X}{R} = \frac{w - R}{h - X} \tag{4}$$

Formula (4) can be converted into a binary quadratic equation:

$$X^2 - hX + wR - R^2 = 0 \tag{5}$$

The solution is:

$$X = \frac{h \pm \sqrt{h^2 - 4(wR - R^2)}}{2} \tag{6}$$

Obviously, $\frac{h}{2} < \frac{h+\sqrt{h^2-4(wr-R^2)}}{2} < h$ and $0 < \frac{h-\sqrt{h^2-4(wR-R^2)}}{2} < \frac{h}{2}$. Therefore, we use $h/2$ as the boundary to define S. When the predicted value of S is 1, $X = \frac{h-\sqrt{h^2-4(wR-R^2)}}{2}$. When the predicted value of S is 0, $X = \frac{h+\sqrt{h^2-4(wR-R^2)}}{2}$.

Finally, the coordinates of the horizontal rectangle ABCD can be obtained according to (x, y, w, h):

$$\begin{cases} A = (x + w/2, y - h/2) \\ B = (x - w/2, y - h/2) \\ C = (x - w/2, y + h/2) \\ D = (x + w/2, y + h/2) \end{cases} \tag{7}$$

and the coordinates of the rotated rectangle EFGH can be obtained according to ABCD and X, R as follows:

$$\begin{cases} E = (x + w/2, y - h/2 + X) \\ F = (x - w/2 + R, y - h/2) \\ G = (x - w/2, y + h/2 - X) \\ H = (x + w/2 - R, y + h/2) \end{cases} \tag{8}$$

The meaning of each parameter is shown in Figure 4:

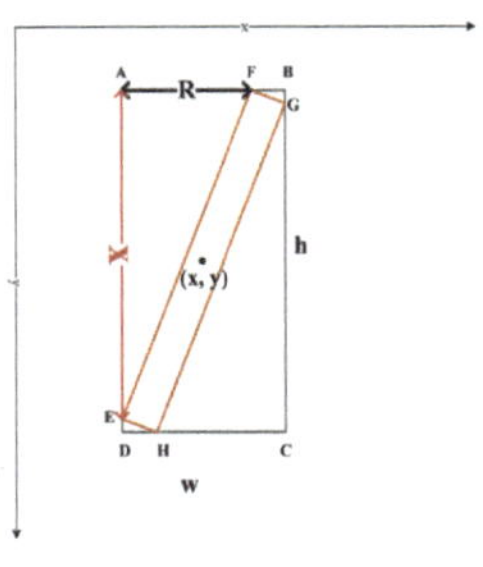

Figure 4. Schematic diagram of the rectangle in the coordinate axis. (x, y, w, h, R, X) are known, and vertex coordinates (A, B, C, D, E, F, G, H) are unknown.

2.2.3. Double-Heads

Full connection is sensitive to spatial location and is more suitable for classification tasks, while convolution has a better effect on object information extraction and is suitable for regression tasks [42]. Because there are both classification parameters and regression parameters in our rotated box coding method, in order to reduce the coupling between the classification and regression tasks, we use two branches to detect classification and regression respectively. The fc-head is used for the confidence and inclination direction S of the ship. We use conv-head to predict horizontal rectangles and R. In Figure 5, we show the specific process of ICR-head. The fc-head structure uses two $1024 \times 7 \times 7$ full connection layers connected in series. The conv-head structure uses 1×1 conv to convert the feature map to $7 \times 7 \times 1024$ and finally avgpool it to $1 \times 1 \times 1024$. The features extracted by classification and regression are different. Using two branches is beneficial for different detection heads to pay more attention to their own tasks.

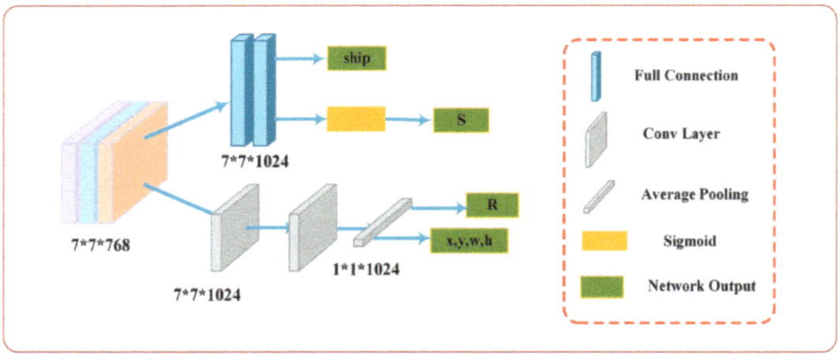

Figure 5. Illustration of the process of the ICR-head module. Each part is marked in the red box on the right.

2.3. Deformable RoI

In the traditional two-stage object detection algorithm, the region proposal output through the RPN network is mapped into the feature map, and RoI will be pooled into 7×7 feature areas using RoI pooling or RoIAlign. However, because the length–width ratio of ships is usually too large, the features of ships cannot be evenly obtained by pooling them into 7×7 feature areas. Therefore, we use RoIAlign to divide the region of interest in the feature map into feature regions with three different dimensions: $7 \times 7, 12 \times 4$ and 4×12 (Figure 6). The reason for this is that the 3:1 ratio can adapt to most ships, so that their features can be extracted evenly. Secondly, the 7×7 grid divides the feature into 49 parts, and the 4×12 grid divides the feature map into 48 parts. We fill a zero pixel behind it to obtain 49 parts, so as to deform it into a 7×7 grid and fuse the three feature maps. In the process of fusion, we use transformation and deformation operations. Transformation uses a self-learning offset matrix to pay attention to the useful information in the feature map. The specific steps are shown in Figure 7a. Deformation is to squeeze 4×12 or 12×4 into one dimension, followed by a pixel, then expand it into 7×7 two-dimensional features (see Figure 7b). The expression of the transformation and deformation operation is as follows:

$$F = \sum_{i} \sigma(w_i \cdot \Delta m_i \cdot F_i + \beta) \tag{9}$$

where Δm_i is self-learning attention matrix, and w_i is the weight ratio of this matrix. i is the index, and $i = 1, 2, 3$. F_i represents the features of different shapes after pooling. The multiplication of these three parameters indicates the transformation operation. β is the supplementary 0 pixel point. $\sigma()$ is a deformation operation designed to rearrange the features of each small part to form a 7×7 size. $\sum$ means concatenate the three features.

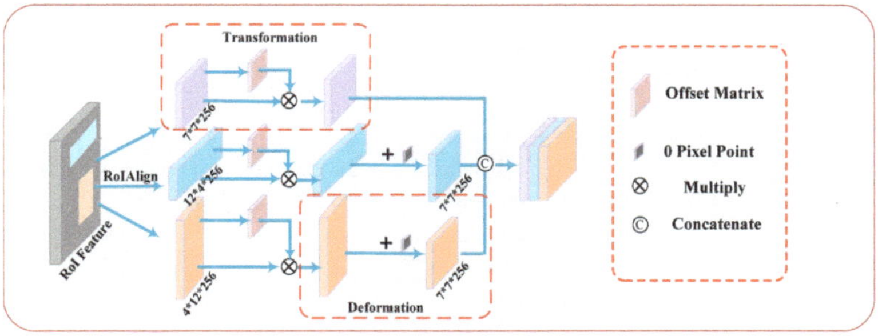

Figure 6. Illustration of process of DRoI module. DRoI module uses three transformation operations and two deformation operations. Transformation operation and deformation operation are circled with red frames.

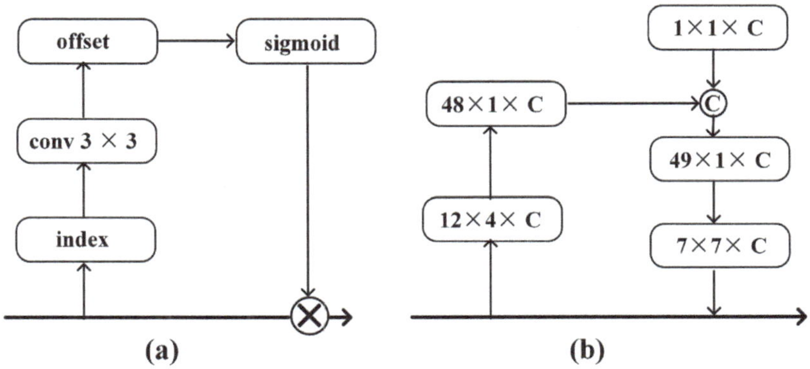

Figure 7. (**a**) The step of transformation operation; (**b**) The process of deformation operation.

2.4. Multitask Loss

OSCD-Net is a two-stage network. The first stage is RPN, which is used to extract candidate regions. The loss here is the same as in [17]:

$$L_{RPN} = \frac{1}{N_{cls}} \sum_i L_{cls}(p_i, p_i^*) + \lambda_1 \frac{1}{N_{reg}} \sum_i p_i^* L_{reg} \tag{10}$$

Here, N_{cls} and N_{reg} are the number of total proposals and positive proposals in a mini-batch fed into the head. i is the index of an anchor and p is the predicted probability. p_i^* is the ground-truth label, the positive sample is 1 and the negative sample is 0. In the second stage, there is the loss of classification and orientation parameters (x, y, w, h, R, S), of which the fc-head loss is:

$$L_{cls}(p, u) = -\log pu \tag{11}$$

$$L_S(s, s^*) = -[s^* log(s) + (1 - s^*) log(1 - s)] \tag{12}$$

where L_{cls} is a classification loss used to judge the confidence of the ships, and L_S is used to identify unique box. s is the predicted value, and s^* is the ground-truth label. The conv-head loss is:

$$L_{reg} = L_h + \lambda_2 L_r \tag{13}$$

$$L_r = smooth_{L_1}(r - r^*) \tag{14}$$

$$L_h = \sum_{i=(x,y,w,h)} smooth_{L_1}(v_i - v_i^*) \tag{15}$$

in which

$$smooth_{L_1}(x) = \begin{cases} 0.5x^2, & |x| \le 1 \\ |x| - 0.5, & otherwise. \end{cases} \tag{16}$$

where L_h is the loss for horizontal box regression, v_i^* is the normalization of the ground-truth of the horizontal box regression and v_i is the predicted value. r^* is the ground-truth of the intersection circle, and r is the predicted value. λ_2 is a super parameter used to balance the total loss.

The total loss is:

$$L_{total} = L_{RPN} + L_{reg} + L_{cls} + \lambda L_S \tag{17}$$

In the experiment, the values of λ_1, λ_2, and λ were 1, 2, and 1.

2.5. Inference

In the inference stage, we first use horizontal NMS to eliminate some candidate boxes, and then use directional NMS to further screen the candidate boxes. In the process of inference, we find that there is no solution for ships placed horizontally. Through analysis, the reason is that $h^2 - 4(wr - r^2) < 0$, which leads to the non-intersection of the predicted circle and the regular rectangle on the vertical edge, as shown in Figure 8a. The blue box is the ground-truth label. At this time, the circumscribed circle of the ground-truth is just tangent to the horizontal rectangle. When the predicted R is slightly larger, the circle and the horizontal rectangle cannot intersect (see the red circle of Figure 8a), resulting in no solution. For this case, we treat $h^2 - 4(wr - r^2)$ as equal to 0. At this time, $X = h/2$. What is drawn here is a parallelogram, as shown in Figure 8b. When calculating the IoU, find its minimum circumscribed rectangle. The entire process of the OSCD-Net is briefly summarized in Algorithm 1.

Algorithm 1 The pseudocode for OSCD-Net.

Input: Ship image X and its labes boxes $B(x_1, y_1, x_2, y_2, x_3, y_3, x_4, y_4)$
Output: The score of the proposals C, the coordinates of the proposals $S(x_1, y_1, x_2, y_2, x_3, y_3, x_4, y_4)$
Initialize: Initialize the pre-training weight w for network;
Decode B into $O(x, y, w, h, R, S)$. Feed X and B into the network to extract feature maps.
Use RPN to get the proposal boxes and map it to the feature maps to get the feature candidate region F.
for $i = 1$ to 3 **do**
  Obtain F_i by using RoIAlign pool F.
  Obtain F_i' by TD operation.
end for
Obtain F' by fusing F_1', F_2', F_3',and get $O'(x', y', w', h', S')$ through ICR-head.
Obtain S_1 and S_2 by decoding O'
if $S' < \frac{h}{2}$ **then**
  $S = S_1$
else
  $S = S_2$
end if

(a) (b)

Figure 8. (**a**) The case where the short side of the horizontal rectangle does not intersect the circle. (**b**) The parallelogram obtained by connecting the midpoint and intersection of the short side of the horizontal rectangle.

3. Experiment

In this section, we first introduce the experimental datasets, evaluation metrics and implementation details in detail. Then, we conducted an ablation study to analyze our ICR-head and DRoI mouble. Finally, we compare other recent methods with several examples.

3.1. Datasets

(1) HRSC2016 [46]: HRSC2016 is a public dataset of remotely sensed ships, including 1070 pictures and 2976 examples. The pictures are collected from Google Earth. There are 2886 ships standing under the port with complex background, including warships, aircraft carriers, cargo ships, fishing boats, etc. The image resolution ranges from 2 m to 0.4 m, and the image size ranges from 300×300 to 1500×900, most of which are larger than 1000×600. In the experiment, we use the HRSC official division of the trainset and the testset. A total of 646 pictures are used for training, and 444 pictures are used for testing.

(2) DOTA-Ship: DOTA [47] is a large-scale dataset with 15 object categories for object detection in remote sensing images. It uses four vertex coordinates for annotation and contains 2806 aerial images from different sensors and platforms. The size of the images ranges from around 800×800 to 4000×4000 pixels. For DOTA-Ship, we manually extract all ship images on the DOTA trainset and verset, with a total of 434 images, including 39028 instances. We randomly divided the trainset and testset according to the ratio of 6:4. During the training, we cut the picture to 1000×1000 pixels.

(3) SSDD+ [14]: SSDD is the first dataset published at home and abroad for ship object detection in SAR images. There are a total of 12 images and 2456 images for each ship. In order to detect ship targets with rotating frame, the label of SSDD is improved, and the rotation angle information is added on the basis of category and position. This data set is called SSDD+. In order to verify the robustness of our algorithm, we also use SAR images to evaluate the algorithm, using SSDD+ with rotation annotation. We randomly divided the trainset and testset according to the ratio of 7:3.

3.2. Evaluation Metrics

For the detection accuracy of ship inspection, we use the average accuracy (mAP) as the evaluation standard, which is consistent with Pascal VOC 2007 (AP(07)) [48] and VOC 2012(AP(12)). AP is the average value of precision based on different recall rates. The form of Precision and Recall is as follows:

$$Precision = \frac{TP}{TP + FP} \tag{18}$$

$$Recall = \frac{TP}{TP + FN} \tag{19}$$

Here, *TP*, *FP* and *FN* represent the number of correct numbers, false positives and false negatives, respectively. *TP* means that the intersection over union (IoU) between the predicted bounding box and the object is higher than 0.5. AP and is defined as:

$$AP = \int_0^1 p(r)dr \tag{20}$$

where p represents precision, r represents recall and p is a function that takes r as a parameter, which is equal to taking the area under the curve. AP(07) is used to select the precision maximum value of 11 points with recall greater than 0–1 and then take the average. Compared with AP(07), the recall of AP(12) is no longer 11 points, but all recall values. In order to highlight the gains achieved by our method over the current existing similar techniques, we added the Precision (P), Recall (R) and $F1$-Score ($F1$) evaluation metrics. $F1$ is the harmonic mean of the Precision and Recall, which is defined as:

$$F1 = \frac{2PR}{P+R} \tag{21}$$

3.3. *Training Details*

In order to verify the effectiveness of our innovation, we used ResNet-101 and FPN in the feature extraction, which can be consistent with other methods. We use ResNet-101 pretrained on ImageNet as the feature extractor. Our network is based on the pytoch framework and trained on two NVIDIA Geforce GTX 2080 Ti GPUs. We use the SGD optimizer [49] to optimize our network. The momentum and weight attenuation are set to 0.9 and 5×10^{-3}, respectively. The initial learning rate is 7.5×10^{-3}, and the learning rate is reduced once every 10 iterations. For the HRSC2016 and DOTA-Ship datasets, our training model is 20 K steps, and the learning rate decays in {14 K, 18 K}. For the SSDD+ dataset, our training model is 12K steps, and the learning rate decays in {8 K, 10 K}.

3.4. *Ablation Study*

In order to verify the effectiveness of our method as a directional object regression method, we designed the same Faster R-CNN + FPN as baseline, using five parameters (Figure 9a), eight parameters (Figure 9b) and our representation (x, y, w, h, r, S) (Figure 9c) as the output parameters of the model to detect rotating ships. Table 1 shows the number of prediction parameters of three different rotated object detection methods and their effects on HRSC2016 dataset. RHC is our representative method, which is to obtain a rotated rectangle through a horizontal rectangle and an intersecting circle. In Faster R-CNN + five parameters (FP), we obtained an AP(07) of 80.9%. Here, the five parameters are (x, y, w, h, θ). (x, y) is the center of the rotating ship, and w and h are the width and length of the ship, respectively. θ is the included angle between the rectangle and the positive half-axis of the X-axis. The range of θ is $(-\frac{\pi}{2}, \frac{\pi}{2})$. For eight parameters (EP), the representation is $(x, y, w, h, \alpha_1, \alpha_2, \alpha_3, \alpha_4)$. (x, y, w, h) are the parameters of the horizontal boxm and $(\alpha_1, \alpha_2, \alpha_3, \alpha_4)$ are the offset of the orientation box from the horizontal box. For specific operations, please refer to [32]. The mAP of Faster R-CNN + EP is 88.2%. Compared with Faster R-CNN + FP, the mAP of Faster R-CNN + EP is improved by 7.3%, so it is usually better to use boundary information than angle information to detect directional objects. However, its parameters have also increased. Faster R-CNN + RHC is our representation. We use five regression variables (x, y, w, h, r) and one classification variable (S) to obtain a orientation box. Compared with method of five parameters, we did not increase the number of regression parameters but only added a classification variable. However, we greatly increase the accuracy, with a 7.7% improvement for mAP. Meanwhile, compared with the Faster R-CNN + EP, we reduce the number of regression parameters from eight to five, and mAP is also increased slightly. Moreover, the accuracy of S is more than 98.9%. Therefore, the increase in S will hardly affect our accuracy. This experiment proves that our method has a good improvement in precision compared with the FP method. At the same time, compared with the EP method, we predict fewer parameters, and the accuracy of our method has also been slightly improved. It is worth noting that we have added a classification parameter, but it is easier to perform the classification task than the regression task.

Table 1. Performance comparison of Faster R-CNN under different rotated box representations. The highest value is indicated by bold numbers (%).

Method	Num	S	AP(07)	AP(12)
Faster R-CNN+FP	5	-	80.9	82.9
Faster R-CNN+EP	8	-	88.20	89.0
Faster R-CNN+RHC (ours)	5	98.9	**88.6**	**89.2**

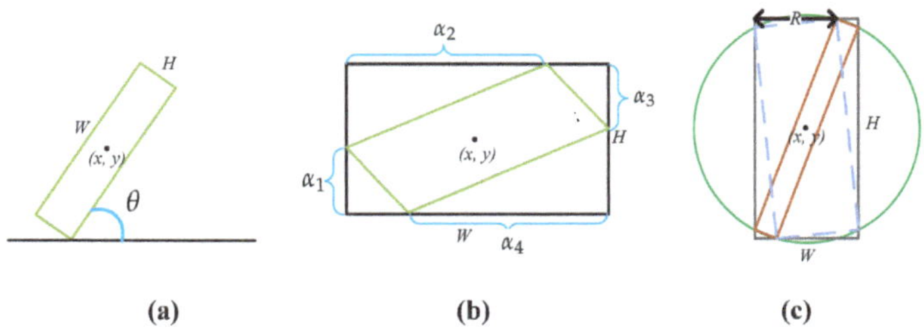

Figure 9. (**a**) The tilt target represented by (x, y, w, h, θ); (**b**) the tilt target represented by $(x, y, w, h, \alpha_1, \alpha_2, \alpha_3, \alpha_4)$; (**c**) the tilt target represented by (x, y, w, h, R, S).

Simultaneously, we also designed the ablation study of each module, which is based on Faster R-CNN + RHC, as shown in Table 2. ICR-head uses the form of double-heads on the basis of Faster-RCNN + RHC. From the experimental results, double-heads improves the AP(12) of our method by 2.89%, which indicates that branch decoupling is more suitable for our detection head. The addition of the DRoI module also further improves the AP. It can be seen from the experiment that both the ICR-head and DRoI modules play a role in the network model, and mAP reaches the highest value when they are used at the same time. In addition, we comparatively show some of the effective detection results in Figure 10. It can be seen from this experiment that Faster-RCNN OBB's angle prediction is not very accurate, leading to the offset of some detection boxes. There are also some small ships missing detection. When we use ICR-head, the inclination angle of the ship is more accurate, and some small ships can be detected. As can be seen from the last column, the boundary of the ship is proposed more accurately after the DRoI module is added. For some slender ships, the IoU of the ground truth and detection results is improved. In this ablation experiment, we found that ICR-head can more accurately capture the position information of the ship, so as to improve the AP. The DRoI module can further refine the boundary information, and is more accurate for ship detection with a large aspect ratio.

Table 2. Ablation study of OSCD-Net on the HRSC2016 dataset. The highest value is indicated by bold numbers (%).

RHC	ICR-Head	DRoI	AP(07)	AP(12)
√			88.60	89.21
√	√		89.17	92.10
√		√	89.21	91.73
√	√	√	**89.98**	**93.52**

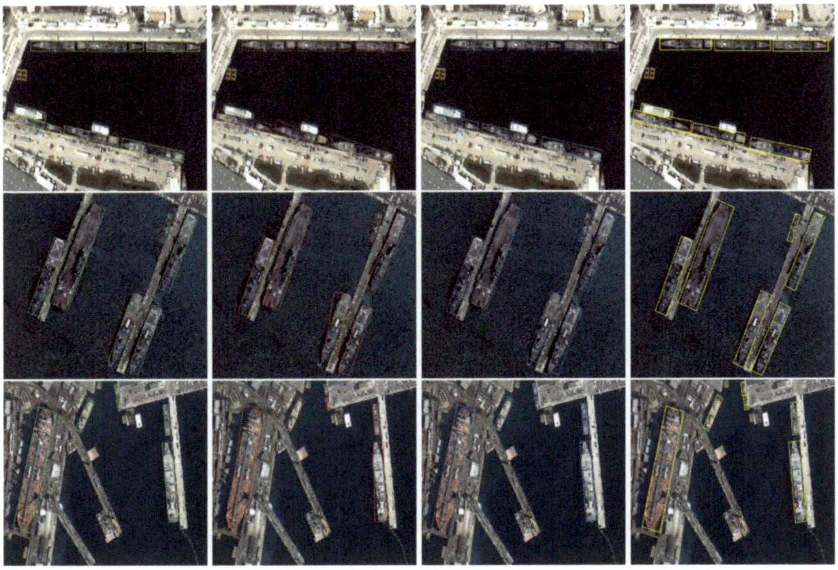

Figure 10. Comparison of ablation study methods. In the first column, we mark the ground truth with green boxes. The second column shows the detection results of Faster-RCNN OBB. The detection result in blue boxes are the effect of our use of ICR-head. In the last column, the yellow box shows the detection results used by both the DRoI module and ICR-head.

3.5. Comparison with Others

In this part, we use AP(07) and AP(12) to compare our method on three datasets, and partial detection results of each dataset. In addition, we also draw the precision–recall curves of the partial comparison method.

1. Experiments on HRSC2016: To further prove the effectiveness of the proposed OSCD-Net, we compared it with other deep learning ship detection methods popular in recent years: Faster R-CNN oriented bounding box (OBB), RetinaNet OBB, R2CNN [26], RRPN [31], RoI transformer [27], Gliding vertex [32], CSL [34] and RITSD [50], as shown in Table 3. Among them, except R2CNN, all use data enhancement. The main data enhancement methods are horizontal flip and vertical flip, random rotation, random clipping, deformation and scaling. It is worth noting that all detectors adopt ResNet-101 pretrained on the ImageNet, except for RRPN adopting VGG-16 [51]. In addition, except for R2CNN and RRPN, FPN structure is used for others. It can be seen from the experiment that the mAP of adding OBB directly on Faster-RCNN and RetinaNet is low. The structures of R2CNN and RRPN are so simple that the mAP is not high. RoI transform proposes a rotated RoI, which makes the AP(07) reach 86.22% and the AP(12) reach 87.18%. Gliding vertex uses the offset of the vertex instead of the angle θ on the basis of Faster R-CNN, and the AP(07) reaches 88.20%. CSL uses angle classification instead of traditional θ, and AP(07) reaches 89.62%. RITSD adds the third stage of rotation on the basis of the traditional second stage, and the AP(07) reaches 89.70% and the AP(12) reaches 92.98%. Our OSCD-Net reaches 89.98% in AP(07) and 93.52% in AP(12). Experiments indicate that, compared with the recently proposed deep learning object detection methods, OSCD-Net has an advanced performance. Except for taking AP as the evaluation metric, we also added the comparison method with Precision, Recall and F1. As shown in Table 4, compared with other methods, OSCD-Net has obvious advantages in Precision and Recall. In Figure 11, we show the results of some comparison methods. Among them, the yellow box is the detection box, and the blue box and the red box are some ships that we manually marked as inaccurate detection and missing detection, respectively. Faster R-CNN OBB and the RoI transformer use FP representation to detect rotating rectangles, and it can be found that there are many

ships with inaccurate detection in these two methods. In the third panel of Figure 11, the RoI transformer and Gliding vertex are missed when the two ships are close to each other. In addition, we also show the precision–recall curves of Faster R-CNN OBB, Retinanet OBB, RoI transformer and Gliding vertex, as shown in Figure 12.

Table 3. Performance evaluation of nine different methods on HRSC2016 dataset. The highest value is indicated by bold numbers (%).

Method	Data Aug	Backbone	FPN	AP(07)	AP(12)
Faster R-CNN OBB	√	ResNet101	√	80.90	82.91
RetinaNet OBB	√	ResNet101	√	79.17	82.10
R2CNN	-	ResNet101	-	72.36	74.35
RRPN	√	VGG16	-	79.60	-
RoI transformer	√	ResNet101	√	86.22	87.18
Gliding vertex	√	ResNet101	√	88.20	89.02
CSL	√	ResNet101	√	89.62	-
RITSD	√	ResNet101	√	89.70	92.98
OSCD-Net (ours)	√	ResNet101	√	**89.90**	**93.52**

Table 4. Performance evaluation with Precision(P), Recall(R) and F1 on the HRSC2016 dataset. The highest value is indicated by bold numbers (%).

Method	P	R	F1
Faster R-CNN OBB	86.24	87.17	86.70
RetinaNet OBB	87.12	85.21	86.15
RoI transformer	88.01	90.15	89.07
Gliding vertex	89.81	92.67	91.22
OSCD-Net (ours)	**90.12**	**94.54**	**92.27**

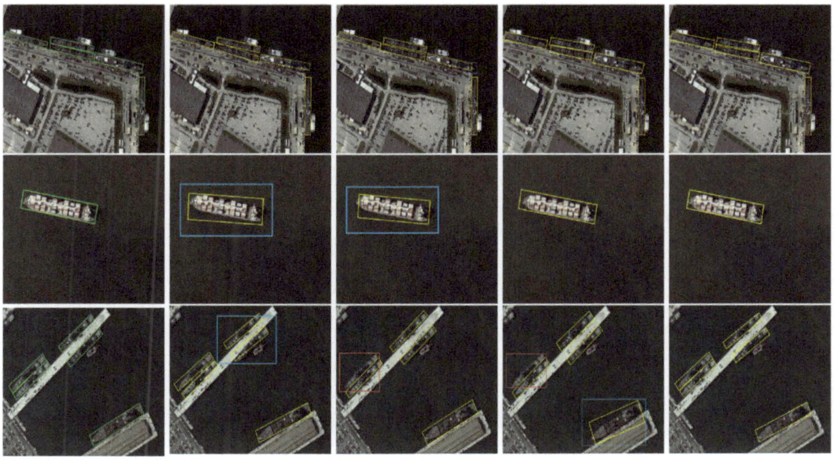

Ground Truth (**a**) (**b**) (**c**) (**d**)

Figure 11. Comparison experiment detection results on HRSC2016 dataset. The green box in the first column represents the ground truth. The yellow box indicates the detection result. The red box indicates the missed detection area, and the blue box indicates the inaccurate detection area. (**a**) The detection result of Faster R-CNN. (**b**) The detection result of RoI transformer. (**c**) The detection result of Gliding vertex. (**d**) The detection result of OSCD-Net.

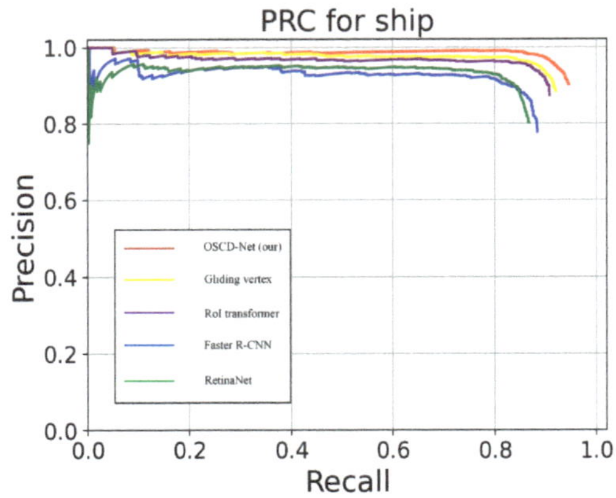

Figure 12. Precision–recall curve of different methods on the HRSC2016 dataset.

2. Experiments on DOTA-Ship: Compared with the HRSC2016 dataset, the ships in DOTA-Ship are smaller and more dense, which makes the detection task more challenging. We randomly divided the DOTA-Ship dataset into a 6:4 trainset and testset. Table 5 shows the comparison and ablation experiments of our method on DOTA-Ship. Here, we compare our method with Faster R-CNN OBB, R2CNN, RRPN, RoI transform and Gliding vertex. These methods all use data enhancement and FPN structure, except R2CNN, and the backbone network is ResNet-101. OSCD-Net adds ICR-head to the baseline and OSCD-Net* adds ICR-head and DRoI module. The results indicate that the mAP of our method is also significantly improved compared with other popular ship detection methods. Compared with baseline (Faster R-CNN OBB), our method is improved by 9.46% in the DOTA-Ship dataset. In Table 6, we conducted a comparative experiment on the DOTA-Ship dataset with Precision, Recall and F1 as evaluation standards. Our method has obvious advantages in Recall and higher Precision than other methods. In Figure 13, we show the comparison of detection results of some methods on DOTA-ship dataset. We can find that the detection boxes of Faster R-CNN OBB have a significant offset in the first two figures, which will lead to the reduction in the IoU. However, using edges to detect rotating ships does not have such a problem. Both the RoI transformer and Gliding vertex have a serious problem of missing detection. OSCD-Net can also miss for partial small and dense ships, but this is less of a problem than other methods. We also show the precision–recall curves on the DOTA-Ship dataset in Figure 14.

Table 5. Performance evaluation of nine different methods on DOTA-Ship dataset. The highest value is indicated by bold numbers (%).

Method	Data Aug	FPN	AP(07)
Faster R-CNN OBB	√	√	58.49
RetinaNet OBB	√	√	56.21
R2CNN	-	-	52.12
RoI transformer	√	√	63.75
Gliding vertex	√	√	66.53
OSCD-Net (ours)	√	√	67.62
OSCD-Net* (ours)	√	√	**67.95**

Table 6. Performance evaluation with Precision(P), Recall(R) and F1 on the DOTA-Ship dataset. The highest value is indicated by bold numbers (%).

Method	P	R	F1
Faster R-CNN OBB	62.92	70.19	66.36
RetinaNet OBB	62.81	68.25	65.42
RoI transformer	64.89	75.10	69.62
Gliding vertex	69.01	77.50	73.01
OSCD-Net (ours)	**69.04**	**78.15**	**73.31**

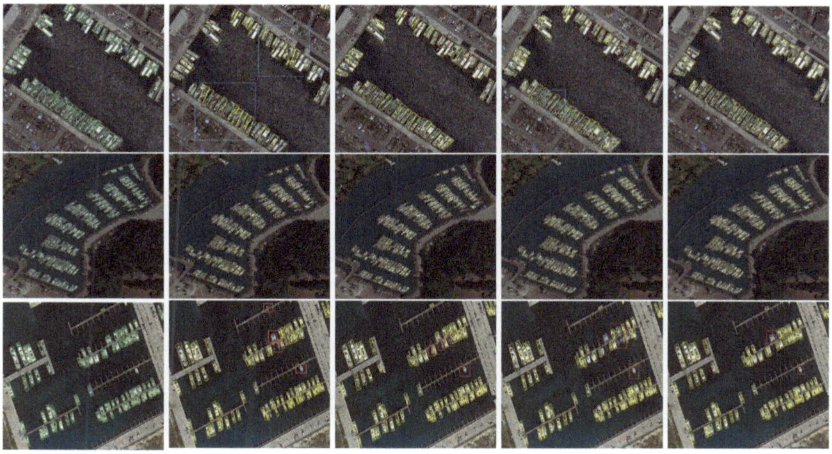

Ground Truth (a) (b) (c) (d)

Figure 13. Comparison of experiment detection results on the DOTA-Ship dataset. The green box in the first column represents the ground truth. The yellow box indicates the detection result. The red box indicates the missed detection area, and the blue box indicates the inaccurate detection area. (**a**) The detection result of Faster R-CNN. (**b**) The detection result of RoI transformer. (**c**) The detection result of Gliding vertex. (**d**) The detection result of OSCD-Net.

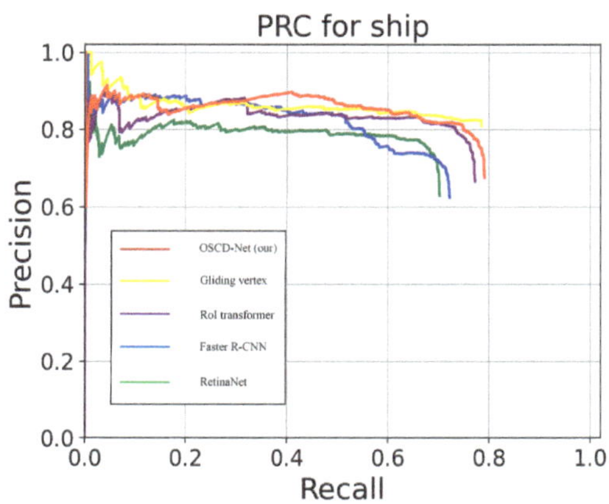

Figure 14. Precision–recall curve of different methods on the DOTA-Ship dataset.

3. Experiments on SSDD+: Because the SSDD+ is an SAR image dataset and the SAR image has the characteristics of inconsistent azimuth and range resolution, prominent target geometric characteristics, strong noise, low definition and large image difference when the pitch angle and azimuth angle change, the SSDD+ dataset tests the robustness

of the model. We adopt the data enhancement strategy of image flipping. We randomly divided the SSDD+ dataset into a 7:3 train set and test set. The comparison methods and experimental parameters here are the same as in the DOTA-Ship dataset. As shown in Table 7, compared with other popular ship detection methods, the mAP of our method is also significantly improved. OSCD-Net adds ICR-head to baseline and OSCD-Net* adds ICR-head and the DRoI module. In Table 8, the recall of OSCD-Net is higher than other methods, and the precision is also higher than other methods except Gliding vertex. In the precision–recall curves on the SSDD+ dataset (Figure 15), the precision of our method is worse than other methods when the recall is low. However, with the improvement in the recall, the precision of our method gradually exceeds that of other methods. Our method is also better than the other methods in terms of the F1, which is the standard to measure the combination of precision and recall. In the resulting images of the comparison (Figure 16), Faster R-CNN OBB and the RoI transformer detect multiple ships as one because many ships are close to each other in the first image. This is due to feature extraction errors. Although the Gliding vertex is able to extract features correctly, there is a missing detection. In the second picture, the missed detection of Faster R-CNN OBB is serious, and the detection of RoI transformer and Gliding vertex targets is not accurate enough (marked with blue box). The OSCD-Net detection results are obviously close to the real marking of ships and less redundant information.

Table 7. Performance evaluation of nine different methods on the SSDD+ dataset. The highest value is indicated by bold numbers (%).

Method	Data Aug	FPN	AP(07)	AP(12)
Faster R-CNN OBB	√	√	75.95	77.45
RetinaNet OBB	√	√	69.17	72.18
R2CNN	-	-	67.78	68.95
RoI transformer	√	√	78.32	79.82
Gliding vertex	√	√	81.22	83.20
OSCD-Net(ours)	√	√	81.01	83.49
OSCD-Net*(ours)	√	√	**82.90**	**84.52**

Table 8. Performance evaluation with Precision(P), Recall(R) and F1 on the SSDD+ dataset. The highest value is indicated by bold numbers (%).

Method	P	R	F1
Faster R-CNN OBB	85.24	83.42	84.32
RetinaNet OBB	84.10	81.93	83.00
RoI transformer	88.23	87.15	87.69
Gliding vertex	**90.04**	88.24	89.13
OSCD-Net (ours)	89.25	**90.45**	**89.85**

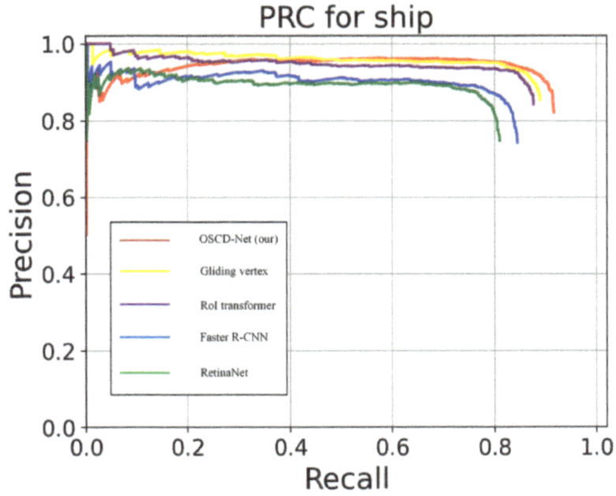

Figure 15. Precision–recall curve of different methods on the SSDD+ dataset.

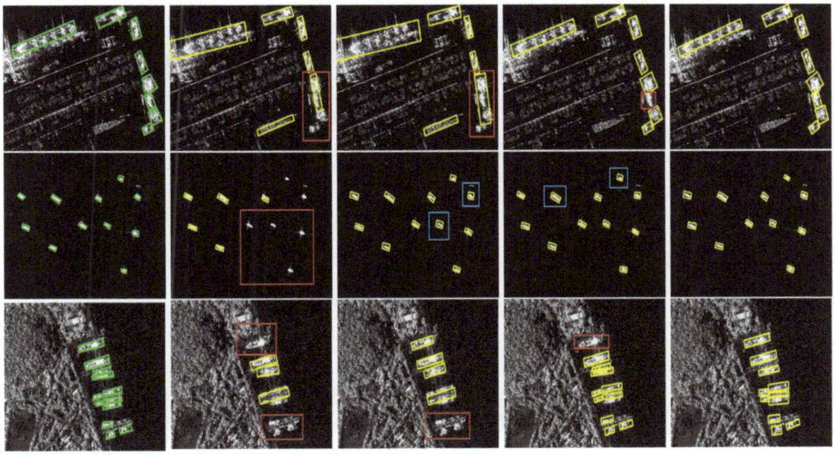

Figure 16. Comparison of experiment detection results on the SSDD+ dataset. The green box in the first column represents the ground truth. The yellow box indicates the detection result. The red box indicates the missed detection area, and the blue box indicates the inaccurate detection area. (**a**) The detection result of Faster R-CNN. (**b**) The detection result of RoI transformer. (**c**) The detection result of Gliding vertex. (**d**) The detection result of OSCD-Net.

It can be seen from the comparative experiment that the accuracy of our model has been improved on the three datasets. In addition, we show the partial testset results of the three datasets in Figure 17. It can be seen from the renderings of the testset that in HRSC2016 and SSDD+, the detection results of our network are relatively good, and most pictures can be detected correctly. However, for dense and small ships in the DOTA-Ship dataset, our network has missed detections. This may be because our network is not very suitable for small and dense ships. We will try our best to overcome this problem in the follow-up work.

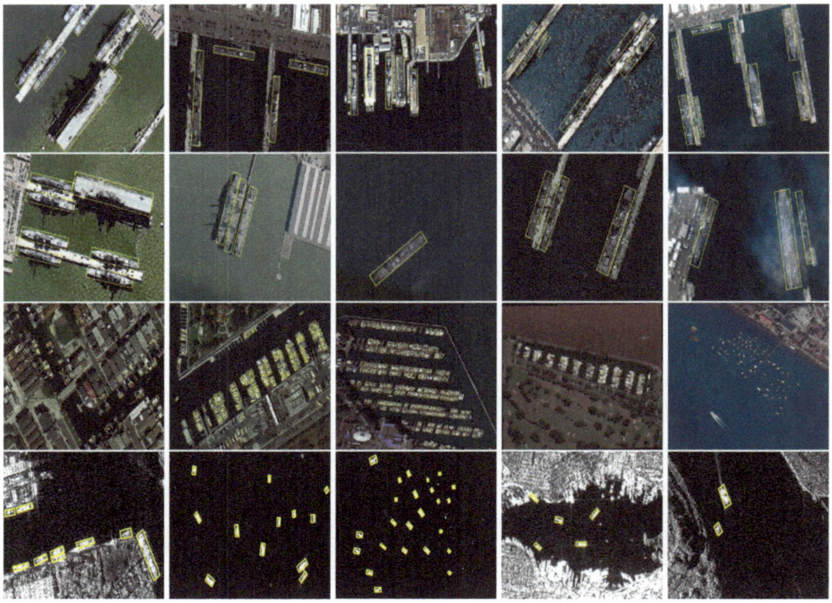

Figure 17. Some detection results on the test datasets.

4. Conclusions

Because deep learning is able to automatically learn the essential features of a large amount of data, it has achieved good results in the field of computer vision and image processing. The same is true for ship detection in the field of remote sensing. In this paper, a novel oriented ship detection method is proposed based on a intersecting circle and a deformable RoI. Ships in remote sensing images have the problems of inclination in any direction and large length–width ratios. To address these problems, ICR-head and a DRoI module were designed by us. ICR-head is a rotated detection head, which can obtain the rotation information of the ship. DRoI pools candidate features into multiple shapes and recombines features through transformation and deformation operations, so as to solve the problem of the large length–width ratio of ships.

Our method focuses on solving the ship's arbitrary tilt and excessive length–width ratio. We are also solving the problem from a geometric point of view. In the future, we will try to solve the two challenges of arbitrary ship tilt and excessive length–width ratio at the feature point of view. Of course, these are not the only challenges of ship detection. The complexity of the background and the size of the ship are also important challenges in ship inspection. This is also one of the main tasks of our future work. With the continuous development of remote sensing technology, whether in optical remote sensing image or SAR remote sensing image, there are a lot of data available for us to use. There is a certain similarity as long as it is the same species. We believe that remote sensing image ship detection has a good prospect in the future.

Author Contributions: All authors have made significant contributions to this work. J.Z. and R.H. proposed the method and wrote the paper; R.H. and B.P. provided suggestions and designed the experiments; Y.L. and B.P. revised the paper. All authors have read and agreed to the published version of the manuscript.

Funding: This work was supported in part by the National Natural Science Foundation of China under the Grant 62001252; the Beijing-Tianjin-Hebei Basic Research Cooperation Project under the Grant F2021203109; the Natural Science Foundation of Hebei Province of China under the Grant F2020202008; and the Scientific and Technological Research Project of Hebei Province Universities and Colleges under the Grant ZD2021311.

Data Availability Statement: No new data were created or analyzed in this study.

Acknowledgments: The author thanks the anonymous reviewers for their valuable comments and the editors for their support of the work.

Conflicts of Interest: The authors declare no conflict of interest.

References

1. Zou, Z.; Shi, Z. Random Access Memories: A New Paradigm for Target Detection in High Resolution Aerial Remote Sensing Images. *IEEE Trans. Image Process.* **2018**, *27*, 1100–1111. [CrossRef] [PubMed]
2. Wang, C.; Bai, X.; Wang, S.; Zhou, J.; Ren, P. Multiscale visual attention networks for object detection in VHR remote sensing images. *IEEE Geosci. Remote Sens. Lett.* **2018**, *16*, 310–314. [CrossRef]
3. Liu, E.; Zheng, Y.; Pan, B.; Xu, X.; Shi, Z. DCL-Net: Augmenting the Capability of Classification and Localization for Remote Sensing Object Detection. *IEEE Trans. Geosci. Remote Sens.* **2021**, *59*, 7933–7944. [CrossRef]
4. Yu, Y.; Wang, J.; Qiang, H.; Jiang, M.; Tang, E.; Yu, C.; Zhang, Y.; Li, J. Sparse anchoring guided high-resolution capsule network for geospatial object detection from remote sensing imagery. *Int. J. Appl. Earth Obs. Geoinf.* **2021**, *104*, 102548. [CrossRef]
5. Yuan, H.; Huang, K.; Ren, C.; Xiong, Y.; Duan, J.; Yang, Z. Pomelo Tree Detection Method Based on Attention Mechanism and Cross-Layer Feature Fusion. *Remote Sens.* **2022**, *14*, 3902. [CrossRef]
6. Dong, X.; Qin, Y.; Gao, Y.; Fu, R.; Liu, S.; Ye, Y. Attention-Based Multi-Level Feature Fusion for Object Detection in Remote Sensing Images. *Remote Sens.* **2022**, *14*, 3735. [CrossRef]
7. Li, Q.; Mou, L.; Liu, Q.; Wang, Y.; Zhu, X.X. HSF-Net: Multiscale Deep Feature Embedding for Ship Detection in Optical Remote Sensing Imagery. *IEEE Trans. Geosci. Remote Sens.* **2018**, *56*, 7147–7161. [CrossRef]
8. Zhou, Q.; Song, F.; Chen, Z.; Zhang, R.; Jiang, P.; Lei, T. Harbor Ship Detection Based on Channel Weighting and Spatial Information Fusion. *J. Phys. Conf. Ser.* **2021**, *1738*, 012057. [CrossRef]
9. Li, X.; Li, D.; Liu, H.; Wan, J.; Chen, Z.; Liu, Q. A-BFPN: An Attention-Guided Balanced Feature Pyramid Network for SAR Ship Detection. *Remote Sens.* **2022**, *14*, 3829. [CrossRef]
10. Zhao, Y.; Zhao, L.; Xiong, B.; Kuang, G. Attention Receptive Pyramid Network for Ship Detection in SAR Images. *IEEE J. Sel. Top. Appl. Earth Obs. Remote Sens.* **2020**, *13*, 2738–2756. [CrossRef]
11. Nie, X.; Duan, M.; Ding, H.; Hu, B.; Wong, E.K. Attention Mask R-CNN for Ship Detection and Segmentation from Remote Sensing Images. *IEEE Access* **2020**, *8*, 9325–9334. [CrossRef]
12. Hu, J.; Zhi, X.; Shi, T.; Zhang, W.; Cui, Y.; Zhao, S. PAG-YOLO: A Portable Attention-Guided YOLO Network for Small Ship Detection. *Remote Sens.* **2021**, *13*, 3059. [CrossRef]
13. Yao, Y.; Jiang, Z.; Zhang, H.; Zhao, D.; Cai, B. Ship detection in optical remote sensing images based on deep convolutional neural networks. *J. Appl. Remote Sens.* **2017**, *11*, 042611. [CrossRef]
14. Li, J.; Qu, C.; Shao, J. Ship detection in SAR images based on an improved faster R-CNN. In Proceedings of the 2017 SAR in Big Data Era: Models, Methods and Applications (BIGSARDATA), Beijing, China, 13–14 November 2017; pp. 1–6. [CrossRef]
15. Tang, G.; Zhuge, Y.; Claramunt, C.; Men, S. N-YOLO: A SAR Ship Detection Using Noise-Classifying and Complete-Target Extraction. *Remote Sens.* **2021**, *13*, 871. [CrossRef]
16. Girshick, R. Fast R-CNN. In Proceedings of the IEEE International Conference on Computer Vision (ICCV), Santiago, Chile, 7–13 December 2015; pp. 1440–1448.
17. Ren, S.; He, K.; Girshick, R.; Sun, J. Faster R-CNN: Towards Real-Time Object Detection with Region Proposal Networks. In *Proceedings of the Advances in Neural Information Processing Systems*; Cortes, C., Lawrence, N., Lee, D., Sugiyama, M., Garnett, R., Eds.; Curran Associates, Inc.: New York, NY, USA, 2015; Volume 28.
18. Redmon, J.; Divvala, S.; Girshick, R.; Farhadi, A. You Only Look Once: Unified, Real-Time Object Detection. In Proceedings of the 2016 IEEE Conference on Computer Vision and Pattern Recognition (CVPR), Las Vegas, NV, USA, 27–30 June 2016; pp. 779–788. [CrossRef]
19. Liu, W.; Anguelov, D.; Erhan, D.; Szegedy, C.; Reed, S.; Fu, C.Y.; Berg, A.C. SSD: Single Shot MultiBox Detector. In *Proceedings of the Computer Vision—ECCV 2016*; Leibe, B., Matas, J., Sebe, N., Welling, M., Eds.; Springer International Publishing: Cham, Switzerland, 2016; pp. 21–37.
20. Lin, T.Y.; Goyal, P.; Girshick, R.; He, K.; Dollar, P. Focal Loss for Dense Object Detection. In Proceedings of the IEEE International Conference on Computer Vision (ICCV), Venice, Italy, 22–29 October 2017.
21. Xie, X.; Li, L.; An, Z.; Lu, G.; Zhou, Z. Small Ship Detection Based on Hybrid Anchor Structure and Feature Super-Resolution. *Remote Sens.* **2022**, *14*, 3530. [CrossRef]
22. Jin, K.; Chen, Y.; Xu, B.; Yin, J.; Wang, X.; Yang, J. A Patch-to-Pixel Convolutional Neural Network for Small Ship Detection With PolSAR Images. *IEEE Trans. Geosci. Remote Sens.* **2020**, *58*, 6623–6638. [CrossRef]
23. Zhu, M.; Hu, G.; Zhou, H.; Wang, S.; Feng, Z.; Yue, S. A Ship Detection Method via Redesigned FCOS in Large-Scale SAR Images. *Remote Sens.* **2022**, *14*, 1153. [CrossRef]
24. Sun, Z.; Dai, M.; Leng, X.; Lei, Y.; Xiong, B.; Ji, K.; Kuang, G. An Anchor-Free Detection Method for Ship Targets in High-Resolution SAR Images. *IEEE J. Sel. Top. Appl. Earth Obs. Remote Sens.* **2021**, *14*, 7799–7816. [CrossRef]

25. Chen, C.; He, C.; Hu, C.; Pei, H.; Jiao, L. MSARN: A Deep Neural Network Based on an Adaptive Recalibration Mechanism for Multiscale and Arbitrary-Oriented SAR Ship Detection. *IEEE Access* **2019**, *7*, 159262–159283. [CrossRef]
26. Jiang, Y.; Zhu, X.; Wang, X.; Yang, S.; Li, W.; Wang, H.; Fu, P.; Luo, Z. R2CNN: Rotational region CNN for orientation robust scene text detection. *arXiv* **2017**, arXiv:1706.09579.
27. Ding, J.; Xue, N.; Long, Y.; Xia, G.S.; Lu, Q. Learning RoI Transformer for Oriented Object Detection in Aerial Images. In Proceedings of the 2019 IEEE/CVF Conference on Computer Vision and Pattern Recognition (CVPR), Long Beach, CA, USA, 16–17 June 2019; pp. 2844–2853. [CrossRef]
28. Han, J.; Ding, J.; Xue, N.; Xia, G.S. ReDet: A Rotation-equivariant Detector for Aerial Object Detection. In Proceedings of the 2021 IEEE/CVF Conference on Computer Vision and Pattern Recognition (CVPR), Long Beach, CA, USA, 16–17 June 2021; pp. 2785–2794. [CrossRef]
29. Yang, X.; Yang, X.; Yang, J.; Ming, Q.; Wang, W.; Tian, Q.; Yan, J. Learning High-Precision Bounding Box for Rotated Object Detection via Kullback-Leibler Divergence. In *Proceedings of the Advances in Neural Information Processing Systems*; Ranzato, M., Beygelzimer, A., Dauphin, Y., Liang, P., Vaughan, J.W., Eds.; Curran Associates, Inc.: New York, NY, USA, 2021; Volume 34, pp. 18381–18394.
30. Li, L.; Zhou, Z.; Wang, B.; Miao, L.; Zong, H. A Novel CNN-Based Method for Accurate Ship Detection in HR Optical Remote Sensing Images via Rotated Bounding Box. *IEEE Trans. Geosci. Remote Sens.* **2021**, *59*, 686–699. [CrossRef]
31. Ma, J.; Shao, W.; Ye, H.; Wang, L.; Wang, H.; Zheng, Y.; Xue, X. Arbitrary-Oriented Scene Text Detection via Rotation Proposals. *IEEE Trans. Multimed.* **2018**, *20*, 3111–3122. [CrossRef]
32. Xu, Y.; Fu, M.; Wang, Q.; Wang, Y.; Chen, K.; Xia, G.S.; Bai, X. Gliding Vertex on the Horizontal Bounding Box for Multi-Oriented Object Detection. *IEEE Trans. Pattern Anal. Mach. Intell.* **2021**, *43*, 1452–1459. [CrossRef] [PubMed]
33. Liu, Q.; Xiang, X.; Yang, Z.; Hu, Y.; Hong, Y. Arbitrary Direction Ship Detection in Remote-Sensing Images Based on Multitask Learning and Multiregion Feature Fusion. *IEEE Trans. Geosci. Remote Sens.* **2021**, *59*, 1553–1564. [CrossRef]
34. Yang, X.; Yan, J. Arbitrary-oriented object detection with circular smooth label. In *Proceedings of the European Conference on Computer Vision*; Springer: Cham, Switzerland, 2020; pp. 677–694.
35. Yang, X.; Hou, L.; Zhou, Y.; Wang, W.; Yan, J. Dense Label Encoding for Boundary Discontinuity Free Rotation Detection. In Proceedings of the 2021 IEEE/CVF Conference on Computer Vision and Pattern Recognition (CVPR), Nashville, TN, USA, 20–25 June 2021; pp. 15814–15824. [CrossRef]
36. Shao, Z.; Zhang, X.; Zhang, T.; Xu, X.; Zeng, T. RBFA-Net: A Rotated Balanced Feature-Aligned Network for Rotated SAR Ship Detection and Classification. *Remote Sens.* **2022**, *14*, 3345. [CrossRef]
37. Cui, Z.; Leng, J.; Liu, Y.; Zhang, T.; Quan, P.; Zhao, W. SKNet: Detecting Rotated Ships as Keypoints in Optical Remote Sensing Images. *IEEE Trans. Geosci. Remote Sens.* **2021**, *59*, 8826–8840. [CrossRef]
38. Yu, Y.; Yang, X.; Li, J.; Gao, X. A Cascade Rotated Anchor-Aided Detector for Ship Detection in Remote Sensing Images. *IEEE Trans. Geosci. Remote Sens.* **2022**, *60*, 1–14. [CrossRef]
39. Zhu, M.; Hu, G.; Zhou, H.; Wang, S.; Zhang, Y.; Yue, S.; Bai, Y.; Zang, K. Arbitrary-Oriented Ship Detection Based on RetinaNet for Remote Sensing Images. *IEEE J. Sel. Top. Appl. Earth Obs. Remote Sens.* **2021**, *14*, 6694–6706. [CrossRef]
40. Pan, Z.; Yang, R.; Zhang, Z. MSR2N: Multi-Stage Rotational Region Based Network for Arbitrary-Oriented Ship Detection in SAR Images. *Sensors* **2020**, *20*, 2340. [CrossRef]
41. Ran, B.; You, Y.; Li, Z.; Liu, F. Arbitrary-Oriented Ship Detection Method Based on Improved Regression Model for Target Direction Detection Network. In Proceedings of the IGARSS 2020—2020 IEEE International Geoscience and Remote Sensing Symposium, Waikoloa, HI, USA, 26 September–2 October 2020; pp. 964–967. [CrossRef]
42. Wu, Y.; Chen, Y.; Yuan, L.; Liu, Z.; Wang, L.; Li, H.; Fu, Y. Rethinking Classification and Localization for Object Detection. In Proceedings of the 2020 IEEE/CVF Conference on Computer Vision and Pattern Recognition (CVPR), Seattle, WA, USA, 13–19 June 2020; pp. 10183–10192. [CrossRef]
43. He, K.; Gkioxari, G.; Dollar, P.; Girshick, R. Mask R-CNN. In Proceedings of the IEEE International Conference on Computer Vision (ICCV), Venice, Italy, 22–29 October 2017.
44. He, K.; Zhang, X.; Ren, S.; Sun, J. Deep residual learning for image recognition. In Proceedings of the IEEE Conference on Computer Vision and Pattern Recognition, Las Vegas, NV, USA, 27–30 June 2016; pp. 770–778.
45. Lin, T.Y.; Dollár, P.; Girshick, R.; He, K.; Hariharan, B.; Belongie, S. Feature Pyramid Networks for Object Detection. In Proceedings of the 2017 IEEE Conference on Computer Vision and Pattern Recognition (CVPR), Honolulu, HI, USA, 21–26 July 2017; pp. 936–944. [CrossRef]
46. Liu, Z.; Wang, H.; Weng, L.; Yang, Y. Ship Rotated Bounding Box Space for Ship Extraction From High-Resolution Optical Satellite Images with Complex Backgrounds. *IEEE Geosci. Remote Sens. Lett.* **2017**, *13*, 1074–1078. [CrossRef]
47. Xia, G.S.; Bai, X.; Ding, J.; Zhu, Z.; Belongie, S.; Luo, J.; Datcu, M.; Pelillo, M.; Zhang, L. DOTA: A Large-Scale Dataset for Object Detection in Aerial Images. In Proceedings of the 2018 IEEE/CVF Conference on Computer Vision and Pattern Recognition, Salt Lake City, UT, USA, 18–23 June 2018; pp. 3974–3983. [CrossRef]
48. Everingham, M.; Van Gool, L.; Williams, C.K.; Winn, J.; Zisserman, A. The pascal visual object classes (voc) challenge. *Int. J. Comput. Vis.* **2010**, *88*, 303–338. [CrossRef]
49. Bottou, L. Stochastic Gradient Descent Tricks. In *Neural Networks: Tricks of the Trade*, 2nd ed.; Montavon, G., Orr, G.B., Müller, K.R., Eds.; Springer: Berlin/Heidelberg, Germany, 2012; pp. 421–436. [CrossRef]

50. Wang, T.; Li, Y. Rotation-Invariant Task-Aware Spatial Disentanglement in Rotated Ship Detection Based on the Three-Stage Method. *IEEE Trans. Geosci. Remote Sens.* **2022**, *60*, 1–12. [CrossRef]
51. Simonyan, K.; Zisserman, A. Very Deep Convolutional Networks for Large-Scale Image Recognition. *arXiv* **2014**, arXiv:1409.1556.

remote sensing

Article

MSCNet: A Multilevel Stacked Context Network for Oriented Object Detection in Optical Remote Sensing Images

Rui Zhang [1,2], Xinxin Zhang [1,2,*], Yuchao Zheng [1,2], Dahan Wang [1,2] and Lizhong Hua [1,2]

[1] College of Computer and Information Engineering, Xiamen University of Technology, Xiamen 361024, China
[2] Fujian Key Laboratory of Pattern Recognition and Image Understanding, Xiamen University of Technology, Xiamen 361024, China
* Correspondence: zhangxinxin@xmut.edu.cn

Abstract: Oriented object detection has recently become a hot research topic in remote sensing because it provides a better spatial expression of oriented target objects. Although research has made considerable progress in this field, the feature of multiscale and arbitrary directions still poses great challenges for oriented object detection tasks. In this paper, a multilevel stacked context network (MSCNet) is proposed to enhance target detection accuracy by aggregating the semantic relationships between different objects and contexts in remote sensing images. Additionally, to alleviate the impact of the defects of the traditional oriented bounding box representation, the feasibility of using a Gaussian distribution instead of the traditional representation is discussed in this paper. Finally, we verified the performance of our work on two common remote sensing datasets, and the results show that our proposed network improved on the baseline.

Keywords: oriented object detection; multilevel stacked context; remote sensing images

Citation: Zhang, R.; Zhang, X.; Zheng, Y.; Wang, D.; Hua, L. MSCNet: A Multilevel Stacked Context Network for Oriented Object Detection in Optical Remote Sensing Images. *Remote Sens.* **2022**, *14*, 5066. https://doi.org/10.3390/rs14205066

Academic Editors: Qian Du, Yanni Dong and Xiaochen Yang

Received: 9 August 2022
Accepted: 3 October 2022
Published: 11 October 2022

1. Introduction

With the rapid development of remote sensing technology, remote sensing data have gradually decreased acquisition costs, enriched data sources, and improved image resolution and quality. Consequently, remote sensing images have gradually become popular and are used in various industries. As a popular image processing method, convolutional neural networks have also become a powerful tool for processing remote sensing data. Compared with traditional manual feature construction, convolutional neural networks directly generate feature representations from the original image pixels, and the deep stacked structure of convolutional neural networks helps extract more abstract semantic features; these results are beyond the reach of the traditional method. For example, in object detection, Ren et al. [1] proposed a classical two-stage object detection network.

To address previous studies on natural images, a series of convolutional neural network object detection methods suitable for remote sensing images are proposed. For example, Li et al. [2] designed a multiangle anchor for a region proposal network (RPN), and proposed a two-channel feature fusion network to learn local and context features. Liu et al. [3] used dilated convolution to extract features and dynamically adjusted the weight of each position in the dilated convolution kernel to consider the detection accuracy of both large and small objects.

However, due to the orthographic characteristics of optical remote sensing images, object detection networks suitable for natural images have difficulty handling objects in arbitrary directions in remote sensing images. Because most of the objects in natural images are usually perpendicular to the ground due to gravity, the influence of direction is smaller than that in the remote sensing image. However, remote sensing images are taken from the upper space of the earth, and objects on the ground usually have different directions. This problem creates new challenges for the object detection task of remote sensing images. For instance, due to the dense distribution of some targets in remote sensing images, multiple targets may crowd in the same region of interest (RoI) proposed by the RPN. As a result, overlapping features will make it difficult to train subsequent classifiers.

To address this challenge, people try to use oriented bounding boxes instead of traditional horizontal detection bounding boxes in remote sensing object detection tasks to improve detection performance. Much progress has also been made in this area, and most oriented bounding boxes have been developed by using a Faster R-CNN with a feature pyramid network (FPN) [4] structure, such as DRBox [5] and CAD-Net [6]. Although these studies have verified the excellent performance of these methods on multiple datasets, there is still room for improvement. For example, Zhang et al. combined a global context module and a local pyramid context module in their CAD-Net to enhance both global and local semantic features. Additionally, the attention mechanism of spatial scale is designed in combination with the FPN structure. This combination of global and local context design is the first approach used in remote sensing image target detection and has been proven to be an effective idea. However, in studies using this approach, the relationship between the global and local context features of each layer is not fully utilized.

Many studies have indicated that in deep convolutional networks, the deeper features of images contain mainly semantic information, while lower features show more boundary details of objects. Additionally, remote sensing images contain more positional relations and semantic connections between objects than natural images due to the particularity of orthophotos. Moreover, due to the local limitations of the convolution operation, it is difficult to model the relationship between two distant objects by using the convolution operation. For example, two different regions in an image have the same target, and the semantic relationship between them may be similar. In addition, the spatial relationship also helps improve the detection performance of the network. For example, bridges and harbors have a similar appearance and are usually located in water, but both ends of a bridge are usually connected to the road, and bridges have a spatial relationship with vehicles; in contrast, only one end of a harbor is usually connected to the road, and harbors have a spatial relationship with ships. Due to the local effective receiving field of the convolution operation, these relationships cannot be effectively modeled by the convolution layer alone.

Thus, it is natural to improve the performance of the detection network by capturing and modeling the connections between objects and contexts, and explicitly modeling long-range relations. Following this idea, we designed a multilevel stacked context network (MSCNet) for remote sensing rotation object detection. First, to effectively capture the relationship between long-range objects, a multilevel stacked semantic capture (MSSC) module is embedded in the network. This module obtains different receiving domains by adding multiple parallels dilated convolutions with different dilated rates at the c5 layer of ResNet and overlays the convolutions in a global-to-local manner to take full advantage of the context relations of different distances. In addition, we perform a multichannel weighted fusion of the RoI in the RPN stage to provide better RoI features for further regression and classification operations.

The main contributions are summarized as follows:

1. An effective multilevel stacked semantic acquisition and enhancement module is proposed to enhance the representation of the FPN on remote sensing images.
2. The Gaussian Wasserstein distance loss is used to alleviate the criticality problems caused by the traditional oriented bounding box representation.
3. An improved RoI allocation strategy is used to enhance FPN multilevel information aggregation, which improves the detection performance of multiscale targets by using a multichannel weighted fusion structure instead of a single layer allocation strategy.

The rest of this paper is organized as follows. Section 2 will introduce the main object detection methods and the related studies on remote sensing images for oriented object detection networks in detail. Section 3 introduces the methods we proposed for MSSCNet. Section 4 provides the experimental results and discussions. Finally, Section 5 gives the conclusions.

2. Related Work

2.1. Anchor-Based Object Detectors

Anchor-based object detection networks focus mainly on anchor generation and secondarily on classification and regression. Anchor-based detectors are divided into two main types according to the number of network stages: one-stage detectors and two-stage detectors. As an example of two-stage detectors, R-CNN and its variants [7,8] generate several proposals for an RoI before training, and then the RoI mapped back to the feature map is further classified and regressed. The main one-stage detectors include SSD [9], RetinaNet [10], RefineDet [11], YOLOV2, and variant YOLOV2 networks [12–14]. Unlike two-stage detectors, which generate a series region proposal by using a Selective Search algorithm or an RPN, one-stage detectors predict category and location information directly through the backbone.

Although anchor-based detectors play a dominant role in object detection tasks, these detectors still have the following shortcomings: (1) The anchor design depends on human presets, and the appropriate anchor size and aspect ratio need to be set for different datasets. (2) A fixed-anchor design is not conducive to detecting extreme scale targets. (3) Generating the proposal requires many anchor samplings; however, having many anchors will lead to sample imbalance. (4) To achieve good performance, some hyperparameters need to be carefully adjusted; these include anchor shape and the IoU threshold of positive and negative samples.

2.2. Anchor-Free Object Detectors

Since anchor-based object detectors are subject to various anchor constraints, some people proposed an anchor-free architecture. Anchor-free detectors are also divided into two forms: key-point-based and center-based detectors.

CornerNet [15], borrowing the idea from algorithms such as human gesture estimation, denotes the detection box by combining the upper left corner and lower right corner points of the target. ExtremeNet [16] defines a key point as an extreme point, predicts four multipeak heatmaps for each target to find the extreme point, and predicts the target center through the center heatmap. The geometric center of the extreme point corresponds to the score of the center heatmap as the grouping condition.

FCOS [17] is similar to FCN [18]; FCOS uses a fully convolutional network to regress the distance from each position of the feature map to the four edges of the targets directly and obtains an effect similar to an anchor-based detector. CenterNet [19] transforms the detection problem into a key point problem and uses a heatmap to directly predict the center point and the size of the target. Each target only predicts one positive sample center point, so non-maximum suppression (NMS) is no longer needed to filter positive and negative samples.

2.3. Arbitrarily Oriented Object Detectors

Since most of the objects to be detected are placed horizontally in the natural image object detection task, early studies were based on horizontal bounding boxes. However, with the gradual deepening of the research, when the object detection scene is extended to text detection, aerial image detection, and 3D target detection, the shortcoming of the horizontal bounding box is gradually exposed; that is, the horizontal bounding box cannot provide accurate positions, and it also affects the performance of the detection network. Therefore, it is necessary to propose an effective rotating detector to generate an oriented instead of a horizontal bounding box for the object.

Among the rotating detectors, a significant problem is how to reasonably represent an oriented target. Different detectors give different definitions. In remote sensing image object detectors, SCRDet, R3Det, CADNet [6,20,21], etc., θ is added into the definition of the detection box as an additional parameter. This method of definition is the most concise and intuitive, and Figure 1 shows two different representations of five parameter-oriented bounding boxes: OpenCV representation and long-side representation. Parameter

$\theta \in [-90°, 0°)$ in the OpenCV representation represents the acute angle or right angle between the bounding box and the x-axis; $\theta \in [-90°, 90°]$ in the long-side representation represents the angle between the long edge of the bounding box and the x-axis.

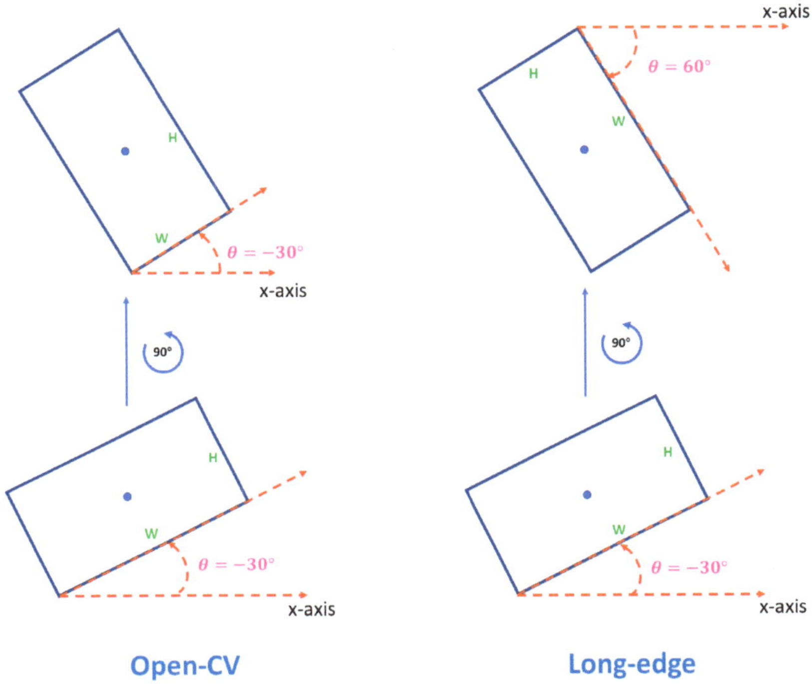

Figure 1. The two representations of oriented bounding boxes. Before rotation occurs, both representations of the bounding boxes are $\{x, y, w, h, \theta = -30°\}$. After the bounding boxes are rotated 90 degrees counter-clockwise, the bounding box is represented in Open-CV as $\{x, y, w' = h, h' = w, \theta = -30°\}$, while the bounding box is represented in Long-edge as $\{x, y, w' = w, h' = h, \theta = 60°\}$.

Unlike the above detectors, Gliding Vertex, Oriented R-CNN [22,23], and others define the oriented bounding box with six parameters $\{x, y, w, h, \theta_1, \theta_2\}$, where $\{x, y, w, h\}$ contains the vertices of the horizontal bounding box of the object; θ_1 and θ_2 are the ratios of the distance to the width and height, respectively, from the vertex of the oriented bounding box to a specific anchor point on the horizontal bounding box; and we usually set the end or midpoint on the width and height of the horizontal bounding box as the anchor point, as shown in Figure 2. Additionally, a few studies still regard the angle of the oriented box as a classification problem; this view is unlike that of most mainstream studies, which regard angle prediction as a regression problem [24].

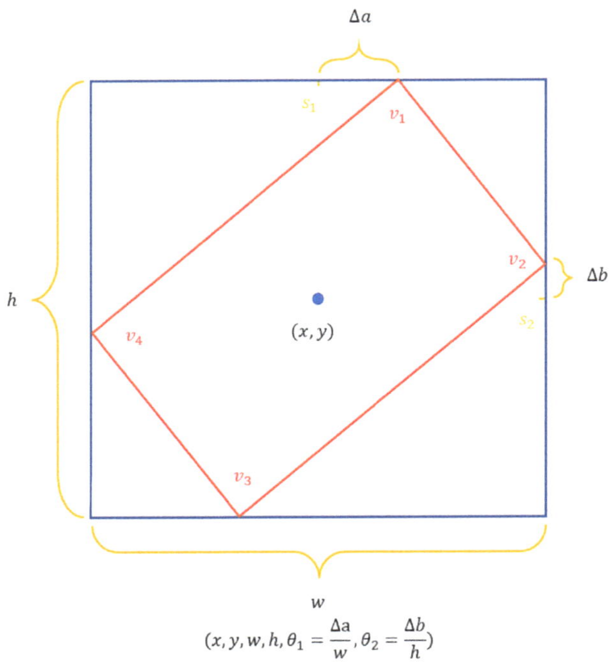

Figure 2. Illustration of six parameters representation for oriented bounding boxes. $s1$ and $s2$ are the midpoints on the width and height of the external horizontal bounding box, Δa and Δb are offsets from the intersecting points to $s1, s2$, or other key points. It adopts $\{x, y, w, h, \theta_1, \theta_2\}$ to represent oriented objects.

3. Methods

3.1. Overview

In this section, we will introduce our proposed network in detail. Figure 3 shows the overall structure of our proposed MSCNet. MSCNet is based on the two-stage detector Faster R-CNN. First, in the feature extraction stage, the long-range relationship capture module is added to enhance the interpretation of the model to the object relationship. For the RPN part, we generate oriented proposals by predicting the anchor offsets and the foreground and background scores. Here, it is necessary to define appropriate rotation parameters when using an oriented bounding box. The simplest way is to add another parameter θ representing the rotation angle, which is expressed as $\{x, y, w, h, \theta\}$. However, this definition will affect detection performance, and we will explain the definition in detail in Section 3.3.1. Thus, to reduce this effect, a Gaussian distribution is used to represent the oriented bounding box, and the Gaussian Wasserstein distance is introduced as a measure of similarity between oriented bounding boxes. Furthermore, in the FPN part, the context information is removed by using the Adaptive RoI Assignment (ARA) module to improve the multiscale expression capability of the FPN. Then, in the R-CNN part, we conduct secondary classification and regression of the proposals in the RPN to obtain the final detection results. Next, we will introduce each part in detail.

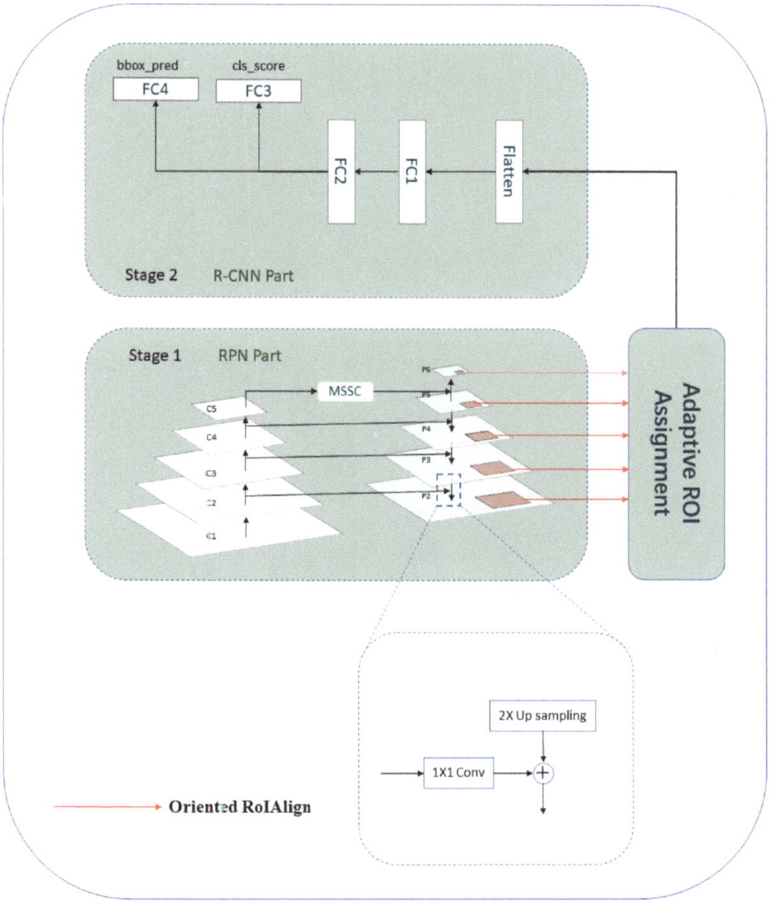

Figure 3. The overall structure of MSCNet.

3.2. MSSC Module

The FPN structure constructs a top-down fusion path to enhance the features of each scale by using a pyramid structure and can improve the detection ability of the network for multiscale targets. Previous studies have shown that the C5 layer in ResNet usually contains rich location and semantic information. Therefore, additional information is added to the $M5$ layer of the FPN by long-range semantic and positional capture of the C5 layer and is added to the bottom layers $\{M4, M3, M2\}$ along with the top-down fusion path. In this paper, an MSSC module is proposed. This module uses a parallel method to obtain receptive fields of different sizes and aggregates them in a stacking way from global receptive fields to local receptive fields. In this way, this module can take full advantage of relationships between multiple contexts and contexts at different levels.

In detail, at the C5 layer, the dilatation convolution with four different dilatation coefficients is used to capture the context information at multiple scales, with $\{3, 6, 12, 18\}$ (see Figure 4).

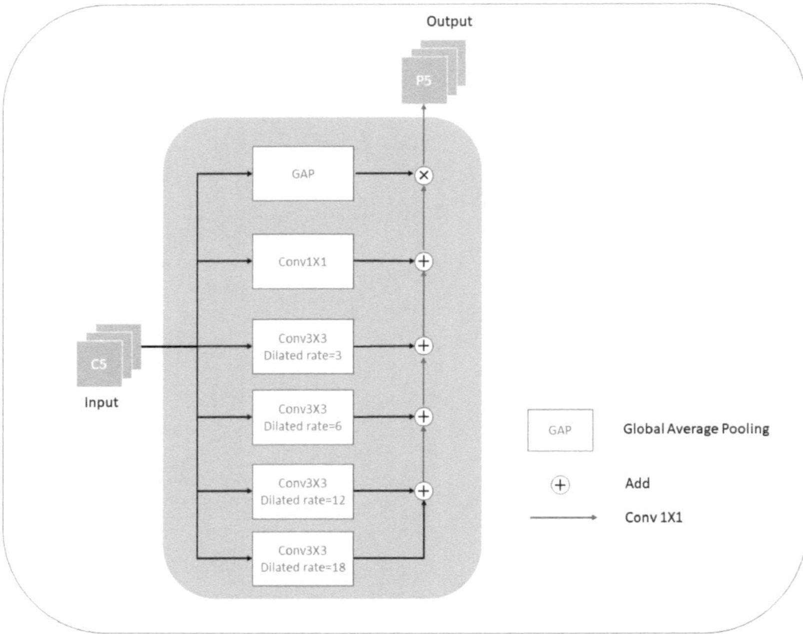

Figure 4. The overall structure of the MSSC module.

3.3. RPN Head with Oriented Bounding Box

To detect oriented objects, we use the RPN with proposing rotating proposals as the first-stage detection head. In detail, we use five levels, P2–P6, which are generated by the FPN with MSSC to generate $a * r$ anchors for each point on each feature map level. Here, $a = \{32^2, 64^2, 128^2, 256^2, 512^2\}$ contains the number of anchor pixel areas at each level, and $r = \{1:2, 1:1, 2:1\}$ represents the number of anchor aspect ratios at each position. Each anchor is represented by a four-dimensional vector $\alpha = (\alpha_x, \alpha_y, \alpha_w, \alpha_h)$, where α_x, α_y is the center point coordinate of the anchor and α_w, α_h are the width and height, respectively, of the anchor.

Each feature output of the FPN is followed by the same detector head, which is used for anchor foreground and background prediction and offset regression. The RPN head consists of a parallel classification branch and regression branch structure. The classification branch proposes an object-ness score for each anchor; this approach is the same as that of Faster RCNN. The regression branch outputs offsets $\Delta = (\Delta x, \Delta y, \Delta w, \Delta h, \Delta\theta)$ based on the anchor for each proposal. To calculate the loss later, it is necessary to decode the offset to obtain the oriented proposal (Figure 5). Here is the decoding formula:

$$\begin{cases} x = \Delta x \cdot \alpha_w + \alpha_x \\ y = \Delta y \cdot \alpha_h + \alpha_y \\ w = \alpha_w \cdot e^{\Delta w} \\ h = \alpha_h \cdot e^{\Delta h} \\ \theta = \Delta\theta \end{cases} \tag{1}$$

where (x, y) is the proposed center point coordinate and w, h are the width and height, respectively, of the proposed oriented bounding box. $\theta \in [-90^\circ, 0^\circ)$ is the proposed deflection angle and indicates an acute or right angle between the bounding box and the x-axis (OpenCV definition).

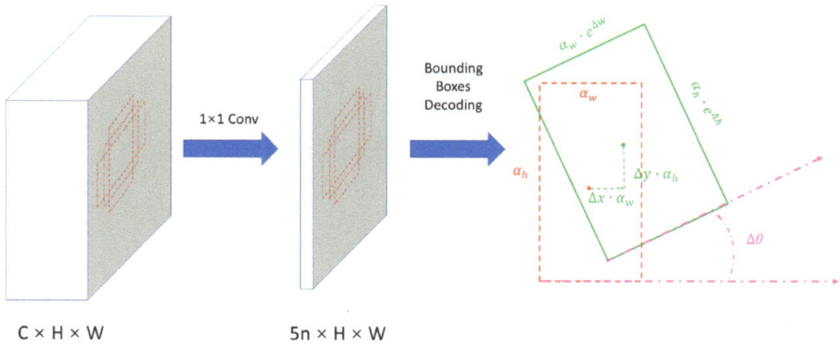

Figure 5. The Region Proposal with Oriented bounding boxes. The prediction of offsets is generated by 1×1 convolution, and n is the number of anchors generated for each position.

3.3.1. RPN Loss Function

To supervise the two prediction branches of the RPN, it is necessary to divide the proposal into positive and negative samples before calculating the loss. The sampling strategy used in this paper is similar to that used by Faster RCNN: (1) an anchor that obtains an IoU overlap higher than 0.7 with any smallest enclosing rectangle of ground-truth is set as a positive sample; (2) an anchor with the maximum IoU is used, and an IoU greater than 0.3 of the smallest ground-truth enclosing rectangle is set as a positive sample; (3) an anchor with an IoU overlap less than 0.3, and any smallest ground-truth enclosing rectangle is set as a negative sample; and (4) an anchor that is neither a positive sample nor a negative sample is set as invalid sampling and does not participate in training.

IoU loss [25] has achieved good results in many object detection tasks, such as those performed by Faster RCNN, FCOS, SCRDet, and YoloV4. However, in the oriented object detection task, IoU loss is not a good measure of the overlap of two rotating bounding boxes. For an object with a large aspect ratio, the angular offset results in a small IoU, as shown in Figure 6. The figure shows two certain angle overlapping bounding boxes. The larger the aspect ratio, the thinner and longer the bounding boxes, and the smaller the IoU, and the decrease rate is particularly obvious when the aspect ratio is in the range of 1 to 4. The larger the overlap angle, the faster the decrease. Additionally, the $L1$ loss also has some shortcomings in the oriented object detection task; they include (1) inconsistency with the measurement and loss function, (2) square-like problems, and (3) boundary discontinuity detection problems [26]. To minimize the influence caused by the above, the Wasserstein distance is adopted as a measure of the similarity between the two oriented bounding boxes, and the Gaussian Wasserstein distance loss (GWD loss) [26] is adopted as the regression loss function of the RPN.

GWD loss, a measurement proposed by Yang et al., reasonably describes the difference between two rotation boxes. GWD loss uses a two-dimensional Gaussian distribution $\mathcal{N}(\mathbf{m}, \mathbf{\Sigma})$ to describe a rotation bounding box and proposes a new regression loss based on the Wasserstein distance to replace the IoU loss. The conversion formula is as follows:

$$
\begin{aligned}
\mathbf{\Sigma}^{1/2} &= \mathbf{RSR}^\top \\
&= \begin{pmatrix} \cos\theta & -\sin\theta \\ \sin\theta & \cos\theta \end{pmatrix} \begin{pmatrix} \frac{w}{2} & 0 \\ 0 & \frac{h}{2} \end{pmatrix} \begin{pmatrix} \cos\theta & \sin\theta \\ -\sin\theta & \cos\theta \end{pmatrix} \\
&= \begin{pmatrix} \frac{w}{2}\cos^2\theta + \frac{h}{2}\sin^2\theta & \frac{w-h}{2}\cos\theta\sin\theta \\ \frac{w-h}{2}\cos\theta\sin\theta & \frac{w}{2}\sin^2\theta + \frac{h}{2}\cos^2\theta \end{pmatrix} \\
\mathbf{m} &= (x, y)
\end{aligned}
\tag{2}
$$

where **R** is the rotation matrix and **S** is the diagonal matrix of the eigenvalues. The Wasserstein distance between Gaussian distributions **W** can be expressed as:

$$\begin{aligned} W_2^2(X,Y) &= \|m_1 - m_2\|_2^2 + \mathrm{tr}\left[\Sigma_1 + \Sigma_2 - 2\left(\Sigma_1^{1/2}\Sigma_2\Sigma_1^{1/2}\right)^{1/2}\right] \\ &= \|m_1 - m_2\|_2^2 + \left\|\Sigma_1^{1/2} - \Sigma_2^{1/2}\right\|_F^2 \\ &= (x_1 - x_2)^2 + (y_1 - y_2)^2 + \frac{(w_1 - w_2)^2 + (h_1 - h_2)^2}{4} \\ &= l_2\text{-norm}\left(\left[x_1, y_1, \frac{w_1}{2}, \frac{h_1}{2}\right]^\top, \left[x_2, y_2, \frac{w_2}{2}, \frac{h_2}{2}\right]^\top\right) \end{aligned} \tag{3}$$

Therefore, the loss function of the RPN is as follows:

$$L1 = \lambda_1 \frac{1}{N}\sum_{i=1}^{N} F_{cls}(p_i, p_{i^*}) + \lambda_2 \frac{1}{N}[\![p_{i^*}]\!]\sum_{i=1}^{N} F_{reg}(t_i, t_{i^*}) \tag{4}$$

$$F_{reg}(b, \hat{b}) = 1 - \frac{1}{\tau + \mathcal{F}(D(b, \hat{b}))}, \tau \geq 1 \tag{5}$$

where λ_1 and λ_2 are the weights of the loss, N is the total number of anchors in a mini-batch, i is the anchor index, p_i is the classification prediction output of the RPN, p_{i^*} is the ground-truth label of the i-th anchor, t_i is the Gaussian distribution by 2D Gaussian transformation of the decoded five-dimensional vector of the i-th positive anchor, t_{i^*} is the Gaussian distribution for the corresponding ground-truth bounding box, F_{cls} is the cross-entropy loss, and F_{reg} is the GWD loss. In Formula (5), τ is a super parameter used to adjust the loss, $D(\cdot)$ is the Wasserstein distance calculation function of two Gaussian distributions, and $\mathcal{F}(\cdot)$ is a nonlinear function used to smooth the Wasserstein distance.

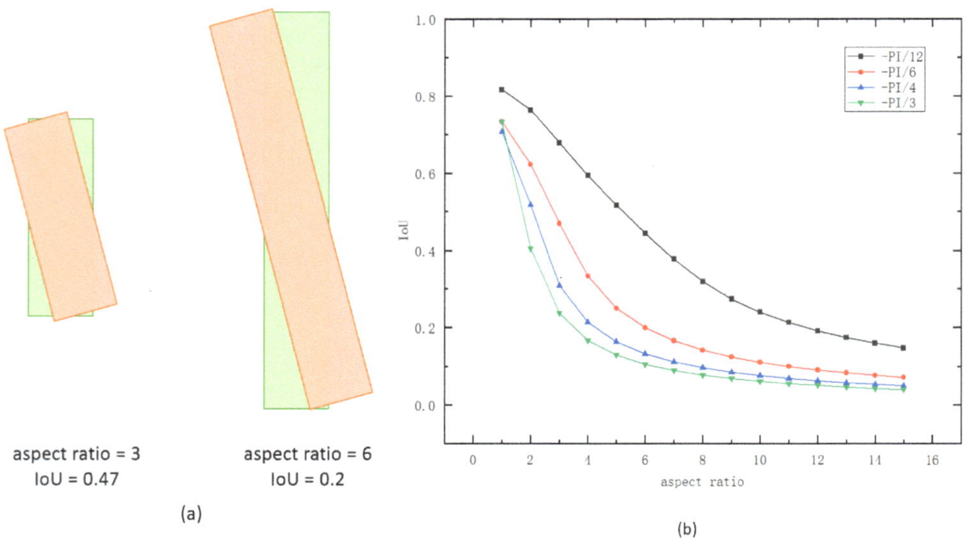

Figure 6. (**a**) indicates the IoU of two bounding boxes with different aspect ratios at an angle of 15°. (**b**) shows the trend of the IoU of the oriented bounding boxes overlapping at different angles as the aspect ratio increases.

3.4. R-CNN Head

For the second stage of the detector, the input consists of $\{P2, P3, P4, P5, P6\}$ provided by the FPN, and a group of RoIs provided by the RPN. For the convenience of the R-CNN head, to facilitate further reasoning and make full use of the FPN information of the different levels, we use oriented RoIAlign to extract fixed-size feature vectors from each layer of the FPN for each RoI, instead of allocating each RoI to a specific layer of the FPN in advance according to the size of the RoI. Then, the feature vectors are fed into the ARA module for adaptive weighted feature fusion. Finally, the fused features are input into two parallel fully connected layers to output the classification score and re-finetune the offset.

3.4.1. Oriented RoIAlign

The oriented proposal generated by the RPN is a five-dimensional vector (x, y, w, h, θ), where $\theta \in [-90°, 0°)$. We mapped the coordinates of the RoIs to the feature map with the stride of $s = [4, 8, 16, 32]$ to extract a fixed-size feature vector. Then, each RoI mapped to the FPN levels was divided into a 7×7 grid, and the feature value in each grid was obtained through bilinear interpolation with a sampling rate of four. This process is the same as RRoI warping [27].

3.4.2. Adaptive RoI Assignment

In the original Faster R-CNN with an FPN, the RoIs generated by the RPN are allocated to different layers of the FPN according to size. Then, fixed-size feature vectors will be extracted by oriented RoIAlign for second-stage reasoning. Typically, larger RoIs are assigned to deeper levels, while smaller RoIs are assigned to shallower levels, but this allocation strategy does not take full advantage of the different levels of information. Therefore, devising a more rational allocation strategy is a natural idea. Specifically, feature vectors from different levels extracted by RoIAlign are first concatenated; then, the weight maps are generated through the ARA module; and the RoI feature vectors of different levels are weighted and fused. The structure of the ARA is shown in Figure 7. The concatenated vectors pass through a global maximum pooling layer, and then two 1×1 convolution layers are used to generate the spatial weight maps.

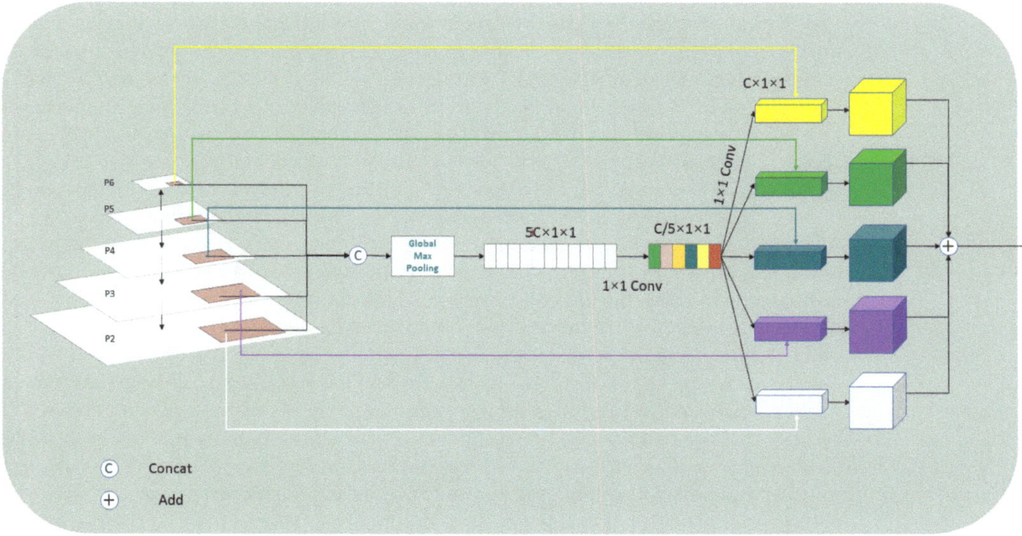

Figure 7. The overall structure of the Adaptive RoI Assignment module. It calculates the weights of each layer of the FPN and fuses the layers to obtain the subsequent features.

4. Experiments

4.1. Datasets and Evaluation Metrics

We test the performance of our proposed method on two common public remote sensing datasets, namely, DOTA [28] and HRSC2016 [29]. All reported results follow standard PASCAL VOC 2007 [30] mean average precision (mAP) metrics.

DOTA is a large-scale public dataset commonly used in remote sensing object detection tasks. It includes 2806 aerial images with sizes ranging from 800 × 800 to 4000 × 4000 pixels. The label contains 188,282 instances of 15 common ground target categories, each marked by its bounding vertices in clockwise order. There are two kinds of detection tasks for the DOTA dataset: detection with oriented bounding boxes (OBB) and detection with horizontal bounding boxes (HBB).

HRSC2016 is a public dataset for oriented ship detection from the Northwest University of Technology. The dataset contains 1061 aerial images with sizes ranging from 300 × 300 to 1500 × 900 pixels. Its annotations are divided into 3 categories and 27 subcategories, with 2976 targets, and use the OBB annotation format.

4.2. Parameter Detail

The experiments of this paper are based on the MMDetection platform [31]. All experiments were finished on a single RTX 3080 with the batch size set to two. We optimize the overall network by using the SGD algorithm with a momentum of 0.9 and a weight decay of 0.0001. We use ResNet50 as our backbone, and it is pretrained on ImageNet [32]. The weight of the RPN loss is set as $\lambda_1 = 1$ and $\lambda_2 = 5$. For DOTA, we used a sliding window with a size of 1024 × 1024 pixels and an overlapping width of 200 pixels for clipping. Twelve epochs were trained in total, the learning rate was set to 0.005, and it was reduced by a ratio of 0.1 after the 8th and 11th epochs. For HRSC2016, we scaled the image to 1333 × 800 before it was sent to the network, and we trained 36 epochs in total. The learning rate was set to 0.005, and it was reduced by a ratio of 0.1 after the 24-th and 33-th epochs.

4.3. Main Results

We compared the proposed method with 17 other oriented object detectors in the DOTA dataset while using the Faster R-CNN network with the rotation bounding boxes' RoI as the baseline, and the results are shown in Table 1. When ResNet-50 is used as the backbone, our method achieves 75.17% mAP, which is 1.38% mAP higher than that obtained by the baseline, and which is superior to that obtained by other detectors. The detection results visualization on DOTA are presented in Figure 8 and the results are presented in Table 1. The table clearly shows that compared with the baseline, the detection accuracies of BD, GTF, SV, RA, and HC have been significantly improved for the following reasons: (1) with the help of the multilevel stacked context module, the recognition accuracies of BD, GTF, and HC with obvious spatial relationship have improved; (2) under the combination of the MSSC and the ARA modules, the features generated by different levels of RPN are fully used to improve the detection accuracy of small targets such as SV; (3) RA shapes are similar to squares, and the introduction of the GWD loss alleviates square-like problems.

We compared the proposed method with 10 other oriented object detectors on HRSC2016, and the results are shown in Table 2. When ResNet-50 is used as the backbone, our method achieves 90.65% AP, which is 0.69% higher than that achieved by the baseline. Some results visualizations are shown in Figure 9, and a comparison of performances is presented in Table 2. This expectation is also consistent with the results of the first experiment; that is, multilevel context stacking helps to improve the network to obtain the characteristics of objects with obvious spatial context relationships. This finding is verified on ships, harbors, and large vehicles, with evidence taken from two datasets.

Table 1. Comparison with state-of-the-art methods on the DOTA dataset. The acronyms for category names are as follows: PL (plane), BD (baseball diamond), BR (Bridge), GTF (Ground field track), SV (Small vehicle), LV (Large vehicle), SH (Ship), TC (Tennis court), BC (Basketball court), ST (Storage tank), SBF (Soccer-ball field), RA (Roundabout), HA (Harbor), SP (Swimming pool), and HC (Helicopter).

	Method	Backbone	Input_size	PL	BD	BR	GTF	SV	LV	SH	TC	BC	ST	SBF	RA	HA	SP	HC	mAP
One-stage	PIoU [33]	DLA-34	512 × 512	80.90	69.70	24.10	60.20	38.30	64.40	64.80	**90.90**	77.20	70.40	46.50	37.10	57.10	61.90	64.00	60.5
	RetinaNet	R-50-FPN	512 × 512	88.67	77.62	41.81	58.17	74.58	71.64	79.11	90.29	82.18	74.32	54.75	60.60	62.57	69.67	60.64	68.43
	DAL [34]	R-50-FPN	800 × 800	88.68	76.55	45.08	66.80	67.00	76.76	79.74	90.84	79.54	78.45	57.71	62.27	69.05	**73.14**	60.11	71.44
	RSDet [35]	R-101-FPN	800 × 800	89.80	82.90	48.60	65.20	69.50	70.10	70.20	90.50	85.60	83.40	62.50	63.90	65.60	67.20	**68.00**	72.2
	P-RSDet [36]	R-101	800 × 800	88.58	77.83	50.44	69.29	71.10	75.79	78.66	90.88	80.10	81.71	57.92	63.03	66.30	69.77	63.13	72.3
	DRN [37]	H-104	1024 × 1024	89.71	82.34	47.22	64.10	76.22	74.43	85.84	90.57	86.18	84.89	57.65	61.93	69.30	69.63	58.48	73.23
	CFC-Net [38]	R-101	800 × 800	89.08	80.41	52.41	70.02	76.28	78.11	87.21	90.89	84.47	85.64	60.51	61.52	67.82	68.02	50.09	73.5
	R3Det	R-101-FPN	800 × 800	88.76	83.09	50.91	67.27	76.23	80.39	86.72	90.78	84.68	83.24	61.98	61.35	66.91	70.63	53.94	73.79
	S^2ANet [39]	R-50-FPN	1024 × 1024	89.11	82.84	48.37	71.11	78.11	78.39	87.25	90.83	84.90	85.64	60.36	62.60	65.26	69.13	57.94	74.12
Two-stage	Faster R-CNN	R-50-FPN	1024 × 1024	88.44	73.06	44.86	59.09	73.25	71.49	77.11	90.84	78.94	83.90	48.59	62.95	62.18	64.91	56.18	69.05
	RoI Transformer	R-101-FPN	1024 × 1024	88.64	78.52	43.44	75.92	68.81	73.68	83.59	90.74	77.27	81.46	58.39	53.54	62.83	58.93	47.67	69.56
	CAD-Net	R-101-FPN	1600 × 1600	87.80	82.40	49.40	73.50	71.10	63.50	76.60	90.90	79.20	73.30	48.40	60.90	62.00	67.00	62.20	69.9
	CenterMap-Net [40]	R-50-FPN	1024 × 1024	88.88	81.24	**53.15**	60.65	**78.62**	66.55	78.10	88.83	77.80	83.61	49.36	66.19	72.10	72.36	58.70	71.74
	SCRDet	R-101-FPN	800 × 800	89.98	80.65	52.09	68.36	68.36	60.32	72.41	90.85	**87.94**	86.86	65.02	66.68	66.25	68.24	65.21	72.61
	FAOD [41]	R-101-FPN	1024 × 1024	**90.21**	79.58	45.49	76.41	73.18	68.27	79.56	90.83	83.40	84.68	53.40	65.42	74.17	69.69	64.86	73.28
	Mask OBB [42]	R-50-FPN	1024 × 1024	89.61	**85.09**	51.85	72.90	75.28	73.23	85.57	90.37	82.08	85.05	55.73	68.39	71.61	69.87	66.33	74.86
	Gliding Vertex	R-101-FPN	1024 × 1024	89.64	85.00	52.26	**77.34**	73.01	73.14	86.82	90.74	79.02	86.81	59.55	**70.91**	72.94	70.86	57.32	75.02
	baseline	R-50-FPN	1024 × 1024	89.95	76.80	50.44	70.84	66.94	**84.40**	**88.74**	90.80	69.87	87.44	**77.42**	67.98	74.89	62.27	48.01	73.79
	MSCNet(ours)	R-50-FPN	1024 × 1024	89.81	79.92	48.62	74.18	68.44	84.28	88.4	90.79	72.49	**87.54**	75.44	69.43	**75.31**	60.07	62.85	**75.17**

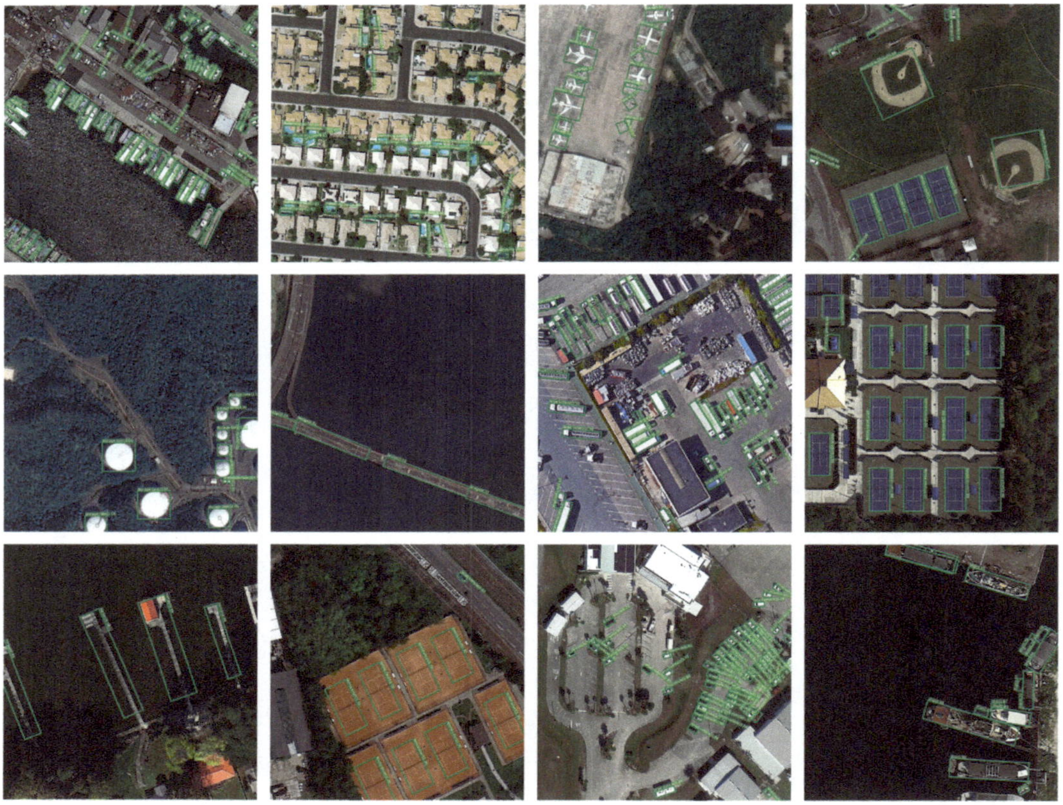

Figure 8. Visualization of detection results obtained by MSCNet on the DOTA dataset. The confidence threshold is set to 0.3 in visualization. Our method can perform well despite different targets with different scales and directions.

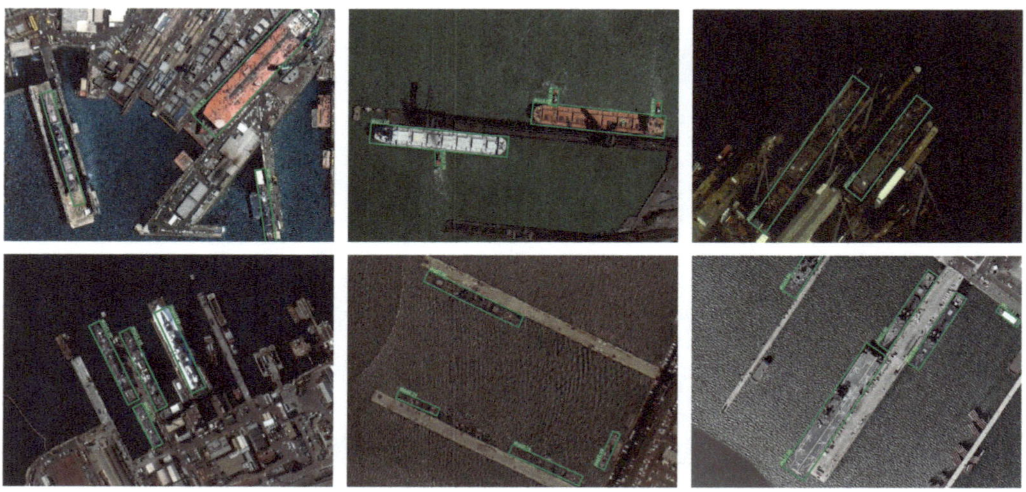

Figure 9. Visualization of detection results obtained by MSCNet on the HRSC2016 dataset.

Table 2. Comparison of network performances on the HRSC2016 dataset.

Model	Backbone	Input_Size	AP_{50}
R2CNN [43]	ResNet101	800 × 800	73.07
RRPN [44]	ResNet101	800 × 800	79.08
RoI-Transformer	ResNet101	512 × 800	86.2
Gliding Vertex	ResNet101	512 × 800	88.2
DAL	ResNet101	416 × 416	88.95
R3Det	ResNet101	800 × 800	89.26
DCL [45]	ResNet101	800 × 800	89.46
CSL	ResNet50	800 × 800	89.62
GWD	ResNet101	800 × 800	89.85
Oriented R-CNN	ResNet101	1333 × 800	90.5
baseline	ResNet50	1333 × 800	89.96
MSCNet(ours)	ResNet50	1333 × 800	**90.65**

4.4. *Ablation Study*

4.4.1. Baseline Setup

To evaluate the individual effects of the different modules, this section takes Faster R-CNN with oriented RoI head and SmoothL1 loss as the baseline and verifies each module on two datasets. The result of the ablation study is shown in Table 3.

Table 3. The respective contributions of the proposed module to the network. MSSC: multilevel stacked semantic capture, GWD: Gaussian Wasserstein distance loss, and ARA: Adaptive RoI Assignment.

Modules			mAP	AP50
MSSC	GWD	ARA	DOTA	HRSC
			73.79	89.96
√			74.27	90.36
	√		74.22	90.21
		√	74.81	90.18
√	√		75.12	90.38
√		√	74.4	90.38
	√	√	74.94	90.47
√	√	√	75.17	90.65

4.4.2. Effect on Each Module

Effect on MSSC: Due to the complexity of remote sensing images, noise information will be introduced in the feature extraction stage, and it is difficult to extract high-quality target features when the target information is mixed with the noise, leading to a decrease in detection ability. In the MSSC module, a top-down fusion path is added to the FPN structure, and a stack semantic capture structure is used to improve the ability to capture multiscale targets in complex scenes. MSSC module improves the baseline performance by 0.48% mAP on DOTA and 0.4% AP on HRSC2016. Because of the aggregation of stacked multilevel semantics, the FPN makes full use of multilevel semantic relations, and experiments also show that this approach is effective. The response of the network to different targets in the scene with and without the MSSC module is shown in Figure 10. It can be concluded that after using the MSSC module, the response of the network to the background-independent noise is suppressed and the response to the target is enhanced.

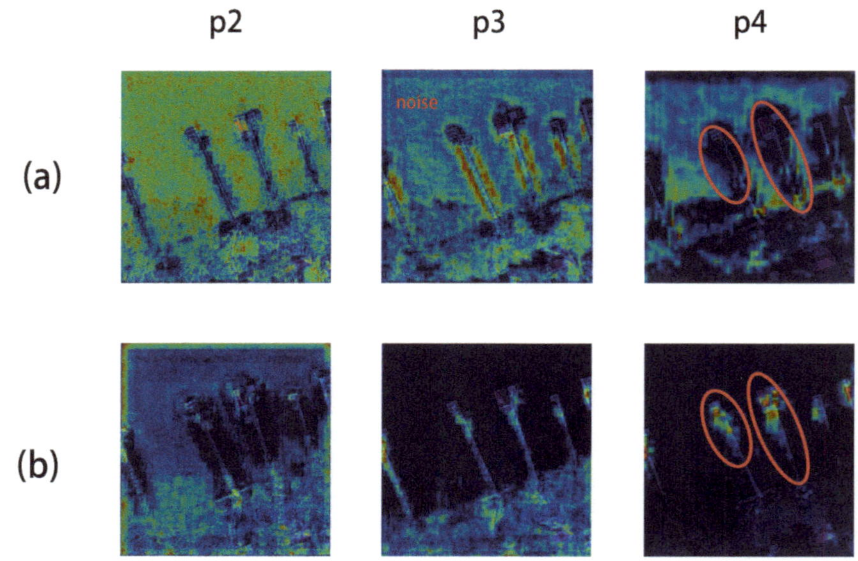

Figure 10. Visualization of the p2-p4 layer outputs of the FPN with and without the MSSC module. (**a**) MSSC is not used. (**b**) MSSC is used. Due to the addition of MSSC, the background noise unrelated to the detection target is suppressed, and the target is enhanced in (**b**).

Effect on GWD: Compared with the traditional five-parameter bounding box representation, the Gaussian distribution representation makes up for some defects of the five-parameter representation. On this basis, a reasonable measurement method is used to describe the similarity between bounding boxes with different aspect ratios and different rotation angles, and the method compensated for the effect between the measurement and reality in oriented object detection tasks. Additionally, the experimental results show that Gaussian distribution representation positively affects the detection task. Using the Gaussian distribution to represent the oriented bounding box and using the Gaussian Wasserstein distance as an overlap metric of two oriented bounding boxes improves the baseline by 0.43% mAP on DOTA and 0.25% AP on HRSC2016. Compared with the performance of HRSC2016, the performance of this method on DOTA is more obvious because DOTA has a variety of objects and many square-like objects; consequently, traditional representation methods are more likely to encounter a bottleneck problem. It can be seen from the ablation experiment in Table 3, after replacing SmoothL1 loss in baseline with GWD loss, the model performance improves by 0.43% mAP on DOTA and by 0.25% AP on HRSC. These experimental results shows that GWD loss has better performance than SmoothL1 loss in oriented object detection tasks to some extent, and the former can better describe the difference between two oriented objects. Figure 11 shows the model convergence ratio and accuracy of GWD loss and SmoothL1 loss when trained on DOTA. It can be seen that at about 750 iterations, the model with SmoothL1 loss showed an oscillation. Compared with the former, the model with GWD loss had smoother convergence and higher accuracy.

Effect on ARA: Adaptive RoI assignment aggregates information from multilevels of the FPN by adaptive weighting, which is especially effective for targets with large-scale variations, especially in DOTA. It improves the baseline performance by 1.02% mAP on DOTA and 0.22% AP on HRSC2016.

Any combination of the two modules can outperform the single module. When all three modules are used in combination, we obtain the highest detection accuracy on both datasets, with improvements of 1.38% mAP and 0.69% AP compared to the mAP and AP obtained by the baseline on DOTA and HRSC2016, respectively. In addition, the inference speed and complexity of MSSCNet are given in Table 4.

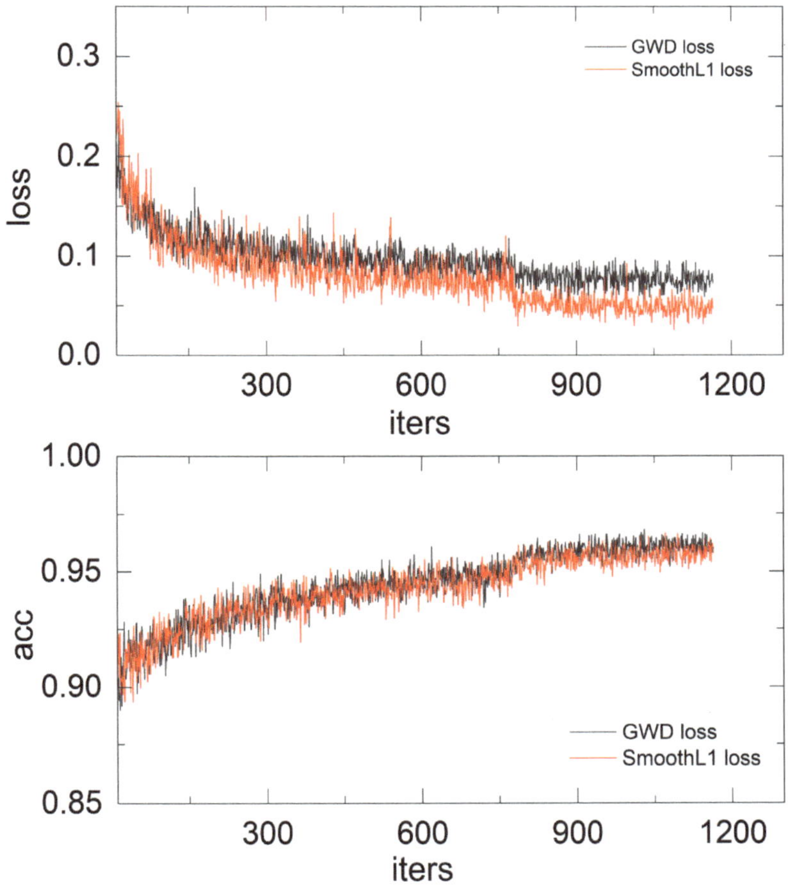

Figure 11. Comparison of model training convergence ratio and accuracy using GWD loss and SmoothL1 loss on DOTA.

Table 4. Comparison of model inference speed and Flops metrics.

	Input Shape	Backbone	Flops	FPS
baseline	$3 \times 1024 \times 1024$	ResNet50	211.3 GFLOPs	23 img/s
MSSCNet	$3 \times 1024 \times 1024$	ResNet50	503.6 GFLOPs	10.6 img/s

5. Conclusions

In this paper, an arbitrarily oriented object detector is proposed according to the characteristics of remote sensing images. Aiming at the deficiency of traditional object detectors in remote sensing, we enhance the modeling ability of the network for the relationship between the objects at different distances by using multilayer stack structures, make full use of the RoI feature of different levels by using a multichannel weighted fusion structure and use a reasonable representation method to represent the oriented object. Moreover, experimental results show that the proposed method outperforms the baseline network on two remote sensing public datasets, and the method performs the best in the relevant oriented object detection networks. Of course, there is still some room for improvement in this work. For example, in the ablation study, when MSSC and ARA were used together, the result obtained on the DOTA dataset was lower than that obtained when ARA was used alone. Additionally, on the DOTA dataset, the proposed model is ineffective in some categories, such as BR and SP. In future work, it is necessary to further explore why these problems occur and find a better combination of MSSC and ARA to create a more effective multilevel feature allocation method.

Author Contributions: R.Z. wrote the manuscript, designed the comparative experiments, designed the architecture and performed the comparative experiments; X.Z. and D.W. supervised the study and revised the manuscript; Y.Z. and L.H. revised the manuscript and gave comments and suggestions to the manuscript. All authors have read and agreed to the published version of the manuscript.

Funding: This work was supported by the National Natural Science Foundation of China (62103345), the Science and Technology Planning Project of Fujian Province (2020H0023, 2020J02160, 2020J01265), the Science and Technology Planning Project of Quanzhou (2020C074), and the Science and Technology Climbing Program of Xiamen University of Technology (XPDKT18030).

Data Availability Statement: Not applicable.

Acknowledgments: We would like to thank the anonymous reviewers for their constructive and valuable suggestions on the earlier drafts of this manuscript.

Conflicts of Interest: The authors declare no conflict of interest.

Abbreviations

ARA	Adaptive RoI Assignment
FPN	Feature Pyramid Network
GWD	Gaussian Wasserstein Distance
IoU	Intersection over Union
mAP	mean Average Precision
MSCNet	Multilevel Stacked Context Network
MSSC	Multilevel Stacked Semantic Capture
NMS	Non-Maximum Suppression
RoI	Region of Interest
RPN	Region Proposal Network
SGD	Stochastic Gradient Descent

References

1. Ren, S.; He, K.; Girshick, R.; Sun, J. Faster r-cnn: Towards real-time object detection with region proposal networks. *arXiv* **2015**, arXiv:1506.01497.
2. Li, K.; Cheng, G.; Bu, S.; You, X. Rotation-insensitive and context-augmented object detection in remote sensing images. *IEEE Trans. Geosci. Remote Sens.* **2017**, *56*, 2337–2348. [CrossRef]
3. Liu, W.; Ma, L.; Wang, J. Detection of multiclass objects in optical remote sensing images. *IEEE Geosci. Remote Sens. Lett.* **2018**, *16*, 791–795. [CrossRef]
4. Lin, T.Y.; Dollár, P.; Girshick, R.; He, K.; Hariharan, B.; Belongie, S. Feature pyramid networks for object detection. In Proceedings of the IEEE Conference on Computer Vision and Pattern Recognition, Honolulu, HI, USA, 21–26 July 2017; pp. 2117–2125.
5. Liu, L.; Pan, Z.; Lei, B. Learning a rotation invariant detector with rotatable bounding box. *arXiv* **2017**, arXiv:1711.09405.
6. Zhang, G.; Lu, S.; Zhang, W. CAD-Net: A context-aware detection network for objects in remote sensing imagery. *IEEE Trans. Geosci. Remote Sens.* **2019**, *57*, 10015–10024. [CrossRef]
7. Girshick, R.; Donahue, J.; Darrell, T.; Malik, J. Rich feature hierarchies for accurate object detection and semantic segmentation. In Proceedings of the IEEE Conference on Computer Vision and Pattern Recognition, Columbus, OH, USA, 23–28 June 2014; pp. 580–587.
8. Girshick, R. Fast r-cnn. In Proceedings of the IEEE International Conference on Computer Vision, Santiago, Chile, 7–13 December 2015; pp. 1440–1448.
9. Liu, W.; Anguelov, D.; Erhan, D.; Szegedy, C.; Reed, S.; Fu, C.Y.; Berg, A.C. Ssd: Single shot multibox detector. In Proceedings of the European Conference on Computer Vision, Amesterdam, The Netherlands, 8 October 2016; Springer: Cham, Switzerland, 2016; pp. 21–37.
10. Lin, T.Y.; Goyal, P.; Girshick, R.; He, K.; Dollár, P. Focal loss for dense object detection. In Proceedings of the IEEE International Conference on Computer Vision, Venice, Italy, 22–29 October 2017; pp. 2980–2988.
11. Lin, G.; Milan, A.; Shen, C.; Reid, I. Refinenet: Multi-path refinement networks for high-resolution semantic segmentation. In Proceedings of the IEEE Conference on Computer Vision and Pattern Recognition, Honolulu, HI, USA, 21–26 July 2017; pp. 1925–1934.

12. Redmon, J.; Farhadi, A. YOLO9000: better, faster, stronger. In Proceedings of the IEEE Conference on Computer Vision and Pattern Recognition, Honolulu, HI, USA, 21–26 July 2017; pp. 7263–7271.
13. Redmon, J.; Farhadi, A. Yolov3: An incremental improvement. *arXiv* **2018**, arXiv:1804.02767.
14. Bochkovskiy, A.; Wang, C.Y.; Liao, H.Y.M. Yolov4: Optimal speed and accuracy of object detection. *arXiv* **2020**, arXiv:2004.10934.
15. Law, H.; Deng, J. Cornernet: Detecting objects as paired keypoints. In Proceedings of the European Conference on Computer Vision (ECCV), Munich, Germany, 8–14 September 2018; pp. 734–750.
16. Zhou, X.; Zhuo, J.; Krahenbuhl, P. Bottom-up object detection by grouping extreme and center points. In Proceedings of the IEEE/CVF Conference on Computer Vision and Pattern Recognition, Long Beach, CA, USA, 15–20 June 2019; pp. 850–859.
17. Tian, Z.; Shen, C.; Chen, H.; He, T. Fcos: Fully convolutional one-stage object detection. In Proceedings of the IEEE/CVF International Conference on Computer Vision, Seoul, Korea, 27 October–2 November 2019; pp. 9627–9636.
18. Long, J.; Shelhamer, E.; Darrell, T. Fully convolutional networks for semantic segmentation. In Proceedings of the IEEE Conference on Computer Vision and Pattern Recognition, Boston, MA, USA, 7–12 June 2015; pp. 3431–3440.
19. Duan, K.; Bai, S.; Xie, L.; Qi, H.; Huang, Q.; Tian, Q. Centernet: Keypoint triplets for object detection. In Proceedings of the IEEE/CVF International Conference on Computer Vision, Seoul, Korea, 27 October–2 November 2019; pp. 6569–6578.
20. Yang, X.; Yang, J.; Yan, J.; Zhang, Y.; Zhang, T.; Guo, Z.; Sun, X.; Fu, K. Scrdet: Towards more robust detection for small, cluttered and rotated objects. In Proceedings of the IEEE/CVF International Conference on Computer Vision, Seoul, Korea, 27 October–2 November 2019; pp. 8232–8241.
21. Yang, X.; Liu, Q.; Yan, J.; Li, A.; Zhang, Z.; Yu, G. R3det: Refined single-stage detector with feature refinement for rotating object. *arXiv* **2019**, arXiv:1908.05612.
22. Xu, Y.; Fu, M.; Wang, Q.; Wang, Y.; Chen, K.; Xia, G.S.; Bai, X. Gliding vertex on the horizontal bounding box for multi-oriented object detection. *IEEE Trans. Pattern Anal. Mach. Intell.* **2020**, *43*, 1452–1459. [CrossRef]
23. Xie, X.; Cheng, G.; Wang, J.; Yao, X.; Han, J. Oriented r-cnn for object detection. In Proceedings of the IEEE/CVF International Conference on Computer Vision, Montreal, BC, Canada, 11–17 October 2021; pp. 3520–3529.
24. Yang, X.; Yan, J. Arbitrary-oriented object detection with circular smooth label. In Proceedings of the sEuropean Conference on Computer Vision, Online, 28 August 2020; Springer: Cham, Switzerland, 2020; pp. 677–694.
25. Yu, J.; Jiang, Y.; Wang, Z.; Cao, Z.; Huang, T. Unitbox: An advanced object detection network. In Proceedings of the 24th ACM International Conference on Multimedia, Amsterdam, The Netherlands, 15–19 October 2016; pp. 516–520.
26. Yang, X.; Yan, J.; Ming, Q.; Wang, W.; Zhang, X.; Tian, Q. Rethinking rotated object detection with gaussian wasserstein distance loss. In Proceedings of the International Conference on Machine Learning, Online, 18–24 July 2021; pp. 11830–11841.
27. Ding, J.; Xue, N.; Long, Y.; Xia, G.S.; Lu, Q. Learning roi transformer for oriented object detection in aerial images. In Proceedings of the IEEE/CVF Conference on Computer Vision and Pattern Recognition, Long Beach, CA, USA, 15–20 June 2019; pp. 2849–2858.
28. Xia, G.S.; Bai, X.; Ding, J.; Zhu, Z.; Belongie, S.; Luo, J.; Datcu, M.; Pelillo, M.; Zhang, L. DOTA: A large-scale dataset for object detection in aerial images. In Proceedings of the IEEE Conference on Computer Vision and Pattern Recognition, Salt Lake City, UT, USA, 18–23 June 2018; pp. 3974–3983.
29. Liu, Z.; Yuan, L.; Weng, L.; Yang, Y. A high resolution optical satellite image dataset for ship recognition and some new baselines. In International conference on pattern recognition applications and methods. *SciTePress* **2017**, *2*, 324–331.
30. Everingham, M.; Van Gool, L.; Williams, C.K.; Winn, J.; Zisserman, A. The pascal visual object classes (voc) challenge. *Int. J. Comput. Vis.* **2010**, *88*, 303–338. [CrossRef]
31. Chen, K.; Wang, J.; Pang, J.; Cao, Y.; Xiong, Y.; Li, X.; Sun, S.; Feng, W.; Liu, Z.; Xu, J.; et al. MMDetection: Open mmlab detection toolbox and benchmark. *arXiv* **2019**, arXiv:1906.07155.
32. Deng, J. A large-scale hierarchical image database. In Proceedings of the IEEE Computer Vision and Pattern Recognition, Miami, FL, USA, 20–25 June 2009.
33. Chen, Z.; Chen, K.; Lin, W.; See, J.; Yu, H.; Ke, Y.; Yang, C. Piou loss: Towards accurate oriented object detection in complex environments. In Proceedings of the European Conference on Computer Vision, Online, 28 August 2020; Springer: Cham, Switzerland, 2020; pp. 195–211.
34. Ming, Q.; Zhou, Z.; Miao, L.; Zhang, H.; Li, L. Dynamic anchor learning for arbitrary-oriented object detection. *arXiv* **2020**, arXiv:2012.04150.
35. Qian, W.; Yang, X.; Peng, S.; Guo, Y.; Yan, J. Learning modulated loss for rotated object detection. *arXiv* **2019**, arXiv:1911.08299.
36. Zhou, L.; Wei, H.; Li, H.; Zhao, W.; Zhang, Y. Objects detection for remote sensing images based on polar coordinates. *arXiv* **2020**, arXiv:2001.02988.
37. Pan, X.; Ren, Y.; Sheng, K.; Dong, W.; Yuan, H.; Guo, X.; Ma, C.; Xu, C. Dynamic refinement network for oriented and densely packed object detection. In Proceedings of the IEEE/CVF Conference on Computer Vision and Pattern Recognition, Seattle, WA, USA, 13–19 June 2020; pp. 11207–11216.
38. Ming, Q.; Miao, L.; Zhou, Z.; Dong, Y. CFC-Net: A critical feature capturing network for arbitrary-oriented object detection in remote-sensing images. *IEEE Trans. Geosci. Remote Sens.* **2021**, *61*, 1–4. [CrossRef]
39. Han, J.; Ding, J.; Li, J.; Xia, G.S. Align deep features for oriented object detection. *IEEE Trans. Geosci. Remote Sens.* **2021**, *60*, 1–11. [CrossRef]
40. Wang, J.; Yang, W.; Li, H.C.; Zhang, H.; Xia, G.S. Learning center probability map for detecting objects in aerial images. *IEEE Trans. Geosci. Remote Sens.* **2020**, *59*, 4307–4323. [CrossRef]

41. Li, C.; Xu, C.; Cui, Z.; Wang, D.; Zhang, T.; Yang, J. Feature-attentioned object detection in remote sensing imagery. In Proceedings of the 2019 IEEE International Conference on Image Processing (ICIP), Taipei, China, 22–25 September 2019; pp. 3886–3890.
42. Wang, J.; Ding, J.; Guo, H.; Cheng, W.; Pan, T.; Yang, W. Mask OBB: A semantic attention-based mask oriented bounding box representation for multi-category object detection in aerial images. *Remote Sens.* **2019**, *11*, 2930. [CrossRef]
43. Jiang, Y.; Zhu, X.; Wang, X.; Yang, S.; Li, W.; Wang, H.; Fu, P.; Luo, Z. R2CNN: Rotational region CNN for orientation robust scene text detection. *arXiv* **2017**, arXiv:1706.09579.
44. Ma, J.; Shao, W.; Ye, H.; Wang, L.; Wang, H.; Zheng, Y.; Xue, X. Arbitrary-oriented scene text detection via rotation proposals. *IEEE Trans. Multimed.* **2018**, *20*, 3111–3122. [CrossRef]
45. Yang, X.; Hou, L.; Zhou, Y.; Wang, W.; Yan, J. Dense label encoding for boundary discontinuity free rotation detection. In Proceedings of the IEEE/CVF Conference on Computer Vision and Pattern Recognition, Nashville, TN, USA, 20–25 June 2021; pp. 15819–15829.

remote sensing

MDPI

Article

A Defect Detection Method Based on BC-YOLO for Transmission Line Components in UAV Remote Sensing Images

Wenxia Bao [1], Xiang Du [1], Nian Wang [1,*], Mu Yuan [1] and Xianjun Yang [2]

1 School of Electronics and Information Engineering, Anhui University, Hefei 230601, China
2 Hefei Institutes of Physical Science, Chinese Academy of Sciences, Hefei 230031, China
* Correspondence: wn_xlb@ahu.edu.cn

Abstract: Vibration dampers and insulators are important components of transmission lines, and it is therefore important for the normal operation of transmission lines to detect defects in these components in a timely manner. In this paper, we provide an automatic detection method for component defects through patrolling inspection by an unmanned aerial vehicle (UAV). We constructed a dataset of vibration dampers and insulators (DVDI) on transmission lines in images obtained by the UAV. It is difficult to detect defects in vibration dampers and insulators from UAV images, as these components and their defective parts are very small parts of the images, and the components vary greatly in terms of their shape and color and are easily confused with the background. In view of this, we use the end-to-end coordinate attention and bidirectional feature pyramid network "you only look once" (BC-YOLO) to detect component defects. To make the network focus on the features of vibration dampers and insulators rather than the complex backgrounds, we added the coordinate attention (CA) module to YOLOv5. CA encodes each channel separately along the vertical and horizontal directions, which allows the attention module to simultaneously capture remote spatial interactions with precise location information and helps the network locate targets of interest more accurately. In the multiscale feature fusion stage, different input features have different resolutions, and their contributions to the fused output features are usually unequal. However, PANet treats each input feature equally and simply sums them up without distinction. In this paper, we replace the original PANet feature fusion framework in YOLOv5 with a bidirectional feature pyramid network (BiFPN). BiFPN introduces learnable weights to learn the importance of different features, which can make the network focus more on the feature mapping that contributes more to the output features. To verify the effectiveness of our method, we conducted a test in DVDI, and its mAP@0.5 reached 89.1%, a value 2.7% higher than for YOLOv5.

Keywords: vibration dampers; insulators; feature fusion; YOLOv5; defect detection; attention mechanism

Citation: Bao, W.; Du, X.; Wang, N.; Yuan, M.; Yang, X. A Defect Detection Method Based on BC-YOLO for Transmission Line Components in UAV Remote Sensing Images. *Remote Sens.* **2022**, *14*, 5176. https://doi.org/10.3390/rs14205176

Academic Editors: Qian Du, Yanni Dong and Xiaochen Yang

Received: 2 September 2022
Accepted: 13 October 2022
Published: 16 October 2022

1. Introduction

Insulators and vibration dampers are exposed to the outdoor environment for long periods of time, are subject to weather and high mechanical tension, and are therefore likely to fail physically [1,2]. The primary function of insulators is to support wires and prevent the current from returning to ground, and when insulators have problems such as fractures or cracks, they are easily pierced, resulting in zero insulation resistance at both ends of the insulator string. Insulation is then lost, leading to power interruptions and outages [3]. The primary function of vibration dampers is to protect power lines from wind-induced periodic vibrations, thus reducing accidents such as transmission line fatigue and segmental strands [4]. When vibration dampers are exposed to the environment for a long period, they tend to rust. Furthermore, rusting may cause the steel strands to loosen, and the hammer head of the vibration dampers will then easily deform, slide, and fall off [5,6], which affects power transmission. The timely detection and replacement of damaged insulators and vibration dampers can therefore guarantee the effective and normal operation of the transmission lines.

Transmission line inspection is an important component of a power system, and the most primitive form is manual inspection. However, traditional manual inspection requires significant human and material resources, and the detection speed is slow. With the continuous development of science and technology, modern power line inspection involves helicopter [7], robot [8], and unmanned aerial vehicle (UAV) inspection methods [9] rather than the original manual methods. Of these, UAV inspection has gradually become an important part of transmission line inspection due to its low cost, high efficiency, and flexibility [10]. However, in the transmission line images taken by UAVs, the vibration dampers are relatively small, and defects such as rust are easily confused with the background color. Although the insulators are much larger than vibration dampers, the location of their self-destruction defect may only be a small piece, so it is also difficult to locate on longer insulation strings in the remote sensing images. Moreover, the insulators and vibration dampers are distributed in complex environments such as forests, houses, and fields, making inspection challenging. The above problems are common in UAV inspection, belonging to small target detection problems in computer vision. The existing methods are insufficient for the accurate detection of insulators and vibration dampers; therefore, this area needs to be further studied. The research we have conducted can be summarized in the following two points:

- We construct a dataset of images of vibration dampers and insulators using remote sensing images taken by UAVs, which we refer to as DVDI. There are three types of insulators (XWP, LXY, and FXBW) and four types of vibration dampers (FD, FDZ, FFH, and FR). Each type of vibration damper or insulator may have defects, or may be normal.
- We propose a defect detection method for insulator and vibration dampers, named BC-YOLO. We introduce the CA module into YOLOv5. This module embeds the location information into the channel attention by decomposing the channel attention. The CA module enhances the network's ability to detect insulators and vibration dampers in complex backgrounds. We use BiFPN instead of the original PANet feature fusion framework to better balance the feature information at different scales by weighting each scale. The BiFPN feature fusion framework enhances the network's ability to detect small targets such as vibration dampers.

The rest of this paper consists of the following chapters. Section 2 presents the relevant literature on target detection. Section 3 introduces our dataset and gives an overview of its pre-processing part. Section 4 introduces our proposed BC-YOLO network structure. Section 5 presents the related experiments and analysis. Section 6 analyzes the confusion matrix generated by training. Section 7 concludes the paper while suggesting future work.

2. Related Work

Tiantian et al. [11] proposed a feature fusion-based insulator detection method, which consisted of a histogram of directional gradient features after principal component analysis dimensionality reduction and local binary pattern features. Support vector machine (SVM) fusion features were used to build the training model. The sliding window method was used to search for candidate regions, and the non-maximum suppression method was applied to fuse the candidate windows. Finally, the location of insulator strings was calculated by linear fitting. In Reddy et al. [12], video monitoring technology is used to obtain insulator images, and discrete orthogonal s-transform and k-means clustering algorithms are used to determine the state of insulators. In Reddy et al. [13], the features of the insulator are extracted by discrete orthogonal transformation, after which the state of the insulator is estimated by SVM. Li et al. [14] based a tilt correction method on principal component analysis, which enabled them to obtain accurate feature extraction curves from insulator images. Five features were extracted from the feature curves, and finally SVM was used to identify insulators with these five features. In Zhao et al. [15], insulators are located by azimuth angle detection and binary shape prior knowledge. The possible orientation angles of the insulator are initially detected, and for each possible angle, the insulator is retained according to its binary shape prior knowledge, and finally, all possible angles are traversed to locate the insulator. Wu et al. [16] proposed a new texture segmentation algorithm to segment complex insulator images into subregions with closed, smooth contours. The texture features of the insulators were extracted using a gray level co-occurrence matrix and calculated using a fast gray level co-occurrence integrated algorithm. In Liao et al. [17], the insulators are located using the local features of the insulators. Firstly, multi-scale and multi-features are introduced to represent the local features of the insulators. After that, the local features of the insulators are trained, and finally, the insulators are localized by a coarse-to-fine strategy. Jabid et al. [18] proposed a new rotation-invariant insulator detection method. Rotational invariance was achieved by an efficient method to estimate the rotation angle of all insulators in an image. Local directional map features based on sliding windows were extracted from the images, and each sliding window was classified using a support vector machine. Zhao et al. [19] proposed a method for representing the appearance of insulator string infrared images based on binary robust invariant scalable key points and vectors of local aggregation descriptors. An SVM-based classification model was integrated into a multi-scale sliding window framework for locating insulator strings in infrared images. Guifeng et al. [20] proposed a swarm optimization clustering algorithm for insulator image segmentation. Their method used an ant colony clustering algorithm for the clustering and segmentation of insulator images. Images taken during a UAV inspection are of very high resolution, but the insulators, and particularly the vibration dampers, on the transmission line are relatively small. The above methods are not effective for the detection of small target objects in high resolution images. In addition, vibration dampers may be easily confused with trees in the forest due to their color, which makes the defect detection difficult.

With the rise of deep learning techniques, more and more people are conducting target detection based on deep learning. The detection methods based on deep learning can be divided into two main categories: single-stage methods (e.g., SSD and YOLO) and two-stage methods (e.g., Faster-RCNN). In general, two-stage methods are not as fast as single-stage methods, and have larger network weight files. Li et al. [21] proposed a method for automatically detecting birds' nests on transmission lines based on Faster-RCNN. The problem of insufficient data samples was solved by zooming in on the bird nest images. Wanguo et al. [22] chose a candidate region-based SSD algorithm for defect localization and identification. The problem of insufficient samples is solved by horizontal mirroring and multi-scale training, and the network parameters are adjusted to select the appropriate network parameters for transmission line defect detection. Wu et al. [23] proposed a CenterNet-based insulator defect detection method. Blurred images are reconstructed using super-resolution in the data pre-processing part to enhance the dataset, while attention mechanisms are used in the net to reduce the interference of the background. Wu et al. [24] proposed an improved YOLOv3 algorithm for the detection of electrical connector defects. First, the K-means clustering algorithm was used to cluster the dataset and improve the detection accuracy of the defective targets. Then, single-scale feature mapping was used for detection rather than multi-scale prediction as in the original network, which not only reduced the computational effort, but also avoided false detection to some extent. Bao et al. [25] proposed a YOLOv4-based PMA-YOLO network, which adds parallel mixed attention (PMA) to the YOLOv4 network to make the network more focused on the target information. In addition, the K-means algorithm is introduced to re-cluster the anchor of vibration dampers. Finally, a multi-stage migration learning strategy is used to improve the training efficiency.

The YOLO series of target detection networks is a popular one-stage object detection method, which has made great progress in target detection by directly predicting the class and location of various objects using separate CNN networks. Compared with Regions with Revolutionary Neural Networks (RCNN) series object detection methods, YOLOv5 achieves a balance between detection speed and accuracy. In particular, YOLOv5 uses the most advanced optimization strategies in the field of CNNs, which is based on an iteration of the original YOLO series. Different degrees of optimization have been applied to data pre-processing, backbone networks, activation functions, training strategies, and anchor clustering, among others. So YOLOv5 is more suitable as a rapid benchmark network for detecting defects in insulators and vibration dampers on transmission lines. However, due to the high resolution of the remote sensing images taken by UAVs, the size of the target represented by the vibration dampers is small, and there is a large difference between the vibration dampers and insulator in size. Therefore, in this paper, a CA module is added to the structure of the backbone, and the original PANet module is replaced with Bi-FPN to improve the detection accuracy of the network. The BC-YOLO network can detect four transmission line defects (normal and damaged insulators, and normal and dislodged vibration dampers), with a mAP@0.5 reaching 89.1%.

3. Dataset

In this paper, we construct a dataset containing insulators and vibration dampers, called DVDI. All of the images in the dataset were taken by UAVs during overhead transmission line inspections by the Chinese Academy of Electric Power. The filming equipment used during the transmission line inspection was a UAV manufactured by DJI, model Phantom 4 Pro V2.0. The UAV maintains a distance of about 10 m from the transmission tower during each power inspection. The UAV model and the image during the inspection are shown in Figure 1.

(**a**) (**b**)

Figure 1. (**a**) Phantom 4 Pro V2.0 UAV; (**b**) UAV inspection shooting data.

3.1. DVDI Dataset

There are many types of insulators and vibration dampers on transmission lines, and the corresponding type is chosen according to the usage scenario. For the most common vibration dampers, there are four types (FD, FDZ, FDY, and FFH), as shown in Figure 2. There are three common types of insulators (FXBW, LXY, and XWP), as shown in Figure 3. The DVDI dataset includes normal and defective vibration dampers as well as normal and defective insulators, as shown in Figure 4.

(a) (b) (c) (d)

Figure 2. Four common types of vibration dampers: (**a**) FD; (**b**) FDZ; (**c**) FDY; (**d**) FFH.

(a) (b) (c)

Figure 3. Three common types of insulators: (**a**) FXBW; (**b**) LXY; (**c**) XWP.

(a)

Figure 4. *Cont.*

(b)

Figure 4. (a) Examples of normal and defective vibration dampers; (b) examples of normal and defective insulators.

After screening, there were a total of 976 UAV remote sensing images, which contained complex backgrounds with areas such as trees, forests, and mountains, as shown in Figure 5. It can be seen from these images that the colors of some insulators and vibration dampers mean that they are easily confused with the background, and they are therefore not easy to detect.

Figure 5. Partial images with vibration dampers and insulators.

3.2. Data Pre-Processing

There are 976 vibration damper and insulator images in the DVDI dataset. In order to improve the generalization ability of the models, the data augmentation techniques are often used in deep convolutional networks. These images are rotated 12° and 180°, and flipped horizontally to obtain the images of defective insulators and vibration dampers from different perspectives. The number of images in the DVDI is increased to 1500 after augmentation. The process of power inspection is affected by the intensity of the light, and insulators and vibration dampers often appear against complex mountainous backgrounds. The dual effects of light intensity and a complex environment can cause false detections and missed detections, and we therefore enhanced the contrast of the detected images to improve the overall quality. For this purpose, we used a gamma transform [26] to enhance the contrast of the image, specifically to correct images with too much or too little gray. The transformation formula is shown in Equation (1), and is a product operation that is applied to each pixel value in the original image. When $\gamma > 1$, this has a stretching effect on the histogram of the gray distribution of the image (making the grayscale stretch toward a high gray value), while a value of $\gamma < 1$ has a shrinking effect on the histogram of the gray distribution of the image (making the grayscale move toward a low gray value). Examples of pre-processed images are shown in Figures 6 and 7.

$$s = cy^{\gamma} \tag{1}$$

(a) (b)

Figure 6. (**a**) Original image; (**b**) horizontal flip.

(a) (b)

Figure 7. (**a**) Original image; (**b**) gamma transform.

3.3. Data Annotation

We used LabelImg as a labeling tool for the insulators and vibration dampers. The labeling rules were as follows: if an insulator was normal, it was labeled as 1, while if there was damage, it was labeled as 2. If a vibration damper was normal without serious bending, it was labeled as 3, while if there was shedding or serious bending, it was labeled as 4. An example of partial image annotation is shown in Figure 8. The labeled data were stored in Pascal VOC [27] format, and the labeled file format was XML. After that, we divided our dataset into a training set, validation set, and test set in the ratio of 6:2:2. The number of annotations for each class is shown in Table 1.

Figure 8. Example of annotation with LabelImg.

Table 1. Label numbers per class.

Classes	Number
1 (normal insulators)	2252
2 (damaged insulators)	632
3 (normal vibration dampers)	3610
4 (damaged vibration dampers)	636

4. Method

4.1. YOLOv5 Network Architecture

YOLOv5 is a single-stage target inspection method that locates and classifies targets by directly regressing the relative positions of candidate boxes. YOLOv5 is the latest network in a series of several iterations of the original YOLO model. Various improvements have been made in terms of data pre-processing, feature extraction, and feature fusion, have been made to greatly improve the detection accuracy of the network. A diagram of the structure of the original YOLOv5 is given in Figure 9. It has three main parts: a backbone, neck, and head. The backbone contains C3, CBL, Focus, and spatial pyramid pooling (SPP) modules [28] to extract features from the input image and pass them to the neck layer. The CBL module is composed of Convolution, Batch Normalization, and Leaky Rule functions. The C3 module splits the input into two branches, one passing through the CBL first and then through the residual structure. After the other branch passes through the CBL, concat combines the outputs of the two branches. The SPP module is a multi-scale feature fusion stage that uses four different sizes of maximum pooling. The Focus module is a slicing operation of the image, preserving more complete information about the down-sampling of the image for subsequent feature extraction. A detailed diagram of the C3, CBL, Focus, SPP module is shown in the last panel of Figure 9. The neck layer exploits the structure of PANet [29] to produce feature pyramids. The algorithm enhances the detection capability of targets at different scales by the bidirectional fusion of low-level semantic space and high-level semantic features. The head layer consists of an anchor frame that is applied to the multi-scale feature map of the neck module that generates detection frames and divides them into corresponding categories, coordinates, and confidence levels.

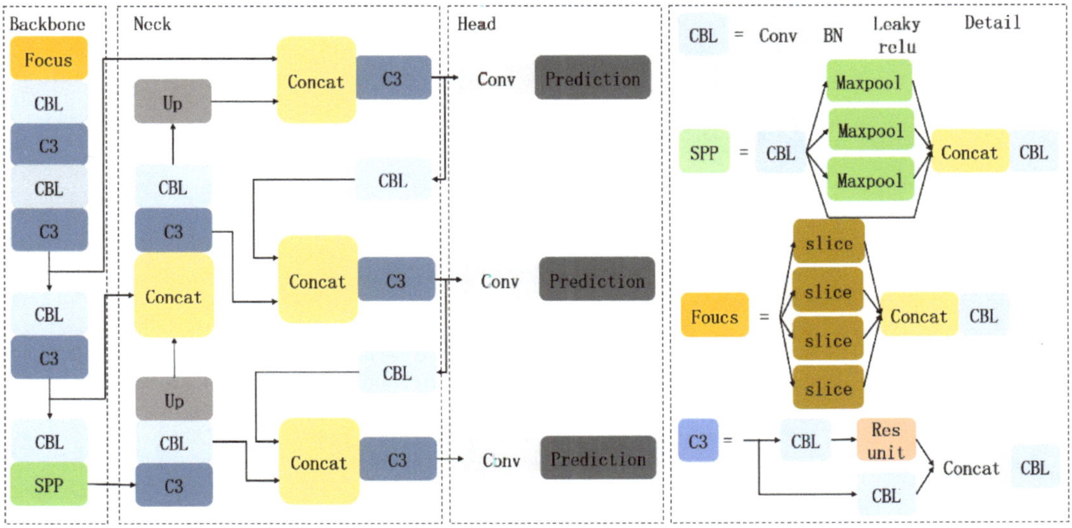

Figure 9. Architecture of the original YOLOv5 network.

4.2. Architecture of the BC-YOLO Network

The remote sensing images taken by UAVs during power inspections have complex backgrounds, and the different types of insulators and vibration dampers may have varying fault shapes and sizes. Existing methods cannot achieve accurate detection and classification of insulators and vibration dampers at the same time. Therefore, we improve on YOLOv5 and propose the BC-YOLO network. In the backbone module, we introduce a CA [30] mechanism, which can help the network to locate the detection target more accurately, thus reducing the interference of target background information. In the neck module, we introduce Bi-FPN [31] to replace the original PANet, which enhances the detection capability of the network for small targets such as vibration dampers by adding weights to each scale to adjust the contribution of each scale. A diagram of the structure of the YOLOv5 is given in Figure 10.

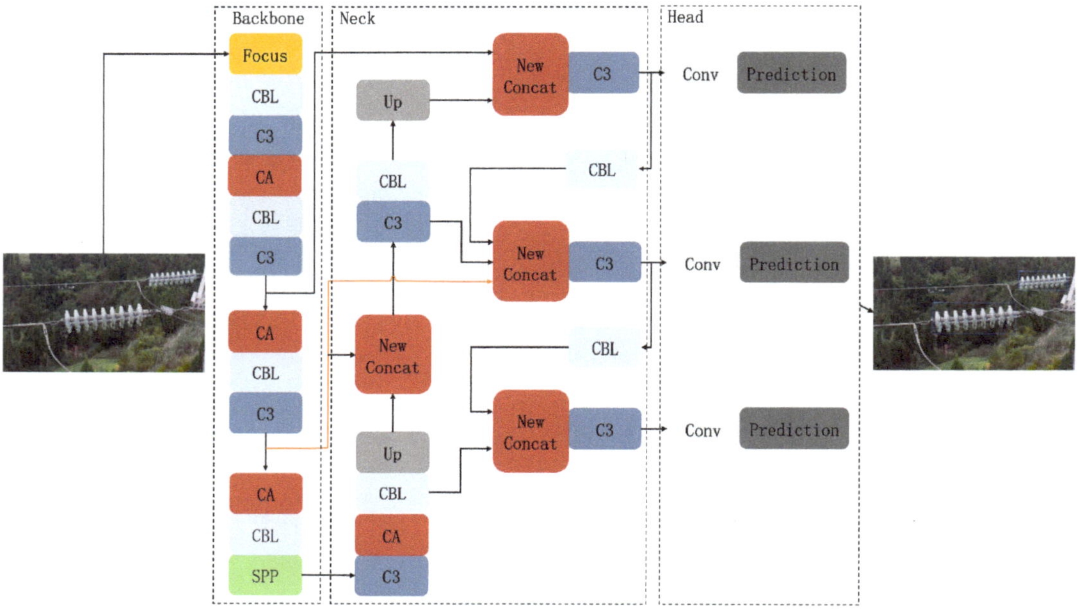

Figure 10. Architecture of the BC-YOLO network.

4.2.1. Attention Mechanism Module

The attention mechanism emerged to make the network more focused on the target information of the current task and less on other information. However, traditional attention mechanisms such as squeeze-and-excitation (SE) [32] only consider the channel information, where location information is equally important in visual targets. In view of this, the convolutional block attention module (CBAM) [33] was derived from SE, and aims to introduce location information through the use of global pooling on the channels. However, this approach can only capture local information, and cannot obtain long-range-dependent information. After several convolution layers, each position of the feature maps contains information about a local area of the original image, and CBAM is used as a weighting factor by taking the maximum and average values of multiple channels for each position, meaning that this weighting only takes into account the information on the local area. In contrast, CA attention avoids the introduction of a larger overhead by embedding the location information into the channel attention, thus allowing the mobile network to obtain information about a larger area. We therefore introduce a CA attention mechanism to improve the detection performance of our network. To solve the problem of position loss due to global average pooling, we efficiently integrate spatial coordinate information by decomposing channel attention into two parallel 1D feature encodings. Specifically, in order to focus attention on the height and width of the image and to encode the precise location information, the input features are divided into two directions, height and width, for global averaging pooling to obtain feature maps in both the height and width directions, respectively, as shown in Equations (2) and (3). The feature maps in the height and width directions of the obtained global perceptual field are then stitched together, after which they are fed into the convolution module with a shared convolution kernel of 1 × 1. Following this, the batch-normalized feature map F1 is fed into the sigmoid activation function to obtain f, as shown in Equation (4). The feature map f is then convolved with a convolution kernel of 1 × 1 according to the original height and width, and the attention weights in the height and width directions are obtained after the activation function, as shown in Equations (5) and (6). Finally, the attention weights in the height and width directions and the original feature map are calculated by multiplying and weighting to obtain feature maps with attention weights in the width and height directions, as shown in Equation (7). The structure of the coordination of the attention mechanism is shown in Figure 11.

$$z_c^h(h) = \frac{1}{W} \sum_{0 \le i \le W} x_c(h, i) \tag{2}$$

$$z_c^w(h) = \frac{1}{H} \sum_{0 \le j \le H} x_c(j, w) \tag{3}$$

$$\mathrm{f} = \delta\left(F_1\left(\left[z^h, z^w\right]\right)\right) \tag{4}$$

$$g^h = \sigma\left(F_h\left(f^h\right)\right) \tag{5}$$

$$g^w = \sigma(F_w(f^w)) \tag{6}$$

$$y_c(i, j) = x_c(i, j) \times g_c^h(i) \times g_c^w(j) \tag{7}$$

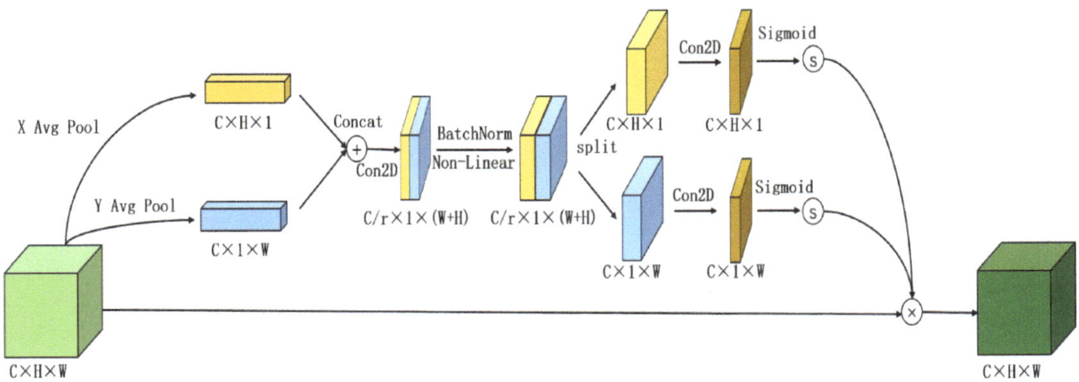

Figure 11. Structure of the coordinate attention mechanism.

4.2.2. Feature Fusion-Enhanced BiFPN

Feature pyramids emerged as a feature fusion framework that was derived to detect objects at different scales. However, the traditional structures of feature pyramids such as FPN (feature pyramid network) [34] and PANet add up different input features when fusing them. Since these different input features have different resolutions, the contributions of these features to the fused features usually vary. To solve this problem, we adopt BiFPN, a feature fusion framework, and replace the original PANet with BiFPN in the neck module. The main improvement made at this stage is to remove any node with only one input edge, since without feature fusion, this will contribute less to the feature network with different feature fusion. Secondly, if the original input node is at the same level as the output node, the amount adds an extra edge between the input node and the output node to fuse more features without adding too much cost.

PANet simply adds up the different features when fusing them, as shown in Equations (8) and (9). Since the resolution of different input features is different, the contribution to the input features is usually also different. To solve this problem, BiFPN adjusts the contribution of each scale by adding weights to each scale feature. We take the features of the middle layer as an example, and describe the two fusion features of BiFPN at level 2, where P_2^{td} is the intermediate feature at level 2 of the top-down path and P_2^{out} is the output feature at level 2 of the bottom-up path, as shown in Equations (10) and (11). All of the other features are constructed similarly. It is worth noting that BiFPN uses a depthwise separable convolution for feature fusion and adds batch normalization and activation after each convolution. The principle of operation of PANet and BiFPN is illustrated in Figure 12.

$$P_2^{td} = Conv\left(P_2^{in} + Resize\left(P_3^{out}\right)\right) \tag{8}$$

$$P_2^{out} = Conv\left(P_2^{td} + Resize\left(P_1^{out}\right)\right) \tag{9}$$

$$P_2^{td} = Conv\left(\frac{\omega_1 P_2^{in} + \omega_2 Resize\left(P_3^{in}\right)}{\omega_1 + \omega_2 + \epsilon}\right) \tag{10}$$

$$P_2^{out} = Conv\left(\frac{\omega_1' P_2^{in} + \omega_2' P_2^{td} + \omega_3' Resize\left(P_1^{out}\right)}{\omega_1' + \omega_2' + \omega_3' + \epsilon}\right) \tag{11}$$

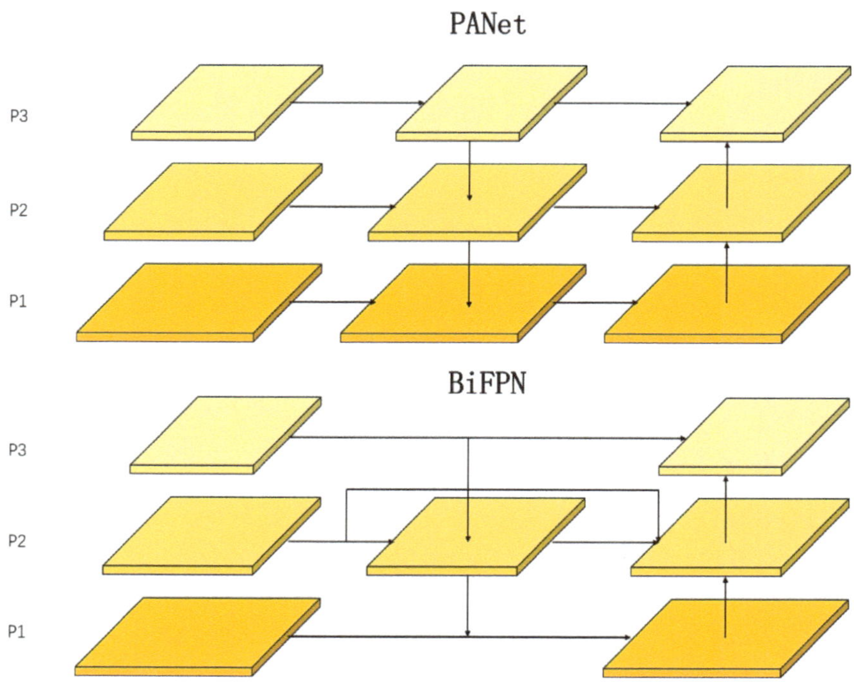

Figure 12. Diagram of the structure of PANet and BiFPN.

4.3. Proposed Framework

Figure 13 illustrates the framework of our transmission line defect detection method. The framework consists of data acquisition, pre-processing, and the BC-YOLO network testing module. We use remote sensing images taken during UAV power inspections to construct a dataset of vibration dampers and insulators called DVDI.

The specific steps in the transmission line defect detection process are as follows:

1. Remote sensing images of transmission lines are taken by the UAV during a power inspection.
2. The images are pre-processed using the gamma transform, and the defect dataset is expanded by rotational mirroring.
3. LabelImg is used to label the dataset, and the categories and boxes of insulators and vibration dampers are saved in an XML file.
4. The dataset is divided into a training set, validation set, and test set in the ratio of 6:2:2, and the resolution of the images is adjusted to 416 × 416 after feeding into the network.
5. The divided dataset is trained using BC-YOLO.
6. The loss function is observed during training, and the network weights are saved when the loss is minimized.
7. The saved network weights are used to detect insulators and vibration dampers with anomalies.

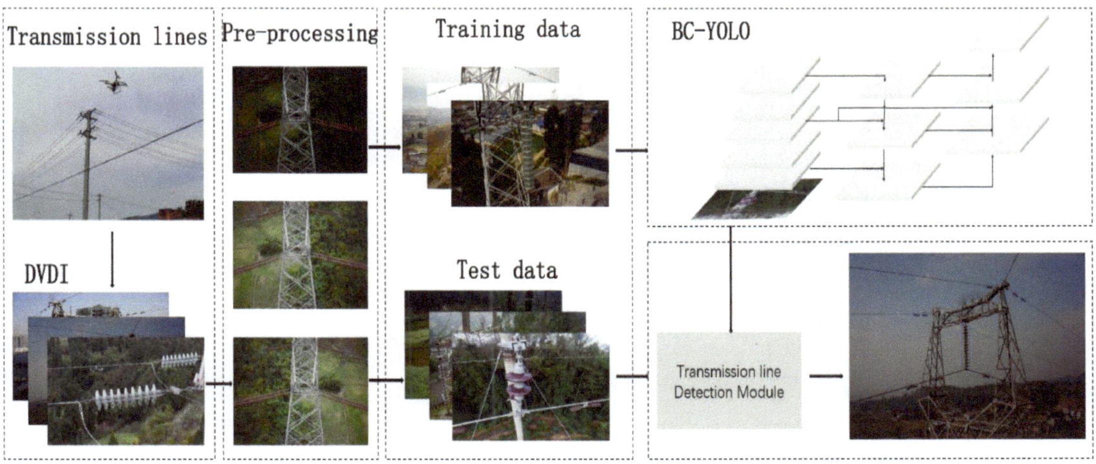

Figure 13. Schematic diagram of the proposed framework for insulator and vibration damper defect detection.

5. Experimental Results and Analysis

5.1. Experimental Environment and Parameters

The software environment and hardware parameters we used during the experiment are shown in Table 2.

Table 2. Software and hardware configuration of the experimental environment.

Platform	Configuration
Integrated development environment	PyCharm
Scripting language	Python3.8
Deep learning frame	PyTorch1.9.1
CPU model	Inter Core i7-9700k
Operating system	Ubuntu18.04 LTS 64-bits
GPU model	NVIDIA GeForce RTX 2080Ti
GPU accelerator	CUDA 10.2
Neural network accelerator	cuDNN7.6.5

The parameters of our experiments in training BC-YOLO are shown in Table 3.

Table 3. BC-YOLO experimental training parameters.

Parameter	Configuration
Neural network optimizer	SGD
Learning rate	0.01
Training epochs	300
Momentum	0.937
Batch size	4
Weight decay	0.0005

5.2. *Performance Evaluation Index*

In this paper, we use the Precision, Recall, Average Precision, and Mean Average Precision as evaluation indices for the network performance. The Recall rate represents how many true positive samples are retrieved by the network compared to the total number of positive instances, and is calculated as shown in Equation (12). Precision indicates how many of the predicted positive samples are positive, and the calculation method is shown in Equation (13). The AP is the area of the region enclosed by the Precision–Recall (P-R) curve and the coordinate axis, which is calculated as shown in Equation (14). The mAP is the average accuracy, which measures the overall detection effect of the network, as shown in Equation (15). The confusion matrix is given in Table 4.

Table 4. Confusion matrix.

	True	False
Positive	TP	FP
Negative	NT	FN

TP is predicted by the network as a positive sample of positive class, FP is predicted by the network as a negative sample of positive class, FN is predicted by the network as a positive sample of negative class, and TN is predicted by the network as a negative sample of negative class.

$$\text{Recall} = \frac{TP}{TP+FP} \tag{12}$$

$$\text{Precision} = \frac{TP}{TP+FN} \tag{13}$$

$$\text{AP} = \int_0^1 P(R)dR \tag{14}$$

$$\text{mAP} = \frac{\sum_{i=0}^{i=n} AP}{n} \tag{15}$$

5.3. *Results and Analysis*

5.3.1. Comparative Experiments on Attention Mechanisms

In this paper, we compare the three common attention mechanisms of SE, CBAM, and CA, which are commonly used in target detection tasks, and the results are shown in Table 4.

The three attention mechanisms are consistent in the positions they add in the YOLOv5 network. There are four versions of the YOLOv5 detection network, namely YOLOv5x, YOLOv5l, YOLOv5m, and YOLOv5s. The model used in this paper is YOLOv5x, which has the highest accuracy among the four versions. Firstly, adding the SE attention mechanism to the backbone improves the Recall by 1.9 points and the mAP by 1.9 points compared to the original YOLOv5x. Secondly, when the CBAM attention mechanism is added, the Recall is improved by 1.9 points and the mAP by 1.6 points. Finally, adding the CA attention mechanism improves the Recall by 4.1 points and the mAP by 2.0 points. From the results shown in the Table 5 below, we see that the effect of CA is the most significant.

Table 5. Effects of different attention mechanisms on network performance.

Baseline	SE	CBAM	CA	Recall(%)	Precision(%)	mAP(%)
YOLOv5x				82.7	89.9	86.4
	√			84.6	88.1	88.3
		√		84.6	87.4	88.0
			√	86.8	88.0	88.4

We use heat maps to visualize the output feature maps for adding different attention mechanisms, as shown in Figure 14. From the results shown in the heat map of Figure 14, the network module with the CA mechanism detects the critical parts of transmission line detection more accurately than with SE or CBAM.

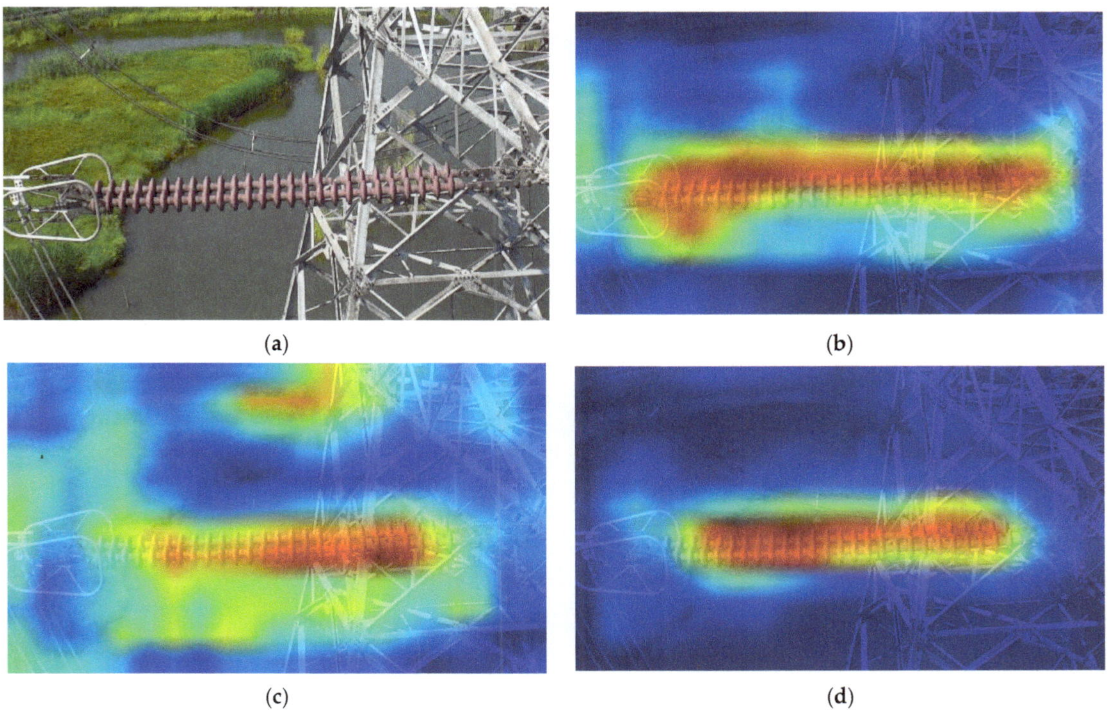

Figure 14. Heat maps for different attention modules: (**a**) original image; (**b**) SE; (**c**) CBAM; (**d**) CA.

5.3.2. Comparison of Experiments That Add Different Modules

In this experiment, we first used YOLOv5x as a baseline and tested the effect of adding the new feature pyramid BiFPN. The results showed that the mAP was improved by 1.8

points, and Recall by 2.7 points. This indicates that the new feature fusion can effectively fuse more feature layers. At the same time, in the process of feature fusion, the network pays more attention to the input features with great contribution, which enhances the learning ability of the model. In the second test, we compared the addition of the CA attention mechanism. The results showed that the mAP increased by two points and Recall by 4.1 points compared to the original baseline, which indicates that adding the attention mechanism to the backbone can effectively help the network to capture important information and long-distance dependencies when extracting features. Finally, the effect of fusing BiFPN and CA was explored, and the results showed that the mAP improved by 2.7 points and Recall by 4.0 points compared to the original baseline, meaning that the fusion of BiFPN and CA can effectively improve the accuracy of transmission line defect detection. The network detection results are presented in Table 6.

Table 6. Effect of each module on network performance.

Baseline	BiFPN	CA	Recall(%)	Precision(%)	mAP(%)
YOLOv5x			82.7	89.9	86.4
	√		85.4	87.3	88.2
		√	86.8	88.0	88.4
	√	√	86.7	86.2	89.1

5.3.3. Comparison of Different Object Detection Networks

To further verify the advantages of our method, we compare it with SSD [35], RetinaNet [36], CenterNet [37], and YOLOv4 [38] methods, and the detection results are shown in Table 7. Our method outperformed the other methods for all classes of mAP, especially for small targets such as vibration dampers. This is due to the small sizes of the numerous vibration damper targets, some of which are easily confused with forests, fields, and dead leaves due to their color, making it difficult for general target detection methods to detect them. In contrast, the methods used in the YOLO series can effectively handle detection tasks involving large differences in target sizes. When the CA mechanism and the new Bi-FPN feature fusion method were added, the accuracy of all categories improved significantly, indicating that this module can effectively improve the detection accuracy of the network. A visualization of the test results for each type of method is shown in Figure 15.

Table 7. Detection results for some mainstream object detection networks.

Method	1(AP)	2(AP)	3(AP)	4(AP)	mAP(%)
SSD	73.0	81.0	22.0	47.0	55.86
RetinaNet	78.0	78.0	26.0	36.0	54.37
CenterNet	85.0	85.0	52.0	59.0	70.33
YOLOv4	87.0	78.0	74.0	83.0	80.26
YOLOv5x	88.4	87.0	82.9	87.4	86.4
Proposed method	90.0	89.2	84.3	92.7	89.1

(**a**)

(**b**)

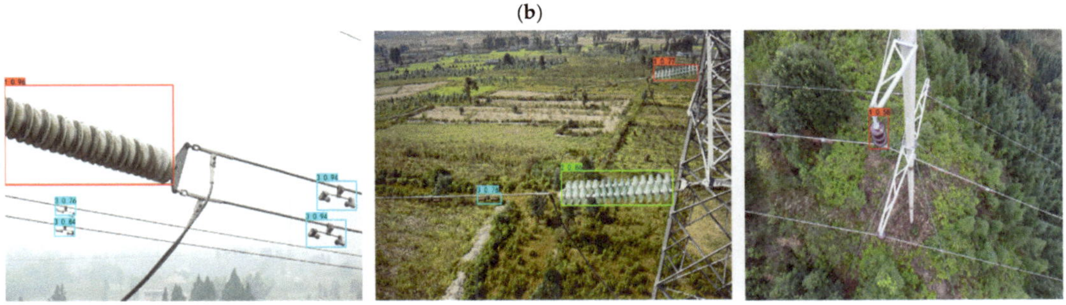

(**c**)

(**d**)

Figure 15. *Cont.*

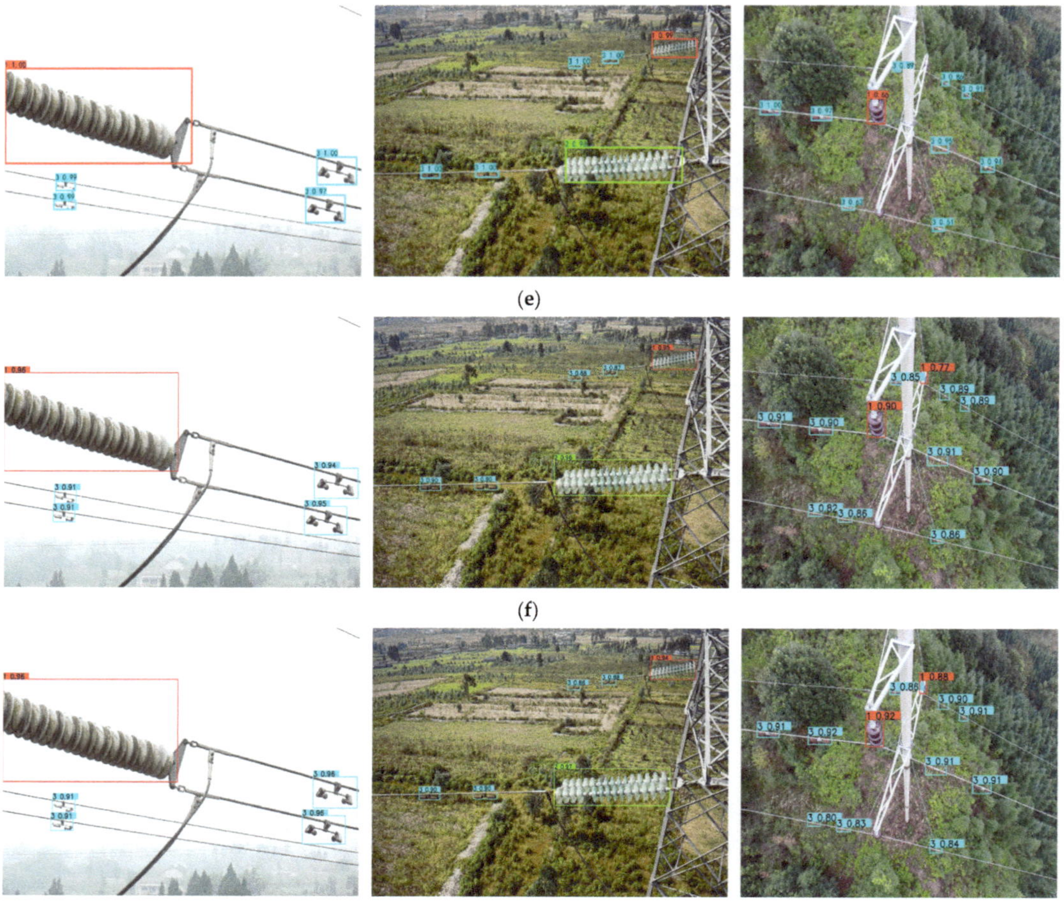

(e)

(f)

(g)

Figure 15. Results on the test dataset: (**a**) original images; (**b**) SSD; (**c**) RetinaNet; (**d**) CenterNet; (**e**) YOLOv4; (**f**) YOLOv5x; (**g**) proposed method.

6. Discussion

The confusion matrix is a standard form of representing the accuracy evaluation, also known as the error matrix, and is represented by an n × n matrix form. A confusion matrix was obtained from our training results after normalization, as shown in Figure 16, where each row represents the true attribution category of the data and each column represents the category predicted by the network. The total amount of data in each row represents the total amount of true data in that category, and the values in each column represent the number of categories predicted by the network for that category.

From the confusion matrix, it can be seen that the detection of defects in both insulators and vibration dampers is affected by the background; the vibration dampers are influenced more strongly by the background than the insulators, as the latter are larger and less likely to be obscured. The color of an insulator is less likely to be confused with the background color, and there is therefore less interference from the background. In contrast, the vibration dampers are easily obscured by transmission towers, transmission lines, and buildings due to their small size. The number of vibration dampers is also large, and since they are easily confused with the background color, this causes certain difficulties in detection. In particular, normal vibration dampers are easily confused with dead leaves, houses, and forests, and the background therefore has more influence on the normal vibration dampers than the defective ones. Due to the excessive number of insulator pieces in some insulator strings, their damage may not be easily detected and therefore may be mistaken for normal insulators.

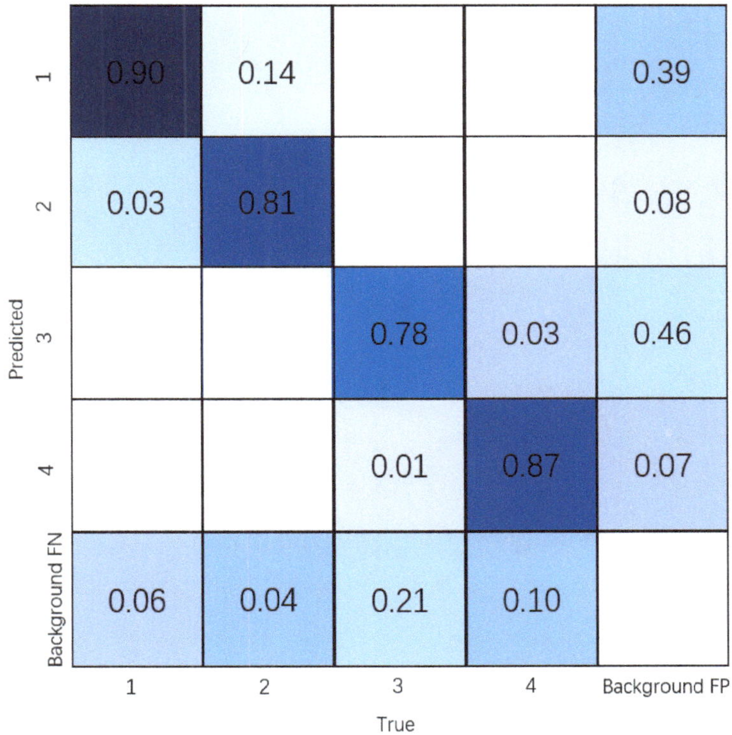

Figure 16. Confusion matrix.

Although the detection accuracy of defective vibration dampers and insulators images is improved using the method proposed in the paper, a few defective vibration dampers were still missed. Some examples are shown in Figure 17. In Figure 17a, the missed detection of the defective vibration dampers is caused by the fact that their color is very close to the color of the land. In Figure 17b, a defective vibration damper is partially occluded, so it is also undetected. Therefore, occlusion and similar color are the main reasons for missed detection, which are also the limitations of the proposed method.

(a)

Figure 17. *Cont.*

(**b**)

Figure 17. Example of partially missed detection: (**a**) Color confusion miss detection; (**b**) Missing detection from masking.

7. Conclusions

To accurately detect insulators and vibration dampers on transmission lines with a complex background, in low light, and with large size differences between targets, we have proposed an optimized version of YOLOv5 through the use of methods such as a CA mechanism and Bi-FPN enhanced feature fusion, to develop the BC-YOLO insulator and vibration damper defect detection method. The findings of this paper can be summarized in the following three points: (i) the datasets used in the experiments contained pictures of multiple types of insulators and vibration dampers, taken during the power inspection process; (ii) a CA mechanism is added to the feature extraction module of YOLOv5, which allows the mobile network to obtain information about a larger area without introducing a larger overhead by embedding the location information into the channel attention; (iii) for the case of insulators and vibration dampers with large differences in size, we introduce Bi-FPN, based on the original feature fusion method, which can effectively improve the efficiency of feature fusion and improve the detection accuracy of small targets. Through an experimental comparison, it was found that the mAP@0.5 for BC-YOLO reached 89.1% on the test set of the DVDI, a value 2.7% higher than for YOLOv5.

In future research, we will collect additional data on defective insulators and vibration dampers taken during UAV inspections, and will expand the types of defects for the insulators and vibration dampers, which will allow the network to be adapted to the detection of insulators and vibration dampers with various types of defects. We will also investigate how to compress the model as a way to increase the detection speed while maintaining the detection accuracy of the model.

Author Contributions: Conceptualization, X.D.; data curation, W.B. and N.W.; formal analysis, W.B. and X.D.; investigation, N.W. and X.Y.; methodology, X.D., W.B., and M.Y.; resources, W.B., N.W., and X.Y.; supervision, W.B. and N.W.; validation, X.D. and M.Y.; writing—original draft, X.D.; writing—review and editing, W.B., X.D., N.W., and M.Y. All authors have read and agreed to the published version of the manuscript.

Funding: This research was funded by the National Key Research and Development Program of China (Grant No. 2020YFF0303803) and Anhui Natural Science Foundation (Grant No. 2208085MC60).

Data Availability Statement: https://github.com/Emp-8/DVDI, accessed on 12 October 2022.

Acknowledgments: The authors sincerely thank the anonymous reviewers for their critical comments and suggestions for improving the manuscript.

Conflicts of Interest: The authors declare no conflict of interest.

References

1. Sadykova, D.; Pernebayeva, D.; Bagheri, M.; James, A. IN-YOLO: Real-time detection of outdoor high voltage insulators using UAV imaging. IEEE Trans. *Power Deliv.* **2019**, *35*, 1599–1601. [CrossRef]
2. Liang, H.; Zuo, C.; Wei, W. Detection and evaluation method of transmission line defects based on deep learning. *IEEE Access* **2020**, *8*, 38448–38458. [CrossRef]
3. Wang, S.; Liu, Y.; Qing, Y.; Wang, C.; Lan, T.; Yao, R. Detection of Insulator Defects With Improved ResNeSt and Region Proposal Network. *IEEE Access* **2020**, *8*, 184841–184850. [CrossRef]
4. Siddiqui, Z.A.; Park, U. A drone based transmission line components inspection system with deep learning technique. *Energies* **2020**, *13*, 3348. [CrossRef]
5. Jiang, S.; Jiang, W.; Huang, W.; Yang, L. UAV-based oblique photogrammetry for outdoor data acquisition and offsite visual inspection of transmission line. *Remote Sens.* **2017**, *9*, 278. [CrossRef]
6. Liu, X.; Lin, Y.; Jiang, H.; Miao, X.; Chen, J. Slippage fault diagnosis of dampers for transmission lines based on faster R-CNN and distance constraint. *Electr. Power Syst. Res.* **2021**, *199*, 107449. [CrossRef]
7. Yang, L.; Fan, J.; Liu, Y.; Li, E.; Peng, J.; Liang, Z. A review on state-of-the-art power line inspection techniques. *IEEE Trans. Instrum. Meas.* **2020**, *69*, 9350–9365. [CrossRef]
8. Katrasnik, J.; Pernus, F.; Likar, B. A survey of mobile robots for distribution power line inspection. *IEEET rans. Power Deliv.* **2009**, *25*, 485–493. [CrossRef]
9. Nguyen, V.N.; Jenssen, R.; Roverso, D. Automatic autonomous vision-based power line inspection: A review of current status and the potential role of deep learning. *Int. J. Electr. Power Energy Syst.* **2018**, *99*, 107–120. [CrossRef]
10. Matikainen, L.; Lehtomäki, M.; Ahokas, E.; Hyyppä, J.; Karjalainen, M.; Jaakkola, A.; Kukko, A.; Heinonen, T. Remote sensing methods for power line corridor surveys.ISPRS J. Photogramm. *Remote Sens.* **2016**, *119*, 10–31.
11. Tiantian, Y.; Guodong, Y.; Junzhi, Y. Feature fusion based insulator detection for aerial inspection. In Proceedings of the 2017 36th Chinese Control Conference (CCC), Dalian, China, 26–28 July 2017; pp. 10972–10977.
12. Reddy, M.J.B.; Chandra, B.K.; Mohanta, D.K. A DOST based approach for the condition monitoring of 11 kV distribution line insulators. In *IEEE Transactions on Dielectrics and Electrical Insulation*; IEEE: Piscataway, NJ, USA, 2011; Volume 18, pp. 588–595. [CrossRef]
13. Reddy, M.J.B.; B, K.C.; Mohanta, D.K. Condition monitoring of 11 kV distribution system insulators incorporating complex imagery using combined DOST-SVM approach. In *IEEE Transactions on Dielectrics and Electrical Insulation*; IEEE: Piscataway, NJ, USA, 2013; Volume 20, pp. 664–674. [CrossRef]
14. Li, B.; Wu, D.; Cong, Y.; Xia, Y.; Tang, Y. A method of insulator detection from video sequence. In Proceedings of the 4th International Symposium on Information Science and Engineering (ISISE), Shanghai, China, 14–16 December 2012; pp. 386–389.
15. Zhao, Z.; Liu, N.; Wang, L. Localization of multiple insulators by orientation angle detection and binary shape prior knowledge. In *IEEE Transactions on Dielectrics and Electrical Insulation*; IEEE: Piscataway, NJ, USA, 2015; Volume 22, pp. 3421–3428. [CrossRef]
16. Wu, Q.; An, J.; Lin, B. A Texture Segmentation Algorithm Based on PCA and Global Minimization Active Contour Model for Aerial Insulator Images. *IEEE J. Sel. Top. Appl. Earth Obs. Remote Sens.* **2013**, *5*, 1509–1518. [CrossRef]
17. Liao, S.; An, J. A robust insulator detection algorithm based on local features and spatial orders for aerial images. *IEEE Geosci. Remote Sens. Lett.* **2015**, *12*, 963–967. [CrossRef]
18. Jabid, T.; Uddin, M.Z. Rotation invariant power line insulator detection using local directional pattern and support vector machine. In Proceedings of the International Conference on Engineering and Technology Innovations 2016 (ICISET), Wuhan, China, 26–24 March 2016; pp. 1–4.
19. Zhao, Z.; Xu, G.; Qi, Y. Representation of binary feature pooling for detection of insulator strings in infrared images. In *IEEE Transactions on Dielectrics and Electrical Insulation*; IEEE: Piscataway, NJ, USA, 2016; Volume 23, pp. 2858–2866. [CrossRef]

20. Guifeng, Z.; Zhiren, T. Swarm optimization clustering for image segmentation of insulators. In Proceedings of the IEEE Conference on Industrial Electronics and Applications (ICIEA), Xi'an, China, 16–21 June 2019; pp. 397–401.
21. Li, F.; Xin, J.; Chen, T.; Xin, L.; Wei, Z.; Li, Y.; Zhang, Y.; Jin, H.; Tu, Y.; Zhou, X.; et al. An Automatic Detection Method of Bird's Nest on Transmission Line Tower Based on Faster_RCNN. *IEEE Access* **2020**, *8*, 164214–164221. [CrossRef]
22. Wanguo, W.; Zhenli, W.; Bin, L.; Yuechen, Y.; Xiaobin, S. Typical Defect Detection Technology of Transmission Line Based on Deep Learning. In Proceedings of the 2019 Chinese Automation Congress (CAC), Hangzhou, China, 22–24 November 2019; pp. 1185–1189. [CrossRef]
23. Wu, C.; Ma, X.; Kong, X.; Zhu, H. Research on insulator defect detection algorithm of transmission line based on CenterNet. *PLoS ONE* **2021**, *16*, e0255135. [CrossRef]
24. Wu, W.; Li, Q. Machine Vision Inspection of Electrical Connectors Based on Improved Yolo v3. *IEEE Access* **2020**, *8*, 166184–166196. [CrossRef]
25. Bao, W.; Ren, Y.; Wang, N.; Hu, G.; Yang, X. Detection of Abnormal Vibration Dampers on Transmission Lines in UAV Remote Sensing Images with PMA-YOLO. *Remote Sens.* **2021**, *13*, 4134. [CrossRef]
26. Liu, B.; Wang, X.; Jin, W.; Chen, Y.; Liu, C. Infrared image detail enhancement based on local adaptive gamma correction. *Chin. Opt. Lett.* **2012**, *10*, 25–29.
27. Everingham, M.; Van Gool, L.; Williams, C.K.; Winn, J.; Zisserman, A. The pascal Visual Object Classes (VOC) challenge. *Int. J. Comput. Vis.* **2010**, *88*, 303–338. [CrossRef]
28. He, K.; Zhang, X.; Ren, S.; Sun, J. Spatial pyramid pooling in deep convolutional networks for visual recognition. *IEEE Trans. Pattern Anal. Mach. Intell.* **2015**, *37*, 1904–1916. [CrossRef]
29. Liu, S.; Qi, L.; Qin, H.; Shi, J.; Jia, J. Path aggregation network for instance segmentation. In Proceedings of the 2018 IEEE/CVF Conference on Computer Vision and Pattern Recognition (CVPR), Salt Lake City, UT, USA, 18–23 June 2018; pp. 8759–8768. [CrossRef]
30. Hou, Q.; Zhou, D.; Feng, J. Coordinate Attention for Efficient Mobile Network Design. In Proceedings of the 2021 IEEE/CVF Conference on Computer Vision and Pattern Recognition (CVPR), Nashville, TN, USA, 20–25 June 2021; pp. 13708–13717. [CrossRef]
31. Tan, M.; Pang, R.; Le, Q.V. EfficientDet: Scalable and Efficient Object Detection. In Proceedings of the 2020 IEEE/CVF Conference on Computer Vision and Pattern Recognition (CVPR), Seattle, WA, USA, 13–19 June 2020; pp. 10778–10787. [CrossRef]
32. Hu, J.; Shen, L.; Albanie, S.; Sun, G.; Wu, E. Squeeze-and-Excitation Networks. In *IEEE Transactions on Pattern Analysis and Machine Intelligence*; IEEE: Piscataway, NJ, USA, 2020; Volume 42, pp. 2011–2023. [CrossRef]
33. Woo, S.; Park, J.; Lee, J.; Kweon, I.S. CBAM: Convolutional block attention module. In Proceedings of the 2018 European Conference on Computer Vision (ECCV), Munich, Germany, 8–14 September 2018; pp. 3–19. [CrossRef]
34. Lin, T.Y.; Dollar, P.; Girshick, R.; He, K.; Hariharan, B.; Belongie, S. Feature pyramid networks for object detection. In Proceedings of the 2017 IEEE Conference on Computer Vision and Pattern Recognition (CVPR), Honolulu, HI, USA, 21–26 July 2017. [CrossRef]
35. Liu, W.; Anguelov, D.; Erhan, D.; Szegedy, C.; Reed, S.; Fu, C.Y.; Berg, A.C. SSD: Single shot multibox detector. *arXiv* **2015**, arXiv:1512.02325.
36. Lin, T.; Goyal, P.; Girshick, R.; He, K.; Dollár, P. Focal loss for dense object detection. In Proceedings of the 2017 IEEE International Conference on Computer Vision (ICCV), Venice, Italy, 22–29 October 2017; pp. 2999–3007. [CrossRef]
37. Zhou, X.Y.; Wang, D.Q.; Krhenbühl, P. Objects as points. *arXiv* **2019**, arXiv:1904.07850.
38. Bochkovskiy, A.; Wang, C.-Y.; Liao, H.-J.M. YOLOv4 optimal speed and accuracy of object detection. *arXiv* **2020**, arXiv:2004.10934.

remote sensing

MDPI

Article

MQANet: Multi-Task Quadruple Attention Network of Multi-Object Semantic Segmentation from Remote Sensing Images

Yuxia Li [1], Yu Si [1], Zhonggui Tong [1], Lei He [2,3,*], Jinglin Zhang [1], Shiyu Luo [1] and Yushu Gong [1]

[1] School of Automation Engineering, University of Electronic Science and Technology of China, Chengdu 611731, China
[2] School of Software Engineering, Chengdu University of Information Technology, Chengdu 610225, China
[3] Sichuan Province Engineering Technology Research Center of Support Software of Informatization Application, Chengdu 610225, China
* Correspondence: helei1978@cuit.edu.cn

Abstract: Multi-object semantic segmentation from remote sensing images has gained significant attention in land resource surveying, global change monitoring, and disaster detection. Compared to other application scenarios, the objects in the remote sensing field are larger and have a wider range of distribution. In addition, some similar targets, such as roads and concrete-roofed buildings, are easily misjudged. However, existing convolutional neural networks operate only in the local receptive field, and this limits their capacity to represent the potential association between different objects and surrounding features. This paper develops a Multi-task Quadruple Attention Network (MQANet) to address the above-mentioned issues and increase segmentation accuracy. The MQANet contains four attention modules: position attention module (PAM), channel attention module (CAM), label attention module (LAM), and edge attention module (EAM). The quadruple attention modules obtain global features by expanding the receptive fields of the network and introducing spatial context information in the label. Then, a multi-tasking mechanism which splits a multi-category segmentation task into several binary-classification segmentation tasks is introduced to improve the ability to identify similar objects. The proposed MQANet network was applied to the Potsdam dataset, the Vaihingen dataset and self-annotated images from Chongzhou and Wuzhen (CZ-WZ), representative cities in China. Our MQANet performs better over the baseline net by a large margin of +6.33 OA and +7.05 Mean F1-score on the Vaihingen dataset, +3.57 OA and +2.83 Mean F1-score on the Potsdam dataset, and +3.88 OA and +8.65 Mean F1-score on the self-annotated dataset (CZ-WZ dataset). In addition, each image execution time of the MQANet model is reduced 66.6 ms compared to UNet. Moreover, the effectiveness of MQANet was also proven by comparative experiments with other studies.

Keywords: deep learning; remote sensing; semantic segmentation; multi-task learning; attention mechanism

Citation: Li, Y.; Si, Y.; Tong, Z.; He, L.; Zhang, J.; Luo, S.; Gong, Y. MQANet: Multi-Task Quadruple Attention Network of Multi-Object Semantic Segmentation from Remote Sensing Images. *Remote Sens.* **2022**, *14*, 6256. https://doi.org/10.3390/rs14246256

Academic Editors: Qian Du, Yanni Dong and Xiaochen Yang

Received: 7 November 2022
Accepted: 5 December 2022
Published: 10 December 2022

1. Introduction

Semantic segmentation tasks classify each pixel in an image into several regions with specific semantic categories and often appear in fields such as human–computer interaction, computer photography, image search engines, and augmented reality. In these applications, the extraction targets are usually clear in semantics and have a small coverage area. However, things are different in remote sensing images. The targets in remote sensing images have a wider range of distribution and more complex features. On the one hand, such characteristics provide richer target detail information for feature detection, such as color, contour, and texture. On the other hand, much more complex interference is introduced into segmentation tasks.

The traditional remote sensing image segmentation methods mainly include support vector machine (SVM) methods [1], superpixel-based methods [2], and semisupervised geodesic-based methods [3]. Traditional methods can achieve better results with a small sample size. Still, with the increase in sample data, the accuracy of traditional methods has not significantly improved to meet the application requirements. Usually, traditional methods are only effective for some specific scenarios, with poor universality.

Focused on the above situation, compared with traditional methods in machine learning, Deep Convolutional Neural Networks (DCNNs), such as FCN [4], have shown excellent feature extraction and object representation abilities [5]. Many approaches have been proposed to increase the receptive field of convolutional neural networks. Unet [6] and SegNet [7] propose skip connection, trying to connect the same-sized feature maps in the encoder and decoder layers. The DeepLab series network models are all based on encoder–decoder architecture. Atrous convolution to expand the receptive field is introduced in DeepLab v1 [8] and DeepLab v2 [9]. Furthermore, Atrous Spatial Pyramid Pooling (ASPP) was proposed to expand the perceptual field further and enhance the spatial feature extraction capability using multi-layer cavity convolution. DeepLab v3 [10] introduced image-level multiscale features in ASPP to further improve the feature extraction capability. DeepLab v3+ [11] used a modified Xception [12] encoder and a lightweight decoder to improve the resolution of segmentation results. FarSeg [13] uses two encoder branches to enhance the extraction of foreground and background, respectively. In addition, some studies have been conducted to improve the deep learning methods according to the characteristics of remote sensing images. EFCNet [14] introduces the separable convolutional module (SCM) to alleviate the problem of numerous parameters for the semantic segmentation of high-resolution remote sensing images. DSPCANet [15] introduces the internal residual block (R2_Block) to enhance the receptive field of the network and learn the ground feature representation from additional DSM images. Sharifi et al. [16] proposed the ResUNet-a model with post-processing. The model contains the output of joint connection, which significantly improves the efficiency and robustness of the farmland extraction model. However, stacking and aggregating convolutional layers perform poorly in covering global receptive fields. Additionally, these methods do not effectively extract global contextual information.

A helpful method for obtaining global contextual information is the self-attention mechanism. Nonlocal [17] proposes a generalized, simple, nonlocal operation operator that can be directly embedded into neural networks. DANet [18] introduces the self-attention mechanism to capture feature dependencies in the spatial and channel dimensions. A2-Nets [19], expectation–maximization attention networks [20], and CBAM [21] introduced a self-attention mechanism to merge global features by different descriptors. LANet [22] proposes a patch attention module and an attention embedding module to merge high-level and low-level features in the model. HMANet [23] proposes a class augmented attention module to obtain class-level information. SSFTT [24] utilizes the self-attention mechanism to construct 3D and 2D convolutional layers to jointly extract shallow spectral and spatial features. However, methods that rely on self-attention only cause the network to pay attention to itself and overlook the spatial context relationship hidden in the labels.

To solve the above-mentioned issues, we suggest a brand new structure termed the Label Attention Module (LAM). LAM fully uses the label's spatial context via the attention module. However, the way to generate attention is different than the self-attention module. LAM optimizes the attention probability map by introducing label information.

Furthermore, we proposed a triple-attention network called TANet, which contains LAM and two self-attention modules presented in DANet: PAM and CAM. TANet can help enhance semantic segmentation accuracy due to the triple attention module's ability to strengthen global information extraction.

Adding LAM can significantly increase the segmentation accuracy of a large range of targets. However, misjudgment problems would appear with respect to some targets presenting similar features, such as impervious surfaces and concrete roof buildings. Due to the competition between different categories, the probability of the misjudgment category affects the real category. In order to reduce the competition between categories, we introduced a multi-task mechanism. The multi-task mechanism converts a multi-category segmentation task into multiple binary-classification segmentation tasks. All categories share an encoder, and each category owns a separate decoder. The multi-task architecture can effectively improve the segmentation accuracy of similar categories. Furthermore, we perform some edge optimization to improve the accuracy of the edge area of different categories. The edge optimization includes two new edge branches and an edge attention module. Combined with the triple attention and multi-task architecture mentioned above, the model contains four attention modules, so we name this network Multi-Task Quadruple Attention Network. We conducted experiments on two public datasets (Potsdam and Vaihingen datasets) and a self-made dataset (CZ-WZ dataset) to demonstrate the effectiveness of the proposed model.

The main contributions of this paper are as follows:

(1) We propose the label attention module (LAM) to learn the spatial contextual information of features from the label instead of information from the network itself.
(2) A Triple Attention Network is designed to obtain global features of large objects. It significantly improves the semantic segmentation accuracy of large objects in remote sensing images.
(3) A Multi-task TANet (MTANet) architecture is proposed to reduce the misjudgment between similar categories.
(4) Based on the MTANet model, A MQANet model is constructed to optimize the edge area of semantic segmentation.

2. Related Search

2.1. Semantic Segmentation

Semantic segmentation divides images into regional blocks with certain semantics and obtains a segmented image with pixel-by-pixel semantic annotation. Traditional segmentation methods focus on designing a feature descriptor for each pixel. However, the specially designed handcrafted feature descriptors are challenging to adapt to other scenarios.

Deep learning methods extract features directly from the data itself. Patch-wise classification achieves semantic segmentation based on the classification results of patches. However, this method is too expensive to compute. FCN [4] proposes a paradigm to obtain semantic segmentation directly from feature upsampling, significantly improving the computational cost. After FCN was proposed, various structures were proposed to improve the segmentation accuracy. Most networks use encoder–decoder architectures, such as Unet [6], SegNet [7], and DeeplabV3+ [11]. Encoders are used for feature extraction, and decoders perform pixel-level classification based on the features obtained by the encoder to obtain semantic segmentation results. Furthermore, Zhou et al. [25] introduced D-LinkNet with multiscale dilation rates to collect contextual information and extract additional global features. GPSNet [26] tries to dynamically aggregate a discriminative semantic context using a comparative feature aggregation module. Furthermore, it can gather free-form semantic context information adaptively. CGNet [27] constructs the joint feature from local features and the surrounding context effectively and efficiently. In addition, CGNet utilizes the global context to improve the joint feature.

2.2. Attention Mechanism

The human visual mechanism inspired the formation of the spatial attention mechanism. When the human eye sees an image, it will automatically give greater attention to critical locations. Therefore, different components of the image feature map should have diverse weights.

Vaswani et al. [28] introduced the self-attention mechanism, which does not use complex models like RNN or CNN, and it only relies on the attention model to parallelize training with global information. Self-attention obtains contextual dependency information at a long distance by capturing the spatial dependencies between any two locations in the feature map. Specifically, assume the standard form of each element in the sequence is (Q, K, V). Q denotes query, K denotes key, and V denotes value. In DCNNs, we generate Q, K, and V from feature maps extracted by the networks' backbones. The attention mechanism calculates attention weight by the similarity between Q and K. Then, attention output comes from the V value and the attention weight's weighted summation. Self-attention differs from standard attention mechanisms: three matrices—the question, key, and value—are identical in the self-attention method. The attention output result is shown in Equation (1). The output of the attention module integrates features from Q, K, and V. A network might benefit from the attention module by catching more global context over local features and enhancing the pixel-level prediction's feature representations.

$$Attention(Q, K, V) = softmax(\frac{QK^T}{\sqrt{d_k}})V \tag{1}$$

where d_k denotes the normalization coefficient, and K^T refers to the transposition of K.

DANet [18] is a classic self-attention network application that introduces a self-attention mechanism to obtain the feature dependence separately in the spatial and channel dimensions. BAM [29] performs better in element-wise summation of spatial and channel attention. PSANet [30] is designed with a bidirectional information dissemination path, where each location aggregates information from all other locations to help predict itself. Unlike previous works, we propose a label attention module to capture global contexts from the label directly, ensuring that the spatial context information can be smoothly transferred to the attention map.

2.3. Multi-Task Learning

Initially introduced by Caruana [31], multi-task learning aims to solve multiple tasks with one architecture. In order to theoretically explain that multi-task learning has better learning effects than single-task learning, relevant papers give theoretical proof in different aspects. For example, the inductive bias learning method proposed by Baxter et al. [32] proved that multi-task learning could achieve better generalization errors than single-task learning in a specific environment. Ben-David [33] proposed a more general concept of connection between tasks, which can be used in many real-life scenarios. This concept proved the strict upper bound of the generalization error of multi-task learning. Maurer [34] applied dictionary learning and sparse coding to multi-task learning and gave the generalization error of multi-task learning by measuring the complexity of the hypothesis. Ando and Zhang et al. [35] assume that all tasks share a standard structure and give reliable estimates of the shared parameters of multiple tasks when there are enough tasks.

Multi-task learning in deep neural networks can be divided into two categories based on whether parameters are shared softly or firmly between tasks [36].

Soft parameter sharing allows for the regularization of different tasks' bottom parameters instead of keeping the bottom parameters the same. For example, Dung et al. [37] apply the L2 norm to restrict the distance between the parameters of different tasks to ensure the relevance between tasks. Yang et al. [38] proposed regularizing bottom parameters with a trace norm. Compared with the complex parameter-constrained multi-task deep learning model, the soft-constrained multi-task learning model has fewer constraints. It can produce better outcomes when the task association is not quite tight. Multi-task deep learning models need to learn a network architecture suitable for multiple tasks simultaneously. Thus, the models are more robust and resistant to overfitting.

The hard sharing method is another multi-task learning method. This method assumes that there is some connection between the different tasks. As in Figure 1, different tasks obtain standard low-level features in the shared bottom layers and confirm the uniqueness of various tasks; each task uses a unique layer to capture high-level features. Our work transforms a multi-classification task into multiple binary classification tasks with high similarity between tasks, so we choose to use hard sharing as an approach.

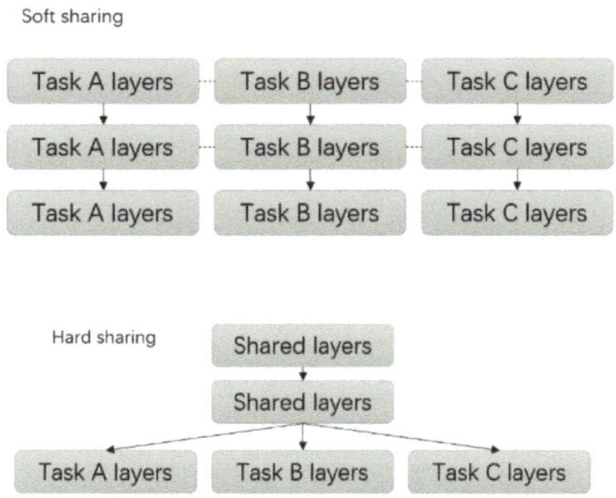

Figure 1. Two different parameter-sharing methods in multi-task learning.

3. Methodology

The proposed network model consists of encoder, decoder, and edge optimization module. The encoder structure is the same as the VGG architecture adopted by UNet. For the decoder, the proposed network converts a multi-category semantic segmentation task into multiple binary-segmentation tasks. The number of decoders is the same as the number of semantic segmentation categories of the ground object. Each decoder contains quadruple attention modules, PAM, CAM, LAM, and EAM. The optimization algorithm for edge extraction consists of three parts, the edge map branch, the distance map branch, and the edge attention module. The architecture of the proposed method is shown in Figure 2.

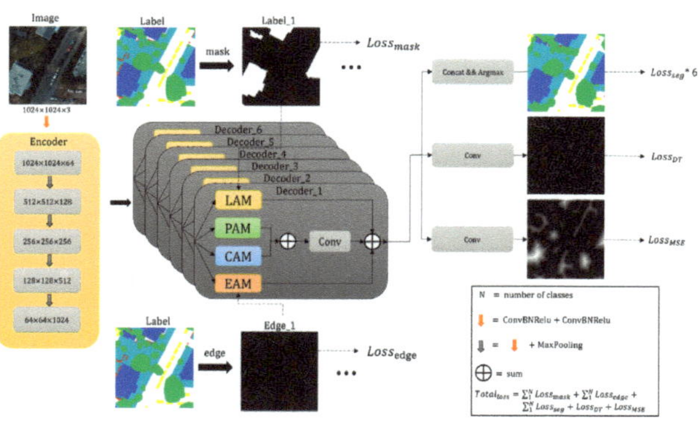

Figure 2. An overview of the Multi-task QANet (MQANet).

Sections 3.1 and 3.2 describe the PAM, CAM, and LAM. In Section 3.3, we introduce the multi-task architecture of this network. In Section 3.4, edge optimization is introduced, including two edge map branches and the EAM.

In Figure 2, all categories share one encoder, and each category owns a separate triple attention decoder. Each decoder contains four attention modules. LAM needs label information, and EAM needs edge labels during training.

3.1. PAM and CAM

Similar to human learning, machine learning is considered attention. The attention mechanism's core goal is to find critical information from various information items for the current task. Position Attention Module (PAM) and Channel Attention Module (CAM) are practical self-attention modules. PAM captures spatial global dependencies, and CAM pays attention to the importance of each channel dimension.

Figure 3 illustrates the structure of the PAM. In the PAM, the input generates two parts of feature maps: one is represented as Q and K to calculate an attention probability map in the shape $(H \times W) \times (H \times W)$, and the other is used as V. Q, K, and V denote query features, key features, and value features. Furthermore, C, H, and W represent the attention probability map's channel, height, and weight. Then, the optimization attention map is reshaped to obtain the final prediction map.

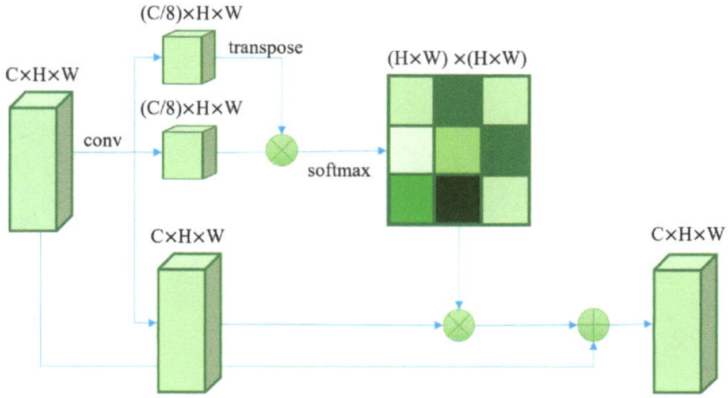

Figure 3. Architecture of Position Attention Module [18].

The overall structure of the PAM is shown in Equation (2).

$$
\begin{aligned}
Att &= Softmax\left(Q_{(HW \times C)} \cdot K_{(C \times HW)}\right) \\
F_{out} &= \left(V_{(C \times HW)} \cdot Att\right) reshape(C \times H \times W) + Input_{(C \times H \times W)}
\end{aligned}
\tag{2}
$$

where Att denotes the attention probability map calculated, and F_{out} is the final output, obtained by summing the input and the optimization attention map. $reshape(C \times H \times W)$ refers to the size of the feature map reshape to $C \times H \times W$.

The width and height of the feature are multiplied by a large value and reduce the number of channels to one-eighth of the original one by using convolution to facilitate operations. Then the dimension of the number of channels is eliminated by matrix multiplication. This operation does not affect the shape of the feature.

PAM uses a spatial attention map to select aggregating contexts. In addition, PAM has a global contextual view. Similar semantic features enhance intra-class compactness and semantic consistency [18].

The CAM is similar in structure to the PAM, but it has a few differences. The structure of the CAM is given in Figure 4. The first difference is that the number of channels is smaller, so there is no need to change the feature map's shape using convolution to reduce the number of operations. The other point is that the shape of the generated attention weight map is changed, and CAM focuses on the connection between the different channels of the features. In the network structure, CAM swaps the position of the location attention module dot product, and the shape of the generated attention weight map is $(C \times C)$, thus establishing the influence relationship between features in different channels.

The overall structure of CAM is shown in Equation (3).

$$
\begin{aligned}
Att &= softmax\left(Q_{(C \times HW)} \cdot K_{(HW \times C)}\right) \\
F_{out} &= \left(Att \cdot V_{(C \times HW)}\right) reshape(C \times H \times W) + Input_{(C \times H \times W)}
\end{aligned}
\tag{3}
$$

where Att denotes the attention probability map calculated, and F_{out} is the final output, obtained by summing the input and the optimization attention map. $reshape(C \times H \times W)$ refers to the size of the feature map reshape to $C \times H \times W$.

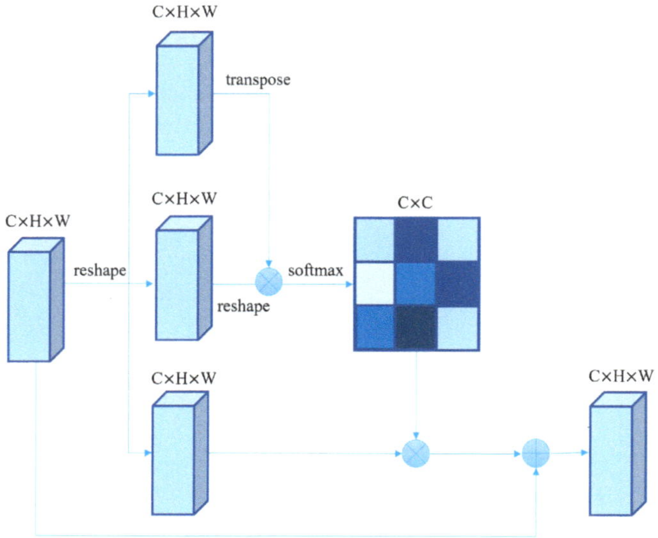

Figure 4. Architecture of Channel Attention Module [18].

3.2. Label Attention Module (LAM)

We built a brand-new attention mechanism at different views, inspired by the effectiveness of attention-based methods. Unlike the self-attention mechanism, we use the label to generate attention probability maps. Thus, LAM can gather more global features. The structure is shown in Figure 5.

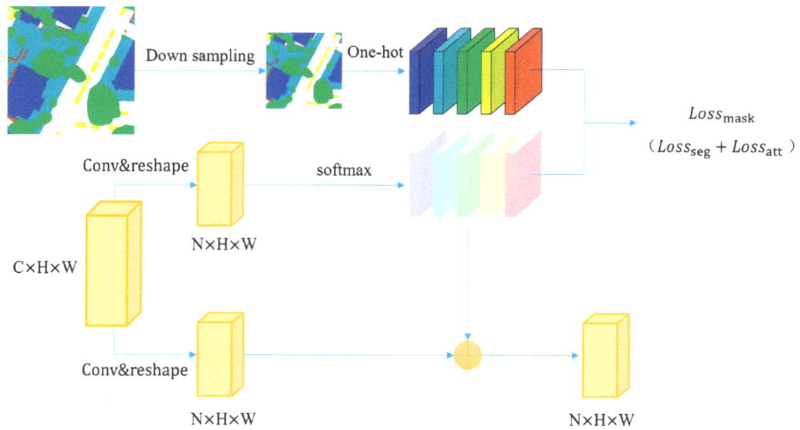

Figure 5. Architecture of Label Attention Module.

The input ($C \times H \times W$) is turned into two parts shaped ($N \times H \times W$) by convolution and reshaping, where N is the number of classifications. LAM's output is the weighted summation of the attention and value parts. In addition, after the SoftMax function, a loss function is computed the attention probability map and a reshaped one-hot label.

$$\begin{gathered} Att = softmax(Conv\left(Input_{(C\times H\times W)}\right)_{(N\times H\times W)}) \\ F_{out} = Conv(Conv(Input_{(C\times H\times W)})_{(N\times H\times W)} + Att)_{(N\times H\times W)} \end{gathered} \quad (4)$$

where Att denotes the attention probability map, and F_{out} denotes the output features of the attention module.

The convolution neural network's backpropagation technique parameter optimization is Equation (5). The loss function is shown in Equation (6).

$$W = W - \frac{\eta}{batch\_size} \sum \frac{\partial L_{seg}}{\partial W} \tag{5}$$

$$\begin{aligned} Loss_{seg} &= CE(predict, label) \\ Loss_{att} &= CE\left(predict, label_{down\_sampling}\right) \\ W &= W - \tfrac{\eta}{batch\_size} \sum \left(\tfrac{\partial L_{seg}}{\partial W} + \tfrac{\partial L_{att}}{\partial W}\right) \end{aligned} \tag{6}$$

where η is the learning rate, and $\frac{\partial L_{seg}}{\partial W}$ is the derivative of the loss function with respect to the parameters of the layer. CE refers to the cross-entropy loss function.

The loss function $Loss_{mask}$ consists of two parts: the segmentation part and the label attention part, which are, respectively, defined as $Loss_{seg}$, $Loss_{att}$. The label attention loss helps LAM generate prediction map prototypes that facilitate the transfer of feature information.

3.3. Multi-Task TANet

The attention mechanism helps improve the semantic segmentation accuracy of large objects. However, with the attention model it is difficult to distinguish some similar features. Some trees without leaves are very similar to low vegetation (Figure 6), and even humans will misunderstand them without carefully checking. Thus, the attention module failed to distinguish the tree area.

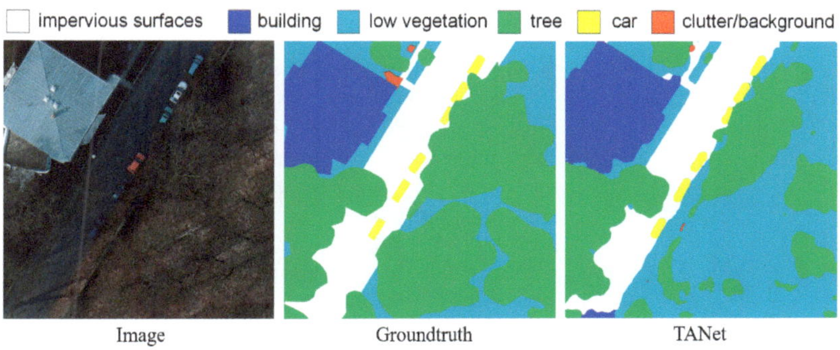

Figure 6. The features of the trees in the red rectangular area are similar to those of low vegetation, and TANet misjudged the area as low vegetation.

In the traditional encoder–decoder structure, one decoder generates probability maps of multiple output results by the Softmax function in Equation (7). For each pixel, the category with the highest probability is determined as the classification result for this point.

$$\begin{aligned} Softmax(Z_i) &= \tfrac{e^{z_i}}{\sum_{c=1}^{C} e^{z_c}} \\ c_i &= argmax(Softmax(Z_i)) \end{aligned} \tag{7}$$

where Z_i denotes the input vector to the softmax function, and C is the number of categories classified.

However, there is inter-class competition between different categories of each pixel. The sum of the probabilities of all categories is 1, and different categories share a decoder to restrict each other. If the probability of a particular point after decoding by the decoder is relatively uniform, it is easy to cause misjudgments. Three solutions are proposed to solve this problem.

The first and most straightforward idea is to change the loss function, using multiple binary-classification sigmoid cross-entropy loss instead of SoftMax cross-entropy loss. We call this model TANet with multiple losses. The loss function is shown in Equation (8).

$$\begin{aligned} p_i &= Sigmoid(Z_i) \\ Loss_{seg} &= \sum_i^n BCE(p_i, label) \\ Loss_{att} &= \sum_i^n BCE\left(att, label_{down\_sampling}\right) \\ Loss &= Loss_{seg} + Loss_{att} \end{aligned} \tag{8}$$

where p_i denotes the output of the inference model, and att denotes the attention map of LAM. BCE denotes the binary cross entropy loss.

The second method is Multi-model TANet. This method is to train a semantic segmentation network for each category and combine all the binary segmentation results. It is not easy to merge the results of multiple models. Here, we use the most intuitive method to combine the predicted probability maps of multiple binary segmentation models, and each pixel takes the category with the highest probability.

Another considerable solution is introducing multi-task learning, which converts a multi-category segmentation task into multiple binary-classification segmentation tasks. We call this method Multi-task TANet. The loss function is the same as TANet with multiple losses. All categories share an encoder, and each category owns a separate decoder. As shown in Figure 2, we use the TANet decoder mentioned above. For each pixel, we use the probability of output for each category as the confidence level and select the category with the highest confidence level as the classification result of the pixel.

For the above three methods, we have conducted experiments to find the best model.

3.4. Edge Optimization

In order to further improve the accuracy of the extracted edges, we made some targeted improvements to the model. Firstly, inspired by [39], we add a new branch of edge extraction. The purpose of this branch is to obtain edge maps of different categories. By computing loss between the edge branch extraction edge result and the edge truth map, the edge bifurcation of the extraction results can be improved.

However, the standard semantic segmentation loss function, like cross-entropy, is unsuitable for computing edge map loss. The extraction result of the edge map may have a slight misalignment with the actual value map, and the standard loss function will produce a significant deviation. Hence, a suitable loss function for the edge map must be selected. We changed the loss function from cross entropy to DT Loss [40].

DT Loss uses distance transform (DT), which transforms an edge map into a distance map. In our task, the DT Loss can be represented in Equation (9).

$$L_{DT} = \sum\nolimits_{(p_{jk} != l_{jk})} I_{DT}(j, k) \tag{9}$$

where I_{DT} denotes distance map, and j, k denotes the pixel coordinates. p_{jk} and l_{jk} represent the predicted edge map and the actual edge map.

Inspired by the distance transform method, we can transform the discrete edge map into a continuous distance map. Since the distance map already contains edge information, we can directly add a new branch to predict the distance map, and the loss function can be the common MSE.

Furthermore, an edge attention module (EAM) is added to each decoder. As shown in Figure 7, the overall architecture of EAM is consistent with of LAM, the difference is that the attention probability map used in EAM is the edge map of labels.

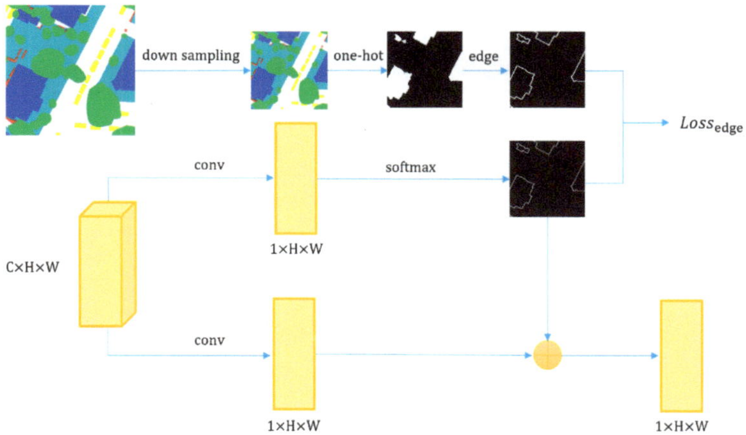

Figure 7. Architecture of Edge Attention Module.

The loss function $Loss_{edge}$ is just the edge map attention part. Structure and loss of EAM is shown as follows.

$$att_{edge} = Softmax(Conv(Input_{C\times H\times W})_{1\times H\times W}) \tag{10}$$

$$F_{out} = Conv(Conv\Big(Input_{(C\times H\times W)})_{(1\times H\times W)} + Att\Big)_{(1\times H\times W)} \tag{11}$$

$$Loss_{edge} = CE\Big(att_{edge}, edgemap_{down\_sampling}\Big) \tag{12}$$

where Att denotes the attention probability map, and F_{out} denotes the output features of the attention module. EAM's output is the weighted summation of the attention and value parts.

Using the boundaries of different feature categories in the labels, an edge map is drawn as the guiding information of the edge attention probability map, and the rest of the model is consistent with all the label attention modules. Since the model contains multiple decoders, each corresponding to a feature class, the edge attention module in each decoder has the same basic structure.

By combining these three improvements, the edges of the extraction results can be further improved.

3.5. Descriptions of Datasets

In order to demonstrate the effectiveness of the model, several remote sensing image semantic segmentation datasets are used in this paper. On the one hand, we use two publicly available ISPRS remote sensing image semantic segmentation datasets. On the other hand, a self-made CZ-WZ multi-category semantic segmentation dataset is used. The differences in resolution and image sensors between the two datasets are significant, so experiments on the three datasets separately can verify the model structure's effectiveness in different data styles.

Firstly, the Potsdam 2D semantic labeling dataset [41] contains 38 patches, each consisting of a true orthophoto (TOP) extracted from a larger TOP mosaic. The label contains six categories: impervious surfaces, buildings, low vegetation, trees, cars, and clutter/background. Four spectral bands exist in each TOP image (red, green, blue, and near-infrared), and we only use the RGB channels in this work.

The second dataset is the Vaihingen dataset [42], which contains 33 TOP patches of different sizes. The ground sampling distance of the TOP is 9 cm. The reference data are divided into the same six categories as the Potsdam dataset. Each TOP image has three spectral bands (red, green, and near-infrared).

The third dataset is the CZ-WZ dataset. The original images come from the Changzhou area in Sichuan Province and the Wuzhen area in Zhejiang Province, China. The spatial resolution of the experimental remote sensing images is 0.51 m. Each TOP image has three spectral bands (red, green, and blue). The label is self-made and classifies the remote sensing image features into five categories: buildings, roads, vegetation, water bodies, and backgrounds.

Each training image is cut into 1024 × 1024 patches. After cutting, the Potsdam dataset contains 1176 training samples and 504 test samples; the Vaihingen dataset contains 221 training samples and 63 test samples; the CZ-WZ dataset contains 1080 training samples and 260 test samples. The training of convolution network models usually requires a large number of samples. Based on the existing dataset, this work increases the network training samples through data enhancement. The original and label images are flipped horizontally or vertically, cut at random positions, and randomly transformed HSV. Data enhancement changes the number of training samples to five times the original.

3.6. Evaluation Metrics

Following the evaluation method used in the literature [22], we evaluate the performance of methods by three metrics: overall accuracy (OA), per-class $F1$ score, and average $F1$ score. OA is the ratio of the number of correct pixels to the total number of pixels. $F1$ score for classification is calculated as the harmonic mean of precision and recall [22].

We calculate the $F1$ score for each foreground category to assess the proposed network's performance. We also calculate the OA for the whole dataset. The calculation formula is as follows.

$$recall = \frac{TP}{TP + FN}, precision = \frac{TP}{TP + FP} \tag{13}$$

$$F1 = \frac{2 \times precision \times recall}{precision + recall}, mean\_F1 = \sum_{i}^{N-1} F1_i / N \tag{14}$$

$$OA = \frac{TP + TN}{TP + TN + FP + FN} \tag{15}$$

with the following terms: True Positive example (TP), False Positive Example (FP), True Negative example (TN), False Negative example (FN).

4. Results and Discussion

In this section, we validate the effectiveness of the proposed attention modules and the multi-task framework. Firstly, we use UNet as our baseline and then utilize an ablation study to show the tests of the proposed triple attention modules. Then, we experimented with the three methods above to find the best model.

4.1. Ablation Study of Triple Attention Modules

The proposed TANet contains PAM, CAM, and LAM, three attention modules. PAM and CAM are self-attention modules proposed by DANet, and LAM is proposed by our paper which is a label attention module. In order to verify the effectiveness of the three attention modules, we replaced UNet's decoder with PAM + CAM, LAM, and TANet (PAM + CAM + LAM).

4.1.1. Experiments Results on Potsdam Datasets

The attention modules focus on targets with a large area and a wide distribution range, such as buildings, low vegetation, and impervious surfaces. As shown in Table 1, after replacing the decoder with PAM and CAM, the F1 scores of impervious surfaces increases by 0.85%, and building F1 score increases by 2.13%. As for LAM, the F1 score for building increases by 1.24% and that for tree by 0.51%. This is because TANet combines the advantages of the previous two models and thus achieves the best classification results. TANet increases F1 of impervious surfaces, buildings, low vegetation, and trees compared with UNet by 1.80%, 2.48%, 0.85%, and 0.72%, respectively. An interesting result is that UNet obtains the highest score for the segmentation result of the car. Since the attention module in this article pays more attention to large-scale and complex targets, there is no noticeable improvement for small targets such as cars.

Table 1. F1 score, Mean F1, and OA index results of multi-object on Potsdam dataset.

Method	Per-Class F1 Score (%)					Mean F1 (%)	OA (%)
	Imp. Surf.	Building	Low Veg.	Tree	Car		
UNet	87.91	91.31	81.76	82.72	**88.91**	86.52	85.48
PAM + CAM	88.76	93.44	82.09	82.21	88.21	86.94	86.41
LAM	88.34	92.45	81.74	83.23	88.01	86.75	86.06
TANet	**89.71**	**93.79**	**82.61**	**83.44**	88.78	**87.67**	**87.23**

4.1.2. Experiment Results on Vaihingen Datasets

We perform experiments on the ISPRS Vaihingen benchmark to assess TANet's performance further. We used the same training and testing setup in the experiments on the Vaihingen dataset. As shown in Table 2, TANet increases the F1 score of impervious surfaces, buildings, low vegetation, trees, and cars compared with UNet by 3.43%, 3.77%, 8.63%, 1.45%, and 9.15%. In this experiment, the performance of using the LAM decoder alone is closer to TANet, which means LAM played a more critical role in this model.

Table 2. F1 score, Mean F1, and OA index results of multi-object on Vaihingen dataset.

Method	Per-Class F1 Score (%)					Mean F1 (%)	OA (%)
	Imp. Surf.	Building	Low Veg.	Tree	Car		
UNet	84.45	87.32	69.77	83.16	63.12	77.56	81.27
PAM + CAM	86.13	88.88	74.43	82.54	71.40	80.68	82.83
LAM	87.85	91.01	78.25	**84.73**	70.85	82.51	85.06
TANet	**87.88**	**91.09**	**78.40**	84.61	**72.27**	**82.85**	**85.26**

4.1.3. Experiments Results on CZ-WZ Datasets

We conducted experiments on CZ-WZ datasets to further evaluate the effectiveness of TANet. As shown in Table 3, in terms of individual category accuracy, it is evident that the improvement is excellent for roads and water bodies. Using UNet as a benchmark, the F1 scores for roads are improved by 9.16%, 12.16%, and 13.41% for PAM + CAM, LAM, and TANet, respectively. This is because UNet cannot effectively learn the global features for this type of distribution with a large range and infrequent sample occurrence, which results in lower accuracy. Moreover, after the attention mechanism is introduced, the extraction accuracy of roads is significantly improved, which shows that the global features are very effective for road extraction. Like roads, water bodies also belong to the category with a larger distribution range and lower frequency of occurrence. They are improved by 6.20%, 6.26%, and 6.59% relative to UNet, PAM + CAM, LAM, and TANet, respectively. Vegetation was also entered as a feature type with a large distribution range. However, vegetation occurred in the sample at a high frequency, so the extraction accuracy of each scheme exceeded 90%. For the vegetation category, PAM + CAM has more improvement than LAM, while LAM has more improvement on roads and water bodies. The two schemes have some overlapping parts for the overall accuracy improvement. However, there are still differences in different categories, combining the advantages of TANet and fusing the two to obtain optimal accuracy.

Table 3. F1 score, Mean F1, and OA index results of multi-object on CZ-WZ dataset.

Method	Per-Class F1 Score (%)				Mean F1 (%)	OA (%)
	Road	Building	Veg.	Water		
UNet	60.75	74.13	90.26	77.32	75.62	81.41
PAM + CAM	69.91	77.22	91.05	83.52	80.43	83.46
LAM	72.91	**77.82**	90.87	83.58	81.30	83.35
TANet	**74.16**	77.31	**91.20**	**83.91**	**81.65**	**83.96**

We compare the results before using the proposed module and after in Figures 8 and 9. It can be clearly observed that the ability to recognize a wide range of object types is enhanced after combining the attention module. For example, in the second row in Figure 8, part of the impervious surface area is covered by shadows. TANet recognizes the impervious surfaces correctly after the global features introduced by the attention module's introduction.

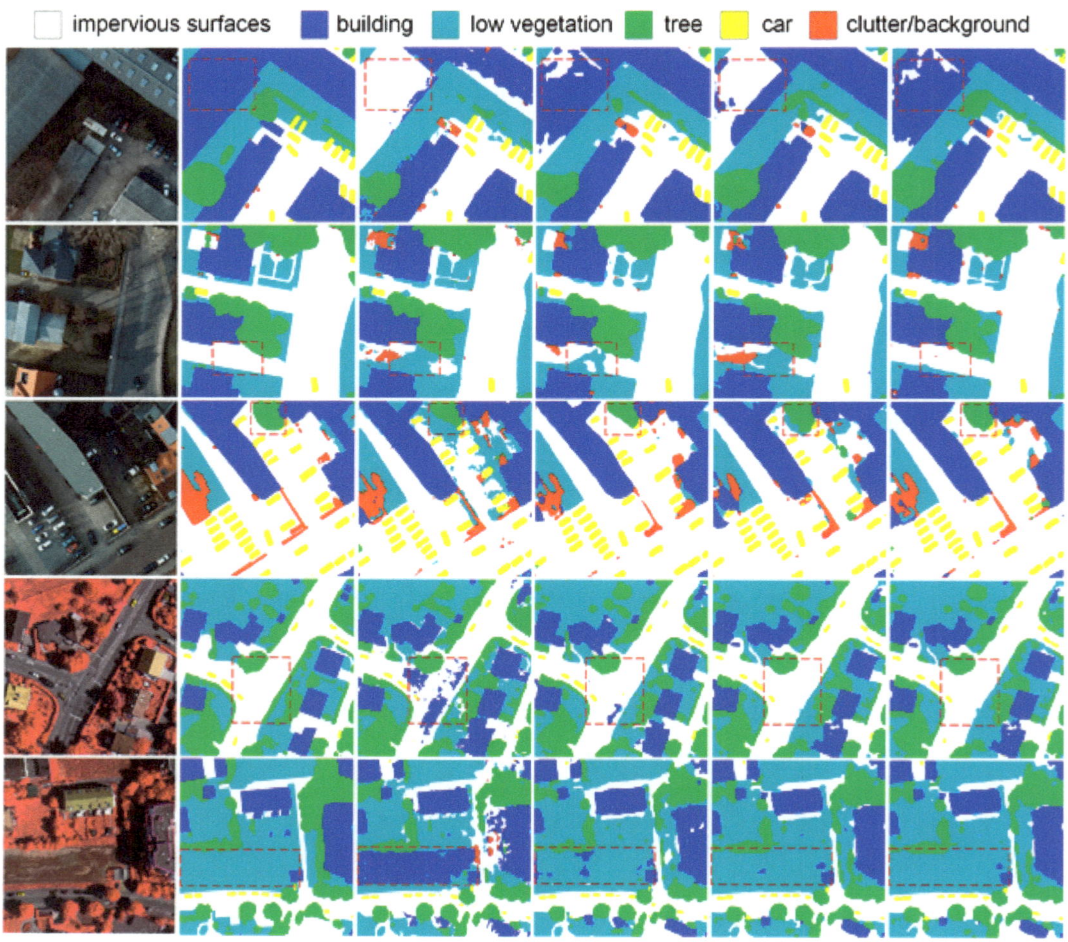

Image Label UNet PAM + CAM LAM TANet

Figure 8. Results on the ablation study of triple attention modules. The first three rows are from the Potsdam dataset and the other is from the Vaihingen dataset.

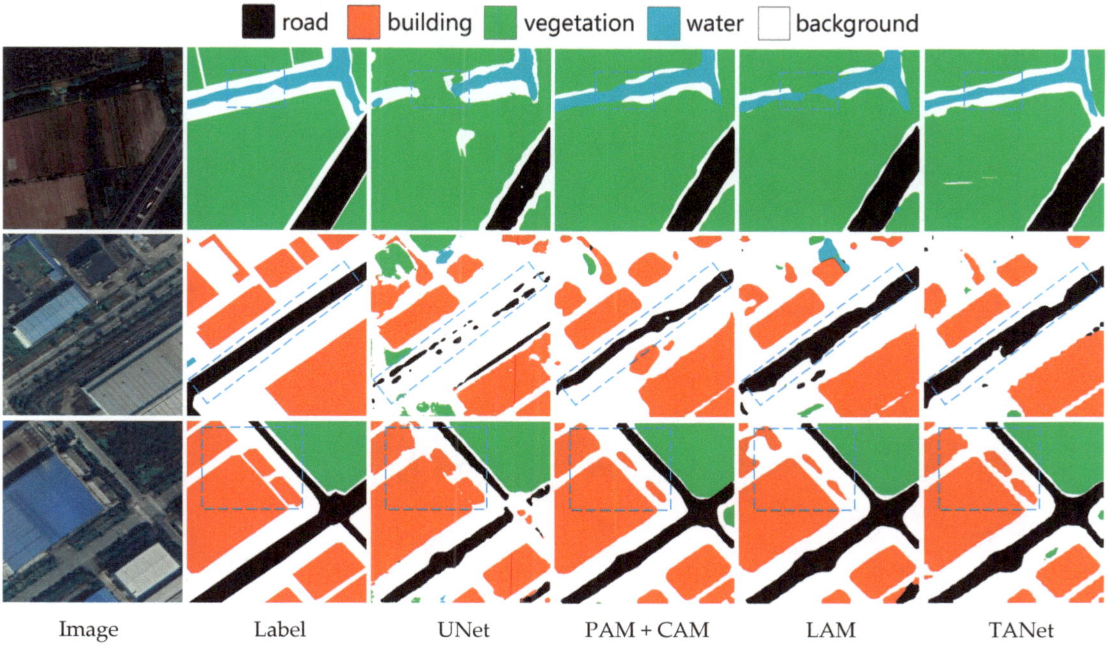

Figure 9. Examples of semantic segmentation results on ablation study of triple attention modules (CZ-WZ dataset).

4.2. Visualization of LAM

We visualize attention maps in LAM to better understand our attention modules. The number of channels of attention maps generated by LAM is the same as the number of segmentation categories, and we overlay the results of each category of the attention map on the original image. Figure 10 shows the attention map results of impervious surfaces, buildings, and cars. The feature response extracted by the attention modules is similar to the segmentation result. Moreover, attention maps of LAM have made some corrections to PAM and CAM extraction results. In the red rectangle, the segmentation results of PAM and CAM are wrong, and with the help of the LAM's attention map, TANet obtains the correct recognition result.

4.3. Ablation Study of Multi-Task Learning

After the introduction of the attention module, the segmentation accuracy of a wide range of targets was improved. Observing the extraction results of the TANet model, it is found that there are still some misjudgments of similar features. We conducted several experiments to solve this problem according to the ideas proposed in Section 3.3.

(1) TANet with multiple losses

We use binary cross-entropy loss for each category instead of softmax cross-entropy loss.

(2) Multi-model TANet

For each category, we trained an individual network. For each input image, we fused the output probability maps of each category to obtain the final output result.

(3) Multi-task TANet

We convert a multi-category segmentation task into multiple binary-classification segmentation tasks. All tasks share one encoder, and each task owns an individual decoder.

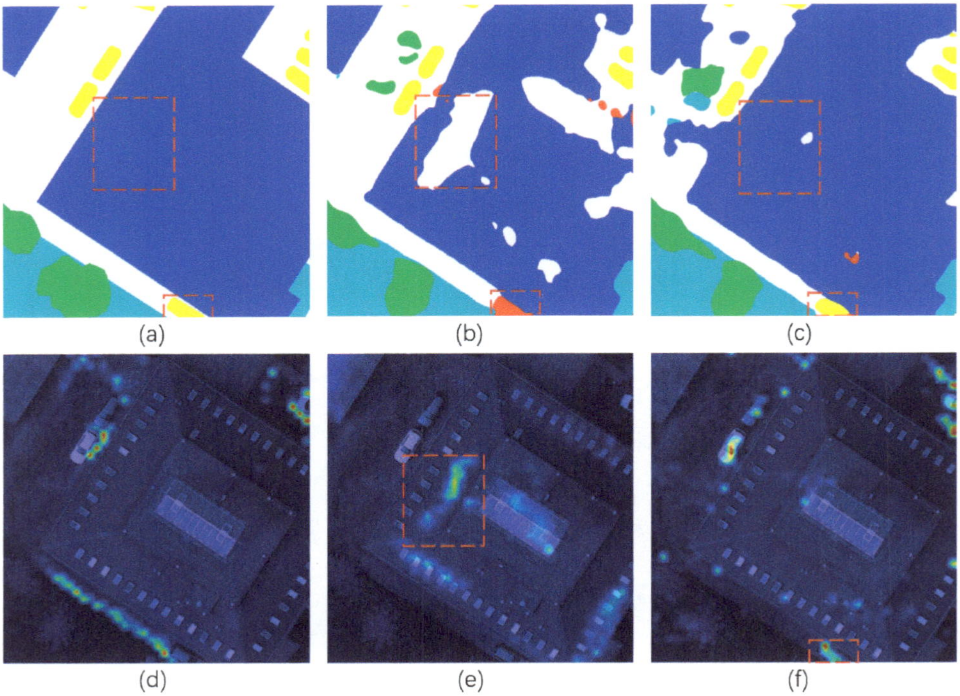

Figure 10. Visualization of LAM attention map. (**a**) label, (**b**) PAM + LAM result, (**c**) TANet result, (**d**) imperious surface attention map, (**e**) building attention map, (**f**) car attention map.

4.3.1. Experiment Results on Potsdam Datasets

As shown in Table 4, Multi-task TANet improves the performance remarkably. Compared with the baseline UNet, they employ a multi-task decoder yielding 88.35% in OA, which brings a 2.87% improvement. The results show that the accuracy can only be improved slightly if the loss function is replaced without increasing the number of decoders. Since all categories share one decoder, and the characteristics of different categories are quite different, it is difficult for one decoder to summarize the global characteristics of all categories. Another finding is that the fusion results of multiple binary segmentation models are poor. Since the models are trained separately, merging the results of multiple models is not an easy task.

Table 4. F1 score, Mean F1, and OA index results of different models on Potsdam dataset.

Method	Per-Class F1 Score (%)					Mean F1 (%)	OA (%)
	Imp. Surf.	Building	Low Veg.	Tree	Car		
UNet	87.91	91.31	81.76	82.72	88.91	86.52	85.48
TANet	89.71	93.79	82.61	83.44	88.78	87.67	87.23
TANet with multiple loss	89.52	94.00	82.91	83.82	88.49	87.72	87.35
Multi-model TANet	86.09	89.16	81.33	81.59	86.81	82.33	83.75
Multi-task TANet	**90.54**	**94.60**	**83.90**	**84.87**	**89.98**	**88.54**	**88.35**

4.3.2. Experiment Results on Vaihingen Dataset

Table 5 reports the quantitative results of the Vaihingen datasets. Compared with the baseline UNet, the methods that combine multi-task ideas achieved higher accuracy. Similar to the performance of the Potsdam dataset, changing the loss function does not significantly improve the model. However, the multi-model fusion performed better on the Vaihingen dataset, especially the tree and car categories, which achieved the highest F1 scores. The performance of multi-model TANet is not stable in different datasets. In terms of large-area features, Multi-task TANet achieved better results. Similar categories are easier to distinguish under the action of multiple decoders.

Table 5. F1 score, Mean F1, and OA index results of different models (Vaihingen dataset).

Method	Per-Class F1 Score (%)					Mean F1 (%)	OA (%)
	Imp. Surf.	Building	Low Veg.	Tree	Car		
UNet	84.45	87.32	69.77	83.16	63.12	77.56	81.27
TANet	87.88	91.09	78.40	84.61	72.27	82.85	85.26
TANet with multiple loss	88.95	92.11	79.49	84.89	71.02	83.29	85.35
Multi-model TANet	88.37	90.77	79.43	**85.65**	**74.83**	83.81	85.67
Multi-task TANet	**89.55**	**92.41**	**80.44**	85.60	74.00	**84.40**	**86.89**

4.3.3. Experiment Results on CZ-WZ Dataset

As shown in Table 6, multi-task TANet significantly improves the performance of image semantic segmentation. Compared to UNet, TANet with multiple losses, Multi-model TANet, and Multi-task TANet improved 5.27%, 5.00%, and 8.04% in F1 score, and 2.14%, 1.88%, and 3.39% in OA score, respectively. Compared to TANet, the extraction accuracy of Multi-loss TANet and Multi-model TANet decreased, and Multi-task TANet achieved the highest accuracy. Different multitasking strategies have different effects on the model. TANet with multiple loss only uses the multitasking loss function without changing the model structure. The model still has parameter competition, and it is still difficult to avoid the problem of misclassification of similar categories. The multi-model strategy used by Multi-model TANet has higher extraction accuracy in a single model. However, it is more difficult to fuse the multi-category results. The method of directly taking the most significant term of probability value does not completely merge the multi-category results, which is caused by the differences in the predicted probability values of different models, so the multitask approach of the multi-model strategy has some instability.

Table 6. F1 score, Mean F1, and OA index results of different models (CZ-WZ dataset).

Method	Per-Class F1 Score (%)				Mean F1 (%)	OA (%)
	Road	Building	Veg.	Water		
UNet	60.75	74.13	90.26	77.32	75.62	81.41
TANet	74.16	77.31	91.20	83.91	81.65	83.96
TANet with multiple loss	73.06	76.64	91.06	82.78	80.89	83.55
Multi-model TANet	72.86	76.44	90.90	82.28	80.62	83.29
Multi-task TANet	**77.02**	**79.44**	**91.54**	**86.67**	**83.66**	**84.80**

Figures 11 and 12 compare the segmented results on different approaches to multitasking ideas. The multi-task mechanism focuses on solving the problem of misidentification of similar features. Multi-task TANet with multiple decoders achieved the best results. Since each category has a dedicated decoder, each decoder is more focused, thereby improving the ability of category discrimination.

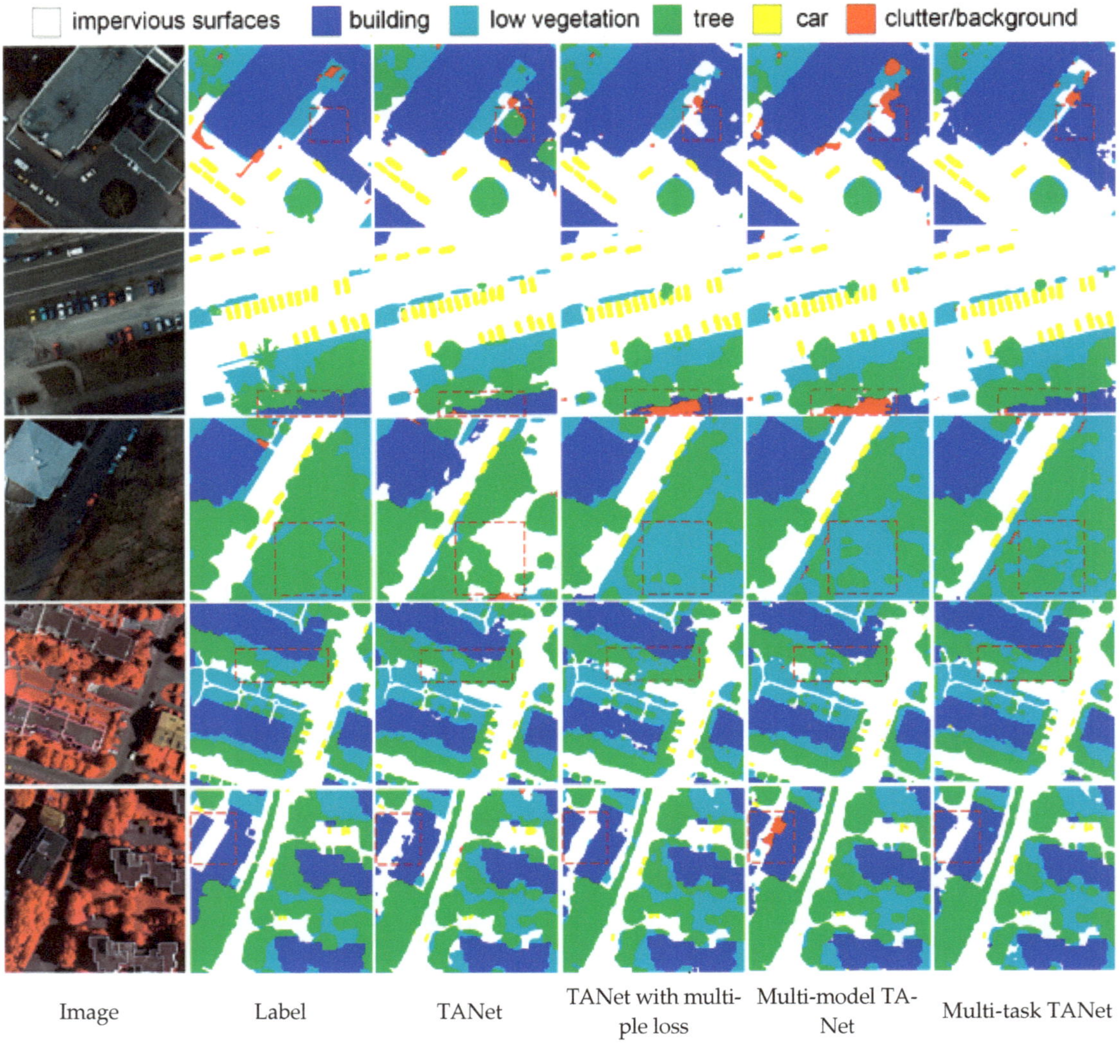

Figure 11. Examples of semantic segmentation results on ablation study of multi-task learning. The first three rows are from the Potsdam dataset and the others are from the Vaihingen dataset.

4.4. Edge Optimization Results and Discussion

The optimization includes the edge map branch (EB), the distance map branch (DB), and EAM. Firstly, to prove the validity of the DT Loss, we introduce the edge map branch and compare the results using DT Loss and cross-entropy (CE). In Table 7, EB_DT denotes the edge map branch with DT Loss, and EB_CE denotes the edge map branch with CE Loss. According to Table 7, the results of different modules of edge optimization on three datasets can be obtained, and the edge optimization of EB_DT and DB is the best result in MQANet.

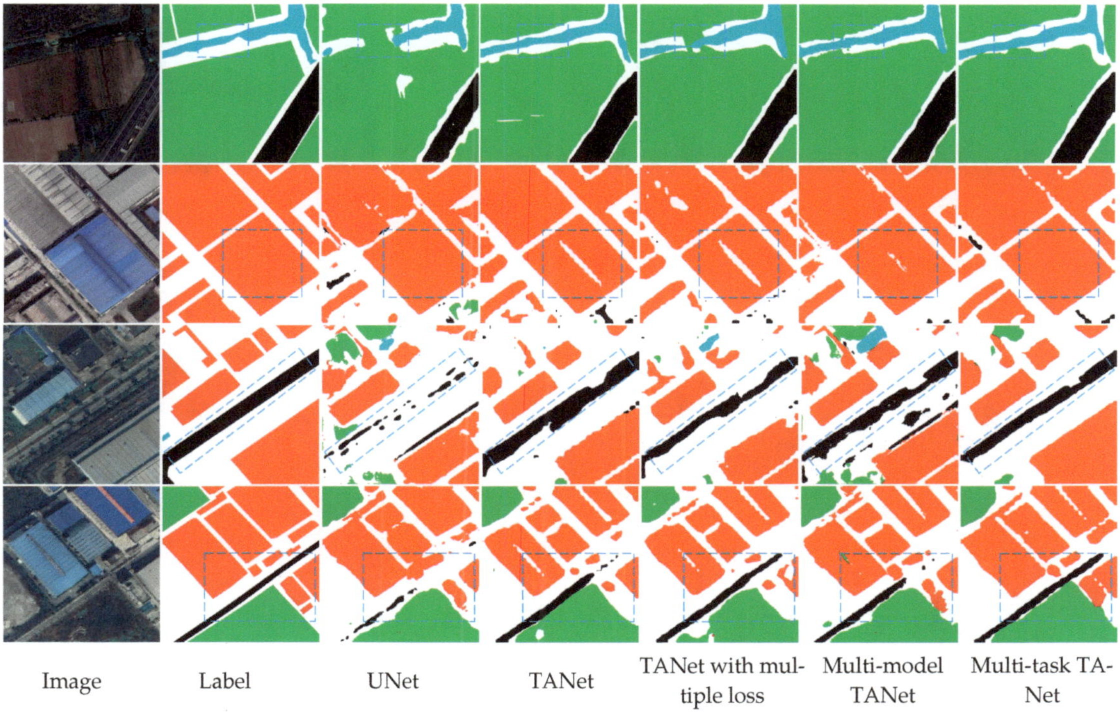

Figure 12. Examples of semantic segmentation results on ablation study of multi-task learning (CZ-WZ dataset).

Table 7. Mean F1 and OA index results on edge optimization.

Method	EB_DT	EB_CE	DB	Potsdam Dataset		Vaihingen Dataset		CZ-WZ Dataset	
				Mean F1 (%)	OA (%)	Mean F1 (%)	OA (%)	Mean F1 (%)	OA (%)
MTANet	×	×	×	88.54	88.35	84.40	86.89	83.66	84.80
MTANet	✓	×	×	89.16	88.77	84.36	86.93	83.94	85.10
MTANet	×	✓	×	89.02	88.50	84.14	86.32	82.48	84.28
MQANet	✓	×	×	89.29	88.96	84.48	87.32	84.10	85.15
MQANet	✓	×	✓	**89.35**	**89.05**	**84.61**	**87.60**	**84.27**	**85.29**

Then we evaluate the effects of the proposed distance branch and EAM. In Table 7, DB denotes the distance branch. MQANet means the previous MTANet combined with edge attention to form a Multi-task Quadruple Attention Network. Table 7 shows that all three modules have improved the segmentation results. In Table 7, MTANet is Multi-mask TANet, and MQANet is Multi-mask QANet.

The two datasets from Potsdam and Vaihingen were applied to tests for Unet, TANet, and Multi-task QANet. The results are shown in Figures 13 and 14. Compared with the three test datasets, the segmentation results of Multi-task QANet present a better effect than Multi-task TANet. The modules, the first line of Figure 13, can clearly show the performance, especially for the edge contour of the target.

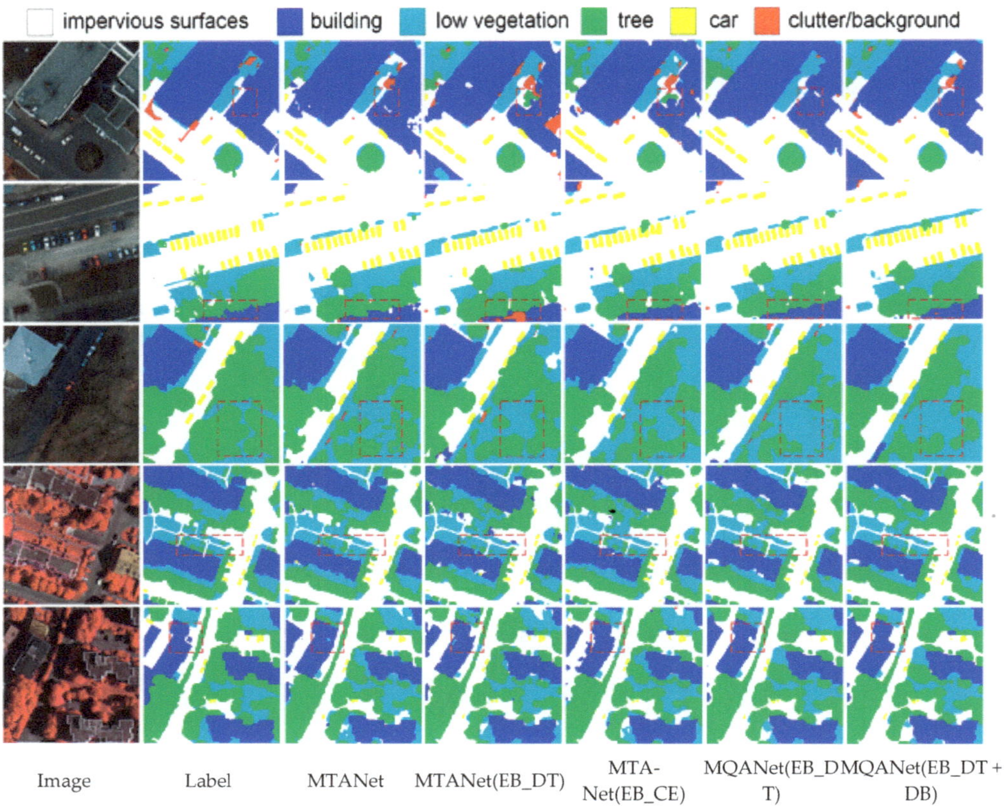

Figure 13. Results on ablation study of edge optimization. The first three rows are from the Potsdam dataset and the others are from the Vaihingen dataset.

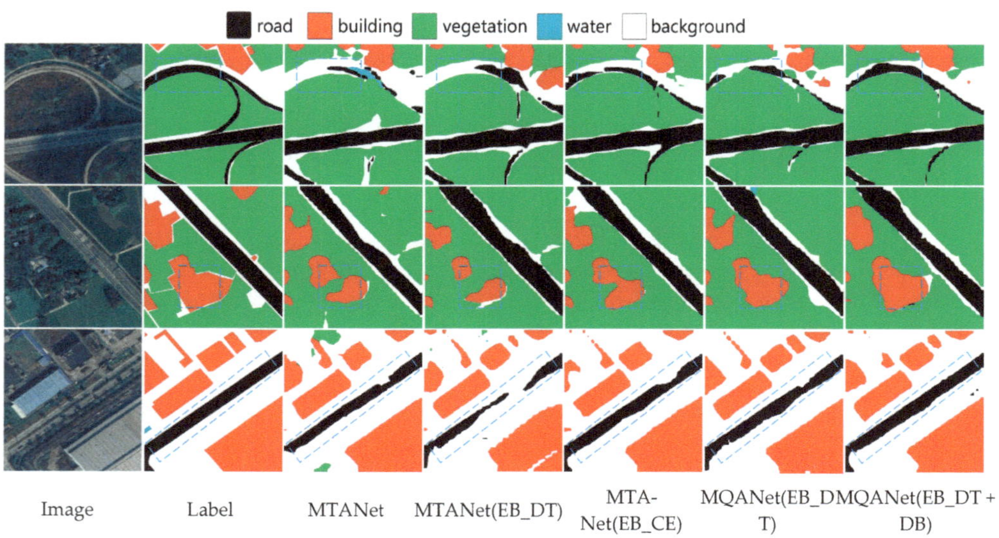

Figure 14. Examples of semantic segmentation results on ablation study of edge optimization (CZ-WZ dataset).

Table 7 shows that after optimization of each module of edge extraction, the overall accuracy of the model is improved to a certain extent, and compared with MTANet (EB_CE), MQANet (EB_DT + DB) shows an improvement of 0.61% and 0.49% in F1 score and OA, respectively. The slight improvement since the edge part accounts for a smaller proportion of the total image area. The optimized edge extraction accuracy contributes less to the overall accuracy improvement. From the results of MTANet (EB_DT) and MQANet (EB_DT), it can be seen that different edge extraction loss functions significantly impact the overall accuracy, and using the cross-entropy loss function to extract edges even reduces the original model accuracy. Moreover, the distance transform loss function can be better applied to the edge extraction task. The results of MQANet (EB_DT + DB) show that the newly introduced distance transform branch and edge attention module can both help to improve the accuracy.

4.5. Discussion of Overall Experimental Results

Table 8 is presented to analyze the results of our methods based on the two public datasets. The Mean F1 and OA of the surface objects have been promoted for the two datasets. For the Potsdam dataset, the 2.83% Mean F1 and 3.57% OA of MQANet are higher than those of Unet. For the Vaihingen dataset, the 7.05% Mean F1 and 6.33% OA of MQANet are higher than Unet.

Table 8. Quantitative results of our methods.

Method	Potsdam Dataset		Vaihingen Dataset		Execution Time/Per Image (ms)
	Mean F1 (%)	OA (%)	Mean F1 (%)	OA (%)	
UNet (baseline)	86.52	85.48	77.56	81.27	125.4
TANet (ours)	87.67	87.23	82.85	85.26	49.8
MTANet (ours)	88.54	88.35	84.40	86.89	55.0
MQANet (ours)	89.35	89.05	84.61	87.60	58.8

In addition, we tested the execution time of our proposed networks and Unet based on the method provided by Dong et al. [43], and the test results is the right column in Table 8. The test method sets the batch size to 1 and lets the network predict 200 images, the final execution time is the average of the total running time. That is the execution time of the network for a single image. From Table 8, it can be seen that the execution time of each image is reduced 66.6 ms compared to UNet, and the execution time of UNet is more than doubled compared to MQANet. Because our proposed multi-tasking mechanism splits a multi-category segmentation task into several binary-classification segmentation tasks, it can effectively reduce execution time and achieve better performance than the UNet baseline.

The experiment's results demonstrate that MQANet obtains optimal accuracy on two public datasets of ISPRS. Our module introduces two attention mechanisms: self-attention and label-attention. Thus, we achieve an enhanced overall accuracy compared to the standard self-attention model. In addition, this paper uses a multi-decoder model to reduce the parameter competition among different classes and performs additional optimization for the edge regions, thus achieving optimal accuracy.

To further describe the model accuracy distribution of different datasets, we count the sample number of each model in different accuracy intervals. We use overall accuracy (OA) as the accuracy evaluation index to draw histograms, as shown in Figure 15.

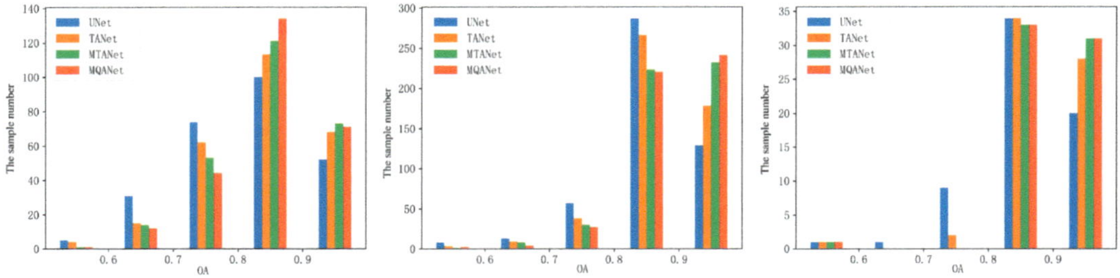

(a) (b) (c)

Figure 15. OA distribution of different models results in three datasets. (**a**) OA distribution map on CZ-WZ data set, (**b**) OA distribution map on Potsdam data set, (**c**) OA distribution map on Vaihingen data set.

As shown in Figure 15, in the Chongzhou–Wuzhen dataset (CZ-WZ dataset), when OA is below 0.8, the number of samples in UNet is significantly higher than that of other models. TANet, MTANet, and MQANet show a decreasing trend. In contrast, when the accuracy is above 0.8, MQANet has the largest number of samples. The above results show that MQANet achieved optimal accuracy in most samples of the CZ-WZ dataset. In the Potsdam and Vaihingen datasets, the regularities are similar to those in the CZ-WZ dataset. When the accuracy is above 0.9, the number of samples of UNet is significantly lower than that of other models, and MQANet has the largest number of samples. This indicates that MQANet also achieves the optimal accuracy in most samples of Potsdam and Vaihingen datasets.

To obtain a more detailed sample distribution, we plotted the extraction accuracy of each sample as a broken line diagram in the ISPRS-Vaihingen test dataset. As shown in Figure 16, the extraction results of MQANet are the highest in four models. TANet, MTANet, and MQANet show an upward trend in the Vaihingen test sample, which further indicates that the proposed model has a high OA index in the Vaihingen test dataset.

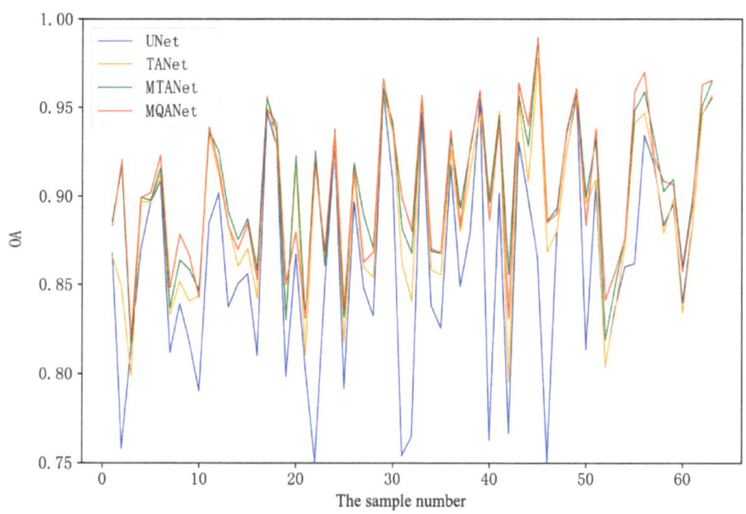

Figure 16. OA Statistical distribution map of model results in Vaihingen dataset.

In addition, we compared the current work on two publicly available datasets to verify the effectiveness of our optimal model, MQANet, still using two evaluation metrics, the average F1 and OA. Among the existing works compared with this paper, the work CBAMNet [21] containing the self-attentive mechanism, SENet [44], and the latest network

Deeplabv3+ [11] with the spatial feature pyramid structure are included, in addition to the latest CRMS [45] network with optimal feature extraction using the multiscale residual module. The experimental results are shown in Table 9. As can be seen from Table 9, the results of our proposed model MQANet show better results on both datasets. On the Vaihingen dataset, our MQANet network has higher Mean F1 and OA than other networks, and on the Potsdam dataset, our MQANet network has higher Mean F1 than other networks, and only OA is 0.08% lower than DSPCANet [15]. Mean F1 score for classification is calculated as the harmonic mean of precision and recall [22], and OA is the ratio of the number of correct pixels to the total number of pixels. This shows that our MQANet has certain advantages in the equalization of various objects identification.

Table 9. Quantitative results of the current work.

Method	Potsdam Dataset		Vaihingen Dataset	
	Mean F1 (%)	**OA (%)**	**Mean F1 (%)**	**OA (%)**
CBAMNet [21]	86.04	85.14	83.77	86.47
Deeplabv3+ [11]	88.01	87.06	83.77	85.71
SENet [44]	87.97	87.63	82.85	85.26
CRMS [45]	89.02	88.92	83.25	86.40
EFCNet [14]	80.17	81.77	81.87	85.46
DSPCANet [15]	87.19	**90.13**	84.46	87.32
MQANet (ours)	**89.35**	89.05	**84.61**	**87.60**

5. Conclusions

A Multi-task Quadruple Attention Network (MQANet) is proposed to improve the accuracy of multi-object semantic segmentation of remote sensing images. We introduce the attention mechanism to obtain more global features and improve the accuracy of the large object area. Furthermore, two self-attention modules are introduced, which are named PAM + CAM, and the OA and Mean F1 are improved. Then, we build a label attention module (LAM) and combine all three attention modules into a triple attention network (TANet). Meanwhile, we proposed three alternative methods to improve the ability to identify similar objects: Multi-task TANet (MTANet). Experiment results show that the multi-task learning method obtains the highest accuracy. Finally, some edge optimizations are made to improve the accuracy of the edge area further, and we combine Multi-task TANet and edge optimizations as the Multi-task QANet (MQANet).

Three datasets were used to verify the accuracy of the proposed model. Compared with the baseline UNet in the Vaihingen dataset, MQANet improved the OA and Mean F1 by 6.33% and 7.05%, respectively. MTANet improved the OA and Mean F1 by 5.48% and 6.84%, respectively. Compared with the baseline UNet in the Potsdam dataset, MQANet improved the OA and Mean F1 by 3.57% and 2.83%, respectively. MTANet improved the OA and Mean F1 by 2.87% and 2.02%, respectively. Compared with the baseline UNet in the CZ-WZ dataset, MQANet improved the OA and Mean F1 by 3.88% and 8.65%, respectively. MTANet improved the OA and Mean F1 by 3.39% and 8.04%, respectively. Through extensive experiments, the proposed MQANet outperforms other methods by a large margin on Vaihingen, Potsdam and self-annotated datasets (CZ-WZ dataset). The results demonstrate that the proposed model (MQANet) has a large accuracy improvement in both F1 and OA indices, and the quadruple attention modules are helpful for large object semantic segmentation of RS images.

The proposed multi-tasking mechanism splits a multi-category segmentation task into several binary-classification segmentation tasks, each of which requires a separate decoder. The types of multi-object semantic segmentation involve 5 or 6 categories in our study, which can achieve better results. However, if the objects are subdivided into dozens or even more categories, the model needs to construct a decoder for each category. A large number of decoders may cause the size expansion of the model, and the applicability of the model may be decreased. In the future, more types of multi-object will be tested to optimize a more robust multi-task semantic segmentation.

Author Contributions: Conceptualization, Y.L. and Y.S.; methodology, Y.L. and L.H.; software, Y.L.; validation, Y.S., Z.T. and J.Z.; formal analysis, Y.S.; investigation, Z.T.; resources, L.H.; data curation, Y.G.; writing—original draft preparation, Y.L.; writing—review and editing, L.H. and S.L.; visualization, Y.S.; supervision, Y.L. and Z.T.; project administration, L.H.; funding acquisition, L.H. All authors have read and agreed to the published version of the manuscript.

Funding: This research was funded by the Key Projects of Global Change and Response of Ministry of Science and Technology of China under Grant 2020YFA0608203, and in part by the fundamental Research Funds for the Central Universities, UESTC (ZYGX2019J064), in part by Major Science and Technology Projects of Sichuan Province under Grant No.2022ZDZX0001, in part by the Science and Technology Support Project of Sichuan Province under Grant 2021YFS0335, in part by China Meteorological Administration Project under Grant FY-APP-2021.0304 and CXFZ2022J031.

Data Availability Statement: The authors would like to thank the team of Potsdam 2-D semantic labeling data and the Vaihingen dataset for the data and experiments.

Acknowledgments: The authors appreciate the reviewers for their constructive comments and kind help to improve the manuscript.

Conflicts of Interest: The authors declare no conflict of interest.

References

1. Ye, Q.; Zhao, H.; Li, Z.; Yang, X.; Gao, S.; Yin, T.; Ye, N. L1-Norm distance minimization-based fast robust twin support vector $ k $-plane clustering. *IEEE Trans. Neural Netw. Learn. Syst.* **2017**, *29*, 4494–4503. [CrossRef]
2. Sun, L.; Ma, C.; Chen, Y.; Shim, H.J.; Wu, Z.; Jeon, B. Adjacent superpixel-based multiscale spatial-spectral kernel for hyperspectral classification. *IEEE J. Sel. Top. Appl. Earth Obs. Remote Sens.* **2019**, *12*, 1905–1919. [CrossRef]
3. Duan, Y.; Huang, H.; Wang, T. Semisupervised feature extraction of hyperspectral image using nonlinear geodesic sparse hypergraphs. *IEEE Trans. Geosci. Remote Sens.* **2021**, *60*, 1–15. [CrossRef]
4. Long, J.; Shelhamer, E.; Darrell, T. Fully convolutional networks for semantic segmentation. In Proceedings of the IEEE Conference on Computer Vision and Pattern Recognition, Boston, MA, USA, 7–12 June 2015; pp. 3431–3440.
5. Gualtieri, J.A.; Cromp, R.F. Support vector machines for hyperspectral remote sensing classification. In *27th AIPR Workshop: Advances in Computer-Assisted Recognition*; SPIE: Bellingham, WA, USA, 1999; pp. 221–232.
6. Ronneberger, O.; Fischer, P.; Brox, T. U-net: Convolutional networks for biomedical image segmentation. In *International Conference on Medical Image Computing and Computer-Assisted Intervention*; Springer: Berlin/Heidelberg, Germany, 2015; pp. 234–241.
7. Badrinarayanan, V.; Kendall, A.; Cipolla, R. Segnet: A deep convolutional encoder-decoder architecture for image segmentation. *IEEE Trans. Pattern Anal. Mach. Intell.* **2017**, *39*, 2481–2495. [CrossRef] [PubMed]
8. Chen, L.-C.; Papandreou, G.; Kokkinos, I.; Murphy, K.; Yuille, A.L. Semantic image segmentation with deep convolutional nets and fully connected crfs. In Proceedings of the International Conference on Learning Representations, Banff, AB, Canada, 14–16 April 2014.
9. Chen, L.-C.; Papandreou, G.; Kokkinos, I.; Murphy, K.; Yuille, A.L. Deeplab: Semantic image segmentation with deep convolutional nets, atrous convolution, and fully connected crfs. *IEEE Trans. Pattern Anal. Mach. Intell.* **2017**, *40*, 834–848. [CrossRef]
10. Chen, L.C.; Papandreou, G.; Kokkinos, I.; Murphy, K.; Yuille, A.L. Rethinking atrous convolution for semantic image segmentation liang-chieh. *IEEE Trans. Pattern Anal. Mach. Intell.* **2018**, *5*.
11. Chen, L.-C.; Zhu, Y.; Papandreou, G.; Schroff, F.; Adam, H. Encoder-decoder with atrous separable convolution for semantic image segmentation. In Proceedings of the European Conference on Computer Vision (ECCV), Munich, Germany, 8–14 September 2018; pp. 801–818.
12. Chollet, F. Xception: Deep learning with depthwise separable convolutions. In Proceedings of the IEEE Conference on Computer Vision and Pattern Recognition (CVPR), Honolulu, HI, USA, 21–26 July 2017.
13. Zheng, Z.; Zhong, Y.; Wang, J.; Ma, A. Foreground-aware relation network for geospatial object segmentation in high spatial resolution remote sensing imagery. In Proceedings of the IEEE/CVF Conference on Computer Vision and Pattern Recognition, Virtual, 13–19 June 2020; pp. 4096–4105.
14. Chen, L.; Dou, X.; Peng, J.; Li, W.; Sun, B.; Li, H. EFCNet: Ensemble Full Convolutional Network for Semantic Segmentation of High-Resolution Remote Sensing Images. *IEEE Geosci. Remote Sens. Lett.* **2021**, *19*, 1–5. [CrossRef]
15. Li, Y.C.; Li, H.C.; Hu, W.S.; Yu, H.L. DSPCANet: Dual-Channel Scale-Aware Segmentation Network With Position and Channel Attentions for High-Resolution Aerial Images. *IEEE J. Sel. Top. Appl. Earth Obs. Remote Sens.* **2021**, *14*, 8552–8565. [CrossRef]
16. Sharifi, A.; Mahdipour, H.; Moradi, E.; Tariq, A. Agricultural field extraction with deep learning algorithm and satellite imagery. *J. Indian Soc. Remote Sens.* **2022**, *50*, 417–423. [CrossRef]

17. Wang, X.; Girshick, R.; Gupta, A.; He, K. Non-local neural networks. In Proceedings of the IEEE Conference on Computer Vision and Pattern Recognition, Salt Lake City, UT, USA, 18–22 June 2018; pp. 7794–7803.
18. Fu, J.; Liu, J.; Tian, H.; Li, Y.; Bao, Y.; Fang, Z.; Lu, H. Dual attention network for scene segmentation. In Proceedings of the IEEE/CVF Conference on Computer Vision and Pattern Recognition, Long Beach, CA, USA, 16–20 June 2019; pp. 3146–3154.
19. Chen, Y.; Kalantidis, Y.; Li, J.; Yan, S.; Feng, J. A^ 2-nets: Double attention networks. *Adv. Neural Inf. Process. Syst.* **2018**, *31*.
20. Li, X.; Zhong, Z.; Wu, J.; Yang, Y.; Lin, Z.; Liu, H. Expectation-maximization attention networks for semantic segmentation. In Proceedings of the IEEE/CVF International Conference on Computer Vision, Seoul, Korea, 27 October–2 November 2019; pp. 9167–9176.
21. Woo, S.; Park, J.; Lee, J.-Y.; Kweon, I.S. Cbam: Convolutional block attention module. In Proceedings of the European Conference on Computer Vision (ECCV), Munich, Germany, 8–14 September 2018; pp. 3–19.
22. Ding, L.; Tang, H.; Bruzzone, L. LANet: Local attention embedding to improve the semantic segmentation of remote sensing images. *IEEE Trans. Geosci. Remote Sens.* **2020**, *59*, 426–435. [CrossRef]
23. Niu, R.; Sun, X.; Tian, Y.; Diao, W.; Chen, K.; Fu, K. Hybrid multiple attention network for semantic segmentation in aerial images. *IEEE Trans. Geosci. Remote Sens.* **2021**, *60*, 1–18. [CrossRef]
24. Sun, L.; Zhao, G.; Zheng, Y.; Wu, Z. Spectral–Spatial Feature Tokenization Transformer for Hyperspectral Image Classification. *IEEE Trans. Geosci. Remote Sens.* **2022**, *60*, 1–14.
25. Zhou, L.; Zhang, C.; Wu, M. D-LinkNet: LinkNet with pretrained encoder and dilated convolution for high resolution satellite imagery road extraction. In Proceedings of the IEEE Conference on Computer Vision and Pattern Recognition Workshops, Salt Lake City, UT, USA, 18–22 June 2018; pp. 182–186.
26. Geng, Q.; Zhang, H.; Qi, X.; Huang, G.; Yang, R.; Zhou, Z. Gated path selection network for semantic segmentation. *IEEE Trans. Image Process.* **2021**, *30*, 2436–2449. [CrossRef]
27. Wu, T.; Tang, S.; Zhang, R.; Cao, J.; Zhang, Y. Cgnet: A light-weight context guided network for semantic segmentation. *IEEE Trans. Image Process.* **2020**, *30*, 1169–1179. [CrossRef]
28. Vaswani, A.; Shazeer, N.; Parmar, N.; Uszkoreit, J.; Jones, L.; Gomez, A.N.; Kaiser, Ł.; Polosukhin, I. Attention is all you need. *Adv. Neural Inf. Process. Syst.* **2017**, *30*.
29. Park, J.; Woo, S.; Lee, J.; Kweon, I. BAM: Bottleneck Attention Module. In Proceedings of the British Machine Vision Conference, Newcastle, UK, 3–6 September 2018.
30. Zhao, H.; Zhang, Y.; Liu, S.; Shi, J.; Loy, C.C.; Lin, D.; Jia, J. Psanet: Point-wise spatial attention network for scene parsing. In Proceedings of the European Conference on Computer Vision (ECCV), Munich, Germany, 8–14 September 2018; pp. 267–283.
31. Caruana, R. Multitask learning. *Mach. Learn.* **1997**, *28*, 41–75. [CrossRef]
32. Baxter, J. A model of inductive bias learning. *J. Artif. Intell. Res.* **2000**, *12*, 149–198. [CrossRef]
33. Ben-David, S.; Schuller, R. Exploiting task relatedness for multiple task learning. In *Learning Theory and Kernel Machines*; Springer: Berlin/Heidelberg, Germany, 2003; pp. 567–580.
34. Maurer, A.; Pontil, M.; Romera-Paredes, B. Sparse coding for multitask and transfer learning. In Proceedings of the International Conference on Machine Learning PMLR, Atlanta, GA, USA, 16–21 June 2013; pp. 343–351.
35. Ando, R.K.; Zhang, T.; Bartlett, P. A framework for learning predictive structures from multiple tasks and unlabeled data. *J. Mach. Learn. Res.* **2005**, 6.
36. Nakamura, A.T.M.; Grassi, V., Jr.; Wolf, D.F. An effective combination of loss gradients for multi-task learning applied on instance segmentation and depth estimation. *Eng. Appl. Artif. Intell.* **2021**, *100*, 104205. [CrossRef]
37. Duong, L.; Cohn, T.; Bird, S.; Cook, P. Low resource dependency parsing: Cross-lingual parameter sharing in a neural network parser. In Proceedings of the 53rd Annual Meeting of the Association for Computational Linguistics and the 7th International Joint Conference on Natural Language Processing (Volume 2: Short Papers), Beijing, China, 26–31 July 2015; pp. 845–850.
38. Yang, Y.; Hospedales, T. Deep Multi-task Representation Learning: A Tensor Factorisation Approach. In Proceedings of the 5th International Conference on Learning Representations, Toulon, France, 24–26 April 2017.
39. Kendall, A.; Gal, Y.; Cipolla, R. Multi-task learning using uncertainty to weigh losses for scene geometry and semantics. In Proceedings of the IEEE Conference on Computer Vision and Pattern Recognition, Salt Lake City, UT, USA, 18–22 June 2018; pp. 7482–7491.
40. Borse, S.; Wang, Y.; Zhang, Y.; Porikli, F. Inverseform: A loss function for structured boundary-aware segmentation. In Proceedings of the IEEE/CVF Conference on Computer Vision and Pattern Recognition, Virtual, 19–25 June 2021; pp. 5901–5911.
41. ISPRS. 2D Semantic Labeling Contest—Potsdam. Available online: https://www.isprs.org/education/benchmarks/UrbanSemLab/2d-sem-label-potsdam.aspx (accessed on 4 September 2018).
42. ISPRS. 2D Semantic Labeling Contest—Vaihingen. Available online: https://www.isprs.org/education/benchmarks/UrbanSemLab/2d-sem-label-vaihingen.aspx (accessed on 4 September 2018).
43. Chu, X.; Chen, L.; Yu, W. NAFSSR: Stereo Image Super-Resolution Using NAFNet. In Proceedings of the IEEE/CVF Conference on Computer Vision and Pattern Recognition (CVPR) Workshops, New Orleand, LA, USA, 19–24 June 2022; pp. 1239–1248.

44. Hu, J.; Shen, L.; Sun, G. Squeeze-and-excitation networks. In Proceedings of the IEEE Conference on Computer Vision and Pattern Recognition, Salt Lake City, UT, USA, 18–22 June 2018; pp. 7132–7141.
45. Liu, Z. Semantic Segmentation of Remote sensing images via combining residuals and multi-scale modules. In Proceedings of the ICMLCA 2021; 2nd International Conference on Machine Learning and Computer Application, Shenyang, China, 17–19 December 2021; pp. 1–4.

remote sensing

MDPI

Article

OrtDet: An Orientation Robust Detector via Transformer for Object Detection in Aerial Images

Ling Zhao [1], Tianhua Liu [1], Shuchun Xie [2,*], Haoze Huang [1] and Ji Qi [1]

[1] School of Geosciences and Info-Physics, Central South University, South Lushan Road, Changsha 410083, China
[2] School of Traffic and Transportation Engineering, Changsha University of Science & Technology, Changsha 410114, China
* Correspondence: xiesc@csust.edu.cn

Abstract: The detection of arbitrarily rotated objects in aerial images is challenging due to the highly complex backgrounds and the multiple angles of objects. Existing detectors are not robust relative to the varying angle of objects because the CNNs do not explicitly model the orientation's variation. In this paper, we propose an Orientation Robust Detector (OrtDet) to solve this problem, which aims to learn features that change accordingly with the object's rotation (i.e., rotation-equivariant features). Specifically, we introduce a vision transformer as the backbone to capture its remote contextual associations via the degree of feature similarities. By capturing the features of each part of the object and their relative spatial distribution, OrtDet can learn features that have a complete response to any direction of the object. In addition, we use the tokens concatenation layer (TCL) strategy, which generates a pyramidal feature hierarchy for addressing vastly different scales of objects. To avoid the confusion of angle regression, we predict the relative gliding offsets of the vertices in each corresponding side of the horizontal bounding boxes (HBBs) to represent the oriented bounding boxes (OBBs). To intuitively reflect the robustness of the detector, a new metric, the mean rotation precision (mRP), is proposed to quantitatively measure the model's learning ability for a rotation-equivariant feature. Experiments on the DOTA-v1.0, DOTA-v1.5, and HRSC2016 datasets show that our method improves the mAP by 0.5, 1.1, and 2.2 and reduces mRP detection fluctuations by 0.74, 0.56, and 0.52, respectively.

Keywords: object detection; rotation-equivariant; self-attention

Citation: Zhao, L.; Liu, T.; Xie, S.; Huang, H.; Qi, J. OrtDet: An Orientation Robust Detector via Transformer for Object Detection in Aerial Images. *Remote Sens.* **2022**, *14*, 6329. https://doi.org/10.3390/rs14246329

Academic Editor: Michael K Ng

Received: 27 October 2022
Accepted: 8 December 2022
Published: 14 December 2022

1. Introduction

Object detectors for remote sensing images are designed to quickly and accurately search for target objects in images, such as vehicles, aircraft, playgrounds, and bridges. They are essential in traffic management, target detection (military), land use, and urban planning [1–3]. As remote sensing images are acquired with a top-view perspective, oriented object detectors use OBBs, which provide more accurate orientation information of the objects than the general object detectors [4–8] that use HBBs. This application enables better detections for objects with dense distributions, large aspect ratios, and arbitrary directions.

Many detectors designed for OBB tasks on remote sensing have reported promising results. In order to enhance the robustness and make the detector capable of detecting objects in arbitrary directions, most are devoted to learning the rotation-equivariant features of the objects [4,8–11]. As analyzed in the experiments of this paper, in the commonly used dataset of remote sensing, many categories not only have a small number of samples but also are concentrated in small angle intervals. Traditional detectors based on CNNs cannot learn rotation equivalence accurately because the convolutional kernels do not follow the directional changes of the objects [9,12]. Existing detectors can detect arbitrary orientation objects by employing larger capacity networks that fit the feature expression of the object under different directions in the training set [13]. As shown in Figure 1a, we found that when the angle in the validation set is rotated to increase its distribution and training set differences, the performance of traditional detectors can differ by nearly 15 AP. Therefore, the robustness of the detector cannot be guaranteed. As shown in Figure 1b, the features maps extracted by traditional detectors are concentrated in the center of the object no matter how they are rotated, and it is difficult to learn the directional information of the object. It is crucial for the detector to encode the rotational equivalence features of the objects more accurately. For example, the CNNs have translation equivalence feature, so the presence of the object at any position does not have a significant impact.

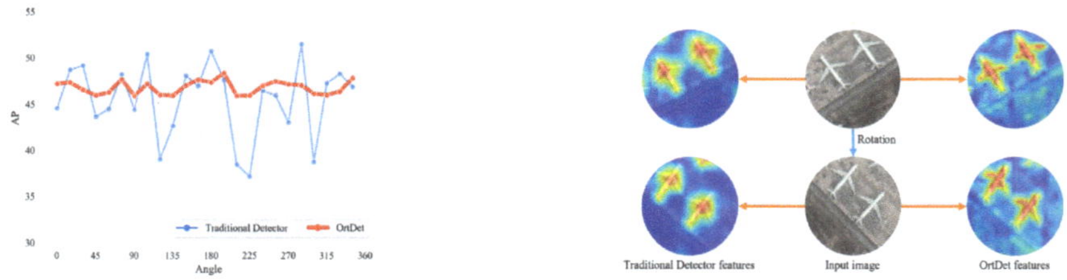

(a) (b)

Figure 1. (**a**) The horizontal coordinate is the objects rotation angle in the validation set, the vertical coordinate is the AP, the red curve is our detector OrtDet, and the blue curve is the traditional detector. (**b**) Illustration of our method. The orange arrow represents the feature heat map, the blue arrow represents the rotation operation. Traditional detectors have difficulty in learning the orientation information of the objects, and their extraction features are focused on the center of the objects regardless of the rotation. Our OrtDet learned the orientation information of the object better.

Recently, to improve the ability of CNNs to encode the rotation equivalence of the object, group convolutions [14] have been proposed to extend the traditional CNNs to obtain the rotation equivalence features in larger groups, which produce feature maps with dedicated orientation channels to record features in different directions of the image [15–17]. The ReDet [13] builds on this approach by designing a rotation-invariant region of interest (RiRoI) structure to obtain features of the object in the directional channel and successfully confer the CNN models with rotation equivalence in detection tasks. However, such methods not only increase the computational complexity of the CNNs but also introduce a large amount of nonessential computations, e.g., most of the remote sensing image regions are composed of a background, and we do not need to obtain features of multiple directions in the background region. Cheng [18] et al. explicitly added a rotation-invariant regularizer to the CNN's features by optimizing a new loss function to force a tight mapping of the feature representations of the training samples before and after rotations to obtain rotation equivalence images. However, such methods force the CNN to learn high-dimensional image-level rotation equivariant/invariant features. In contrast, instance level rotation-equivalence features are more noteworthy in the object detection task.

In addition, the existing detectors [8,19–21] measure the precision using the mean average precision (mAP) metric. There is no specific metric for measuring the robustness of the object direction in the oriented object detection task. To a certain extent, the ability to learn more rotation equivalence features results in a higher mAP metric. However, the higher mAP does not equate to better rotation equivalence features of the object. The training and validation sets are usually divided randomly on the dataset, so there are independent and identically distributed [22–24] in direction distribution. Therefore, the detectors can achieve a high mAP by fitting the feature expression in each direction of the train set. It is equivalent to informing the detector in advance of the test set's object in the corresponding direction feature, and it cannot learn rotation equivalence features. In particular, for remote sensing dataset bridges, courses and ports, and other objects, due to the limitation of the number of satellites and the limitation of the overhead perspective, the number of objects is not only small but also more than 50% of objects directions are concentrated in the range of $0 \pm 15°$. When the angle distributions of the training and test dataset are inconsistent, the mAP will be substantially reduced. Thus, the mAP metric can only reflect the accuracy of the detector for the current angle distribution dataset and does not reflect the robustness of the detector.

In this paper, we propose a novel OrtDet architecture for remote sensing images to improve robustness by learning the rotation equivalence features of the objects. Firstly, we introduce the vision transformer architecture as the backbone. As shown in Figure 1b, compared with the traditional detector, which focuses on the central part of the object regardless of how the object is rotated, and the extracted feature is very coarse, the OrtDet can regress the overall features of the object (e.g., nose, fuselage, wings, etc.) more finely and accurately, so we can identify the orientation of the object by its features, which is more suitable for the oriented object detection task. Unlike CNNs with fixed and limited perceptual fields, it relies on an intrinsic self-attentive mechanism as the main module to capture its remote contextual associations via the degree of feature similarity. The feature similarity region can change adaptively according to the object direction. As shown in Figure 2a, it can be seen that the feature similarity regions obtained by self-attention at each point are treated as nondifferentiated at the beginning of training. When the model is trained, as shown in Figure 2b, the model learns that the similarity region of the features at each point in the object converges adaptively from the full image range to the region where the object is located, and Figure 2c shows that this similarity region is consistent with the change in the object's orientation, thus demonstrating that this approach can better learn rotational equivalence features. This direction also follows the ViT [25] philosophy to reduce the "induction bias" while pursuing generalized features. Self-attention [26] computations have fewer inductive bias than CNNs, such as translation equivalence and localization, but it can still learn translation equivalence and scale equivalence features under certain datasets and supervised/self-supervised tasks [25,27], so it can also learn object rotation equivalence under the corresponding supervised tasks, i.e., oriented object detection tasks. Secondly, the transformer's architecture is a plain, nonhierarchical architecture that maintains a single-scale feature map [25]. We use the Tokens Concatenation Layer (TCL) layer, which reduces its spatial resolution with the increase in network depth and thus generates a multiscale features map.Then, we leverage the Feature Pyramid Network (FPN) [28] module to assign feature levels according to the scale of objects. To reduce the computational complexity of the model, we use the window mechanism and shift window mechanism to reduce the computational overhead of the model while maintaining the relationship between the windows of the images [29].

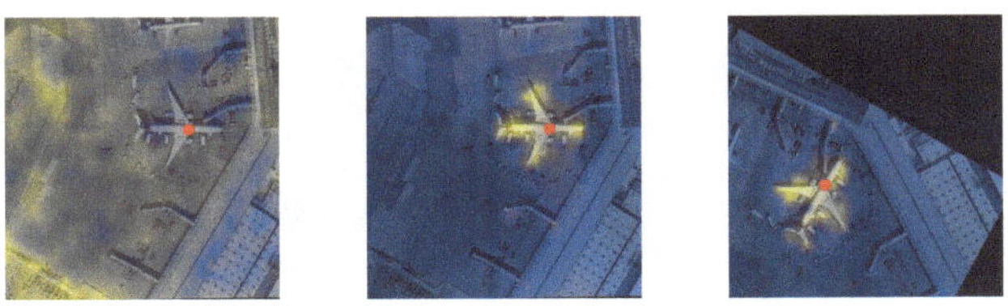

(a) (b) (c)

Figure 2. The relative importance of a single token (red point) with respect to the other tokens in the image is visualized. The shade of yellow color represents the degree of importance. (**a**) represents the beginning of the transformer's backbone training. (**b**) represents the state after the transformer's backbone is trained. (**c**) shows the object at different angles in the image.

In addition, we use the vertex offset method to accurately describe an oriented object by indirectly regressing the glide of each vertex offset of the object with the smallest HBBs on each corresponding side. The principle is outlined in Section 3.4. This representation can further alleviate the instability of the regression's results due to the change in the object's direction. Furthermore, to quantitatively measure the ability of the detector to learn the object rotation equivalence features, we propose a new evaluation metric: mean Rotation Precision (mRP). It measures the fluctuation of the accuracy of the detector for the same dataset at different angle distributions. A smaller mRP means that the accuracy is less affected by the object direction in the dataset and is more robust. The effectiveness of the proposed method is demonstrated by extensive experiments on the remote sensing datasets DOTA-v1.0, DOTA-v1.5, and HRSC2016.

The significant contributions of this paper can be summarized as follows:

- We propose an Orientation Robust Detector for remote sensing images. It adaptively captures remote contextual associations via the degree of feature similarity. The rotation equivalence feature can be learned more accurately for objects at different angles.
- We use windowing mechanism and TCL strategy to reduce the computational complexity of transformer and generate multi-scale features. A more efficient and robust method is used to represent the oriented objects to reduce the regression's confusion problem.
- We propose a new metric, mRP, compared to the traditional mAP; it reflects the ability to learn object rotation equivalences by quantitatively describing the fluctuation of the accuracy in the test set for different object angle distributions.
- We carry out experiments on the DOTA-V1.0, DOTA-V1.5, and HRSC2016 datasets to demonstrate the effectiveness of our method, which involves significant improvements in detection performances as measured by both mAP and mRP metrics.

2. Related Work

2.1. Oriented Object Detection

Unlike most HBB object detectors [30–35], OBB object detectors [4,8,13,19,20] are more suitable for the overhead perspective of the remote sensing images. To detect objects in arbitrary directions, the method [36] determines OBBs by directly introducing angular parameters based on the Faster RCNN [31] regression center coordinates, width, and height. Rotated RPN anchors [37–39] with different angles, aspect ratios, and scales are defined manually to perform the regression of the object, while RRoI pooling is used to rotate the object features for normalization. However, the above method dramatically increases the extra calculation burden in the training and testing phases. The RoI-Transformer [4] interposes a light-weighting module between RPN and RCNN. A lightweight module is inserted to transform horizontal RoIs (HRoIs) into rotation RoIs (RRoIs), thus avoiding the generation of a large number of anchors and reducing the computational complexity. To avoid the problems of abrupt boundary variability and angular periodicity, the gliding vertex [19] first extracts the external horizontal bounding boxes of the object and then introduces scale factors on each side of the HBBs to locate the vertices of the OBBs. CSL [20] used angle classification instead of regression and designed soft labels to cope with the boundary problem, which achieved good results, but angle classification also led to an excessive amount of output parameters. R3Det [8] and S2A-Net [10] designed a feature refinement module using feature interpolation to achieve better spatial alignments with oriented objects. Some complex and uncommon methods have recently been developed to represent OBBs using the middle line [7] and polar coordinates [40]. Zhang et al. [41] propose a novel oriented object detector based on CNN with adaptive object orientation features. All the above methods aim to improve the representation of the OBBs or feature representation of the object. The former can address the loss of abruptness caused by the angle periodicity and the interaction between the length and width of the edges. The latter can extract the exact features of the corresponding rotated object.

2.2. Rotation Equivalence Networks

To alleviate the deficiency that CNNs have a poor representation of rotational equivalence features, the group convolution method [14] is proposed to incorporate a 4-fold approximation with respect to the rotational equivalence feature. HexaConv [17] used a hexagonal raster expansion relative to a 6-fold rotation equivalence and obtained rotation-equivalence features at more angles. The ReDet [13] approach builds on this by modifying the RRoI structure in the detector to allow the cyclic switching of the direction channels and the interpolation of the corresponding features to obtain the rotation-equivalence features of the object. STN [42] and DCN [43] are widely used in oriented object detection tasks by explicitly modelling the rotation variation in the detector. The RICNN [44] optimizes a new loss function by introducing regularization. Cheng [18] et al. explicitly add a rotation-invariant regulariser to the CNN's features by optimizing a new loss function to force a tight mapping of the feature representations to achieve rotational equivalence features. OSIm [45] combines the rotation-equivalence channel features constructed in the frequency domain with the original spatial channel features to cope with the problem. Although these methods approximate rotation-equivalence features, they all require many training samples and a large number of parameters. Secondly, these methods all focus on image-level rotation-equivalence features instead of instance-level rotation-equivalence features.

2.3. Transformer-Based Networks

In contrast to the above methods based on CNN backbone methods, our approach employs a transformer structure based on the self-attention mechanism [26] as the backbone of the detector. The advantage of the transformer lies in using self-attention for capturing global contextual information to establish a long-range dependence. ViT [25] applied a standard transformer directly to image classification by splitting the image into patches and not focusing on pixels, leading to very competitive results with CNNs. Since the architecture is a single-scale feature map with high computational complexity, it cannot be directly used as a backbone for detection tasks [27]. The swin transformer [29] was proposed to reduce the high number of computations by introducing the window and shift window mechanism. By limiting self-attention computation to the local window in the image, it is able to achieve higher accuracies than all existing CNN-based state-of-the-art methods. DETR [34] reduces the computational complexity of the transformer by combining the transformer and CNN, applying transformer architecture to object detection for the first time. Deformable DETR [46] borrows the idea of a variable convolution neural network that enables the transformer's ability to detect multiscale objects. Adaptive clustering transformers [47] further reduce the computational complexity of the model by clustering and accelerate the convergence process. SMCA-DETR [48] introduced a spatially modulated Gaussian mechanism to achieve fast convergence speeds compared with DETR. In the field of remote sensing, the transformer is also used in the field of object detection [49] and instance segmentatio [50,51]. In addition, the CRTransSar [52] structure has also been proposed for applications in SAR images for the object detection of ships. Zhang et al. [53] proposed a TRS structure to apply transformer for a remote sensing scene-classification task. The above methods use transformers as the backbone in the model, when combining them with CNNs to reduce their computational complexity, and there are no allocated supervised tasks for mining the rotational equivalence features of the detector.

3. Method

3.1. Overview

This section presents the principle of the proposed OrtDet to encode rotation-equivalence features and the box regression mechanism. The overall architecture of OrtDet is shown in Figure 3. Firstly, the image is split into tokens by patch partition and is fed into the backbone of the transformer's architecture. Then, it produces multiscale feature expressions relative to the image by the TCL layer. Secondly, the features corresponding to the smallest

HBBs of the object are extracted by RPN and RoI Align. Finally, the OBBs are determined by regressing the four vertices offset of the HBBs by the FC layer.

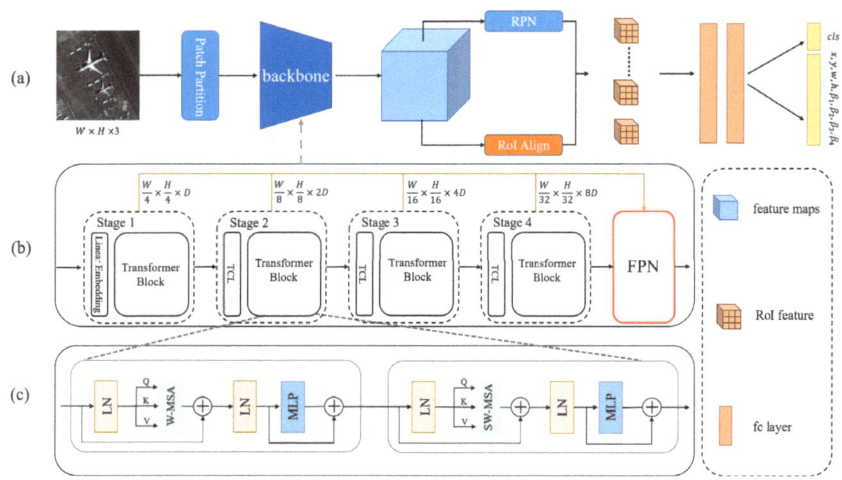

Figure 3. Overview of the proposed method. (**a**) Main structure of rotating object detection. (**b**) Main structure of Transformer model backbone. The brown arrows represent the features at each scale in the FPN. (**c**) Structure of the transformer block for feature extraction, where LN represents layer norm operations, W-MSA represents self-attention operations in the rule window, and SW-MAS represents self-attention operations in the offset window.

3.2. Feature Learning

We introduce the transformer architecture as the backbone. To enrich the spatial neighborhood information of each pixel and reduce the computation, input image I is first divided into $H \times W$ patches, $x_i \in \mathbb{R}^{p^2 * C}, i \in \{1, \cdots, N\}$ by a patch partition module, where P is the size of each patch, C is the channel number of the image (in RGB image usually 3), and N is the total number of patches (Figure 4). The Transformer uses a constant latent vector size D via all of its layers, so the tokens are projected to an arbitrary channel dimension D by using linear embedding layer U and each patch is mapped to a token that enriches the feature's information. The token after patch partition is denoted by z^0, as shown in Equation (1).

$$z^0 = [x_1 U, x_2 U, \cdots, x_N U], U \in \mathbb{R}^{(p^2 \cdot C) \cdot D}, z^0 \in \mathbb{R}^{N \cdot D} \quad (1)$$

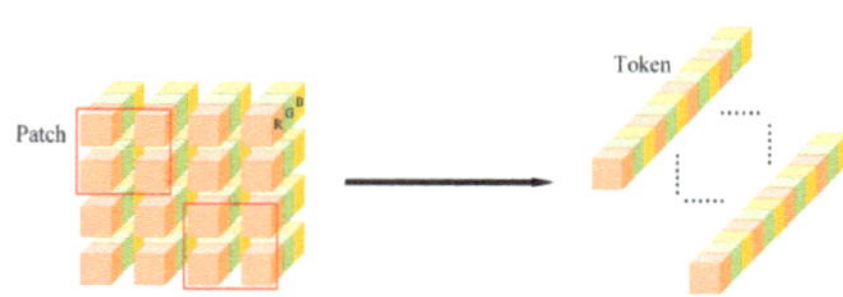

Figure 4. The original image is divided into nonoverlapping Patches(red boxes), and the original spectral features in the Patches are concatenated in the spatial dimension to be output as tokens of the original image. The dotted lines represent the rest of the Tokens that are omitted

Our model uses multi-head self-attention (MSA) [26] to find the relative importance of a single token with respect to the other tokens in the image, which is measured by calculating the attention weight between each token. To obtain the attention weight, the Q (query), K (key), and V (value) vectors are generated by multiplying the element against three learned matrices U_{QKV}. (Equation (2)). The vector for Q, K and V dimensions is denoted by D_K. Vector Q and K contain the similarity of each token in the image obtained by the dot product of Q and K. Next, the output similarity matrix is weighted by multiplying the V vector to obtain the final results. Since vectors Q, K, and V are generated by tokens, the method is called self-attention.

$$[Q, K, V] = Z^i U_{QKV}, U_{QKV} \in \mathbb{R}^{D*3D_K} \tag{2}$$

To determine the relevance between tokens, the dot products of the Q of the focused token and the K of other tokens are obtained. The dot product's result is normalized by D_K and fed into a softmax to obtain attention weight AT (Equation (3)). The AT determines the relative importance of every pair of tokens in the image.

$$AT = Softmax(\frac{QK^T}{\sqrt{D_K}}), AT \in \mathbb{R}^{N*N} \tag{3}$$

Each row of the AT represents the relative importance of a single token with respect to the other tokens, so the V of each token is multiplied by the output of the softmax. The result can be interpreted as the most closely associated tokens, thus extracting its corresponding feature Since the highest score is usually used to represent itself, it is necessary to use MSA to extract different relevant features.

$$ST = AT \cdot V \tag{4}$$

The MSA block computes the scaled dot-product attention separately for N heads using the previous operation, but each of the different heads is projected to a different Q-K-V vector using a different U_{QKV} matrix. The output of each of the N heads is concatenated and projected to the specified feature dimension using a feedforward layer with learnable weights W.

$$MSA(z) = Concat(SA_1(z_i); SA_2(z_i), \cdots, SA_N(z_i)) \cdot W, W \in \mathbb{R}^{nD_k*D} \tag{5}$$

To further reduce the computational complexity of the transformer architecture, we adopt a windowing mechanism. Instead of performing self-attention with all the remaining tokens in the image, we divide the tokens into separate nonoverlapping windows by region, as shown in Figure 5a. Moreover, only the self-attention between each token in each window is calculated, as shown in Equation (6). W_MSA represent the MSA operations in windows.

$$\hat{z}^l = W\_MSA(LN(z^{l-1})) + z^{l-1} \tag{6}$$

$$z^l = MLP(LN(\hat{z}^l)) + \hat{z}^l \tag{7}$$

Secondly , to prevent the restrictive isolation of tokens in each window, as shown in Figure 5b, the shift windows operation is introduced to allow information interactions between adjacent and nonoverlapping local windows. It significantly enhances the feature representation capability of the model, as shown in Equation (8). SW_MSA represent MSA operations in shift windows. Each transformer block followed by a two-layer MLP and LayerNorm(LN) layer is applied before each MSA module and each MLP module, as shown in Equations (7) and (9).

$$\hat{z}^{l+1} = SW\_MSA(LN(z^l)) + z^l \tag{8}$$

$$z^{l+1} = MLP(LN(\hat{z}^{l+1})) + \hat{z}^{l+1} \tag{9}$$

(a) (b)

Figure 5. Self-attention in windows portioning (**a**) and Shift windows partitioning (**b**) operation. The orange, yellow, green and blue colors represent Tokens in each window in (**a**), the red boxed pixel in (**b**) can be used to perform a self-attention operation with the four adjacent pixels in the previous windows

3.3. TCL Layer

We use the TCL layer to transform nonhierarchical, single-scale features into hierarchical, multiscale features and conduct the task of multiscale object detection in remote sensing images. The detailed operation procedures are as follows.

Specifically, we perform a concatenation operation on the 2×2 neighbor tokens in the feature map, which reduce the feature scale of tokens to half of the original one and change the channel dimension from D-dimension to 4D-dimension. To prevent the dimension from growing too fast and to render the merged features smoother, we use linear layer U_d to reduce the feature dimension from $4D$ to $2D$. (Equation (10))

$$z'_{\frac{j}{2},\frac{i}{2}} = Concat(z_{(i,j)}, z_{(i+1,j)}, z_{(i,j+1)}, z_{(i+1,j+1)}) \cdot U_d, U_d \in \mathbb{R}^{4D*2D}, z' \in \mathbb{R}^{\frac{N}{2}*2D} \quad (10)$$

In stage 1, we do not use the TCL layer to reduce the feature scale because the original image input feature dimension is expanded to the D dimension using linear embedding. In stage 2, the feature scale is reduced from $\frac{W}{4} \times \frac{H}{4}$ to $\frac{W}{8} \times \frac{H}{8}$ using the TCL layer, and in stage 3 and stage 4, the feature scale is reduced to $\frac{W}{16} \times \frac{H}{16}$ and $\frac{W}{32} \times \frac{H}{32}$, respectively. As shown in Figure 6, the model is designed as a hierarchical pyramid structure via the TCL layer to produce a hierarchical feature representation. Stage 1 and Stage 2 downsampled the images by $4\times$ and $8\times$, respectively, which produced shallow features with strong spatial and geometric features, but lacked semantic features for the images. Stage 3 and Stage 4 downsampled the images by $16\times$ and $32\times$, respectively, which produced deep features that responded to high-level semantic features of the objects, but lacked spatial geometric features. Therefore, we use the FPN [28] module to fuse the shallow and deep features of the image and perform the object detection task independently in each level by top-down and skip connections. Thus, we can satisfy multiscale object detection features in remote sensing images.

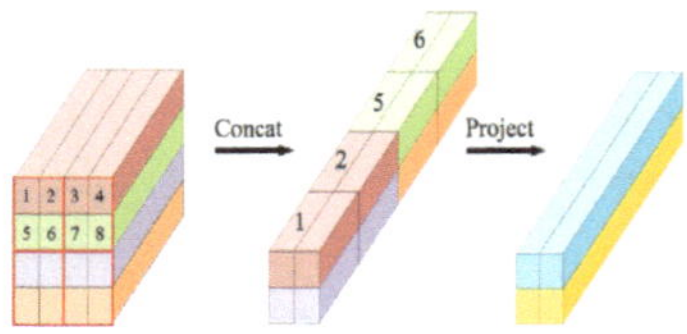

Figure 6. To describe this operation more visually, we use different colors for the features of different rows. The original tokens in the red box are concatenated with their feature vectors in the spatial dimension(e.g. 1, 2, 5 and 6 tokens), then projected to the specified feature dimension.

3.4. Rotated Object Box Regression

The direct regression of the angle parameters is challenging for the detector. Therefore, we use the regression of HBB corner point offsets to determine the OBB. An intuitive illustration of the method is depicted in Figure 7. We first calculate the HBB (black box B_h

in Figure 7) of the OBB (blue box B_o in Figure 7). B_h is represented using the conventional methods (x, y, h, w), with (x, y) representing the center point and (h, w) representing the width and height, respectively. In addition to the HBB's representation, four additional parameters (β_1,β_2,β_3 and β_4) are introduced to describe the OBB of the object. The formula for $\beta_i, i \in \{1, 2, 3, 4\}$ is shown in (Equation (11))

$$\begin{aligned} \beta_i &= \frac{l_i}{w}, i \in \{1, 3\} \\ \beta_i &= \frac{l_i}{h}, i \in \{2, 4\} \end{aligned} \tag{11}$$

where l_i represents the offset of each corner point in the OBB from the corresponding corner point in the HBB. The normalization operation allows the model to converge quickly to the target object.

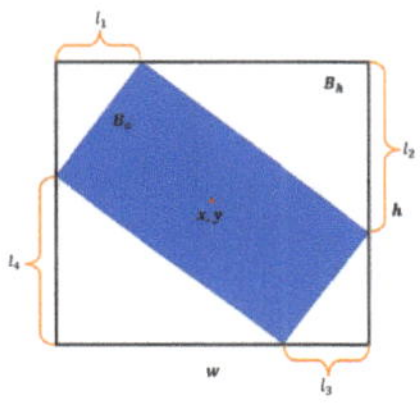

Figure 7. First, determine black box B_h (x, y, w, and h) of the object's border, and the red dot represents the coordinates of the object center point. Then calculate the offset of each corner point(β_1, β_2, β_3 and β_4) in the HBB to describe the blue box, B_o, of the object.

The input image is first fed into the backbone to extract features and generate HBBs proposals with RPN. Then, the regional features extracted via RoIAlign on the proposals are passed through the FC layers to generate final results, including a HBB (x, y, h, and w) and four variables (β_1, β_2, β_3, and β_4) defining the OBB. The loss function of the model is similar to that of the Faster RCNN [31]. The difference in the regression loss of object bounding box L_{reg} is calculated according to Equation (12).

$$\begin{aligned} L_{reg} &= \lambda_1 L_h + \lambda_2 L_\beta \\ L_\beta &= \sum_{i=0}^{4} smooth_{L_1}(\beta_i - \widehat{\beta}_i) \end{aligned} \tag{12}$$

where L_h is the HBBs loss of the object, which is calculated in the same way as in the Faster RCNN, and$\widetilde{\beta}_i$ represents the true offset of each corner point of the rotating border. λ_1 and λ_2 are hyperparameters to balance the importance of each part of the loss.

3.5. Rotated Precision Metric

To address the defect that the mAP does not accurately measure the model's ability to learn the rotation equivalence of the object, we propose a new mean rotational precision (mRP) as an evaluation metric. Firstly, all images in test set I containing K categories are randomly rotated by N sets of different angles to form an extended test set $\{I^1, I^2, I^3, \ldots, I^n\}$. Then, the accuracy of AP_j^i is evaluated for each object class in the test set I_j^i $(1 \leq i \leq n, 1 \leq j \leq k)$, and the variance value, RP_j, of AP_j^i accuracies for each category in the rotated test set is calculated as follows:

$$RP_j = \frac{\sum_{i=1}^{n}(AP_i^j - M_j)^2}{k} \tag{13}$$

where M_j represents the average accuracy of each category under different angle distributions.

$$M_j = \frac{\sum_{n}^{i=1} AP_i^j}{n} \tag{14}$$

Finally, the mRP results are obtained by averaging various types of objects:

$$mRP = \frac{\sum_{j=1}^{k} (RP_j - M)^2}{k} \tag{15}$$

where M represents the average of the various types of RP_j accuracies.

$$M = \frac{\sum_{j=1}^{k} RP_j}{k} \tag{16}$$

RP_j represents the degree to which the accuracy is affected by the angle for each category, with smaller values representing the model's ability to learn the rotation equivalence feature of the object better. mRP is the robustness of the mAP for each dataset, and a smaller mRP means a more robust model, leading to fewer errors due to changes in the object's direction.

4. Experiments and Analysis

This section evaluates the proposed OrtDet on three public datasets designed for remote sensing oriented object detection task. We first introduce the dataset used in the experiment (Section 4.1). Then, we describe the implementation details and hyperparameter design (Section 4.2). Finally, we analyze our model and present our experimental results on different datasets (Sections 4.3–4.5).

4.1. Dataset Description

The experiments evaluate the proposed OrtDet on three publicly available datasets designed for remote sensing oriented object detection.

DOTA-v1.0 [36] is one of the largest datasets for oriented object detection in the remote sensing field released by Wuhan University in China. It contains 2806 remote sensing images with sizes ranging from 800 × 800 to 4000 × 4000 and containing 188,282 object instances and 15 object categories, which includes the following: Plane (PL), Baseball diamond (BD), Bridge (BR), Ground track field (GTF), Small vehicle (SV), Large vehicle (LV), Ship (SH), Tennis court (TC), Basketball court (BC), Storage tank (ST), Soccer-ball field (SBF), Roundabout (RA), Harbor (HA), Swimming pool (SP), and Helicopter (HC).

DOTA-v1.5 is a new dataset released in the 2019 DOTA Challenge that adds to the DOTA-v1.0 dataset, mainly by adding a new category: Container Crane (CC) and giving many small object instances smaller than 10 pixels for specific annotations, especially on the SV category for more detailed annotation. Compared to DOTA-v1.0, the DOTA-v1.5 dataset is more challenging, but the model training can be more stable.

HRSC2016 [54] is a challenging ship detection dataset with OBBs annotation produced by Northwestern Polytechnical University in China in 2016, which contains 1061 aerial images that range in size from 300 × 300 to 1500 × 900. It contains 436 training sets, 181 validation sets, and 444 test sets. Similarly to DOTA, we used separate training and validation set. All images were resized to (800,512) without changing the aspect ratio, and a random horizontal flip was applied during training.

For a fair comparison, we process the dataset similarly to the baseline method. When evaluating the mAP metric in the DOTA dataset, the training and validation sets are used for training, the test set is used for testing, and the accuracy is evaluated using the official code. To calculate the rotation precision of the mRP, we need to rotate the image and label information simultaneously to evaluate the accuracy of the dataset at different angular distributions. As we do not have access to the label information of the test set, we only use the training set for training and rotate the validation set for testing. We cropped the original image to 1024 × 1024 for image processing, set the step size to 824, and used 0 to paddle images smaller than 1024 after cropping. The sample distribution in each category in the DOTA-v1.5 dataset is shown in Figure 8a. The number of samples in each category is shown in Figure 8b. As depicted in the figure, it can be seen that in the remote sensing dataset, ground nonfixed objects mainly include targets such as vehicles, planes and ships, and ground-fixed objects mainly include targets such as courts, bridges, and ports. The number of objects for ground nonfixed objects is usually much larger than that of ground-fixed objects. Furthermore, the object directions of ground-fixed objects are concentrated in a few angle intervals. On the other hand, due to the arbitrary nature of the direction in ground nonfixed objects, the object's directions are evenly distributed in each interval, so it is consistent with the previous analysis of this paper, which proves that the proposed model has a wide range of application scenarios in the field of remote sensing.

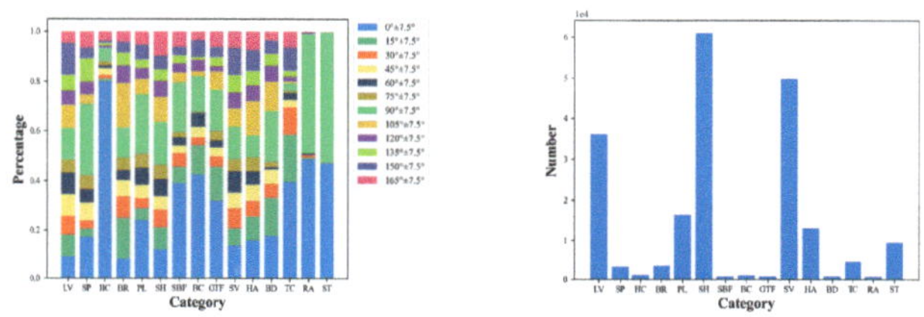

(a) (b)

Figure 8. (**a**) The direct distribution of each category in the DOTA-V1.0 dataset by cropping the images at each angle. (**b**) The number of samples in each category after cropping the images in the DOTA-V1.0 dataset.

4.2. Implementation Details and Hyperparameter Design

The proposed method is mainly experimented on the RTX3090 using the MMDetection framework [55]. The initial learning rate is 7.5×10^{-3} with a batch size of 4 for training and divided by 10 in each learning rate decay step. For the DOTA-v1.0 and DOTA-v1.5 datasets, the training epoch is set to 12, and the learning rate decayed at 8 and 11 epochs. We cropped the original image to 1024 × 1024 with a step size of 824. The data augmentation strategy is used to balance the categories in the training sample. For the HRSC2016 dataset, the training epoch was set to 36, and the learning rate decayed at 24 and 33 epochs. In all experiments, the network is trained by the SGD optimizer with momentum and weight decay set to 0.9 and 5×10^{-4}, respectively. The window size is set to 7 × 7 by default. The parameter, D, of the channel number is set to 96, and the expansion layer of each MLP is 4 for all experiments (e.g., the MLP input dimension is 96 and, the hidden layer's dimension is 384). In multi-head self-attention, the number of heads gradually increased as the transformer block deepened. In stage 1, stage 2, stage 3 and stage 4, the number of heads is 3, 6, 12, and 24, and the numbers of layers is set to 2, 2, 6, and 2, respectively. In addition, the hyperparameters λ_1 and λ_2 in Equation (12) are both set to 1. Data augmentation was performed by only using random horizontal and vertical flips. When evaluating the mRP's metrics, to ensure the authenticity of the results, we rotate the validation set by multiple angles to ensure that the operation does not change the size of the object, and we use a rotation followed by a cropping sequence for both the images and the annotation files. Moreover, we do not use rotated augmentation in the training process for the mRP metric.

4.3. Detection Precision

We compare the OrtDet model proposed in this paper with representative state-of-the-art methods on the three benchmark dataset in Tables 1–3. We indicate in boldface and highlight the best results on each column for the different methods. Our method shows the best performance on either dataset, achieving 74.09% mAP on the DOTA-v1.0 dataset, 64.4% mAP on DOTA-v1.5, and 89.6% mAP on the HRSC2016 dataset. In addition, compared to the same vertex offset regression of the Gliding Vertex method, our OrtDet, can significantly improve the accuracy for small objects categories, such as Large Vehicle, Small Vehicle, Ship, and to a certain extent for other categories. The reason is that small objects are more sensitive to angular regression, and the slight deviation for small objects is more likely to reduce the IoU between the ground truth and the regression bounding box. In contrast, our OrtDet can obtain rotation-equivalence features better than CNNs, so the improvement in accuracy for such objects is noticeable. Moreover, in the DOTA-v1.5 dataset, the Small Vehicle category incorporates a large number of labels below 10 pixels. Thus, vertex offset regression is unsuitable for describing such tiny rotating objects, resulting in lower accuracies for the Small Vehicle category than other oriented object detection models. However, our detector can significantly alleviate this deficiency.

Table 1. Mean average precision (%) of the different methods for the DOTA-v1.0 dataset.

Methods	PL	BD	BR	GTF	SV	LV	SH	TC	BC	ST	SBF	RA	HA	SP	HC	mAP
RRPN [37]	88.52	71.20	31.66	59.30	51.85	56.19	57.25	90.80	72.84	67.38	56.69	52.84	53.08	51.94	53.58	61.01
RoI-Transformer [4]	88.64	78.52	43.44	**75.92**	68.81	73.68	83.59	90.74	77.27	81.46	58.39	53.54	62.83	58.93	47.67	69.56
Gliding-Vertex [19]	88.58	**84.94**	49.87	73.90	70.80	75.62	79.21	90.60	78.06	**86.26**	58.11	**70.11**	**75.88**	**72.81**	47.82	73.51
R^3Det [8]	**89.59**	81.17	50.53	66.10	70.92	**78.66**	78.21	90.81	**85.26**	84.23	**61.81**	63.77	68.16	69.83	**67.17**	73.74
OrtDet (Ours)	89.02	84.82	**50.68**	71.45	**72.06**	74.98	**87.64**	**90.85**	78.76	85.81	58.97	68.95	67.29	70.49	59.61	**74.09**

Table 2. Mean average precision (%) of the different methods for the DOTA-v1.5 dataset.

Methods	PL	BD	BR	GTF	SV	LV	SH	TC	BC	ST	SBF	RA	HA	SP	HC	CC	mAP
RetinaNet-O [33]	71.4	77.6	42.1	64.7	44.5	56.8	73.3	90.8	76.0	60.0	46.9	69.2	59.6	64.5	48.1	0.8	59.2
FR-O [36]	71.9	74.5	44.5	59.9	51.3	69.0	79.4	**90.8**	**77.4**	67.5	47.7	69.7	61.2	65.3	**60.5**	1.54	62.0
Mask R-CNN [56]	76.8	73.5	49.9	57.8	51.3	71.3	79.7	90.5	74.2	66.0	46.2	70.6	63.1	64.5	67.8	9.42	62.7
Gliding Vertex [19]	72.1	82.4	47.5	**73.4**	34.6	68.0	80.3	90.7	71.1	69.1	**49.8**	68.1	**68.5**	66.9	57.8	**14.6**	63.3
HTC [57]	**77.8**	73.7	**51.4**	64.0	**51.5**	**73.3**	80.3	90.5	75.1	67.3	48.5	70.6	64.8	64.4	55.9	5.15	63.4
OrtDet (Ours)	72.3	**84.5**	48.7	71.1	42.1	73.2	**80.7**	90.7	71.2	**69.5**	47.8	**71.3**	67.9	**67.4**	58.5	13.8	**64.4**

Table 3. Mean average precision (%) and mean rotation precision of the different methods for the HRSC2016 dataset.

Methods	RRPN [37]	RoI-Transformer [4]	Gliding-Vertex [19]	OrtDet (Ours)
mAP	79.6	86.2	87.4	**89.6**
mRP	2.55	2.37	2.28	**1.76**

4.4. Rotation Equivalence

In order to quantitatively assess the ability of baseline and the OrtDet to learn object rotation-equivalence features, the experiments are conducted to measure the extent to which the accuracy was affected by the angles distribution of each category in three datasets using the RP metric. The mRP is further used to quantitatively measure the overall fluctuation degree of the detector. The results are shown in Tables 3–5. We indicate in boldface and highlight the best results on each column for the different methods.

Table 4. Mean rotation precision (%) of the different methods for the DOTA-v1.0 dataset.

Methods	PL	BD	BR	GTF	SV	LV	SH	TC	BC	ST	SBF	RA	HA	SP	HC	mRP
FR-O [36]	0.28	3.75	2.06	3.98	3.37	2.18	0.47	5.07	4.20	2.58	4.18	2.05	4.72	2.95	8.79	3.45
HTC [57]	0.26	1.96	1.71	3.80	3.60	2.10	0.30	2.09	4.26	2.05	4.04	2.32	3.28	2.44	8.84	2.87
RoI-Transformer [4]	0.42	2.51	**1.65**	2.82	2.88	2.62	**0.23**	0.22	4.15	2.27	2.92	2.40	3.91	2.30	10.2	2.82
R^3Det [8]	0.34	**1.50**	2.00	4.08	2.28	1.55	0.26	3.03	5.24	1.71	2.63	**1.63**	3.05	2.13	10.22	2.77
Gliding-Vertex [19]	0.13	3.63	1.83	2.97	2.70	0.66	0.35	0.21	3.20	0.84	2.92	2.23	3.82	3.75	7.64	2.38
OrtDet (Ours)	**0.12**	2.84	2.18	**2.23**	**2.25**	**0.65**	0.31	**0.09**	**1.23**	**0.25**	**1.02**	1.98	**2.83**	**1.76**	**5.06**	**1.64**

Table 5. Mean rotation precision (%) of the different methods for the DOTA-v1.5 dataset.

Methods	PL	BD	BR	GTF	SV	LV	SH	TC	BC	ST	SBF	RA	HA	SP	HC	CC	mRP
FR-O [36]	0.66	3.07	1.53	3.73	3.15	2.23	0.27	5.47	5.40	0.41	3.17	1.54	4.86	0.77	8.23	**0.74**	2.83
Mask R-CNN [56]	0.57	2.46	1.54	3.74	2.53	1.78	0.16	4.76	4.78	0.33	3.58	2.15	2.47	0.82	9.49	1.68	2.68
HTC [57]	0.36	2.77	**0.87**	3.51	1.93	0.97	0.49	4.21	4.03	0.36	3.94	1.61	2.94	1.48	7.92	3.06	2.53
RoI-Transformer [4]	0.40	3.00	1.44	2.82	1.60	0.90	0.53	1.77	3.78	0.48	2.44	2.11	3.69	1.14	8.65	3.85	2.41
Gliding-Vertex [19]	0.67	2.68	2.27	3.13	**0.30**	0.94	0.07	1.74	3.14	0.21	1.29	**1.03**	2.25	0.95	7.73	1.82	1.89
OrtDet (Ours)	**0.23**	**1.35**	1.23	**2.11**	0.76	**0.80**	**0.05**	**0.63**	**1.60**	**0.12**	**0.72**	1.11	**1.58**	**0.90**	**5.01**	0.99	**1.20**

The experimental results in Tables 5 and 6 show that the RP of certain categories for all the detectors are significantly smaller than others, such as aircraft, ships, and vehicles, for which the directional characteristics are apparent. The accuracy variation of plane and ship with their angular distribution is further shown in Figure 9. The distribution of these categories in the training set is shown in Figure 8a. As these categories are nonground fixed objects, they can be distributed at any angle in the overhead perspective of the remote sensing image. Hence, their distribution in each angle interval in the dataset is approximately balanced. Figure 10a,b show the plane and ship categories distribution on each angle interval in detail. In addition, the number of objects in these categories is much larger than the number of objects in other categories from Figure 8b. Thus, the CNN-based backbone has sufficient data for learning the features of these objects in different directions. Thus, it does not need to learn the rotation-equivalence features of the object and can still accurately regress the OBBs of these categories. The OrtDet model can further reduce RP fluctuations in these categories by learning rotation-equivalence features.

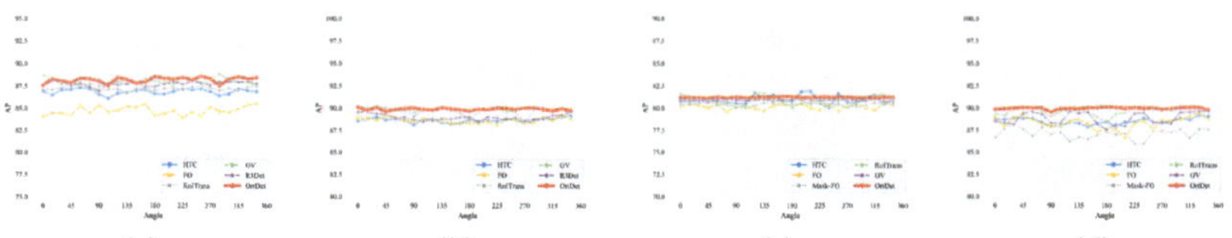

(a) (b) (c) (d)

Figure 9. (**a**,**b**) show the fluctuation of the detection accuracy of the ship and plane categories in the DOTA-V1.0 dataset with the rotation angle of the dataset, respectively. (**c**,**d**) show the fluctuation of the detection accuracy of the ship and plane categories in the DOTA-V1.5 dataset with the rotation angle of the dataset, respectively.

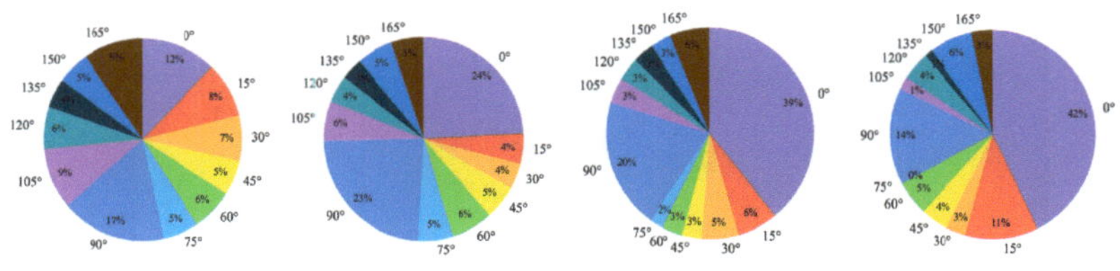

(a) (b) (c) (d)

Figure 10. (**a**–**d**) represent the distribution of ship, plane, soccer-ball fields and basketball courts in each angle interval, respectively. We count the number of objects belonging to each angle interval from 0° to 165° in 15° intervals. From the figure, we can find that the distribution of plane and ship in each angle interval is much more balanced than that of basketball courts and soccer-ball fields.

Table 6. Performance comparisons of the backbone on OrtDet.

Methods	Backbone	DOTA-v1.0		DOTA-v1.5		Size (Mb)
		mAP	mRP	mAP	mRP	
OrtDet	ResNet	70.89	2.38	59.89	1.75	60.4
	Transformer	72.32	1.64	61.50	1.19	52.9

For the roundabout and storage tank categories, although they are distributed within a specific range of angles, their accuracy does not fluctuate particularly significantly regardless of the rotation of the dataset due to the approximate circular shape of the objects. The main reason for the slight change is that the detector's result is always horizontal no matter how the circular object is rotated. However, to make the labels follow the rotation of the object, we rotate the ground truth of the dataset accordingly, thus resulting in a slight deviation of the result of the circular object from the ground truth after rotation.

The categories Basketball-court, Soccer ball field, Ground track field, and Harbor are ground-fixed objects. The angles do not change flexibly in the overhead perspective of remote sensing images, resulting in a relatively concentrated distribution of the object's angles. For example, the angle of the basketball court category in Figure 10c is concentrated around 0° and 90°, and the training samples of these categories are tiny compared with those of the plane, ship, and vehicle, as shown in Figure 8b. Therefore, the detector in the baseline does not has sufficient data to learn the features of these categories in other directions. Therefore, when the object angles of these categories in the validation set are inconsistent with the training data, the accuracy will drop sharply, proving that they cannot really learn the rotation equivalence feature. Contrarily, the accuracy of the OrtDet is much less affected by the angle distribution than that of the baseline. The model can rely on training data with a more concentrated angle distribution to obtain robustness results in the validation set with a completely different angle distribution. Figure 11 shows that the experimental results demonstrate that the method is much more capable of learning rotation-equivalence features than the baseline. As depicted in Figure 12, we visualize and evaluate the Basketball-Court, Helicopter, Plane, Soccer-ball-field, and Tennis-Court categories using a confidence threshold of 0.7. The OrtDet model proposed in this paper does not change significantly with a change in the object's direction when the accuracy is high, while baseline methods are prone to missing detections and false detections due to the change in the object's angle.

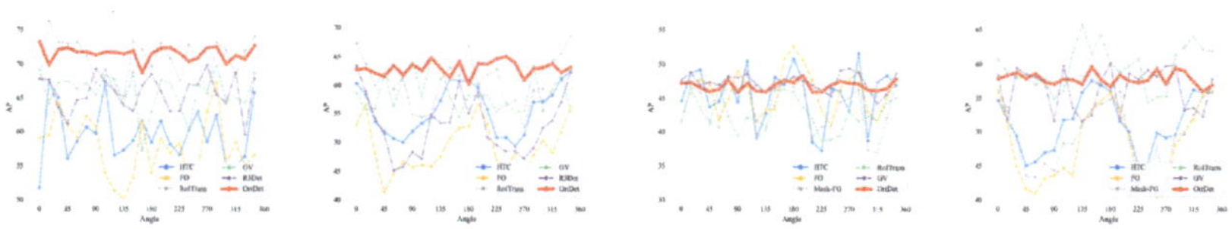

(a) (b) (c) (d)

Figure 11. (**a**,**b**) show the fluctuation of the detection accuracy of the Soccert-ball-field and Basketball-court categories in the DOTA-V1.0 dataset with the rotation angle of the dataset, respectively. (**c**,**d**) show the fluctuation of the detection accuracy of the Soccert-ball-field category and the Basketball-court category with the rotation angle of the dataset in the DOTA-V1.5 dataset, respectively.

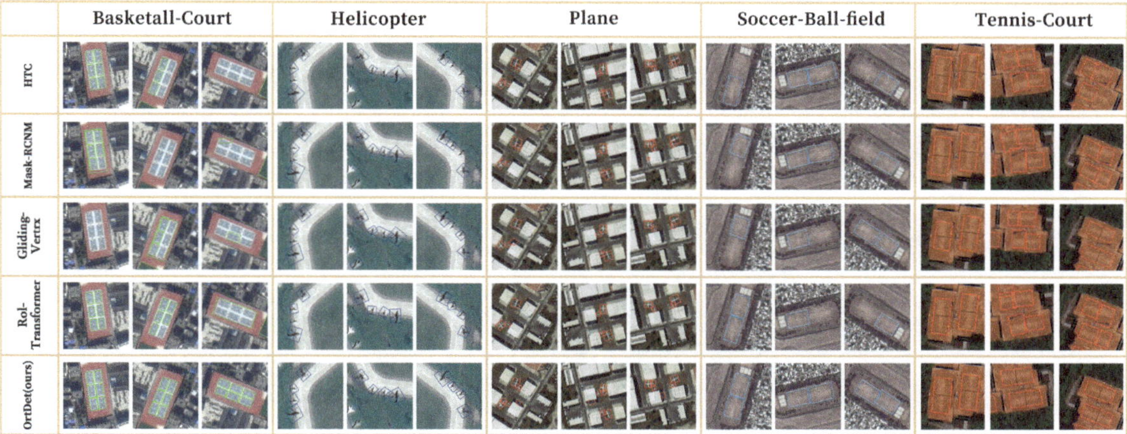

Figure 12. Qualitative comparison with baseline methods in detecting the identical objects categories of different orientations in the DOTA-v1.5 dataset by rotating an input image with different angles. The bounding box is the result of the detection for different categories. The OrtDet architecture is subject to much smaller angular images of the object than the baseline.

Figure 13 shows that in the DOTA-v1.0 and DOTA-v1.5 datasets, the OrtDet has a much smaller overall fluctuation than baselines. At the same time, the mAP of each angular distribution is slightly higher or similar to that of the baseline. Therefore, the OrtDet has a more significant advantage in encoding rotation equivalence features, and its learning of rotation equivalence features makes the detector more robust. In particular, in remote sensing images, due to the limitation of the satellite trajectory and overhead perspective, generating a large number of training samples with a more uniform angle distribution for fixed ground features is challenging. Thus, the robust detector is more beneficial to the task of the oriented detector in remote sensing images.

4.5. Ablation Study

In this section, we perform a series of ablation studies on the DOTA dataset to validate the effectiveness of the proposed method in this paper.

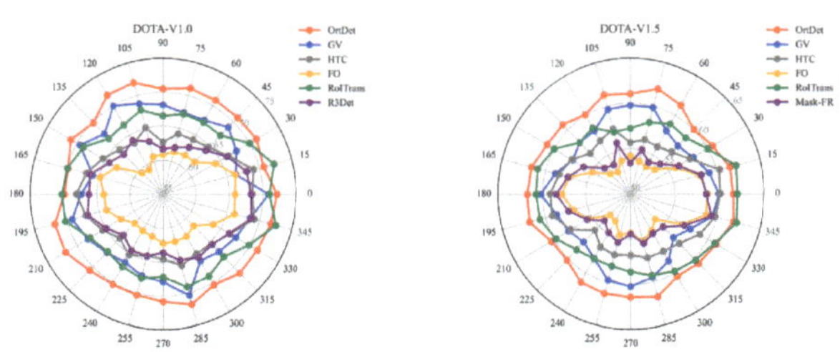

(a) (b)

Figure 13. A higher roundness of the curve indicates a higher robustness of the detector, and a smaller distance between the curve and the outer circle indicates higher accuracies. (**a**) Variation of the detection accuracy with the rotation angle of the validation set on the DOTA-v1.0 dataset. (**b**) The variation of the detection accuracy with the rotation angle of the validation set on the DOTA-v1.5 dataset.

Transformer Architecture. To verify the effectiveness of the transformer architecture, we compared the ResNet backbone based on CNN. The experimental results are shown in Table 6. Compared with the ResNet, the transformer architecture can obtain a higher mAP with further reductions in the number of parameters and significantly reduce the degree of mRP fluctuations. Thus, the transformer can better learn rotation equivalence features, and the rotation equivalence features of the object can improve the accuracy.

Vertex offset. In order to verify that the vertex offset method can accurately regress the OBBs, we compare the vertex offset method used in OrtDet with the most commonly used direct regression of the object angle method. The experimental results are shown in Table 7. The direct regression of the object angle will reduce the accuracy due to the imprecision of the angle regression. In contrast, the method of angle vertex offsets will slightly increase the number of parameters of the model due to the regression. However, this method can reduce the deviation of the regression results due to the change in the object angle based on the acquisition of the object rotation equivalence features and further improve the accuracy of the OBBs. Moreover, this method can improve the accuracy and reduce the fluctuation of the results.

Table 7. Performance comparisons of the box regression on OrtDet.

Methods	Box Reg	DOTA-v1.0		DOTA-v1.5		Size (Mb)
		mAP	mRP	mAP	mRP	
OrtDet	RBox reg	71.88	2.04	61.33	1.88	50.8
	Vertex offset	72.32	1.64	61.50	1.19	52.9

5. Conclusions

In this paper, we propose the Orientation Robust Detector to improve the robustness relative to the object's direction in the oriented object detection task. Introducing the transformer's architecture alleviates the deficiency observed in traditional CNNs in learning object rotation equivalence. Moreover, its accuracy is robust relative to inconsistent angle distribution problems in the test set with the training set and/or the problem where data are insufficient in each angle range. Hence, the false and missed detections caused by the object direction variations are significantly reduced. We used the TCL layer to generate the transformer's pyramidal feature hierarchy to address the large-scale differences of objects in remote sensing. We also reduced the computational complexity of the model by introducing a windowing mechanism and a shift windowing mechanism. In addition, we predict the HBBs of the object and the relative gliding offset to represent the OBBs, which avoids the deviation of the OBBs regression due to the change in the object's direction. Finally, the mRP metric is proposed to quantitatively reflect the degree of accuracy fluctuation when the object orientation of the dataset changes as a measure of the model's learning ability for object rotation equivalence features. Experimental results fully demonstrate the effectiveness and robustness of our OrtDet. Nevertheless, the windowing mechanism introduced to reduce the complexity of the model leads to a limited interaction range between image regions. Our future work will focus on expanding the image's interaction range in the transformer architecture as much as possible while being computationally tractable.

Author Contributions: Conceptualization, L.Z.; Methodology, T.L.; Formal analysis, S.X.; Writing—original draft, L.Z.; Writing—review and editing, T.L., H.H. and J.Q.; Visualization, H.H.; Project administration, S.X., L.Z.; Funding acquisition, S.X. All authors have read and agreed to the published version of the manuscript.

Funding: This research was funded by National Natural Science Foundation of China(42171458); Natural Science Foundation of Hunan Province, China(2021JJ30818); Supported by the Young Teacher Development Program of Changsha University of Science and Technology(2019QJCZ006)

Data Availability Statement: The data that support the findings of this study are available from the author upon reasonable request. This data is available at https://captain-whu.github.io/DOTA/index.html (10 October 2022).

Conflicts of Interest: The authors declare no conflict of interest.

References

1. Li, K.; Wan, G.; Cheng, G.; Meng, L.; Han, J. Object detection in optical remote sensing images: A survey and a new benchmark. *ISPRS J. Photogramm. Remote Sens.* **2020**, *159*, 296–307. [CrossRef]
2. Cheng, G.; Han, J. A survey on object detection in optical remote sensing images. *ISPRS J. Photogramm. Remote Sens.* **2016**, *117*, 11–28. [CrossRef]
3. Wu, J.; Cao, C.; Zhou, Y.; Zeng, X.; Feng, Z.; Wu, Q.; Huang, Z. Multiple Ship Tracking in Remote Sensing Images Using Deep Learning. *Remote Sens.* **2021**, *13*, 3601. [CrossRef]
4. Ding, J.; Xue, N.; Long, Y.; Xia, G.S.; Lu, Q. Learning roi transformer for oriented object detection in aerial images. In Proceedings of the IEEE/CVF Conference on Computer Vision and Pattern Recognition, Long Beach, CA, USA, 15–20 June 2019; pp. 2849–2858.
5. Azimi, S.M.; Vig, E.; Bahmanyar, R.; Körner, M.; Reinartz, P. Towards multi-class object detection in unconstrained remote sensing imagery. In Proceedings of the Asian Conference on Computer Vision, Perth, Australia, 2–6 December 2018; Springer: Berlin/Heidelberg, Germany, 2018; pp. 150–165.
6. Yang, X.; Yang, J.; Yan, J.; Zhang, Y.; Zhang, T.; Guo, Z.; Sun, X.; Fu, K.S. Towards more robust detection for small, cluttered and rotated objects. In Proceedings of the IEEE International Conference on Computer Vision, Seoul, Republic of Korea, 27 October–2 November 2019; Volume 27, pp. 8232–8241.
7. Wei, H.; Zhang, Y.; Chang, Z.; Li, H.; Wang, H.; Sun, X. Oriented objects as pairs of middle lines. *ISPRS J. Photogramm. Remote Sens.* **2020**, *169*, 268–279. [CrossRef]
8. Yang, X.; Yan, J.; Feng, Z.; He, T. R3det: Refined single-stage detector with feature refinement for rotating object. *Proc. Aaai Conf. Artif. Intell.* **2021**, *35*, 3163–3171. [CrossRef]
9. Kalra, A.; Stoppi, G.; Brown, B.; Agarwal, R.; Kadambi, A. Towards Rotation Invariance in Object Detection. In Proceedings of the IEEE/CVF International Conference on Computer Vision, Montreal, QC, Canada, 10–17 October 2021; pp. 3530–3540.
10. Han, J.; Ding, J.; Li, J.; Xia, G.S. Align deep features for oriented object detection. *IEEE Trans. Geosci. Remote Sens.* **2021**, *60*, 5602511. [CrossRef]
11. Ming, Q.; Miao, L.; Zhou, Z.; Yang, X.; Dong, Y. Optimization for arbitrary-oriented object detection via representation invariance loss. *IEEE Geosci. Remote Sens. Lett.* **2021**, *19*, 8021505. [CrossRef]
12. Buslaev, A.; Iglovikov, V.I.; Khvedchenya, E.; Parinov, A.; Druzhinin, M.; Kalinin, A.A. Albumentations: Fast and flexible image augmentations. *Information* **2020**, *11*, 125. [CrossRef]
13. Han, J.; Ding, J.; Xue, N.; Xia, G.S. Redet: A rotation-equivariant detector for aerial object detection. In Proceedings of the IEEE/CVF Conference on Computer Vision and Pattern Recognition, Nashville, TN, USA, 20–25 June 2021; pp. 2786–2795.
14. Cohen, T.; Welling, M. Group equivariant convolutional networks. In Proceedings of the International Conference on Machine Learning, PMLR, New York, NY, USA, 20–22 June 2016; pp. 2990–2999.
15. Worrall, D.E.; Garbin, S.J.; Turmukhambetov, D.; Brostow, G.J. Harmonic networks: Deep translation and rotation equivariance. In Proceedings of the IEEE Conference on Computer Vision and Pattern Recognition, Honolulu, HI, USA, 21–26 July 2017; pp. 5028–5037.
16. Cohen, T.S.; Geiger, M.; Köhler, J.; Welling, M. Spherical CNNs. In Proceedings of the International Conference on Learning Representations, Vancouver, BC, Canada, 30 April–3 May 2018.
17. Hoogeboom, E.; Peters, J.W.; Cohen, T.S.; Welling, M. Hexaconv. *arXiv* **2018**, arXiv:1803.02108.
18. Cheng, G.; Han, J.; Zhou, P.; Xu, D. Learning rotation-invariant and fisher discriminative convolutional neural networks for object detection. *IEEE Trans. Image Process.* **2018**, *28*, 265–278. [CrossRef]
19. Xu, Y.; Fu, M.; Wang, Q.; Wang, Y.; Chen, K.; Xia, G.S.; Bai, X. Gliding vertex on the horizontal bounding box for multi-oriented object detection. *IEEE Trans. Pattern Anal. Mach. Intell.* **2020**, *43*, 1452–1459. [CrossRef] [PubMed]
20. Yang, X.; Yan, J. Arbitrary-oriented object detection with circular smooth label. In Proceedings of the European Conference on Computer Vision, Glasgow, UK, 23–28 August 2020; Springer: Berlin/Heidelberg, Germany, 2020; pp. 677–694.
21. Zhang, F.; Wang, X.; Zhou, S.; Wang, Y.; Hou, Y. Arbitrary-oriented ship detection through center-head point extraction. *IEEE Trans. Geosci. Remote Sens.* **2021**, *60*, 5612414. [CrossRef]
22. Shen, Z.; Liu, J.; He, Y.; Zhang, X.; Xu, R.; Yu, H.; Cui, P. Towards out-of-distribution generalization: A survey. *arXiv* **2021**, arXiv:2108.13624.
23. Shen, Z.; Cui, P.; Zhang, T.; Kunag, K. Stable learning via sample reweighting. *Proc. AAAI Conf. Artif. Intell.* **2020**, *34*, 5692–5699. [CrossRef]
24. Kuang, K.; Xiong, R.; Cui, P.; Athey, S.; Li, B. Stable prediction with model misspecification and agnostic distribution shift. *Proc. AAAI Conf. Artif. Intell.* **2020**, *34*, 4485–4492. [CrossRef]
25. Dosovitskiy, A.; Beyer, L.; Kolesnikov, A. An image is worth 16x16 words: Transformers for image recognition at scale. *arXiv* **2020**, arXiv:2010.11929.
26. Vaswani, A.; Shazeer, N.; Parmar, N.; Uszkoreit, J.; Jones, L.; Gomez, A.N.; Kaiser, Ł.; Polosukhin, I. Attention is all you need. In Proceedings of the Advances in Neural Information Processing Systems, Long Beach, CA, USA, 4–9 December 2017; Volume 30.
27. Li, Y.; Mao, H.; Girshick, R.; He, K. Exploring plain vision transformer backbones for object detection. *arXiv* **2022**, arXiv:2203.16527.
28. Lin, T.Y.; Dollár, P.; Girshick, R.; He, K.; Hariharan, B.; Belongie, S. Feature pyramid networks for object detection. In Proceedings of the IEEE Conference on Computer Vision and Pattern Recognition, Honolulu, HI, USA, 21–26 July 2017; pp. 2117–2125.

29. Liu, Z.; Lin, Y.; Cao, Y.; Hu, H.; Wei, Y.; Zhang, Z.; Lin, S.; Guo, B. Swin transformer: Hierarchical vision transformer using shifted windows. In Proceedings of the IEEE/CVF International Conference on Computer Vision, Montreal, QC, Canada, 10–17 October 2021; pp. 10012–10022.
30. Liu, W.; Anguelov, D.; Erhan, D.; Szegedy, C.; Reed, S.; Fu, C.Y.; Berg, A.C. Ssd: Single shot multibox detector. In Proceedings of the European Conference on Computer Vision, Amsterdam, The Netherlands, 11–14 October 2016; Springer: Berlin/Heidelberg, Germany, 2016; pp. 21–37.
31. Ren, S.; He, K.; Girshick, R.; Sun, J. Faster r-cnn: Towards real-time object detection with region proposal networks. In Proceedings of the Advances in Neural Information Processing Systems, Montreal, QC, Canada, 7–12 December 2015; Volume 28.
32. Redmon, J.; Divvala, S.; Girshick, R.; Farhadi, A. You only look once: Unified, real-time object detection. In Proceedings of the IEEE Conference on Computer Vision and Pattern Recognition, Las Vegas, NV, USA, 27–30 June 2016; pp. 779–788.
33. Lin, T.Y.; Goyal, P.; Girshick, R.; He, K.; Dollár, P. Focal loss for dense object detection. In Proceedings of the IEEE International Conference on Computer Vision, Venice, Italy, 22–29 October 2017; pp. 2980–2988.
34. Carion, N.; Massa, F.; Synnaeve, G.; Usunier, N.; Kirillov, A.; Zagoruyko, S. End-to-end object detection with transformers. In Proceedings of the European Conference on Computer Vision, Glasgow, UK, 23–28 August 2020; Springer: Berlin/Heidelberg, Germany, 2020; pp. 213–229.
35. Zhou, X.; Wang, D.; Krähenbühl, P. Objects as points. *arXiv* **2019**, arXiv:1904.07850.
36. Xia, G.S.; Bai, X.; Ding, J.; Zhu, Z.; Belongie, S.; Luo, J.; Datcu, M.; Pelillo, M.; Zhang, L. DOTA: A large-scale dataset for object detection in aerial images. In Proceedings of the IEEE Conference on Computer Vision and Pattern Recognition, Salt Lake City, UT, USA, 18–23 June 2018; pp. 3974–3983.
37. Ma, J.; Shao, W.; Ye, H.; Wang, L.; Wang, H.; Zheng, Y.; Xue, X. Arbitrary-oriented scene text detection via rotation proposals. *IEEE Trans. Multimed.* **2018**, *20*, 3111–3122. [CrossRef]
38. Zhang, Z.; Guo, W.; Zhu, S.; Yu, W. Toward arbitrary-oriented ship detection with rotated region proposal and discrimination networks. *IEEE Geosci. Remote Sens. Lett.* **2018**, *15*, 1745–1749. [CrossRef]
39. Liu, L.; Pan, Z.; Lei, B. Learning a rotation invariant detector with rotatable bounding box. *arXiv* **2017**, arXiv:1711.09405.
40. Zhou, L.; Wei, H.; Li, H.; Zhao, W.; Zhang, Y.; Zhang, Y. Arbitrary-oriented object detection in remote sensing images based on polar coordinates. *IEEE Access* **2020**, *8*, 223373–223384. [CrossRef]
41. Dong, Z.; Wang, M.; Wang, Y.; Liu, Y.; Feng, Y.; Xu, W. Multi-Oriented Object Detection in High-Resolution Remote Sensing Imagery Based on Convolutional Neural Networks with Adaptive Object Orientation Features. *Remote Sens.* **2022**, *14*, 950. [CrossRef]
42. Jaderberg, M.; Simonyan, K.; Zisserman, A.; et al. Spatial transformer networks. In Proceedings of the Advances in Neural Information Processing Systems, Montreal, QC, Canada, 7–12 December 2015; Volume 28.
43. Dai, J.; Qi, H.; Xiong, Y.; Li, Y.; Zhang, G.; Hu, H.; Wei, Y. Deformable convolutional networks. In Proceedings of the IEEE International Conference on Computer Vision, Montreal, QC, Canada, 10–17 October 2017; pp. 764–773.
44. Cheng, G.; Zhou, P.; Han, J. Learning rotation-invariant convolutional neural networks for object detection in VHR optical remote sensing images. *IEEE Trans. Geosci. Remote Sens.* **2016**, *54*, 7405–7415. [CrossRef]
45. Wu, X.; Hong, D.; Tian, J.; Chanussot, J.; Li, W.; Tao, R. ORSIm detector: A novel object detection framework in optical remote sensing imagery using spatial-frequency channel features. *IEEE Trans. Geosci. Remote Sens.* **2019**, *57*, 5146–5158. [CrossRef]
46. Zhu, X.; Su, W.; Lu, L. Deformable detr: Deformable transformers for end-to-end object detection. *arXiv* **2020**, arXiv:2010.04159.
47. Zheng, M.; Gao, P.; Zhang, R.; Li, K.; Wang, X.; Li, H.; Dong, H. End-to-end object detection with adaptive clustering transformer. *arXiv* **2020**, arXiv:2011.09315.
48. Gao, P.; Zheng, M.; Wang, X.; Dai, J.; Li, H. Fast convergence of detr with spatially modulated co-attention. In Proceedings of the IEEE/CVF International Conference on Computer Vision, Montreal, QC, Canada, 10–17 October 2021; pp. 3621–3630.
49. Li, Q.; Chen, Y.; Zeng, Y. Transformer with Transfer CNN for Remote-Sensing-Image Object Detection. *Remote Sens.* **2022**, *14*, 984. [CrossRef]
50. Xu, X.; Feng, Z.; Cao, C.; Li, M.; Wu, J.; Wu, Z.; Shang, Y.; Ye, S. An improved swin transformer-based model for remote sensing object detection and instance segmentation. *Remote Sens.* **2021**, *13*, 4779. [CrossRef]
51. Panboonyuen, T.; Jitkajornwanich, K.; Lawawirojwong, S.; Srestasathiern, P.; Vateekul, P. Transformer-Based Decoder Designs for Semantic Segmentation on Remotely Sensed Images. *Remote Sens.* **2021**, *13*, 5100. [CrossRef]
52. Xia, R.; Chen, J.; Huang, Z.; Wan, H.; Wu, B.; Sun, L.; Yao, B.; Xiang, H.; Xing, M. CRTransSar: A Visual Transformer Based on Contextual Joint Representation Learning for SAR Ship Detection. *Remote Sens.* **2022**, *14*, 1488. [CrossRef]
53. Zhang, J.; Zhao, H.; Li, J. TRS: Transformers for remote sensing scene classification. *Remote Sens.* **2021**, *13*, 4143. [CrossRef]
54. Liu, Z.; Yuan, L.; Weng, L.; Yang, Y. A high resolution optical satellite image dataset for ship recognition and some new baselines. In Proceedings of the International Conference on Pattern Recognition Applications and Methods, Porto, Portugal, 24–26 February 2017; SciTePress: Setubal, Portugal, 2017; Volume 2, pp. 324–331.
55. Chen, K.; Wang, J.; Pang, J.; Cao, Y.; Xiong, Y.; Li, X.; Sun, S.; Feng, W.; Liu, Z.; Xu, J.; et al. MMDetection: Open mmlab detection toolbox and benchmark. *arXiv* **2019**, arXiv:1906.07155.

56. He, K.; Gkioxari, G.; Dollár, P.; Girshick, R. Mask r-cnn. In Proceedings of the IEEE International Conference on Computer Vision, Venice, Italy, 22–29 October 2017; pp. 2961–2969.
57. Chen, K.; Pang, J.; Wang, J.; Xiong, Y.; Li, X.; Sun, S.; Feng, W.; Liu, Z.; Shi, J.; Ouyang, W.; et al. Hybrid task cascade for instance segmentation. In Proceedings of the IEEE/CVF Conference on Computer Vision and Pattern Recognition, Seoul, Republic of Korea, 27 October–2 November 2019; pp. 4974–4983.

remote sensing

Article

Object Counting in Remote Sensing via Triple Attention and Scale-Aware Network

Xiangyu Guo [1], Marco Anisetti [2], Mingliang Gao [1,*] and Gwanggil Jeon [1,3]

[1] School of Electrical and Electronic Engineering, Shandong University of Technology, Zibo 255000, China; 21404020512@stumail.sdut.edu.cn (X.G.); gjeon@inu.ac.kr (G.J.)
[2] Department of Computer Science, Universit'a degli Studi di Milano, 20133 Milano, Italy; marco.anisetti@unimi.it
[3] Department of Embedded Systems Engineering, Incheon National University, Incheon 22012, Republic of Korea
* Correspondence: mlgao@sdut.edu.cn

Abstract: Object counting is a fundamental task in remote sensing analysis. Nevertheless, it has been barely studied compared with object counting in natural images due to the challenging factors, e.g., background clutter and scale variation. This paper proposes a triple attention and scale-aware network (TASNet). Specifically, a triple view attention (TVA) module is adopted to remedy the background clutter, which executes three-dimension attention operations on the input tensor. In this case, it can capture the interaction dependencies between three dimensions to distinguish the object region. Meanwhile, a pyramid feature aggregation (PFA) module is employed to relieve the scale variation. The PFA module is built in a four-branch architecture, and each branch has a similar structure composed of dilated convolution layers to enlarge the receptive field. Furthermore, a scale transmit connection is introduced to enable the lower branch to acquire the upper branch's scale, increasing the output's scale diversity. Experimental results on remote sensing datasets prove that the proposed model can address the issues of background clutter and scale variation. Moreover, it outperforms the state-of-the-art (SOTA) competitors subjectively and objectively.

Keywords: remote sensing; object counting; attention mechanism; scale variation; background clutter

Citation: Guo, X.; Anisetti, M.; Gao, M.; Jeon, G. Object Counting in Remote Sensing via Triple Attention and Scale-Aware Network. *Remote Sens.* **2022**, *14*, 6363. https://doi.org/10.3390/rs14246363

Academic Editor: Fridon Shubitidze

Received: 18 October 2022
Accepted: 13 December 2022
Published: 15 December 2022

1. Introduction

With the rapid development of remote sensing technologies and satellite platforms, high-quality and quantity remote sensing images are provided for implementing specific tasks, e.g., object detection [1,2], image classification [3] and image super-resolution [4]. Compared with the aforementioned tasks which have been extensively investigated, object counting in remote sensing images has been barely explored due to the challenging factors, e.g., background clutter and scale variation.

The purpose of object counting is to infer the number of instances existing in images. It plays an essential and fundamental role in urban planning [5], environment management [6], monitoring system [7] and public safety [8]. Object counting has drawn much attention, and various approaches have been proposed which can be classified in three categories, i.e., detection-based methods [9,10], regression-based methods [11] and density estimation-based methods [12,13]. The detection-based methods generate bounding boxes by a designed object detector, and they sum the bounding boxes to obtain the object counts. These methods are suitable for counting large objects in sparse scenes. Once the scene becomes crowded and the object size is small, the counting performance will deteriorate dramatically. To address this issue, the regression-based methods are proposed to directly learn a mapping from the image to counts. They can deal with the problem of occlusion and complex background while ignoring the spatial information. Nowadays, benefiting from the strong feature representation ability of convolutional neural network (CNN), density estimation-based methods have outperformed the aforementioned two methods and become the mainstream in the domain of object counting [14]. The key idea of the density estimation-based method is to regress a density map by a CNN, and then, the pixels of the map are summed to generate the final count value [14–16]. Nevertheless, the scale variation and background clutter still constrain the accuracy and robustness of the object counting, especially in remote sensing images.

Some examples of objects with scale variation and background clutter in remote sensing images are depicted in Figure 1. The first column shows the challenges of background clutter. Remote sensing images are taken from the sky, and some vehicles may be sheltered under trees (enclosed by red circles), so it is necessary to accurately identify targets. The second column displays the problem of scale variation (the ships marked in red bounding boxes). One can see that the size varies obviously in the images due to the different types of ships. Thus, the counting model should possess the capacity of capturing large-range scale information.

Figure 1. Examples of objects with scale variation and background clutter in remote sensing.

To mitigate the background clutter, a simple yet effective tool is the visual attention mechanism, which can adjust the weights between the foreground and background. Gao et al. [17] introduced a dual attention module containing channel and spatial units to identify the foreground. Moreover, SwinCounter [13] was proposed to capture global context information based on the Swin Transformer. Specifically, the SwinCounter leverages several shifted windows to compute self-attention with others, which can highlight the tiny object region. Similarly, the PSGCNet [16] deployed a global context module to overcome the background noise by building dependencies between different channels.

To cope with the problem of scale variation, most solutions adopt multiple columns networks to capture multiscale information. For example, MCNN [15] is built to extract multiscale information with a three-branch architecture for crowd counting. Different filters are employed in each column to acquire various receptive fields. Furthermore, several different mechanisms, e.g., dilated convolution [12] and spatial pyramid pooling [18], are employed to boost the scale diversity. Chen et al. [19] leveraged four parallel dilated convolution layers to capture pyramid features to tackle the scale variation. Dai et al. [12] utilized some dense blocks with dense connections to obtain different features with multiple scales. Gao et al. [16] deployed a two-path module, where a local path with a pyramid architecture is built to capture multiscale information, and a global path is developed to capture the features of the large objects.

In this paper, a triple attention and scale-aware network (TASNet) is built to simultaneously address the problems of background clutter and scale variation in remote sensing images. First, a TVA module is proposed to address the problem of background clutter. It is built in a three-branch pattern, with each branch generating a refined feature by performing an attention operation across different dimensions. Then, the optimized features are averaged to generate the output tensor. In this way, the TVA module can boost the dependencies between the channel and spatial dimensions, which helps distinguish the foreground from the background. To relieve the adverse effect of scale variation, a PFA module is built with a four-branch structure. The four branches possess a similar configuration, which adopts dilated convolution to extract the multiscale information. Moreover, a scale transmit mechanism is introduced, which enables the upper branch to pass the scale cue to the lower branch to increase the scale diversity. Overall, the contributions are as follows:

1. A triple attention scale-aware network (TASNet) is built in a divide-and-conquer manner to address the problem of background clutter and scale variation for object counting in remote sensing images.
2. A TVA module, which executes attention operations on features in three views, is built to deal with the background clutter. A PFA module adopting a four-branch architecture is proposed to capture multiscale information.
3. Extensive experiments are carried out to verify the performance of object counting in challenging remote sensing scenarios. Meanwhile, detailed ablation studies are conducted to prove the effectiveness of the different compound modes, backbone networks and the multiscale feature fusion mechanisms within the proposed model.

2. Related Literature

In this section, we present the related work in background clutter and scale variation, which are the two inevitable challenges in object counting in remote sensing.

2.1. Solutions for Background Clutter

Rich context information can guide the model to emphasize the foreground region and suppress background noise. The attention mechanism fits the bill and has been a powerful tool in object counting [20,21]. For example, Gao et al. [22] deployed two parallel attention modules (channel and spatial attention) to build a crowd counting network. Specifically, channel attention aims to alleviate the error estimation of complex backgrounds, while spatial attention can guide the model to perceive global information. Zhu et al. [23] built a dual path network consisting of a density map path for predicting a coarse map, and an attention map path to generate a head probability map. With the assistance of the probability map, the final density map can distinguish the head region from the background. Gao et al. [17] introduced a concatenated attention module to fuse multiscale features and restrain the background clutter. It first employs a channel unit to recognize the object area and then adopts a spatial unit to divide different density levels for encoding a wide range of dependencies. Jiang et al. [24] built an attention scaling network to recognize different density regions, which are generated by multiplying attention masks and scaling factors.

In addition to the assistance of the attention mechanism, many semantic segmentation methods are adopted to address the problem of background clutter. For example, Khan et al. [25] directly employed a scene segmentation framework for crowd counting. It is composed of three components, namely a classification module, semantic scene segmentation (SSS) module and density estimation module. Specifically, the SSS module can produce a segmentation map to highlight the head region. Meng et al. [26] proposed a regularized surrogate task based on binary segmentation. With the help of binary segmentation, the network can generate a hard uncertainty map and a soft uncertainty map to suppress the background noise. Gao et al. [27] designed a foreground and background segmentation (FBS) module to recognize the interest region. The FBS module is built with an encoder–decoder structure to produce a segmentation map, which is transmitted to the density head branch to emphasize the crowd region. Liu et al. [28] built several inter-related segmentation maps to assist predicting the high-quality density map. Specifically, for each segmentation map, the predicted values higher than the given threshold value are set to one; otherwise, they are set to zero.

However, the attention modules adopted in these networks still have weaknesses in transmitting information to each other, leading to weak dependencies between global and local perspectives. To this end, we build a TVA module into the network to enhance the dependencies among different dimensions.

2.2. Solutions for Scale Variation

Scale variation is another inevitable challenge limiting the improvement of counting performance. Many attempts have been devoted to addressing this problem. Zhang et al. [15] proposed the first multi-column network with diverse filter kernels to address this problem. Cao et al. [29] employed different filters to take scale features and then aggregate them as the input of the next layer. In addition to using a multi-column architecture, dilated convolution has been adopted to capture multiscale information. Li et al. [30] made full use of the dilated convolution and built a congested scene recognition (CSR) network to enlarge the receptive fields while saving the computation parameters. Liu et al. [31] introduced a structured feature representation learning mechanism to promote the complementarity between different scale feature maps. Meanwhile, a hierarchically structured loss function was proposed to boost the local correlation of different regions. Chen et al. [19] built a scale pyramid module (SPM) which contains four convolution layers with different dilated rates (2, 4, 8, 12) to extract multiscale features. Liu et al. [32] designed a contextual aware network (CAN) to enhance the counting performance. It encodes local scale feature maps and perceives different head sizes. Zhu et al. [33] proposed a multi-level features aggregation network. It consists of a key component termed scale and level aggregation module (SLAM). The SLAM first utilizes a four-branch scale aggregation (SA) to extract the multiscale information and then utilizes a channel attention (CA) to adjust the channel weights of each feature. Finally, the outputs of SA and CA are summed to generate multiscale features. Dai et al. [12] deployed three successive residual dense blocks to learn local and global context information. Duan et al. [34] embedded a context aggregation module (CAM) into the network to fuse multiscale context information. Moreover, the proposed module can retain tiny details by several pixel attention modules subsequently. Han et al. [35] designed a tree-like scale diversity module, which contains nine different sizes of receptive fields to extract scale features. In addition, a cross-scale communication is introduced to boost the complementary scale information fusion. Chen et al. [36] proposed a multiscale semantic refining module to solve the scale variation.

Unfortunately, the aforementioned methods have some limitations and weaknesses. For example, the dependencies of the global and local information should be built considering the independence of each column in a multi-column architecture. To this end, we introduce a modified multi-column module to enhance the dependencies of each column.

3. Methodology

3.1. Network Architecture

The architecture of the proposed TASNet is depicted in Figure 2. It comprises four parts, i.e., a backbone for basic feature extraction, a TVA module for identifying the object region, a PFA module for suppressing the scale variation, and a decoder for generating the final prediction.

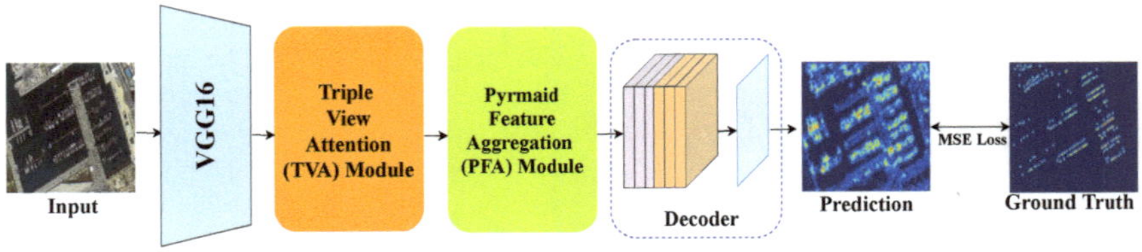

Figure 2. Architecture of the TASNet for object counting in remote sensing.

Specifically, the tailored VGG16 [37] (with ten convolution layers and three max pooling layers) is selected as the backbone. This way, the output feature size is reduced to 1/4 of the input. Subsequently, the TVA module is built to mitigate the background clutter with a three-branch architecture. Afterward, the PFA module is utilized to extract the scale information through a multi-column structure. Lastly, a decoder including three successive convolution layers (3×3 kernel size with 512 channels), three deformable convolution layers (3×3 kernel sizes with 512, 256, and 128 channels), and a 1×1 convolution are equipped to generate the final density map. The prediction is supervised by the ground truth density map generated by the Gaussian kernel, and the MSE loss is applied to measure the l_2 distance between the two maps.

3.2. TVA Module

The complex background in remote sensing images may mislead the network to identify the object incorrectly, which weakens the counting performance. Therefore, we propose a triple view attention (TVA) module to deal with the challenge. Figure 3 exhibits the architecture of the TVA module.

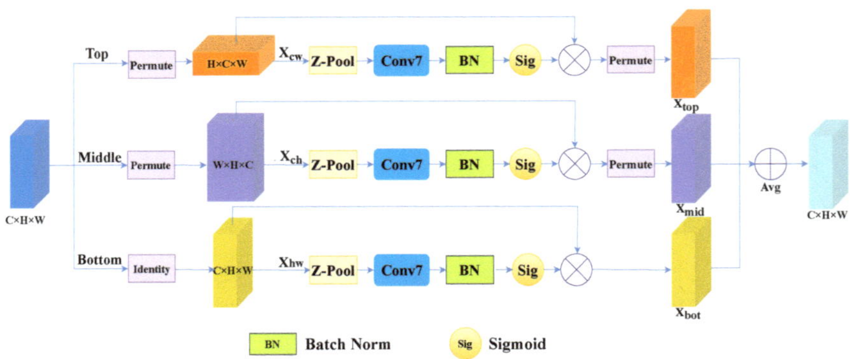

Figure 3. Architecture of the TVA module.

For an input $\mathbf{X} \in \mathbb{R}^{C \times H \times W}$, three branches are established to perform attention operations on different dimensions and output three intermediate features. The three intermediate outputs are averaged to generate the final results.

Specifically, the goal of the top branch is to build the dependencies between the width dimension spatially and the channel dimension. First, the input is viewed as a new shape $\mathbf{X_{cw}} \in \mathbb{R}^{H \times C \times W}$. Then, $\mathbf{X_{cw}}$ is performed through a Z-pool operation [38], which aims to retain rich details and squeeze the depth of the height dimension spatially. The Z-pool is executed by combining the max pooled and average pooled features and reduces the zeroth dimension to 2. Mathematically, it is formulated as

$$Z\text{-}pool(\mathbf{X}) = \mathrm{Cat}(\mathrm{Maxpool}(\mathbf{X}), \mathrm{Avgpool}(\mathbf{X})), \tag{1}$$

where Cat means the concatenate operation. Maxpool and Avgpool denote the operators of max pooling and average pooling, respectively. Afterward, the compressed tensor is fed into a standard convolution layer with a kernel size of k (set to 7 in this paper). Next, a batch normalization layer is added to compress the zeroth dimension to 1. The optimized weights can be obtained by a sigmoid function and are applied to $\mathbf{X}_{cw}$. Finally, the output $\mathbf{X}_{top} \in \mathbb{R}^{C\times H\times W}$ of the top branch is acquired and re-viewed as the shape of input $\mathbf{X} \in \mathbb{R}^{C\times H\times W}$. In sum, the top branch can be represented as,

$$\mathbf{X}_{top} = \mathbf{X}_{cw} \odot \mathrm{Sig}(\mathrm{BN}(\mathrm{Conv}(z^{th}_{pool}(\mathbf{X}_{cw})))), \tag{2}$$

where BN and Sig are the batch normalization and sigmoid function, respectively. $\odot$ is a dot product operation.

Like the top branch, the middle branch aims to build the dependencies between the spatial height dimension and channel dimension. It reshapes the input into $\mathbf{X}_{ch} \in \mathbb{R}^{W\times H\times C}$ and then performs the same operation as the top branch. Overall, it is formulated as,

$$\mathbf{X}_{mid} = \mathbf{X}_{ch} \odot \mathrm{Sig}(\mathrm{BN}(\mathrm{Conv}(z^{th}_{pool}(\mathbf{X}_{ch})))), \tag{3}$$

The input of the bottom branch directly executes the attention operation compared with the upper branch. Following that, the same procedure is performed to obtain the attention result $\mathbf{X}_{hw} \in \mathbb{R}^{C\times H\times W}$. Then, the bottom branch can be formulated as,

$$\mathbf{X}_{bot} = \mathbf{X}_{hw} \odot \mathrm{Sig}(\mathrm{BN}(\mathrm{Conv}(z^{th}_{pool}(\mathbf{X}_{hw})))), \tag{4}$$

The final feature $\mathbf{X}_{tva} \in \mathbb{R}^{C\times H\times W}$ is generated by averaging the outputs of three branches, which is formulated as,

$$\mathbf{X}_{tva} = \frac{1}{3}(\mathbf{X}_{top} + \mathbf{X}_{mid} + \mathbf{X}_{bot}). \tag{5}$$

3.3. PFA Module

The scale variation in remote sensing images restrains the CNN from capturing large-range information of different sizes, which seriously degrades the estimated performance. To address this problem, a PFA module is built to extract wide-range scale features. The framework of the PFA module is depicted in Figure 4. It consists of a four-branch structure and a feature aggregation layer. The purpose of the four-branch structure is to capture multiscale information. Meanwhile, the feature aggregation layer aims to fuse the previously generated maps and compress the number of channels to be the same as the input.

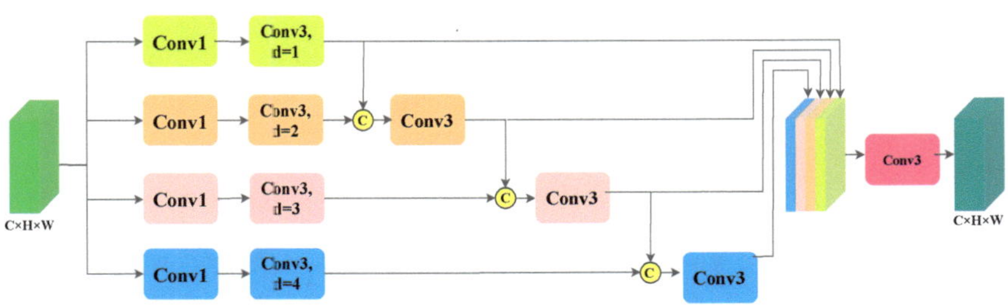

Figure 4. Architecture of the PFA module.

Specifically, given an input $\mathbb{R}^{C\times H\times W}$, it is fed into the four-branch structure. In each branch, a 1×1 convolution layer is first utilized to compress the number of channels to 1/4 of the input feature. Afterwards, dilated convolution layers with dilated rates 1, 2, 3 and 4 are employed to enlarge the receptive field and form the pyramid features. Different from the previous multi-column networks [15,19], which extract information in each column separately, the proposed PFA module can transmit scale information from the upper branch to the lower branch. Particularly, the features of each branch are concatenated with the subsequent branch features. Then, a convolutional layer with a kernel size of

3×3 is subsequently deployed to fuse the features with diverse scales. Each branch is formulated as,

$$\begin{cases} f_1 = \text{Conv}_3^1(\text{Compress}(\mathbf{X})), \\ f_2 = \text{Conv}_3^1(\text{Cat}(f_1, \text{Conv}_3^2(\text{Compress}(\mathbf{X})))), \\ f_3 = \text{Conv}_3^1(\text{Cat}(f_2, \text{Conv}_3^3(\text{Compress}(\mathbf{X})))), \\ f_4 = \text{Conv}_3^1(\text{Cat}(f_3, \text{Conv}_3^4(\text{Compress}(\mathbf{X})))), \end{cases} \tag{6}$$

where $f_1 \sim f_4$ represents four intermediate features. $Conv_k^d$ denotes a convolution layer with filter size k and dilated rate d. Cat and Compress are the concatenate and compress operations, respectively.

The transition connection enables the lower branch to perceive multiscale information. Locally, the lower three branches can extract multiscale features. Globally, the PFA module enables the network to capture a wider scale range. Subsequently, a concatenate operation is executed on the generated feature map of each branch to aggregate them and generate a fused feature. Eventually, the output $\hat{\mathbb{R}}^{C \times H \times W}$ is obtained by a 3×3 convolution layer. The PFA module is formulated as,

$$\mathbf{X}_{pfa} = \text{Conv}3(\text{Cat}(f_1, f_2, f_3, f_4)), \tag{7}$$

3.4. Density Map Generation

To generate the ground truth density map, the dot map is filleted by a Gaussian function, which is the common criteria in the domain of object counting [15,39]. Supposing that the location of an object is at pixel x_i; then, it is defined by an impulse function $\delta(x - x_i)$. In this case, given a remote sensing image with N annotations, it can be defined as,

$$I(x) = \sum_{x=1}^{N} \delta(x - x_i), \tag{8}$$

Nevertheless, the density map generated is not continuous and cannot be employed to train the model. Therefore, the density map is converted into a continuous density function by convolving $I(x)$ with a Gaussian kernel $G_{\sigma_i}(x)$. The final density map can be formulated as,

$$D(x) = \sum_{i=1}^{N} \delta(x - x_i) * G_{\sigma_i}(x), \tag{9}$$

where σ_i denotes the standard deviation of the Gaussian kernel. Following the previous work [17], we set σ to 15.

3.5. Loss Function

The widely used "Mean Squared Error (MSE)" loss is adopted to train the model, which optimizes the model by minimizing the Euclidean distance between prediction and ground truth density maps. It is formulated as follows,

$$l(\theta) = \frac{1}{N} \sum_{i=1}^{N} \left\| X_i^{Est}(\theta) - X_i^{GT} \right\|_2^2, \tag{10}$$

where N represents the batch size, and θ denotes the parameters to be trained. X_i^{GT} and X_i^{Est} denote the ground truth map and estimated density map, respectively.

4. Experiments

4.1. Datasets

Remote Sensing Object Counting (RSOC) dataset: The RSOC dataset [17] is the largest remote sensing counting benchmark. Precisely, it consists of 3057 satellite images with a total of 286,539 annotations. It is further divided into four subdatasets based on the type of objects: buildings, small-vehicle, large-vehicle, and ship.

Car Parking Lot (CARPK) dataset: The CARPK dataset [40] includes 1448 drone-view images from four parking lots, of which 89,777 cars are labeled. The training and test sets contain 989 and 459 images, respectively.

Pontifical Catholic University of Paranà+ (PUCPR+) dataset: The PUCPR+ dataset [40] is a large-scale vehicle-counting dataset. The images are captured in parking lot scenes under different weather conditions, i.e., rainy, cloudy and sunny. In particular, it consists of 125 images with 16,456 annotations, among which 100 images are used for training and 25 images are used for testing, respectively. More details of the three datasets are listed in Table 1.

Table 1. Detailed information of the RSOC, CARPK and PUCPR+ datasets.

Datasets	Platform	Images	Train/Test	Size (Avg.)	Annotation Format
RSOC_Building	Satellite	2468	1205/163	512 × 512	Center point
RSOC_Large-vehicle	Satellite	172	108/64	1552 × 1573	Bounding box
RSOC_Small-vehicle	Satellite	280	222/58	2473 × 2339	Bounding box
RSOC_Ship	Satellite	137	97/40	2558 × 2668	Bounding box
CARPK	Drone	1448	989/459	720 × 1280	Bounding box
PUCPR+	Camera	125	100/25	720 × 1280	Bounding box

4.2. Implementation Details

All experiments are implemented based on the PyTorch framework and executed on an NVIDIA 3090Ti GPU. During the training phase, the SGD optimizer with the initial learning rate of 1×10^{-7} and the decay rate of 5×10^{-4} is adopted to optimize the model. Furthermore, for high-resolution datasets, i.e., large-vehicle, small-vehicle, and ship, we resize them to 1024 × 768 in order to avoid out-of-memory.

For data augmentation, the crop size of the Building subdataset is set to 256 × 256, and the other three subdatasets in RSOC are set to 512 × 384. Then, we adopt mirror flipping to double them. Finally, the batch size is set to 32 for the Building dataset and 1 for the other datasets.

4.3. Evaluation Protocols

The widely used Mean Average Error (MAE) and Root Mean Squared Error (RMSE) are adopted [15,22] to evaluate the counting accuracy and robustness. They are formulated using,

$$MAE = \frac{1}{T}\sum_{i=1}^{T}\left|V_{est}^{i} - V_{gt}^{i}\right|, \tag{11}$$

$$RMSE = \sqrt{\frac{1}{T}\sum_{i=1}^{T}\left|V_{est}^{i} - V_{gt}^{i}\right|^{2}}, \tag{12}$$

where T is the number of the test sample. V_{est}^{i} and V_{gt}^{i} denotes the estimated and ground truth value of the *i-th* image, respectively.

Meanwhile, we adopt the indicators of GFLOPs and runtime [27,41] to evaluate the efficiency of the models.

4.4. Experiments on the RSOC Dataset

The objective comparison results between the proposed method and SOTA methods on the RSOC dataset are presented in Table 2.

Table 2. Comparative results on the RSOC dataset. The best and the second-best results are highlighted in red and blue, respectively.

Method	Building		Small-Vehicle		Large-Vehicle		Ship	
	MAE	RMSE	MAE	RMSE	MAE	RMSE	MAE	RMSE
MCNN [15]	13.65	16.56	488.65	1317.44	36.56	55.55	263.91	412.30
CMTL [42]	12.78	15.99	490.53	1321.11	61.02	78.25	251.17	403.07
SANet [29]	29.01	32.96	497.22	1276.66	62.78	79.65	302.37	436.91
CSRNet [30]	8.00	11.78	443.72	1252.22	34.10	46.42	240.01	394.81
SCAR [22]	26.90	31.35	497.22	1276.65	62.78	79.64	302.37	436.92
SPN [19]	7.74	11.48	455.16	1252.92	36.21	50.65	241.43	392.88
CAN [32]	9.12	13.38	457.36	1260.39	34.56	49.63	282.69	423.44
SFCN [43]	8.94	12.87	440.70	1248.27	33.93	49.74	240.16	394.81
DSNet [12]	8.93	12.61	405.83	1254.16	27.44	42.38	206.25	346.96
SFANet [23]	8.18	11.75	435.29	1284.15	29.04	47.01	201.61	332.87
ASPDN [17]	7.54	10.52	433.23	1238.61	18.76	31.06	193.83	318.95
TASNet (Ours)	7.63	11.25	394.89	1196.83	22.75	37.13	191.82	278.17

One can see that the proposed TASNet achieves the best performance on the small-vehicle and ship datasets. Meanwhile, it achieves the second-best results on building and large-vehicle datasets. The possible reasons are that the building and large-vehicle datasets have more low-density scenarios than the small-vehicle and ship datasets. The proposed PFA module pays much more attention to the dense scenes, resulting in a suboptimal counting performance for sparse regions. Particularly, on the most challenging small-vehicle dataset, the TASNet improves the MAE by 20.5% and 9.3%, and RMSE by 6.3% and 4.1%, compared with SCAR [22] and SFANet [23], both of which adopt an attention mechanism to suppress the background clutter. On the ship dataset, the TASNet outperforms the ASPDN [17] in both MAE and RMSE. Although the results on building and large-vehicle datasets are the second best, the results are still competitive and convincing, especially compared with the multi-column networks, e.g., SANet [29] and SPN [19]. Notably, the small-vehicle and ship datasets are more challenging than the other two datasets in terms of background clutter and scale variation. Experimental results prove that the proposed method can well address these two problems and presents superior counting performance. Even so, there still exists a large improvement room on small-vehicle and ship datasets in accuracy and robustness.

The visualization results on the four subdatasets of the RSOC dataset are presented in Figures 5, 6, 7 and 8, respectively. Within these four figures, from top to bottom are the input remote sensing images, the corresponding ground truth density maps and the predicted density maps, where 'GT' and 'Est' denote the real value and estimated value, respectively. From the visualizations results, one can see that the estimated counting number and the density map are very close to ground truth.

Some visual comparison of TASNet with other models (SFAN [23], DSNet [12] and ASPDN [17]) on the RSOC_ship dataset are illustrated in Figure 9. It proves that the TASNet can better handle the background clutter and scale variation than the other three methods. In addition, the estimated values are closer to ground truth compared with other methods.

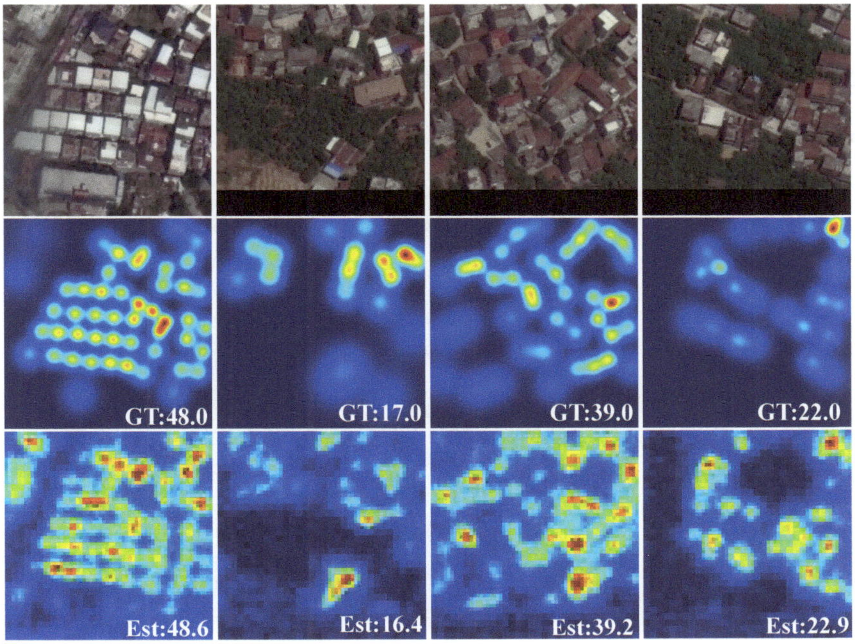

Figure 5. Subjective results on the RSOC_building dataset.

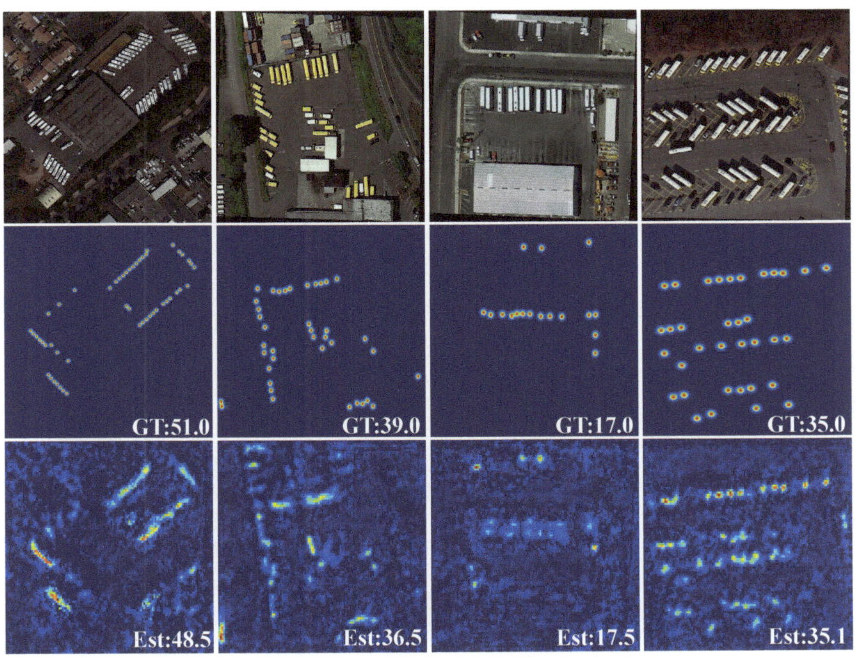

Figure 6. Subjective results on the RSOC_large-vehicle subdataset.

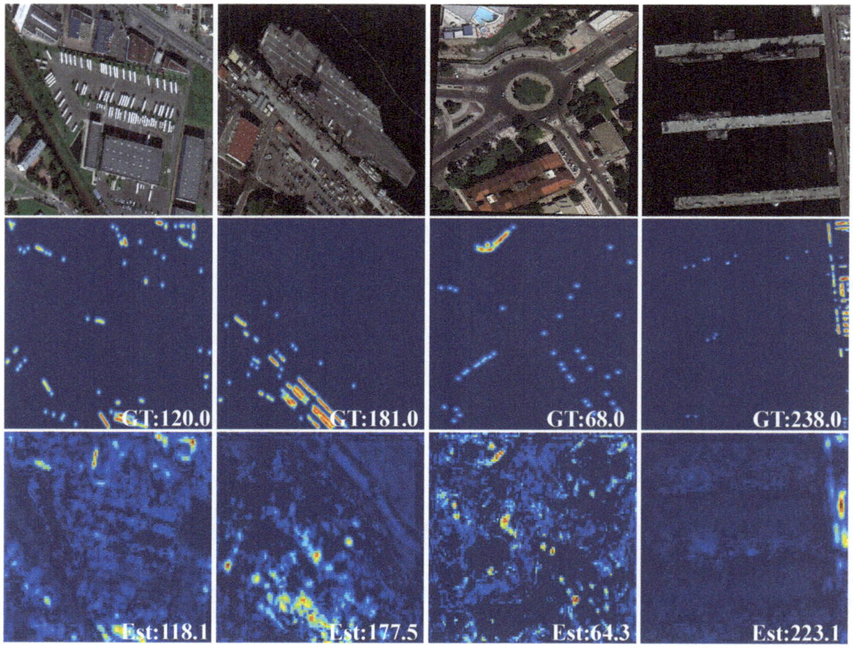

Figure 7. Subjective results on the RSOC_small-vehicle subdataset.

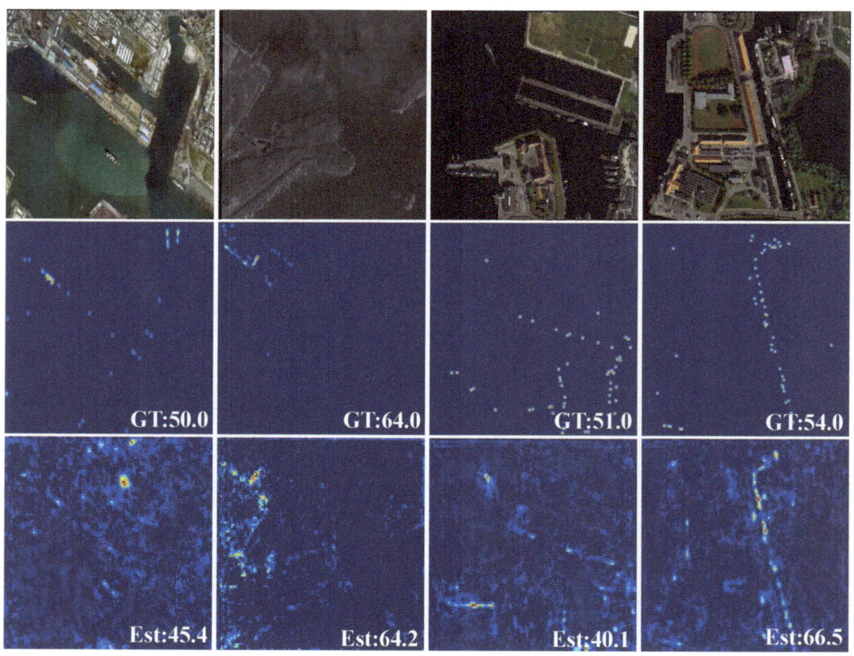

Figure 8. Subjective results on the RSOC_ship subdataset.

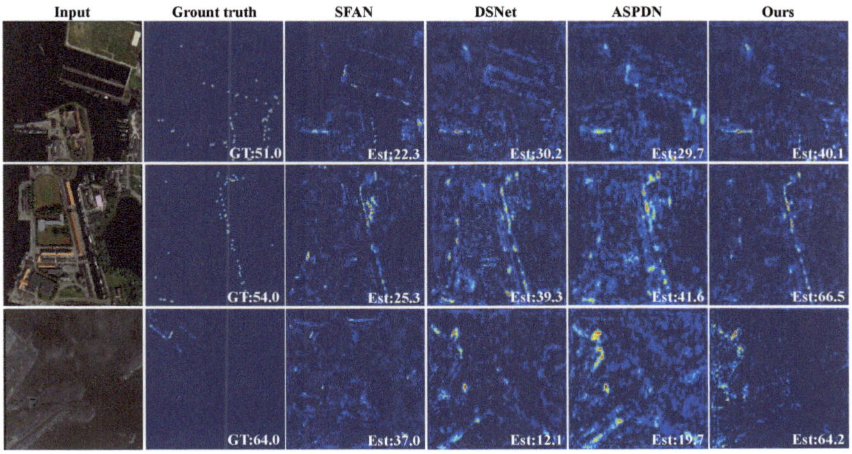

Figure 9. Subjective comparisons of different models on the RSOC_ship subdataset.

4.5. Experiments on the CARPK Dataset

The objective comparison results on the CARPK dataset are shown in Table 3.

Table 3. Experimental results on the CARPK and PUCPR+ datasets. The best and the second-best results are highlighted in red and blue, respectively.

Methods	CARPK		PUCPR+	
	MAE	RMSE	MAE	RMSE
YOLO [9]	102.89	110.02	156.72	200.54
FRCN [44]	103.48	110.64	156.76	200.59
LEP [45]	51.83	-	15.17	-
LPN [40]	23.80	36.79	22.76	34.46
SSD [46]	37.33	42.32	119.24	132.22
RetinaNet [47]	16.62	22.30	24.58	33.12
One-Look Regression [48]	59.46	66.84	21.88	36.73
MCNN [15]	39.10	43.30	21.86	29.53
CSRNet [30]	11.48	13.32	8.65	10.24
BL [49]	9.58	11.38	6.54	8.13
PSGCNet [16]	8.15	10.46	5.24	7.36
TASNet (Ours)	7.16	10.23	5.16	6.76

The comparison methods fall into three categories, i.e., detection-based methods (from the first row to sixth row), regression-based methods (the seventh row) and deep learning-based methods (from the eighth row to eleventh row). Obviously, there is still a big margin between the first two methods and the deep learning-based methods. In addition, the counting performance of the regression-based method [48] is worse than that of most detection-based methods [40,45].

Compared with other deep learning methods, the proposed TASNet performs best in MAE and RMSE. Specifically, the TASNet achieved a score of 7.16 and 10.23 in MAE and RMSE, which are far ahead of the competitors. In particular, it improves the MAE by 12.1% compared with the PSGCNet [16], which is built specifically to cope with the problems of background clutter and scale variation. Figure 10 shows some visualized results on the CARPK dataset. One can see that the estimated maps are very close to the ground truth.

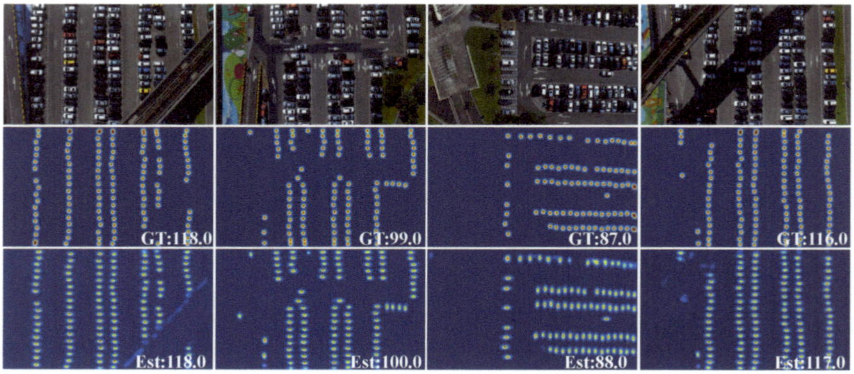

Figure 10. Subjective results on the CARPK dataset.

4.6. Experiments on the PUCPR+ Dataset

The objective comparison results on the PUCPR+ dataset are also shown in Table 3. Different from the CARPK dataset, the PUCPR+ dataset has a severe problem of weather variation, which is a big challenge for object counting. The experimental results show that the TASNet scores 5.16 in MAE and 6.76 in RMSE, which both outperform the competitors. Specifically, it reduces by 40.3% and 33.9% compared with CSRNet [30], which is proposed for addressing scale variation. Figure 11 shows some visualization results on the PUCPR+ datasets. It also verifies the effectiveness of the proposed TASNet in generating the density map and estimating the object counting.

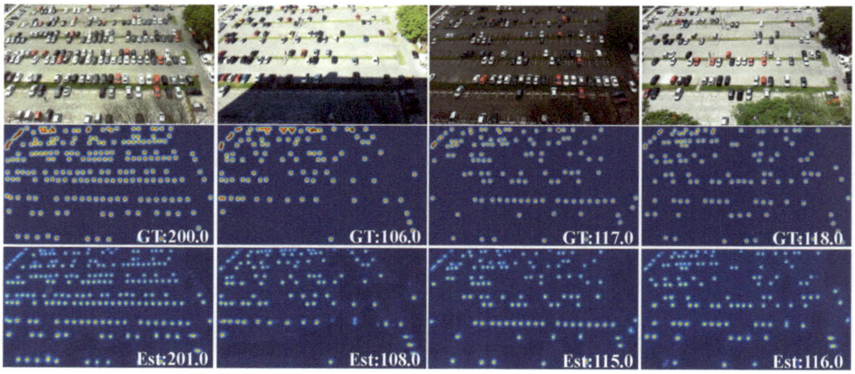

Figure 11. Subjective results on the PUCPR+ dataset.

4.7. Ablation Studies

In this section, we conduct a series of ablation studies to verify the different compound modes, the backbone networks, and the multiscale information fusion mechanisms.

4.7.1. Ablation Studies on the Modules

To trial the effectiveness of each module and the effect of different compound modes, we set up a series of ablation studies. The objective comparison results are listed in Table 4. The configuration information and analysis are described below.

Table 4. Comparative results of the baseline with different compound modes on the ship dataset. The best performances are highlighted in **bold**.

Methods	MAE	RMSE
Baseline	240.01	394.81
Baseline + TVA	237.88	373.97
Baseline + PFA	223.11	362.42
Baseline + PFA + TVA	214.22	312.35
Baseline+Parallel (TVA & PFA)	205.54	291.72
Baseline + TVA + PFA	**191.82**	**278.17**

a. **Baseline:** The baseline is considered as the pre-trained VGG16 with the decoder. It shows that the output results of the baseline are the worst.
b. **Baseline + TVA:** The combination is to insert the TVA module between VGG16 and the decoder. One can see that the TVA module is beneficial in boosting the counting performance.
c. **Baseline + PFA:** The group embeds the PFA module into the baseline. It proves that the PFA module is also conducive to the estimated performance.
d. **Baseline + PFA + TVA:** Insert PFA and TVA modules successively in the baseline. It can be observed that the MAE and RMSE are improved by 10.7% and 20.9% compared with the baseline, respectively. It reveals that the combination of PFA and TVA modules is better than that of a single module.
e. **Baseline + Parallel (TVA and PFA):** Connect the PFA and TVA modules in parallel and then add them to the baseline. Again, the results show that the performance improves more than the aforementioned compound modes.
f. **Baseline + TVA + PFA:** Embed the TVA and PFA modules successively in the baseline. Intuitively, it exhibits the best performance in MAE and RMSE compared with all the configurations mentioned above.

The visualization results of the baseline with different components are shown in Figure 12. It shows that both the TVA and PFA module are helpful to boost the counting performance. The proposed TVA module can effectively suppress the background noise. The PFA module can capture more large-scale information, but it incorrectly estimates other objects. Moreover, the problems of background clutter and scale variation are alleviated by adding the two modules to the baseline, but the compound mode of "Baseline + TVA + PFA" (TASNet) achieves the best results.

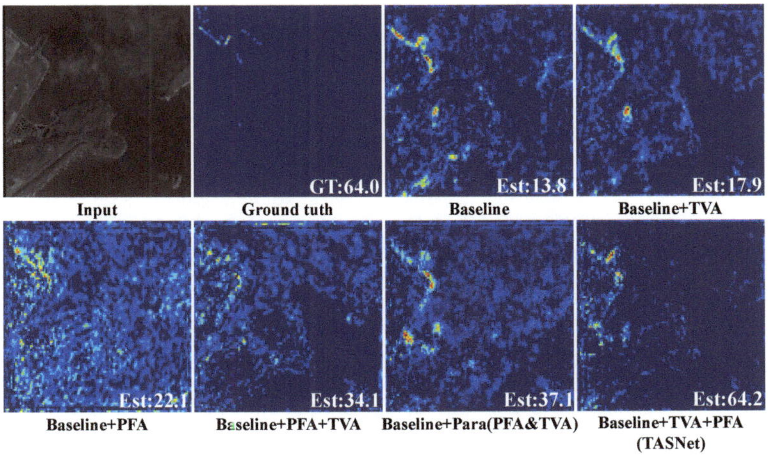

Figure 12. The qualitative comparison of the baseline with different components.

4.7.2. Ablation Studies on Backbone Networks

In addition to exploring the effectiveness of the proposed modules, ablation studies on backbone networks are also explored. Three networks, ResNet-50 [50], ResNeXt [51] and VGG-16, are adopted as backbone. The comparative results are shown in Table 5. It proves that adopting VGG-16 as the backbone network obtains the best performance. In fact, VGG is generally adopted as the backbone for feature extraction in most counting tasks [12,17,30] because of its strong generalization ability.

Table 5. Comparative results of different backbones on the ship dataset. The best performances are highlighted in **bold**.

Methods	MAE	RMSE
TASNet (Resnet-50)	215.39	346.81
TASNet (ResneXt)	197.83	327.66
TASNet (VGG-16)	**191.82**	**278.17**

4.7.3. Ablation Studies on Multiscale Feature Fusion Mechanisms

In order to further prove the superiority of the proposed PFA module in multiscale feature fusion, three multiscale feature fusion mechanisms, namely scale pyramid module (SPM) [19], dense scale dilated convolution block (DDCB) [12] and pyramid scale module (PSM) [16], are compared. The comparative results are depicted in Table 6. From the table, one can see that the PFA module is superior to the other three modules in multiscale information fusion. Specifically, compared with SPM and PSM which are built with pyramid structures, the proposed PFA module improves by 0.45% and 13.03% in MAE, and 15.51% and 15.89% in RMSE. Meanwhile, compared with DDCB, which is a single block with multiple dense connections, the PFA module reduces by 1.37% and 5.53% in MAE and RMSE. The advantage of the PFA module is that it introduces an information transmit mechanism in the pyramid structure which can capture more scale features.

Table 6. Comparative results of different multiscale feature fusion mechanisms on the ship dataset. The best performances are highlighted in **bold**.

Methods	MAE	RMSE
Baseline + TVA + SPM	192.69	329.45
Baseline + TVA+DDCB	194.48	294.46
Baseline + TVA + PSM	220.58	330.71
Baseline + TVA + PFA	**191.82**	**278.17**

4.8. *Efficiency Comparison*

To evaluate the efficiency of the proposed TASNet, we conduct comparative experiments to test the computational complexity and inferring time on an RTX 3090Ti GPU. The comparison results are reported in Table 7. One can see that the proposed TASNet scores 20.5 M and 377.04 in parameters and GFLOPs both achieving the second-best results among the competitors. Although the SFANet [23] ranks first in parameters and GFLOPs, it performs worse in counting accuracy and robustness than the TASNet (as shown in Table 2).

For the inferring time, the TASNet achieve 33.6, outperforming all competitors. Specifically, compared with the ASPDN [17], which performs best in the RSOC_building and RSOC_large-vehicle (as shown in Table 2), the TASNet improves by 40.2% in inferring time, which demonstrates that the TASNet is more efficient to ASPDN. In addition, the TASNet decreases by 1.4% compared with SFANet [23] with the least number of parameters, as it has a more bloated model.

Table 7. Comparison results of the TASNet and other models in parameters, GFLOPs and inferring time. The input size is set to 1024 × 768.

Methods	Params (M)	GFLOPs	Inferring Time (ms)
SFANet [23]	17.0	312.83	34.1
DSNet [12]	20.7	379.15	34.5
ASPDN [17]	22.7	455.45	56.2
PSGCNet [16]	27.5	385.59	33.7
TASNet (Ours)	20.5	377.04	33.6

4.9. Failure Cases

Although the proposed TASNet is capable of achieving the superior counting performance against other counting methods, it still has some failure cases, especially on the RSOC dataset. Some failure cases are visualized in Figure 13. One can see that there is a large gap between the estimated value and the ground truth, especially for tiny object counting in dense scenarios.

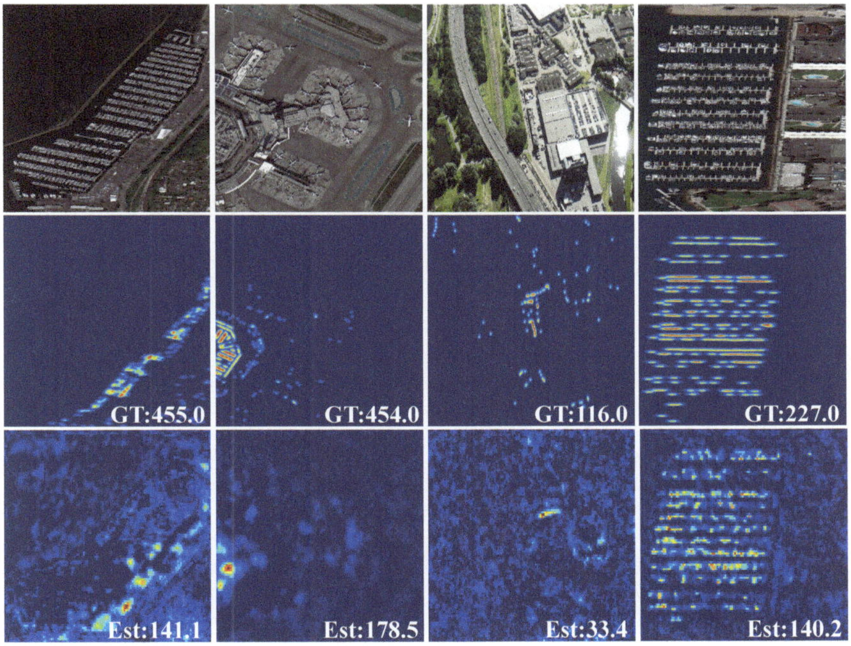

Figure 13. Failure cases on the RSOC dataset.

Tiny object counting is very valuable for real-world vision application and differs from general counting tasks. For instance, since the objects are tiny while the whole input image has a relatively large field-of-view, there is much less information from the targeting objects and much more from background distractions. The distinctions make tiny object counting a uniquely challenging task [52]. It indicates that there is still a large amount of room for tiny object counting.

5. Conclusions and Future Work

In this paper, we present the TASNet to address the problems of background clutter and scale variation for object counting in remote sensing. The proposed TASNet is characterized by two key components, i.e., a TVA module and a PFA module. Specifically, the TVA module is built to capture the dependencies of features across the spatial and channel dimensions. It can emphasize the object region and suppress the background noise. Meanwhile, the PFA module is introduced to solve the scale variation by extracting multiscale information with a four-branch architecture. Experimental results on extensive remote sensing datasets verified the effectiveness and superiority of the proposed TASNet. In the future, more efforts are expected to tiny object counting in dense scenarios, as the tiny objects have fewer detailed information and are highly susceptible to background interference.

Author Contributions: Conceptualization, X.G., M.G. and G.J.; methodology, X.G. and M.G.; software, X.G. and G.J.; validation, X.G. and M.G.; formal analysis, X.G. and M.G.; investigation, X.G. and M.A.; resources, M.G. and M.A.; writing, X.G.; supervision, M.G., G.J. and M.A.; project administration, X.G., G.J. and M.G. All authors have read and agreed to the published version of the manuscript.

Funding: This work is supported in part by the National Natural Science Foundation of China (No. 61801272).

Data Availability Statement: Not applicable.

Acknowledgments: This work is supported in part by the National Natural Science Foundation of China (No. 61801272). Many thanks to Abdellah Chehri in the Department of Mathematics and Computer Science at the Royal Military College of Canada for his help in proofreading and polishing the language.

Conflicts of Interest: The authors declare no conflict of interest.

References

1. Zhang, Q.; Cong, R.; Li, C.; Cheng, M.M.; Fang, Y.; Cao, X.; Zhao, Y.; Kwong, S.T.W. Dense Attention Fluid Network for Salient Object Detection in Optical Remote Sensing Images. *IEEE Trans. Image Process.* **2021**, *30*, 1305–1317. [CrossRef]
2. Gadamsetty, S.; Ch, R.; Ch, A.; Iwendi, C.; Gadekallu, T.R. Hash-Based Deep Learning Approach for Remote Sensing Satellite Imagery Detection. *Water* **2022**, *14*, 707. [CrossRef]
3. Bazi, Y.; Bashmal, L.; Rahhal, M.M.A.; Dayil, R.A.; Ajlan, N.A. Vision Transformers for Remote Sensing Image Classification. *Remote. Sens.* **2021**, *13*, 516. [CrossRef]
4. Zhang, S.; Yuan, Q.; Li, J.; Sun, J.; Zhang, X. Scene-Adaptive Remote Sensing Image Super-Resolution Using a Multiscale Attention Network. *IEEE Trans. Geosci. Remote. Sens.* **2020**, *58*, 4764–4779. [CrossRef]
5. Rathore, M.M.; Ahmad, A.; Paul, A.; Rho, S. Urban planning and building smart cities based on the Internet of Things using Big Data analytics. *Comput. Netw.* **2016**, *101*, 63–80. [CrossRef]
6. Grinias, I.; Panagiotakis, C.; Tziritas, G. MRF-based segmentation and unsupervised classification for building and road detection in peri-urban areas of high-resolution satellite images. *Isprs J. Photogramm. Remote. Sens.* **2016**, *122*, 145–166. [CrossRef]
7. Benedek, C.; Descombes, X.; Zerubia, J. Building Development Monitoring in Multitemporal Remotely Sensed Image Pairs with Stochastic Birth-Death Dynamics. *IEEE Trans. Pattern Anal. Mach. Intell.* **2012**, *34*, 33–50. [CrossRef]
8. Fan, Y.; Wen, Q.; Wang, W.; Wang, P.; Li, L.; Zhang, P. Quantifying Disaster Physical Damage Using Remote Sensing Data—A Technical Work Flow and Case Study of the 2014 Ludian Earthquake in China. *Int. J. Disaster Risk Sci.* **2017**, *8*, 471–488. [CrossRef]
9. Redmon, J.; Divvala, S.K.; Girshick, R.B.; Farhadi, A. You Only Look Once: Unified, Real-Time Object Detection. In Proceedings of the 2016 IEEE Conference on Computer Vision and Pattern Recognition (CVPR), Las Vegas, NV, USA, 27–30 June 2016; pp. 779–788. [CrossRef]
10. Girshick, R.B. Fast R-CNN. In Proceedings of the 2015 IEEE International Conference on Computer Vision (ICCV), Santiago, Chile, 7–13 December 2015; pp. 1440–1448. [CrossRef]
11. Pham, V.Q.; Kozakaya, T.; Yamaguchi, O.; Okada, R. COUNT Forest: CO-Voting Uncertain Number of Targets Using Random Forest for Crowd Density Estimation. In Proceedings of the 2015 IEEE International Conference on Computer Vision (ICCV), Santiago, Chile, 7–13 December 2015; pp. 3253–3261. [CrossRef]
12. Dai, F.; Liu, H.; Ma, Y.; Cao, J.; Zhao, Q.; Zhang, Y. Dense Scale Network for Crowd Counting. In Proceedings of the 2021 International Conference on Multimedia Retrieval, Tokyo, Japan, 22–24 March 2021. [CrossRef]
13. Gao, J.; Gong, M.; Li, X. Global Multi-Scale Information Fusion for Multi-Class Object Counting in Remote Sensing Images. *Remote. Sens.* **2022**, *14*, 4026. [CrossRef]
14. Gao, G.; Gao, J.; Liu, Q.; Wang, Q.; Wang, Y. CNN-based Density Estimation and Crowd Counting: A Survey. *arXiv* **2020**, arXiv:2003.12783.

15. Zhang, Y.; Zhou, D.; Chen, S.; Gao, S.; Ma, Y. Single-Image Crowd Counting via Multi-Column Convolutional Neural Network. In Proceedings of the IEEE Conference on Computer Vision and Pattern Recognition (CVPR), Las Vegas, NV, USA, 27–30 June 2016; pp. 589–597. [CrossRef]
16. Gao, G.; Liu, Q.; Hu, Z.; Li, L.; Wen, Q.; Wang, Y. PSGCNet: A Pyramidal Scale and Global Context Guided Network for Dense Object Counting in Remote-Sensing Images. *IEEE Trans. Geosci. Remote. Sens.* **2022**, *60*, 1–12. [CrossRef]
17. Gao, G.; Liu, Q.; Wang, Y. Counting From Sky: A Large-Scale Data Set for Remote Sensing Object Counting and a Benchmark Method. *IEEE Trans. Geosci. Remote. Sens.* **2021**, *59*, 3642–3655. [CrossRef]
18. Lan, M.; Zhang, Y.; Zhang, L.; Du, B. Global context based automatic road segmentation via dilated convolutional neural network. *Inf. Sci.* **2020**, *535*, 156–171. [CrossRef]
19. Chen, X.; Bin, Y.; Sang, N.; Gao, C. Scale Pyramid Network for Crowd Counting. In Proceedings of the 2019 IEEE Winter Conference on Applications of Computer Vision (WACV), Waikoloa, HI, USA, 7–11 January 2019; pp. 1941–1950. [CrossRef]
20. Guo, X.; Gao, M.; Zhai, W.; Shang, J.; Li, Q. Spatial-Frequency Attention Network for Crowd Counting. *Big Data* **2022**, *10*, 453–465. [CrossRef] [PubMed]
21. Zhai, W.; Gao, M.; Anisetti, M.; Li, Q.; Jeon, S.; Pan, J. Group-split attention network for crowd counting. *J. Electron. Imaging* **2022**, *31*, 41214. [CrossRef]
22. Gao, J.; Wang, Q.; Yuan, Y. SCAR: Spatial-/channel-wise attention regression networks for crowd counting. *Neurocomputing* **2019**, *363*, 1–8. [CrossRef]
23. Zhu, L.; Zhao, Z.; Lu, C.; Lin, Y.; Peng, Y.; Yao, T. Dual Path Multi-Scale Fusion Networks with Attention for Crowd Counting. *arXiv* **2019**, arXiv:1902.01115.
24. Jiang, X.; Zhang, L.; Xu, M.; Zhang, T.; Lv, P.; Zhou, B.; Yang, X.; Pang, Y. Attention Scaling for Crowd Counting. In Proceedings of the IEEE Conference on Computer Vision and Pattern Recognition (CVPR), Seattle, WA, USA, 13–19 June 2020; pp. 4705–4714. [CrossRef]
25. Khan, K.; Khan, R.; Albattah, W.; Nayab, D.; Qamar, A.M.; Habib, S.; Islam, M. Crowd Counting Using End-to-End Semantic Image Segmentation. *Electronics* **2021**, *10*, 1293. [CrossRef]
26. Meng, Y.; Zhang, H.; Zhao, Y.; Yang, X.; Qian, X.; Huang, X.; Zheng, Y. Spatial Uncertainty-Aware Semi-Supervised Crowd Counting. In Proceedings of the 2021 IEEE/CVF International Conference on Computer Vision (ICCV), Montreal, QC, Canada, 10–17 October 2021; pp. 15529–15539. [CrossRef]
27. Gao, J.; Wang, Q.; Li, X. PCC Net: Perspective Crowd Counting via Spatial Convolutional Network. *IEEE Trans. Circuits Syst. Video Technol.* **2020**, *30*, 3486–3498. [CrossRef]
28. Liu, Y.; Liu, L.; Wang, P.; Zhang, P.; Lei, Y. Semi-Supervised Crowd Counting via Self-Training on Surrogate Tasks. *arXiv* **2020**, arXiv:2007.03207.
29. Cao, X.; Wang, Z.; Zhao, Y.; Su, F. Scale Aggregation Network for Accurate and Efficient Crowd Counting. In Proceedings of the European Conference on Computer Vision (ECCV), Munich, Germany, 8–14 September 2018; ._45. [CrossRef]
30. Li, Y.; Zhang, X.; Chen, D. CSRNet: Dilated Convolutional Neural Networks for Understanding the Highly Congested Scenes. In Proceedings of the IEEE Conference on Computer Vision and Pattern Recognition (CVPR), Salt Lake City, UT, USA, 18–23 June 2018; pp. 1091–1100. [CrossRef]
31. Liu, L.; Qiu, Z.; Li, G.; Liu, S.; Ouyang, W.; Lin, L. Crowd Counting With Deep Structured Scale Integration Network. In Proceedings of the 2019 IEEE/CVF International Conference on Computer Vision (ICCV), Seoul, Republic of Korea, 27 October–2 November 2019; pp. 1774–1783. [CrossRef]
32. Liu, W.; Salzmann, M.; Fua, P. Context-Aware Crowd Counting. In Proceedings of the IEEE Conference on Computer Vision and Pattern Recognition (CVPR), Long Beach, CA, USA, 15–20 June 2019; pp. 5094–5103. [CrossRef]
33. Zhu, F.; Yan, H.; Chen, X.; Li, T.; Zhang, Z. A multi-scale and multi-level feature aggregation network for crowd counting. *Neurocomputing* **2021**, *423*, 46–56. [CrossRef]
34. Duan, Z.; Wang, S.; Di, H.; Deng, J. Distillation Remote Sensing Object Counting via Multi-Scale Context Feature Aggregation. *IEEE Trans. Geosci. Remote. Sens.* **2022**, *60*, 1–12. [CrossRef]
35. Han, K.; Wang, Y.; Tian, Q.; Guo, J.; Xu, C.; Xu, C. GhostNet: More Features From Cheap Operations. In Proceedings of the 2020 IEEE/CVF Conference on Computer Vision and Pattern Recognition (CVPR), Seattle, WA, USA, 13–19 June 2020; pp. 1577–1586. [CrossRef]
36. Chen, J.; Wang, K.; Su, W.; Wang, Z. SSR-HEF: Crowd Counting With Multiscale Semantic Refining and Hard Example Focusing. *IEEE Trans. Ind. Inform.* **2022**, *18*, 6547–6557. [CrossRef]
37. Simonyan, K.; Zisserman, A. Very Deep Convolutional Networks for Large-Scale Image Recognition. *arXiv* **2015**, arXiv:1409.1556.
38. Misra, D.; Nalamada, T.; Arasanipalai, A.U.; Hou, Q. Rotate to Attend: Convolutional Triplet Attention Module. In Proceedings of the 2021 IEEE Winter Conference on Applications of Computer Vision (WACV), Waikoloa, HI, USA, 3–8 January 2021; pp. 3138–3147. [CrossRef]
39. Zhai, W.; Gao, M.; Souri, A.; Li, Q.; Guo, X.; Shang, J.; Zou, G. An attentive hierarchy ConvNet for crowd counting in smart city. *Clust. Comput.* **2022**. [CrossRef]
40. Hsieh, M.R.; Lin, Y.L.; Hsu, W.H. Drone-Based Object Counting by Spatially Regularized Regional Proposal Network. In Proceedings of the International Conference on Computer Vision (ICCV), Venice, Italy, 22–29 October 2017; pp. 4165–4173. [CrossRef]

41. Wang, P.; Gao, C.; Wang, Y.; Li, H.; Gao, Y. MobileCount: An efficient encoder-decoder framework for real-time crowd counting. *Neurocomputing* **2020**, *407*, 292–299. [CrossRef]
42. Sindagi, V.; Patel, V. CNN-Based cascaded multi-task learning of high-level prior and density estimation for crowd counting. In Proceedings of the IEEE International Conference on Advanced Video and Signal Based Surveillance (AVSS), Lecce, Italy, 29 August–1 September 2017; pp. 1–6. [CrossRef]
43. Wang, Q.; Gao, J.; Lin, W.; Yuan, Y. Learning From Synthetic Data for Crowd Counting in the Wild. In Proceedings of the IEEE Conference on Computer Vision and Pattern Recognition (CVPR), Long Beach, CA, USA, 15–20 June 2019; pp. 8190–8199. [CrossRef]
44. Ren, S.; He, K.; Girshick, R.B.; Sun, J. Faster R-CNN: Towards Real-Time Object Detection with Region Proposal Networks. *TPAMI* **2015**, *39*, 1137–1149. [CrossRef] [PubMed]
45. Stahl, T.; Pintea, S.L.; Gemert, J.C.V. Divide and Count: Generic Object Counting by Image Divisions. *IEEE Trans. Image Process.* **2019**, *28*, 1035–1044. [CrossRef]
46. Liu, W.; Anguelov, D.; Erhan, D.; Szegedy, C.; Reed, S.E.; Fu, C.Y.; Berg, A.C. SSD: Single Shot MultiBox Detector. In Proceedings of the European Conference on Computer Vision (ECCV), Amsterdam, The Netherlands, 11–14 October 2016; pp. 21–37. [CrossRef]
47. Lin, T.Y.; Goyal, P.; Girshick, R.B.; He, K.; Dollár, P. Focal Loss for Dense Object Detection. *IEEE Trans. Pattern Anal. Mach. Intell.* **2020**, *42*, 318–327. [CrossRef]
48. Mundhenk, T.N.; Konjevod, G.; Sakla, W.A.; Boakye, K. A Large Contextual Dataset for Classification, Detection and Counting of Cars with Deep Learning. In Proceedings of the European Conference on Computer Vision (ECCV), Amsterdam, The Netherlands, 11–14 October 2016. [CrossRef]
49. Ma, Z.; Wei, X.; Hong, X.; Gong, Y. Bayesian Loss for Crowd Count Estimation With Point Supervision. In Proceedings of the 2019 IEEE/CVF International Conference on Computer Vision (ICCV), Seoul, Republic of Korea, 27 October–2 November 2019; pp. 6141–6150. [CrossRef]
50. He, K.; Zhang, X.; Ren, S.; Sun, J. Deep Residual Learning for Image Recognition. In Proceedings of the 2016 IEEE Conference on Computer Vision and Pattern Recognition (CVPR), Las Vegas, NV, USA, 27–30 June 2016; pp. 770–778. [CrossRef]
51. Xie, S.; Girshick, R.B.; Dollár, P.; Tu, Z.; He, K. Aggregated Residual Transformations for Deep Neural Networks. In Proceedings of the 2017 IEEE Conference on Computer Vision and Pattern Recognition (CVPR), Honolulu, HI, USA, 21–26 July 2017; pp. 5987–5995. [CrossRef]
52. Yu, X.; Han, Z.; Gong, Y.; Jan, N.; Zhao, J. The 1st Tiny Object Detection Challenge: Methods and Results. In Proceedings of the 2020 ECCV Workshops, Glasgow, UK, 23–28 August 2020. [CrossRef]

remote sensing

Article

A Spatial Cross-Scale Attention Network and Global Average Accuracy Loss for SAR Ship Detection

Lili Zhang [1], Yuxuan Liu [1,*], Lele Qu [1], Jiannan Cai [1] and Junpeng Fang [2]

[1] School of Electronical and Information Engineering, Shenyang Aerospace University, Shenyang 110136, China
[2] School of Integrated Circuits, Tsinghua University, Beijing 100084, China
* Correspondence: liuyuxuan1@stu.sau.edu.cn

Abstract: A neural network-based object detection algorithm has the advantages of high accuracy and end-to-end processing, and it has been widely used in synthetic aperture radar (SAR) ship detection. However, the multi-scale variation of ship targets, the complex background of near-shore scenes, and the dense arrangement of some ships make it difficult to improve detection accuracy. To solve the above problem, in this paper, a spatial cross-scale attention network (SCSA-Net) for SAR image ship detection is proposed, which includes a novel spatial cross-scale attention (SCSA) module for eliminating the interference of land background. The SCSA module uses the features at each scale output from the backbone to calculate where the network needs attention in space and enhances the features of the feature pyramid network (FPN) output to eliminate interference from noise, and land complex backgrounds. In addition, this paper analyzes the reasons for the "score shift" problem caused by average precision loss (AP loss) and proposes the global average precision loss (GAP loss) to solve the "score shift" problem. GAP loss enables the network to distinguish positive samples and negative samples faster than focal loss and AP loss, and achieve higher accuracy. Finally, we validate and illustrate the effectiveness of the proposed method by performing it on SAR Ship Detection Dataset (SSDD), SAR-ship-dataset, and High-Resolution SAR Images Dataset (HRSID). The experimental results show that the proposed method can significantly reduce the interference of background noise on the ship detection results, improve the detection accuracy, and achieve superior results to the existing methods.

Keywords: object detection; deep learning; synthetic aperture radar (SAR); spatial cross-scale attention (SCSA); global average accuracy loss (GAP loss)

Citation: Zhang, L.; Liu, Y.; Qu, L.; Cai, J.; Fang, J. A Spatial Cross-Scale Attention Network and Global Average Accuracy Loss for SAR Ship Detection. *Remote Sens.* **2023**, *15*, 350. https://doi.org/10.3390/rs15020350

Academic Editor: João Catalão Fernandes

Received: 11 November 2022
Revised: 31 December 2022
Accepted: 3 January 2023
Published: 6 January 2023

1. Introduction

Synthetic aperture radar (SAR) is a microwave sensor based on the scattering characteristics of electromagnetic waves for imaging, which has a certain cloud and ground penetration capability to detect hidden targets. This characteristic makes it suited for marine monitoring, mapping, and military applications. With the continuous exploitation of marine resources, the monitoring of marine vessels based on SAR images has received increasing attention. SAR image object detection aims to automatically locate and identify specific targets from images and has important application prospects in defense and civil fields such as target identification, target detection, marine development, and terrain classification [1–4].

The development of SAR image ship detection methods can be divided into two stages [5]: traditional detection methods represented by constant false alarm rate (CFAR) algorithms and deep learning methods represented by convolutional neural networks. The constant false alarm rate algorithm adaptively adjusts the threshold value using statistical models and sample selection strategies. It has been widely used due to its constant false alarm rate and low complexity. For example, the bilateral CFAR algorithm [6] takes into account the intensity distribution and spatial distribution when selecting thresholds to improve the accuracy of detection. The two-parameter CFAR [7] uses log-normal as the statistical model with more accurate parameter estimation and achieves better detection performance in a multi-target environment. The modified CFAR algorithm proposed in the paper [8] uses variable guard windows to improve its detection performance for the multi-scale scene. Li et al. [9] propose an adaptive CFAR method based on intensity and texture feature fusion attention contrast mechanism to suppress clutter background and speckle noise with better performance in complex backgrounds and multi-target marine environments. The above CFAR algorithms rely on the land-segmentation algorithm in the nearshore scene, while the adaptive superpixel-level CFAR algorithm [10] is based on the superpixel segmentation method, which does not require additional land-segmentation algorithm, and its detection performance is still reliable in near-shore scenarios. However, these algorithms are based on the modeling of clutter statistical features, which are more sensitive to the complex shoreline, ocean clutter, and coherent scattering noise, with low detection accuracy and poor generalization [5]. Currently, deep learning is the mainstream research direction due to its excellent feature extraction capability and end-to-end training process, which is gradually replacing the traditional methods.

In the field of target detection, deep learning methods based on convolutional neural networks have developed rapidly due to the translation invariance and weight-sharing properties of convolution. The existing target detection methods based on convolutional neural networks can be divided into two-stage and one-stage methods. The two-stage network first generates the proposed region in the first stage and then performs classification and correction of the proposed region in the second stage. This method has higher detection accuracy and better detection for small targets, such as region-convolutional neural network (R-CNN) [11], Fast R-CNN [12], Faster R-CNN [13], Sparse R-CNN [14], etc. The one-stage network does not need to generate proposed regions and directly outputs the classification and location results of the target. Although the accuracy of the one-stage network is generally lower than that of the two-stage network, they have higher detection speed and simple training steps, such as YOLO series [15–22], single shot detection (SSD) [23], RetinaNet [24], fully convolutional one-stage object detection (FCOS) [25], etc.

Since the release of the SAR ship detection dataset (SSDD) by Li et al. [26] in 2017, SAR image ship detection based on convolutional neural networks developed rapidly, and subsequently in, 2019 and 2020, Wang et al. [27] and Wei et al. [28] released the SAR-Ship-Dataset and High-Resolution SAR Images Dataset (HRSID), respectively. Additionally, in 2021, Zhang et al. [29] corrected the mislabeling of the initial version of the SSDD and used a standardized format, which further facilitated the development of the field. In recent years, Shi et al. [30] proposed a feature aggregation enhancement pyramid network (FAEPN) to enhance the extraction capability of the network for multi-scale targets, taking into account the quality of classification and regression when assigning positive and negative samples. Cui et al. [31] proposed a dense attention pyramid network (DAPN), which used dense connectivity in the feature pyramid network (FPN) to obtain richer semantic information about each scale and used a convolutional block attention module (CBAM) to enhance the features. Zhang et al. [32] proposed a quad-feature pyramid network (Quad-FPN) to extract complex multi-scale features, and then replaced non-maximal suppression (NMS) with soft non-maximal suppression (soft-NMS) to improve the detection of densely aligned targets. Li et al. [4] proposed an attention-guided balanced feature pyramid network (A-BFPN), which contained an enhanced refinement module (ERM) for reducing background interference and a channel attention-guided fusion network for reducing confounding effects in mixed feature maps. Different from the above-mentioned anchor-base network, Zhu et al. [33] proposed a novel anchor-free network based on FCOS for SAR image target detection, in which a deformable convolution (Dconv) and an improved residual network (IRN) were added to optimize the feature extraction capability of the network. Wu et al. [34] proposed the instance segmentation-assisted ship detection network (ISASDNet), with a global reasoning module, which achieved good performance in instance segmentation and object detection. Wang et al. [35] used intersection over union (IoU) K-means to solve the extreme aspect ratio problem and embedded a soft threshold attention module (STA) in the network to suppress the effects of noise and complex backgrounds. Tian et al. [36] employed an object characteristic-driven image enhancement (OCIE) module to enhance the variety of datasets and explored the dense feature reuse (DFR) module and receptive field expansion (RFE) module to increase the network receptive field and strengthen the transmission of information flow. Wei et al. [37] adopted a high-resolution feature pyramid network (HRFPN) to make full use of the information in the high-resolution and low-resolution features and proposed a high-resolution ship detection network (HR-SDNet) on this basis.

In order to further improve the detection accuracy of the network, in this paper, a new anchor-free spatial cross-scale attention network (SCSA-Net) is proposed, which contains a novel spatial cross-scale attention (SCSA) module that enhances features at different scales while mitigating the interference generated by complex backgrounds such as land. In addition, to solve the "score shift" problem of average accuracy loss (AP loss), this paper improves AP loss and proposes global average accuracy loss (GAP loss). Three widely used datasets are used to experimentally validate the proposed methods and confirm their effectiveness.

The main contributions of our work are as follows:

1. A novel cross-scale spatial attention module is proposed, which consists of a cross-scale attention module and a spatial attention redistribution module. The former dynamically adjusts the position of network attention by combining information from different scales. The latter redistributes spatial attention to mitigate the influence of complex backgrounds and make the ship more distinctive.
2. We analyze the reasons why AP loss generates the "score shift" problem and propose a global average accuracy loss (GAP loss) to solve it. Compared to traditional methods using focus loss as the classification loss, training with GAP loss allows the network to optimize directly with the average precision (AP) as the target and to distinguish between positive and negative samples more quickly, achieving better detection results.
3. We propose an anchor-free spatial cross-scale attention network (SCSA-Net) for ship detection in SAR images, which reached 98.7% AP on the SSDD, 97.9% AP on the SAR-Ship-Dataset and 95.4% AP on the HRSID, achieving state-of-the-art performance.

The remainder of this paper is organized as follows: the second section introduces the proposed method; the third section describes the experiments and analysis of the results; the fourth section discusses and suggests future work; finally, the fifth section summarizes this paper.

2. Methods

This section describes SCSA-Net for SAR ship detection, which contains the overall architecture of SCSA-Net, the spatial cross-scale attention module, and GAP loss.

2.1. Overall Architecture of SCSA-Net

The main framework of SCSA-Net is depicted in Figure 1. FCOS [25] is chosen as the baseline, which is one of the most representative one-stage anchor-free detectors. Following the FCOS, the ResNet-50 is used as the backbone of SCSA-Net for feature extraction and the FPN is employed to solve the multi-scale target detection problem. Additionally, a new spatial cross-scale spatial attention module is added to the network, which dynamically enhances the FPN output features based on the backbone output features. The enhanced feature is then forwarded to the class-box subnet for classification and regression.

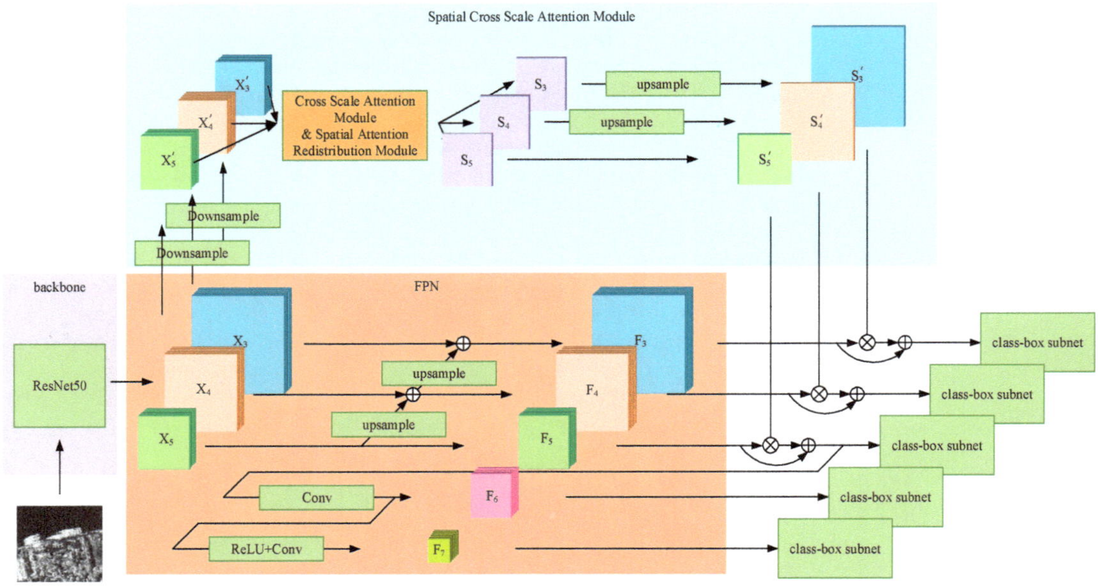

Figure 1. Overview of our proposed SCSA-Net. It consists of four sections: the backbone of the SCSA-Net, feature pyramid network (FPN), spatial cross-scale attention module, and class-box subnet.

2.2. Spatial Cross-Scale Attention Module

In the offshore scenes of SAR images, the network usually performs well because of the large distinction between sea level clutter and ship targets. However, in the near-shore scenes of SAR images, the land portion of the background is complex and the network easily confuses ship targets with smaller islands and land backgrounds having similar ship shapes. This leads to false alarms and missed detections. The spatial cross-scale attention module is proposed to reduce the interference of land-to-ship targets. It consists of two parts, the cross-scale attention module, and the spatial attention redistribution module. The cross-scale attention module determines the regions requiring attention at each scale, while the spatial attention redistribution module redistributes the attention regions at each scale spatially to locate more accurate attention regions.

2.2.1. Cross-Scale Attention Module

The structure of the cross-scale attention module is shown in Figure 2. The multi-head attention module is used to extract the attention regions of each scale more accurately. To be more intuitive, the formula in [38] is quoted as:

$$\begin{gathered} MultiHead(Q,K,V) = Concat(head_1,\ldots,head_n)W^O \\ where\ head_i = soft\max\left(\frac{\left(QW_i^Q\right)\left(KW_i^K\right)^T}{\sqrt{d_k}}\right)\left(VW_i^V\right) \end{gathered} \quad (1)$$

where $W_i^Q \in \mathbb{R}^{d_{model} \times d_k}$, $W_i^K \in \mathbb{R}^{d_{model} \times d_k}$, $W_i^V \in \mathbb{R}^{d_{model} \times d_v}$, and $W^O \in \mathbb{R}^{d_v \times d_{model}}$ are the projection matrices, where d_k and d_v are the dimensions of K and V respectively, and d_{model} is the dimension of the multi-head attention module. In this paper, the multi-head attention inputs of Q, K, and V are the same, so we simplify $MultiHead(Q,K,V)$ to $MultiHead(X)$ when $Q = K = V = X$.

The features from the last residual block of i-th stage in ResNet-50 are $M_i \in \mathbb{R}^{C_i \times H_i \times W_i}$, $i = 3,4,5$ where C_i, H_i, W_i indicate the channel number, spatial height, and width, respectively. First, the convolution of size 1 is used in M_i to obtain $X_i \in \mathbb{R}^{256 \times H_i \times W_i}, i = 3,4,5$ in order to reduce the channel dimension. Then, the maximum pooling layer is used to downsample X_i to obtain X_i' to reduce the computing cost $X_i^v = V\left(X_i'\right) \in \mathbb{R}^{256\times(H_5\cdot W_5)}, i = 3,4,5$ denotes the vectorization function, and $X_i' = V^{-1}(X_i^v) \in \mathbb{R}^{256\times H_5\times W_5}$ denotes its inverse function. The enhanced feature $A_i \in \mathbb{R}^{256\times H_5\times W_5}, i = 3,4,5$ are calculated as:

$$A_i = \begin{cases} V^{-1}(\ MultiHead\ (V(X_i'))), & i=5 \\ V^{-1}(\ MultiHead\ (V(X_i' + A_{i+1}))), & i=3,4 \end{cases} \tag{2}$$

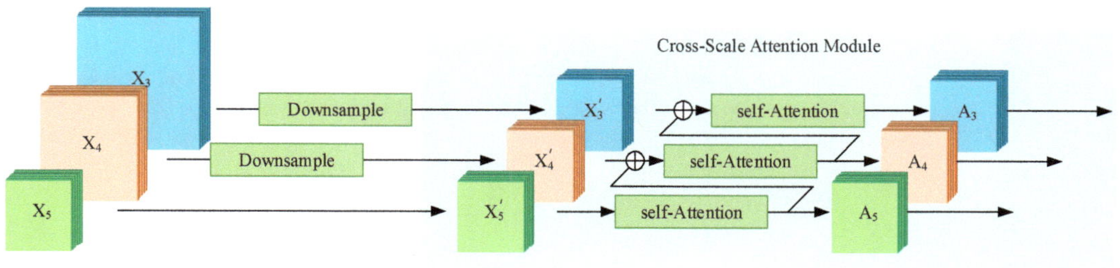

Figure 2. The structure of the cross-scale attention module.

2.2.2. Spatial Attention Redistribution Module

The structure of the spatial redistribution model is shown in Figure 3. First, using a deformable convolution and concatenation on A_i to obtain $A_c \in \mathbb{R}^{(256\cdot 3)\times H_5\times W_5}$. Then A_c is fed into the convolution layer and SoftMax activation layer to obtain $A_s \in \mathbb{R}^{5\times H_5\times W_5}$. The $A_{s_j} \in \mathbb{R}^{1\times H_5\times W_5}$ denotes the feature map of the j-th channel in the A_s, where $A_{s_j}, j = 3, 4, 5$ are considered as attention weights only on the j-th scale, except that some regions possibly will be interesting or suppressed at all scales, such as common features of ship targets or land areas that do not contain targets. For this reason, A_{s_1} and A_{s_2} are respectively used as the weights that require attention and suppression at all scales. Then, the redistributed attention weight $S_i \in \mathbb{R}^{1\times H_5\times W_5}$ can be formulated as:

$$S_i = A_{s_i} + A_{s_1} - A_{s_2} \tag{3}$$

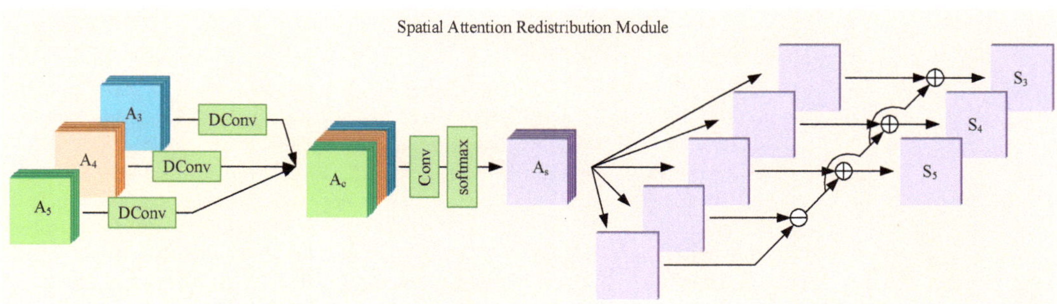

Figure 3. The structure of the spatial attention redistribution module.

Finally, S_i are reshaped by upsampling into $S_i' \in \mathbb{R}^{1\times H_i\times W_i}$ used to enhance features $F_i, i = 3,4,5$ from FPN. Following the idea of FCOS, to further improve the detection capability of large-scale targets, we use a series of convolutions of stride 2 and size 3 for the enhanced features of F_5 to obtain feature maps F_6 and F_7 with larger receptive fields.

$$F_6 = Conv(F_5 \cdot BD(S_5)) \tag{4}$$

$$F_7 = Conv(ReLU(F_6)) \tag{5}$$

where $BD()$ means broadcast operation, $Conv()$ means a convolution with a kernel size of 3, a stride of 2, and a padding of 1.

2.3. Global Average Precision Loss

2.3.1. Average Precision Loss

In the object detection task, average precision (AP) is an important criterion for the detection result. The range of AP is $[0,1]$, the closer the AP is to 1, the more accurate the detection result is and the closer the AP is to 0, the worse the detection result is. The main optimization goal of the object detection task is to make the AP as close to 1 as possible. Based on this goal, AP loss was proposed by Chen et al. [39]. With a large number of samples, AP loss can achieve better performance than focal loss because it allows the network to be optimized directly for the AP, and is not affected by a large number of true negative samples [39]. To be more intuitive, the formula of AP loss in [39] is quoted as:

$$\begin{aligned} \mathcal{L}_{AP} &= \frac{1}{|P|}\sum_{i\in P}\frac{\sum_{j\in N} H(s_j-s_i)}{1+\sum_{j\in P, j\neq i} H(s_j-s_i)+\sum_{j\in N} H(s_j-s_i)} \\ where\ H(x) &= \begin{cases} 0, & x<-\delta \\ \frac{x}{2\delta}+0.5, & -\delta\le x\le\delta \\ 1, & \delta<x \end{cases} \end{aligned} \tag{6}$$

where s_i is the classification score of the box b_i output by the network without the sigmoid function, and each b_i will be assigned a label $t_i \in \{-1,0,1\}$ (label -1 for not counted into the AP loss). $P=\{i|t_i=1\}$ and $N=\{i|t_i=0\}$ are the set of positive and negative samples, respectively, and $|P|$ is the size of set P. Further, Chen et al. [39] give a formula for the gradient g_i of s_i to address the problem that $\mathcal{L}_{AP}$ is non-differentiable in the backpropagation process.

$$\begin{aligned} g_i &= \sum_j L_{ji}\cdot y_{ji} - \sum_j L_{ij}\cdot y_{ij} \\ where\ L_{ij} &= \frac{H(s_j-s_i)}{1+\sum_{j\in P, j\neq i} H(s_j-s_i)+\sum_{j\in N} H(s_j-s_i)} \end{aligned} \tag{7}$$

where $\forall i,j, y_{ij}=1_{t_i=\mathbf{1},t_j=0}$ and $\mathbf{1}$ is an indicator function that equals to 1 only if the subscript condition holds (i.e., $t_i=1, t_j=0$), otherwise 0.

2.3.2. Global Average Precision Loss

Chen et al. [39] point out that training with AP loss has the problem of "score shift". That is, although the AP of each of the two images is high, the AP may become lower if the two images are put together, because the scores of the positive or negative samples of the two images may be in different ranges, and the score of negative samples on one of the images may be higher than the score of positive samples in the other image. Chen et al. [39] avoid this situation through minibatch training but it still exists in two different batches. The main reason for the "score shift" is that AP loss only deals with those positive samples that have smaller scores than the negative samples and those negative samples that have larger scores than the positive samples. If the score of a positive sample is larger than all the negative samples, then its gradient g_i will be 0. Similarly, if the score of a negative sample is smaller than all the positive samples, then its gradient will also be 0. This leads to the fact that once the score of a sample jumps out of the range $r_k = \left(\max_{i\in N_k}(s_{i,k}), \min_{j\in P_k}\left(s_{j,k}\right)\right)$ in the k-th batch, then its gradient of AP loss will become 0 unless the region changes so that it is included again, and when $\min_{j\in P_k}\left(s_{j,k}\right) > \max_{i\in N_k}(s_{i,k})$, the gradient of AP loss for all samples will become 0. Thus, the greater the difference in r_k, the more serious the problem of "score shift". In addition, in the early stage of the training, the scores of positive and negative samples are very close to each other, which will result in a small range of r_k.

Therefore, training with AP loss can only produce a very small gap in the positive and negative samples, leading to poor results.

To solve the above problem, we propose a global average precision loss (GAP loss). First, to make the range r_k of each batch similar, we calculate the mean value of positive and negative sample scores for multiple batches and take them into account separately when calculating L_{ij}. Thus, the formula for L_{ij} becomes as follows, while for more intuition we divide L_{ij} into L_{ij}^P for $\{i, j|t_i = 1, t_j = 0\}$ and L_{ij}^N for $\{i, j|t_i = 0, t_j = 1\}$.

$$L_{ij}^N = \frac{H(s_j - s_i)}{1 + \sum\limits_{j \in P^{(k)}, j \neq i} H(s_j - s_i) + \sum\limits_{j \in N^{(k)}} H(s_j - s_i)} + \frac{H\left(s_j - s_P^{(k)}\right)}{1 + \sum\limits_{j \in P^{(k)}, k \neq i} H\left(s_j - s_P^{(k)}\right) + \sum\limits_{j \in N^{(k)}} H\left(s_j - s_P^{(k)}\right)} \tag{8}$$

$$L_{ij}^P = \frac{H(s_j - s_i) + \left|N^{(k)}\right| \cdot H\left(s_N^{(k)} - s_i\right)}{1 + \sum\limits_{j \in P^{(k)}, j \neq i} H(s_j - s_i) + \sum\limits_{j \in N^{(k)}} H(s_j - s_i) + \left|N^{(k)}\right| \cdot H\left(s_N^{(k)} - s_i\right)} \tag{9}$$

where $P^{(k)}$ and $N^{(k)}$ are the set of positive and negative samples in the k-th batch, respectively. $\left|P^{(k)}\right|$ and $\left|N^{(k)}\right|$ is the size of set $P^{(k)}$ and $N^{(k)}$. $s_P^{(k)}$ and $s_N^{(k)}$ are the exponentially weighted averages of the positive and negative sample scores of multiple batches and are calculated as follows:

$$s_P^{(k)} = \beta s_P^{(k-1)} + (1 - \beta) \frac{1}{\left|P^{(k)}\right|} \sum_{i \in P^{(k)}} s_i \tag{10}$$

$$s_N^{(k)} = \beta s_N^{(k-1)} + (1 - \beta) \frac{1}{\left|N^{(k)}\right|} \sum_{i \in N^{(k)}} s_i \tag{11}$$

Compared with Equation (7), the extra term in the Equation (8) is equivalent to adding $\left|P^{(k)}\right|$ additional positive samples with a score of $s_P^{(k)}$ to the set of positive samples. Likewise, Equation (9) adds $\left|N^{(k)}\right|$ additional negative samples with score $s_N^{(k)}$ to the set of negative samples. In this way, the r_k of each batch will contain both $s_N^{(k)}$ and $s_P^{(k)}$, thus alleviating the problem of "score shift". We then added additional terms to the gradient g_i to allow the sample to still obtain a non-zero gradient even if it jumps out of r_k, and the gradient of GAP loss is calculated as follows:

$$g_i' = \sum_j \left(L_{ji}^N + \left(\left(L_{ji}^N\right)^\gamma + 1\right) \cdot Sigmoid(s_i)^\gamma\right) \cdot y_{ji} - \sum_j \left(L_{ij}^P + \left(\left(L_{ij}^P\right)^\gamma + 1\right) \cdot (1 - Sigmoid(s_i))^\gamma\right) \cdot y_{ij} \tag{12}$$

Taking positive samples as an example, positive samples with small activation values and large L_{ij}^P will obtain a larger value of g_i' when L_{ij}^P is not zero compared to g_i, while as L_{ij}^P is equal to zero, positive samples will obtain a non-zero gradient based on their activation values, thus allowing the network to focus on those difficult positive samples with small activation values and large L_{ij}^P, while not "ignoring" those simple samples that jump out at r_k.

3. Experiment

In this section, we first introduce the experimental datasets, evaluation metrics, and implementation details. Then, we conducted an ablation study to analyze SCSA-Net. Finally, we compare other recent methods with several examples.

3.1. Datasets

1. Official-SSDD (SSDD): Currently, the SSDD [26] dataset published in 2017 is the most widely used in the SAR ship detection field. Subsequently, Zhang et al. published the

updated official SSDD dataset of the SSDD dataset in 2021 [29], which corrected the wrong labels in SSDD and provided richer label formats. The official SSDD dataset contains complex backgrounds and multi-scale offshore and inshore targets. Most of the images are 500 pixels wide, and the SSDD has a variety of SAR image samples with resolutions ranging from 1 m to 15 m from different sensors of RadarSat-2, Terra SAR-X, and Sentinel-1. The average size of ships in SSDD is only ~35 × 35 pixels. In this paper, we refer to Official SSDD as SSDD for convenience.
2. SAR-Ship-Dataset: SAR-Ship-Dataset was released by Wang et al. [27] in 2019. It contains 43,819 images with 256 × 256 image sizes, mainly from Sentinel-1 and Gaofen-3. SAR ships in SAR-Ship-Dataset are provided with resolutions from 5 m to 20 m, and HH, HV, VV, and VH polarizations. Same to their original reports in [27], the entire dataset is randomly divided into training (70%), validation (20%), and test dataset (10%).
3. High-Resolution SAR Images Dataset (HRSID): The HRSID proposed by Wei et al. [28] is constructed by using original SAR images from the Sentinel-1B, TerraSAR-X, and TanDEM-X satellites. The HRSID contains 5604 images of 800 × 800 size and 16,951 ship targets. These images have various polarization rates, imaging modes, imaging conditions, etc. As in its original reports in [28], the ratio of the training set and the test set is 13:7 according to its default configuration files.

3.2. Evaluation Metrics

In order to evaluate the detection performance quantitatively, the evaluation criteria we used include the precision rate, recall rate, and average precision (AP). The calculation method is as follows:

$$\text{Precision} = \frac{TP}{TP + FP} \tag{13}$$

$$\text{Recall} = \frac{TP}{TP + FN} \tag{14}$$

For a predicted box, if the IoU between it and the corresponding ground truth box is greater than 0.5, then it is defined as true positive (TP), otherwise, it is false positive (FP). For a ground truth box, if there is no predicted box with its IoU greater than 0.5, it is defined as a false negative (FN). AP is defined as:

$$\text{AP} = \int_0^1 p(r)dr \tag{15}$$

where p and r represents precision and recall at an IoU threshold of 0.5, and $p()$ is a function that takes r as a parameter, which is equal to taking the area under the curve.

3.3. Training Details

The ResNet-50 pre-trained on ImageNet is carried out to initialize our backbone. The initial learning rate of the SGD optimizer is 0.01, which is divided by 10 at each decay step. The number of image iterations per epoch is 16 K. All training processes are carried out on a Tesla V100 GPU (32 G) server.

3.4. Ablation Study

To visualize the effect of GAP loss, we counted the mean curves of the classification scores of positive and negative samples predicted by the network during the training phase, as shown in Figures 4–6. The curves are the output of the network trained by focal loss, AP loss, and GAP loss, respectively. To be more intuitive, the classification scores are normalized by a sigmoid function to convert them into probabilities, which represents the probability that the network considers the sample to be a ship. For positive samples, it should be as close to 1 as possible, and for negative samples, it should be as close to 0 as possible. In the initial stage of training, the positive and negative sample scores of the network prediction are in the range of 0.5. No matter what kind of loss is used for training, the negative sample scores can drop to 0 quickly, but only the positive sample scores corresponding to the GAP loss can quickly increase to about 1. For the network trained with focal loss and AP loss, the output positive sample score will first drop from 0.5 to approximately 0.17 and 0.04 before gradually increasing. Additionally, since the gradient of AP loss for samples out of r_k is 0, the average of the positive sample score corresponding to the AP loss at the end of training can only reach about 0.1, such that the gap of scores between positive and negative samples is relatively small, leading to poorer detection results. The focal loss can gradually increase the positive sample score of the network to about 0.9, but it is still lower than the positive sample score corresponding to the GAP loss.

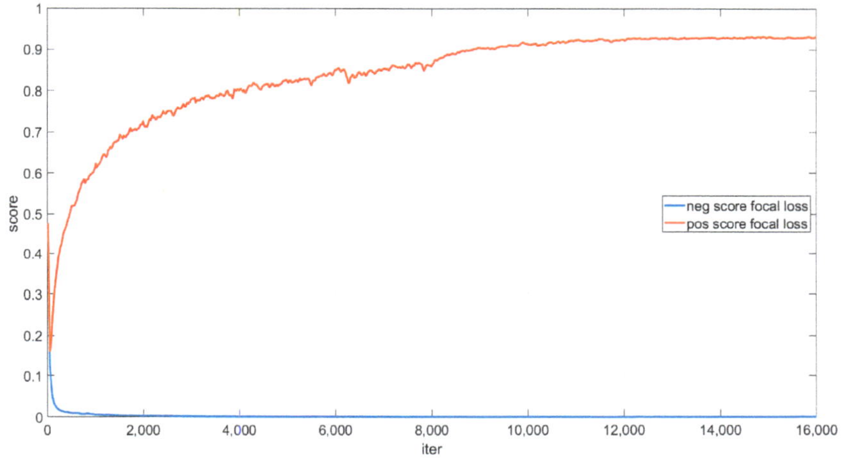

Figure 4. Curves of the average classification scores of positive and negative samples output from the network trained by focal loss during the training phase.

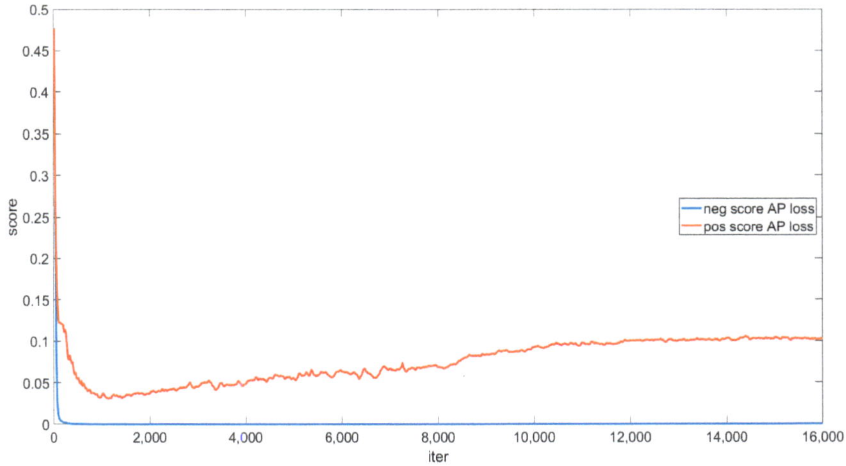

Figure 5. Curves of the average classification scores of positive and negative samples output from the network trained by AP loss during the training phase.

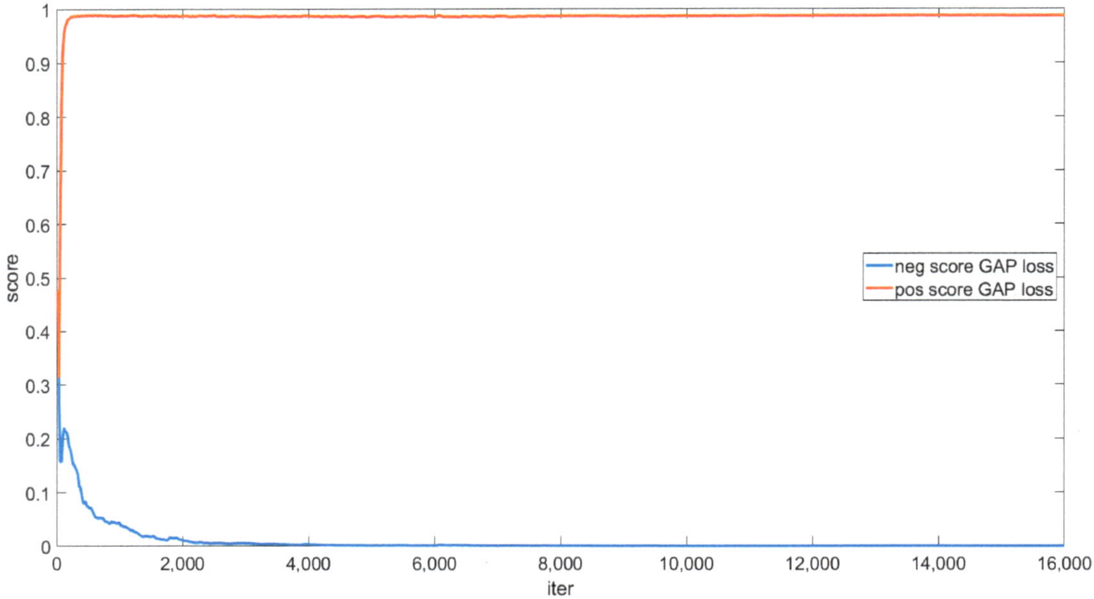

Figure 6. Curves of the average classification scores of positive and negative samples output from the network trained by GAP loss during the training phase.

To evaluate the effectiveness of the SCSA module and GAP loss, we performed ablation studies on SSDD, SAR-Ship-Dataset, and HRSID with different settings of SCSA-Net, respectively. The results of the ablation study are shown in Table 1 and Figure 7. Firstly, by adding the SCSA module, the performance of the network can be improved regardless of the loss used in training. Second, since AP loss results in a smaller gap between positive and negative sample scores, replacing the focal loss with AP loss results in a decrease in AP, with or without the SCSA module. In contrast, direct replacement of focal loss with GAP loss can improve AP with or without the SCSA module. Finally, using both the GAP loss and SCSA modules achieved the best results on all three datasets, 98.7% AP on the SSDD dataset, 97.9% AP on the SAR-Ship-Dataset dataset, and 95.4% AP on the HRSID dataset, respectively.

Table 1. Ablation study of SCSA-Net on SSDD, SAR-Ship-Dataset, and HRSID.

	Baseline	Different Settings of SCSA-Net				
Focal loss	√			√		
AP loss		√			√	
GAP loss			√			√
SCSA				√	√	√
SSDD	96.5	94.5	98.0	97.2	95.5	**98.7**
SAR-Ship-Dataset	96.1	95.4	97.0	96.3	95.8	**97.9**
HRSID	92.0	89.9	94.2	93.2	90.9	**95.4**

√ indicates that the corresponding item is used. The best results are bolded.

Some visualization results of the ablation experiments on SSDD, SAR-Ship-Dataset, and HRSID are shown in Figures 8–10, where Figure 8a–d–Figure 10a–d show the ground truths, detection results of FCOS trained by focal loss, detection results of FCOS trained by GAP loss, and detection results of FCOS with SCSA module trained by GAP loss, respectively. The samples include different background states in the offshore, deep sea, and moored in port, respectively. Since some of the ships have large noise and more complex nearshore backgrounds, the detection results of FCOS are easily affected by them, leading to missed detections and false alarms, as shown in Figures 8, 9 and 10b. By training FCOS with GAP loss, the scores of negative samples can be effectively reduced and the scores of positive samples can be increased, resulting in better AP, as shown in Figures 8, 9 and 10c. However, for some difficult samples, even training with GAP loss still cannot achieve good results due to the limited expressiveness of the network. By adding the SCSA module, the feature extraction capability of the network is improved, and the problem of missed detection and false alarm in some near-shore scenes is effectively solved, as shown in Figures 8, 9 and 10d.

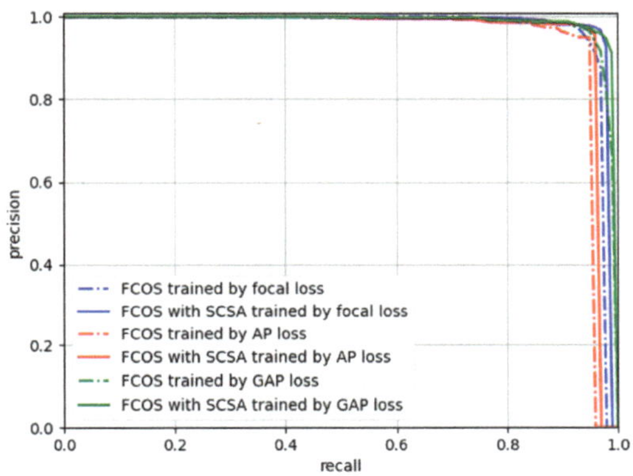

(**a**)

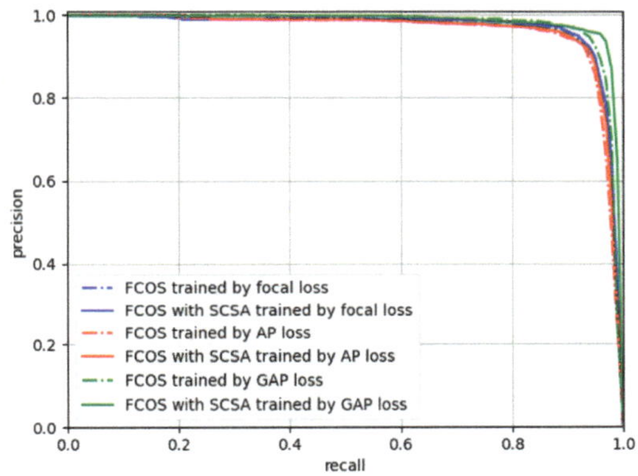

(**b**)

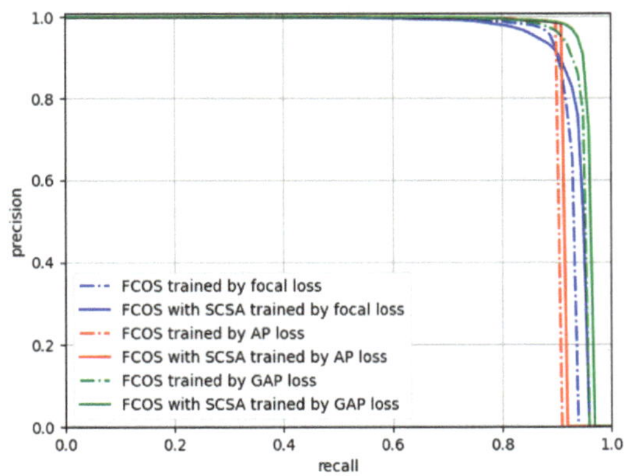

(**c**)

Figure 7. Precision-Recall curves (PR curves) of SCSA-Net with different settings. (**a**) PR curves of SCSA-Net with different settings on SSDD. (**b**) PR curves of SCSA-Net with different settings on SAR-Ship-Dataset. (**c**) PR curves of SCSA-Net with different settings on HRSID.

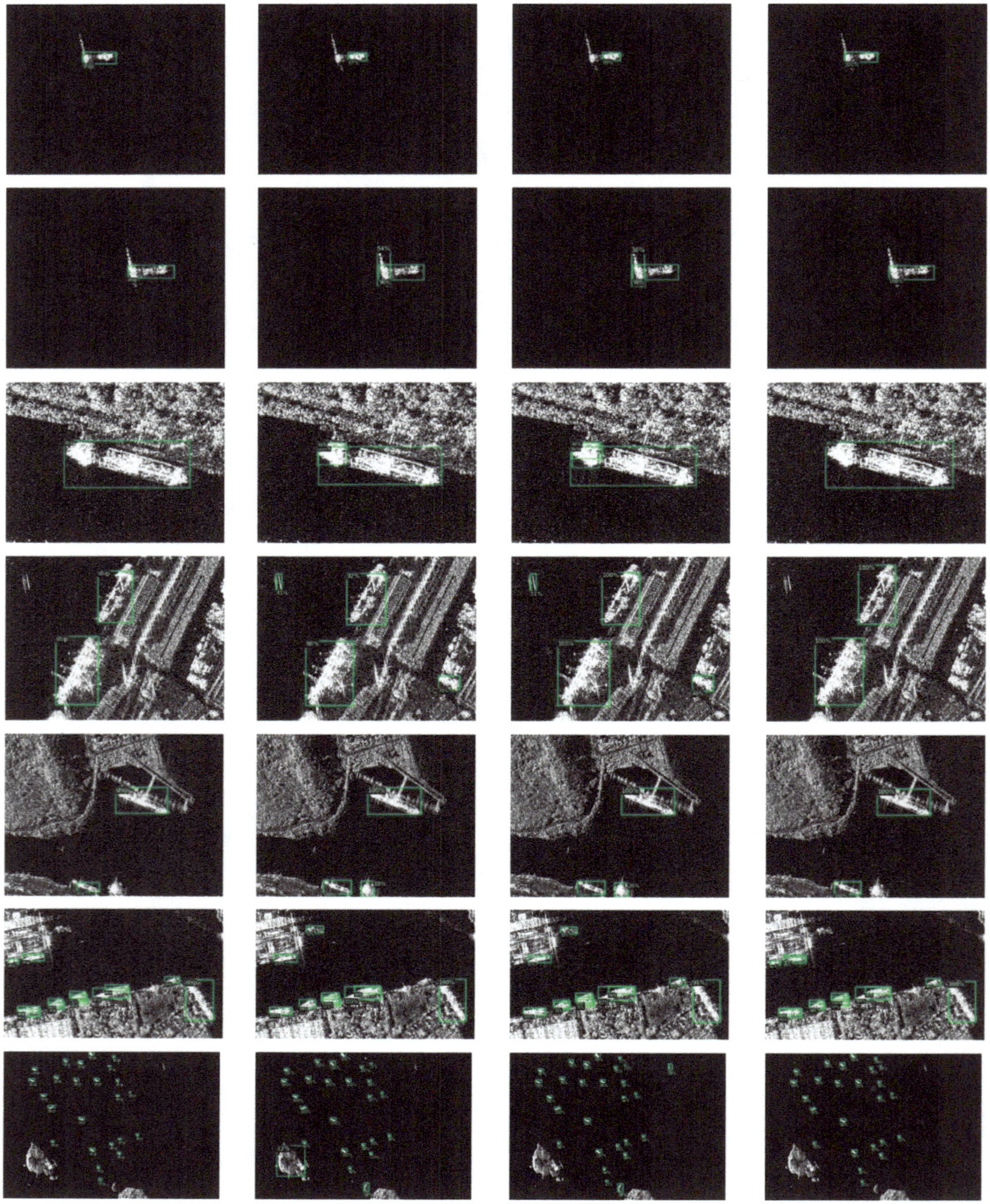

Figure 8. *Cont.*

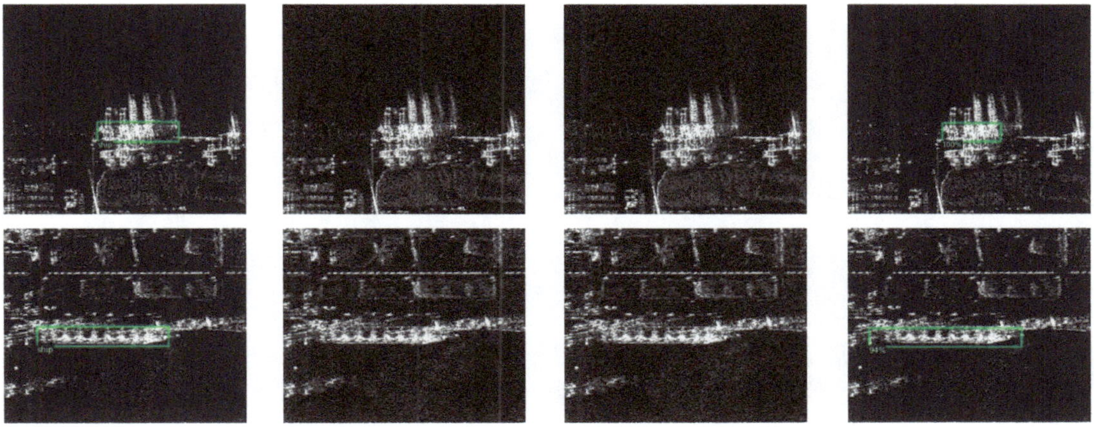

(**a**) (**b**) (**c**) (**d**)

Figure 8. Visualization results of ablation experiments on the SSDD. (**a**) Ground truths. (**b**) Detection results of FCOS trained by focal loss. (**c**) Detection results of FCOS trained by GAP loss. (**d**) Detection results of FCOS with SCSA module trained by GAP loss.

Figure 9. *Cont.*

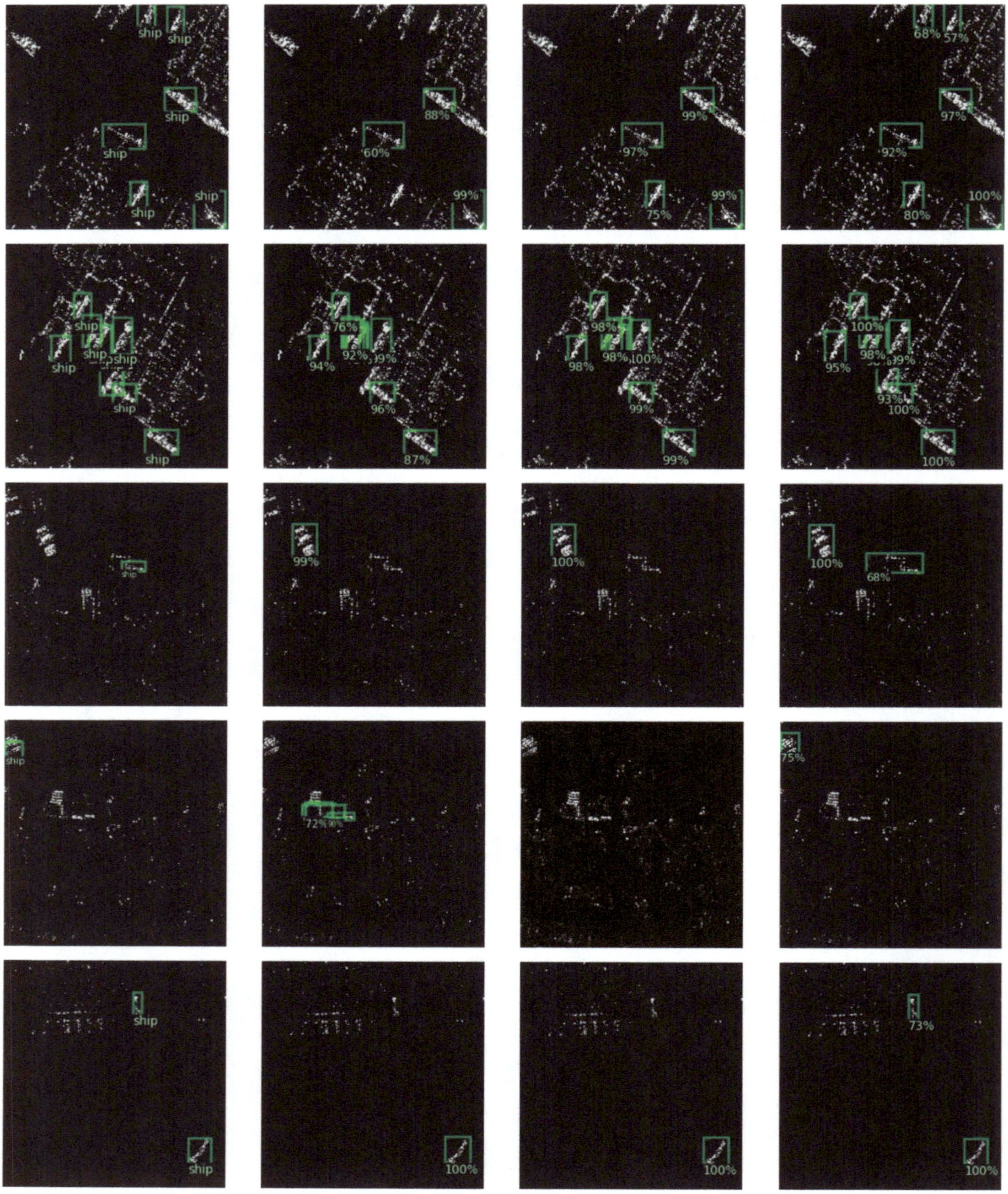

Figure 9. *Cont.*

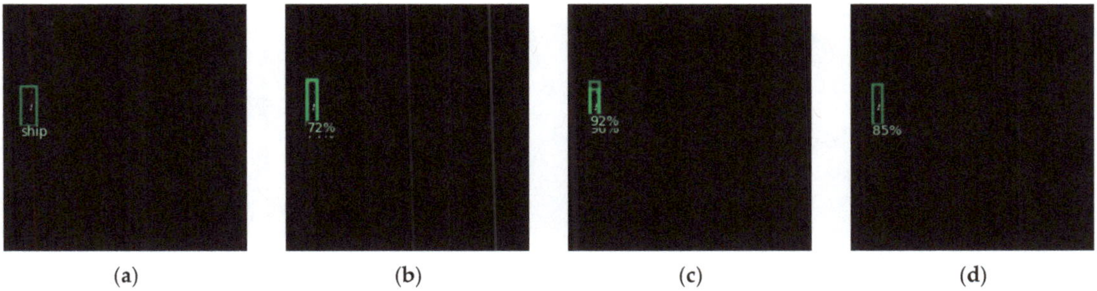

Figure 9. Visualization results of ablation experiments on the SAR-Ship-Dataset. (**a**) Ground truths. (**b**) Detection results of FCOS trained by focal loss. (**c**) Detection results of FCOS trained by GAP loss. (**d**) Detection results of FCOS with SCSA module trained by GAP loss.

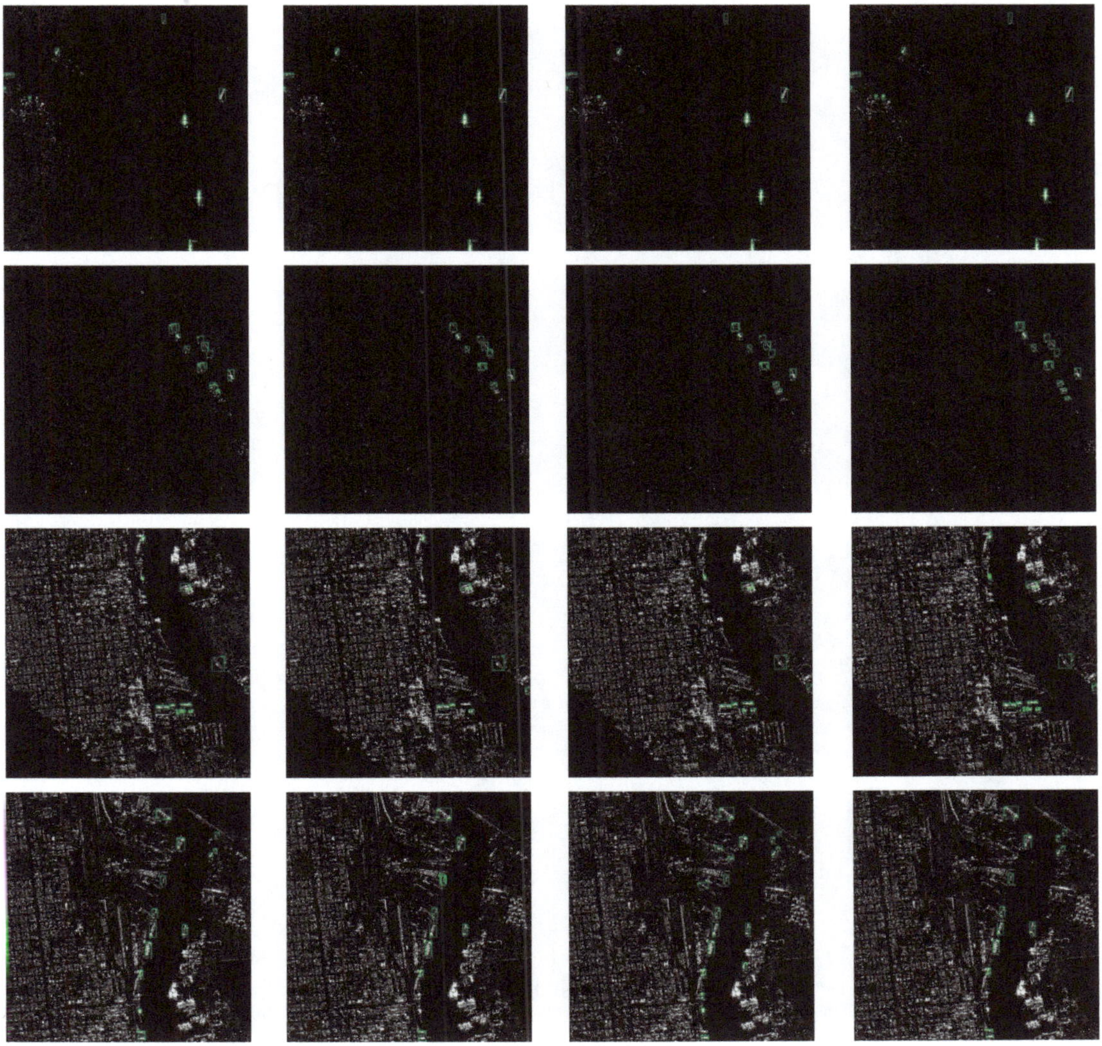

Figure 10. *Cont.*

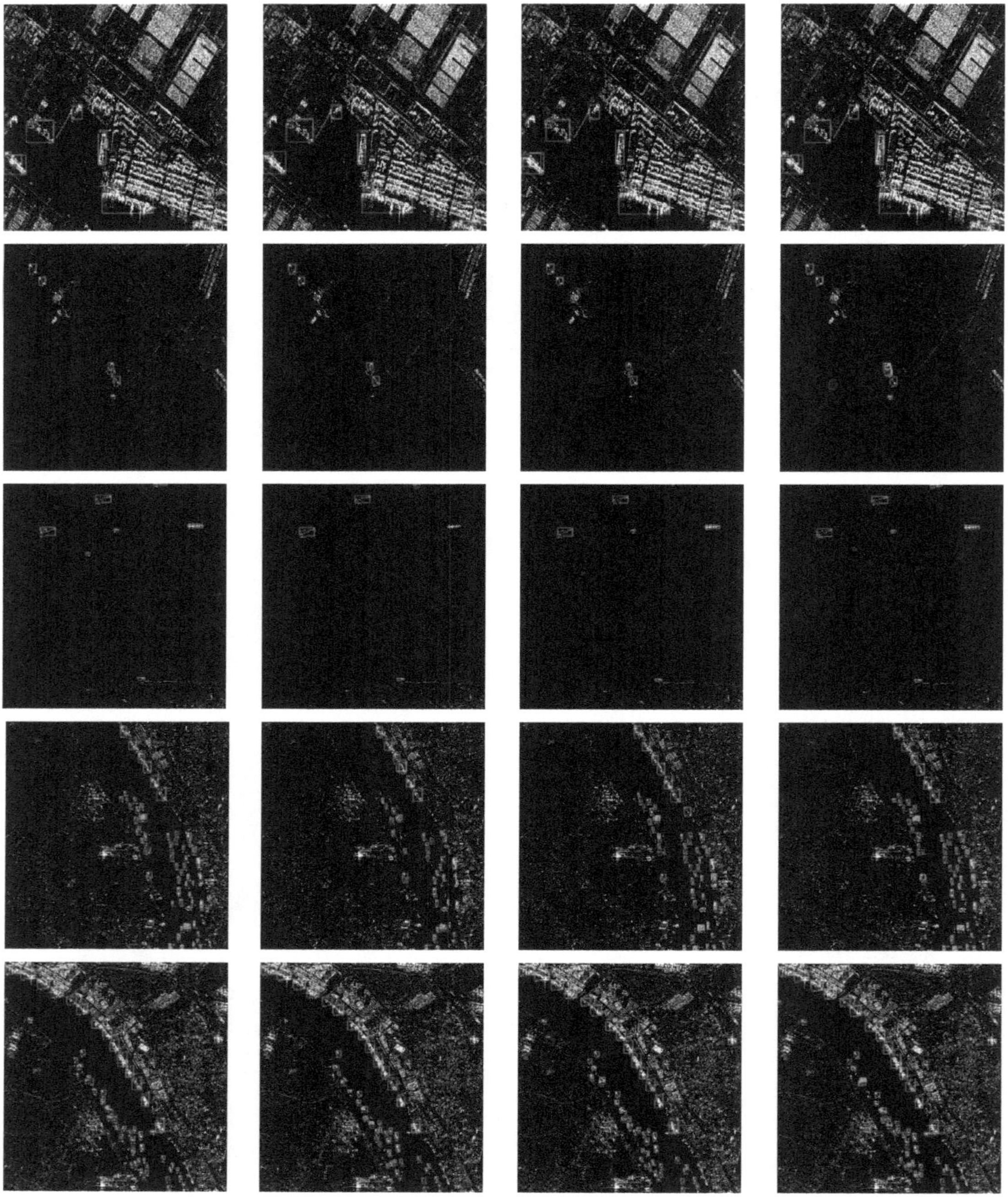

(a) (b) (c) (d)

Figure 10. Visualization results of ablation experiments on the HRSID. (**a**) Ground truths. (**b**) Detection results of FCOS trained by focal loss. (**c**) Detection results of FCOS trained by GAP loss. (**d**) Detection results of FCOS with SCSA module trained by GAP loss.

To further confirm the effectiveness of the SCSA module, we use different configurations of the SCSA module in the network and train the network using GAP loss. The detection results on SSDD, SAR-Ship-Dataset, and HRSID are shown in Table 2. The difference in settings is whether A_{s_1} and A_{s_2} are considered when calculating S_i in the Equation (3). The AP values of the results can be improved by using either A_{s_1} or A_{s_2} on the three different datasets. By using A_{s_1} additionally in the calculation of S_i, 0.096%, 0.187%, and 0.157% AP are improved on the SSDD, SAR-Ship-Dataset, and HRSID, respectively. By using A_{s_2} in the calculation of S_i, 0.116%, 0.195%, and 0.184% AP are improved on the SSDD, SAR-Ship-Dataset, and HRSID, respectively. Additionally, the best detection results can be achieved by using A_{s_1} and A_{s_2} together, with an improvement of 0.270%, 0.311%, and 0.346% AP on the SSDD, SAR-Ship-Dataset, and HRSID, respectively.

Table 2. Results of different settings of SCSA module on SSDD, SAR-Ship-Dataset, and HRSID.

	Different Settings of SCSA Module			
A_{s_1}		√		√
A_{s_2}			√	√
SSDD	98.475	98.571	98.591	**98.745**
SAR-Ship-Dataset	97.571	97.758	97.766	**97.882**
HRSID	95.046	95.203	95.230	**95.392**

√ indicates that the corresponding item is used. The best results are bold.

3.5. Comparison with the Latest SAR Ship Detection Methods

To further demonstrate the advancement and superiority of our proposed method, we experimentally validated with the latest SAR ship detection method using the SSDD, SAR-Ship-Dataset, and HRSID, as shown in Tables 3–5. The proposed method achieves the best results on all three widely used datasets. On the SSDD, ISASDNet+r101 [34] achieves the highest 96.8% AP in the two-stage network, while the proposed SCSA-Net can achieve 98.7% AP, which is 1.9% AP improvement compared to ISASDNet+r101 [34] and 0.3% AP improvement compared to the anchor-free network proposed by Zhu et al. [33] (Table 3). On SAR-Ship-Dataset, SCSA-Net achieves 97.9% AP, which is 2.1% AP improvement compared to the highest 95.8% AP in the two-stage network (Table 4). SCSA-Net achieves 95.4% AP on HRSID (Table 5). As a one-stage network, SCSA-Net not only has the best performance in the one-stage network but also exceeds that of the two-stage network.

Table 3. Comparison with the latest SAR ship detection methods on SSDD.

	Methods	AP
Two-stage	Faster R-CNN [13]	90.8
	SER Faster R-CNN [40]	91.5
	ISASDNet+r50 [34]	95.4
	ISASDNet+r101 [34]	96.8
	STANet-50+FPN [35]	95.7
	Mask-RCNN(OCIE-DFR-RFE) [36]	92.1
One-stage	ResNet-50+Quad-FPN [32]	96.6
	YOLOV3 (OCIE-DFR-RFE) [36]	68.8
	ASAFE [30]	95.2
	A-BFPN [4]	96.8
	HR-SDNet [37]	89.4
	Unnamed method * [33]	98.4
	ours	**98.7**

* indicates that the authors of this paper did not name the proposed overall method. The best result is bold.

Table 4. Comparison with the latest SAR ship detection methods on SAR-Ship-Dataset.

	Methods	AP
Two-stage	Faster R-CNN [13]	91.7
	SER Faster R-CNN [40]	92.2
	ISASDNet+r50 [34]	95.3
	ISASDNet+r101 [34]	95.8
One-stage	ResNet-50+Quad-FPN [32]	94.4
	HR-SDNet [37]	92.3
	ours	**97.9**

The best result is bold.

Table 5. Comparison with the latest SAR ship detection methods on HRSID.

	Methods	AP
Two-stage	Faster R-CNN [13]	80.7
	SER Faster R-CNN [40]	81.5
One-stage	ResNet-50+Quad-FPN [32]	90.9
	HR-SDNet [37]	85.9
	ours	**95.4**

The best result is bold.

It should be noted that some SAR ship detection algorithms compared in Tables 3–5 cannot be reproduced using the same experimental equipment, experimental environment, and experimental parameters because there is no open-source code, so we directly quote the relevant performance indicators in the corresponding papers. Our proposed method achieves higher AP detection performance than the latest SAR ship detection methods.

To visualize the detection performance of different methods, Figure 11 shows the comparison results of one-stage networks on SSDD (first three rows), SAR-Ship-Dataset (rows 4 to 6), and HRSID (rows 7 to 9), where Figure 11a–d show the ground truths, detection results of HR-SDNet [37], detection results of ResNet-50+Quad-FPN [32], and detection results of SCSA-Net proposed in this paper, respectively. Additionally, Figure 12 shows the corresponding precision-recall curves (PR curves). As shown in Figure 11a–c, in the near-shore scene, the network detects a low target score due to the interference of the background, and it is easy to generate false alarms and missed detections. Compared with other methods, SCSA-Net can alleviate this problem with higher classification scores of targets in nearshore scenes, reducing false alarms and missed detections, as well as obtaining better detection performance, as shown in Figure 11d. Meanwhile, HR-SDNet and ResNet-50+Quad-FPN cannot effectively handle densely arranged targets in nearshore scenes and are prone to miss-detection and treat multiple targets as one, as shown in rows 1, 2, and 5 of Figure 11b,c, while SCSA-Net possesses a stronger ability to handle such targets than them, as shown in rows 1, 2, and 5 of Figure 11d.

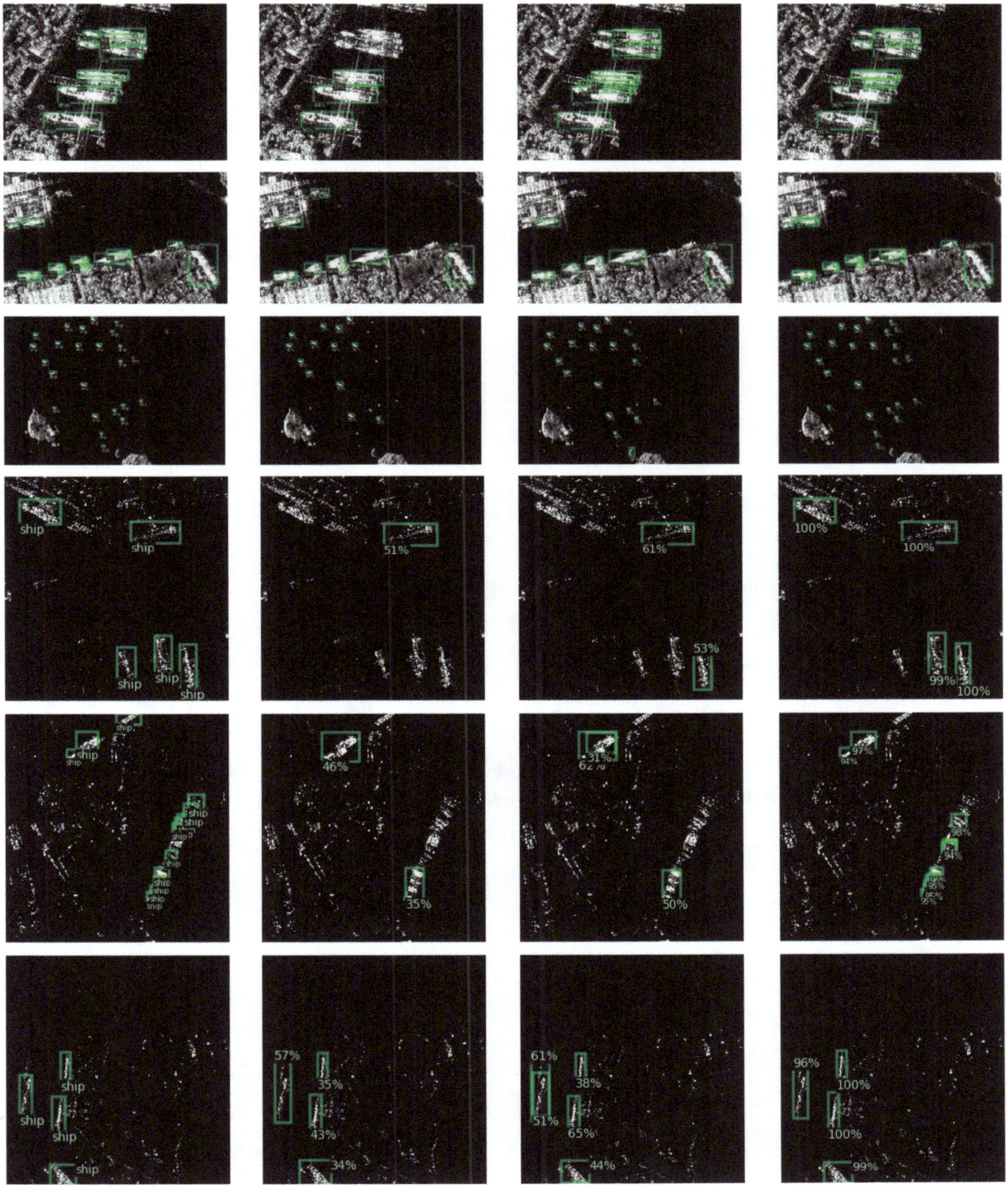

Figure 11. *Cont.*

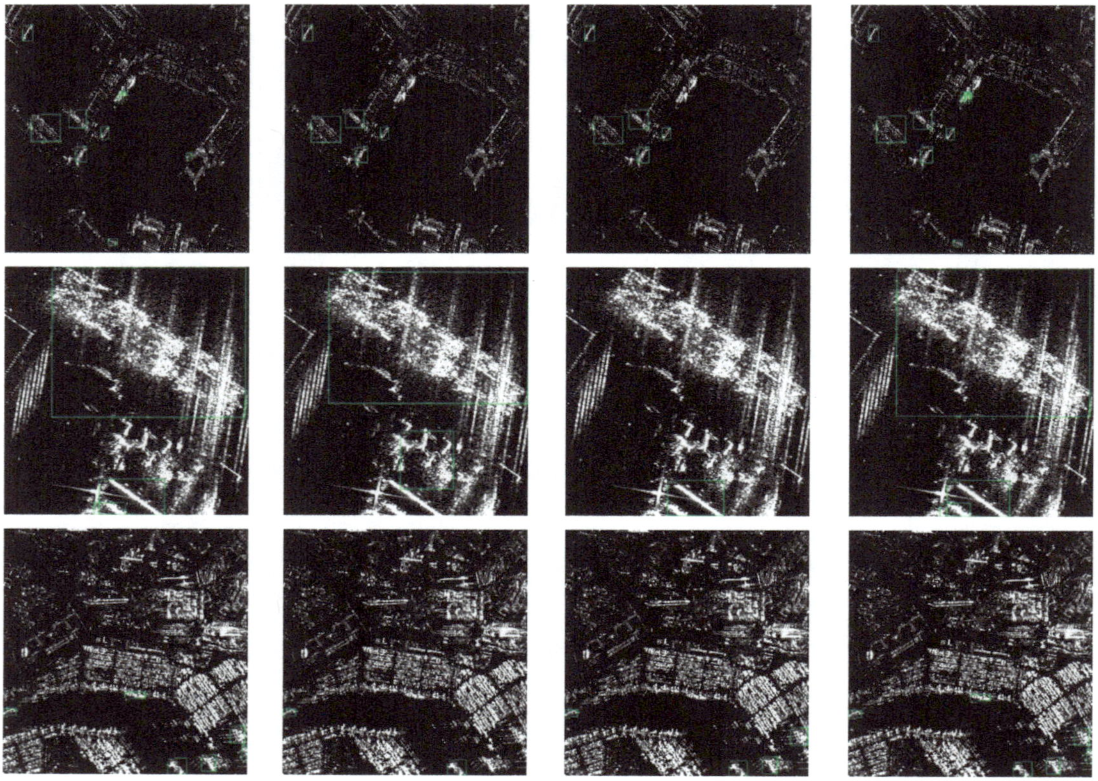

(**a**) (**b**) (**c**) (**d**)

Figure 11. Comparison of the results of different methods. (**a**) Ground truths. (**b**) Detection results of HR-SDNet. (**c**) Detection results of ResNet-50+Quad-FPN. (**d**) Detection results of SCSA-Net.

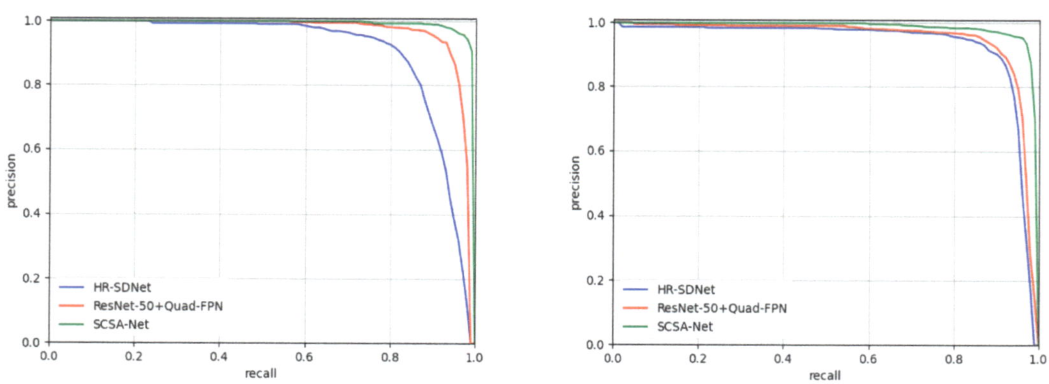

(**a**) (**b**)

Figure 12. *Cont.*

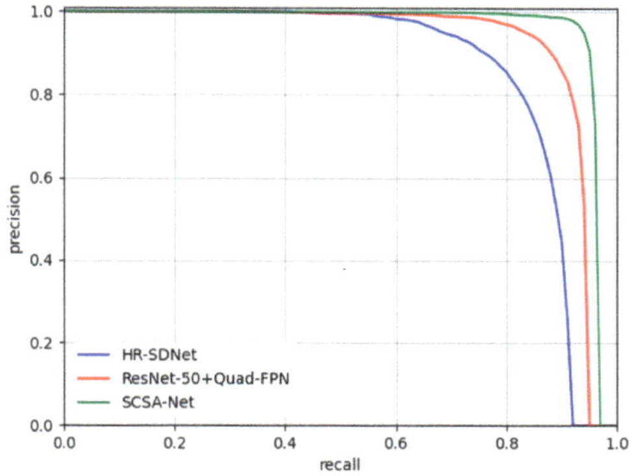

(c)

Figure 12. PR curves of different methods. (**a**) PR curves of different methods on SSDD. (**b**) PR curves of different methods on SAR-Ship-Dataset. (**c**) PR curves of different methods on HRSID.

4. Discussion

The experimental results on SSDD, SAR-Ship-Dataset, and HRSID validate the effectiveness of the proposed method in this paper. However, the horizontal rectangular box detection method used by the proposed method cannot obtain the angle information of the ship. The rotatable rectangular box can locate the ship target more accurately to reduce the background information contained in the box and help to obtain the ship heading and aspect ratio information. Therefore, the subsequent research direction is how to obtain the rotation angle information of the ship to obtain a more accurate rotatable box and further improve the detection performance. Additionally, as shown in Table 6, the SCSA module added to the SCSA-Net accounted for 23.1% of the overall parameters, resulting in a 26.6% drop in FPS. Although the SCSA module is effective in improving accuracy, its impact on the amount of network computation and test speed cannot be ignored. Further, although the one-stage network is less computationally intensive compared to the two-stage network, it is still computing-heavy and difficult to deploy for some embedded devices. Therefore, it is also the goal of our future work to accurately locate and remove useless parameters from the network to reduce the parameter size and computation of the model without affecting the accuracy.

Table 6. The number of parameters of each module and the testing speed of the network.

	Backbone (ResNet-50)	FPN	SCSA Module	Class-Box Subnet	Total Param(M)	FPS
Param(M)	23.45	3.87	9.61	4.74	-	-
FCOS	√	√		√	32.06	30
SCSA-Net	√	√	√	√	41.67	22

√ indicates that the corresponding item is used.

5. Conclusions

In this work, a one-stage anchor-free SCSA-Net is developed to accurately detect ship targets in SAR images. To improve the feature extraction ability of the network and reduce the interference of nearshore background to the ship target, an SCSA module is proposed to dynamically enhance the features in space. In addition, a GAP loss is proposed, which enables the network to optimize directly with AP as the target, and solves the "score shift" problem in AP loss, so that it can effectively improve the score of the predicted tar and promote the detection accuracy of the network. We can conclude the experimental results on SSDD, SAR-Ship-Dataset, and HRSID: (1) by using the SCSA module, the accuracy can be improved on all three datasets regardless of whether the training is performed using focal loss, AP loss, or GAP loss. Its effectiveness is confirmed; (2) by using GAP loss for training, the gap between the average classification scores of positive and negative samples can be effectively widened, and the accuracy can be improved with or without the SCSA module; (3) compared with other methods, the SCSA-Net proposed in this paper has higher detection accuracy on all three datasets.

Future work: our future work will focus on reducing the computing scale of the network without affecting its accuracy, and studying the application of arbitrarily oriented objects.

Author Contributions: Conceptualization, L.Z.; methodology, L.Z.; software, L.Z.; validation, L.Z. and Y.L.; formal analysis, L.Z. and Y.L.; investigation, L.Z.; resources, L.Z. and Y.L.; data curation, L.Z.; writing-original draft preparation, Y.L.; writing-review and editing, L.Z., Y.L., L.Q., J.C. and J.F.; visualization, L.Z. All authors have read and agreed to the published version of the manuscript.

Funding: This work was supported in part by the Xingliao Talents Program of Liaoning Province under Grant XLYC1907134, in part by the Scientific Research Project of the Department of Education of Liaoning Province under Grant LJKZ0174, and in part by Liaoning BaiQianWan Talents Program under Grant 2018B21.

Data Availability Statement: Not applicable.

Acknowledgments: We gratefully appreciate the publishers of the SSDD dataset, SAR-Ship-Dataset, and HRSID, and the editors and reviewers for their efforts and contributions.

Conflicts of Interest: The authors declare no conflict of interest.

References

1. Zhang, P.; Xu, H.; Tian, T.; Gao, P.; Li, L.; Zhao, T.; Zhang, N.; Tian, J. SEFEPNet: Scale Expansion and Feature Enhancement Pyramid Network for SAR Aircraft Detection with Small Sample Dataset. *IEEE J. Sel. Top. Appl. Earth Obs. Remote Sens.* **2022**, *15*, 3365–3375. [CrossRef]
2. Hong, Z.; Yang, T.; Tong, X.; Zhang, Y.; Jiang, S.; Zhou, R.; Han, Y.; Wang, J.; Yang, S.; Liu, S. Multi-Scale Ship Detection from SAR and Optical Imagery Via A More Accurate YOLOv3. *IEEE J. Sel. Top. Appl. Earth Obs. Remote Sens.* **2021**, *14*, 6083–6101. [CrossRef]
3. Du, L.; Dai, H.; Wang, Y.; Xie, X.; Wang, Z. Target discrimination based on weakly supervised learning for high-resolution SAR images in complex scenes. *IEEE Trans. Geosci. Remote Sens.* **2020**, *58*, 461–472. [CrossRef]
4. Li, X.; Li, D.; Liu, H.; Wan, J.; Chen, Z.; Liu, Q. A-BFPN: An Attention-Guided Balanced Feature Pyramid Network for SAR Ship Detection. *Remote Sens.* **2022**, *14*, 3829. [CrossRef]
5. Li, S.; Fu, X.; Dong, J. Improved Ship Detection Algorithm Based on YOLOX for SAR Outline Enhancement Image. *Remote Sens.* **2022**, *14*, 4070. [CrossRef]
6. Leng, X.; Ji, K.; Yang, K.; Zou, H. A Bilateral CFAR Algorithm for Ship Detection in SAR Images. *IEEE Geosci. Remote Sens. Lett.* **2015**, *15*, 1536–1540. [CrossRef]
7. Ai, J.; Yang, X.; Song, J.; Dong, Z.; Jia, L.; Zhou, F. An Adaptively Truncated Clutter-Statistics-Based Two-Parameter CFAR Detector in SAR Imagery. *IEEE J. Ocean. Eng.* **2018**, *43*, 267–279. [CrossRef]
8. Dai, H.; Du, L.; Wang, Y.; Wang, Z. A Modified CFAR Algorithm Based on Object Proposals for Ship Target Detection in SAR Images. *IEEE Geosci. Remote Sens. Lett.* **2016**, *13*, 1925–1929. [CrossRef]
9. Li, N.; Pan, X.; Yang, L.; Huang, Z.; Wu, Z.; Zheng, G. Adaptive CFAR Method for SAR Ship Detection Using Intensity and Texture Feature Fusion Attention Contrast Mechanism. *Sensors* **2022**, *22*, 8116. [CrossRef]
10. Li, M.; Cui, X.; Chen, S. Adaptive Superpixel-Level CFAR Detector for SAR Inshore Dense Ship Detection. *IEEE Geosci. Remote Sens. Lett.* **2022**, *19*, 4010405. [CrossRef]
11. Girshick, R.; Donahue, J.; Darrell, T.; Malik, J. Rich Feature Hierarchies for Accurate Object Detection and Semantic Segmentation. In Proceedings of the 2014 IEEE Conference on Computer Vision and Pattern Recognition, Columbus, OH, USA, 23–28 June 2014; pp. 580–587. [CrossRef]
12. Girshick, R. Fast R-CNN. In Proceedings of the 2015 IEEE International Conference on Computer Vision (ICCV), Santiago, Chile, 7–13 December 2015; pp. 1440–1448. [CrossRef]
13. Ren, S.; He, K.; Girshick, R.; Sun, J. Faster R-CNN: Towards Real-Time Object Detection with Region Proposal Networks. *IEEE Trans. Pattern Anal. Mach. Intell.* **2017**, *39*, 1137–1149. [CrossRef] [PubMed]
14. Sun, P.; Zhang, R.; Jiang, Y.; Kong, T.; Xu, C.; Zhan, W.; Tomizuka, M.; Li, L.; Yuan, Z.; Wang, C.; et al. Sparse R-CNN: End-to-End Object Detection with Learnable Proposals. In Proceedings of the IEEE/CVF Conference on Computer Vision and Pattern Recognition (CVPR), Nashville, TN, USA, 20–25 June 2021; pp. 14454–14463.

15. Redmon, J.; Divvala, S.; Girshick, R.; Farhadi, A. You Only Look Once: Unified, Real-Time Object Detection. In Proceedings of the 2016 IEEE Conference on Computer Vision and Pattern Recognition (CVPR), Las Vegas, NV, USA, 27–30 June 2016; pp. 779–788. [CrossRef]
16. Redmon, J.; Farhadi, A. YOLO9000: Better, Faster, Stronger. In Proceedings of the 2017 IEEE Conference on Computer Vision and Pattern Recognition (CVPR), Honolulu, HI, USA, 21–26 July 2017; pp. 6517–6525. [CrossRef]
17. Redmon, J.; Farhadi, A. YOLOv3: An Incremental Improvement. *arXiv* **2018**, arXiv:1804.02767. [CrossRef]
18. Bochkovskiy, A.; Wang, C.; Liao, H. YOLOv4: Optimal Speed and Accuracy of Object Detection. *arXiv* **2020**, arXiv:2004.10934. [CrossRef]
19. Tang, G.; Liu, S.; Fujino, I.; Claramunt, C.; Wang, Y.; Men, S. H-YOLO: A Single-Shot Ship Detection Approach Based on Region of Interest Preselected Network. *Remote Sens.* **2020**, *12*, 4192. [CrossRef]
20. Tang, G.; Zhuge, Y.; Claramunt, C.; Men, S. N-YOLO: A SAR Ship Detection Using Noise-Classifying and Complete-Target Extraction. *Remote Sens.* **2021**, *13*, 871. [CrossRef]
21. Xie, F.; Lin, B.; Liu, Y. Research on the Coordinate Attention Mechanism Fuse in a YOLOv5 Deep Learning Detector for the SAR Ship Detection Task. *Sensors* **2022**, *22*, 3370. [CrossRef]
22. Zhu, H.; Xie, Y.; Huang, H.; Jing, C.; Rong, Y.; Wang, C. DB-YOLO: A Duplicate Bilateral YOLO Network for Multi-Scale Ship Detection in SAR Images. *Sensors* **2021**, *21*, 8146. [CrossRef]
23. Liu, W.; Anguelov, D.; Erhan, D.; Szegedy, C.; Reed, S.E.; Fu, C.Y.; Berg, A.C. SSD: Single Shot MultiBox Detector. In Proceedings of the European Conference on Computer Vision, Amsterdam, The Netherlands, 8–16 October 2016; pp. 21–37. [CrossRef]
24. Lin, T.Y.; Goyal, P.; Girshick, R.; He, K.; Dollár, P. Focal loss for dense object detection. In Proceedings of the IEEE International Conference on Computer Vision, Venice, Italy, 22–29 October 2017; pp. 2980–2988. [CrossRef]
25. Tian, Z.; Shen, C.; Chen, H.; He, T. FCOS: Fully Convolutional One-Stage Object Detection. In Proceedings of the 2019 IEEE/CVF International Conference on Computer Vision (ICCV), Seoul, Republic of Korea, 27–28 October 2019; pp. 9626–9635. [CrossRef]
26. Li, J.; Qu, C.; Shao, J. Ship detection in SAR images based on an improved faster R-CNN. In Proceedings of the 2017 SAR in Big Data Era: Models, Methods and Applications (BIGSARDATA), Beijing, China, 13–14 November 2017; pp. 1–6. [CrossRef]
27. Wang, Y.; Wang, C.; Zhang, H.; Dong, Y.; Wei, S. A SAR Dataset of Ship Detection for Deep Learning under Complex Backgrounds. *Remote Sens.* **2019**, *11*, 765. [CrossRef]
28. Wei, S.; Zeng, X.; Qu, Q.; Wang, M.; Su, H.; Shi, J. HRSID: A High-Resolution SAR Images Dataset for Ship Detection and Instance Segmentation. *IEEE Access.* **2020**, *8*, 120234–120254. [CrossRef]
29. Zhang, T.; Zhang, X.; Li, J.; Xu, X.; Wang, B.; Zhan, X.; Xu, Y.; Ke, X.; Zeng, T.; Su, H.; et al. SAR Ship Detection Dataset (SSDD): Official Release and Comprehensive Data Analysis. *Remote Sens.* **2021**, *13*, 3690. [CrossRef]
30. Shi, H.; Fang, Z.; Wang, Y.; Chen, L. An Adaptive Sample Assignment Strategy Based on Feature Enhancement for Ship Detection in SAR Images. *Remote Sens.* **2022**, *14*, 2238. [CrossRef]
31. Cui, Z.; Li, Q.; Cao, Z.; Liu, N. Dense Attention Pyramid Networks for Multi-Scale Ship Detection in SAR Images. *IEEE Trans. Geosci. Remote Sens.* **2019**, *57*, 8983–8997. [CrossRef]
32. Zhang, T.; Zhang, X.; Ke, X. Quad-FPN: A Novel Quad Feature Pyramid Network for SAR Ship Detection. *Remote Sens.* **2021**, *13*, 2771. [CrossRef]
33. Zhu, M.; Hu, G.; Li, S.; Zhou, H.; Wang, S.; Feng, Z. A Novel Anchor-Free Method Based on FCOS + ATSS for Ship Detection in SAR Images. *Remote Sens.* **2022**, *14*, 2034. [CrossRef]
34. Wu, Z.; Hou, B.; Ren, B.; Ren, Z.; Wang, S.; Jiao, L. A Deep Detection Network Based on Interaction of Instance Segmentation and Object Detection for SAR Images. *Remote Sens.* **2021**, *13*, 2582. [CrossRef]
35. Wang, R.; Shao, S.; An, M.; Li, J.; Wang, S.; Xu, X. Soft Thresholding Attention Network for Adaptive Feature Denoising in SAR Ship Detection. *IEEE Access* **2021**, *9*, 29090–29105. [CrossRef]
36. Tian, L.; Cao, Y.; He, B.; Zhang, Y.; He, C.; Li, D. Image Enhancement Driven by Object Characteristics and Dense Feature Reuse Network for Ship Target Detection in Remote Sensing Imagery. *Remote Sens.* **2021**, *13*, 1327. [CrossRef]
37. Wei, S.; Su, H.; Ming, J.; Wang, C.; Yan, M.; Kumar, D.; Shi, J.; Zhang, X. Precise and Robust Ship Detection for High-Resolution SAR Imagery Based on HR-SDNet. *Remote Sens.* **2020**, *12*, 167. [CrossRef]
38. Ashish, V.; Noam, S.; Niki, P.; Jakob, U.; Llion, J.; Aidan, N.G.; Lukasz, K.; Illia, P. Attention Is All You Need. In Proceedings of the Advances in Neural Information Processing Systems, Long Beach, CA, USA, 4–9 December 2017.
39. Chen, K.; Li, J.; Lin, W.; See, J.; Wang, J.; Duan, L.; Chen, Z.; He, C.; Zou, J. Towards Accurate One-Stage Object Detection With AP-Loss. In Proceedings of the IEEE/CVF Conference on Computer Vision and Pattern Recognition (CVPR), Long Beach, CA, USA, 15–20 June 2019; pp. 5119–5127.
40. Lin, Z.; Ji, K.; Leng, X.; Kuang, G. Squeeze and Excitation Rank Faster R-CNN for Ship Detection in SAR Images. *IEEE Geosci. Remote Sens. Lett.* **2019**, *16*, 751–755. [CrossRef]

remote sensing

Article

Improved Neural Network with Spatial Pyramid Pooling and Online Datasets Preprocessing for Underwater Target Detection Based on Side Scan Sonar Imagery

Jinrui Li [1,2,†], Libin Chen [1,†], Jian Shen [3,†], Xiongwu Xiao [2,*], Xiaosong Liu [4], Xin Sun [4], Xiao Wang [5] and Deren Li [2]

1 College of Resources and Environment, Yangtze University, Wuhan 430100, China
2 State Key Laboratory of Information Engineering in Surveying, Mapping and Remote Sensing, Wuhan University, Wuhan 430072, China
3 School of Computer Science, Wuhan University, Wuhan 430072, China
4 Coal Geological Geophysical Exploration Surveying & Mapping Institute of Shanxi Province, Jinzhong 030621, China
5 School of Marine Technology and Geomatics, Jiangsu Ocean University, Lianyungang 222000, China
* Correspondence: xwxiao@whu.edu.cn; Tel.: +86-186-0276-2010
† These authors contributed equally to this work.

Abstract: Fast and high-accuracy detection of underwater targets based on side scan sonar images has great potential for marine fisheries, underwater security, marine mapping, underwater engineering and other applications. The following problems, however, must be addressed when using low-resolution side scan sonar images for underwater target detection: (1) the detection performance is limited due to the restriction on the input of multi-scale images; (2) the widely used deep learning algorithms have a low detection effect due to their complex convolution layer structures; (3) the detection performance is limited due to insufficient model complexity in training process; and (4) the number of samples is not enough because of the bad dataset preprocessing methods. To solve these problems, an improved neural network for underwater target detection—which is based on side scan sonar images and fully utilizes spatial pyramid pooling and online dataset preprocessing based on the You Look Only Once version three (YOLO V3) algorithm—is proposed. The methodology of the proposed approach is as follows: (1) the AlexNet, GoogleNet, VGGNet and the ResNet networks and an adopted YOLO V3 algorithm were the backbone networks. The structure of the YOLO V3 model is more mature and compact and has higher target detection accuracy and better detection efficiency than the other models; (2) spatial pyramid pooling was added at the end of the convolution layer to improve detection performance. Spatial pyramid pooling breaks the scale restrictions when inputting images to improve feature extraction because spatial pyramid pooling enables the backbone network to learn faster at high accuracy; and (3) online dataset preprocessing based on YOLO V3 with spatial pyramid pooling increases the number of samples and improves the complexity of the model to further improve detection process performance. Three-side scan imagery datasets were used for training and were tested in experiments. The quantitative evaluation using Accuracy, Recall, Precision, mAP and F1-Score metrics indicates that: for the AlexNet, GoogleNet, VGGNet and ResNet algorithms, when spatial pyramid pooling is added to their backbone networks, the average detection accuracy of the three sets of data was improved by 2%, 4%, 2% and 2%, respectively, as compared to their original formulations. Compared with the original YOLO V3 model, the proposed ODP+YOLO V3+SPP underwater target detection algorithm model has improved detection performance through the mAP qualitative evaluation index has increased by 6%, the Precision qualitative evaluation index has increased by 13%, and the detection efficiency has increased by 9.34%. These demonstrate that adding spatial pyramid pooling and online dataset preprocessing can improve the target detection accuracy of these commonly used algorithms. The proposed, improved neural network with spatial pyramid pooling and online dataset preprocessing based on the YOLO V3 method achieves the highest scores for underwater target detection results for sunken ships, fish flocks and seafloor topography, with mAP scores of 98%, 91% and 96% for the above three kinds of datasets, respectively.

Citation: Li, J.; Chen, L.; Shen, J.; Xiao, X.; Liu, X.; Sun, X.; Wang, X.; Li, D. Improved Neural Network with Spatial Pyramid Pooling and Online Datasets Preprocessing for Underwater Target Detection Based on Side Scan Sonar Imagery. *Remote Sens.* **2023**, *15*, 440. https://doi.org/10.3390/rs15020440

Academic Editor: Andrzej Stateczny

Received: 1 December 2022
Revised: 29 December 2022
Accepted: 7 January 2023
Published: 11 January 2023

Keywords: underwater target detection; spatial pyramid pooling; online datasets preprocessing; YOLO V3; side scan sonar imagery

1. Introduction

In the field of underwater target detection, there are many cases of effective object detection and recognition. For example, balancing learning (BL) for high-quality Synthetic Aperture Radar (SAR) ship detection [1], two-stage faster region-based convolutional neural network [2], novel quad feature pyramid network [3] and hybrid task cascade plus (HTC+) for SAR ship instance segmentation [4] have been applied in the synthetic aperture radar learning approach for target detection and recognition from remote sensing images. Taken together, these experimental results demonstrate improvements in the recognition and the detection of underwater targets when utilizing a high-speed focusing algorithm for circular synthetic aperture radar [5], high-speed and high-accurate detector (H2Det) with SAR images [6], HyperLi-Net for high-accuracy and high-speed SAR ship detection [7]. Exploiting a balance scene learning mechanism for offshore and inshore target detection [8] improves detection accuracy and efficiency.

In the underwater target detection field, side scan sonar has been widely used in marine research [9], seabed terrain classification [10] and oceanographic surveys [11] because it can acquire a large number of images within the sonar measurement range [12], which is also a classification of detection capability of side-scan sonar and the need for the coverage of a specified sweeping area [13]. But with the processing of the side-scan sonar datasets, the traditional method is cumbersome and inefficient; it does not work very well. The traditional machine learning method is cumbersome and inefficient, cannot completely realize high-accuracy end-to-end target detection and is limited by the structure of the models influencing the detection time [14]. At present, the application of deep learning methods in many fields has achieved excellent results and, in recent years, has also been successful [15].

At present, the methods of target detection in side scan sonar images can be classified as contour extraction and image segmentation methods. Wang et al. [16] used a particle swarm optimization algorithm to improve the threshold segmentation method to segment the vector and improve the accuracy and efficiency of underwater target segmentation. Gong et al. [17] combined the fast fuzzy c-means clustering algorithm (SCFFCM) with MRF to effectively improve the ability of noise suppression in sonar image segmentation. Dzieciuch et al. [18] reported on the preliminary methods and results of applying a non-linear classification method, convolutional neural networks (CNN) to mine detection in noisy sonar imagery. Rhinelander et al. [19] adopted extra feature extraction, and data engineering can result in better classification performance compared to parameter optimization alone for target classification with side scan sonar. Song et al. [20] proposed using deep CNN to segment images of side scan sonar into three parts: highlight areas with objects, regions of shadow and sea-bottom reverberation areas. Zhu et al. [21] proposed and tested an approach on a set of sonar images obtained by a UUV equipped with side scan sonar. Automatic target detection was achieved through the use of matched filters, while target classification is achieved with the trained SVM classifier based on features extracted by the CNN. Einsidler et al. [22] demonstrated the application of deep learning for ATR in side scan sonar imagery; in particular, supervised-learning Region-based CNN (R-CNN), for detecting objects in sonar images. Kim et al. [23] proposed a convolutional neural network algorithm based on Faster R-CNN (Region based Convolutional Neural Networks) learning based on the region of interest in which the details of the neural network are self-organized to fit the data. Wu et al. [24] proposed a novel Convolutional neural network with an Efficient Convolutional Network (ECNet) architecture to implement semantic segmentation of side scan sonar images. Wang et al. [25] proposed a real-time semantic segmentation network RT-Seg for side scan sonar images. Yu et al. [26] employed the R2CNN module to obtain accurate sonar image features, which reduces errors and improves accuracy. Additionally, a self-guidance module was introduced to ensure the network's stability and to optimize the segmentation results. Wang et al. [27] proposed an underwater object detection method based on the YOLO V3 network, demonstrating that the YOLO V3 network is an effective way to improve the accuracy of underwater object detection. Li et al. [28] studied the problems of low accuracy, low efficiency and low missed detection rate in sonar image target detection. The You Only Look Once version 3 (YOLO V3) model is used to perform one-time neural network processing on side scan sonar images applied to sonar image detection in this paper. Jin et al. [29] proposed inputting sonar images directly into the improved convolutional neural network (CNN) for seabed sediment classification without extracting image features.

As demonstrated by this research, deep learning has been widely used in side scan sonar image detection. However, deep learning networks cannot be applied to different resolutions and different scales, while the sample input scales and scales of the network processing create inconsistency issues because of the limited inputting scales of the images. The usual approach is to compress, expand or crop the images [30], but this will ultimately limit the accuracy of object detection because of the loss of image information [31] and add distortion to the entire image [32]. Some methods of side scan sonar image target detection and some conventional neural networks, such as Alexnet, GoogleNet, ResNet, etc. have low accuracy [33] and add complexity to network structures in the process of side scan sonar image processing [34].

Because of limiting the scales of the input images and classical target detection methods, some strategies and solutions that we can adopt in typical target detection based on side scan sonar images are as follows: (1) For the problem that the existing network is difficult to apply to different resolutions and different scales, and the inconsistency between the sample input scales [35] and the network processing scales. We propose to add an algorithm to the model to break the limiting of the input scales on the detection process. Spatial pyramid pooling can use pooling to divide the CNN features into several fixed sizes [36], which can solve the problems in the process of image detection in a certain extent. (2) To solve the problem of low accuracy of side scan sonar image detection using some classical convolutional neural network methods due to the structure of the model, we can use the YOLO network for further network optimization and improvement. Besides, we also adopted online dataset preprocessing (ODP) for improving the experimental results and achieving high detection accuracy.

In this paper, we propose an improved deep learning network based on spatial pyramid pooling and online dataset preprocessing for side scan sonar image object detection to resolve the multi-scale inputting problem [37]. Our proposed approach uses online dataset processing, with random flips to increasing the number of samples and enhancing the complexity of the model [38], so as to improve detection accuracy. We use the data expansion method to preprocess the data [39], thus meeting the pre-data preparation requirements for image classification. Our approach uses spatial pyramid pooling in which multiple pools of different sizes with the size of the output feature vector [40] are fixed to realize the input operations with images of different sizes. The spatial pyramid pooling network is used for target detection as it can adapt to multi-dimensional inputting [41], automatically extracting missing input vector features. Spatial pyramid pooling has successfully achieved accuracy improvements in image target detection as it is added between the feature extraction module and the fully connected layer [42]. To further address the limited multi-scale inputting problem, we also adopted the YOLO V3 as this is an output network that uses multi-scale features for object detection and adjusts the basic network structure to realize efficient high-accuracy detection performance.

The improvement of the lightweight network algorithm based on YOLO V3 has had a lot of fruitful research. Yue Li et al. [43] proposed an improved YOLO-SC (YOLO-Submarine Cable) detection method, based on the YOLO V3 algorithm, to build a testing environment for submarine cables and to create a submarine cable image dataset. Moran

Ju et al. [44] proposed a multi-scale target detection algorithm involving the improvement of YOLO V3 by a mathematical derivation method based on Intersection over Union (IOU). Xu et al. [45] adopted DenseNet (Densely Connected Network) based on the YOLO V3 in detecting remote sensing targets to enhance feature extraction capability. What this research has in common with the above proof applications is the improved YOLO V3 algorithm. What distinguishes this research from similar applications is that we propose an online data preprocessing method to increase the amount of data from the data input; thus, the complexity of the model can be improved, and better experimental results can be obtained.

A large number of systematic experiments indicate that, by adding the spatial pyramid pooling, the evaluation of side scan image target detection with the commonly used networks can be significantly improved. Three kinds of datasets were used for quantitative evaluation in the experiments, including the Sunken ship, Fish flock and Seafloor topography. The evaluation indicators, [46] including Accuracy, Recall, Precision, and F1 Score, measured the overall accuracy of the model prediction. These evaluation experiments assessed the efficiency of the network model as modified in the experiment compared with the unimproved networks. Based on the spatial pyramid pooling, the YOLO V3 used in this paper for three kinds of targets shows greater efficiency than AlexNet, GoogleNet, VGGNet and ResNet neural networks. This result shows that spatial pyramid pooling effectively deals with the problem of missing data, improves the efficiency of the model and can have higher detection accuracy.

The results show that the proposed approach effectively realizes higher detection accuracy. The proposed online dataset preprocessing approach incorporating random flip, based on the YOLO V3 model with spatial pyramid pooling, increased the number of samples and enhanced the complexity of the model, thus improving detection accuracy. This study can provide theoretical and practical value for improved detection when using side scan sonar images. The rest of this paper is organized as follows. Section 2 presents the principle methodology for the paper. Experiment details, including datasets description and preprocessing, training model process, results and discussion, etc., are shown in Section 3. Conclusions are drawn in Section 4.

2. Methodology

2.1. Improved Neural Networks with Spatial Pyramid Pooling and Online Datasets Preprocessing for Underwater Target Detection

As shown in Figure 1, we start with dataset preprocessing. In the experiment, five kinds of neural network frameworks for target detection were adopted, and then spatial pyramid pooling was added into the five neural network structures, respectively. Online dataset preprocessing (ODP) is input to the network that is our proposed target detection method based on the YOLO V3 algorithm with spatial pyramid pooling and online dataset preprocessing for side scan sonar underwater images.

Finally, five kinds of neural networks were compared between those with spatial pyramid pooling added and those without it added to verify the effects of spatial pyramid pooling; based on spatial pyramid pooling with YOLO V3, the performance was compared between the online datasets preprocessing being adopted and without it being adopted to verify the effectiveness of the online datasets preprocessing.

2.2. Datasets Preprocessing

Seafloor conditions are complex and changeable [47], affected by irregular water flow and light [48], and the imaging quality is spotty. In addition, the characteristics of sonar images, such as blurred edges and low contrast [49], will eventually cause serious problems, such as image blurring and color distortion. Therefore, preprocessing should improve the quality of side scan sonar images, which can make the target contour more prominent. Preprocessing permits effective identification of observation targets. The preprocessing scheme of datasets is as follows.

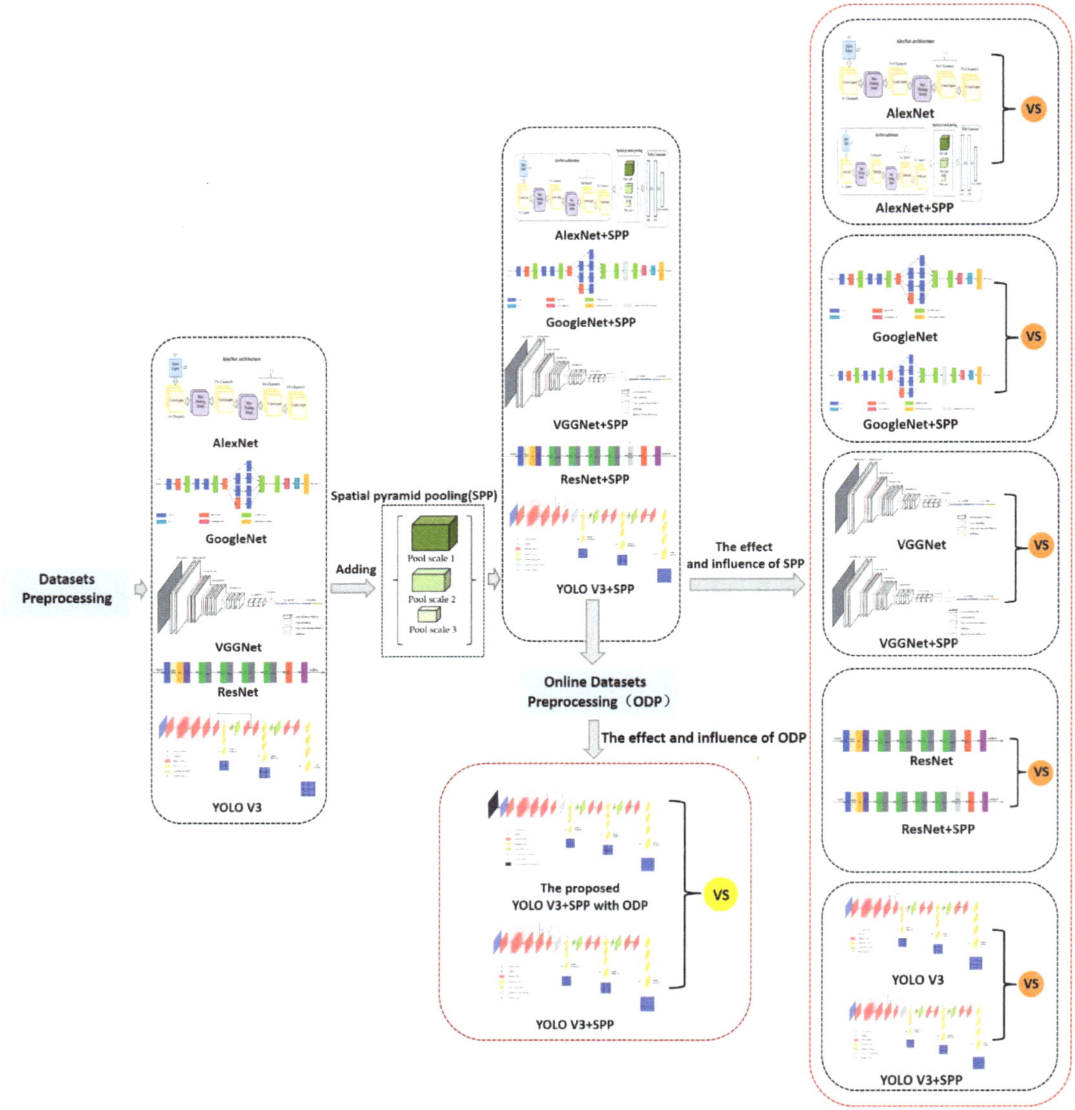

Figure 1. The overall experiment flow chart and our proposed ODP+YOLO V3+SPP method.

(1) The expanded datasets

The original dataset consisted of three categories (sunken ship, fish flock and seafloor topography), each with 50 images and a total of 150 images and with 250 images and a total of 750 images (Figure 2).

(2) Data Augmentation

The original sample data dataset used in this paper is small, and thus subject to overfitting [50] in the absence of more training samples. The number of different types of images in the training set is often not balanced enough [51], leading to bias phenomena [52] in the trained model representation to effectively train the network. The original dataset is augmented to improve the detection accuracy and generalization ability of the model.

By introducing a correlate specifying the directory where the original data is located, the augmentation method is set to geometrically rotate (to the left or the right) with a probability of 0.7 to within 25 degrees of geometric rotation. Sampling 250 images, i.e., augmentation to 250 images per category, for a total of 750 images (Table 1).

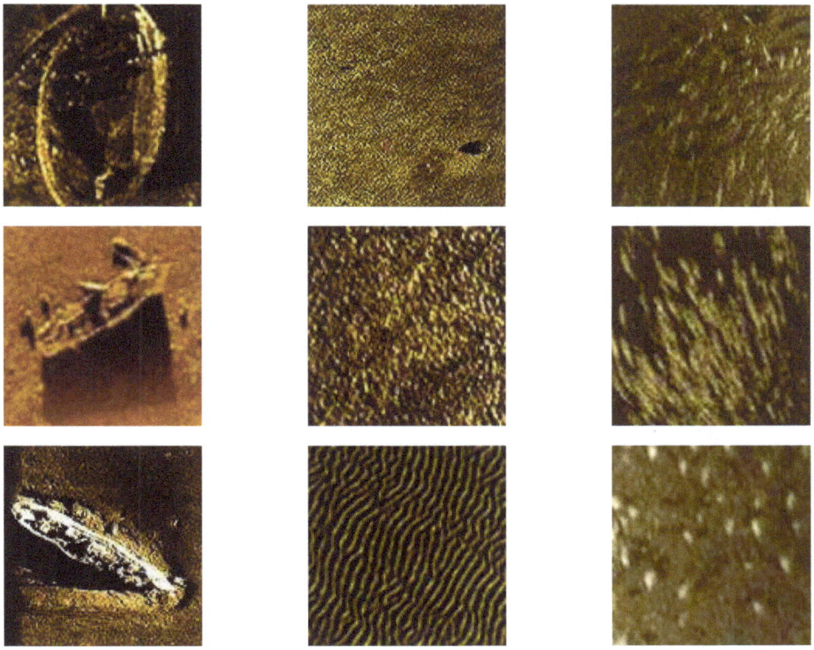

(**a**) Sunken ship (**b**) Fish flock (**c**) Seafloor topography

Figure 2. The demonstration of the original side scan sonar images from three kinds of datasets including Sunken Ship, Fish Flock, Seafloor Topography.

Table 1. Detailed description of the datasets.

Dataset Class	Sunken Ship	Fish Flock	Seafloor Topography
Image Quantity	250	250	250

(3) Partition the datasets

The newly expanded post-dataset was divided into the training set and the validation set in a 7:3 ratio, that is, 175 for training and 75 for testing in each category.

(4) YOLO dataset annotation

Image datasets are annotated and stored in a text file with the same name as the image. Example: 00.50.51.01.0. The first integer represents the object class, with three categories (0,1,2); the second and third values represent the scale of the object's central location to the entire image; the fourth and fifth values represent the scale of the object's width and height to the entire image. Since there is only one object in each image, the following four numbers are the same, except for the first number, which is the category of the object.

2.3. *Using Online Dataset Preprocessing to Improve Target Detection Performance*

2.3.1. DataLoader Platform

The DataLoader is a tool in Pytorch for processing model input data [53]. It combines datasets and samplers by providing single or multi-threading of the object [54] on the datasets. The goal is to do an initialization of the data [55] and divide the training data into groups. This function is used in the training model to break up the training data into groups [56]. During the training process, the DataLoader encapsulates the custom Dataset into batch-sized tensors according to batch size, shuffle, and so on operation process.

2.3.2. Random Flip

As a kind of data enhancement method, the random flip is adopted, which includes horizontal and vertical flips according to the actual target [57], and is usually realized by modifying the configuration file [58]. Random flips will randomly produce numbers between 0 and 1.

Random flips randomly produce numbers ranking between 0 and 1 (Algorithm 1). The main effect of horizontal flipping on labels is the x coordinate, that is, x = 1 − x0, labels [: , 1] = 1 − labels [: , 1]. The main effect of vertical flipping on labels is the y coordinate, which is y = 1 − y0, labels [: , 2] = 1 − labels [: , 2].

Algorithm 1: Random Flip Algorithm

Input: All initialized Datasets
Based platform: Dataloader
Output: Updated Random Flip Datasets
Listing 1: Random Flip Process

```
→ dataset = create_dataloader
→ mlc = int(np.concatenate (dataset.labels, 0)[:,0].max()) #max label class
→ nb = len (train_loader) #number of batches
→→ assert mlc < nc Possible class labels are 0-{nc—1}
→→ if Rank in [-1,0]:
→→→val_loader =create_dataloader
→→ if not resume:
→→→labels = np.concatenate (datasets.labels, 0)
→→→→if plots:
→→→→→plot_labels (labels, names, save_dir)
```

End procedure

2.4. *Networks Adopted in the Experiments*

The convolution backbone network acts as a feature extractor and is related to accuracy and speed performance, playing an important role in object detection. Convolution Neural Networks (CNNs) are a deep learning model widely used in sonar image classification. CNNs can realize advanced solutions for object detection and classification in the field of computer vision [59]. The core feature of CNN is the capture of local features and the automatic acquisition of semantic information at different levels of abstraction [60]. Because of its relatively fast training speed, it has also achieved very good results in image based target detection and has been widely used. For example, HOG-ShipLSNet [61] for ship classification, Polarization fusion with geometric feature embedding for SAR ship classification [62], and there have been many mature experiments. CNN can pool and map the initial image to extract higher-level semantic information [63] layer by layer. The specific convolution process is below:

$$z_j^m = f\left(\sum_{i\in T_j} k_{i.j}^m * z_i^{m-1} + b_j^m\right) \tag{1}$$

where f is the activation function; $\Updownarrow$ is the number of layers; $k_{i,j}$ is filtering; b_j is the bias; $*$ is the convolution operation; and T_j represents a collection to store input matrices.

Through years of development, CNNs have gradually achieved remarkable performance and accuracy by increasing the number of layers of the neural network; thus, they have also increased the number of samples for learning and training. In its process, a back-propagation algorithm feeds back the error into the model layer by layer [64], gradually training it by forwarding feedback, finally causing the neural network model to converge.

2.4.1. AlexNet

AlexNet is used to input images constrained to a fixed size, randomly truncated into 224 × 224 blocks [65]. Different feature maps are passed to subsequent convolutional layers [66] after a series of operations. This network uses five convolutional layers as the base layer and adds another three layers as a fully connected network structure [67], which acts as a classifier. Through layers of the network, more computing resources are calculated; the two convolutional layers will communicate with each other; the feature dimension [68] will be added at the same time; and the calculation amount will be superimposed in convolutional layers with almost no size limit on images.

2.4.2. GoogleNet

The Google team came up with a web architecture called GoogLeNet [69] for image detection and classification, based on the Hebbian model, which consists of nine Inception modules stacked on top of each other [70]. This architecture increases the depth and width of the network, improves the utilization of the internal computing resources [71] and constructs the mathematical model of the GoogLeNet network. The success of the GoogLeNet mainly depends on the Inception module, which can deal with the problems of over-fitting, large amounts of computation and low accuracy.

2.4.3. VGGNet

VGGNet network structure is simple, regular and efficient. Compared with VGGNet, AlexNet and GoogleNet replace a 5 × 5 convolutional kernel with two 3 × 3 small convolutional cores [72] and reduce the parameters. The outstanding contribution of VGGNet is to demonstrate that very small convolutions can be effectively improved by increasing network depth. The VGGNet network structure has a total of 16 layers including 13 convolution layers [73], 5 pooling layers and three full connection layers.

2.4.4. ResNet

When the VGGNet propagates backwards, there will be a loss in the transmission process, and even a gradient explosion and a gradient disappearance will occur when the transmission reaches a deeper layer. In 2015, Kaiming He proposed deep residual networks (ResNet) for target detection and classification, target localization [74] and segmentation [75]. In contrast to VGGNet, ResNet has skip connections between different layers [76], adding shallow information directly to the convolution layer and jumping some gradients directly to the shallow layer to mitigate the loss problem.

The deeper the network layer is, the more abundant the features that can be extracted [77] and the more semantic the image can reflect, but the simple stack will lead to the disappearance of network gradients [78]. ResNet introduces residual learning to deal with the problem that the deep learning network is difficult to be optimized. Its essence is to use the optimal mapping and use the stack nonlinear layer to fit another mapping.

2.4.5. YOLO V3

The classification network is the backbone network, and target detection and image match the sub-direction of the classification task. From AlexNet to ResNet, the backbone network is getting better. We add the regression head of the predicted position and the estimation of the NMS maximum a series of content and operations [79] to form the model of YOLO V3.

YOLO is an abbreviation for You Only Look Once. The YOLO versions are in the process of upgrading iterations. Fast speed is the most outstanding core characteristic of the YOLO algorithm in its small size [80]. YOLO shows great ability in generalization due to its generalized features [81] and good performance transferring to others. YOLO has many lower sampling layers [82], improving the effectiveness of the network because the target features are not exhaustive. The original YOLO uses global information and reduces detection errors when using the background as an object. YOLO V2 includes two-stage training, which adds Anchor with K-means. In YOLO V3, the network adopts FPN [83] for multi-scale detection.

Inspired by ResNet, YOLO V3 uses Darknet-53 as the backbone feature extraction [84] network, with better robustness and deeper network layers to increase the number of convolutional layers; the residual network is used in the convolution layer many times to make the network more easily converge and to enhance the adaptability of the model to the complex scene.

2.5. Multi-Scale Inputting Based on the Spatial Pyramid Pooling

In recent years, the study of multi-dimension input based on CNN has been one of the hot fields in deep learning, and two main classes of multi-dimension object detection methods have been developed [85]. To detect the multi-feature graph extracted from different layers of networks independently [86], the graph is fused [87]. To improve the precision of multi-scale detection, the multi-feature map is fused with the feature map and receptive fields of different sizes.

Based on the CNN model, the Spatial Pyramid Pooling Network method was proposed in 2014, which enables the non-fixed input image size to map arbitrary-sized features into fixed-sized feature vectors [88]; it shows the powerful performance of detecting multi-scale object fusion.

The process of the Spatial Pyramid Pooling Network is to extract features of different sizes to effectively utilize global information. If an image is an input, it will be divided into different sizes, and each block will extract a feature [89]. In max pooling, the value of the largest block in the image is calculated, and then the corresponding neuron output is obtained in turn.

The Spatial Pyramid Pooling Network has many pools of different sizes to ensure that the outputting feature vector size [90] is fixed to achieve the pooling goal. The CNN and its characteristics greatly limit the input of image size. If the images are processed, it will lead to image quality shrinkage and information loss [91]. To solve this problem, the Spatial Pyramid Pooling Network method is added to the CNN.

2.6. Networks with Spatial Pyramid Pooling Adopted Experiments in the Models

Based on the ability of the spatial pyramid pool to break through the restriction of input size, we add it to the traditional classical algorithms AlexNet, GoogleNet, VGGNet and ResNet network models, respectively. The effect, however, is restricted by the structure of the above network models; the detection accuracy has some limitations [92]. Therefore, we add a spatial pyramid module in YOLO V3 to realize the function of stitching multi-scale region features. Multi-scale aggregation is used to extract features and to converge the network in the spatial pyramid pool.

Therefore, we compare YOLO V3 network with other networks vertically, using the same networks with and without the pyramid pool model, using the qualitative evaluation by evaluation indicators, to highlight the high performance of the YOLO V3 Model based on spatial pyramid pooling.

2.6.1. AlexNet + SPP

Spatial pyramid pooling is a method of combining input images through three pool modules and convolving different input images to extract the features of each module, solving the problem of training multiple images after a selective search. Adding Spatial Pyramid Pooling to AlexNet (Figure 3) eliminates the crop/warp image normalization process and solves the information loss and storage problems caused by image distortion. We used Spatial Pyramid Pooling to fix the convolution layer in the process of fine-tuning the AlexNet network structure (Algorithm 2), replacing the last Pooling layer before the full connection layer, thus the CNN network can train the role of different sizes of images. Finally, the original image can be directly sent to the network for training, accelerating the network to achieve the end-to-end process of object detection.

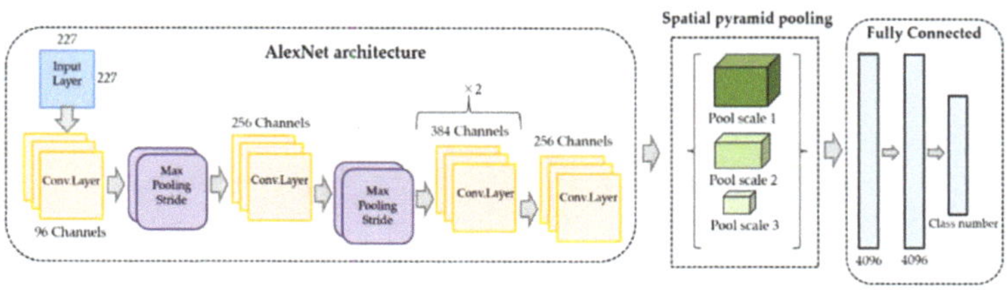

Figure 3. The model of AlexNet + SPP.

Algorithm 2: AlexNet + SPP Algorithm

Input: SPP algorithm input
Based network: AlexNet algorithm
Output: AlexNet + SPP algorithm output
Listing 2: SPP Process
→ class SPPLayer (nn.Module):
→ **def** __init__(self, num_levels, pool_type = 'max_pool'):
→ → super (SPPLayer, self).__init__()
→ → self.num_levels = num_levels
→ → → self.pool_type = pool_type
End procedure

2.6.2. GoogleNet + SPP

The success of the GoogleNet is largely due to the Inception module, which can be seen as a stack of Inception modules throughout the GoogLeNet mainframe. The spatial pyramid algorithm structure in GoogleNet is added after the Inception module (Figure 4). The Inception Module mainly adds a convolution layer to CNN (Algorithm 3), using ReLU as the activation function, and its main function is to reduce the dimension of network features and to reduce a lot of computation without sacrificing the performance of the network model. The Spatial pyramid pooling structure in GoogleNet is added after the Inception template.

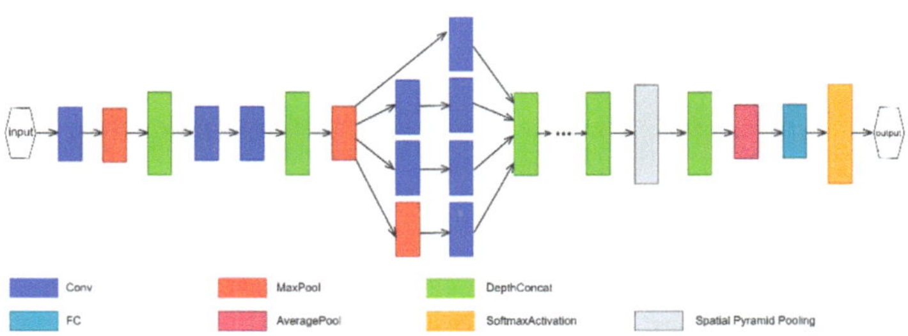

Figure 4. The model of GoogleNet + SPP.

Algorithm 3: GoogleNet + SPP Algorithm

Input: SPP algorithm input
Based network: GoogleNet algorithm
Output: GoogleNet + SPP algorithm output
Listing 3: SPP Process

```
→ class SPPLayer (nn.Module):
→ def __init__(self, num_levels, pool_type = 'max_pool'):
→ → super (SPPLayer, self).__init__()
→ → self.num_levels = num_levels
→ → → self.pool_type = pool_type
```

End procedure

2.6.3. VGGNet + SPP

VGGNet is compact, uses maximum pooling connections and uses ReLU as the activation function between the hidden layers. Therefore, the VGGNet is often used to extract image features, increase the deep of the convolutional neural network and use small convolutional cores, which play a great role in the final detection of the network (Figure 5). We add the spatial pyramid pooling to the last layer (Algorithm 4) of the VGGNet's convolution layer.

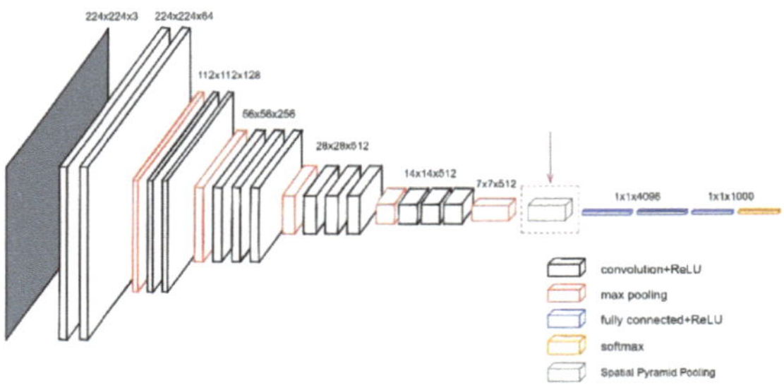

Figure 5. The model of VGGNet + SPP.

Algorithm 4: VGGNet + SPP Algorithm

Input: SPP algorithm input
Based network: VGGNet algorithm
Output: VGGNet + SPP algorithm output
Listing 4: SPP Process

```
→ class SPPLayer (nn.Module):
→ def __init__(self, num_levels, pool_type = 'max_pool'):
→ → super (SPPLayer, self).__init__()
→ → self.num_levels = num_levels
→ → → self.pool_type = pool_type
```

End procedure

2.6.4. ResNet + SPP

The residual module is very important in the deep neural network, which is often used in the process of modeling. The success of ResNet is that it solves the problem of network degradation with the help of a residual module, which improves the depth of the network, gets the feature of stronger expression ability and has higher accuracy (Figure 6).

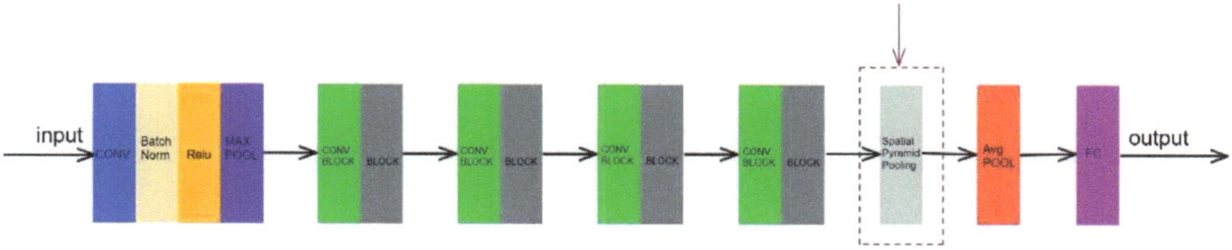

Figure 6. The model of ResNet + SPP.

The spatial pyramid pooling structure in ResNet is added between ResNet and the full connection layer (Algorithm 5), and the deep residual network without activating the normalized layer is designed. The main goal is to achieve different input sizes while ensuring the fixed-size network, which is now well-optimized and fully utilized. It is helpful to increase the depth of the network while avoiding gradient disappearance.

Algorithm 5: ResNet + SPP Algorithm

Input: SPP algorithm input
Based network: ResNet algorithm
Output: ResNet + SPP algorithm output
Listing 5: SPP Process

```
→ class SPPLayer (nn.Module):
→ def __init__(self, num_levels, pool_type = 'max_pool'):
→ → super (SPPLayer, self).__init__()
→ → self.num_levels = num_levels
→ → → self.pool_type = pool_type
return solution
```

2.6.5. YOLO V3 + SPP

YOLO V3 outputs three feature maps of different sizes, from top to bottom, corresponding to the features of the deep layer, middle layer and shallow layer, respectively. In contrast to the small size and large receptive field of the deep feature map, the shallow feature map is more convenient for detecting small-scale objects, which is similar to the FPN structure.

With the addition of a spatial pyramid pooling structure, YOLO V3 can increase performance with little time consumption (Figure 7). The fusion of local features and global features is realized by the SPP module, in which the SPP part is added in the first Set block (Algorithm 6). By maximizing the pool of different receptive fields and finally stitching the dimensions, the fusion information of different scales can be obtained, thus improving the performance of the model.

Algorithm 6: YOLO V3 + SPP Algorithm

Input: SPP algorithm input
Based network: YOLO V3 algorithm
Output: YOLO V3 Net + SPP algorithm output
Listing 6: SPP Process

```
→ class SPP (nn.Module):
→ def __init__(self, c1, c2, k = (5, 9, 13)):
→ → super ()._init_()
→ → self.cv1 = Conv (c1, c_, 1, 1)
→ → self.cv2 = Conv (c_ *(len(k) + 1), c2, 1, 1)
→ → → self.m = nn.ModuleList ([nn.MaxPool])
```

End procedure

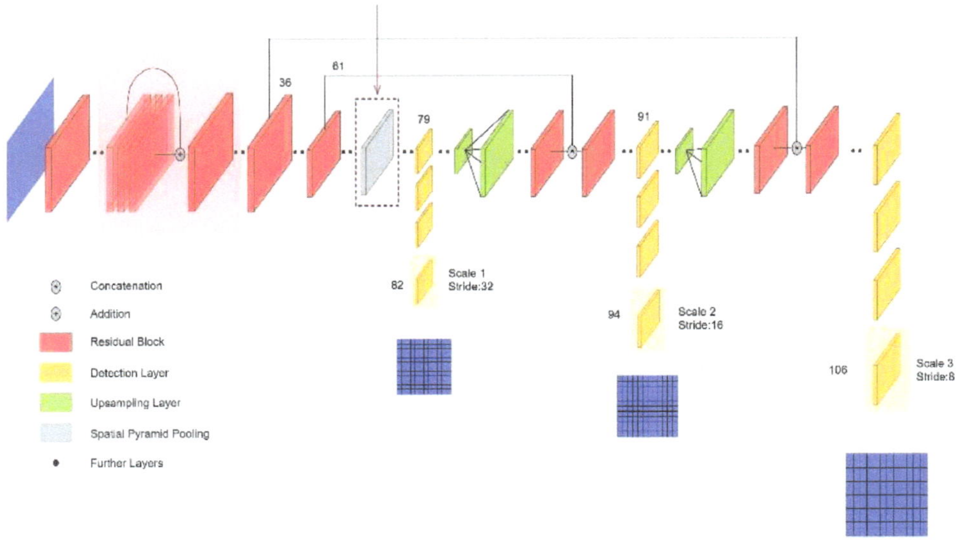

Figure 7. The model of YOLO V3 + SPP.

2.6.6. Online Dataset Preprocessing with YOLO V3 + SPP

We adopt online dataset preprocessing (ODP) to improve detection performance based on the DataLoader platform in Section 2.3. The input part of the improved neural network with spatial pyramid pooling is based on the YOLO V3 method (YOLO V3 + SPP model with Online Dataset Preprocessing) to form an experimental group (Figure 8). It is used to compare with the YOLO V3 model with spatial pyramid pooling in Section 2.6.5 to test the effect of random data flipping before the input part (Algorithm 7) to the impact on the evaluation of identification.

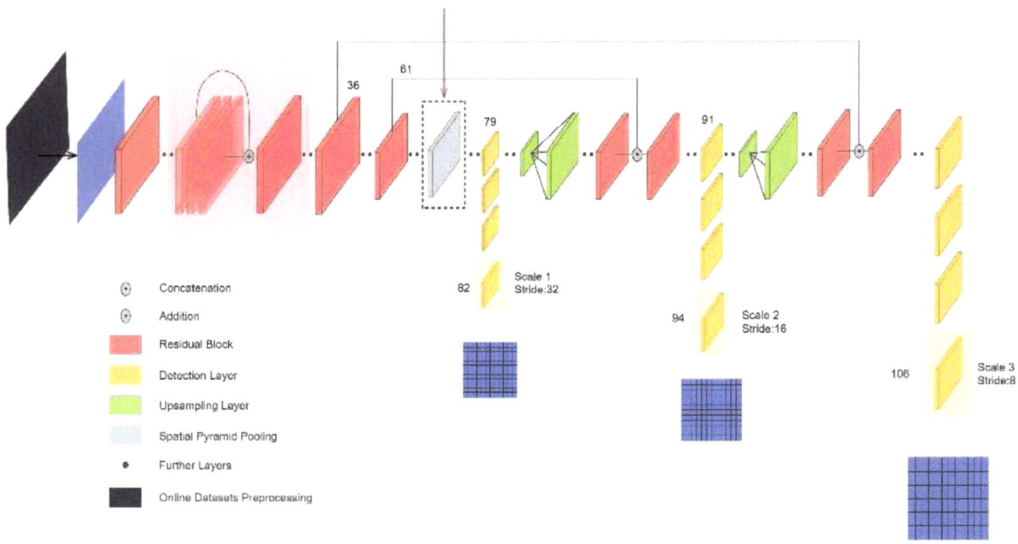

Figure 8. The model of Online Datasets preprocessing with YOLO V3 + SPP.

Algorithm 7: Online Datasets Preprocessing YOLO V3 + SPP Algorithm

```
Input: SPP algorithm input
Based network: YOLO V3 algorithm
Output: YOLO V3 Net + SPP algorithm output
Listing 7: SPP Process
1: → class SPP (nn.Module):
2: → def __init__(self, c1, c2, k = (5, 9, 13)):
3: → → super ()._init_()
4: → → self.cv1 = Conv (c1, c_, 1, 1)
5: → → self.cv2 = Conv (c_ *(len(k) + 1), c2, 1, 1)
6: → → → self.m = nn.ModuleList ([nn.MaxPool])
End procedure
```

2.6.7. Performance Comparisons

Through the above experiments, we made four groups of comparison, testing the influence and effect of adding spatial pyramid pooling on four kinds of traditional classical deep learning network methods and then, through a group of experimental comparison, testing the effect of spatial pyramid pooling on YOLO V3 deep learning network method. A set of experiments was carried out to compare the effects of spatial pyramid pooling on YOLO V3 deep learning network method, to test the effect of on-line preprocessing adopting random flip on YOLO V3 with high accuracy and high speed, which has been added to spatial pyramid pooling.

3. Experiments

3.1. Datasets

3.1.1. Datasets Description

To verify the validity of sonar image based target detection by using pyramid pooling improved convolution neural network, the original raw sample of the dataset used in this paper comes from 150 sonar images with 50 images in each category collected by individuals divided into three categories, including sunken ships, fish flocks and seafloor topography (Figure 2), to form a dataset for the research of the underwater target detection algorithm. The description of the Dataset is shown in Table 1.

The amount of data is far from sufficient; it is difficult to train the network, and the network will be overfitting, affecting the accuracy. To efficiently train the network, the original dataset is processed as follows.

3.1.2. Datasets Preprocessing

Data expansion makes the training data as close as possible to the test data, thus improving the prediction accuracy. At the same time, the network is forced to learn more robust features so that the model has stronger generalization ability. After data expansion, the sample of the training set can be increased, which can effectively alleviate the over-fitting of the model and also can bring stronger generalization ability to the model. According to the preprocessing method of Section 2.2, we extend the original dataset and introduce the correlation package, specify the directory in which the original data resides and geometrically rotate the original data (to the left or the right) with a probability of 0.7 to within 25 degrees. The original label of the image should be kept unchanged when the data expansion operation is carried out, and the data expansion should be carried out without changing the label; 250 images were added to each category, for a total of 750 images. The newly expanded post-dataset is divided into a training set and verification set in a 7:3 ratio, that is, 175 training sets and 75 test sets.

3.2. Experimental Preparation

3.2.1. Experimental Condition

In this experiment, the training and testing environments were conducted on a Ubuntu 18.04 PC with an Intel® Xeon® CPU, 31.2 GB memory and an NVIDIA-SMI 470.82.00 GPU with 250.9 GB memory. The procedure was operated on the PyCharm 11.0.13 platform using python language. The deep learning framework written by PyCharm for the underlying language has high flexibility and strong code execution ability. Based on this, the following experiments are designed.

3.2.2. Experiment Model Training

The deep learning framework written by torch for the underlying language has high flexibility and strong code execution ability. Based on this, the following four sets of experiments are designed (Figure 9); the learning rate is set to 0.0001; and a total of 60 steps (epoch) are trained. Firstly, AlexNet, GoogleNet, VGGNet and ResNet neural networks were used as initialization model group to preprocess three kinds of datasets: sunken ships, fish flocks and seafloor topography.

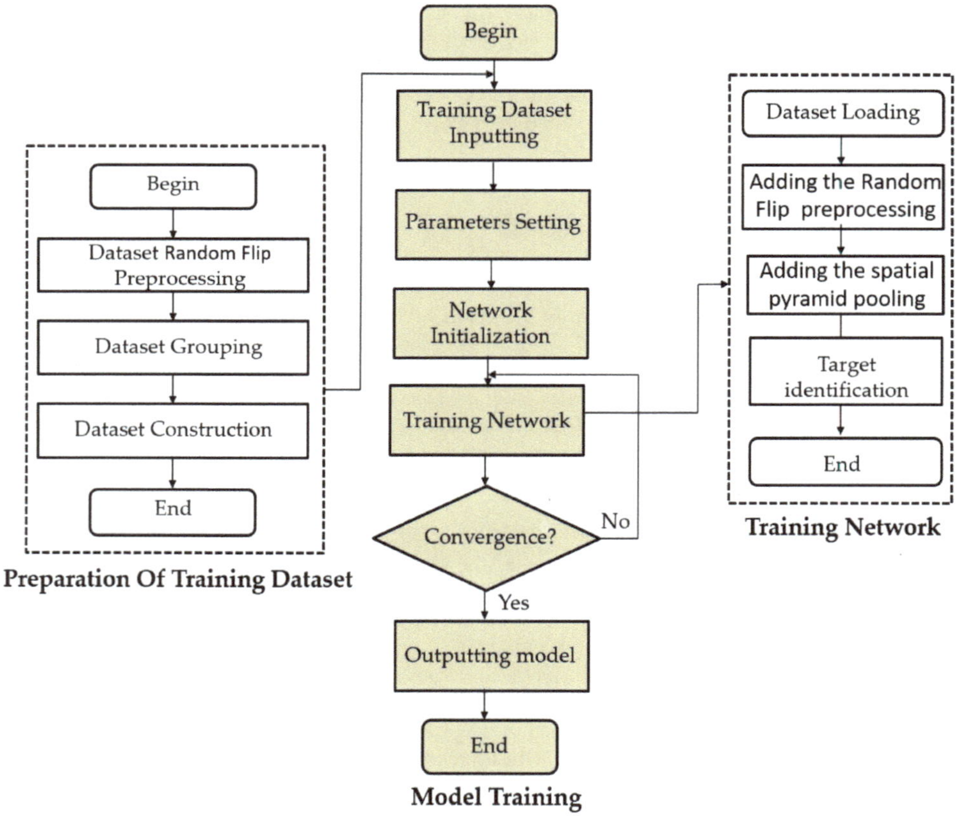

Figure 9. Flow chart of the target detection processes.

Afterwards, then the spatial pyramid pooling algorithm structure is added to the above network deep learning training in contrast to the group without the addition of the spatial pyramid pooling, for proving the effectiveness and reliability of the spatial pyramid pooling algorithm. The second step is to define some variables required for training and placeholders for inputting images and to import the prepared dataset into networks. The third step is to collect and summarize the data obtained from the experiment, reading all the pictures in all subfolders under the path, and storing them in a list. Finally, when loading the data, the data is reordered; the training set and the verification set are loaded in turn; and the data results are displayed through the samples.

In the experiment of the YOLO model, the datasets are enhanced, and the training set is labeled and input into the network. Then the model configuration file is constructed according to the network structure; the configuration file is analyzed through the script; and the YOLO V3 network is constructed by adding the pyramid pool layer by layer. YOLO V3 is used to train three kinds of experimental data, and then, based on YOLO V3, a spatial pyramid pool algorithm is added to construct the network; the spatial pyramid pool network layer is divided into four branches and sampled by the different maximum pool, and three parallel pool layers are added to the network layer 75 to 77, the spatial pyramid structure can achieve the fusion of different scale features, especially for small target detection, which has a good performance and enhances the ability of feature extraction. We input the YOLO V3 model with randomly flipped dataset preprocessing online and compared them to verify the dataset preprocessing effect.

3.3. Evaluation Indicators

In the process of data training, there will be overfitting. To avoid it as much as possible without adding training samples to make the dataset more complex, performing various transformations on the original image [81] can greatly improve the numerical amount of learning and training in the experiment. According to the progress of the experiment, the model parameters were analyzed and adjusted to improve the detection accuracy of the model [82].

To compare, analyze and evaluate accurately, the different network models used in the experiment, their respective detection accuracy, selecting accuracy, precision, recall, mAP and Time were used as indicators. There are four situations that describe the classification results. These situations are:

The true positive (TP): positive samples predicted by the model as positive classes;

The true negative (TN): negative samples predicted as negative classes by the model;

The false positive (FP): positive samples predicted as negative classes by the model;

The false negative (FN): negative samples predicted by the model as positive classes.

The accuracy rate is the percentage of the number of samples that predicts and judges correct data in the model to the total number of data samples. Precision and Recall are mutually constrained. Their calculations follow:

$$OA = \frac{(TP + TN)}{(TP + TN + FP + FN)} \tag{2}$$

$$precision = \frac{TP}{TP + FP} \tag{3}$$

$$recall = \frac{TP}{TP + FN} \tag{4}$$

$$mAP = \frac{\sum_n precision}{n} \tag{5}$$

3.4. Experimental Results

This section presents the analysis of the data results, to verify the improvement effect of spatial pyramid pooling and to verify the effectiveness of adopting online dataset preprocessing (ODP) random flip in the field of sonar image based target detection. To improve the reliability and stability of the experiment, the amount of data in the training set is much larger than that in the test set, and the size of the training set is more than twice that of the test set. After 60 iterations and repeated experiments, a comparison of experimental results is shown in Table 2 for the AlexNet Group, the AlexNet + SPP Group, the GoogleNet Group, the GoogleNet +SPP Group, the ResNet Group and ResNet + SPP Group. This table reports the performance results of eight groups from the number of iterations from three scenes. In order to display the experimental results efficiently, we adopted the median accuracy of the verification set of the multi-round model after training for a more straightforward presentation of the results.

Table 2. The comparison of evaluation indicators on classic network performance.

Networks	Testing Time Used (Mins)	Accuracy	Precision	Recall	F1-Score
AlexNet	5.10	0.76	0.79	0.74	0.71
AlexNet + SPP	4.27	0.78	0.80	0.73	0.74
GoogleNet	5.81	0.71	0.71	0.82	0.76
GoogleNet + SPP	3.64	0.75	0.75	0.73	0.78
VGGNet	14.34	0.84	0.81	0.97	0.88
VGGNet + SPP	8.25	0.86	0.89	0.86	0.89
ResNet	4.39	0.87	0.89	0.87	0.87
ResNet + SPP	4.18	0.89	0.91	0.86	0.88

In the first experiment comparing (+SPP), our proposed adding SPP method yielded higher detection accuracy and faster detection efficient performance (Table 2). For example, In the AlexNet groups, the AlexNet + SPP method shows time used 0.83 min faster than the AlexNet method; the AlexNet + SPP method shows 2% accuracy higher than the AlexNet method; the AlexNet + SPP method shows 1% precision higher than the AlexNet method; the AlexNet + SPP method shows 1% recall lower than the AlexNet method; the AlexNet + SPP method shows 3% F1-score higher than the AlexNet method, indicating that adding SPP has a positive influence on improving detection accuracy and detection efficient performance.

In the second experiment comparison (Table 3), our proposed ODP+YOLO V3+SPP method yielded the highest detection accuracy and the fastest processing efficiency performance. From a comparison with Table 3, the proposed ODP+YOLO V3+SPP outperformed the other tested networks in the YOLO group comparisons. In the mAP evaluation, the Online Datasets Preprocessing YOLO V3 + SPP group shows 4% higher than the YOLO V3 + SPP group; the Online Datasets Preprocessing YOLO V3 + SPP group shows 6% higher than the YOLO V3 group. In Precision evaluation respect, the Online Datasets Preprocessing YOLO V3 + SPP group shows 2% higher than the YOLO V3 + SPP group; the Online Datasets Preprocessing YOLO V3 + SPP group shows 13% higher than the YOLO V3 group. In Recall evaluation respect, the Online Datasets Preprocessing YOLO V3 + SPP group shows 5% lower than the YOLO V3 + SPP group; the Online Datasets Preprocessing YOLO V3 + SPP group shows 3% higher than the YOLO V3 group. In Time used evaluation respect, the Online Datasets Preprocessing YOLO V3 + SPP group shows 0.02 h faster than the YOLO V3 + SPP group; the YOLO V3 + SPP group shows 0.08 h faster than the YOLO V3 group; the Online Datasets Preprocessing YOLO V3 + SPP group shows 0.1 h faster than the YOLO V3 group, that is to say, the detection efficiency has increased by 9.34%. The above experimental results show that compared with the original YOLO V3 model, the proposed ODP+YOLO V3+SPP underwater target detection algorithm model has improved detection performance through the mAP qualitative evaluation index has increased by 6%, the Precision qualitative evaluation index has increased by 13%, and the detection efficiency has increased by 9.34%. The results also indicate our proposed ODP+YOLO V3+SPP method has a positive improving effect on detection accuracy and detection efficiency.

Table 3. Comparison of evaluation indicators on YOLO Network performance.

Networks	Testing Time Used (Hours)	mAP	Precision	Recall
YOLO V3	1.17	0.89	0.78	0.8
YOLO V3 + SPP	1.09	0.91	0.89	0.88
The proposed ODP + YOLO V3 + SPP	1.07	0.95	0.91	0.83

In Figure 10, we can see that the AlexNet network added to the spatial pyramid pooling has higher detection accuracy and better detection performance than without the spatial pyramid pooling. It improves the spatial pyramid pooling showing a positive effect on the image detection of side scan sonar imagery.

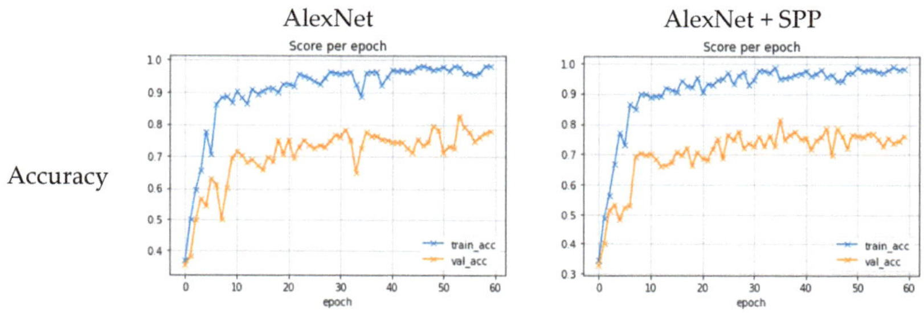

Figure 10. Performance comparison analysis between AlexNet and AlexNet + SPP.

In the GoogleNet groups (Table 2), the GoogleNet + SPP method shows time used 2.17 mins faster than the GoogleNet method; the GoogleNet + SPP method shows 4% accuracy higher than the GoogleNet method; the GoogleNet + SPP method shows 4% precision higher than the GoogleNet method; the GoogleNet + SPP method shows 9% recall lower than the GoogleNet method; the GoogleNet + SPP method shows 2% F1-score higher than the AlexNet method.

In Figure 11, we can see that the GoogleNet network added to the spatial pyramid pooling has higher detection accuracy and better detection performance than without adding spatial pyramid pooling. It improves the spatial pyramid pooling showing a positive effect on the image detection of side scan sonar imagery.

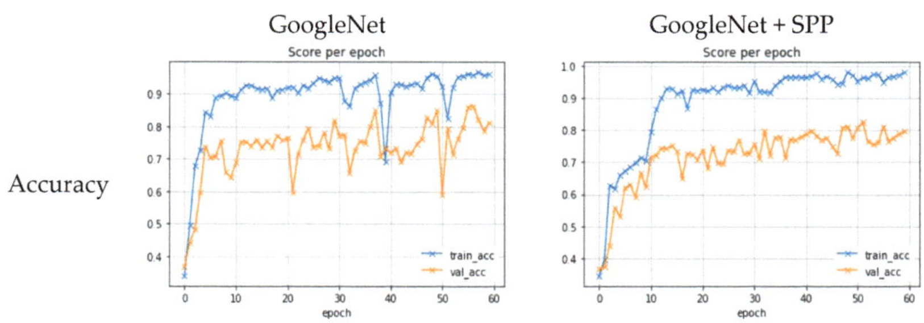

Figure 11. Performance comparison analysis between GoogleNet and GoogleNet + SPP.

In the VGGNet groups (Table 2), the VGGNet + SPP method shows time used 6.09 mins faster than the VGGNet method; the VGGNet + SPP method shows 2% accuracy higher than the VGGNet method; the VGGNet + SPP method shows 8% precision higher than the VGGNet method; the VGGNet + SPP method shows 11% recall lower than the VGGNet method; the VGGNet + SPP method shows 1% F1-score higher than the VGGNet method.

In Figure 12, we can see that the VGGNet network added to the spatial pyramid pooling has higher detection accuracy and better detection performance than without the spatial pyramid pooling. It improves the spatial pyramid pooling, showing a positive effect on the image detection of side scan sonar imagery.

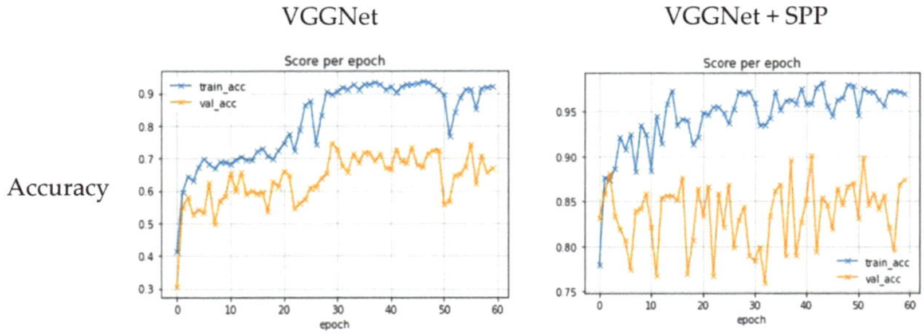

Figure 12. Performance comparison analysis between VGGNet and VGGNet + SPP.

In the ResNet groups (Table 2), the ResNet + SPP method shows time used 0.21 mins faster than the ResNet method; the ResNet + SPP method shows 2% accuracy higher than the ResNet method; the ResNet + SPP method shows 2% precision higher than the ResNet method; the ResNet + SPP method shows 1% recall lower than the ResNet method; the ResNet + SPP method shows 1% F1-score higher than the VGGNet method.

In Figure 13, we can see that the ResNet network added to the spatial pyramid pooling has higher detection accuracy and better detection performance than without the spatial pyramid pooling. It improves the spatial pyramid pooling, showing a positive effect on the image detection of side scan sonar imagery.

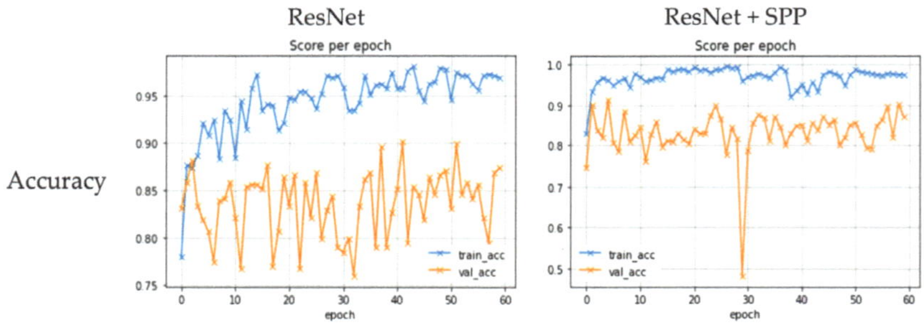

Figure 13. Performance comparison analysis between ResNet and ResNet + SPP.

Above all, the experimental results show that: for AlexNet, GoogleNet, VGGNet, ResNet algorithms, after spatial pyramid pooling is added to their backbone networks in this paper, the average detection accuracy of the three sets of data is increased by 2%, 4%, 2% and 2%, respectively, comparing with the original algorithm. This proves that adding spatial pyramid pooling can improve the target detection accuracy of these commonly used algorithms.

In Figure 14, we can see that the YOLO V3 network added to the spatial pyramid pooling (YOLO V3 + SPP) has higher detection accuracy and better detection performance than without the spatial pyramid pooling. Besides, our proposed improved neural networks with spatial pyramid pooling and online datasets preprocessing (ODP + YOLO V3 + SPP) show the best detection performance in the figure. It improves the spatial pyramid pooling and online datasets preprocessing, showing a positive influence on the image detection of side scan sonar imagery. The detection accuracy has been improved through our proposed method.

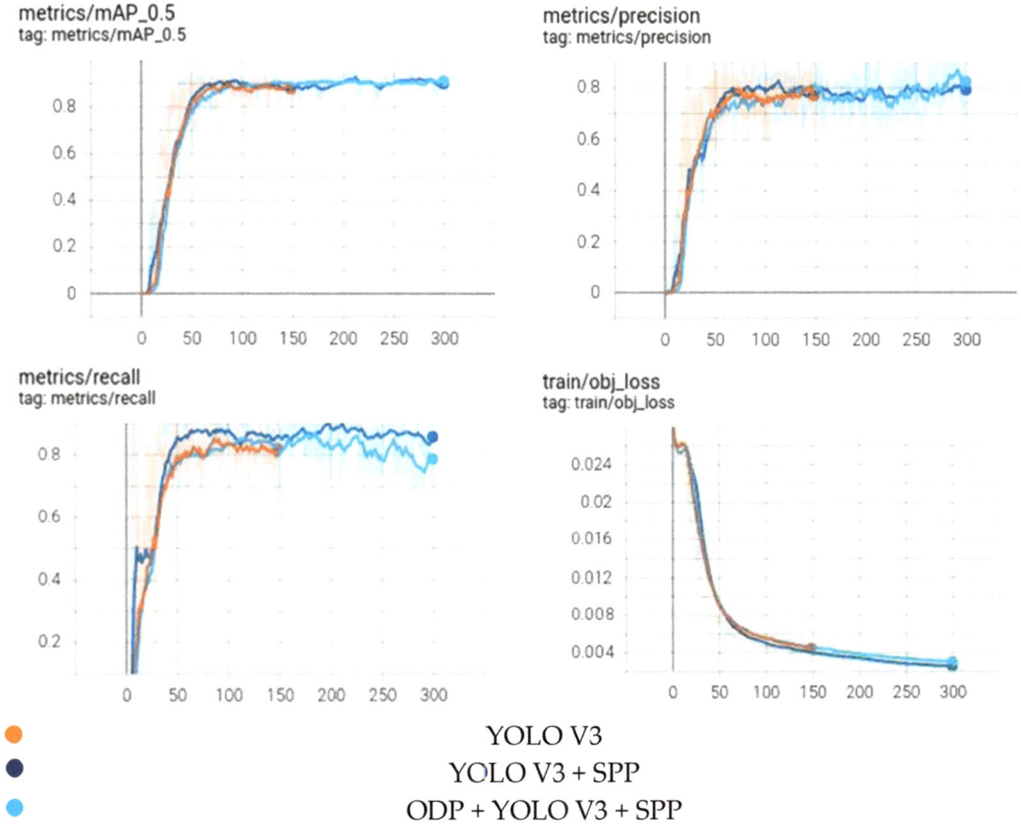

Figure 14. The performance of YOLO groups comparison analysis.

In Table 4 the evaluation indicators for three kinds of side scan sonar datasets based on the proposed YOLO V3 + SPP with online datasets preprocessing method shows that the mAP is 98% for the sunken ship, 91% for the fish flock, 96% for seafloor topography; the Precision is 89% for the sunken ship, 98% for the fish flock, 82% for seafloor topography; the Recall is 88% for the sunken ship, 77% for the fish flock, 84% for seafloor topography. In Figure 15, it shows some example target detection results of the proposed YOLO V3 + SPP with online datasets preprocessing (ODP) method for the three sets of side scan sonar datasets including Sunken Ship, Fish Flock and Seafloor Topography.

Table 4. The evaluation indicators for three kinds of side scan sonar datasets based on the YOLO V3 + SPP with online datasets preprocessing method.

Target Classes	mAP	Precision	Recall
All	0.95	0.9	0.83
Sunken ship	0.98	0.89	0.88
Fish flock	0.91	0.98	0.77
Seafloor topography	0.96	0.82	0.84

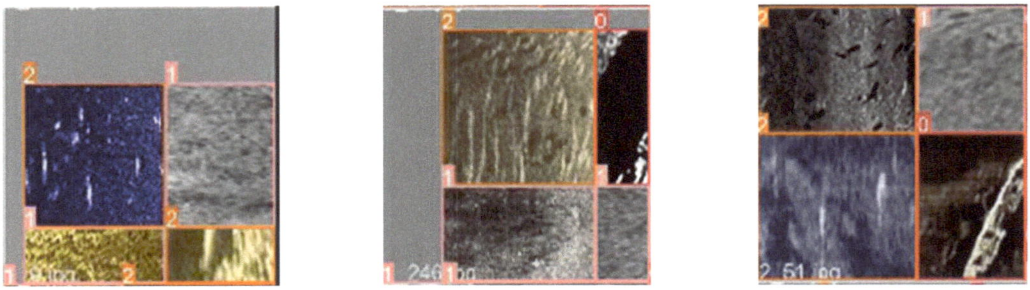

Figure 15. Some example target detection results of the proposed ODP+YOLO V3 + SPP method for the three sets of side scan sonar datasets, including Sunken Ship, Fish Flock, Seafloor Topography: label 0 refers to Sunken Ship; label 1 refers to Fish Flock; and label 2 refers to Seafloor Topography.

To test the efficiency of the ODP + YOLO V3 + SPP method proposed in this paper with the above four traditional classical neural networks and four improved algorithms (a total of eight algorithms) after adding SPP respectively. Considering the length of the paper and other issues, this paper will compare the efficiency of the ODP + YOLO V3 + SPP method proposed in this paper with the fastest method among the 8 traditional classical neural networks (Google + SPP). The overall detection time of all images and the average detection time of each image in the validation set were used to evaluate the efficiency of the method. The overall detection time for all images is the median of the overall detection time for all images in the validation set. The average detection time of each image is obtained by dividing the total detection time of all images in the validation set by the total number of images to obtain the median of the results.

In Table 5, we compare the total detection time of all images and the average detection time of each image in the traditional classical algorithm, the Google + SPP method with the shortest time and the highest efficiency, the Proposed ODP + YOLO V3 + SPP is used for comparison with The Proposed ODP + YOLO V3 + SPP. As mentioned in the 3.1 Datasets section, there are three types of validation sets for the entire dataset, namely sunken ship, fish flock and seafloor topography. The number of images in each type of validation set is 75, and the total number of images in the three types of validation set is 225. We then carry on the experimental result analysis according to this.

Table 5. Comparing the average detection time between the original classic Google + SPP and the proposed ODP + YOLO V3 + SPP in the seconds for each image.

Methods	The Overall Detection Time for All Images	The Average Detection Time for Each Image
Google + SPP	218.25	0.97
The Proposed ODP + YOLO V3 + SPP	128.25	0.57

According to Table 5, on the NVIDIA-SMI 470.82.00, the average detection time for each image with the Google + SPP method was 0.97 s and the overall detection time for all images was 218.25 s. The average detection time of The Proposed ODP + YOLO V3 + SPP for each image was 0.57 s and the overall detection time for all images was 128.25 s. Our method proposed ODP + YOLO V3 + SPP method is faster than the Google + SPP method. The efficiency for all datasets by The Proposed ODP + YOLO V3 + SPP method is improved by 41%. Therefore, The Proposed ODP + YOLO V3 + SPP method in this paper shows the best detection efficiency.

From the above experimental results, our method an improved neural network with the spatial pyramid pooling and online dataset preprocessing based on the YOLO V3 shows good performance in improving detection accuracy and improving detection speed. Compared with Google+SPP group, which is the best detection efficient model in the several classic networks (Table 2), the proposed ODP+YOLO V3+SPP model's detection efficiency has increased by 41% (Table 5). Compared with the original YOLO V3 model, the proposed ODP+YOLO V3+SPP underwater target detection algorithm model has improved detection performance through the mAP qualitative evaluation index has increased by 6%, the Precision qualitative evaluation index has increased by 13%, and the detection efficiency has increased by 9.34% (Table 3). Our approach shows the best detection performance, and it is conducive to improvement in the detection of typical side scan sonar datasets.

3.5. Discussion

The method proposed in this paper combines the spatial pyramid pooling algorithm and the online datasets preprocessing based on the dataLoader platform to improve the detection performance (e.g., detection accuracy and efficiency) of the side scan sonar images. After conducting a comparison of the proposed approach in the experiments, the results show our approach yielded higher detection performance without adopting the Spatial Pyramid Pooling Method, without adopting the Online Dataset Preprocessing Method.

The results prove in the process of target detection, the detection evaluation indicators of the same network with the spatial pyramid pooling compared with that of the original network show better performance. It has a certain stability, which verifies that the pools of different sizes of the spatial pyramid pooling algorithm can make the fixed size of the output feature vector helps to improve the detection accuracy of sonar images. For example, compared with the AlexNet Group, the AlexNet + SPP Group has less time used and higher accuracy, etc. Spatial pyramid pooling better preserves more feature information of images, which is beneficial to improve detection accuracy and efficiency.

This result shows the performances of YOLO group comparisons on detection. We can find that YOLO V3 with adding spatial pyramid pooling has better performance than YOLO V3 without adding spatial pyramid pooling. The detection accuracy of the YOLO V3 network can be improved by using spatial pyramid pooling. Besides, online datasets preprocessing YOLO V3 with adding spatial pyramid pooling has the best performance in the three groups, which demonstrates the good performance of YOLO models in the design of experiments, especially which includes spatial pyramid pooling and online datasets preprocessing, showing better performance than other network methods in the paper.

The method we proposed can better realize the balance of improving accuracy and efficiency than other comparison methods in the paper, which is an improved neural network with spatial pyramid pooling and online dataset preprocessing based on the YOLO V3 method for typical underwater target detection.

4. Conclusions

In this paper, we propose an improved neural network with spatial pyramid pooling and online dataset preprocessing based on the YOLO V3 method for underwater target detection, realizing the balance of improving accuracy and efficiency. The algorithm proposed in this paper delivers high detection accuracy, but more importantly, it can achieve an optimal balance between improved detection accuracy and improved detection speed when tackling underwater targets. Compared with the original YOLO V3 model, the proposed ODP+YOLO V3+SPP underwater target detection algorithm model has improved detection performance through the mAP qualitative evaluation index has increased by 6%, the Precision qualitative evaluation index has increased by 13%, and the detection efficiency has increased by 9.34%.

The spatial pyramid pooling algorithm was added to improve sonar image detection to break the limitations of the input image size. The limited size of input data is a common problem; therefore, the overall quality of input data may be lacking. The traditional methods to solve this problem often result in missing information, and furthermore, simple data processing methods often cause information uncertainty. In this research, we adopted methods that improved the integrity of the information and thus increased the accuracy of the classification model. YOLO V3 with spatial pyramid pooling. As compared with other tested models, this approach uses a more streamlined structure that enhances feature extraction and reduces vanishing gradients by incorporating a spatial pyramid structure. Results show that the application of spatial pyramid pooling can improve the accuracy of side scan sonar image based target detection. Using online dataset preprocessing with random flips improved detection performance. Lots of experiments are needed to find the optimal convolution kernel parameters to reduce the dependence on neural network parameters and data volume.

For the typical target detection based on side scan sonar images, the detection accuracy of common neural network algorithms was improved over the original algorithms after adding spatial pyramid pooling. Therefore, (1) the detection accuracy of typical target detection based on side scan sonar image can be notably improved by using spatial pyramid pooling as proposed in this paper; (2) in this paper, online dataset preprocessing flip helps improve underwater target detection accuracy based on side scan sonar images.

In the future, our research will further focus on the deep learning based feature extraction methods, and more effective methods will be integrated. The proposed model can be used in the future to improve the accuracy of side scan image classification tasks. At the same time, our framework will be enhanced with more advanced deep learning architectures to make further efforts to improve model efficiency.

Author Contributions: J.L., X.X., L.C., J.S. and D.L. proposed the idea and wrote the manuscript; J.L., X.X., J.S., X.L., X.S. and X.W. designed and performed experiments; X.X., L.C. and D.L. helped to revise the manuscript. All authors have read and agreed to the published version of the manuscript.

Funding: The work was supported by the National Natural Science Foundation of China (Grant No. 42101449), the Natural Science Foundation of Hubei Province, China (Grant Nos. 2022CFB773, 2020CFA001), the Key Research and Development Project of Jinzhong City, China (Grant No. Y211006), the Science and Technology Program of Southwest China Research Institute of Electronic Equipment (Grant No. JS20200500114), the Chutian Scholar Program of Hubei Province, the Yellow Crane Talent Scheme and LIESMARS Special Research Funding. We sincerely thank the editors and anonymous reviewers for their outstanding comments.

Data Availability Statement: The data are not publicly available due to the restrictions of containing information that could compromise the privacy of research participants. So participants of this study did not agree for their data to be shared publicly, so supporting data is not available.

Acknowledgments: The authors wish to thank the editor and reviewers for their suggestions, and thanks to Xiao Wang at the Jiangsu Ocean University for providing us with several sets of experimental data. We are thankful to Stephen C. McClure for providing us English editing of the manuscript freely.

Conflicts of Interest: The authors declare no conflict of interest.

References

1. Zhang, T.; Zhang, X.; Liu, C.; Shi, J.; Wei, S.; Ahmad, I.; Zhou, X.; Pan, D.; Li, J.; Su, H. Balance learning for ship detection from synthetic aperture radar remote sensing imagery. *ISPRS J. Photogramm. Remote Sens.* **2021**, *182*, 190–207. [CrossRef]
2. Siradjuddin, I.A.; Muntasa, A. Faster Region-based Convolutional Neural Network for Mask Face Detection. In Proceedings of the 2021 5th International Conference on Informatics and Computational Sciences (ICICoS), Semarang, Indonesia, 24–25 November 2021; IEEE: Piscataway, NJ, USA, 2021; pp. 282–286.
3. Zhang, T.; Zhang, X.; Ke, X. Quad-FPN: A novel quad feature pyramid network for SAR ship detection. *Remote Sens.* **2021**, *13*, 2771. [CrossRef]
4. Zhang, T.; Zhang, X. HTC+ for SAR Ship Instance Segmentation. *Remote Sens.* **2022**, *14*, 2395. [CrossRef]
5. Bara, M.; Sagues, L.; Paniagua, F.; Broquetas, A.; Fàbregas, X. High-speed focusing algorithm for circular synthetic aperture radar (C-SAR). *Electron. Lett.* **2000**, *36*, 1. [CrossRef]
6. Zhu, M.; Hu, G.; Zhou, H.; Wang, S. H2Det: A high-speed and high-accurate ship detector in SAR images. *IEEE J. Sel. Top. Appl. Earth Obs. Remote Sens.* **2021**, *14*, 12455–12466. [CrossRef]
7. Zhang, T.; Zhang, X.; Shi, J.; Wei, S. HyperLi-Net: A hyper-light deep learning network for high-accurate and high-speed ship detection from synthetic aperture radar imagery. *ISPRS J. Photogramm. Remote Sens.* **2020**, *167*, 123–153. [CrossRef]
8. Zhang, T.; Zhang, X.; Shi, J.; Wei, S.; Wang, J.; Li, J.; Su, H.; Zhou, Y. Balance scene learning mechanism for offshore and inshore ship detection in SAR images. *IEEE Geosci. Remote Sens. Lett.* **2020**, *19*, 1–5. [CrossRef]

9. Shang, X.; Zhao, J.; Zhang, H. Automatic overlapping area determination and segmentation for multiple side scan sonar images mosaic. *IEEE J. Sel. Top. Appl. Earth Obs. Remote Sens.* **2021**, *14*, 2886–2900. [CrossRef]
10. Qin, X.; Luo, X.; Wu, Z.; Shang, J. Optimizing the sediment classification of small side-scan sonar images based on deep learning. *IEEE Access* **2021**, *9*, 29416–29428. [CrossRef]
11. Zhu, M.; Song, Y.; Guo, J.; Feng, C.; Li, G.; Yan, T.; He, B. PCA and kernel-based extreme learning machine for side-scan sonar image classification. In Proceedings of the 2017 IEEE Underwater Technology (UT), Busan, Republic of Korea, 21–24 February 2017; IEEE: Piscataway, NJ, USA, 2017; pp. 1–4.
12. Fallon, M.F.; Kaess, M.; Johannsson, H.; Leonard, J.J. Efficient AUV navigation fusing acoustic ranging and side-scan sonar. In Proceedings of the 2011 IEEE International Conference on Robotics and Automation, Shanghai, China, 9–13 May 2011; IEEE: Piscataway, NJ, USA, 2011; pp. 2398–2405.
13. Wu, Z.; Yang, F.; Tang, Y. Side-scan sonar and sub-bottom profiler surveying. In *High-Resolution Seafloor Survey and Applications;* Springer: Singapore, 2021; pp. 95–122.
14. Al-Qatf, M.; Lasheng, Y.; Al-Habib, M.; Yu, L. Deep Learning Approach Combining Sparse Autoencoder With SVM for Network Intrusion Detection. *IEEE Access* **2018**, *6*, 52843–52856. [CrossRef]
15. Bertasius, G.; Shi, J.; Torresani, L. Deepedge: A multi-scale bifurcated deep network for top-down contour detection. In Proceedings of the IEEE Conference on Computer Vision and Pattern Recognition, Boston, MA, USA, 7–12 June 2015; pp. 4380–4389.
16. Wang, Y.; Sun, Y.; Lv, P.; Wang, H. Detection of line weld defects based on multiple thresholds and support vector machine. *NDT E Int.* **2008**, *41*, 517–524. [CrossRef]
17. Gong, M.; Su, L.; Jia, M.; Chen, W. Fuzzy clustering with a modified MRF energy function for change detection in synthetic aperture radar images. *IEEE Trans. Fuzzy Syst.* **2013**, *22*, 98–109. [CrossRef]
18. Dzieciuch, I.; Gebhardt, D.; Barngrover, C.; Parikh, K. Non-Linear Convolutional Neural Network for Automatic Detection of Mine-Like Objects in Sonar Imagery. In Proceedings of the International Conference on Applications in Nonlinear Dynamics, Rome, Italy, 21–25 May 2017; Springer: Cham, Switzerland, 2017; pp. 309–314.
19. Rhinelander, J. Feature extraction and target classification of side-scan sonar images. In Proceedings of the 2016 IEEE Symposium Series on Computational Intelligence (SSCI), Athens, Greece, 6–9 December 2016.
20. Song, Y.; Zhu, Y.; Li, G.; Feng, C.; He, B.; Yan, T. Side scan sonar segmentation using deep convolutional neural network. In Proceedings of the OCEANS 2017-Anchorage, Anchorage, AK, USA, 18–21 September 2017; IEEE: Piscataway, NJ, USA, 2017; pp. 1–4.
21. Zhu, P.; Isaacs, J.; Fu, B.; Ferrari, S. Deep learning feature extraction for target recognition and classification in underwater sonar images. In Proceedings of the 2017 IEEE 56th Annual Conference on Decision and Control (CDC), Melbourne, Australia, 12–15 December 2017; IEEE: Piscataway, NJ, USA, 2017; pp. 2724–2731.
22. Einsidler, D.; Dhanak, M.; Beaujean, P.P. A deep learning approach to target recognition in side scan sonar imagery. In Proceedings of the OCEANS 2018 MTS/IEEE Charleston, Charleston, SC, USA, 22–25 October 2018; IEEE: Piscataway, NJ, USA, 2018; pp. 1–4.
23. Kim, J.; Choi, J.W.; Kwon, H.; Oh, R.; Son, S. The application of convolutional neural networks for automatic detection of underwater object in side scan sonar images. *J. Acoust. Soc. Korea* **2018**, *37*, 118–128.
24. Wu, M.; Wang, Q.; Rigall, E.; Li, K.; Zhu, W.; He, B. ECNet: Efficient convolutional networks for side scan sonar image segmentation. *Sensors* **2019**, *19*, 2009. [CrossRef] [PubMed]
25. Wang, Q.; Wu, M.; Yu, F.; Feng, C.; Li, K.; Zhu, Y.; Rigall, E.; He, B. Rt-seg: A real-time semantic segmentation network for side scan sonar images. *Sensors* **2019**, *19*, 1985. [CrossRef] [PubMed]
26. Yu, F.; He, B.; Li, K.; Yan, T.; Shen, Y.; Wang, Q.; Wu, M. Side scan sonar images segmentation for AUV with recurrent residual convolutional neural network module and self-guidance module. *Appl. Ocean Res.* **2021**, *113*, 102608. [CrossRef]
27. Wang, Y.; Liu, J.; Yu, S.; Wang, K.; Han, Z.; Tang, Y. Underwater Object Detection based on YOLO-v3 network. In Proceedings of the 2021 IEEE International Conference on Unmanned Systems (ICUS), Beijing, China, 15–17 October 2021; IEEE: Piscataway, NJ, USA, 2021; pp. 571–575.
28. Li, J.W.; Cao, X. Target Recognition and Detection in Side-scan Sonar Images based on YOLO v3 Model. In Proceedings of the 2022 41st Chinese Control Conference (CCC), Hefei, China, 25–27 July 2022; IEEE: Piscataway, NJ, USA, 2022; pp. 7186–7190.
29. Jin, S.; Zhang, N.; Bian, G.; Cui, Y. A seabed sediment classification model based on PSO-AlexNet. In Proceedings of the 2nd International Conference on Signal Image Processing and Communication (ICSIPC 2022), Qingdao, China, 20–22 May 2022; SPIE: Bellingham, WA, USA, 2022; Volume 12246, pp. 349–361.
30. Jia, X.; Wei, X.; Cao, X.; Foroosh, H. Comdefend: An efficient image compression model to defend adversarial examples. In Proceedings of the IEEE/CVF conference on computer vision and pattern recognition, Long Beach, CA, USA, 15–20 June 2019; pp. 6084–6092.
31. Yang, X.; Yan, J. Arbitrary-oriented object detection with circular smooth label. In Proceedings of the European Conference on Computer Vision, Glasgow, UK, 23–28 August 2020; pp. 677–694.
32. Sahin, F.E.; Tanguay, A.R. Distortion optimization for wide-angle computational cameras. *Optics Express* **2018**, *26*, 5478–5487. [CrossRef]
33. Zhang, X.; Karaman, S.; Chang, S.F. Detecting and simulating artifacts in gan fake images. In Proceedings of the 2019 IEEE International Workshop on Information Forensics and Security (WIFS), Delft, The Netherlands, 9–12 December 2019; IEEE: Piscataway, NJ, USA, 2019; pp. 1–6.

34. Sheikh, H.R.; Bovik, A.C. Image information and visual quality. *IEEE Trans. Image Process.* **2006**, *15*, 430–444. [CrossRef]
35. Zhang, W.; Ma, K.; Yan, J.; Deng, D.; Wang, Z. Blind image quality assessment using a deep bilinear convolutional neural network. *IEEE Trans. Circuits Syst. Video Technol.* **2018**, *30*, 36–47. [CrossRef]
36. Canziani, A.; Paszke, A.; Culurciello, E. An analysis of deep neural network models for practical applications. *arXiv* **2016**, arXiv:1605.07678.
37. Zhou, Z.; Cui, Z.; Zang, Z.; Meng, X.; Cao, Z.; Yang, J. UltraHi-PrNet: An Ultra-High Precision Deep Learning Network for Dense Multi-Scale Target Detection in SAR Images. *Remote Sens.* **2022**, *14*, 5596. [CrossRef]
38. Han, B.; Hu, Z.; Su, Z.; Bai, X.; Yin, S.; Luo, J.; Zhao, Y. Mask_LaC R-CNN for measuring morphological features of fish. *Measurement* **2022**, *203*, 111859. [CrossRef]
39. Wang, H.; Shi, Y.; Yue, Y.; Zhao, H. Study on freshwater fish image recognition integrating SPP and DenseNet network. In Proceedings of the 2020 IEEE International Conference on Mechatronics and Automation (ICMA), Osaka, Japan, 24–26 April 2020; IEEE: Piscataway, NJ, USA, 2020; pp. 564–569.
40. Guo, H.; Gao, H.; Guo, C.; Lu, J.; Lin, Y. Dock detection method in remote sensing images based on improved YOLOv4. In Proceedings of the Fourteenth International Conference on Digital Image Processing (ICDIP 2022), Wuhan, China, 20–23 May 2022; SPIE: Bellingham, WA, USA, 2022; Volume 12342, pp. 105–112.
41. Han, Q.; Yin, Q.; Zheng, X.; Chen, Z. Remote sensing image building detection method based on Mask R-CNN. *Complex Intell. Syst.* **2022**, *8*, 1847–1855. [CrossRef]
42. Le, Y.; Yang, X. Tiny imagenet visual recognition challenge. *CS 231n* **2015**, *7*, 3.
43. Li, Y.; Zhang, X.; Shen, Z. YOLO-Submarine Cable: An Improved YOLO-V3 Network for Object Detection on Submarine Cable Images. *J. Mar. Sci. Eng.* **2022**, *10*, 1143. [CrossRef]
44. Ju, M.; Luo, H.; Wang, Z.; Hui, B.; Chang, Z. The application of improved YOLO V3 in multi-scale target detection. *Appl. Sci.* **2019**, *9*, 3775. [CrossRef]
45. Xu, D.; Wu, Y. Improved YOLO-V3 with DenseNet for multi-scale remote sensing target detection. *Sensors* **2020**, *20*, 4276. [CrossRef] [PubMed]
46. Karathanassi, V.; Kolokousis, P.; Ioannidou, S. A comparison study on fusion methods using evaluation indicators. *Int. J. Remote Sens.* **2007**, *28*, 2309–2341. [CrossRef]
47. Shang, R.; Wang, J.; Jiao, L.; Rustam, S.; Hou, B.; Li, Y. SAR targets classification based on deep memory convolution neural networks and transfer parameters. *IEEE J. Sel. Top. Appl. Earth Obs. Remote Sens.* **2018**, *11*, 2834–2846. [CrossRef]
48. Pang, Y.; Zhao, X.; Zhang, L.; Lu, H. Multi-scale interactive network for salient object detection. In Proceedings of the IEEE/CVF Conference on Computer Vision and Pattern Recognition, Seattle, WA, USA, 13–19 June 2020; pp. 9413–9422.
49. Sriram, S.; Vinayakumar, R.; Sowmya, V.; Aamoun, A.; Soman, K. Multi-scale learning based malware variant detection using spatial pyramid pooling network. In Proceedings of the IEEE INFOCOM 2020-IEEE Conference on Computer Communications Workshops (INFOCOM WKSHPS), Toronto, ON, Canada, 6–9 July 2020; IEEE: Piscataway, NJ, USA, 2020; pp. 740–745.
50. Li, S.; Yuan, S.; Liu, S.; Wen, J.; Huang, Q.; Zhang, Z. Characteristics of Low-Frequency Acoustic Wave Propagation in Ice-Covered Shallow Water Environment. *Appl. Sci.* **2021**, *11*, 7815. [CrossRef]
51. Sebens, K.P.; Witting, J.; Helmuth, B. Effects of water flow and branch spacing on particle capture by the reef coral Madracis mirabilis (Duchassaing and Michelotti). *J. Exp. Mar. Biol. Ecol.* **1997**, *211*, 1–28. [CrossRef]
52. Wang, X.; Wang, L.; Li, G.; Xie, X. A Robust and Fast Method for Sidescan Sonar Image Segmentation Based on Region Growing. *Sensors* **2021**, *21*, 6960. [CrossRef] [PubMed]
53. Ying, X. An overview of overfitting and its solutions. *J. Phys. Conf. Ser.* **2019**, *1168*, 022022. [CrossRef]
54. Pecorelli, F.; Di Nucci, D.; De Roover, C.; De Lucia, A. A large empirical assessment of the role of data balancing in machine-learning-based code smell detection. *J. Syst. Softw.* **2020**, *169*, 110693. [CrossRef]
55. Osetsky, Y.; Barashev, A.V.; Zhang, Y. Sluggish, chemical bias and percolation phenomena in atomic transport by vacancy and interstitial diffusion in NiFe alloys. *Curr. Opin. Solid State Mater. Sci.* **2021**, *25*, 100961. [CrossRef]
56. Paszke, A.; Gross, S.; Massa, F.; Lerer, A.; Chintala, S. Pytorch: An Imperative Style, High-Performance Deep Learning Library. In *Advances in Neural Information Processing Systems*; MIT Press: Cambridge, MA, USA, 2019; Volume 32.
57. Moura, P.; Crocker, P.; Nunes, P. High-level multi-threading programming in logtalk. In Proceedings of the International Symposium on Practical Aspects of Declarative Languages, San Francisco, CA, USA, 7–8 January 2008; Springer: Berlin/Heidelberg, Germany, 2008; pp. 265–281.
58. He, K.; Girshick, R.; Dollár, P. Rethinking imagenet pre-training. In Proceedings of the IEEE/CVF International Conference on Computer Vision, Seoul, Republic of Korea, 27 October–2 November 2019; pp. 4918–4927.
59. Jaikrishnan, S.V.J.; Chantarakasemchit, O.; Meesad, P. A breakup machine learning approach for breast cancer prediction. In Proceedings of the 2019 11th International Conference on Information Technology and Electrical Engineering (ICITEE), Pattaya, Thailand, 10–11 October 2019; IEEE: Piscataway, NJ, USA, 2019; pp. 1–6.
60. Chu, H.; Xiong, X.; Gao, Y.J.; Luo, J.; Jing, H. Diffuse reflection and reciprocity-protected transmission via a random-flip metasurface. *Sci. Adv.* **2021**, *7*, eabj0935. [CrossRef]
61. Li, Z.; Peng, C.; Yu, G.; Zhang, X.; Deng, Y.; Sun, J. Detnet: A backbone network for object detection. *arXiv* **2018**, arXiv:1804.06215.

62. Albawi, S.; Mohammed, T.A.; Al-Zawi, S. Understanding of a convolutional neural network. In Proceedings of the 2017 International Conference on Engineering and Technology (ICET), Antalya, Turkey, 21–23 August 2017; IEEE: Piscataway, NJ, USA, 2017; pp. 1–6.
63. Wang, Z.J.; Turko, R.; Shaikh, O.; Park, H.; Das, N. CNN explainer: Learning convolutional neural networks with interactive visualization. *IEEE Trans. Vis. Comput. Graph.* **2020**, *27*, 1396–1406. [CrossRef]
64. O'Shea, K.; Nash, R. An introduction to convolutional neural networks. *arXiv* **2015**, arXiv:1511.08458.
65. Zhang, T.; Zhang, X.; Ke, X.; Liu, C.; Xu, X. HOG-ShipCLSNet: A novel deep learning network with hog feature fusion for SAR ship classification. *IEEE Trans. Geosci. Remote Sens.* **2021**, *60*, 1–22. [CrossRef]
66. Wang, Y.; Wang, C.; Zhang, H. Ship classification in high-resolution SAR images using deep learning of small datasets. *Sensors* **2018**, *18*, 2929. [CrossRef] [PubMed]
67. Nguyen, D.P.T.; Matsuo, Y.; Ishizuka, M. Exploiting syntactic and semantic information for relation extraction from wikipedia. In Proceedings of the IJCAI Workshop on Text-Mining & Link-Analysis (TextLink 2007), Hyderabad, India, 6–12 January 2007.
68. Sacramento, J.; Ponte Costa, R.; Bengio, Y.; Senn, W. Dendritic cortical microcircuits approximate the backpropagation algorithm. In *Advances in Neural Information Processing Systems*; MIT Press: Cambridge, MA, USA, 2018; Volume 31.
69. Alom, M.Z.; Taha, T.M.; Yakopcic, C.; Westberg, S.; Asari, V.K. The history began from alexnet: A comprehensive survey on deep learning approaches. *arXiv* **2018**, arXiv:1803.01164.
70. Alippi, C.; Disabato, S.; Roveri, M. Moving convolutional neural networks to embedded systems: The alexnet and VGG-16 case. In Proceedings of the 2018 17th ACM/IEEE International Conference on Information Processing in Sensor Networks (IPSN), Porto, Portugal, 11–13 April 2018; IEEE: Piscataway, NJ, USA, 2018; pp. 212–223.
71 Yuan, Z.W.; Zhang, J. Feature extraction and image retrieval based on AlexNet. In Proceedings of the Eighth International Conference on Digital Image Processing (ICDIP 2016), Chengu, China, 20–22 May 2016; SPIE: Bellingham, WA, USA, 2016; Volume 10033, pp. 65–69.
72. Ballester, P.; Araujo, R.M. On the performance of GoogLeNet and AlexNet applied to sketches. In Proceedings of the Thirtieth AAAI Conference on Artificial Intelligence, Phoenix, AZ, USA, 12–17 February 2016.
73. Anand, R.; Shanthi, T.; Nithish, M.S.; Lakshman, S. Face recognition and classification using GoogleNET architecture. In *Soft Computing for Problem Solving*; Springer: Singapore, 2020; pp. 261–269.
74. Salavati, P.; Mohammadi, H.M. Obstacle detection using GoogleNet. In Proceedings of the 2018 8th International Conference on Computer and Knowledge Engineering (ICCKE), Mashhad, Iran, 25–26 October 2018; IEEE: Piscataway, NJ, USA, 2018; pp. 326–332.
75. Muhammad, U.; Wang, W.; Chattha, S.P.; Ali, S. Pre-trained VGGNet architecture for remote-sensing image scene classification. In Proceedings of the 2018 24th International Conference on Pattern Recognition (ICPR), Beijing, China, 20–24 August 2018; IEEE: Piscataway, NJ, USA, 2018; pp. 1622–1627.
76. Sathish, K.; Ramasubbareddy, S.; Govinda, K. Detection and localization of multiple objects using VGGNet and single shot detection. In *Emerging Research in Data Engineering Systems and Computer Communications*; Springer: Singapore, 2020; pp. 427–439.
77. Wu, Z.; Shen, C.; Van Den Hengel, A. Wider or deeper: Revisiting the ResNet model for visual recognition. *Pattern Recognit.* **2019**, *90*, 119–133. [CrossRef]
78. Li, S.; Jiao, J.; Han, Y.; Weissman, T. Demystifying ResNet. *arXiv* **2016**, arXiv:1611.01186.
79. Chen, Z.; Xie, Z.; Zhang, W.; Xu, X. ResNet and Model Fusion for Automatic Spoofing Detection. In Proceedings of the Interspeech, Stockholm, Sweden, 20–24 August 2017; pp. 102–106.
80. Khan, R.U.; Zhang, X.; Kumar, R.; Aboagye, E.O.; Kumar, R. Evaluating the performance of ResNet model based on image recognition. In Proceedings of the 2018 International Conference on Computing and Artificial Intelligence, Chengdu, China, 12–14 March 2018; pp. 86–90.
81. He, F.; Liu, T.; Tao, D. Why ResNet works? Residuals generalize. *IEEE Trans. Neural Netw. Learn. Syst.* **2020**, *31*, 5349–5362. [CrossRef]
82. Redmon, J.; Farhadi, A. Yolov3: An incremental improvement. *arXiv* **2018**, arXiv:1804.02767.
83. Zhao, L.; Li, S. Object detection algorithm based on improved YOLOV3. *Electronics* **2020**, *9*, 537. [CrossRef]
84. Won, J.H.; Lee, D.H.; Lee, K.M.; Lin, C.H. An improved YOLOv3-based neural network for de-identification technology. In Proceedings of the 2019 34th International Technical Conference on Circuits/Systems, Computers and Communications (ITC-CSCC), Jeju, Republic of Korea, 23–26 June 2019; IEEE: Piscataway, NJ, USA, 2019; pp. 1–2.
85. Lang, P.; Fu, X.; Martorella, M.; Dong, J.; Xie, M. A comprehensive survey of machine learning applied to radar signal processing. *arXiv* **2020**, arXiv:2009.13702.
86. Lee, Y.H.; Kim, Y. Comparison of CNN and YOLO for Object Detection. *J. Semicond. Disp. Technol.* **2020**, *19*, 85–92.
87. He, K.; Zhang, X.; Ren, S.; Sun, J. Spatial pyramid pooling in deep convolutional networks for visual recognition. *IEEE Trans. Pattern Anal. Mach. Intell.* **2015**, *37*, 1904–1916. [CrossRef]
88. Yue, J.; Mao, S.; Li, M. A deep learning framework for hyperspectral image classification using spatial pyramid pooling. *Remote Sens. Lett.* **2016**, *7*, 875–884. [CrossRef]
89. Huang, Z.; Wang, J.; Fu, X.; Yu, T.; Wang, R. DC-SPP-YOLO: Dense connection and spatial pyramid pooling based YOLO for object detection. *Inf. Sci.* **2020**, *522*, 241–258. [CrossRef]

90. Liu, Y.; Gross, L.; Li, Z.; Li, X.; Qi, W. Automatic building extraction on high-resolution remote sensing imagery using deep convolutional encoder-decoder with spatial pyramid pooling. *IEEE Access* **2019**, *7*, 128774–128786. [CrossRef]
91. Zhang, X.; Wang, W.; Zhao, Y.; Xie, H. An improved YOLOv3 model based on skipping connections and spatial pyramid pooling. *Syst. Sci. Control Eng.* **2021**, *9* (Suppl. 1), 142–149. [CrossRef]
92. Xu, F.; Wang, H.; Sun, X.; Fu, X. Refined marine object detector with attention-based spatial pyramid pooling networks and bidirectional feature fusion strategy. *Neural Comput. Appl.* **2022**, *34*, 14881–14894. [CrossRef]

remote sensing

Article

Contrastive Domain Adaptation-Based Sparse SAR Target Classification under Few-Shot Cases

Hui Bi [1,2,*], Zehao Liu [1,2], Jiarui Deng [1,2], Zhongyuan Ji [1,2] and Jingjing Zhang [1,2]

[1] College of Electronic Engineering, Nanjing University of Aeronautics and Astronautics, Nanjing 211106, China
[2] The Key Laboratory of Radar Imaging and Microwave Photonics, Ministry of Eduction, Nanjing University of Aeronautics and Astronautics, Nanjing 211106, China
* Correspondence: bihui@nuaa.edu.cn

Abstract: Due to the imaging mechanism of synthetic aperture radar (SAR), it is difficult and costly to acquire abundant labeled SAR images. Moreover, a typical matched filtering (MF) based image faces the problems of serious noise, sidelobes, and clutters, which will bring down the accuracy of SAR target classification. Different from the MF-based result, a sparse image shows better quality with less noise and higher image signal-to-noise ratio (SNR). Therefore, theoretically using it for target classification will achieve better performance. In this paper, a novel contrastive domain adaptation (CDA) based sparse SAR target classification method is proposed to solve the problem of insufficient samples. In the proposed method, we firstly construct a sparse SAR image dataset by using the complex image based iterative soft thresholding (BiIST) algorithm. Then, the simulated and real SAR datasets are simultaneously sent into an unsupervised domain adaptation framework to reduce the distribution difference and obtain the reconstructed simulated SAR images for subsequent target classification. Finally, the reconstructed simulated images are manually labeled and fed into a shallow convolutional neural network (CNN) for target classification along with a small number of real sparse SAR images. Since the current definition of the number of small samples is still vague and inconsistent, this paper defines few-shot as less than 20 per class. Experimental results based on MSTAR under standard operating conditions (SOC) and extended operating conditions (EOC) show that the reconstructed simulated SAR dataset makes up for the insufficient information from limited real data. Compared with other typical deep learning methods based on limited samples, our method is able to achieve higher accuracy especially under the conditions of few shots.

Keywords: synthetic aperture radar (SAR); convolutional neural network (CNN); contrastive domain adaptation (CDA); few-shot learning; target classification

Citation: Bi, H.; Liu, Z.; Deng, J.; Ji, Z.; Zhang, J. Contrastive Domain Adaptation-Based Sparse SAR Target Classification under Few-Shot Cases. *Remote Sens.* **2023**, *15*, 469. https://doi.org/10.3390/rs15020469

Academic Editors: Qian Du, Yanni Dong and Xiaochen Yang

Received: 12 December 2022
Revised: 5 January 2023
Accepted: 10 January 2023
Published: 13 January 2023

1. Introduction

Different from a conventional optical system, synthetic aperture radar (SAR) is able to work under all-day and all-weather conditions, and hence has been widely used in several military and civilian fields [1,2]. Automatic target recognition (ATR) is one of the important research topics in SAR image processing [3,4]. It consists of three modules, target detection, target discrimination, and target classification, whose key process is target classification. However, the limitation of feature extraction technique hinders the development of target classification. The appearance of deep learning can automatically extract target features instead of manual extraction, which promotes the development of image processing. In 2006, Hinton et al. introduced the concept of deep learning, indicating that a multi-layer convolutional neural network (CNN) has great potential for learning target features [5]. In 2012, Krizhevsky et al. proposed the first deep CNN model called AlexNet for image classification and won the ImageNet competition with its top-5 error being 17.0%. It made deep learning become popular in the field of image processing [6,7]. Since then, various networks such as GoogLeNet [8], ResNet [9], DenseNet [10] have appeared. Nowadays, deep learning methods represented by CNN have been successfully applied to optical image processing. However, compared with abundant labeled datasets in optical images, acquiring labeled SAR images is difficult and expensive. Lack of sufficient labeled SAR data limits the development of deep learning in SAR target classification. In addition, those models that work for optical images could not be used directly for SAR images. Therefore, how to use limited SAR data to improve the performance of SAR target classification has become a research hot spot in the recent years.

Research on a small sample can be separated into three areas, i.e., data augmentation, transfer learning, and model design. Data augmentation mainly focuses on the input data. Transfer learning aims to utilize information from datasets of other domains to support the target domain. Model design aims to propose a new framework suitable for SAR image processing. In 2016, Ding et al. proposed to use three types of data augmentation technologies to alleviate the problem of target translation, speckle noise, and pose missing, respectively [11]. In the same year, Chen et al. introduced the new classification framework all-convolutional networks (A-ConvNet), which only contains convolutional layers and down-sampling layers, greatly reducing the parameters in the network [12]. Experimental results based on the MSTAR dataset demonstrate that its classification accuracy reaches 99.13% when all training samples are input data. In 2017, Hansen et al. revealed the feasibility of transfer learning between a simulated dataset and a real SAR image. It shows that transfer learning enables CNNs to achieve faster convergence and better performance [13]. Lin et al. imitated the principle of ResNet and designed a convolutional highway unit (CHU) to solve the gradient-vanishing problem caused by training deep networks with limited SAR data [14]. In 2019, Zhong et al. proposed to transfer the convolutional layers of the model which is pretrained on ImageNet and compress the model using a filter-based pruning method. As the training data increase to 2700 images per class, the classification accuracy can reach 98.39% [15]. Wang and Xu proposed a novel concatenated rectified linear unit (CReLU) instead of rectified linear unit (ReLU) in order to preserve negative phase information and obtained double feature maps of the previous layer. Experiments on the MSTAR dataset show that the accuracy can reach 88.17% with only 20% of training data [16]. Considering the limited labeled SAR images, Zhang et al. proposed to train a generative adversarial network (GAN) using abundant unlabeled SAR images to apply general characteristics of SAR images to SAR target recognition [17]. Then, Zheng et al. proposed a multi-discriminator GAN (MD-GAN) based on GAN to promote the framework to distinguish generated pseudo-samples from real input data, ensuring the quality of generated images [18]. In 2020, Guo et al. proposed a compact convolutional autoencoder (CCAE) to learn target characteristics. A typical convolutional autoencoder (CAE) fails to enhance the compactness of targets between the same class. CCAE adopts a pair of samples from the same class as input so as to minimize the distance between intra-class samples. Experiments on target classification demonstrate that CCAE can reach 98.59% accuracy [19]. Huang et al. evaluated the efficiency of transfer learning in 2020 and introduced a multi-source domain method to reduce the difference between data of source and target domains to improve the performance of transfer learning. Results on the OpenSARShip dataset validate that the smaller difference gap, the better transfer learning performs [20]. In 2021, Guo et al. verified the feasibility of cross-domain learning from an optical image to a SAR image, and applied the idea to object detection by adding a domain adaptation module to the Faster R-CNN model [21]. In 2022, Yang et al. proposed a dynamic joint correlation alignment network (DJ-CORAL) used for SAR ship classification. It firstly transforms heterogeneous features from two domains into a public subspace to eliminate the heterogeneity, then conducts heterogeneous domain adaptation (HDA) to realize domain shift minimization. Compared with existing semi-supervised transfer learning methods, DJ-CORAL presents better performance [22].

Different from optical images, phase information is unique to SAR images; research on how to make full use of SAR images has made progress. In 2017, Zhang et al. first proposed an architecture called complex-valued CNN (CV-CNN) oriented for SAR target classification.

Experimental results show that CV-CNN achieves higher classification accuracy with phase information being part of input data [23]. Coman et al. divided phase information into real-part and imaginary-part information, and adopted amplitude-real-imaginary three-layer data to form the input data. Results on MSTAR present about 90% accuracy, which alleviates the over-fitting problem caused by the lack of training data [24]. On the basis of CV-CNN, Yu et al. improved the network by introducing a complex-valued fully convolutional neural network (CV-FCNN) in 2020. CV-FCNN only consists of convolutional layers in order to avoid the overfitting problem caused by large amount of parameters. Experiments on MSTAR validate that CV-FCNN presents better performance than CV-CNN [25].

Many achievements in SAR target classification are on the basis of matched filtering (MF) recovered images. The introduction of sparse SAR imaging technology solves the problem that the quality of MF-based SAR images is severely affected by noise, sidelobes, and clutters. Typical sparse recovery algorithms such as iterative soft thresholding (IST) [26,27] and orthogonal matching pursuit (OMP) [28,29] are able to improve the image quality by protruding the target while suppressing background clutter. In 2018, a novel sparse recovery algorithm based on IST was proposed, named as BiIST [30,31]. Compared with OMP and IST, BiIST can enhance target characteristics while remaining an image statistical distribution. Then, in 2021, Bi et al. combined a sparse SAR dataset with typical detection frameworks, i.e., YOLOv3 and Faster R-CNN, showing that higher accuracy can be achieved when using a sparse SAR image instead of an MF-based image as input data [32]. In 2022, Deng et al. proposed a novel sparse SAR target classification framework called amplitude-phase CNN (AP-CNN) to utilize both magnitude and phase of a sparse SAR image recovered by BiIST. Classification network comparison shows that the combination of the sparse SAR dataset and AP-CNN achieves optimal performance [33].

In this paper, we propose a novel sparse SAR target classification method based on contrastive domain adaptation (CDA) for a few shots. In the proposed method, we firstly use the BiIST algorithm to improve the MF-based SAR image performance, and hence construct a novel sparse SAR dataset. Then, the simulated dataset and constructed sparse dataset are sent into an unsupervised domain adaptation framework, so as to transfer target features and minimize distribution difference. After the CDA operation, the reconstructed simulated SAR image which has a similar background distribution and target features to real SAR image will be manually labeled for subsequent classification. Finally, the reconstructed simulated SAR image will be trained in a shallow CNN along with a few real samples. Experimental results based on the MSTAR dataset under standard operating conditions (SOC) show that compared to several typical deep learning methods for SAR target classification, the proposed method presents similar or better performance with enough real samples. With the decrease of the number of real SAR images, the proposed method achieves higher accuracy than other methods, showing great potential for application in real scenes under the extreme shortage of labeled real images.

The main contributions of this paper can be concluded as follows. 1. This paper first applies an unsupervised contrastive domain adaptation framework to minimize the distribution difference between a simulated SAR dataset and real sparse SAR images so as to facilitate transfer learning in sparse SAR image classification. 2. In view of the serious shortage of labeled SAR images in practical applications, the proposed method improves classification accuracy with the real sparse SAR images being reduced to single digits per class.

The rest of this paper is organized as follows. Section 2 introduces the procedure of constructing a sparse real dataset by the BiIST algorithm and the principle of the common transfer learning method. The key process of the proposed algorithm, i.e., CDA, is discussed in Section 3, including components of the unsupervised domain adaptation framework and theories of minimizing the distribution difference between the two datasets. The model of CNN used for target classification is described in Section 4. Experiments based on the sparse MSTAR dataset and the reconstructed simulated dataset under different conditions are shown in Section 5. Section 6 analyzes the experimental results in detail. At last, Section 7 concludes this work.

2. Methodology

2.1. BiIST-Based Sparse SAR Dataset Construction

Although the SAR image recovered by the MF algorithm is more convenient to obtain, it has serious noise with ambiguous target features, which affects its further application in target classification. Therefore, we want to obtain higher-quality images to support subsequent applications. Using the MF-based image $\mathbf{X}_{\mathrm{MF}}$ as input, the sparse SAR imaging model based on complex image data can be expressed as [22,23]

$$\mathbf{X}_{\mathrm{MF}} = \mathbf{X} + \mathbf{N} \tag{1}$$

where $\mathbf{X}$ is the backscattering coefficient of the interested area to be recovered, $\mathbf{N}$ represents the difference between $\mathbf{X}_{\mathrm{MF}}$ and $\mathbf{X}$. Then, we can reconstruct $\mathbf{X}$ by solving

$$\hat{\mathbf{X}} = \min_{\mathbf{X}}\left\{\|\mathbf{X}_{\mathrm{MF}} - \mathbf{X}\|_{\mathrm{F}}^2 + \lambda\|\mathbf{X}\|_1\right\} \tag{2}$$

where λ is the regularization parameter and subscript F denotes the Frobenius norm. As demonstrated in [22,23], BiIST is able to output two kinds of sparse SAR images called sparse solution ($\mathbf{X}_{\mathrm{SP}}$) and non-sparse solution ($\mathbf{X}_{\mathrm{NSP}}$). Similar to OMP and IST, $\mathbf{X}_{\mathrm{SP}}$ improves the image quality with less noise but damages the background distribution and phase information. Compared with $\mathbf{X}_{\mathrm{SP}}$, $\mathbf{X}_{\mathrm{NSP}}$ not only obtains improved SAR image while remaining the background distribution, offering more information for subsequent CDA and target classification. In the following, the real sparse SAR image refers to $\mathbf{X}_{\mathrm{NSP}}$ recovered by BiIST.

2.2. Transfer Learning

The idea of transfer learning (TL) in the field of computer vision (CV) aims to imitate human learning behavior. Firstly, the deep learning frameworks are pretrained on large amounts of data to learn the general features, then these frameworks are applied to specific tasks to reduce calculation and improve performance. Benefiting from the abundant optical image datasets, transfer learning is widely used in practical applications. The traditional transfer learning method is shown in Figure 1. It can be seen that after pretraining the framework with abundant data in source domain, only the last fully connected layer is randomly initialized while parameters in the rest of the framework are fixed. The number of classes in the dataset used for pretraining the network is then changed to the number of classes in the target dataset. However, this method fails to operate in SAR image classification and can be explained from two aspects. Firstly, different from abundant optical image datasets, the lack of labeled SAR image datasets cannot afford the pretraining of the network. In addition, between A and B due to the imaging mechanism of SAR, the same kind of target under different imaging modes or imaging parameters will be different, which may hinder the development of transfer learning in the field of SAR image processing.

Therefore, in this paper, we take above problems into consideration and propose a novel transfer learning method used for sparse SAR target classification, i.e., the contrastive domain adaptation-based sparse SAR target classification. In the proposed method, the training process is a two-step procedure. Firstly, a simulated SAR dataset and real SAR data are simultaneously sent into an unsupervised domain adaptation framework to reduce the distribution difference and obtain the reconstructed simulated SAR images for subsequent target classification. Secondly, the reconstructed simulated images and a few real sparse SAR images form the input of CNN for further target classification.

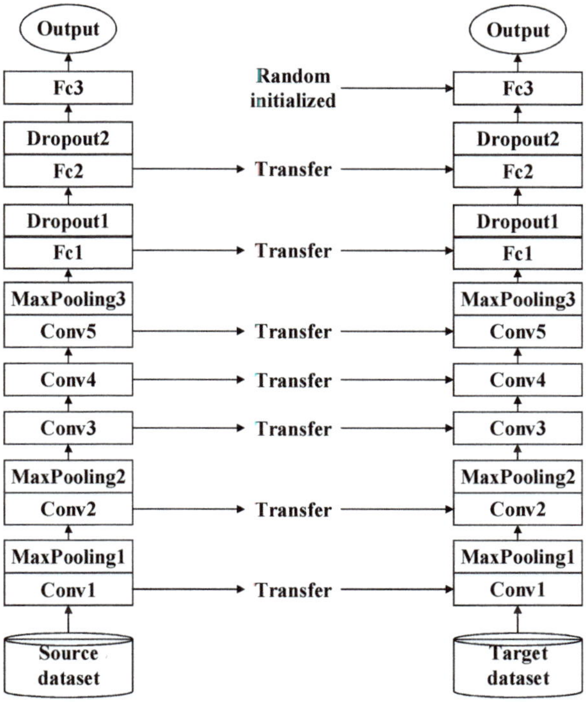

Figure 1. Details of the traditional transfer learning method.

3. Feature Learning-Based Contrastive Domain Adaptation

The key process of the proposed method is described in detail in this section. The flowchart of the unsupervised domain adaptation framework is shown in Figure 2, whose input consists of the simulated SAR dataset **A** and $\mathbf{X}_{NSP}$ dataset **B**. Two image reconstruction models and two discrimination models form a closed loop so as to make the simulated SAR dataset **A** and $\mathbf{X}_{NSP}$ dataset **B** learn characteristics from each other. In Figure 3, the image reconstruction model G_{AB} represents the transition from **A** to **B**. G_{BA} is the opposite, which has the same structure as G_{AB}. Discrimination model D_B is used to identify whether the input data are a reconstructed simulated SAR target or a real sparse SAR target, and D_A shows the opposite. Similarly, D_A and D_B have the same structure as well. The principles of CDA will be explained in the following.

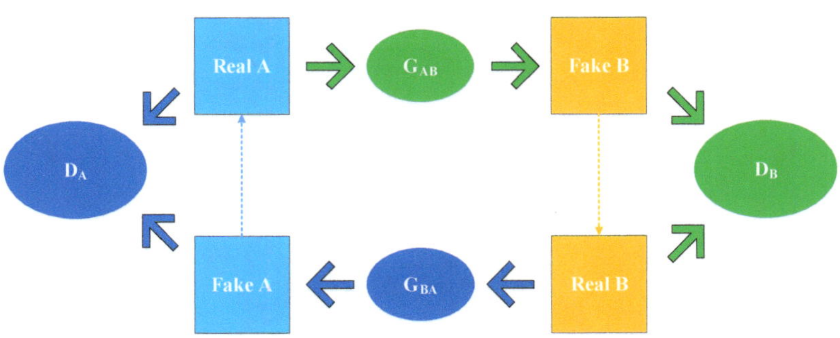

Figure 2. Overview of the contrastive domain adaptation.

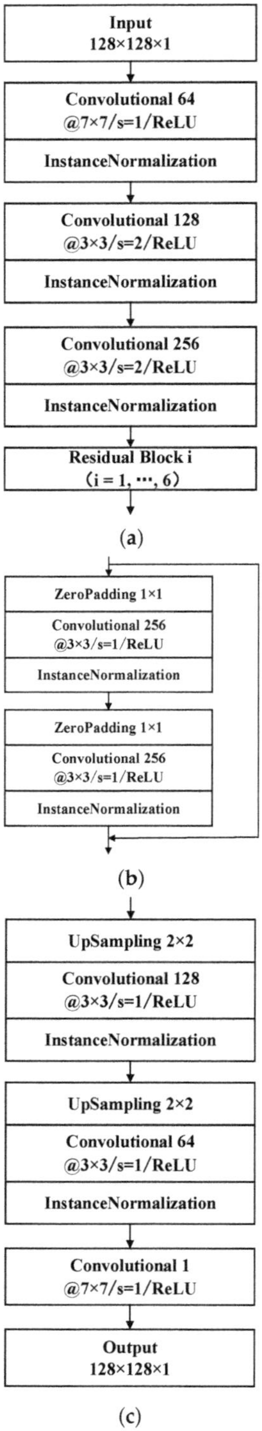

Figure 3. Architecture of the image reconstruction model. (**a**) Feature extraction module. (**b**) Details of the residual block. (**c**) Feature restoration module.

3.1. Feature Learning

Take the transition from the simulated SAR dataset **A** to the $\mathbf{X}_{NSP}$ dataset **B** as an example. Firstly, **A** goes through G_{AB} to generate an image which has a similar distribution to **B**. The architecture of G_{AB} is made up of two modules, i.e., feature extraction module and feature restoration module, as shown in Figure 3. The feature extraction module is depicted in Figure 3a,b. It firstly uses several convolutional layers to extract target features and reduce the dimension of the feature map to 32 × 32 × 256. Six residual blocks are then followed after the convolutional layer to deepen the network. It should be noted that the residual block is only used to learn the target features and does not change the size of the feature map to facilitate the subsequent feature restoration.

3.2. Feature Restoration

The feature restoration module is shown in Figure 3c, which forms the second part of the image reconstruction model. It is found that through multiple up-sampling layers and convolutional layers, the final output is an image of size 128 × 128 × 1, i.e., the same as the input simulated image. The output feature map of up-sampling layer can be written as

$$\begin{cases} h_{out} &= h_{in} \times f \\ w_{out} &= w_{in} \times f \\ c_{out} &= c_{in} \end{cases} \tag{3}$$

where h_{out}, w_{out}, c_{out} denote the height, width, and number of channels of the output feature map. Similarly, h_{in}, w_{in}, c_{in} represent those of the input feature map. f is the kernel size of the up-sampling layer and is set to 2 in this paper. Moreover, it should be noted that the number of channels of the convolutional layer in the feature restoration module corresponds to the one in the feature extraction module, e.g., the channels of the first convolutional layer in the feature restoration module are the same as those of the second convolutional layer in the feature extraction module to keep the same number of the feature map.

3.3. Image Discrimination

Then, a discrimination model is used to identify whether the input data are the reconstructed simulated SAR image or sparse real SAR image, whose structure is shown in Figure 4. In order to maximize the potential of the discriminant model, all the information in each pixel of the input image should be considered and used. Therefore, an activation function named Leaky ReLU is adopted in each convolutional layer, which can be defined as

$$\text{LeakyReLU} = \begin{cases} x_{ij}, x_{ij} &\geqslant 0 \\ \\ \alpha x_{ij}, x_{ij} &< 0 \end{cases} \tag{4}$$

where x_{ij} denotes the value in the pixel of i-th row and j-th column, α is a constant which is set to 0.2 in this paper to retain negative values and avoid the loss of information in the negative axis. It should be noted that instance normalization (IN) is used after each convolutional layer, whether it is an image reconstruction model or a discriminative model. Different from batch normalization (BN) that processes elements of all images in a batch, IN only focuses on the elements of an image so that it can maintain the unique characteristics of each sample. After four convolution operations, the last convolution layer reduces the dimension of the feature map to 8 × 8 × 1, i.e., 64 pixels. The output value of each pixel is zero or one. Zero means that the input data are a reconstructed simulated SAR image generated through the above G_{AB} network. One represents that the input is a real sparse SAR target. The final result is obtained by comprehensive evaluation of 64 output values. During the discrimination process, mean squared error (MSE) is used to measure the difference between reconstructed simulated image and $\mathbf{X}_{\mathrm{NSP}}$, defined as

$$\mathrm{MSE} = \frac{1}{N_{\mathrm{B}}} \sum_{n=1}^{N_{\mathrm{B}}} (x_n - \hat{x}_n)^2 \tag{5}$$

where N_{B} denotes the number of images in the $\mathbf{X}_{\mathrm{NSP}}$ dataset $\mathbf{B}$, x_n is the n-th sample of $\mathbf{B}$, and $\hat{x}_n$ is the corresponding reconstructed sample. After multiple epochs of iterative training, the G_{AB} model is able to generate an image that is closer to $\mathbf{X}_{\mathrm{NSP}}$ (see Figure 5).

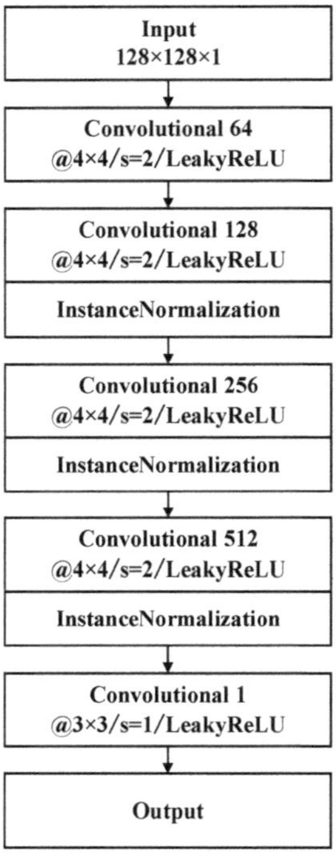

Figure 4. Structure of the discrimination model.

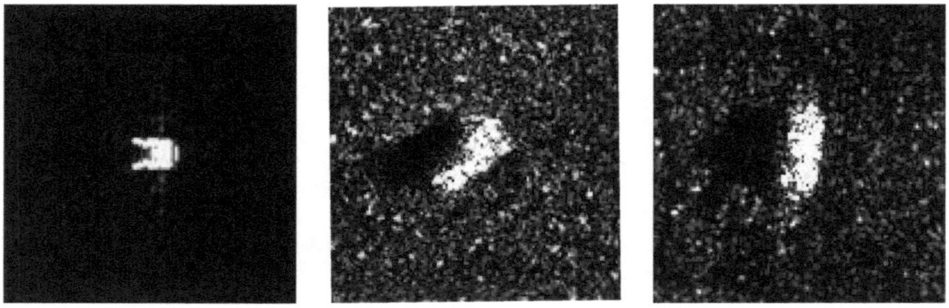

(**a**) (**b**) (**c**)

Figure 5. Presentation of three kinds of SAR images. (**a**) Original simulated SAR image. (**b**) Reconstructed simulated SAR image. (**c**) $\mathbf{X}_{\mathrm{NSP}}$.

Since the two networks used to generate images have the same structure, the simulated SAR images and the sparse SAR images are not paired, i.e., the number of samples in the two datasets is different. In order to ensure that the input samples of each epoch learn the correct mapping from each other during the training process, a constraint condition is added into the framework, which is defined as

$$L_{cons} = E[\|G_{BA}(G_{AB}(x_a)) - x_a\|_1] + E[\|G_{AB}(G_{BA}(x_b)) - x_b\|_1] \quad (6)$$

where $G_{BA}(G_{AB}(x_a))$ indicates the simulated image, x_a generates an image through G_{AB} and then restores it by G_{BA}, $E[\|G_{BA}(G_{AB}(x_a)) - x_a\|_1]$ is used to tell the difference between the restored image and the original x_a. Similarly, $G_{AB}(G_{BA}(x_b))$ indicates the simulated image, x_b generates an image through G_{BA} and then restores it by G_{AB} while $E[\|G_{AB}(G_{BA}(x_b)) - x_b\|_1]$ measures the difference between the restored image and the original x_b. $\mathbf{X}_{NSP}$ will also learn to simulate the target features of the simulated SAR image through the G_{BA} model and D_A, thereby accelerating the reduction of the distribution between **A** and **B**. The principle of transition from **B** to **A** is the same as that described above.

4. CNN-Based Sparse SAR Target Classification

4.1. Principle

Due to the advantages of weight sharing and local connections, CNN has a strong ability in extracting target features without increasing the complexity of the target classification framework. Therefore, in this section, a shallow CNN model is used to classify the sparse SAR target after acquiring a large number of reconstructed simulated SAR images. The flowchart of the SAR target classification is shown in Figure 6. After contrastive domain adaptation, we acquire abundant reconstructed simulated targets which have similar distributions to real SAR images. It is found that the reconstructed simulated SAR images and few real SAR images form the input of CNN and go through multiple convolutional layers and max pooling layers to extract target features. Finally, the softmax activation function is chosen to classify targets. The detailed components of the classification model are presented in Figure 7. As shown in Figure 7, the feature extraction framework consists of five convolutional layers, three max pooling layers, and two fully connected layers. The convolutional layer is different from the traditional neural network. It contains several convolution kernels to extract image characteristics one by one, greatly reducing the parameters in the network. This calculation is described as

$$y_{ij} = \sum_{i,j=1}^{k} (x_{ij} * w_{ij}) + b \quad (7)$$

where k means that the size of the kernel is $k \times k$, w_{ij} is the weight of i-th row and j-th column in the kernel, x_{ij} represents the value of i-th row and j-th column in the region covered by the convolution kernel, y_{ij} denotes the element of i-th row and j-th column in the output feature map, and b means bias which is set to 0 in many cases so as to reduce the computational complexity. The max pooling layer reduces the dimension of the feature map to further decrease the calculation. The size of the output feature map is

$$\begin{cases} h_{out} = \frac{(h_{in} - m)}{s} + 1 \\ w_{out} = \frac{(w_{in} - m)}{s} + 1 \\ c_{out} = c_{in} \end{cases} \quad (8)$$

where s determines stride, and m shows the size of the pooling box. The fully connected layer is responsible for integrating the extracted features and preparing them for classification. To alleviate the overfitting problem caused by limited real targets, two dropout layers are placed after the fully connected layer to randomly prevent some neurons from participating in the training of each epoch and to improve the robustness of the classification framework.

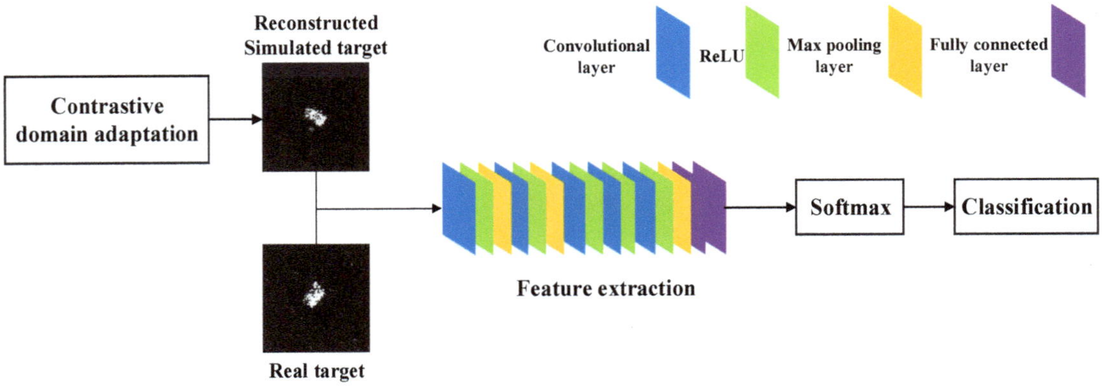

Figure 6. Flowchart of the target classification.

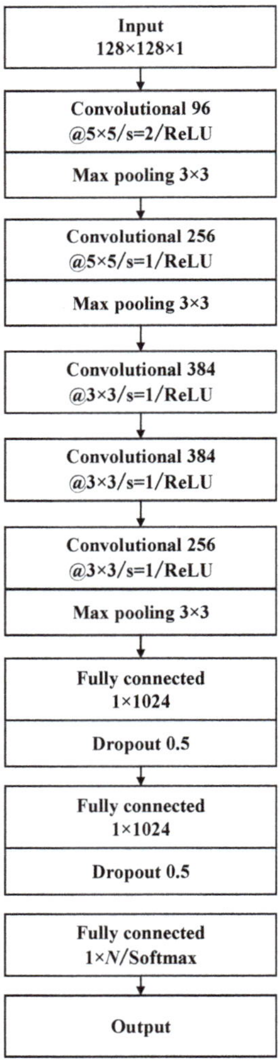

Figure 7. Structure of the classification framework.

4.2. Training Details

During the training process, the learning rate and epoch are set to 0.0005 and 200, respectively. The input dataset is usually separated into training set, validation set, and testing set. The training set is responsible for updating parameters in the network. The validation set separated from the training set is used to monitor the network training. While the testing set does not join the training process and is utilized to evaluate the robustness of the model. It is noted that in this paper, the reconstructed simulated data and a few sparse SAR images form the training set, but the validation set only consists of real SAR data in the training set. Each category in the training set is made up of 190 reconstructed simulated images and a few real sparse SAR images ranging from 1 to 50.

5. Experimental Results Based on MSTAR Dataset

In this section, we take MSTAR as our basic dataset, which is commonly used in SAR image classification. In order to validate the proposed method, we divide the experiments into three main parts. First, experiments under four different situations are conducted to validate whether CDA promotes the classification performance when real samples are insufficient. In the second part of experiments, several typical deep learning methods are used to compare with the proposed method. Finally, experiments under EOC are carried out in which characteristics in the training set and testing set differ a lot, bringing great difficulty for SAR target classification.

5.1. Dataset

In the experiments, the $\mathbf{X}_{NSP}$ dataset is constructed based on MSTAR. MSTAR is a MF-based dataset including 10 kinds of military vehicles. The detailed data description of MSTAR under SOC is shown in Table 1, including serial number, number of samples in the training set and testing set. Data in MSTAR are collected at two depression angles ranging from 0° to 360° azimuth angle. All the experiments are conducted under SOC. The samples of 17° are for the training set, while the images of 15° become the testing set. The optical images and corresponding SAR images of the ten category vehicles are shown in Figure 8. In the experiments, we suppose that only few labeled real sparse SAR images are available, so the samples at 17° are randomly selected to be 1, 5, 10, 20, 30, 40, and 50 per class.

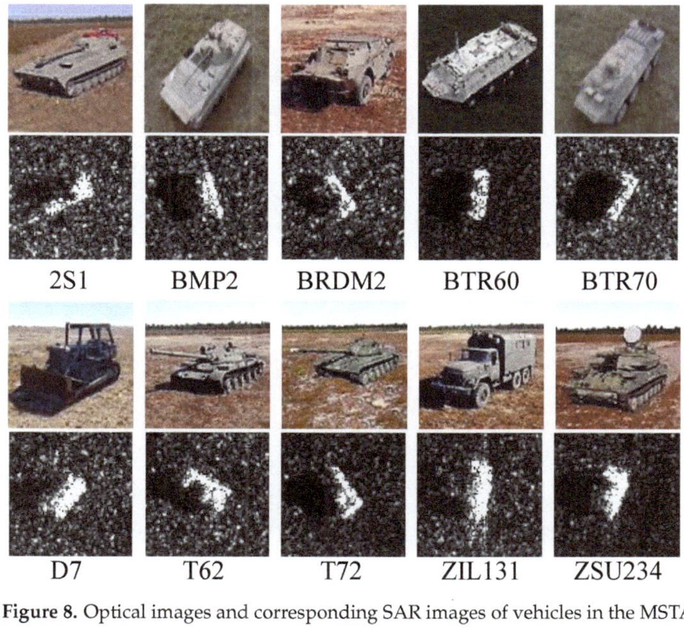

Figure 8. Optical images and corresponding SAR images of vehicles in the MSTAR dataset.

Table 1. Data description for SOC.

Class	Serial No.	Training Set (Depression 17°)	Testing Set (Depression 15°)
2S1	B01	299	274
BMP2	SN9563	233	196
BRDM2	E-71	298	274
BTR60	Kloyt7532	256	195
BTR70	C71	233	196
D7	92v13015	299	274
T62	A51	299	273
T72	SN132	232	196
ZIL131	E12	299	274
ZSU234	D08	299	274
Total		2747	2426

5.2. Study Based on Different Input Data

In order to support our viewpoints, the experiments based on both MF-recovered dataset and sparse SAR dataset are carried out simultaneously. As shown in Figure 9, experimental results of different numbers of training samples under four situations are taken into consideration. It can be concluded from two aspects that without CDA, CNN reaches similar accuracy when training data are sufficient, but fails to present similar performance with training samples decreasing. In addition, compared with the MF-based dataset, the $\mathbf{X}_{NSP}$ dataset achieves higher classification accuracy no matter the number of training samples. Take 5 samples of each class as example, the combination of CDA and sparse SAR dataset outperforms others by 20.6%, 28.0%, and 38.3%, respectively. The confusion matrix in Table 2 shows the classification result of the combination of CDA and sparse dataset, including both correctly and misclassified targets. It is seen that when only 20 training samples per class are available, the accuracy rate can reach more than 80%, and even targets such as T62 and ZSU234 can achieve more than 95%.

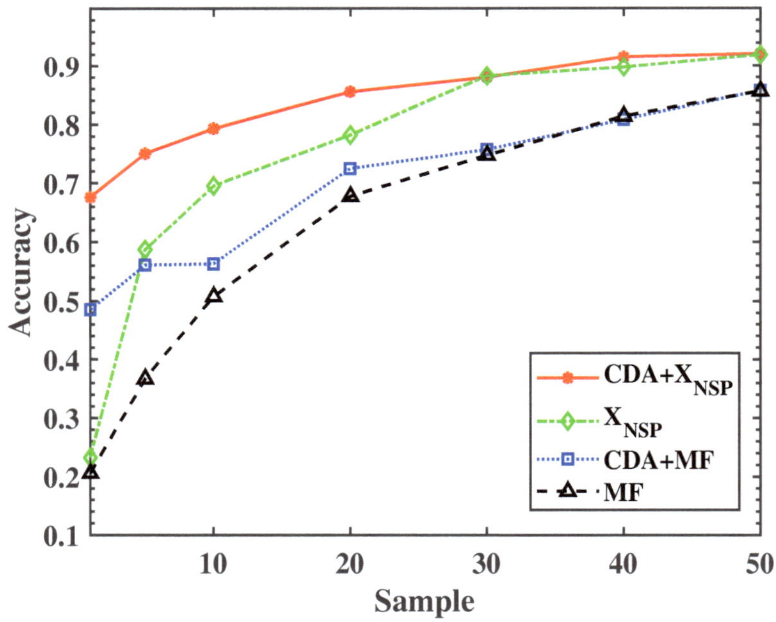

Figure 9. Experimental results of different training samples based on MF-based image and sparse image under SOC.

Table 2. Confusion matrix of classification accuracy based on the CDA and sparse SAR image via 20 samples per class.

Class	2S1	BMP2	BRDM2	BTR60	BTR70	D7	T62	T72	ZIL131	ZSU234	Total
2S1	218	2	4	19	5	2	12	9	2	1	
BMP2	14	108	4	15	35	0	2	16	0	2	
BRDM2	0	5	251	7	4	1	2	0	2	2	
BTR60	15	9	5	145	14	0	0	5	1	1	
BTR70	19	1	4	14	158	0	0	0	0	0	
D7	1	0	0	0	0	242	1	0	17	13	
T62	1	0	0	0	0	1	266	5	0	0	
T72	3	2	0	5	0	0	1	169	0	16	
ZIL131	0	0	0	0	0	24	2	0	248	0	
ZSU234	0	0	0	0	0	5	0	0	0	269	
Accuracy (%)	79.56	55.10	91.61	74.36	80.61	88.32	97.44	86.22	90.51	98.18	85.49

5.3. Comparison between Different Methods

To further validate the proposed method, several deep learning methods used for SAR target classification via few shots are adopted to be comparative experiments. As shown in Table 3, it can be seen that the proposed method achieves higher classification accuracy than CHU and A-ConvNet under all sets of training samples. In addition, when real SAR images are sufficient, the accuracy of M-Net is close to the proposed method. However, when it comes to the small sample case, i.e., with less than 30 samples per class, the proposed method shows better classification performance. While when the number of real SAR images decreases to 20 samples per class, the proposed method achieved 85.5% classification accuracy, respectively outperforming CHU, A-ConvNet, and M-Net by 44.5%, 19.5%, and 8.8%. Moreover, it should be noted that even if only 5 samples per class are available, our method can still achieve more than 75% accuracy. The performance gap with other methods is even more obvious.

Table 3. Classification accuracy of different methods based on different number of training samples.

Samples	CHU [15] *	A-ConvNet [13]	M-Net [34]	Proposed
5	<10.0%	40.0%	-	75.1%
10	<20.0%	51.4%	-	79.3%
20	41.0%	66.0%	76.7%	85.5%
30	47.0%	77.4%	83.3%	88.0%
40	55.0%	84.5%	86.7%	91.6%
50	76.0%	85.7%	91.0%	92.1%

*: Results can be found from Figure 4 of [15].

5.4. Verification under EOC

Considering the generalization of the proposed method, experiments under EOC are carried out as well, in which samples in the training set and testing set differ a lot. Data description for EOC is shown in Table 4, including each category's serial number and number of samples. It can be seen that some military vehicles have different types in the training set and testing set, e.g., the type of T72 in the training set is SN132, but A64 in the testing set, which brings more difficulty for SAR target classification. Experimental results demonstrate the combination of the $\mathbf{X}_{\mathrm{NSP}}$ dataset and CDA achieves better performance than the $\mathbf{X}_{\mathrm{NSP}}$ dataset only, as shown in Figure 10. With real samples decreasing, the performance gap becomes more obvious. Table 5 presents the confusion matrix of classification accuracy based on CDA and the sparse SAR image via 5 samples per class, in which the misclassified targets and targets being correctly classified are recorded in detail. As shown in the table, although ZSU234 reaches a relatively low accuracy of 40.97%, targets such as T72 and BRDM2 achieve around 80% accuracy with only 5 real SAR targets, even 2S1 still achieves over 90% classification accuracy. In Table 6, more indicators are selected to evaluate the experiments. The precision rate denotes the correct proportion of model predictions in all predicted samples. The recall rate is close to classification accuracy which means that the proportion of samples is correctly predicted to all the real samples. F1-score which is written as $F1 = 2 \times \text{precision} \times \text{recall}/(\text{precision} + \text{recall})$ represents a comprehensive evaluation of precision and recall. Tables 5 and 6 give a comprehensive evaluation of experimental results via 5 samples per class.

Table 4. Data description for EOC.

Class	Serial No.	Training Set (Depression 17°)	Testing Set (Depression 30°)
2S1	B01	299	288
BRDM2	E71	298	287
T72	SN132/A64	299	288
ZSU234	D08	299	288
Total		1195	1151

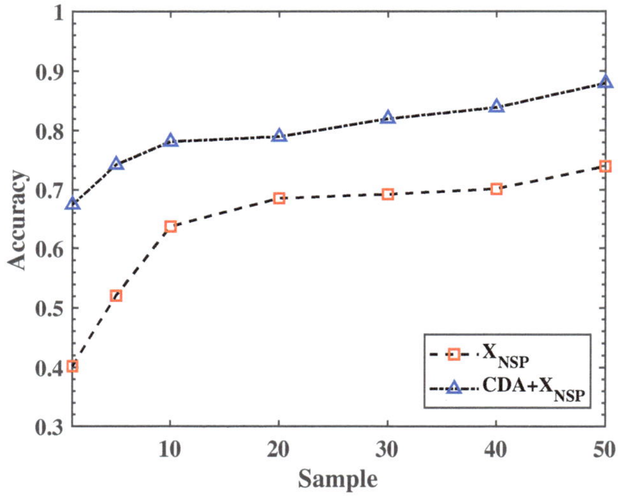

Figure 10. Experimental results of different training samples based on sparse image under EOC.

Table 5. Confusion matrix of classification accuracy based on CDA and sparse SAR image via 5 samples per class.

Class	2S1	BRDM2	T72	ZSU234	Total
2S1	261	24	2	1	
BRDM2	24	250	12	1	
T72	44	1	226	17	
ZSU234	87	0	83	118	
Accuracy (%)	90.63	87.11	78.47	40.97	74.28

Table 6. Classification results based on CDA and sparse SAR image via 5 samples per class.

Class	Precision	Recall	F1-Score	Test Samples
2S1	0.63	0.91	0.74	288
BRDM2	0.91	0.87	0.89	287
T72	0.70	0.78	0.74	288
ZSU234	0.86	0.41	0.56	288
Avg/total	0.77	0.74	0.73	1151

In this section, several experiments on the basis of the MF-based dataset and BiIST-based dataset $\mathbf{X}_{\mathrm{NSP}}$ are conducted to validate the proposed method. From the above analysis, it can be concluded that the CDA-based sparse SAR target classification method can improve the performance or achieve optimal accuracy in the comparison with others under each set of training samples, especially under few-shot cases.

6. Experimental Analysis

Experimental results on SAR target classification can mainly be divided into two scenarios. Under SOC, we carry out four comparative experiments under different sets of training samples ranging from 1 per class to 50 per class. On the one hand, the combination of $\mathbf{X}_{\mathrm{NSP}}$ and CDA achieves similar performance to $\mathbf{X}_{\mathrm{NSP}}$ when the real samples are sufficient, but outperforms $\mathbf{X}_{\mathrm{NSP}}$ with real sparse SAR images reducing. On the other hand, no matter the number of training data, the combination of $\mathbf{X}_{\mathrm{NSP}}$ and CDA always reaches higher classification accuracy compared with MF and the combination of CDA and MF, showing the great potential of non-sparse SAR images. Under EOC, targets in training set and testing set differ a lot; hence, it can be found that CDA is able to improve classification performance especially in the situation of fewer training samples. When the number of training samples comes to 5 per category, the combination of $\mathbf{X}_{\mathrm{NSP}}$ and CDA can still reach over 70% accuracy, which shows the great potential of applying the proposed method in real scenes.

7. Conclusions

In this paper, a novel sparse SAR target classification method on the basis of CDA is proposed for few-shot cases. Experimental results show that compared to typical classification methods via small samples, the proposed method has similar accuracy when there are sufficient real SAR samples. However, with the used real SAR images decreasing to 5 or 10 samples of each category, the accuracy gap between the proposed method and others is more obvious. In addition, it should be also noted that the combination of $\mathbf{X}_{\mathrm{NSP}}$ and CDA obtains optimal performance under SOC and EOC, proving the efficiency and generalization in SAR applications. To further reduce real SAR images while improving the accuracy, we will explore how to integrate phase information in the future.

Author Contributions: Conceptualization, H.B. and J.Z.; methodology, Z.L., J.D. and Z.J.; validation, Z.L. and J.D.; writing—original draft preparation, Z.L. and J.D.; writing—review and editing, Z.J., H.B. and J.Z. All authors have read and agreed to the published version of the manuscript.

Funding: This work was supported in part by the National Natural Science Foundation of China under Grant 62271248 and 61901213, in part by Guangdong Basic and Applied Basic Research Foundation under Grant 2020B1515120060, in part by the Aeronautical Science Foundation of China under Grant 201920052001, and in part by the University Joint Innovation Fund Project of CALT under Grant CALT2021-11.

Data Availability Statement: The complex SAR image data used in this paper is the MSTAR dataset, which can be found at https://www.sdms.afrl.af.mil/index.php?collection=mstar (accessed on 3 August 2022).

Conflicts of Interest: The authors declare no conflict of interest.

Abbreviations

The following abbreviations are used in this paper:

DA	Domain Adaptation
CDA	Contrastive Domain Adaptation
HDA	Heterogeneous Domain Adaptation
SAR	Synthetic Aperture Radar
CV	Computer Vision

MF	Matched Filtering
ATR	Automatic Target Recognition
CHU	Convolutional Highway Unit
A-ConvNets	All-convolutional Networks
CV-CNN	Complex-valued Convolutional Neural Network
CV-FCNN	Complex-valued Fully Convolutional Neural Network
IST	Iterative Soft Thresholding
OMP	Orthogonal Matching Pursuit
CNN	Convolutional Neural Network
ReLU	Rectified Linear Unit
CReLU	Concatenated Rectified Linear Unit
DL	Deep Learning
TL	Transfer Learning
GAN	Generative Adversarial Network
MD-GAN	Multi-discriminators Generative Adversarial Network
CAE	Convolutional Autoencoder
CCAE	Compact Convolutional Autoencoder
DJ-CORAL	Dynamic Joint Correlation Alignment Network
Fc	Fully Connected Layer
RCNN	Regional Convolutional Neural Network
YOLO	You Only Look Once
AP-CNN	Amplitude-phase Convolutional Neural Network
IN	Instance Normalization
BN	Batch Normalization
MSE	Mean Squared Error
MSTAR	Moving and Stationary Target Acquisition and Recognition
SOC	Standard Operating Conditions
EOC	Extended Operating Conditions

References

1. Curlander, J.C.; Mcdonough, R.N. *Synthetic Aperture Radar: Systems and Signal Processing*; Wiley: New York, NY, USA, 1991.
2. Henderson, F.M.; Lewis, A.J. *Principle and Application of Imaging Radar*; John Wiley and Sons: New York, NY, USA, 1998.
3. Dugeon, D.E.; Lacoss, R.T. An overview of automatic target recognition. *Linc. Lab. J.* **1993**, *6*, 3–10.
4. Kreithen, D.E.; Halversen, S.D.; Owirka, G.J. Discriminating targets from clutter. *Linc. Lab. J.* **1993**, *6*, 25–51.
5. Hinton, G.E.; Salakhutdinov, R.R. Reducing the dimensionality of data with neural networks. *Science* **2006**, *313*, 504–507. [CrossRef] [PubMed]
6. Krizhevsky, A.; Sutskever, I.; Hinton, G.E. Imagenet classification with deep convolutional neural networks. *Commun. ACM* **2012**, *60*, 84–90. [CrossRef]
7. Russakovsky, O.; Deng, J.; Su, H.; Krause, J.; Satheesh, S.; Ma, S.; Huang, Z.; Karpathy, A.; Khosla, A.; Bernstein, M.; et al. Imagenet large scale visual recognition challenge. *Int. J. Comput. Vis.* **2015**, *115*, 211–252. [CrossRef]
8. Szegedy, C.; Liu, W.; Jia, Y.; Sermanet, P.; Reed, S.; Anguelov, D.; Erhan, D.; Vanhoucke, D.; Rabinovich, A. Going deeper with convolutions. In Proceedings of the 2015 IEEE Conference on Computer Vision and Pattern Recognition (CVPR), Boston, MA, USA, 7–12 June 2015.
9. He, K.; Zhang, X.; Ren, S.; Sun, J. Delving deep into rectifiers: Surpassing human-level performance on ImageNet classification. In Proceedings of the 2016 IEEE Conference on Computer Vision and Pattern Recognition (CVPR), Santiago, Chile, 26 June–1 July 2016.
10. Huang, G.; Liu, Z.; der Maaten, L.; Weinberger, K.Q. Densely connected convolutional networks. In Proceedings of the 2017 IEEE Conference on Computer Vision and Pattern Recognition (CVPR), Honolulu, HI, USA, 21–26 July 2017.
11. Ding, J.; Chen, B.; Liu, H.; Huang, M. Convolutional neural network with data augmentation for SAR target recognition. *IEEE Geosci. Remote Sens. Lett.* **2016**, *13*, 364–368. [CrossRef]
12. Chen, S.; Wang, H.; Xu, F.; Qiu, Y. Target classification using the deep convolutional networks for SAR images. *IEEE Trans. Geosci. Remote Sens.* **2016**, *54*, 4806–4817. [CrossRef]
13. Hansen, D.M.; Kusk, A.; Dall, J.; Nielsen, A.A.; Engholm, R.; Skriver, H. Improving SAR automatic target recognition models with transfer learning from simulated data. *IEEE Geosci. Remote Sens. Lett.* **2017**, *14*, 1484–1488. [CrossRef]
14. Lin, Z.; Ji, K.; Kang, M.; Leng, X.; Zou, H. Deep convolutional highway unit network for SAR target classification with limited labeled training data. *IEEE Geosci. Remote Sens. Lett.* **2017**, *14*, 1091–1095. [CrossRef]
15. Zhong, C.; Mu, X.; He, X.; Wang, J.; Zhu, M. SAR target image classification based on transfer learning and model compression. *IEEE Geosci. Remote Sens. Lett.* **2019**, *16*, 412–416. [CrossRef]
16. Wang, Z.; Xu, X. Efficient deep convolutional neural networks using CReLU for ATR with limited SAR images. *J. Eng.* **2019**, *2019*, 7615–7618. [CrossRef]

17. Zhang, W.; Zhu, Y.; Fu, Q. Deep transfer learning based on generative adversarial networks for SAR target recognition with label limitation. In Proceedings of the 2019 IEEE International Conference on Signal, Information and Data Processing (ICSIDP), Chongqing, China, 11–13 December 2019.
18. Zheng, C.; Jiang, X.; Liu, X. Semi-Supervised SAR ATR via multi-discriminator generative adversarial network. *IEEE Sens. J.* **2019**, *19*, 7525–7533. [CrossRef]
19. Guo, J.; Wang, L.; Zhu, D.; Hum C. Compact convolutional autoencoder for SAR target recognition. *IET Radar Sonar Navig.* **2020**, *14*, 967–972. [CrossRef]
20. Huang, Z.; Pan, Z.; Lei, B. What, where, and how to transfer in SAR target recognition based on deep CNNs. *IEEE Trans. Geosci. Remote Sens.* **2020**, *58*, 2324–2336. [CrossRef]
21. Guo, Y.; Du, L.; Lyu, G. SAR target detection based on domain adaptive Faster R-CNN with small training data size. *Remote Sens.* **2021**, *13*, 4202. [CrossRef]
22. Yang, G.; Lang, H. Semisupervised heterogeneous domain adaptation via dynamic joint correlation alignment network for ship classification in SAR imagery. *IEEE Geosci. Remote Sens. Lett.* **2022**, *19*, 1–5. [CrossRef]
23. Zhang, Z.; Wang, H.; Xu, F.; Jin, Y.-Q. Complex-valued convolutional neural network and its application in polarimetric SAR image classification. *IEEE Trans. Geosci. Remote Sens.* **2017**, *55*, 7177–7188. [CrossRef]
24. Coman, C.; Thaens, R. A deep learning SAR target classification experiment on MSTAR dataset. In Proceedings of the 19th International Radar Symposium (IRS), Bonn, Germany, 20–22 June 2018.
25. Yu, L.; Hu, Y.; Xie, X.; Lin, Y.; Hong, W. Complex-valued full convolutional neural network for SAR target classification. *IEEE Geosci. Remote Sens. Lett.* **2020**, *17*, 1752–1756. [CrossRef]
26. Daubechies, I.; Defriese, M.; De Mol, C. An iterative thresholding algorithm for linear inverse problems with a sparsity constraint. *Commun. Pure Appl. Math.* **2004**, *57*, 1413–1457. [CrossRef]
27. Bi, H.; Bi, G. Performance analysis of iterative soft thresholding algorithm for L_1 regularization based sparse SAR imaging. In Proceedings of the 2019 Radar Conference (RadarConf), Boston, MA, USA, 22–26 April 2019.
28. Pati, Y.C.; Rezaiifar, R.; Krishnaprasad, P.S. Orthogonal matching pursuit: Recursive function approximation with applications to wavelet decomposition. In Proceedings of the 27th Asilomar Conference on Signals, Systems and Computers, Pacific Grove, CA, USA, 1–3 November 1993.
29. Donoho, D.L.; Tsaig, Y.; Drori, I.; Starck, J. Sparse solution of underdetermined systems of linear equations by stagewise orthogonal matching pursuit. *IEEE Trans. Inf. Theory* **2012**, *58*, 1094–1121. [CrossRef]
30. Bi, H.; Bi, G.; Zhang, B.; Hong, W. A novel iterative thresholding algorithm for complex image based sparse SAR imaging. In Proceedings of the 12th European Conference on Synthetic Aperture Radar (EUSAR), Aachen, Germany, 4–7 June 2018.
31. Bi, H.; Bi, G. A novel iterative soft thresholding algorithm for L_1 regularization based SAR image enhancement. *Sci. China Inf. Sci.* **2019**, *62*, 49303. [CrossRef]
32. Bi, H.; Deng, J.; Yang, T.; Wang, J.; Wang, L. CNN-based target detection and classification when sparse SAR image dataset is available. *IEEE J. Sel. Top. Appl. Earth Obs. Remote Sens.* **2021**, *14*, 5815–6826. [CrossRef]
33. Deng, J.; Bi, H.; Zhang, J.; Liu, Z.; Yu, L. Amplitude-phase CNN-based SAR target classification via complex-calued sparse image. *IEEE J. Sel. Top. Appl. Earth Obs. Remote Sens.* **2022**, *15*, 5214–5221. [CrossRef]
34. Shang, R.; Wang, J.; Jiao, L.; Stolkin, R.; Hou, B.; Li, Y. SAR targets classification based on deep memory convolution neural networks and transfer parameters. *IEEE J. Sel. Top. Appl. Earth Obs. Remote Sens.* **2018**, *11*, 2834–2846. [CrossRef]

remote sensing

Article

Semantic Segmentation of Mesoscale Eddies in the Arabian Sea: A Deep Learning Approach

Mohamad Abed El Rahman Hammoud [1], Peng Zhan [1,2,3], Omar Hakla [4], Omar Knio [1,2] and Ibrahim Hoteit [1,2,*]

1 Physical Science and Engineering, King Abdullah University of Science and Technology, Thuwal 23955, Saudi Arabia
2 Department of Ocean Science and Engineering, Southern University of Science and Technology, Shenzhen 518055, China
3 Southern Marine Science and Engineering Guangdong Laboratory, Guangzhou 510000, China
4 Maroun Semaan Faculty of Engineering and Architecture, American University of Beirut, Beirut 1107, Lebanon
* Correspondence: ibrahim.hoteit@kaust.edu.sa

Abstract: Detecting mesoscale ocean eddies provides a better understanding of the oceanic processes that govern the transport of salt, heat, and carbon. Established eddy detection techniques rely on physical or geometric criteria, and they notoriously fail to predict eddies that are neither circular nor elliptical in shape. Recently, deep learning techniques have been applied for semantic segmentation of mesoscale eddies, relying on the outputs of traditional eddy detection algorithms to supervise the training of the neural network. However, this approach limits the network's predictions because the available annotations are either circular or elliptical. Moreover, current approaches depend on the sea-surface height, temperature, or currents as inputs to the network, and these data may not provide all the information necessary to accurately segment eddies. In the present work, we have trained a neural network for the semantic segmentation of eddies using human-based—and expert-validated—annotations of eddies in the Arabian Sea. Training with human-annotated datasets enables the network predictions to include more complex geometries, which occur commonly in the real ocean. We then examine the impact of different combinations of input surface variables on the segmentation performance of the network. The results indicate that providing additional surface variables as inputs to the network improves the accuracy of the predictions by approximately 5%. We have further fine-tuned another pre-trained neural network to segment eddies and achieved a reduced overall training time and higher accuracy compared to the results from a network trained from scratch.

Keywords: eddy detection; deep neural network; semantic segmentation; Arabian Sea

Citation: Hammoud, M.A.E.R.; Zhan, P.; Hakla, O.; Knio, O.; Hoteit, I. Semantic Segmentation of Mesoscale Eddies in the Arabian Sea: A Deep Learning Approach. *Remote Sens.* **2023**, *15*, 1525. https://doi.org/10.3390/rs15061525

Academic Editors: Xiaochen Yang, Qian Du and Yanni Dong

Received: 26 January 2023
Revised: 19 February 2023
Accepted: 22 February 2023
Published: 10 March 2023

1. Introduction

Energetic swirls in the ocean, known as eddies, are essential for the transport of heat and material [1] and for energy conversion among dynamics at different scales [2]. In oceanography, eddies with diameters ranging between 10 km and 500 km are known as mesoscale eddies [3]. Mesoscale eddies regulate oceanic mixing and climate by promoting the transport of salt, heat, and carbon on both local and global scales [4,5], and by inducing a strong shear that generates sub-mesoscale processes to produce prominent vertical fluxes in the upper ocean [6,7]. Detecting and tracking mesoscale eddies are therefore important for investigating their properties, transport mechanisms, and their contributions to air–sea coupling processes [8].

Established eddy detection methods depend either on physical or on geometric criteria to classify these mesoscale features from the mainstream flow. Physical criteria involve the determination of dynamical properties such as the magnitude of the sea-level anomaly [9], the velocity gradient [10,11], or the Okubo–Weiss (OW) parameter [12]. Generally, these methods define an eddy based on closed contours of these quantities above a certain threshold value; however, they are limited by the fact that no single threshold is optimal for the varying oceanic conditions, and the OW parameter is extremely susceptible to noise in sea-surface height (SSH) data [13,14]. To overcome these weaknesses, Chelton et al. 2011 [13] proposed a "threshold-free" method, partitioning the SSH field using a range of monotonically changing thresholds. They further proposed smoothing the SSH field spatially before processing these data to reduce the noise, which may affect the OW parameter.

Unlike the physical criteria, geometric criteria are based on selecting circular or closed streamlines from the quasi-circular flow patterns. One of the most straightforward methods for identifying eddies based on geometric criteria is to identify instantaneous streamlines mapped onto a plane normal to the vortex core [15]. A more sophisticated implementation is to use the winding-angle (WA) method [16], which has been demonstrated to be more efficient and accurate than the Okubo–Weiss method for identifying mesoscale eddies [17]. Nevertheless, a major drawback of established eddy detection techniques is the geometric restriction on the predicted eddies, where these eddies have to be circular or elliptical by construction of the algorithm.

Recently, deep learning techniques have been advancing rapidly, capitalizing on the ability of deep neural networks (DNNs) to learn sophisticated relationships between input and output fields. These advances have been utilized to solve pressing issues in computer vision, pattern recognition, and physics, to name a few examples. For instance, deep learning has been employed to produce high-quality realistic images [18–20], to help with medical assessments of cancer patients [21,22], and to emulate earth systems [23–25]. Deep learning has also been used to detect objects and segment images [26,27] and videos [28]. In particular, semantic segmentation (e.g., Ref. [29]) and instance segmentation (e.g., Refs. [30,31]) aim to delineate all features or individual features, respectively. In segmentation approaches, a DNN is used to output a classification map, of the same size as the input image, which describes the object or class to which a given pixel belongs [32].

Machine learning techniques for detecting eddies were first introduced by Castelani, 2006 [33], who trained a densely connected neural network to detect eddies from sea surface temperature (SST) data. Deep learning techniques were then employed to help improve the detection accuracy. For instance, both Franz et al., 2018 [34], and Lguensat et al., 2018 [35], depended on SSH data to train a convolutional neural network (CNN) to detect eddies. Both groups of investigators employed semantic segmentation approaches to generate classification maps describing the locations and extents of cyclonic and anticyclonic eddies. Moschos et al., 2020 [36], then proposed using SST data rather than SSH data to train a CNN to produce classification maps that identify the locations of cyclonic and anticyclonic eddies. Similarly, Duo et al., 2019 [37], Xu et al., 2019 [38], and Nian et al., 2019 [39], utilized sea-surface anomalies as inputs for the detection and semantic segmentation of mesoscale eddies. More recently, a combination of SST and SSH data and surface speeds were used as inputs to a deep CNN, which predicted the locations and shapes of individual eddies in an instance-segmentation approach [40]. These previously mentioned methods relied on model-generated eddy labels, which limited the predicted shapes of the mesoscale eddies to be either circular or elliptical. In reality, however, the shapes of these eddies may be distorted, necessitating the requirement for more sophisticated labels [41].

In this study, we have trained a DNN to perform the semantic segmentation of cyclonic and anticyclonic eddies in the Arabian Sea by relying on human-labeled and expert-validated annotated data for mesoscale eddies in that sea. Human-based annotations allow for predictions of complex eddy geometries allowing to identify eddies that are neither circular nor elliptical, which is one of the major limitations of non-learning-based algorithms. In particular, our training dataset includes annotations of different shapes that are not necessarily elliptical, which enables the DNN to predict such complex geometries. The annotations are made openly accessible to promote additional developments in learning-based eddy detection. The employed deep neural network comprises a U-Net [29] *backbone* for feature extraction, with a Pyramid Pooling Module (PPM) [42] as the *decode head* or task-performing network that predicts the output classification maps. We trained this network using as input the SSH, SST, sea-surface salinity (SSS), and zonal and meridional velocity components from the Copernicus Marine Environment Monitoring Service (CMEMS), a global ocean-eddy-resolving reanalysis product [43]. The usage of different surface variables as inputs to the network provides more information about the surface state of the sea and is expected to improve the prediction accuracy compared to the use of a single ocean-surface field. We compared the results obtained by using this trained model to those obtained from another model with the same architecture that we trained using only SSH, SST, or SSS data as the single input, as is typically performed in the literature. We then analyzed the results by examining the sensitivity of the network to different combinations of the input ocean-surface fields. Moreover, to investigate the computational benefits of using "off-the-shelf" networks [44], we compared the results obtained using this DNN to those obtained using a pre-trained ResNet50–FCN [45], which was fine-tuned to predict eddies using a transfer-learning methodology. We also evaluate the performance of a model trained on the Arabian Sea dataset for predicting eddies in the Red Sea to investigate its generalizability for the semantic segmentation of eddies in different basins. This study is the first to analyze the impact of different combinations of ocean-surface fields on the semantic segmentation of eddies, and to examine the utilization of transfer learning for the semantic segmentation of mesoscale eddies.

The remainder of the paper is structured as follows. Section 2 describes the circulation in the Arabian Sea and the data used for training and validating the DNN. The U-Net–PPM, trained from scratch, and the pre-trained ResNet50–FCN are described in Section 3, where the training setup and methodology are presented. The results are presented in Section 4. Finally, Section 5 summarizes the main results and presents the concluding remarks and a discussion of future research directions.

2. Ocean Data

We used the global ocean eddy-resolving reanalysis product from the Copernicus Marine Environment Monitoring Service (CMEMS), which provides reliable and up-to-date information of the global ocean with a comprehensive collection of oceanographic and environmental data and includes information on a wide range of parameters such as temperature, salinity, sea level, ocean currents, and biogeochemical properties. The reanalysis dataset used in this study is available from 1993 to 2020 with a horizontal resolution of $1/12^{\circ}$, or approximately 8 km [43]. This ocean reanalysis was generated using version 3.1 of the Nucleus for European Models of the Ocean (NEMO) model [46], driven at the surface by the ERA-Interim climate reanalysis dataset [47] of the European Centre for Medium-Range Weather Forecasts (ECMWF). This model assimilates *in situ* data from various sources such as satellites and *in situ* measurements using a reduced-order Kalman filter, including along-track altimeter data, remotely sensed SST data, and *in situ* T/S vertical profiles, and the reanalysis data set was subject to rigorous quality control and validation procedures to ensure its accuracy and reliability, and is continually updated as new observations become available. In this study, daily averaged SSH, SST, SSS, and surface current fields are extracted. The spatial and temporal resolution of the CMEMS product are fine enough to resolve the mesoscale eddies in the Arabian Sea [8]. Figure 1 illustrates a snapshot of the surface currents superimposed on the SSH for a day during the summer season.

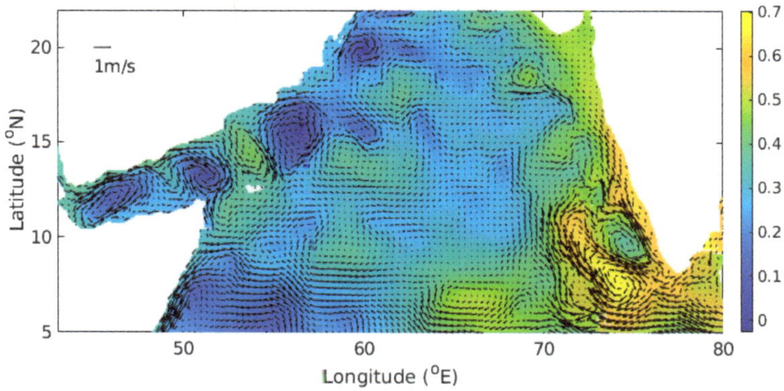

Figure 1. Plot illustrating the SSH and surface currents in the Arabian Sea at a snapshot representative of a typical day during the summer season.

The Somali current along the Somalia coast off northwest Africa makes the Arabian Sea unique compared to other basins with strong western boundary-current systems, since it is not continuous in time but reverses seasonally [48]. This pronounced seasonality is linked to the annually reversing Indian Ocean monsoon. During the southwest monsoon season in summer, anticyclonic oceanic circulation in the Arabian Sea and the northward western-boundary Somali current follow the appearance of a negative curl of the wind stress over the central basin, while the opposite scenario applies during the northeast monsoon season in winter [49]. In addition, the Arabian Sea exhibits pronounced intraseasonal and spatial variability in eddy activity, with enhanced eddy kinetic energy (EKE) during summer, primarily in the western basin. Mesoscale eddies play a key role in modulating the circulation [50] and biological processes [51] in the Arabian Sea—east of the Somalia coast in particular—and they exhibit EKE and eddy-induced transport, which are comparable in magnitude in both the horizontal and vertical directions to other eddy-rich basins in the global oceans [8,52,53].

3. Neural Network

3.1. Neural Network Architecture

In this section, we examine the use of a U-Net–PPM neural-network architecture to segment cyclonic and anticyclonic eddies, and we compare its performance, in terms of prediction accuracy and training time, to that of a pre-trained and fine-tuned Resnet50–FCN. Both network architectures are composed of a CNN backbone for extracting features, followed by a decode head for predicting the classification maps. The first network comprises a U-Net [29] backbone and a PPM [42] decode head, and was trained from randomly initialized model parameters. The second network is a pre-trained CNN that comprises a 50-layer ResNet [54] backbone equipped with a fully convolutional network (FCN) [45] decode head, and we fine-tuned the initial model parameters by training it on our dataset. Both neural network architectures and pipelines are presented in Figure 2, which illustrates the architecture of each DNN along with the convolution blocks used to develop them. This figure portrays the pipeline, showing the input fields, the individual layers/blocks, and the output classification maps.

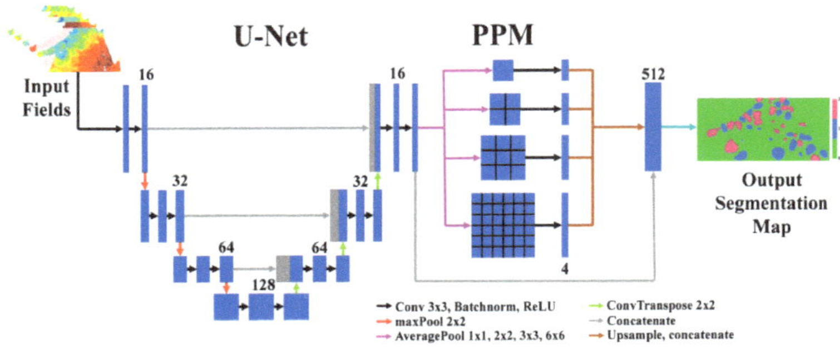

(a) U-Net with PPM and convolution blocks.

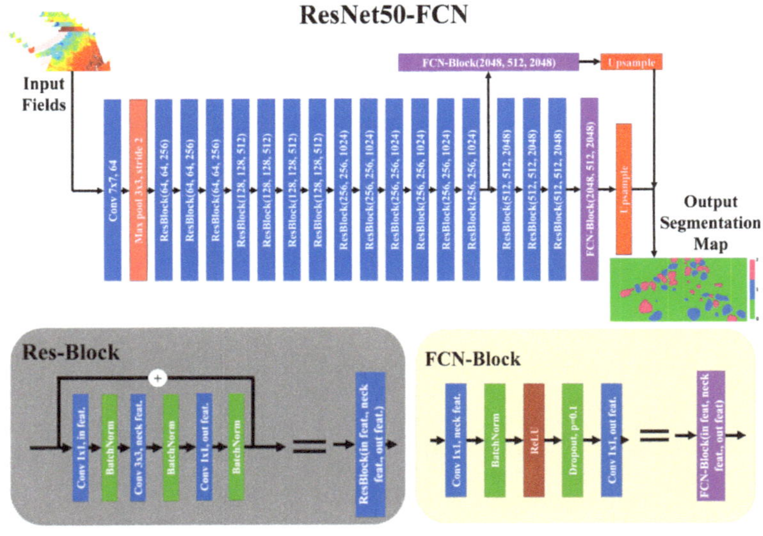

(b) Resnet50-FCN.

Figure 2. Neural−Network Architectures: Schematics of (**a**) the U-Net backbone with the PPM decode head, along with the Res- and FCN-blocks, and (**b**) the ResNet50–FCN that was pre-trained on a subset of the COCO dataset and is available in PyTorch's torchvision. The deep neural network in (**a**) is trained by starting from randomized initial parameters, whereas that of (**b**) was trained from randomly initialized and pre-trained model parameters that are fine-tuned by training on the Arabian Sea dataset.

Figure 2a shows the backbone of the first network, which is a U-Net that comprises the following sequence: a convolutional encoder, bottleneck layers, and an expanding convolutional decoder that mirrors the encoder. Specifically, the U-Net's encoder includes the sequence of a convolution layer with a 3×3 kernel size and a batchnorm followed by ReLU activation function. This sequence is present twice at each level of the encoder, and a maxpooling layer is employed to transition to the next level. The decoder is a reflection of the U-Net encoder, with transposed convolutions and nearest-neighbor upsampling substituted for the convolutions and max-pooling layers, respectively. Finally, a skip connection concatenates the feature maps from the outputs of each encoder layer to the input of the decoder layer at the same level; this has been shown to promote strong gradient propagation and to preserve information from the encoder layers to subsequent layers in the network [29,54].

The second network consists of a 50-layer ResNet CNN backbone formed by a sequence of Res-Blocks, as shown in Figure 2b. The Res-Blocks [54] consist of three consecutive convolution layers, each followed by a BatchNorm layer, with a skip connection between the input and the output of the Res-Block. The convolution layers have convolution kernels of sizes 1×1, 3×3 and 1×1, respectively. To achieve 50 layers, this network employs 16 consecutive Res-Blocks along with an initial convolution layer that is followed by a max pooling layer with a stride of 2. The FCN is merged with the 50-layer ResNet by connecting one FCN-block to the 15th Res-Block and another FCN-block to the last Res-Block. The outputs of these FCN-blocks are upsampled and then joined to predict the classification maps. We used a pre-trained model that is readily available in PyTorch's *torchvision* [55] in a transfer-learning [44] setting, with layers that fit the task-dependent dataset substituted for the first and last layers before training the network on the dataset of interest. This approach is commonly adopted in deep learning applications to reduce the training time, where these pre-trained models were extensively trained on large datasets to provide a large set of features that could be utilized in future tasks [56].

In the present work, we have analyzed the performance of both DNNs in segmenting eddies in the Arabian Sea using different combinations of ocean-surface variables as inputs to the DNN. We trained the first DNN using human-annotated labels, which allows for complex geometrical predictions that differ from the outputs of established eddy-detection algorithms, which are limited to circles and ellipses. Importantly, this study differs from that of Fan et al., 2020 [40], who relied on combining the SSH and SST data into one channel, producing a physically inconsistent quantity. Moreover, their analysis does not report the spatial resolution of the input fields nor the geographical location of the region of interest. Finally, the human-based labels used by Fan et al., 2020 [40], are based solely on SSH data, making them susceptible to larger annotation errors than the ones reported here because they miss the circulation information gained by using the velocities.

3.2. Training and Validation Datasets

We inputted the ocean-surface fields to the DNN to predict the probability that a pixel belongs to a cyclonic eddy, an anticyclonic eddy, or neither (i.e., to the background). These three categories comprise the output classes for which the network predicts a softmax probability that a pixel belongs to a given class. Each ocean-surface variable is represented by an input channel to the network, where the total number of input channels is equal to the number of input variables propagated through the DNN. The input fields are propagated through the DNN, which classifies the image point-by-point according to the class with the highest softmax probability, as further discussed in the following section. Finally, the output classification map is a 2D spatial map with the same resolution as the input, and it indicates whether a grid cell is classified to belong to the background, a cyclonic eddy, or an anticyclonic eddy.

The training and validation datasets rely on the CMEMS ocean reanalysis datasets described in Section 2. In particular, we extracted the ocean-surface fields from the years 2001 to 2010 to generate the training dataset and used those from 2011 as the validation dataset. All surface fields—including the velocity-vector fields—were human-labeled and later validated by an expert to confirm the quality of the labels. This is important since a DNN trained using eddy-detection-based data—cf. [35]—can only represent a surrogate of the eddy-detection model. On the other hand, training a DNN using hand-labeled data that contains labels of different shapes and sizes and which are not constrained to be ellipses or circles, is a surrogate for an arbitrary model described by the human-based annotations. Finally, we note that we have removed the last 10 days of 2010 and the first 10 days of 2011 from the training and validation datasets to avoid any leakage of information between the training and validation datasets that may arise from high-frequency correlations that could bias the performance-metric estimates. This results in training and validation datasets comprising 3642 and 354 samples, respectively.

3.3. Loss Function

We trained both networks by minimizing the cross entropy between the predicted and true masks. The cross entropy is commonly used for multi-class semantic-segmentation problems, cf. [57,58]. In multi-class classification, the cross entropy computes a loss for each class label, and the total loss is reduced by taking the mean or sum of all the loss terms. The cross entropy for C unique classes can be expressed as follows:

$$\mathcal{L}(\boldsymbol{x};\theta) = -\frac{1}{n_x n_y}\sum_{i=1}^{n_x}\sum_{j=1}^{n_y}\sum_{k=1}^{n_c} w_k y_t(\boldsymbol{x}) log[\hat{y}_\theta(\boldsymbol{x})], \tag{1}$$

where n_x, n_y, and n_c are the numbers of horizontal grid cells, vertical grid cells, and classes, respectively. The quantities y_t are the reference annotated labels and $\hat{y}_\theta$ are the θ-parameterized labels predicted by the network at a given grid cell for an input image $\boldsymbol{x} \in \mathbb{R}^{n_x,n_y,n_f}$, where n_f is the number of input ocean fields. The loss employs a class-specific weighting w_k, with $\sum_k w_k = 1$, to mitigate issues that may arise from class imbalance [58], since at any given instance the areas covered by cyclonic and anticyclonic eddies are not equal. The classes are weighted by the inverse of percentage of occurrence of a given class throughout the entire training dataset. During training, we omitted the loss for the background class to avoid overestimating the accuracy of the network because it is considerably easier for the network to predict the background class label, leading to an artificial boost in the overall loss without improving the segmentation accuracy.

3.4. Accuracy Metrics

In addition to the loss, we assessed the performance of the trained model by computing a number of accuracy metrics that are commonly used in semantic segmentation. These metrics include the pixel-wise accuracy ($\mathcal{A}$), the Jaccard index (i.e., the intersection over the union: IoU), the precision ($\mathcal{P}$), and the recall ($\mathcal{R}$) [28,45]. If we define a true positive (TP) as an outcome in which the model correctly predicts a positive class, then a true negative (TN) is an outcome in which the model correctly predicts a negative class. Similarly, a false positive (FP) is an outcome in which the model incorrectly predicts a positive class, and a false negative (FN) is an outcome in which the model incorrectly predicts a negative class.

Using this notation, we can define $\mathcal{A}$ as follows:

$$\mathcal{A} = \frac{TP + TN}{TP + TN + FP + FN}, \tag{2}$$

a measure of the fraction of pixels in the network labeled correctly. The IoU measures the overlap of the target mask with the predicted output; this is given by

$$\text{IoU} = \frac{|A \cap B|}{|A \cup B|}, \tag{3}$$

where A is the set of positive-class pixels from the prediction and B is the set of positive-class pixels from the true labels. The precision $\mathcal{P}$ is the ratio of TP to all positive examples, which is a measure of the exactness of the model's classification of a positive class; it is given by

$$\mathcal{P} = \frac{TP}{TP + FP}. \tag{4}$$

Finally, $\mathcal{R}$ is a measure of the percentage of detected ground-truth labels, expressed as follows:

$$\mathcal{R} = \frac{TP}{TP + FN}. \tag{5}$$

3.5. Training Setup

We trained the DNN eddy detection models for a total of 20 epochs, each of which randomly passes over 2500 sampled instances of the input fields and the corresponding labeled classification maps, which totals to 50,000 random samples during training. Training depends on a stochastic-gradient-descent (SGD) optimization algorithm to minimize the loss function, with momentum 0.9 and weight decay 10^{-4}. We performed an initial parameter search to identify the best learning rate (LR) to maximize the model accuracy. In particular, we trained the DNN for a batch size of 10 and LRs of 0.0001, 0.001, 0.01, 0.02, 0.03, 0.04, and 0.1. We selected the model that maximizes the average accuracy at segmenting cyclonic and anticyclonic eddies as the best model and used it to report our results in Section 4. We performed the same parameter search for each case corresponding to different input surface fields, and we present the eddy-segmentation results from the model with the highest accuracy. We utilized a learning-rate scheduler to reduce the LR by a factor of 10 if the accuracy did not increase for at least three epochs; this enables us to take finer optimization steps in efforts to build an improved model.

4. Experiments and Results

4.1. Semantic Segmentation Using Five Ocean-Surface Fields

In this experiment, the outputs of two U-Net–PPM models—one trained using all five surface ocean fields as an input to the DNN while the other is trained using SSH data only—are contrasted qualitatively against the reference annotations. The parameters of the U-Net–PPM model were randomly initialized, and training was performed using the dataset described in Section 3.2. Figure 3 shows the ground-truth labels and the predicted segmentation maps for randomly selected samples for the case that used all five ocean-surface variables to train the neural network. The results suggest that the trained model is able to predict the location of mesoscale eddies quite comparably to the ground-truth labels. The network also predicts additional eddies that were either missed during annotation or are artifacts or "ghost" eddies, as named by Lguensat et al., 2018 [35]. Moreover, these results suggest that the network smooths out the boundaries of the mesoscale predictions by following the natural variations in the surface fields. This is an interesting takeaway because the labels are based on polygons [59], which are not as smooth as the input fields.

Similarly, Figure 4 shows that the eddy locations predicted using only SSH data as input are also visually comparable to the ground truth labels. The shapes of the mesoscale features, however, are not as accurate as those predicted by the DNN trained on all the ocean-surface variables. In particular, this model enlarges and over-smooths the predicted regions that correspond to mesoscale eddies. The trained model's predictions also merge multiple labels together and exhibit some artifacts in the predicted segmentation masks. Compared to the case in which all the surface variables were used as input, these predictions are less accurate, as indicated by the prediction masks, inaccurately predicting extremely small eddies or overestimating the corresponding prediction areas. While this section presents a qualitative analysis of the network's performance at segmenting eddies using different input variables, the following section quantifies the performance boost achieved from using multiple input fields as opposed to using only SSH data.

4.2. Sensitivity of Segmentation Accuracy to the Input Surface Fields

The results of the previous section indicates that the predicted eddies are better captured qualitatively when all five surface variables are used as input to the DNN in comparison to only using the SSH fields. In this section, the performance of the U-Net–PPM DNN model, with randomly initialized trainable parameters, is quantified for different combinations of input ocean-surface fields. The input layer of the network is the only component of the DNN that is modified in this numerical experiment, varying the number of input channels to correspond to the number of input fields. Note that the training parameters, including the LR and batch size, are held fixed throughout these experiments to allow a fair comparison between different models.

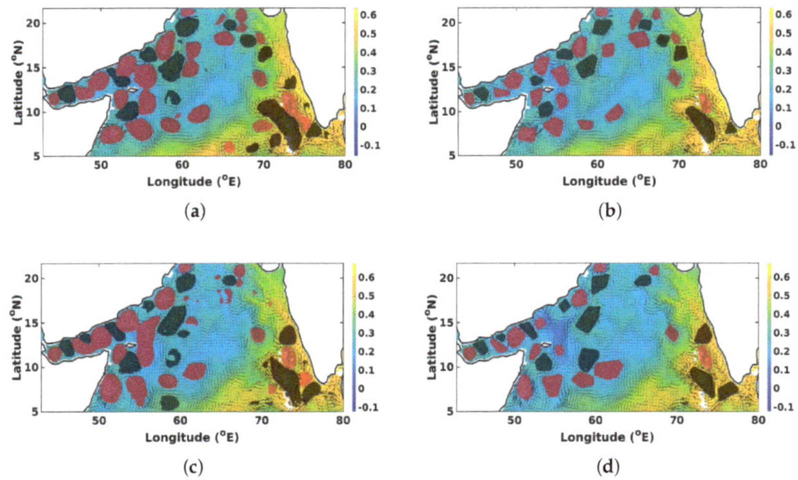

Figure 3. Ground−truth and predicted segmentation masks for the U-Net–PSPnet deep neural network using all five ocean-surface fields. The segmentation maps superpose the SSH data and the surface-velocity vectors. The plots demonstrate that the network captures all the labeled eddies with smoother boundaries. Here, the cyclonic eddies are colored red and the anticyclonic eddies are colored black. The color bars at the sides of each panel indicate the SSH values in meters; (**a**) Prediction—Sample 1; (**b**) Ground Truth—Sample 1; (**c**) Prediction—Sample 2; (**d**) Ground Truth—Sample 2.

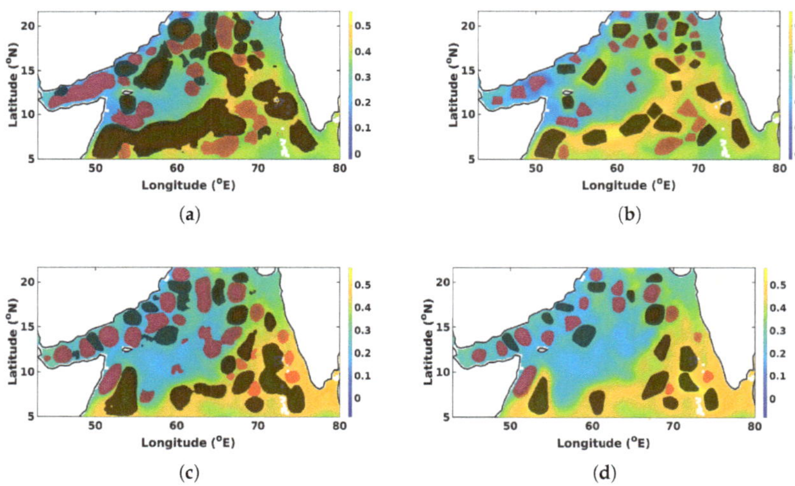

Figure 4. Ground-truth and predicted segmentation masks for the U-Net–PSPnet deep neural network using only the SSH data as input. The segmentation maps are superposed on the SSH data only to show the input to the network alongside the output masks. The plots demonstrate that the network captures the labeled eddies with smoother boundaries than those in the ground-truth plots but lumps some labels together. Here, the cyclonic eddies are colored red and the anticyclonic eddies are colored black. The color bars at the sides of each panel indicate the SSH values in meters; (**a**) Prediction—Sample 1; (**b**) Ground Truth—Sample 1; (**c**) Prediction—Sample 2; (**d**) Ground Truth—Sample 2.

Table 1 reports the accuracy metrics for the different combinations of the input variables. The results suggest that using all five ocean-surface variables yields a 2% boost in $\mathcal{A}$, in comparison to using only SSH as input, and a 5% increase in $\mathcal{A}$ in comparison to using SST or SSS as input. In addition, using SSH or the velocities alone yields a value of $\mathcal{A}$ that is roughly 2% greater than using either—or even both—SSS or SST as input. This indicates that both SSH and the velocities are sufficient to achieve the semantic segmentation of eddies with high $\mathcal{A}$ values; however, using all the ocean variables as input provides the neural network with a rich flow of information, which achieves the highest $\mathcal{A}$ values. The mean IoU (mIoU) varies in a way similar to $\mathcal{A}$, with the largest mIoU obtained when all the ocean-surface fields are used as inputs. Note that the values of the mIoU are expected to be smaller than the $\mathcal{A}$ values because the mIoU penalizes wrong predictions, whereas $\mathcal{A}$ does not.

Table 1. Accuracy Metrics for different inputs to the neural network: The inputs to the U-Net–PSPnet model are varied systematically, and the corresponding accuracy metrics are evaluated and reported in the table. The table shows that the accuracy metrics are improved by approximately 5% when all five ocean-surface variables are used, as opposed to using only the SSH data.

	Accuracy		Mean IoU		$\mathcal{P}$		$\mathcal{R}$	
Input Variables	Cyclonic	Anticyclonic	Cyclonic	Anticyclonic	Cyclonic	Anticyclonic	Cyclonic	Anticyclonic
SSH	0.92871	0.93942	0.45984	0.42608	0.47306	0.44126	0.94267	0.92531
SST	0.89841	0.90285	0.35974	0.30395	0.37709	0.31804	0.88659	0.87279
SSS	0.89623	0.90300	0.34649	0.28808	0.36821	0.30932	0.85457	0.80754
SSH and SST	0.92809	0.93197	0.45975	0.40122	0.47049	0.41150	0.95272	0.94140
SSH and SSS	0.93942	0.94252	0.49387	0.43301	0.51596	0.45322	0.92022	0.90661
SSS and SST	0.90210	0.90361	0.36783	0.30508	0.38596	0.31913	0.88679	0.87388
Velocities	0.92511	0.92935	0.44640	0.38892	0.45945	0.40091	0.94016	0.92856
SSH, SSS, and SST	0.92341	0.92849	0.44366	0.38731	0.45447	0.39805	0.94913	0.93487
SST, SSS, and velocities	0.92524	0.93283	0.44927	0.40000	0.46097	0.41388	0.94653	0.92264
SSH, SSS, and velocities	0.92783	0.92855	0.45525	0.38625	0.46985	0.39846	0.93610	0.92648
SST, SSH, and velocities	0.91893	0.92540	0.43280	0.37998	0.44072	0.38907	**0.96017**	**0.94211**
SSH, SSS, SST, and velocities	**0.94340**	**0.94855**	**0.49602**	**0.45173**	**0.49095**	**0.44603**	0.93994	0.92783

Finally, this table shows that the network is able to not only achieve large values of $\mathcal{A}$ but also to achieve $\mathcal{R}$ values that are much larger than the $\mathcal{P}$ values. Combining these results with those of Figures 3 and 4 demonstrates that the network predicts the majority of the positive classes; i.e., it predicts the eddies correctly. However, the areal coverage of the eddies is overestimated, which adversely affects the $\mathcal{P}$ values, since the eddies are not perfectly covered by the prediction. These results are also in agreement with the predicted semantic-segmentation maps, in which the regions of the predicted labels are almost always larger than those of the true labels. The $\mathcal{P}$ values may also be lower than the $\mathcal{R}$ values because of class imbalance, where the background covers a larger area than the eddies. Moreover, a high $\mathcal{R}$ value with a lower $\mathcal{P}$ value may also be attributable to the network's predictions of eddies that were not labeled; i.e., either missed during annotation or non-existent. Finally, the results also indicate that the effects of the surface variables on $\mathcal{P}$ and $\mathcal{R}$ are similar to their effects on $\mathcal{A}$ and mIoU.

Figure 5 presents the neural-network predictions for each combination of the input variables considered in this sensitivity study. This figure also includes the reference human labels to enable straightforward comparisons of the predictions with the reference for each case. When only a single ocean variable is used as input to the DNN, several fictitious eddies are predicted, as indicated by the artifacts, and the areas of the predicted eddies are overestimated. This is similar to the results of Lguensat et al., 2018 [35]. As the number of input variables is increased, the network predictions improve, as indicated by the predicted regions better covering the actual eddies, as given by the labels. In all but the last case, the DNN prediction does not split the eddy boundaries correctly, as is apparent from the continuous predictions that lump several of the labeled eddies into a single predicted eddy. This problem was largely mitigated when the SSH, SST, SSS, and the velocities were used as input to the network. In that case, the predictions better delineate the eddy boundaries, with a smaller numbers of artifacts than in the other cases.

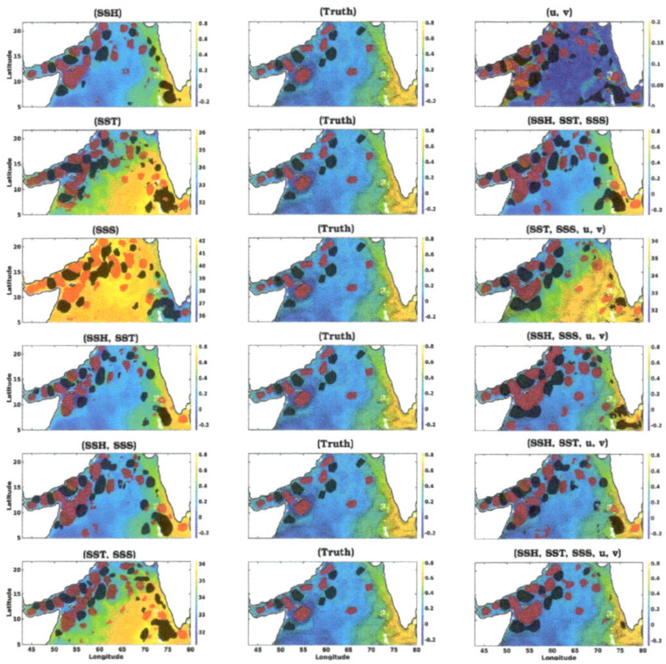

Figure 5. Illustration of the neural network predictions (**left** and **right** columns) for DNNs trained using the input variables indicated by the titles of each subplot, along with the true human labels (**center** column). We used the same input field for all the predictions, and the same truth is repeated in each row to make the comparison easier for each case. Here, cyclonic eddies are colored red and the anticyclonic eddies are colored black. Note that the background and the color bars correspond to the first ocean−surface variable indicated by the subfigure title. The velocity vector field is overlaid on top of the scalar field whenever applicable.

4.3. Transfer Learning Using a Pre-Trained Model

Various pre-trained models are readily available for deployment for a variety of tasks, including object detection [60], human-pose determination [61], and semantic segmentation [28]. These models are generally trained extensively on large datasets—for example, on ImageNet [62]—which helps the network to learn a rich collection of features. These pre-trained models can be further tuned for a particular task using a specific dataset by—for example, in classification problems—reinitializing the input and output layers to accommodate the input dimensions and the output classes required by the new task of interest. Here, we explored the use of a pre-trained ResNet50–FCN [45,63] for the semantic segmentation of mesoscale features in the Arabian Sea using the five ocean-surface fields employed above. The model was pre-trained on a subset of *COCO train2017* using 20 categories that also occur in the *Pascal VOC* dataset. This model achieves an $\mathcal{A}$ of 91.4% and mIoU of 60.5% on *COCO val2017*. We reinitialized the input layer to accommodate five channels, corresponding to the five ocean-surface fields. The classifier was reinitialized to predict three channels, which correspond to the cyclonic-eddy, anticyclonic-eddy, and background classes. The model parameters were fine-tuned by training the model using our human-annotated Arabian Sea ocean-eddies dataset. Furthermore, the same ResNet50–FCN model was trained starting from randomly initialized trainable parameters using the same training dataset, and the resulting predictions were contrasted to the reference annotations and the predictions of the U-net–PPM and fine-tuned ResNet50–FCN.

Figure 6 shows the reference and predicted segmentation masks for the U-net–PPM, ResNet50–FCN, and fine-tuned ResNet50–FCN. The figure shows that all the predictions perform comparably at detecting the mesoscale eddies. The U-net–PPM predicts several "ghost" eddies that are not present in the ResNet50–FCN predictions. In contrast, both ResNet50–FCN predictions are smoother and have less discontinuous predictions. This could be attributed to the ResNet50–FCN model having ∼35 million trainable parameters, far more than the ∼1 million of the U-net–PPM. The figure indicates that the ResNet50–FCN predictions have less "ghost" eddies for the case of the pre-trained model as opposed to the one trained starting from randomly initialized trainable parameters. The predictions are smoother than the labels; however, small artifacts that do not resemble eddies were noticeable in a few samples. Nevertheless, the accuracy of the model is greater than the accuracy of the U-Net–PPM trained with all five ocean-surface variables. This indicates that the pre-trained model possesses a number of features that can be fine-tuned to improve the predictability of eddies.

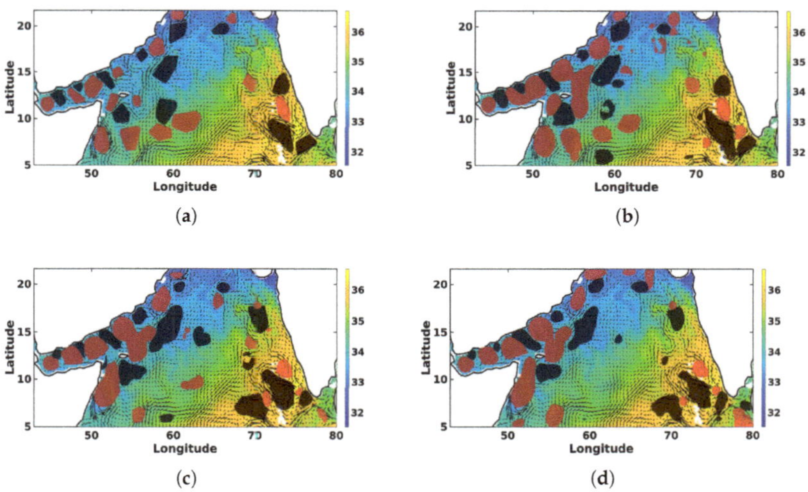

Figure 6. Transfer Learning: Predicted and ground-truth segmentation maps for some randomly selected samples. We obtained these predictions using the fine-tuned ResNet50–FCN model, with all five ocean-surface variables as inputs to the trained model. Cyclonic eddies are colored red and anticyclonic eddies are colored black. The color bars at the sides of each panel indicate the SSH values in meters; (**a**) Reference; (**b**) U-Net–PPM; (c) ResNet50–FCN; (**d**) Pretrained ResNet50–FCN.

Figure 7 outlines the evolution of $\mathcal{A}$ for both cyclonic (C) and anticyclonic (AC) eddies over time, where each marker represents the value of $\mathcal{A}$ at the end of an epoch. The results indicate that the pre-trained ResNet50–FCN model is able to achieve an average $\mathcal{A}$ higher than that of U-Net–PPM for all time steps. Specifically, the pre-trained ResNet50–FCN reaches an average $\mathcal{A}$ of approximately 96% in less than 30 min of training, whereas the U-Net–PPM model achieves an average $\mathcal{A}$ of about 95% in approximately 200 min. This demonstrates that fine-tuning a pre-trained model can help to reduce the consumption of computational resources for such tasks, as the computational time was reduced more than six-fold by fine-tuning the pre-trained network. Furthermore, the randomly initialized ResNet50–FCN was also able to achieve accuracy values larger than those of the U-net–PPM, however, such accuracy values were achieved after about 235 min of training, which is expected from using a large model.

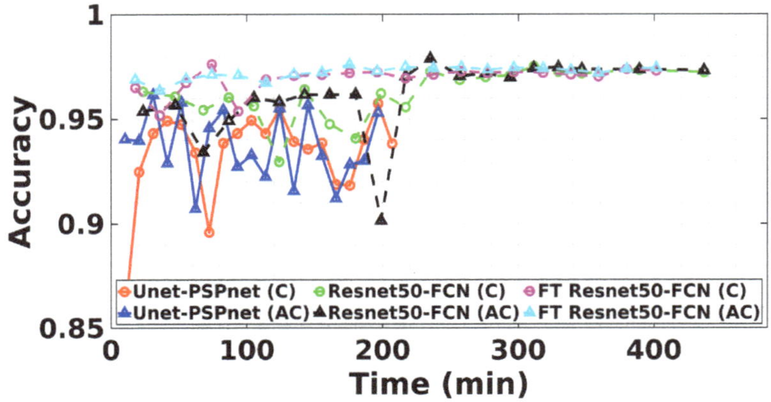

Figure 7. Comparison Learning Curves: Evolution of the pixel accuracy of the U−Net–PSPnet model trained from a random initialization and of a pre−trained ResNet50–FCN model that has been fine-tuned. The curves show that the pre−trained Resnet50-FCN models achieve an accuracy above 95% much faster than the U−Net–PSPnet model. **Note:** FT refers to fine-tuned.

4.4. Generalization to the Red Sea

We finally investigated the performance of the pre-trained model in the semantic segmentation of eddies in a different eddy-rich sea, the Red Sea, using the DNN trained with data labeled for the Arabian Sea oceanic variables. In general, the model's performance is expected to degrade when the inputs are changed to a different sea because of the change in the input distributions. To detect eddies in the Red Sea, we used the fine-tuned ResNet50–FCN using all five ocean-surface variables (SSH, SST, SSS, and the velocities) as input, as they include the richest set of features and achieve the highest accuracy, as discussed in previous sections.

Figure 8 shows the detected-eddy segmentation masks superimposed on the SSH data and the velocity-vector field. Since no ground-truth labels are available for the Red Sea, we performed a qualitative visual assessment of the performance of the model. Specifically, this figure shows that the model is able to identify most, if not all, large-scale mesoscale eddies in the Red Sea. In comparison to those in the Arabian Sea, eddies in the Red Sea are generally smaller and are not greatly deformed from circular shapes. The detected eddies flank each other in the central and northern Red Sea, and their sizes can become as large as the whole width of the basin. This is due to the effect of the large meridional density gradient that energizes the mesoscale dynamics under the topographic constraints [64].

While some eddies are not detected by this neural network, we suspect that this may have occurred because the resolution of the data may be too coarse to enable eddy segmentation in the narrow Red Sea basin [2]. Nevertheless, the network is able to identify several large and semi-persistent eddies that are known to exist in the central Red Sea [2,65–67]. We expect that better performance could be achieved by retraining the DNN using additional data labeled for the Red Sea. Another approach to mitigate the effects of this domain shift is to perform domain adaptation to push the input distribution closer to the one on which the DNN was trained [68,69].

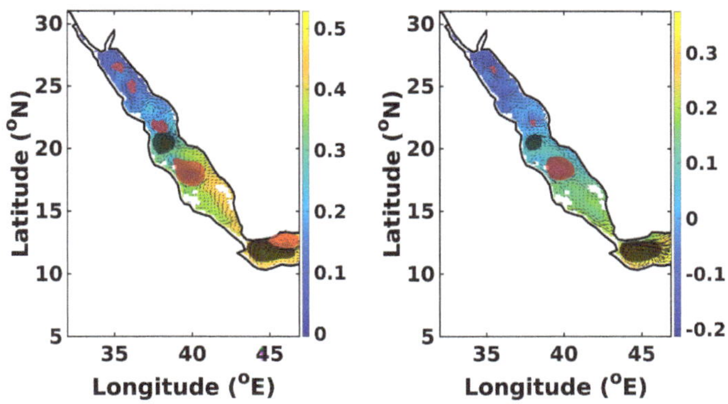

(a) (b)

Figure 8. Red Sea Predictions: Predicted semantic-segmentation maps obtained using the fine-tuned ResNet50–FCN model. Since no ground-truth labels are available, we assessed the performance of the model without using any domain generalization to improve the predictions. Despite the facts that the network was not trained for the Red Sea and that the data for this narrow basin are coarse, the network is able to predict the rough locations of the mesoscale eddies. The plot shows the cyclonic eddies colored red and the anticyclonic eddies colored black. The scalar field shown in the plots corresponds to the SSH values in meters; (**a**) Sample 1; (**b**) Sample 2.

5. Discussion and Conclusions

We have utilized human-annotated labels for cyclonic and anticyclonic eddies in the Arabian Sea to train a deep neural network for the semantic segmentation of these mesoscale features. We first trained a deep neural network (DNN) with a U-Net feature extractor and a PPM decode head starting from random initial parameters. We explored using only the SSH data as input to the DNN, as is commonly done in the literature, in contrast to using multiple surface-state variables (SSH, SST, SSS, and the velocities). The results demonstrate that using all five ocean-surface variables yields a higher accuracy for eddy detection than using only the SSH data as input.

Accordingly, we varied the input variables systematically to assess the sensitivity of the DNN to these inputs. Our numerical experiments showed that the accuracy of the DNN predictions is most sensitive to the SSH and velocities, and we obtained a boost of 2–3% in accuracy when we included these variables as inputs, as opposed to using only the SST or SSS data. We also explored the use of a transfer-learning technique for the semantic segmentation of eddies. In particular, a ResNet50 backbone with an FCN decode head that were pre-trained on a subset of the COCO dataset were fine-tuned by training them on the Arabian Sea eddies dataset. The results of this experiment demonstrated that using a pre-trained model improves the prediction accuracy by approximately 2%. In addition, a pre-trained model reduced the training time by more than half, by simply fine-tuning the learned convolution kernels to the dataset at hand.

Finally, we evaluated the performance of the eddy detection DNN in qualitatively predicting the eddies in the Red Sea. We used the ResNet50–FCN model that was fine-tuned on the Arabian Sea dataset to predict the eddies in the Red Sea. The results show that the model predicts the larger mesoscale eddies but fails to predict the smaller eddies. The performance of this DNN could be improved by using data labeled for the Red Sea, using domain adaptation and generalization techniques, or using higher-resolution data to train the DNN.

This work could be extended further to track or even predict the locations and shapes of eddies. In addition, one might be interested in applying domain-adaptation and generalization techniques to construct a model to predict eddies on a global scale. Moreover, one could consider examining the performance of these models when remote-sensing data is used as input or when higher-resolution input fields are used to capture smaller features in the ocean. Finally, we plan to investigate an instance-segmentation approach to detect and delineate individual eddies in the ocean.

Author Contributions: Conceptualization, M.A.E.R.H., P.Z., O.K., I.H.; methodology, M.A.E.R.H. and P.Z.,; software, M.A.E.R.H.; validation, P.Z., O.K., I.H.; formal analysis, M.A.E.R.H., P.Z., O.K., I.H.; investigation, M.A.E.R.H.; resources, O.K. and I.H.; data curation, M.A.E.R.H., P.Z., O.H.; writing—original draft preparation, M.A.E.R.H. and P.Z.; writing—review and editing, M.A.E.R.H., P.Z., O.H., O.K., I.H.; visualization, M.A.E.R.H. and P.Z.; supervision, O.K. and I.H.; project administration, O.K. and I.H.; funding acquisition, O.K. and I.H. All authors have read and agreed to the published version of the manuscript.

Funding: The research reported in this publication was supported by the Virtual Red Sea Initiative Award #REP/1/3268-01-01.

Institutional Review Board Statement: Not applicable.

Informed Consent Statement: Not applicable.

Data Availability Statement: The datasets used in this study are publicly available in https://doi.org/10.5061/dryad.ht76hdrm3.

Acknowledgments: We thank Issam Lakkis for his helpful discussions. We also thank Jad Bhamdouni, Louis Youssef, Nour Qaraqira, and Omar AlLahham for their help with acquiring the annotation.

Conflicts of Interest: The authors declare no conflict of interest.

References

1. Kundu, P.K.; Cohen, I.M.; Dowling, D.R. Chapter 10—Boundary Layers and Related Topics. In *Fluid Mechanics*, 6th ed.; Kundu, P.K., Cohen, I.M., Dowling, D.R., Eds.; Academic Press: Boston, MA, USA, 2016; pp. 469–532.
2. Zhan, P.; Subramanian, A.C.; Yao, F.; Kartadikaria, A.R.; Guo, D.; Hoteit, I. The eddy kinetic energy budget in the Red Sea. *J. Geophys. Res. Ocean.* **2016**, *121*, 4732–4747. [CrossRef]
3. Tansley, C.E.; Marshall, D.P. Flow past a Cylinder on a β Plane, with Application to Gulf Stream Separation and the Antarctic Circumpolar Current. *J. Phys. Oceanogr.* **2001**, *31*, 3274–3283. [CrossRef]
4. McWilliams, J.C. The nature and consequences of oceanic eddies. In *Ocean Modeling in an Eddying Regime*; American Geophysical Union: Washington, DC, USA, 2008; pp. 5–15.
5. Sommer, J.L.; d'Ovidio, F.; Madec, G. Parameterization of subgrid stirring in eddy resolving ocean models. Part 1: Theory and diagnostics. *Ocean. Model.* **2011**, *39*, 154–169. [CrossRef]
6. Su, Z.; Wang, J.; Klein, P.; Thompson, A.F.; Menemenlis, D. Ocean submesoscales as a key component of the global heat budget. *Nat. Commun.* **2018**, *9*, 1–8. [CrossRef]
7. Zhan, P.; Guo, D.; Krokos, G.; Dong, J.; Duran, R.; Hoteit, I. Submesoscale Processes in the Upper Red Sea. *J. Geophys. Res. Ocean.* **2022**, *127*, e2021JC018015. [CrossRef]
8. Zhan, P.; Guo, D.; Hoteit, I. Eddy-Induced Transport and Kinetic Energy Budget in the Arabian Sea. *Geophys. Res. Lett.* **2020**, *47*, e2020GL090490. [CrossRef]
9. Chaigneau, A.; Eldin, G.; Dewitte, B. Eddy activity in the four major upwelling systems from satellite altimetry (1992–2007). *Prog. Oceanogr.* **2009**, *83*, 117–123. [CrossRef]
10. Isern-Fontanet, J.; García-Ladona, E.; Font, J. Identification of marine eddies from altimetric maps. *J. Atmos. Ocean. Technol.* **2003**, *20*, 772–778. [CrossRef]
11. Kurian, J.; Colas, F.; Capet, X.; McWilliams, J.C.; Chelton, D.B. Eddy properties in the California current system. *J. Geophys. Res. Ocean.* **2011**, *116*. [CrossRef]
12. Chelton, D.B.; Schlax, M.G.; Samelson, R.M.; de Szoeke, R.A. Global observations of large oceanic eddies. *Geophys. Res. Lett.* **2007**, *34*. [CrossRef]
13. Chelton, D.B.; Schlax, M.G.; Samelson, R.M. Global observations of nonlinear mesoscale eddies. *Prog. Oceanogr.* **2011**, *91*, 167–216. [CrossRef]
14. Zhan, P.; Subramanian, A.C.; Yao, F.; Hoteit, I. Eddies in the Red Sea: A statistical and dynamical study. *J. Geophys. Res. Ocean.* **2014**, *119*, 3909–3925. [CrossRef]
15. Robinson, S.K. Coherent motions in the turbulent boundary layer. *Annu. Rev. Fluid Mech.* **1991**, *23*, 601–639. [CrossRef]
16. Sadarjoen, I.A.; Post, F.H. Detection, quantification, and tracking of vortices using streamline geometry. *Comput. Graph.* **2000**, *24*, 333–341. [CrossRef]
17. Chaigneau, A.; Gizolme, A.; Grados, C. Mesoscale eddies off Peru in altimeter records: Identification algorithms and eddy spatio-temporal patterns. *Prog. Oceanogr.* **2008**, *79*, 106–119. [CrossRef]

18. Abdal, R.; Qin, Y.; Wonka, P. Image2StyleGAN: How to Embed Images Into the StyleGAN Latent Space? In Proceedings of the 2019 IEEE/CVF International Conference on Computer Vision (ICCV), Seoul, Republic of Korea, 27–28 October 2019; pp. 4431–4440. [CrossRef]
19. Karras, T.; Aittala, M.; Laine, S.; Härkönen, E.; Hellsten, J.; Lehtinen, J.; Aila, T. Alias-Free Generative Adversarial Networks. In Proceedings of the 35th Conference on Neural Information Processing Systems (NeurIPS), online, 6–14 December 2021.
20. Rombach, R.; Blattmann, A.; Lorenz, D.; Esser, P.; Ommer, B. High-resolution image synthesis with latent diffusion models. In Proceedings of the IEEE/CVF Conference on Computer Vision and Pattern Recognition, New Orleans, LA, USA, 18–24 June 2022; pp. 10684–10695.
21. Wang, G.; Li, W.; Ourselin, S.; Vercauteren, T. Automatic brain tumor segmentation using cascaded anisotropic convolutional neural networks. In Proceedings of the International MICCAI Brainlesion Workshop, Quebec City, QC, Canada, 14 September 2017; pp. 178–190.
22. Wu, N.; Phang, J.; Park, J.; Shen, Y.; Huang, Z.; Zorin, M.; Jastrzkebski, S.; Févry, T.; Katsnelson, J.; Kim, E.; et al. Deep Neural Networks Improve Radiologists' Performance in Breast Cancer Screening. *IEEE Trans. Med. Imaging* **2019**. [CrossRef]
23. Rasp, S.; Pritchard, M.S.; Gentine, P. Deep learning to represent subgrid processes in climate models. *Proc. Natl. Acad. Sci. USA* **2018**, *115*, 9684–9689. [CrossRef]
24. Beucler, T.; Pritchard, M.; Rasp, S.; Ott, J.; Baldi, P.; Gentine, P. Enforcing Analytic Constraints in Neural Networks Emulating Physical Systems. *Phys. Rev. Lett.* **2021**, *126*. [CrossRef]
25. Keisler, R. Forecasting Global Weather with Graph Neural Networks. *arXiv* **2022**, arXiv:2202.07575.
26. Noh, H.; Hong, S.; Han, B. Learning deconvolution network for semantic segmentation. In Proceedings of the IEEE International Conference on Computer Vision, Santiago, Chile, 7–13 December 2015; pp. 1520–1528.
27. Yurtkulu, S.C.; Şahin, Y.H.; Unal, G. Semantic segmentation with extended DeepLabv3 architecture. In Proceedings of the 2019 27th Signal Processing and Communications Applications Conference (SIU), Sivas, Turkey, 24–26 April 2019; pp. 1–4.
28. Contributors, M. MMSegmentation: OpenMMLab Semantic Segmentation Toolbox and Benchmark. 2020. Available online: https://github.com/open-mmlab/mmsegmentation (accessed on 23 February 2023).
29. Ronneberger, O.; Fischer, P.; Brox, T. U-Net: Convolutional Networks for Biomedical Image Segmentation. In Proceedings of the MICCAI 2015: 18th International Conference, Munich, Germany, 5–9 October 2015. [CrossRef]
30. He, K.; Gkioxari, G.; Dollár, P.; Girshick, R. Mask r-cnn. In Proceedings of the IEEE International Conference on Computer Vision, Venice, Italy, 22–29 October 2017; pp. 2961–2969.
31. Bochkovskiy, A.; Wang, C.Y.; Liao, H.Y.M. Yolov4: Optimal speed and accuracy of object detection. *arXiv* **2020**, arXiv:2004.10934.
32. Peng, C.; Zhang, X.; Yu, G.; Luo, G.; Sun, J. Large Kernel Matters – Improve Semantic Segmentation by Global Convolutional Network. In Proceedings of the IEEE Conference on Computer Vision and Pattern Recognition (CVPR), Honolulu, HI, USA, 21–26 July 2017; pp. 4353–4361.
33. Castellani, M. Identification of eddies from sea surface temperature maps with neural networks. *Int. J. Remote. Sens.* **2006**, *27*, 1601–1618. [CrossRef]
34. Franz, K.; Roscher, R.; Milioto, A.; Wenzel, S.; Kusche, J. Ocean Eddy Identification and Tracking Using Neural Networks. In Proceedings of the IGARSS 2018—2018 IEEE International Geoscience and Remote Sensing Symposium, Valencia, Spain, 23–27 July 2018; pp. 6887–6890. [CrossRef]
35. Lguensat, R.; Sun, M.; Fablet, R.; Tandeo, P.; Mason, E.; Chen, G. EddyNet: A Deep Neural Network For Pixel-Wise Classification of Oceanic Eddies. In Proceedings of the IGARSS 2018—2018 IEEE International Geoscience and Remote Sensing Symposium, Valencia, Spain, 23–27 July 2018; pp. 1764–1767. [CrossRef]
36. Moschos, E.; Schwander, O.; Stegner, A.; Gallinari, P. Deep-SST-Eddies: A Deep Learning Framework to Detect Oceanic Eddies in Sea Surface Temperature Images. In Proceedings of the ICASSP 2020—2020 IEEE International Conference on Acoustics, Speech and Signal Processing (ICASSP), Barcelona, Spain, 4–8 May 2020; pp. 4307–4311. [CrossRef]
37. Duo, Z.; Wang, W.; Wang, H. Oceanic Mesoscale Eddy Detection Method Based on Deep Learning. *Remote. Sens.* **2019**, *11*, 1921. [CrossRef]
38. Xu, G.; Cheng, C.; Yang, W.; Xie, W.; Kong, L.; Hang, R.; Ma, F.; Dong, C.; Yang, J. Oceanic Eddy Identification Using an AI Scheme. *Remote. Sens.* **2019**, *11*, 1349. [CrossRef]
39. Nian, R.; Cai, Y.; Zhang, Z.; He, H.; Wu, J.; Yuan, Q.; Geng, X.; Qian, Y.; Yang, H.; He, B. The Identification and Prediction of Mesoscale Eddy Variation via Memory in Memory With Scheduled Sampling for Sea Level Anomaly. *Front. Mar. Sci.* **2021**, *8*, 1689. [CrossRef]
40. Fan, Z.; Zhong, G.; Li, H. A Feature Fusion Network for Multi-modal Mesoscale Eddy Detection. In Proceedings of the 27th International Conference on Neural Information Processing (ICONIP2020), Bangkok, Thailand, 23–27 November 2020; pp. 51–61.
41. Chen, G.; Han, G.; Yang, X. On the intrinsic shape of oceanic eddies derived from satellite altimetry. *Remote. Sens. Environ.* **2019**, *228*, 75–89. [CrossRef]
42. Zhao, H.; Shi, J.; Qi, X.; Wang, X.; Jia, J. Pyramid Scene Parsing Network. In Proceedings of the IEEE Conference on Computer Vision and Pattern Recognition (CVPR), Honolulu, HI, USA, 21–26 July 2017.
43. Ferry, N.; Parent, L.; Garric, G.; Barnier, B.; Jourdain, N.C. Mercator global Eddy permitting ocean reanalysis GLORYS1V1: Description and results. *Mercat.-Ocean. Q. Newsl.* **2010**, *36*, 15–27.
44. Weiss, K.; Khoshgoftaar, T.M.; Wang, D. A survey of transfer learning. *J. Big Data* **2016**, *3*, 1–40. [CrossRef]

45. Long, J.; Shelhamer, E.; Darrell, T. Fully convolutional networks for semantic segmentation. In Proceedings of the IEEE Conference on Computer Vision and Pattern Recognition, Boston, MA, USA, 7–12 June 2015; pp. 3431–3440.
46. Madec, G.; Bourdallé-Badie, R.; Bouttier, P.A.; Bricaud, C.; Bruciaferri, D.; Calvert, D.; Chanut, J.; Clementi, E.; Coward, A.; Delrosso, D.; et al. NEMO Ocean Engine. 2017. Available online: https://www.nemo-ocean.eu/doc/ (accessed on 23 February 2023).
47. Hersbach, H.; Bell, B.; Berrisford, P.; Hirahara, S.; Horányi, A.; Muñoz-Sabater, J.; Nicolas, J.; Peubey, C.; Radu, R.; Schepers, D.; et al. ERA5 hourly data on single levels from 1979 to present. Copernicus Climate Change Service (C3S) Climate Data Store (CDS). 2018. Available online: https://cds.climate.copernicus.eu/cdsapp#!/dataset/reanalysis-era5-single-levels?tab=overview (accessed on 23 February 2023).
48. Schott, F.A.; Xie, S.P.; McCreary, J.P, Jr. Indian Ocean circulation and climate variability. *Rev. Geophys.* **2009**, *47*. [CrossRef]
49. Wang, H.; McClean, J.L.; Talley, L.D.; Yeager, S. Seasonal cycle and annual reversal of the Somali Current in an eddy-resolving global ocean model. *J. Geophys. Res. Ocean.* **2018**, *123*, 6562–6580. [CrossRef]
50. Fischer, A.S.; Weller, R.A.; Rudnick, D.L.; Eriksen, C.C.; Lee, C.M.; Brink, K.H.; Fox, C.A.; Leben, R.R. Mesoscale eddies, coastal upwelling, and the upper-ocean heat budget in the Arabian Sea. *Deep. Sea Res. Part II Top. Stud. Oceanogr.* **2002**, *49*, 2231–2264. [CrossRef]
51. Sergey, P. Mesoscale Eddies of Arabian Sea: Physical-biological Interactions. *Int. J. Mar. Sci.* **2012**, *2*, 51–56. [CrossRef]
52. Scharffenberg, M.G.; Stammer, D. Seasonal variations of the large-scale geostrophic flow field and eddy kinetic energy inferred from the TOPEX/Poseidon and Jason-1 tandem mission data. *J. Geophys. Res.* **2010**, *115*. [CrossRef]
53. Roullet, G.; Capet, X.; Maze, G. Global interior eddy available potential energy diagnosed from Argo floats. *Geophys. Res. Lett.* **2014**, *41*, 1651–1656. [CrossRef]
54. He, K.; Zhang, X.; Ren, S.; Sun, J. Deep Residual Learning for Image Recognition. In Proceedings of the IEEE Conference on Computer Vision and Pattern Recognition (CVPR), Las Vegas, NV, USA, 27–30 June 2016.
55. Paszke, A.; Gross, S.; Massa, F.; Lerer, A.; Bradbury, J.; Chanan, G.; Killeen, T.; Lin, Z.; Gimelshein, N.; Antiga, L.; et al. PyTorch: An Imperative Style, High-Performance Deep Learning Library. In Proceedings of the Annual Conference on Neural Information Processing Systems 2019 (NeurIPS 2019), Vancouver, BC, Canada, 8–14 December 2019; Volume 32.
56. Wang, K.; Gao, X.; Zhao, Y.; Li, X.; Dou, D.; Xu, C.Z. Pay attention to features, transfer learn faster CNNs. In Proceedings of the International Conference on Learning Representations, New Orleans, LA, USA, 6-9 May 2019.
57. Garcia-Garcia, A.; Orts-Escolano, S.; Oprea, S.; Villena-Martinez, V.; Martinez-Gonzalez, P.; Garcia-Rodriguez, J. A survey on deep learning techniques for image and video semantic segmentation. *Appl. Soft Comput.* **2018**, *70*, 41–65. [CrossRef]
58. Jadon, S. A survey of loss functions for semantic segmentation. In Proceedings of the 2020 IEEE Conference on Computational Intelligence in Bioinformatics and Computational Biology (CIBCB), online, 27–29 October 2020; pp. 1–7.
59. Torralba, A.; Russell, B.C.; Yuen, J. Labelme: Online image annotation and applications. *Proc. IEEE* **2010**, *98*, 1467–1484. [CrossRef]
60. Zhou, Y.; Yang, X.; Zhang, G.; Wang, J.; Liu, Y.; Hou, L.; Jiang, X.; Liu, X.; Yan, J.; Lyu, C.; et al. MMRotate: A Rotated Object Detection Benchmark using PyTorch. In Proceedings of the 30th ACM International Conference on Multimedia, Lisbon, Portugal, 10–14 October 2022.
61. Contributors, M. OpenMMLab Pose Estimation Toolbox and Benchmark. 2020. Available online: https://github.com/open-mmlab/mmpose (accessed on 23 February 2023).
62. Deng, J.; Dong, W.; Socher, R.; Li, L.J.; Li, K.; Fei-Fei, L. ImageNet: A large-scale hierarchical image database. In Proceedings of the 2009 IEEE Conference on Computer Vision and Pattern Recognition, Miami, FL, USA, 20–1 June 2009; pp. 248–255. [CrossRef]
63. Marcel, S.; Rodriguez, Y. Torchvision the Machine-Vision Package of Torch. In Proceedings of the 18th ACM International Conference on Multimedia, Florence, Italy, 25–29 October 2010; pp. 1485–1488.
64. Zhan, P.; Krokos, G.; Guo, D.; Hoteit, I. Three-dimensional signature of the Red Sea eddies and eddy-induced transport. *Geophys. Res. Lett.* **2019**, *46*, 2167–2177. [CrossRef]
65. Yao, F.; Hoteit, I.; Pratt, L.J.; Bower, A.S.; Köhl, A.; Gopalakrishnan, G.; Rivas, D. Seasonal overturning circulation in the Red Sea: 2. Winter circulation. *J. Geophys. Res. Ocean.* **2014**, *119*, 2263–2289. [CrossRef]
66. Yao, F.; Hoteit, I.; Pratt, L.J.; Bower, A.S.; Zhai, P.; Köhl, A.; Gopalakrishnan, G. Seasonal overturning circulation in the Red Sea: 1. Model validation and summer circulation. *J. Geophys. Res. Ocean.* **2014**, *119*, 2238–2262. [CrossRef]
67. Zhan, P.; Gopalakrishnan, G.; Subramanian, A.C.; Guo, D.; Hoteit, I. Sensitivity Studies of the Red Sea Eddies Using Adjoint Method. *J. Geophys. Res. Ocean.* **2018**, *123*, 8329–8345. [CrossRef]
68. Sun, B.; Feng, J.; Saenko, K. Return of Frustratingly Easy Domain Adaptation. In Proceedings of the 30th AAAI Conference on Artificial Intelligence, Phoenix, AR, USA, 12–17 February 2016.
69. Sun, B.; Saenko, K. Deep CORAL: Correlation Alignment for Deep Domain Adaptation. In Proceedings of the 14th European Conference on Computer Vision, Amsterdam, The Netherlands, 8–16 October 2016.

remote sensing

Article

An Effective Task Sampling Strategy Based on Category Generation for Fine-Grained Few-Shot Object Recognition

Shifan Liu, Ailong Ma *, Shaoming Pan and Yanfei Zhong

State Key Laboratory of Information Engineering in Surveying, Mapping and Remote Sensing, Wuhan University, Wuhan 430072, China
* Correspondence: maailong007@whu.edu.cn

Abstract: The recognition of fine-grained objects is crucial for future remote sensing applications, but this task is faced with the few-shot problem due to limited labeled data. In addition, the existing few-shot learning methods do not consider the unique characteristics of remote sensing objects, i.e., the complex backgrounds and the difficulty of extracting fine-grained features, leading to suboptimal performance. In this study, we developed an improved task sampling strategy for few-shot learning that optimizes the target distribution. The proposed approach incorporates broad category information, where each sample is assigned both a broad and fine category label and converts the target task distribution into a fine-grained distribution. This ensures that the model focuses on extracting fine-grained features for the corresponding broad category. We also introduce a category generation method that ensures the same number of fine-grained categories in each task to improve the model accuracy. The experimental results demonstrate that the proposed strategy outperforms the existing object recognition methods. We believe that this strategy has the potential to be applied to fine-grained few-shot object recognition, thus contributing to the development of high-precision remote sensing applications.

Keywords: fine-grained; few-shot; object recognition; sampling strategy

Citation: Liu, S.; Ma, A.; Pan, S.; Zhong, Y. An Effective Task Sampling Strategy Based on Category Generation for Fine-Grained Few-Shot Object Recognition. *Remote Sens.* **2023**, *15*, 1552. https://doi.org/10.3390/rs15061552

Academic Editor: Gwanggil Jeon

Received: 3 February 2023
Revised: 9 March 2023
Accepted: 10 March 2023
Published: 12 March 2023

1. Introduction

The field of remote sensing has seen rapid development in the last few decades, with applications in areas, such as environmental resources, hydrology, and geology. Object recognition, which involves mapping visual features to object classes, is a crucial problem in remote sensing image analysis and has diverse civil applications, such as geographic information system (GIS) mapping, transportation planning, and navigation [1–7]. With the release of more remote sensing datasets and the increasing spatial and spectral resolutions, identifying fine-grained object classes, such as the detailed models of airplanes has now become possible. Fine-grained object recognition in computer vision requires the identification of the subtle differences in the local regions of an object. The current methods of fine-grained object recognition include deep convolutional neural network (DCNN)-based methods, localization/recognition-based methods, network integration based methods, and higher-order feature-coding-based methods. The part-based R-CNNs [8] are a representative approach that use a region-based convolutional neural network (R-CNN) to detect objects and their discriminative parts. Meanwhile, attention mechanisms have been applied to learn models without bounding box labels [9–11]. The network-integration-based methods learn multiple models and then integrate the predictions [12–14]. High-level feature encoding combines feature maps using outer products [15–17], with the bilinear model being the most representative. It is, however, unclear how these methods perform with remote sensing objects, but some studies have attempted to apply them. For example, Sumbul et al. [18] introduced multiple sources of data to address the low inter-class variance, small training set size for rare classes, and large class imbalance, while Aygünes et al. [19,20] used weakly supervised location-based recognition for fine-grained tree recognition.

The fine-grained object recognition task, which involves identifying the types of objects among a large number of closely related subcategories, differs from the traditional object identification tasks commonly studied in the remote sensing literature in at least two ways:

- The difficulty of accumulating a large number of similar subcategory samples can greatly limit the size of the training set for certain subcategories;
- The class imbalance can lead the traditional supervised formulations to prefer fitting to classes that occur frequently, while ignoring classes with a limited sample size.

In the data-driven models, these differences can create recognition difficulties, especially when there are limited labeled data, which is known as the few-shot learning (FSL) problem. This is a significant issue in the fine-grained object recognition task. The current FSL methods adapt models learned from auxiliary datasets with large amounts of data to new data with limited labeled data. During training, the concept of a "task" is introduced, where a small set of class-balanced samples are iteratively sampled from the auxiliary dataset, and the model learns the task distribution. New data with limited annotation can be considered a few-shot task and adapted using the task distribution learned from the auxiliary dataset.

Most of the current FSL methods can be divided into three main categories: (1) generation-based methods; (2) initialization-based methods; and (3) metric-based learning methods. The generation-based methods aim to learn a generator that can augment the data for new classes [21–23], while the initialization-based approaches [24] focus on learning a good initialization for the model parameters. The metric-based methods involve measuring the distance between a query set and a support set in the embedding space to determine the predicted categories [25,26]. For the FSL problem of remote sensing object recognition, there is not much current research. However, Fu et al. [27] did introduce an initialization-based few-shot approach to synthetic aperture radar (SAR) object recognition, while noting that the few-shot classification of remote sensing objects is more difficult than that of natural images. Some works have improved the effect of FSL on SAR data by modifying the network and metric functions [28,29]. There have also been some studies that have focused on addressing the few-shot transfer learning problem in SAR data [30,31].

However, the above FSL methods are not able to deal with remote sensing objects very effectively. The diverse sources of remote sensing data make the learned distribution inaccurate, and the complex background of remote sensing objects, the high image noise, and the small size of the objects make the features more difficult to detect with the model, which leads to a reduction of the FSL effect.

The current few-shot methods rely on a meta-learning training paradigm, i.e., learning the task distributions by learning them on sampled tasks. Factors, such as the source of the data and the way the samples are composed in the task, affect the task distribution, while the latter can be considered to be determined by the sampling strategy. Naturally, in order to make the learned task distribution closer to the real task distribution, we start from the perspective of improving the effectiveness of the sampling strategy. We therefore propose an improved sampling strategy for FSL tasks:

- We improve the sampling strategy for the fine-grained object recognition task. The task sampling strategy is improved so that each sampled task contains only subcategory samples belonging to the same broad category, e.g., fine-grained airplane samples belonging to the airplane category. This sampling strategy is used to keep each task-specific model focused on mining fine-grained features and solving the problem of the difficult extraction of remote sensing object features.
- The category synthesis method keeps the sampling task length consistent. For each task sampling operation, for the case where the number of fine-grained categories is less than the number of sampling operations, a category synthesis method is proposed to fill the gaps, to maintain the same task length for each sampling operation, and to ensure that an accurate task distribution is learned instead of a mixed distribution.

2. Background

2.1. Task Sampler

Meta-learning is a popular paradigm for few-shot learning methods. It involves sampling tasks $\mathcal{T}$ on auxiliary datasets $\mathcal{D}_{aux}$, adapting models on the tasks, and learning task distributions $p(\mathcal{T})$. The learned task knowledge $p(\mathcal{T})$ is then transferred to new tasks on few-shot datasets $\mathcal{D}_{new}$. To ensure that it is task knowledge that is transferred, the new data categories and the categories in the auxiliary dataset do not overlap. Tasks are sampled through category-balanced sampling of samples, where each task contains N categories, and each category has $K_{support} + K_{query}$ samples. Since $K_{support}$ and K_{query} are usually small values, each task $\mathcal{T}_i$ is a small-size set. The optimization objective for few-shot learning is shown in the following equation:

$$\arg\min_{\theta} \mathbb{E}_{\mathcal{T}\sim p(\mathcal{T})}\mathcal{L}(\theta, \mathcal{T}) \rightarrow \arg\min_{\theta} \sum_{i}^{n} \frac{1}{n}\mathcal{L}(\theta, \mathcal{T}_i)$$

where θ is model parameters, $\mathcal{L}$ is the loss function, and $p(\mathcal{T})$ is a task distribution.

To solve the few-shot problem using the meta-learning paradigm, the new task should have an identical distribution to the task distribution on the auxiliary data, i.e., $\mathcal{T}_{new} \sim p(\mathcal{T})$. In remote sensing, a task distribution is related to the data source region $\mathcal{R}$, the number of sampling categories N, and the number of samples per category $K_{support}$, K_{query}. Therefore, to ensure the effectiveness of the new task, at least the number of categories and the number of samples per category should be consistent with the auxiliary tasks.

2.2. Few-Shot Learning on Tasks

In FSL, the model learns by storing information about the task distribution among the meta-parameters. This stored information is called meta-information, and it can include the entire model parameters, a portion of the model parameters, or a specific hyperparameter. The optimization direction of the model is guided by the query set in the task during the training process. The overall FSL process is shown in Figure 1.

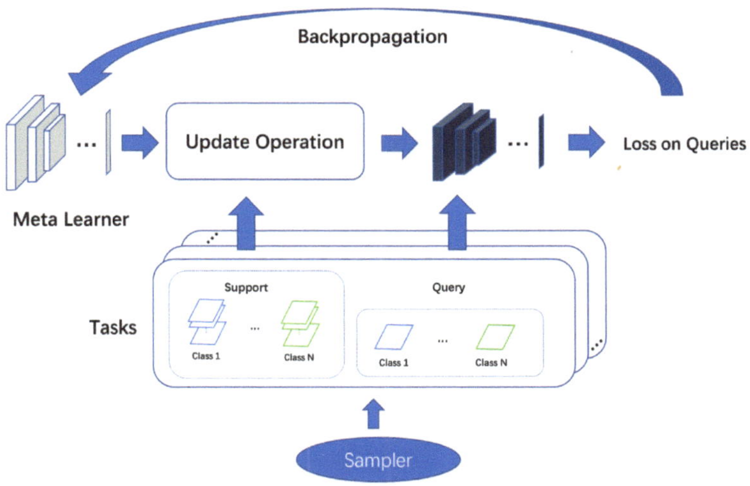

Figure 1. The main process of FSL involves adjusting the meta-parameters based on the loss on the query set using the task-specific parameters, which are fine-tuned from the meta-parameters through update operations on the support set generated by the task sampler. The update operations and loss calculation can vary depending on the specific FSL method being used.

3. Methodology

The overall system for fine-grained object recognition in remote sensing applications is illustrated in Figure 2, showing the input and output schematic. In the meta-learning step, the initialized model is transformed into a meta-model by learning on the auxiliary dataset through the tasks generated by the task sampler. The specific learning process is described in Figure 1. In the inference phase, we sample the few-shot task, fine-tune the meta-model on the support set, and finally output the prediction results for the query set.

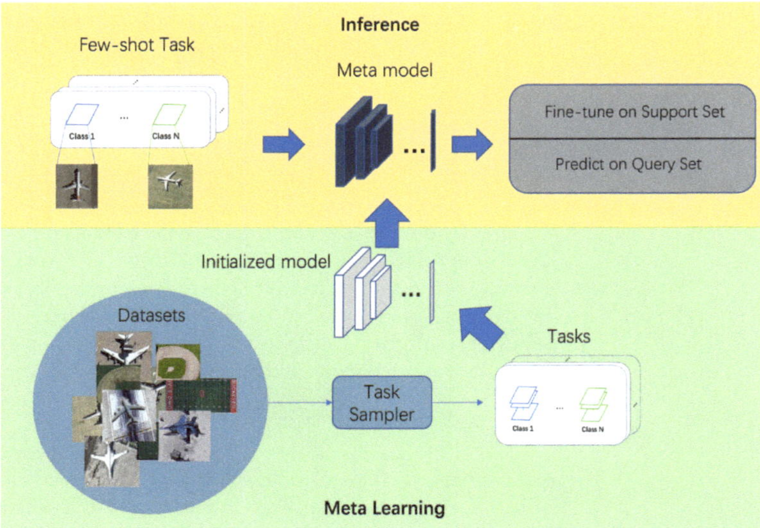

Figure 2. Schematic diagram of the input and output of the whole system. The generation of the meta-model and the subsequent output on the new data are inseparable from the task sampler.

3.1. Problem Formulation

3.1.1. Fine-Grained Problem

In Section 2.1, we discussed the task sampling paradigm for FSL training. However, this sampling strategy does not have a way to solve the problem of the difficult extraction of fine-grained remote sensing object features because the sampled tasks are not focused on mining fine-grained features, and in each task sampling operation, the sampled categories are:

$$\mathcal{C}^N = \{C_i^N | 1 \le i \le N, C_i^N \in \mathcal{C}_{aux}\}$$

We explicitly define each category C_i^N to belong to some broad category S_i^N and have $C_i^N \in S_i^N$. It is seen that $S_i^N = S_j^N$ does not always hold, and under extreme conditions there are even $S_i^N \cap S_j^N = \emptyset$ for any i, j, which means that the task is concerned with distinguishing differences between S_i^N, S_j^N rather than fine-grained C_i^N and C_j^N.

Further, such a sampling approach has a bias for some S_i^N, since

$$p(C_i^N \in S_i^N | \mathcal{T}) = \frac{||\{C_i | C_i \in S_i^N\}||}{||\mathcal{C}_{aux}||}$$

for each sampled category C_i^N from $\mathcal{C}_{aux}$, and the bias on S_i^N depends on the molecular $||\{C_i | C_i \in S_i^N\}||$.

Although the original intention of using this sampling paradigm is for the categories to be balanced in each task, each task will still have a bias toward certain broad categories, which is detrimental to the model learning of a generic large few-shot model.

3.1.2. Validity Issues of Task Distribution Learning

In Sections 2.1 and 2.2, we showed that the purpose of the model optimization is to eventually learn a task distribution. For the remote sensing recognition task, due to the diverse sample sources, the number of certain categories may not reach the required number of samples during the task sampling process, resulting in such a task being considered as coming from a different task distribution, and thus the learned distribution is a mixed distribution, which brings inaccuracy to the model learning.

3.2. *Improved Task Sampler*

In order to solve the problem of the difficult extraction of fine-grained features in the few samples of remote sensing imagery, we improve the task sampling strategy to make each task change from a few-shot task to a fine-grained few-shot task and let the model learn fine-grained few-shot features. Specifically, the sample categories in the unimproved task are from different broad categories, and such a task is called a recognition task. However, such a task cannot focus on fine-grained features because the two samples are from different broad categories which differ significantly, such as aircraft and ships(as discussed in Section 3.1.1).

We improve the sampling strategy by including in each task only samples sampled from a particular large class so that the model will focus on the intra-class differences within the classes in the task, i.e., the fine-grained features. Such improved tasks, which we call fine-grained tasks, end up with a fine-grained task distribution that the model learns. It is worth noting that introducing additional broad category information does not require additional annotation of the original dataset. In fact, for remote sensing data, although there are many fine-grained categories, the broad categories these fine-grained categories belong to are indeed very limited. We do not need additional annotation on the data. We only need to add appropriate judgments in the programming.

At the same time, a category generation technique is designed for the situation of this sampling strategy having an insufficient number of categories to be sampled. We use the idea of mix-up for category generation by sampling the beta distribution to obtain a factor λ, where, for any two category samples i, j considered to generate a new category, this new category sample is obtained by pixel-level fusion as $\lambda i + (1 - \lambda)j$. Specifically, lambda belongs to the distribution:

$$f(\lambda) = \frac{\lambda^{-0.5}(1-\lambda)^{-0.5}}{\int_0^1 u^{-0.5}(1-u)^{-0.5}du} = Beta(\lambda; 0.5, 0.5)$$

We set the λ to a random variable instead of a static parameter value in order to avoid duplication of the resulting new category. For example, if there are only two categories and five categories are sampled, if lambda is a constant parameter, then the new categories generated are bound to be duplicated, so there is no guarantee that the length of the picked task is five. We chose the beta distribution because the distribution is symmetric about 0.5 when the two parameters of the beta distribution agree, and the range of the random variables is 0 to 1. Such a distribution is suitable for fusing pairs of samples.

3.3. *Framework*

The improvement methods discussed previously are integrated into a unified process. The overall flow is shown in Algorithm 1.

Algorithm 1 Fine-Grained Task Sampling Process

Input:

The auxiliary dataset $\mathcal{D}_{aux}$;
Collection of categories in the auxiliary dataset $\mathcal{C}_{aux}$;
Collection of broad categories in the auxiliary dataset $\mathcal{S}_{aux}$;
Number of categories in a task N;

Number of samples per category in a task $K_{support}, K_{query}$.
Output:
Tasks sampled from auxiliary datasets $\mathcal{T}$.
for iteration i = 1,2,3... **do**
Sample a broad category from $\mathcal{S}_{aux}$, denoted as S_i
Calculate $R_i = N \bmod \|\{C_i|C_i \in S_i\}\|$, $N_i = N \mid \|\{C_i|C_i \in S_i\}\|$
for j=1,2,3,..N_i **do**
Sample two category C_j^1, C_j^2 where $C_j^1, C_j^2 \in S_i$ and $C_j^1, C_j^2 \in \mathcal{C}_{aux}$
Sample $K_{support}$ samples belonging to C_j^1, C_j^2 from $\mathcal{D}_{aux}$, denoted as S_j^1, S_j^2
Sample K_{query} samples belonging to C_j^1, C_j^2 from $\mathcal{D}_{aux}$, denoted as Q_j^1, Q_j^2
Sample λ_j from Beta Distribution $\beta(.2, .2)$
For each image i, j in S_j^1, S_j^2 fused as $\lambda_j i + (1 - \lambda_j) j$, denoted as S_j
For each image i, j in Q_j^1, Q_j^2 fused as $\lambda_j i + (1 - \lambda_j) j$, denoted as Q_j
Define $\mathcal{T}_{ij} = (S_j, Q_j)$
end for
for $C_j \in S_i$ **do**
Sample $K_{support}$ samples belonging to C_j from $\mathcal{D}_{aux}$, denoted as S_j
Sample K_{query} samples belonging to C_j from $\mathcal{D}_{aux}$, denoted as Q_j
Define $\mathcal{T}_{ij}^r = (S_j, Q_j)$
end for
Define $\mathcal{T}_i = \{\mathcal{T}_{ij}|j \leq N\}$, $\mathcal{T}_i^r = \{\mathcal{T}_{ij}^r|C_j \in S_i\}$
end for
return $\mathcal{T} = \{\mathcal{T}_i, \mathcal{T}_i^r\}$

We consider the fusion sample $\lambda_j i + (1 - \lambda_j) j$ a brand new category sample. Since λ is a random variable, each fusion sample category is considered as different. As can be seen from the algorithm flow, the category sampled by each task belongs to the broad category S_i. Therefore, each task focuses on the small differences in the fine-grained categories of broad category S_i, forcing the model to mine fine-grained features. This sampling strategy, which further specifies the tasks as fine-grained few-shot tasks, provides an artificial prior to improve the mining ability for fine-grained features. Figure 3 briefly illustrates the difference between a task sampled by the proposed approach and a task sampled by the general policy.

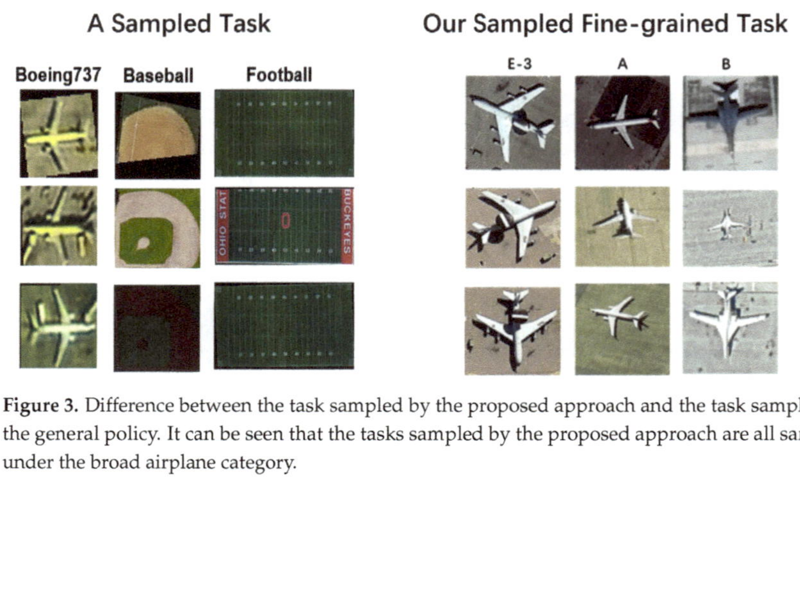

Figure 3. Difference between the task sampled by the proposed approach and the task sampled by the general policy. It can be seen that the tasks sampled by the proposed approach are all samples under the broad airplane category.

4. Datasets

4.1. FAIR1M Fine-Grained Dataset

The different models were trained and evaluated on the FAIR1M [32] dataset.There are three reasons for adopting this dataset:

- In terms of quantity, it is much larger than the other existing object datasets, and it can be divided into auxiliary datasets;
- It provides more rich fine-grained category information;
- It provides better image quality due to careful data cleaning procedures.

The FAIR1M dataset is unique in its focus on fine-grained object recognition in high-resolution remote sensing imagery. The dataset contains more than 15,000 images from over 100 locations worldwide, covering 37 fine-grained categories for object detection. The categories include ten types of airplanes, eight types of ships, nine types of vehicles, four types of courts, and three types of roads. The airplane categories consist of the most common types in civil aviation, such as *Boeing737* and *Airbus A320*. The ship categories are defined by their functions, such as *Passenger Ships* and *Engineering Ships*. The vehicle categories are also defined by their functions, such as *Small Cars* and *Excavators*. There are also categories for courts and roads. The dataset also includes categories for objects that do not belong to the specific types, labeled as *Other-Airplane*, *Other-Ship*, and *Other-Vehicle*. The distribution of instances per category reflects the authenticity and challenge of the dataset.

4.2. MTARSI Datasets

We also evaluated the models on the muti-type aircraft of remote sensing images (MTARSI) dataset [33]. There are two reasons for adopting this dataset:

- The dataset is an aircraft dataset containing 20 fine-grained aircraft categories, which is convenient for evaluating the effect of a model trained on FAIR1M;
- The objects are carefully selected from many airports around the world and are calibrated with correct and high-quality category labeling.

The MTARSI dataset is a valuable resource for researchers working on aircraft recognition in remote sensing images, as it offers a wide range of images with diverse aircraft types and environmental conditions. With careful labeling and data augmentation, the dataset provides an excellent opportunity for developing and evaluating algorithms for fine-grained object recognition in high-resolution remote sensing imagery.

4.3. Object Cropping and Subcategory Division

4.3.1. FAIR1M

Although the dataset is intended for object detection, we carefully cropped the objects according to the annotations, and some examples can be seen in Figure 4. The modified dataset has 5 categories—*Ship* , *Vehicle*, *Airplane*, *Court*, and *Road*—and 38 subcategories, which are shown in Figure 5 and Table 1 The samples of each subcategory show a long-tail distribution in Figure 5. and the cropped objects are suitable for forming a few-shot and fine-grained dataset. We formed the subcategories with a high number of samples into an auxiliary dataset for the meta-training phase, and the remaining samples were incorporated into a few-shot dataset for the testing phase. The detailed division is shown in Tables 2 and 3.

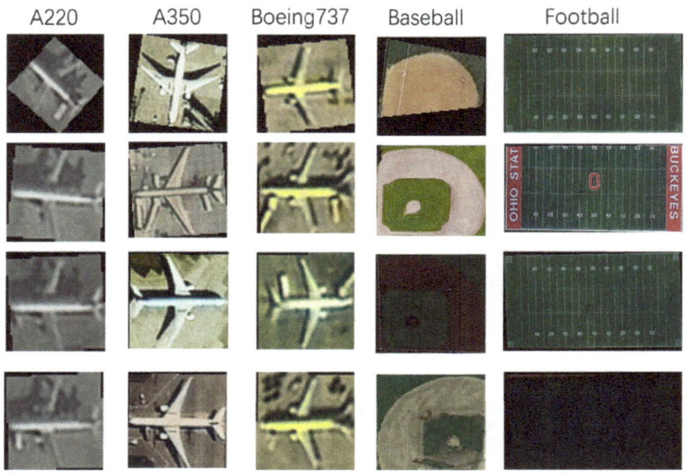

Figure 4. Cropped instances from the FAIR1M dataset.

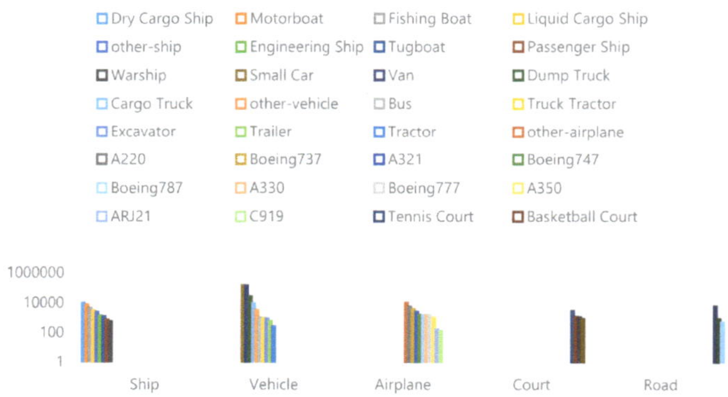

Figure 5. The distribution of the number of cropped subcategories per category.

Table 1. The 5 categories (*Ship*, *Vehicle*, *Airplane*, *Court*, and *Road*) and 38 subcategories in the FAIR1M dataset.

Category	Subcatogory
Ship	*Motorboat, Fishing Boat, Liquid Cargo Ship, Other-Ship, Dry Cargo Ship* *Passenger Ship, Tugboat, Warship, Engineering Ship*
Vehicle	*Small Car,Van* *Bus, Cargo Truck, Dump Truck, other-vehicle, Trailer, Tractor, Excavator, Truck Tractor*
Airplane	*Boeing737, A220, Other-Airplane* *Boeing747, Boeing777, Boeing787, ARJ21, C919, A321, A330, A350*
Court	*Baseball Field, Basketball Court, Tennis Court, Basketball Court* *Football Field*
Road	*Intersection* *Roundabout, Bridge*

Table 2. The subcategories in this table form the auxiliary dataset for training.

Category	Base Subcatogory
Ship	*Motorboat, Fishing Boat, Liquid Cargo Ship, other-ship, Dry Cargo Ship*
Vehicle	*Small Car,Van*
Airplane	*Boeing737, A220, Other-Airplane*
Court	*Baseball Field, Basketball Court, Tennis Court, Basketball Court*
Road	*Intersection*

Table 3. The subcategories in this table form the few-shot dataset for testing.

Category	Noval Subcatogory
Ship	*Passenger Ship, Tugboat, Warship, Engineering Ship*
Vehicle	*Bus, Cargo Truck, Dump Truck, other-vehicle, Trailer, Tractor, Excavator, Truck Tractor*
Airplane	*Boeing747, Boeing777, Boeing787, ARJ21, C919, A321, A330, A350*
Court	*Football Field*
Road	*Roundabout, Bridge*

4.3.2. MTARSI

Since the MTARSI dataset is released for object recognition, each sample is an image containing an object and a background, and no additional processing of the data is required. We apply this dataset directly in the experiments without any modification.

5. Experiments

5.1. General Overview of the Process

Broadly speaking, the experimental procedure was as follows:

- Determine base subcategories and novel subcategories (not overlapping);
- Use the FSL methods to learn from the base subcategories (#epoch: early stopping);
- Fine-tune the obtained model to the novel subcategories with the limited labeled data (only one task).

5.1.1. Input Data

The original images were of various sizes, so they were resized to a fixed size of 224 × 224 pixels for training.

5.1.2. Training

All the models were trained with the Adam optimizer in the inner gradient update and stochastic gradient descent (SGD) in the outer gradient update on a single GPU. The learning rate was set to 10^{-3} for the Adam optimizer and 1 for SGD. We used a meta-learning paradigm for the training, i.e., learning on a sampling task. Overall, the experiments were set to simulate the few sample case by setting the number of samples per class k in a task to 1, 5, and 10. The FSL methods were then applied to learn from the tasks.

5.1.3. Fine-Tuning

Only one task was sampled from the few-shot dataset, and it had the same parameters (N and $K_{support}$) as the task sampled from the auxiliary dataset. The model parameters were fine-tuned on the support set for that task, and the query set was used to evaluate the performance. We sampled 20 times and evaluated the average performance of the model on the 20 tasks. Note that to reduce the variance, we increased the number of K_{query} to 50.

5.1.4. Evaluation Metrics

In this paper, to evaluate the performance of the fine-grained visual classification (FGVC) algorithms, we use the classification accuracy as the metric for this task. $N_{correct}$ and N_{all} denote the number of correctly predicted instances and total instances, respectively. The definition of the classification accuracy is as follows:

$$Acc = \frac{N_{correct}}{N_{all}}$$

5.2. Network Structure

We used the most common network structure for the FSL methods, where the network consisted of four blocks in series. Each block contained a 3 × 3 convolutional layer with 32 channels, a batch normalization layer, a maximum pooling layer of width 2, and a rectified linear unit (ReLU) activation layer.

5.3. Comparing Methods

We used first-order model-agnostic meta-learning(FOML) [24], Reptile [34], and Prototypical network(Proto.) [25] as comparison methods. The first two of these methods are initialization-based methods, and the last one is a metric-learning-based method. The proposed method is based on the Prototypical network with the replacement of the proposed task sampling strategy. All the methods used the same network structure described in the previous section.

5.4. Experiments Results

5.4.1. Experiments on FAIR1M

We performed FSL on the base category divisions in Table 2 and tested on the novel categories in Table 3 to obtain the following results, which are listed in Table 4.

Table 4. Only one task was sampled from the new classes. The model was then tuned on the support set and the precision was tested on the query set. The sampling was conducted 50 times, and the average performance accuracy on these tasks is reported.

ALL	5 Way 1 Shot	5 Way 5 Shot	5 Way 10 Shot
FOML	34.11% ± 9.32%	51.96% ± 10.11%	62.74% ± 10.08%
Reptile	37.24% ± 7.78%	54.32% ± 8.66%	63.36% ± 10.07%
Proto.	37.81% ± 7.67%	59.33% ± 9.45%	65.34% ± 9.13%
Ours	**38.30% ± 8.05%**	**61.41% ± 8.32%**	**66.73% ± 8.98%**

It can be seen from Table 4 that even if the base and novel categories do not overlap, a model trained on the base categories and fine-tuned on a small amount of labeled data from the novel categories can accurately predict the remaining data, demonstrating the effectiveness of the few-shot methods. Consistent with the academic research findings, the current FSL methods based on metric learning have a higher accuracy. Furthermore, the accuracy of each method improves as the number of shots increases. It is quite clear that as the number of samples increases, the distribution learned by the model becomes more accurate. The proposed method, which uses an improved fine-grained task sampling strategy, achieves a higher accuracy and larger performance improvement. Notably, the proposed task sampling strategy was consistent with the other methods during testing, suggesting that learning on the base categories with this strategy can lead to better few-shot results.

Furthermore, in the novel categories, we singled out broad categories for the model tuning and testing, and we picked out the *Aircraft, Ship,* and *Vehicle* categories. The results of these experiments are listed in Tables 5–7. As can be seen, compared to the experimental results in Table 4 obtained on all the categories, the proposed approach shows an improvement, but not by much. Our analysis suggests that when a single broad category is picked for the testing, it is a simple case, such as the large category of aircraft, where the FSL model has a strong generalization ability due to the absence of interference from samples belonging to other broad categories, and the model is fine-tuned to have a higher accuracy, narrowing the gap with the proposed method. However, in most cases, the proposed method still obtains a higher accuracy, which is due to the better fine-grained task distribution learned on the auxiliary dataset. It can therefore be concluded that the key to the effectiveness of FSL lies in the efficiency of learning on the auxiliary dataset.

Table 5. The broad category of ship was picked from the base categories for the experiment, and the broad category of ship from the novel categories was used for the testing.

Ship	5 Way 1 Shot	5 Way 5 Shot	5 Way 10 Shot
FOML	40.34% ± 8.70%	52.26% ± 5.61%	58.49% ± 4.68%
Reptile	41.49% ± 5.98%	55.18% ± 4.96%	58.62% ± 3.45%
Proto.	43.64% ± 5.63%	56.63% ± 4.43%	60.28% ± 3.72%
Ours	**43.98% ± 6.54%**	**56.82% ± 4.18%**	**61.12% ± 3.70%**

Table 6. The broad category of vehicle was picked from the base categories for the experiment, and the broad category of vehicle from the novel categories was used for the testing.

Vehicle	5 Way 1 Shot	5 Way 5 Shot	5 Way 10 Shot
FOML	27.34% ± 5.52%	36.87% ± 5.44%	41.25% ± 5.07%
Reptile	26.45% ± 4.80%	35.54% ± 4.92%	42.38% ± 5.19%
Proto.	29.78% ± 6.06%	42.55% ± 5.62%	47.22% ± 4.99%
Ours	**30.02% ± 6.48%**	41.34% ± 5.57%	46.38% ± 5.29%

Table 7. The broad category of airplane was picked from the base category for the experiment, and the broad category of airplane from the novel categories was used for the testing.

Airplane	5 Way 1 Shot	5 Way 5 Shot	5 Way 10 Shot
FOML	22.33% ± 2.30%	35.68% ± 6.20%	45.39% ± 8.38%
Reptile	27.29% ± 3.83%	37.90% ± 5.21%	46.36% ± 8.23%
Proto.	27.75% ± 5.07%	40.68% ± 6.79%	47.02% ± 8.17%
Ours	**28.16% ± 5.23%**	**41.64% ± 6.53%**	**47.67% ± 7.55%**

5.4.2. Experiments on MTARSI

FSL focuses on learning the task distribution instead of the sample distribution, so it does not require the same classification system for the auxiliary and new datasets. To analyze FSL on a remote sensing dataset, we applied the model trained on the FAIR1M dataset to the MTARSI dataset. All the categories of the MTARSI dataset were used for the testing, with one task sampled for the model tuning and 50 tasks sampled for testing on the remaining data. The results are listed in Table 8.

As can be seen from the results, the effect of the aircraft recognition performance on the FAIR1M dataset (Table 7) is similar to the performance on the MTARSI dataset. Furthermore, the accuracy is about the same, implying that the final performance effect difference does not come from the sample difference between the sampled MTARSI and FAIR1M datasets but depends on the learned task distribution on the auxiliary dataset. The good news is that this implies that the model trained on the auxiliary dataset is highly generalizable, giving the application a high migration adaptation capability.

Table 8. Only one task was sampled from the new classes. The model was then tuned on the support set, and the precision was tested on the query set. The sampling was conducted 50 times, and the average performance accuracy on these tasks is reported.

ALL(Airplane)	5 Way 1 Shot	5 Way 5 Shot	5 Way 10 Shot
FOML	25.30% ± 4.88%	37.22% ± 5.63%	46.06% ± 5.90%
Reptile	27.54% ± 4.25%	40.86% ± 6.35%	48.23% ± 6.36%
Proto.	28.97% ± 5.10%	42.34% ± 6.16%	46.56% ± 6.70%
Ours	**29.06% ± 5.27%**	**44.95% ± 6.77%**	**48.69% ± 6.70%**

6. Conclusions

The remote sensing fine-grained object recognition task has a few-shot problem that is difficult to get around, i.e., the lack of fine-grained object data. Although some few-shot methods have been proposed to address learning from a small number of data samples, these methods have not been adapted to remote sensing objects, where the object context is complex and fine-grained features are difficult to extract. In this study, we first investigated the FSL methods and found that only a small amount of work has been conducted in the field of remote sensing. This inspired us to design a few-shot approach for remote sensing tasks. In this paper, we reviewed the general paradigm of training FSL methods on sampled tasks. We then affirmed the paradigm's usefulness for FSL, but pointed out its shortcomings for remote sensing data. The problem is that the task has a preference for certain broad categories, the task may not be fine-grained, and the model has difficulty extracting the already hard-to-extract fine-grained features of remote sensing imagery, which can adversely affect the model learning. To solve this problem, we proposed an improved task sampling strategy: the fine-grained task sampling strategy. The fine-grained task sampling strategy ensures that the samples in each sampled task come from a certain broad category, which ensures that the model focuses on the small differences in the fine-grained features on that task. This reinforces the model to learn fine-grained features. We tested the effectiveness of this sampling strategy on the FAIR1M and MTARSI datasets. We found that the FSL method can achieve a large accuracy improvement using the improved sampling strategy. In addition, from the performance effect on the two datasets, we pointed out that the performance of FSL depends on the learning effect on the auxiliary datasets. In our future work, we will attempt to further improve the existing network structure to make it more suitable for remote sensing tasks. We will also bring the few-shot approach to other tasks in remote sensing, such as remote sensing object detection.

Author Contributions: S.L.: methodology, software, validation, formal analysis, investigation, data curation, writing—original draft, visualization; A.M.: conceptualization, writing—review and editing, project administration, supervision; S.P.: supervision; Y.Z.: supervision, funding acquisition. All authors have read and agreed to the published version of the manuscript

Funding: This research received no external funding.

Data Availability Statement: All the data come from public datasets, and the data used in the paper can be downloaded from https://www.gaofen-challenge.com/benchmark accessed on 22 January 2022 and https://zenodo.org/record/3464319#.Y9pg6nBBxEY accessed on 25 January 2022.

Acknowledgments: The authors would like to thank the Editor, Associate Editor, and anonymous reviewers for their helpful comments and suggestions that improved this article. This work was supported in part by the National Key Research and Development Program of China under Grant No. 2022YFB3903404, in part by the National Natural Science Foundation of China under Grant No. 42171336

Conflicts of Interest: The authors declare no conflict of interest.

References

1. Zhang, L.; Zhang, L.; Du, B.J.I.G.; Magazine, R.S. Deep learning for remote sensing data: A technical tutorial on the state of the art. *IEEE Geosci. Remote. Sens. Mag.* **2016**, *4*, 22–40.
2. Fatima, S.A.; Kumar, A.; Pratap, A.; Raoof, S.S. Object recognition and detection in remote sensing images: A comparative study. In Proceedings of the 2020 International Conference on Artificial Intelligence and Signal Processing (AISP), Amaravati, India, 10–12 January 2020; pp. 1–5.
3. Jiang, B.; Li, X.; Yin, L.; Yue, W.; Wang, S. Object recognition in remote sensing images using combined deep features. In Proceedings of the 2019 IEEE 3rd Information Technology, Networking, Electronic and Automation Control Conference (ITNEC), Chengdu, China, 15–17 March 2019; pp. 606–610.
4. Yang, C.; Dong, Y.; Du, B.; Zhang, L. Attention-Based Dynamic Alignment and Dynamic Distribution Adaptation for Remote Sensing Cross-Domain Scene Classification. *IEEE Trans. Geosci. Remot. Sens.* **2022**, *60*, 1–13.
5. Chen, W.; Ouyang, S.; Yang, J.; Li, X.; Zhou, G.; Wang, L. JAGAN: A framework for complex land cover classification using Gaofen-5 AHSI images. *IEEE J. Sel. Top. Appl. Earth Obs. Remot. Sens.* **2022**, *15*, 1591–1603. [CrossRef]
6. Chen, W.; Ouyang, S.; Tong, W.; Li, X.; Zheng, X.; Wang, L. GCSANet: A global context spatial attention deep learning network for remote sensing scene classification. *IEEE J. Sel. Top. Appl. Earth Obs. Remot. Sens.* **2022**, *15*, 1150–1162. [CrossRef]
7. Liu, Q.; Dong, Y.; Zhang, Y.; Luo, H. A Fast Dynamic Graph Convolutional Network and CNN Parallel Network for Hyperspectral Image Classification. *IEEE Trans. Geosci. Remot. Sens.* **2022**, *60*, 1–15. [CrossRef]
8. Zhang, N.; Donahue, J.; Girshick, R.; Darrell, T. Part-based R-CNNs for fine-grained category detection. In *Proceedings of the European Conference on Computer Vision, Zurich, Switzerland, 6–12 September 2014*; Springer: Berlin/Heidelberg, Germany, 2014; pp. 834–849.
9. Sermanet, P.; Frome, A.; Real, E. Attention for fine-grained categorization. *arXiv* **2014**. [CrossRef]
10. Sun, M.; Yuan, Y.; Zhou, F.; Ding, E. Multi-attention multi-class constraint for fine-grained image recognition. In Proceedings of the European Conference on Computer Vision (ECCV), Glasgow, UK, 23–28 August 2020; pp. 805–821.
11. Oliveau, Q.; Sahbi, H. Learning attribute representations for remote sensing ship category classification. *IEEE J. Sel. Top. Appl. Earth Obs. Remot. Sens.* **2017**, *10*, 2830–2840. [CrossRef]
12. Ge, Z.; Bewley, A.; McCool, C.; Corke, P.; Upcroft, B.; Sanderson, C. Fine-grained classification via mixture of deep convolutional neural networks. In Proceedings of the 2016 IEEE Winter Conference on Applications of Computer Vision (WACV), Lake Placid, NY, USA, 7–10 March 2016; pp. 1–6.
13. Ge, Z.; McCool, C.; Sanderson, C.; Corke, P. Subset feature learning for fine-grained category classification. In Proceedings of the IEEE Conference on Computer Vision and Pattern Recognition Workshops, Virtual, 19–25 June 2021; pp. 46–52.
14. Wang, D.; Shen, Z.; Shao, J.; Zhang, W.; Xue, X.; Zhang, Z. Multiple granularity descriptors for fine-grained categorization. In Proceedings of the IEEE International Conference on Computer Vision, Cambridge, MA, USA, 20–23 June 1995; pp. 2399–2406.
15. Zhang, X.; Zhou, F.; Lin, Y.; Zhang, S. Embedding label structures for fine-grained feature representation. In Proceedings of the IEEE Conference on Computer Vision and Pattern Recognition, Miami, FL, USA, 20–25 June 2009; pp. 1114–1123.
16. Kong, S.; Fowlkes, C. Low-rank bilinear pooling for fine-grained classification. In Proceedings of the IEEE Conference on Computer Vision and Pattern Recognition, Miami, FL, USA, 20–25 June 2009; pp. 365–374.
17. Yu, C.; Zhao, X.; Zheng, Q.; Zhang, P.; You, X. Hierarchical bilinear pooling for fine-grained visual recognition. In Proceedings of the European Conference on Computer Vision (ECCV), Glasgow, UK, 23–28 August 2020; pp. 574–589.
18. Sumbul, G.; Cinbis, R.G.; Aksoy, S.; Sensing, R. Multisource region attention network for fine-grained object recognition in remote sensing imagery. *IEEE Trans. Geosci. Remot. Sens.* **2019**, *57*, 4929–4937. [CrossRef]
19. Aygüneş, B.; Aksoy, S.; Cinbiş, R.G. Weakly supervised deep convolutional networks for fine-grained object recognition in multispectral images. In Proceedings of the IGARSS 2019–2019 IEEE International Geoscience and Remote Sensing Symposium, Yokohama, Japan, 28 July–2 August 2019; pp. 1478–1481.
20. Aygunes, B.; Cinbis, R.G.; Aksoy, S.J.I.J.O.P.; Sensing, R. Weakly supervised instance attention for multisource fine-grained object recognition with an application to tree species classification. *Isprs J. Photogramm. Remot. Sens.* **2021**, *176*, 262–274. [CrossRef]
21. Schwartz, E.; Karlinsky, L.; Shtok, J.; Harary, S.; Marder, M.; Feris, R.; Kumar, A.; Giryes, R.; Bronstein, A.M. Delta-encoder: An effective sample synthesis method for few-shot object recognition. *arXiv* **2018**. . [CrossRef]
22. Gao, H.; Shou, Z.; Zareian, A.; Zhang, H.; Chang, S.F. Low-shot learning via covariance-preserving adversarial augmentation networks. In Proceedings of the NIPS'18: Proceedings of the 32nd International Conference on Neural Information Processing Systems, Red Hook, NY, USA, 3–8 December 2018.
23. Pfister, T.; Charles, J.; Zisserman, A. Domain-adaptive discriminative one-shot learning of gestures. In *Proceedings of the European Conference on Computer Vision, Glasgow, UK, 23–28 August 2020*; Springer: Berlin/Heidelberg, Germany, 2020; pp. 814–829.
24. Finn, C.; Abbeel, P.; Levine, S. Model-agnostic meta-learning for fast adaptation of deep networks. In Proceedings of the International Conference on Machine Learning, PMLR, Sydney, Australia, 6–11 August 2017; pp. 1126–1135.
25. Snell, J.; Swersky, K.; Zemel, R. Prototypical networks for few-shot learning. In Proceedings of the 31st Conference on Neural Information Processing Systems (NIPS 2017), Long Beach, CA, USA, 4–9 December 2017.
26. Vinyals, O.; Blundell, C.; Lillicrap, T.; Wierstra, D. Matching networks for one shot learning. In Proceedings of the 30th Conference on Neural Information Processing Systems (NIPS 2016), Barcelona, Spain, 5–10 December 2016.

27. Fu, K.; Zhang, T.; Zhang, Y.; Wang, Z.; Sun, X.; Sensing, R. Few-shot SAR target classification via metalearning. *IEEE Trans. Geosci. Remot. Sens.* **2021**, *60*, 1–14. [CrossRef]
28. Tang, J.; Zhang, F.; Zhou, Y.; Yin, Q.; Hu, W. A fast inference networks for SAR target few-shot learning based on improved siamese networks. In Proceedings of the IGARSS 2019–2019 IEEE International Geoscience and Remote Sensing Symposium, Yokohama, Japan, 28 July–2 August 2019; pp. 1212–1215.
29. Gao, F.; Xu, J.; Lang, R.; Wang, J.; Hussain, A.; Zhou, H. A Few-Shot Learning Method for SAR Images Based on Weighted Distance and Feature Fusion. *Remot. Sens.* **2022**, *14*, 4583. [CrossRef]
30. Tai, Y.; Tan, Y.; Xiong, S.; Sun, Z.; Tian, J. Few-shot transfer learning for sar image classification without extra sar samples. *IEEE J. Sel. Top. Appl. Earth Obs. Remot. Sens.* **2022**, *15*, 2240–2253. [CrossRef]
31. Rostami, M.; Kolouri, S.; Eaton, E.; Kim, K. Deep transfer learning for few-shot SAR image classification. *Remot. Sens.* **2019**, *11*, 1374. [CrossRef]
32. Sun, X.; Wang, P.; Yan, Z.; Xu, F.; Wang, R.; Diao, W.; Chen, J.; Li, J.; Feng, Y.; Xu, T.; et al. FAIR1M: A benchmark dataset for fine-grained object recognition in high-resolution remote sensing imagery. *Isprs J. Photogramm. Remot. Sens.* **2022**, *184*, 116–130. [CrossRef]
33. Wu, Z.Z.; Wan, S.H.; Wang, X.F.; Tan, M.; Zou, L.; Li, X.L.; Chen, Y.J.A.S.C. A benchmark data set for aircraft type recognition from remote sensing images. *Appl. Soft Comput.* **2020**, *89*, 106132. [CrossRef]
34. Nichol, A.; Schulman, J. Reptile: A scalable metalearning algorithm. *OpenAI* **2018**, *2*, 4.

remote sensing

Article

MUREN: MUltistage Recursive Enhanced Network for Coal-Fired Power Plant Detection

Shuai Yuan [1], Juepeng Zheng [2,3], Lixian Zhang [2,3], Runmin Dong [2,3], Ray C. C. Cheung [1,*] and Haohuan Fu [2,3]

1 Department of Electrical Engineering, City University of Hong Kong, Hong Kong 999077, China; shuai.yuan@my.cityu.edu.hk
2 Ministry of Education Key Laboratory for Earth System Modeling, Department of Earth System Science, Tsinghua University, Beijing 100084, China; zjp19@mails.tsinghua.edu.cn (J.Z.); zhanglx18@mails.tsinghua.edu.cn (L.Z.); drm@mail.tsinghua.edu.cn (R.D.); haohuan@tsinghua.edu.cn (H.F.)
3 Department of Earth System Science, Tsinghua University—Xi'an Institute of Surveying and Mapping Joint Research Center for Next-Generation Smart Mapping, Beijing 100084, China
* Correspondence: r.cheung@cityu.edu.hk

Abstract: The accurate detection of coal-fired power plants (CFPPs) is meaningful for environmental protection, while challenging. The CFPP is a complex combination of multiple components with varying layouts, unlike clearly defined single objects, such as vehicles. CFPPs are typically located in industrial districts with similar backgrounds, further complicating the detection task. To address this issue, we propose a MUltistage Recursive Enhanced Detection Network (MUREN) for accurate and efficient CFPP detection. The effectiveness of MUREN lies in the following: First, we design a symmetrically enhanced module, including a spatial-enhanced subnetwork (SEN) and a channel-enhanced subnetwork (CEN). SEN learns the spatial relationships to obtain spatial context information. CEN provides adaptive channel recalibration, restraining noise disturbance and highlighting CFPP features. Second, we use a recursive construction set on top of feature pyramid networks to receive features more than once, strengthening feature learning for relatively small CFPPs . We conduct comparative and ablation experiments in two datasets and apply MUREN to the Pearl River Delta region in Guangdong province for CFPP detection. The comparative experiment results show that MUREN improves the mAP by 5.98% compared with the baseline method and outperforms by 4.57–21.38% the existing cutting-edge detection methods, which indicates the promising potential of MUREN in large-scale CFPP detection scenarios.

Keywords: coal-fired power plant detection; composite object detection; deep learning; carbon neutrality

Citation: Yuan, S.; Zheng, J.; Zhang, L.; Dong, R.; Cheung, R.C.C.; Fu, H. MUREN: MUltistage Recursive Enhanced Network for Coal-Fired Power Plant Detection. *Remote Sens.* **2023**, *15*, 2200. https://doi.org/10.3390/rs15082200

Academic Editors: Yanni Dong, Xiaochen Yang and Qian Du

Received: 29 March 2023
Revised: 13 April 2023
Accepted: 15 April 2023
Published: 21 April 2023

1. Introduction

China has emerged as a key force in anthropogenic carbon emissions and reduction as a result of the incredible development of China's economy and society in recent decades [1–3]. During this process, the surge in electrical power consumption has contributed to the result that the burning of fossil fuels (coal) has played the main part in anthropogenic carbon emissions [4].

The Pearl River Delta (PRD) region has had the fastest pace of development [5], while its ever-increasing energy consumption has led to a significant surge in air pollutant emissions [6]. Since a coal-fired power plant (CFPP) is the primary source of carbon dioxide emissions in thermal power, awareness of the precise location of CFPPs and their working status is important for carbon dioxide emissions monitoring and air pollution management. Identifying a combined complex with multiple objects and a nonrigid layout, compounded by the challenges posed by complicated backgrounds and blurred boundaries of CFPPs, is an exceedingly difficult research problem [7–10].

High-resolution satellite images (HRSI) can offer us an orthographic image of CFPPs, which could provide their location and working status from their latitude, longitude, and whether they are smoking or not. However, manual manipulation for searching and distinguishing in enormous pixels is time- and labor-consuming. Deep learning methods are naturally adapted to this issue because of their high efficiency and accuracy. In recent years, with the availability of many well-labeled datasets, such as ImageNet [11] and PASCAL VOC [12], there now exist outstanding deep-learning-based object detection methods such as SSD [13], RetinaNet [14] , YOLO [15], Fast R-CNN [16], Faster R-CNN [17], Dynamic R-CNN [18], Libra R-CNN, [19] and Cascade R-CNN [20] . Specifically, SSD adopts multiscale feature maps and prior boxes with different sizes to conduct multiscale object detection. RetinaNet with Focal Loss can effectively handle sample imbalance in training. Faster R-CNN adds a Region Proposal Network (RPN) into the structure to realize true real-time object detection. Dynamic R-CNN points out the strategy of dynamically adjusting training based on the change in sample distribution based on the training characteristics of target detection. Cascade R-CNN consists of a sequence of detectors to avoid problems of overfitting. These methods have been widely used in many applications such as cancer detection [21,22], animal detection [23,24], traffic light detection [25,26], and face detection [27–29], as well as many regular remote sensing applications, such as tree detection [30–32], ship detection [33,34], car detection [35,36], building detection [37,38], etc.

Nevertheless, using existing vanilla object detection methods for CFPP detection remains challenging. First, the complex nature of the target, which consists of different nonrigid components with blurred boundaries and complex composite spatial relationships, leads to inaccurate detection and imprecise location. Compared with cars and ships, CFPP is more challenging to detect due to similar and complex texture surroundings , resulting in false positives. Figure 1a,b illustrates a CFPP's blurred boundaries and the spatial interrelationship of components, in contrast with the clear boundaries and lack of spatial relationship between cars and ships in Figure 1c,d. The strong contrasts between cars and ships make them easier to detect, while the similar or complex backgrounds of CFPPs hinder accurate detection. Identifying the working status of CFPPs requires correct identification of smoke, further complicating the task.

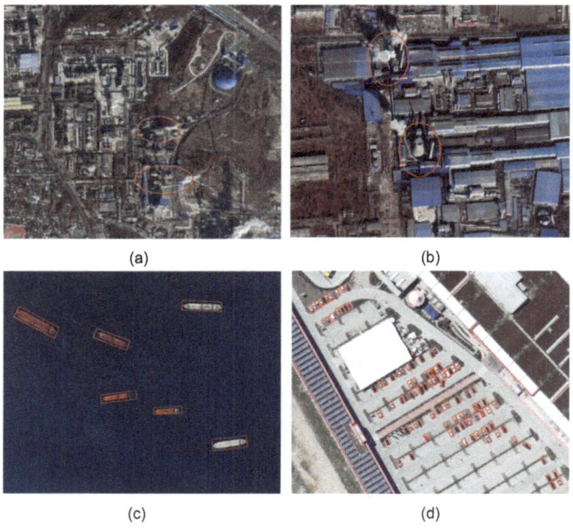

Figure 1. The differences between CFPPs and other objects, i.e., cars and ships, in optical HRSIs. (**a**,**b**) are the optical images of CFPPs, while (**c**,**d**) are the optical images of ships and cars. Compared with CFPP, cars and ships have clearer boundaries and simpler backgrounds. Additionally, their regular shapes make them easier to detect.

To address these problems in CFPP detection, we propose a (MUltistage Recursive Enhanced Detection Network) to improve CFPP detection performance and apply it in large-scale scenarios. Our contributions can be summarized as follows:

(a) We construct a new dataset of CFPPs, including the location and working status of over 300 CFPPs collected from Google Earth at 1-meter resolution. The dataset is published in https://github.com/yuanshuai0914/MUREN (accessed on: 14 April 2023).

(b) We design two enhancement mechanisms, i.e., a channel-enhanced subnetwork and a spatial-enhanced subnetwork embedded into the backbone of our detection method. CEN enhances feature representation for CFPPs and restrains the effects of noise for better training and testing performance. SEN learns the spatial relationship of components in CFPP and enriches the semantic and context information for better localization.

(c) We integrate the recursive connections and improved Atrous Spatial Pyramid Pooling (ASPP) module into the Feature Pyramid Network (FPN). FPN fuses multilevel semantic and spatial information for small object detection. Recursive connections and the ASPP module make FPN receive features twice, boosting feature learning for small and irregular CFPPs.

The remainder of this paper is organized as follows. First, we briefly introduce the related work on object detection methods applied to remote sensing and coal-fired power plant detection in Section 2. After that, we explain our MUREN in detail in Section 3, then show the datasets and study area in Section 4. In Sections 5 and 6, we analyze and discuss the CFPP detection performance of MUREN, including a comparative study and ablation experiments. Finally, we conclude this paper and present future work.

2. Related Works

This section briefly introduces CFPP detection-related works. We introduce object detection in remote sensing from dataset and algorithm aspects and introduce the recent progress in CFPP detection.

2.1. *Object Detection in Remote Sensing*

Object detection has always been a crucial issue in remote sensing because of its valuable and irreplaceable interpretation on optical images [39,40]. However, object detection in remote sensing images (RSIs) often faces many difficulties, such as the variations of sensors on satellites, different application scenarios, complex backgrounds, etc. [41]. To handle these problems, a lot of effort has been made to develop compatible methods for object detection in RSIs, including algorithms and datasets.

2.1.1. Algorithms in Related Works

In the last one or two decades, considerable work has been conducted to develop compatible algorithms for object detection in RSIs. We briefly divide them into two parts: traditional algorithms and machine learning algorithms. In traditional methods, for example, Peng and Liu [42] developed a shadow–context model to extract buildings in 229 dense urban aerial images by combining shadow information with context to verify building regions. Chaudhuri et al. [43] proposed a semiautomatic approach for road detection, whose customized operators include directional morphological enhancement, segmentation, and thinning. Eikvil et al. [44] proposed an automatic approach for vehicle detection consisting of a segmentation step followed by two stages of object classification that utilized multispectral images, panchromatic images, and road networks. In the machine learning field, for example, Yao et al. [45] proposed a computational model to detect airports in optical RSIs, which contained a target-oriented saliency model and a learned conditional random field (CRF) model. earned condition random field Zhu et al. [46] focused on ship detection and proposed a novel hierarchical complete and operational SDSOI approach based on shape and texture features, which is considered a sequential coarse-to-fine elimination process of false alarms. However, the aforementioned algorithms all utilize low-feature or middle-feature representation for object detection [47], which only achieve low accuracy or limited application scenarios.

As a part of machine learning, deep learning has made remarkable progress recently. With its development, the accuracy and efficiency of object detection has made significant improvement. Compared with traditional methods, deep-learning-based object detection algorithms can be quicker, more robust, and more generalized when encountering a huge amount of data, complex backgrounds, and scenarios. Since these advantages of deep learning methods are in accordance with the needs of remote sensing scenarios, remote sensing images based object detection applications using deep learning methods have also increased quickly [30,31,39,41,48–50]. For example, Li et al. [51] presented a rotation-insensitive Region Proposal Network (RPN) by introducing multiangle anchors into the existing RPN based on the Faster R-CNN pipeline, which can effectively handle the problem of geospatial object rotation variations. Liu et al. [52] designed a framework-based YOLOv2 for detecting arbitrarily oriented ships, which can directly predict rotationally oriented bounding boxes. Li et al. [53] designed a feature-attentioned end-to-end framework for object detection in remote sensing imagery, which performed well on the DOTA and UCAS-AOD datasets. Sun et al. [54] proposed a unified part-based convolutional neural network called PBNet for composite object detection in remote sensing imagery, containing a part localization module and a context refinement module.

2.1.2. Datasets in Related Works

Unlike natural scene images, optical RSIs usually provide overlook, sight, and profile information, with larger scales and more spatial information. As a result, it is difficult to directly translate trained recognition from natural scenes to RSIs. With the efforts in research in object detection in RSIs, several object detection datasets of optical RSIs have been collected and constructed. For example, Cheng and Han [39] developed a dataset named NWPU-VHR10, which contains ten geospatial object classes, including airplane, basketball court, baseball diamond, bridge, harbor, ground track field, ship, storage tank, tennis court, and vehicle. It consists of 715 RGB images and 85 pan-sharpened color infrared images, of which the resolution varies from 0.08 m to 2 m. Zhu et al. [55] collected a typical dataset for airplane and vehicle detection, i.e., UCAS-AOD, which consists of 600 images with 3210 airplanes and 310 images with 2819 vehicles. Zhang and Deng [7] constructed a CFPP dataset named BUAA-FFPP60, which contains over 60 CFPPs, including not only locations and class labels but also working status labels. The aforementioned datasets are widely used in object detection in remote sensing research [53,56–59] and make outstanding contributions to the Earth observation community.

2.2. Coal-Fired Power Plant Detection

CFPP monitoring has always been a crucial issue in emission estimates and air pollution management. However, precise CFPP detection has just received attention for a half decade. Before the emergence of deep learning algorithms and the availability of HRSI datasets, related work could only focus on qualitative and experiential analysis [60–62]. Along with the development of algorithms and HRSI datasets, progress has been made on the issue of precise CFPP detection. For instance, Zhang and Deng [7] collected a dataset named BUAA-FFPP60 and compared eight deep learning models in CFPP detection, comprehensively analyzing their performance on accuracy, speed, and hardware cost. Yao et al. [10] presented a Faster-R-CNN-based network to reap chimney and condensing tower integrated detection. Han et al. [8] firstly trained a Faster R-CNN model to detect CFPP, and then used two spatial analysis methods, the digital terrain model (DTM) filtering and main direction test, to remove the false results and add missed targets. Deng and Zhang [9] further increased the scale and level of feature pyramids based on the vanilla Feature Pyramid Network (FPN) to improve the performance in CFPP detection. Yin et al. [63] proposed a CFPP detection network designed for nonrigid components, including a one-stage detector, a context attention multiscale feature extraction subnetwork, and a part-based attention module. The comparative detection results evaluated by the average precision are shown in Figure 2.

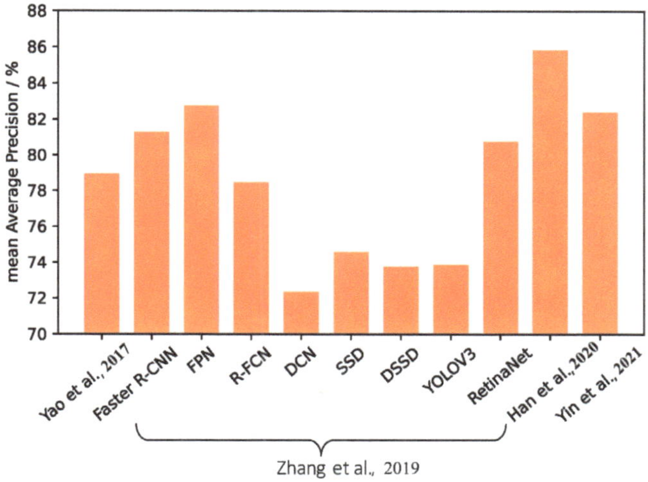

Figure 2. The mAPs of the previous studies on CFPP detection on the BUAA-FFPP60 datasets [7,8,10,63].

Deep learning algorithms are widely used for CFPP detection, but efficiency and accuracy are limited due to the complex nature of CFPPs, including multiple objects with variable layouts and similar background patterns, as well as a lack of a generalized end-to-end workflow for practical applications. To address these issues, we propose MUREN, a multistage end-to-end CFPP detection algorithm that enhances accuracy and efficiency for large-scale RSIs.

3. MUREN

3.1. Overview of Our Method

MUREN is a multistage detection network that functions in an end-to-end manner. Designed for large-scale CFPP detection, MUREN first handles the enormous input pixels. The massive input images are first cropped from 8000 × 8000 pixels to a much smaller size, e.g., 600 × 600 pixels, with 20% overlap of each slice to reduce time and computation consumption, and in the meantime, to save complete individual object information. Our goal is to achieve accurate large-scale composite CFPP detection. Because a typical CFPP consists of various components with nonrigid spatial relationships and blurred boundaries, as well as similar complex textures surrounding it, it is necessary to enhance the feature representation and channel–spatial relationships of these objects and cripple the noise disturbance. As a result, our proposed detection framework includes (as shown in Figure 3) a channel-enhanced sub-network, a spatial-enhanced sub-network, a recursive connection added in the Feature Pyramid Network, and a multistage detector-based Cascade R-CNN.

(1) A channel-enhanced subnetwork (CEN) for tackling the similarity of background patterns. In parallel with ResNet-50 [64], we add a channel-enhanced subnetwork consisting of a global average pooling layer, a global max pooling layer, two fully connected layers, and a batch normalization layer followed by an activation layer, which reaps adaptive channel recalibration and improves the object feature representation.

(2) A spatial-enhanced subnetwork (SEN) for tackling the spatial interrelationship of CFPPs' complex components. In addition to CEN and ResNet-50, we propose a symmetrical spatial-enhanced subnetwork consisting of a global average pooling layer, a global max pooling layer, and a convolutional layer followed by an activation layer.

(3) A recursive connection is added in FPN to strengthen the global feature and receptive field. FPN constructs the feature pyramid to gain different scale features and build connections between the same-scale feature map. We use a recursive connection from the FPN layers to the backbone layers. This connection gives feedback received from the FPN to the previous backbone to strengthen the object feature extraction.
(4) A multistage detector after Region of Interests (RoI) Pooling. We adopt a Cascade R-CNN-based multistage detector in the end, containing three detectors with different Intersection of Union (IoU) thresholds trained sequentially, using the output of a detector as the training set for the next.

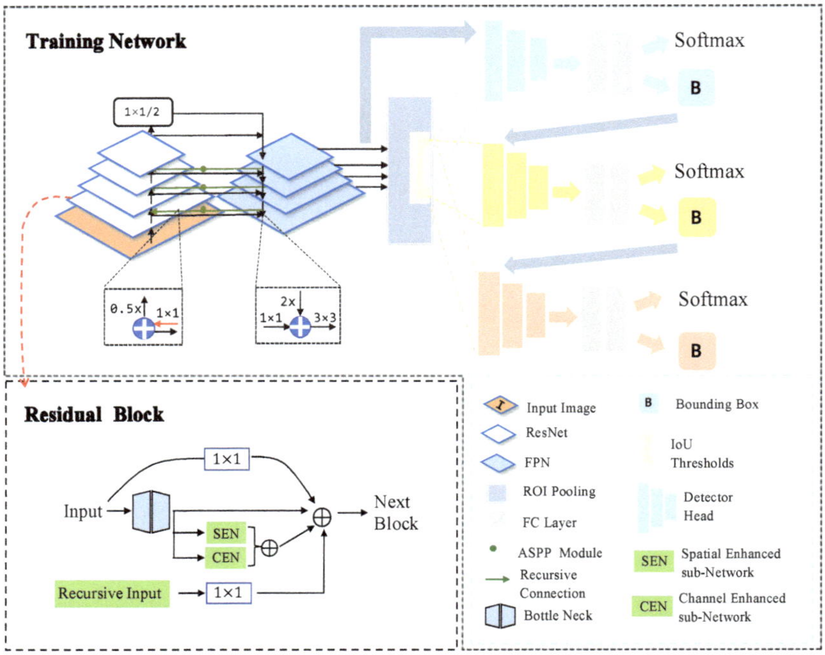

Figure 3. The architecture of our proposed MUREN: the GREEN parts denote our main contributions, including a channel-enhanced subnetwork, a spatial-enhanced subnetwork, and a recursive connection with an improved ASPP module embedded in the Feature Pyramid Network.

3.2. *Symmetrically Enhanced Network*

As a composite target, CFPP owns various components with a nonrigid spatial relationship and blurred boundaries. In addition, the complex backgrounds can create massive obstructions when extracting features and generating RoIs, as well as increase false positives and decrease accuracy. To address this problem, it is essential to strengthen the features of every part and weaken noise disturbance. Here, we propose a symmetrically enhanced network, containing both a channel-enhanced subnetwork and a spatial-enhanced subnetwork, symmetrically located beside the backbone. The structure is shown in the Residual Module in Figure 3.

3.2.1. Channel-Enhanced Subnetwork

As we can see, the channel-enhanced subnetwork has a squeeze-and-excitation block [65]. This subnetwork learns the relationship and importance of each input channel and represents the level of importance as different weights. Then, the weights are applied to matched channels to achieve adaptive channel recalibration by strengthening the object feature and constraining the noise.

In the squeeze section, we first need to separate each channel and perform feature compression on each channel, switching each two-dimensional channel feature information into a real number, representing a global context to some extent. This procedure can be achieved by a global average pooling layer (GAP) and a global max pooling layer (GMP) [66]. GAP computes one real number to represent the global feature in one channel, at the cost of blurred feature information after the average operation. GMP preserves the most prominent information, with the possibility of noise rather than the object feature. As a result, we maintain both GAP and GMP to keep both the global feature and the most prominent signal. Specifically, the input feature map $\mathbb{X} \in \mathbb{R}^{C \times W \times H}$ represents $W \times H$ features with C channels. The output feature map $\mathbb{O}_{\supset} \in \mathbb{R}^{C \times 1 \times 1}$ after global average pooling represents the global average context information. The output feature map $\mathbb{O}_{>} \in \mathbb{R}^{C \times 1 \times 1}$ after global maximum pooling represents the global maximum context information. $y_k^a \in \mathbb{R}$ represents the GAP output $\mathbb{O}_{\supset}$ of the kth feature map in channel dimension. $y_k^m \in \mathbb{R}$ represents the GMP output $\mathbb{O}_{>}$ of the kth feature map in channel dimension. $y_k \in \mathbb{R}$ presents the addition of y_k^m and y_k^a. $x_k(i,j)$ identify the element located at (i,j) in the kth feature map. $\mathbf{Y} \in \mathbb{R}^{C \times 1 \times 1}$ represents the output queued up in channel order. Then, we can define the squeeze operation as follows:

$$y_k^a = \frac{1}{W \times H} \sum_{i=1}^{H} \sum_{j=1}^{W} x_k(i,j) \tag{1}$$

$$y_k^m = max\{x_k(i,j), i \in H, j \in W\} \tag{2}$$

$$y_k = y_k^a \oplus y_k^m \tag{3}$$

$$\mathbf{Y} = array[y_1, y_2, \ldots, y_C] \tag{4}$$

The y_k^a can be identified as the typical representative of the entire feature map and can also be used to extract global context information, which serves as a guide to further enhance network performance. The global average pooling output may eliminate some specific information by averaging; so, we also employ the global max pooling layer and combine it in an addition manner, resulting in y_k.

After the squeeze operation, we need to learn the relationship and importance in a channel-wise manner, gain the weights of each channel, and match these weights with the global average pooling output y_k of the squeeze operation. In the excitation section, we employ a bottleneck-like construction consisting of two fully connected layers, a batch normalization layer followed by a ReLU activation layer, and a Sigmoid activation layer, as shown in Figure 4. Two fully connected layers construct channel weights by two 1×1 convolutions via input feature map fusion and generate a same-sized output as input $\mathbf{Y}$. The batch normalization layer is a feature normalization technique that is inserted into deep learning architectures as a trainable process to reduce internal covariant shift [67], in which the distribution of features often continuously changes over the training iterations. The ReLU activation layer is used because of its capability of preventing gradient vanishing [68]. After the second fully connected layer, we set a Sigmoid activation layer to obtain a normalization value in a range [0,1].

We define $\mathbf{e} \in \mathbb{R}^{C \times 1 \times 1}$ as a real number queue in (0,1) in channel order. σ is the Sigmoid function. $\mathbf{F_1} \in \mathbb{R}^{C \times 1 \times 1}$ and $\mathbf{F_2} \in \mathbb{R}^{C \times 1 \times 1}$ refer to 2 fully connected layers. B is the BN layer. δ represents the ReLU activation function, and $*$ is the convolution operation. The whole procedure can be described as follows:

$$\mathbf{e} = \sigma(\mathbf{F_2}\delta(B * \mathbf{F_1} * \mathbf{Y})) \tag{5}$$

where **e** indicates the importance of each channel, learned from the fully connected and nonlinear layers before. Then, we conduct a channel-wise multiplication to achieve the end-to-end adaptive channel recalibration.

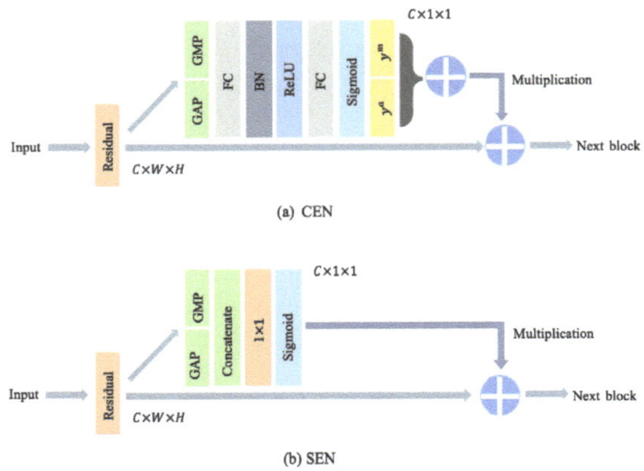

Figure 4. The detailed architectures of CEN and SEN. Both of them are embedded into the residual block of ResNet50.

3.2.2. Spatial-Enhanced Subnetwork

Symmetric with the CEN, the spatial-enhanced subnetwork lies beside the identity shortcut in the backbone. Different from the CEN, this subnetwork focuses on the inter-object spatial relationships of features. Different components of CFPPs in satellite images contribute to the CFPP detection task differently, and each component has special spatial interrelationships with one another. As a result, we still use GAP and GMP, but replace the fully connected layers with convolutional layers. As shown in Figure 4, we first employ GAP and GMP to obtain the global context from the average and the maximum perspectives. Then, we concatenate both of them to generate a new spatial information descriptor. This concatenation can use information from both the average and the maximum perspectives, which has been proven to be efficient in previous efforts [69]. Afterward, we adopt a convolutional layer to extract the feature map from the concatenation descriptor, and the output can indicate the "whereabouts" of the object. Finally, three feature maps, the feature map from the backbone, new feature map from CEN, and new feature map from SEN, are aggregated together by multiplication.

Similarly, we can describe the SEN procedure as follows:

$$\mathbf{F_s} = \sigma(f[\mathbb{O}_{\supset}, \mathbb{O}_{>}]) \tag{6}$$

$\mathbf{F_s} \in \mathbb{R}^{C \times W \times H}$ represents the output feature map of SEN. σ is the Sigmoid function, and f refers to the convolutional layer. $\mathbb{O}_{\supset}, \mathbb{O}_{>}$ represent the output features of GAP and GMP, respectively.

3.3. Recursive Connection in FPN

The recursive connection in FPN is designed for enhancing feature representations in the feature extraction step. There is a reasonable assumption that human visual perception does not process the entire image at once and needs feedback to capture more details in one image [70]. Similarly, it is likely to miss important information in one shot when processing images in neural networks. Therefore, designing a feedback mechanism in neural networks can be efficient in object classification and location. Here, we follow Qiao et al. [71] to employ a Recursive Feature Pyramid (RFP)-based structure, boosting training speed and performance. This structure contains FPN, recursive connections, and an improved Atrous Spatial Pyramid Pooling (ASPP) module.

3.3.1. Feature Pyramid Network

FPN brings significant progress to multiscale object detection performance [72]. This top-down, bottom-up structure with lateral connections can fuse features from different levels to obtain abundant semantic information, achieving good performance in small object detection. In the same way as Lin et al. [72], we define C_i as the ith feature level in the bottom-up pathway, M_i as the ith feature level in the top-down pathway, and P_i as the output feature map. f is the convolution operation. We can describe FPN as follows in Formulas (7)–(9):

$$C_{i+1} = f(C_i) \tag{7}$$

$$M_i = f(M_{i+1}) \oplus f^{1\times1}(C_i) \tag{8}$$

$$P_i = f^{3\times3}(M_i) \tag{9}$$

3.3.2. Recursive Connection

As shown in Figure 3, the recursive connections are added into FPN. These connections bring feedback received from FPN to the backbone, enhancing the FPN representation capability. Here, we denote R_i as the feature transformation before transmitting from FPN to the backbone, and we can obtain the recursive connection Formula (10):

$$C_i = f(C_{i-1}, R_i(M_i)) \tag{10}$$

In order to be compatible with the recursive connections, we redesign the backbone, i.e., ResNet50, on the basis of the two aforementioned enhanced subnetworks. We additionally design a branch road for recursive features in the first residual block in each stage, as shown in Figure 3. To use the feature $R(M)$, we add a 1×1 convolutional layer before the merge.

3.3.3. Improved ASPP Module

As stated in Qiao et al. [71], ASPP [73] is used for the implementation of connections between the FPN and backbone, in which there are four parallel branches taking M_i as their inputs. Outputs denoted as R_i are the concatenation of four feature maps generated from four branches, respectively. Since the 3×3 convolution in the ASPP module will learn some redundant information, the number of parameters is large, so it will take a long time to train. Here, we decompose the 3×3 atrous convolutional layers in ASPP into a 3×1 and a 1×3 layer to maintain its atrous rate [74]. This improved ASPP module can efficiently reduce the parameter amount and boost training speed. Figure 5 shows the improved ASPP module architecture.

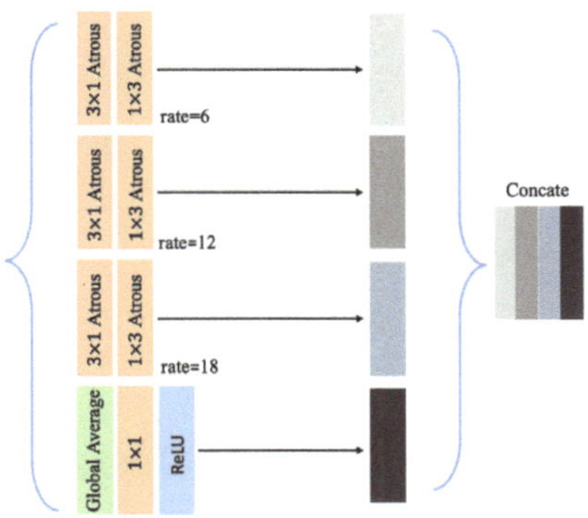

Figure 5. The architecture of the improved ASPP module. Four outputs generated by four branches are concatenated along the channel dimension.

3.4. Multistage Detector

Cascade R-CNN has proved its effectiveness in object detection, addressing the overfitting and inference–time mismatch problems and improving the bbox quality significantly [20]. Here, we employ this cascade architecture as our detector. There are three R-CNN stages using outputs from the current stage to train the next. The three stages are performed in a sequential manner and have their own IoU thresholds, i.e., (0.5, 0.6, and 0.7).

4. Datasets and Study Area

To the best of our knowledge, Yao et al. [10] collected the first dataset for CFPP detection. Zhang and Deng [7] added 101 images to this dataset and named it BUAA-FFPP60. This dataset is so far the most comprehensive and most popular dataset for CFPP detection. However, this dataset is collected in the Beijing–Tianjin–Hebei Region and only contains 318 original pictures, which is still not enough for large-scale applications in other locations for both training and testing. To improve the diversity of dataset resources, we collected and constructed a new dataset called HPGC-CFPP, which contains 300 original CFPP HRSIs collected in the Guangzhou–Foshan–Zhaoqing region from 2015 to 2019, covering over 25,000 km^2. These 300 images contain over 70 CFPPs with 1 m resolution in Google Earth, ranging in size from 500 $\times$ 500 to 1000 $\times$ 1500 pixels. We manually labeled and interpreted four types of CFPPs, including two types of objects, i.e., chimney and condensing tower, and tower working status, i.e., working and nonworking. By pairing one type of object and one status, there are four labels in our datasets, i.e., working chimney, nonworking chimney, working condensing tower, and nonworking condensing tower. As Figure 6 shows, a CFPP's location may contain several different components, including chimneys and condensing towers with different working statuses. Different constructions belonging to one CFPP are included in one labeled image. In this paper, we use both BUAA-FFPP60 and HPGC-CFPP to realize robustness and avoid overfitting. After rotation, mirroring, and blur operations, we obtained 1200 images as our training dataset and 400 images as our test dataset, of which 200 images are for model evaluation, and the 200 images collected in the Guangzhou–Foshan–Zhaoqing region are for large-scale application evaluation.

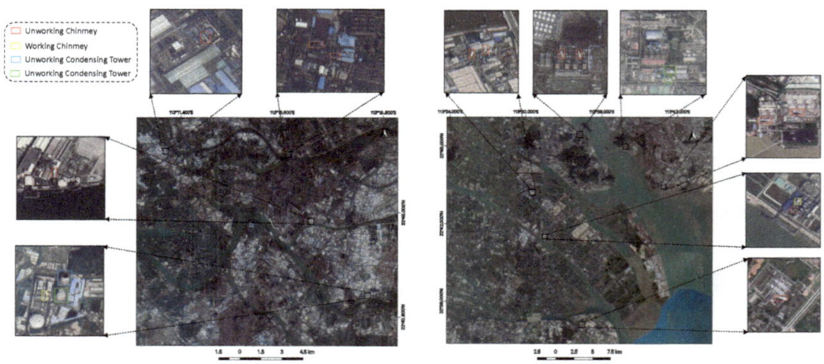

Figure 6. Examples of our datasets, including working chimney, nonworking chimney, working condensing tower, and nonworking condensing tower.

Our study area is the Guangzhou–Foshan–Zhaoqing region, located in the Pearl River Delta (PRD) region, Guangdong province, China, consisting of Guangzhou and Foshan, and Zhaoqing (See Figure 7). Images for large-scale detection are Google Earth's smallest outsourcing rectangles, containing the Guangzhou city, Foshan city, and Zhaoqing city boundaries, with 1-meter resolution . The 200 images collected in this region are used for the evaluation of this large-scale application.

In this section, we discuss our experiment details and results. First, we introduce our experiment parameter settings, platform, and evaluation metric in Section 5.1. Then, we elaborate on the experimental results of MUREN in Sections 5.2 and 5.3. Section 5.4 is for discussion of the comparative study.

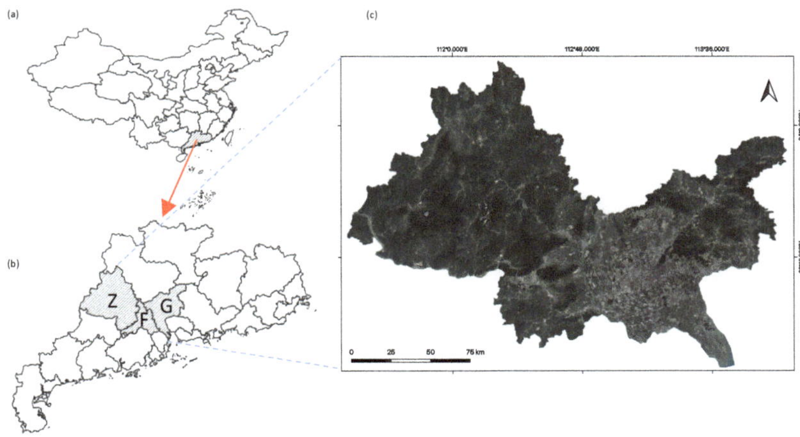

Figure 7. (**a**) is the map of China, (**b**) is the map of Guangdong province and locations of the Guangzhou–Foshan–Zhaoqing region, and (**c**) is our study area.

5. Experimental Results

5.1. Parameter Settings and Evaluation Metric

We conducted our experiment with the PyTorch deep learning framework [75] and MMDetection framework [76], with four NVIDIA GeForce RTX 2080 Ti GPUs. The batch size is set as 4 and the learning rate is 0.005. We use mini-batch stochastic gradient descent (SGD) [77] as the optimizer for classifier training and set a momentum of 0.9 and a weight decay of 0.0005. For the anchor scale, we set 8, and the ratio we set is [0.1, 0.2, 0.5, 1, 2, 5, 10], considering the shape of the chimney and condensing tower. Additionally, we also use multiscale training with the long edge set to 2000 and the short edge randomly sampled from [400, 1400], as well as online hard example mining (OHEM) [78] to handle hard example learning. The loss function for the classifier is CrossEntropy Loss and the loss function for bbox is SmoothL1 Loss. Additionally, we adopt CLAHE and IAASharpen in Albumentations [79] to conduct online image augmentation, with a possibility of 0.5. Soft-NMS (Soft Nonmaximum Suppression) is also used [80] to eliminate overlapping proposals.

There are four parts in the evaluation metric: recall, precision, average precision (AP), and mAP. Recall indicates the model's capability of detecting ground-truth CFPPs. Precision represents the model's capability of detecting CFPPs correctly. AP (average precision) represents the average precision in the Precision–Recall (PR) curve, measuring model performance more comprehensively. mAP is the mean AP, averaging all APs for corresponding label types. They can be calculated from the following formulas:

$$recall = \frac{TP}{TP + FN} \tag{11}$$

$$precision = \frac{TP}{TP + FP} \tag{12}$$

$$AP^{50} = \int_0^1 p(r)dr \tag{13}$$

$$mAP^{50} = \frac{1}{N}\sum_{i=1}^{N} AP_i \tag{14}$$

TP means true positives, denoting the number of CFPP detected correctly; FP means false positives, denoting the number of others detected as CFPPs by mistake; and FN means false negatives, denoting the number of ground truths that are missing in detection results. $p(r)$ means the point on the PR curve. N means the total number of label types. AP^{50} represents that when the IoU metric value between the detected CFPP and a ground-truth CFPP is greater than or equal to 0.5, a CFPP construction will be called as correctly detected.

5.2. MUREN Detection Results

We conduct our experiments on the datasets mentioned above, 1200 images for training and 400 images for testing, and three cities' Google Earth images for prediction. Table 1 shows the detection results of our MUREN, concerning TP, FP, FN, precision, recall, AP, and mAP. We found that our MUREN achieved an mAP of 92.46%, with APs of 88.92%, 86.32%, 98.1%, and 96.49%, respectively, for working chimney, nonworking chimney, working condensing tower, and nonworking condensing tower. Among these, the values of the indexes of the "chimney" category are basically lower than those of the"condensing tower" category; the values of the indexes of the "nonworking" category are basically lower than those of the "working" one. The reasons for these two phenomenons are that first, chimneys have much smaller features than condensing towers, and there is more similarity confusion with chimneys than condensing towers, and second, the "working" status uses smoke, which presents more features than the "nonworking" status, meaning it is easier to detect. Appendix A displays some examples of detection results. We can figure out that our MUREN displays strong ability in CFPP detection, both in dispersed and densely clustered scenarios.

Table 1. The detection results evaluation of MUREN.

Index	Chimney		Condensing Tower	
	Working	Nonworking	Working	Nonworking
recall	88.17%	83.02%	98.32%	97.28%
precision	86.92%	80.46%	99%	92.23%
AP	88.92%	86.32%	98.10%	96.49%
mAP	92.46%			

5.3. Large-Scale Applications

We applied MUREN on a large-scale application in three cities, Guangzhou, Foshan, and Zhaoqing, with 1-meter Google Earth satellite images, and evaluated the detection performance on the 200 labeled images collected in this region. We can find from Table 2 that MUREN had good performance in the large-scale application, achieving an mAP of 87.58%. However, all APs have some decrease compared with Table 1. We discuss this decrease in Section 6.3. Figure 8 shows the visualization of the large-scale application. As we can see, there are fifty-two CFPPs detected by MUREN in total. Additionally, considering that a complete CFPP location must consist of chimneys and condensing towers, we set a distance threshold of 100 m—which is 100 pixels in this image—between a chimney and a condensing tower to identify whether this is a CFPP location. If the distance between a chimney and a condensing tower is less than 100 pixels, we recognize it as a complete CFPP location, denoted by a red point; otherwise, it is an incomplete CFPP location, denoted by a blue point. It turns out that there are nine complete CFPPs and forty-three incomplete

CFPPs in this region. Additionally, CFPPs tend to be located in more urbanized areas with more industries and rivers. The time required for this investigation is about five hours, which is very fast and efficient in large-scale applications. This efficiency guarantees our future work on time series large-scale CFPP detection.

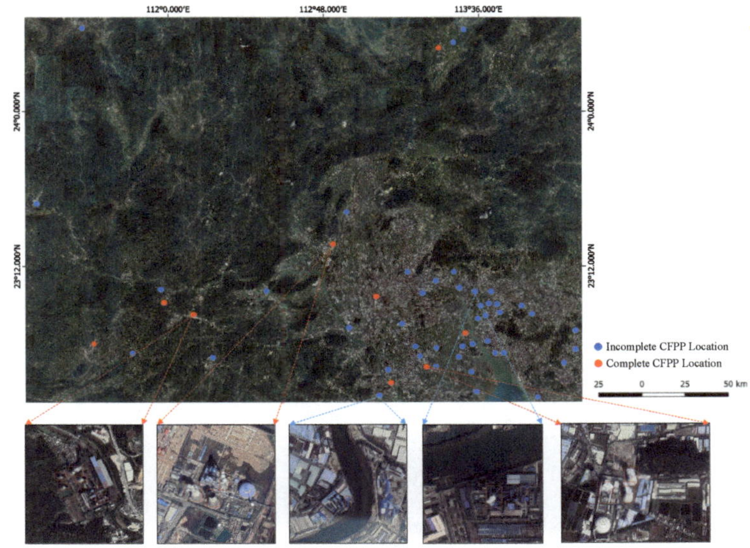

Figure 8. The visualization of large-scale detection in the Guangzhou–Foshan–Zhaoqing region. The blue point denotes an incomplete CFPP location, and the red point denotes a complete CFPP location. There are nine complete CFPPs and forty-three incomplete CFPPs in this region.

Table 2. The large-scale evaluation results of MUREN.

Index	Chimney		Condensing Tower	
	Working	Nonworking	Working	Nonworking
TP	66	89	43	57
FP	13	19	6	7
FN	11	14	1	4
precision	83.43%	82.41%	87.76%	89.06%
recall	85.71%	86.41%	97.73%	93.44%
AP	84.67%	82.72%	92.75%	90.19%
mAP	87.58%			

5.4. Comparative Study between MUREN and Other Object Detection Methods

Precision comparison. We conducted comparative studies between our MUREN and other state-of-the-art object detection methods, including Fast R-CNN, Faster R-CNN [16], Cascade R-CNN [20], Dynamic R-CNN [18], Libra Faster R-CNN [19], SSD [13], and RetinaNet [14]. We list the comparative results in Table 3 with respect to AP and mAP. More detection results can be found in Appendix A. We find that our MUREN reaches an mAP of 92.5%, which improves the mAP by 5.98% compared with the baseline (Cascade R-CNN) and outperforms by 4.57–21.38% the other six deep-learning-based object detection methods in respect to mAP. Specifically, MUREN takes the lead for the working chimney, nonworking chimney, and working condensing tower, yet it is a little behind Libra Faster R-CNN in the nonworking condensing tower. Our MUREN improves the most on chimney detection performance, which presents with fewer pixels in RSIs and is more difficult to detect. This indicates MUREN has good capabilities in small and dense object detection in RSIs. SSD, RetinaNet, and Fast R-CNN perform similarly, with an mAP lower than 75%. Dynamic R-CNN and Faster R-CNN achieve improvements compared with the last three algorithms, with an mAP of around 80%. Cascade R-CNN and Libra Faster R-CNN displayed great performance, achieving an mAP of about 87%. Figures 9 and 10 show the detection results of this comparative study between seven deep-learning-based object detection methods and our MUREN, showing the better performance of MUREN compated with other state-of-the-art methods. We can observe that MUREN stands out mainly in chimney detection and nonworking condensing tower detection. More detection results are listed in Appendix A.

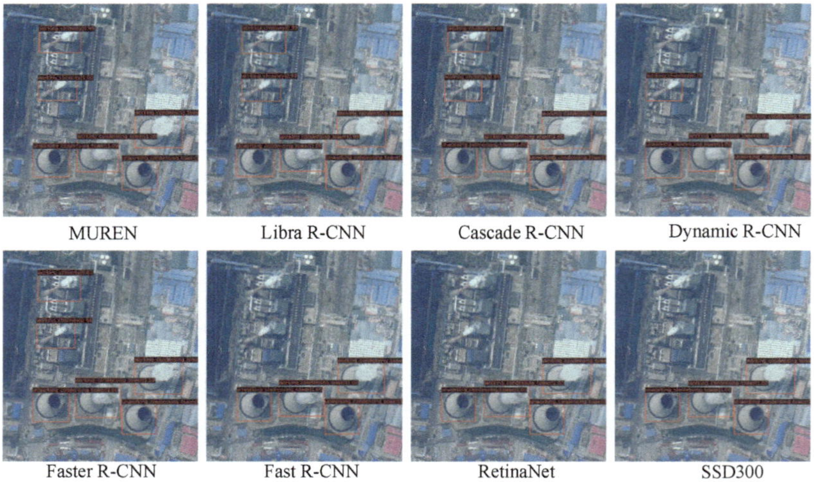

Figure 9. The detection results of eight methods. The box denotes the location, and the text denotes the category and working status. The number denotes the confidence value.

Table 3. The precision comparison between other methods and MUREN.

Method	Working Chimney	Nonworking Chimney	Working Condensing Tower	Nonworking Condensing Tower	mAP
SSD	55.31%	56.38%	88.41%	84.23%	71.08%
RetinaNet	59.78%	58.19%	89.90%	84.21%	73.02%
Fast R-CNN	62.02%	61.56%	90.67%	84.67%	74.73%
Dynamic R-CNN	71.29%	72.76%	92.31%	91.07%	81.85%
Faster R-CNN	70.73%	66.24%	93.44%	89.41%	79.95%
Cascade R-CNN	80.34%	79.92%	93.76%	91.93%	86.48%
Libra Faster R-CNN	82.23%	78.45%	94.32%	**96.67%**	87.89%
MUREN(Ours)	**88.92%**	**86.32%**	**98.10%**	96.49%	**92.46%**

Complexity comparison. We also compare the computational complexity regarding the parameter amount, space occupancy, and training/testing time. These variables can reflect the computational complexity of trained models. As can be seen in Table 4, SSD has a comparatively small parameter amount and space occupancy. The two-stage methods, i.e., Fast R-CNN, Faster R-CNN, Dynamic R-CNN, Cascade R-CNN, and Libra Faster R-CNN, have more parameter amounts and space occupancy. Among these, Dynamic R-CNN has the most amount of parameters and space occupancy. Compared with these methods, our MUREN has a comparatively average parameter amount and space occupancy, with a slight increase compared with the baseline (Cascade R-CNN). The training and testing time are compared in Figure A2. MUREN has one of the longest training times, but the testing time is comparatively short.

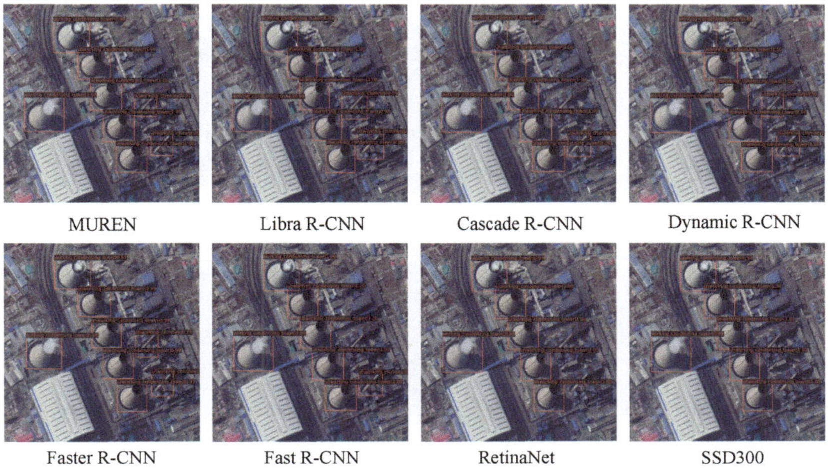

Figure 10. The detection results of eight methods. The box denotes the location, and the text denotes the category and working status. The number denotes the confidence value.

Table 4. The complexity comparison between other methods and MUREN.

Method	Parameter Amount	Space Occupancy
SSD	35 million	224 MB
RetinaNet	40 million	633 MB
Fast R-CNN	42 million	428 MB
Dynamic R-CNN	47 million	631 MB
Faster R-CNN	42 million	437 MB
Cascade R-CNN	44 million	552 MB
Libra Faster R-CNN	45 million	575MB
MUREN(Ours)	45 million	587 MB

6. Discussion

In this part, we reap ablation studies for MUREN, discussing the effectiveness of each strategy in our proposed MUREN, and discuss the limitations of MUREN by offering missed and false detections. Typically, we first assessed the effect of the symmetrically enhanced network in Section 6.1 . Then, we evaluated the contributions of recursive connections in Section 6.2, with the improved ASPP module. These ablation studies explain the effectiveness of every single strategy in our proposed MUREN and offer a deeper sight into it. After that, we discuss the limitations of MUREN and offer some examples of error detection, misdetection, and corresponding illustrations.

6.1. Ablation Study of the Symmetrically Enhanced Network

In this section, we evaluate the effectiveness of the symmetrically enhanced network from two aspects: detection results and feature map visualization. As can be seen from Figure11, we visualize the feature map extracted from the backbone and ResNet50, with and without CEN and SEN. It is clear that ResNet50 with CEN can filter most noise disturbance, but in the meantime, CEN can filter target features falsely to some extent.

Detection results. Table 5 displays the detection results' evaluation. ResNet50 with CEN outperformed ResNet50 without CEN by 1.35% in mAP. Especially, ResNet50 with CEN achieved great improvements in chimney detection. The APs of working chimney detection and nonworking chimney detection increased by 1.96% and 2.23%, respectively, compared with those detected by ResNet50 without CEN.

SEN contributes to the improvement in mAP by 0.4% against the baseline, with APs of 0.99% on working chimneys and 0.52% on nonworking chimneys, respectively. When combining the two enhanced networks, the mAP achieves a number of 89.27%, which outperforms Libra Faster R-CNN. this indicates that both enhanced spatial and channel networks can bring benefits to small object detection. Additionally, this confirms that the richer and more accurate the semantic information, the better the detection performance.

Table 5. The ablation study of CEN and SEN.

Strategy	Working Chimney	Nonworking Chimney	Working Condensing Tower	Nonworking Condensing Tower	mAP
Baseline	80.34%	79.92%	95.76%	92.93%	87.23%
Baseline+CEN	82.30%	82.15%	96.12%	93.78%	88.58%
Baseline+SEN	81.33%	80.45%	**96.25%**	92.67%	87.63%
Baseline+CEN+SEN	**83.91%**	**82.93%**	96.17%	**94.07%**	**89.27%**

Feature map visualization. As can be seen from Figure 11, we visualized the feature map extracted from the backbone and ResNet50, with and without CEN and SEN. It is clear that ResNet50 with CEN can filter most noise disturbance, but in the meantime, CEN can filter target features falsely to some extent. Moreover, it is clear to see from the SEN column of Figure 11 that by adding SEN into the backbone, the spatial interrelationship between components in CFPPs shows up, and more semantic information with richer and accurate location details from the complex remote sensing scenario is extracted. CEN with SEN makes the boundaries of components clear and the location accurate.

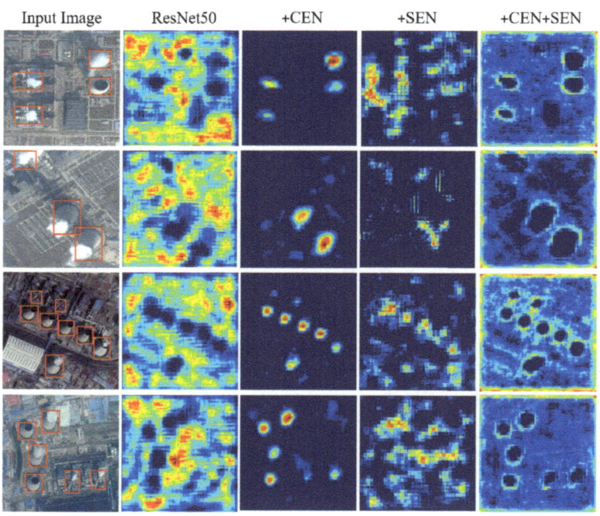

Figure 11. The visualization of feature maps from ResNet50, ResNet50+CEN, ResNet50+SEN, and ResNet50+CEN+SEN. We visualized the feature maps of these four satellite images. Vanilla ResNet50 is seriously affected by surrounding textures, and the features of CFPP are vague and blurred. By adding CEN into the backbone, most noise is eliminated and the features of CFPP are emphasized, but some useful features are also removed. SEN finds more spatial relationship between components in CFPPs. After employing SEN, the features of CFPPs become more comprehensive with clearer boundaries and more accurate locations.

6.2. Ablation Study of Recursive Connections

In the section, we show the ablation study of the recursive connections we propose in Section 3.3. Table 6 shows the results. First, we tested the performance of the baseline (Cascade R-CNN) with ResNet50 and vanilla FPN as its backbone. Then, we added recursive connections and an improved ASPP module in the backbone, evaluating their contributions to the improvements. As we can see, recursive connections can bring >2% improvements in mAP compared with the baseline. The improved ASPP module outperforms the vanilla ASPP module by about 0.5% mAP. Specifically, the APs of working and nonworking chimneys increased by 3.86% and 2.69%, which is higher than those of working and nonworking condensing towers—1.56% and 2.4%, respectively. We can find that recursive connections with the improved ASPP module boost the model ability of small and densely clustered object detection.

Table 6. The ablation study of recursive connections.

Strategy	Working Chimney	Nonworking Chimney	Working Condensing Tower	Nonworking Condensing Tower	mAP
Baseline	80.34%	79.92%	95.76%	92.93%	87.23%
Recursive Connections	82.38%	81.41%	96.83%	94.21%	88.71%
Recursive Connections + Vanilla ASPP	83.02%	82.19%	97.11%	94.76%	89.27%
Recursive Connections + Improved ASPP	**83.98%**	**82.61%**	**97.32%**	**95.33%**	**89.81%**

6.3. Limitations

Even though the superiority of MUREN has been validated through comparative studies and ablation studies, there exist limitations. First, as shown in Section 5.3, the APs all decrease compared with the local evaluation, which indicates the generalization ability of MUREN should be strengthened. We recognize the training data are not that sufficient to support good generalization ability, and this may lead to overfitting to a certain extent. Because the number of CFPPs in specific regions is not as great as other objects, such as trees and vehicles, there is a limit to collecting the same amount of data. Additionally, some the training data were collected in the Beijing–Tianjin–Hebei region, where the different scenarios, backgrounds, and features from the Guangzhou–Foshan–Zhaoqing region may harm the generalization ability. Second, in Figure 12, we show some examples of false detection and misdetection in MUREN results. The red rectangles denote the false detection, and the green rectangles denote the misdetection. As we can see, the main factors of misdetection and false detection are complex backgrounds and similar textures, including pipes, snow, clouds, etc. These factors indicate that although MUREN is designed for CFPP detection in complex scenarios, severe obstructions and similar surroundings can still affect performance.

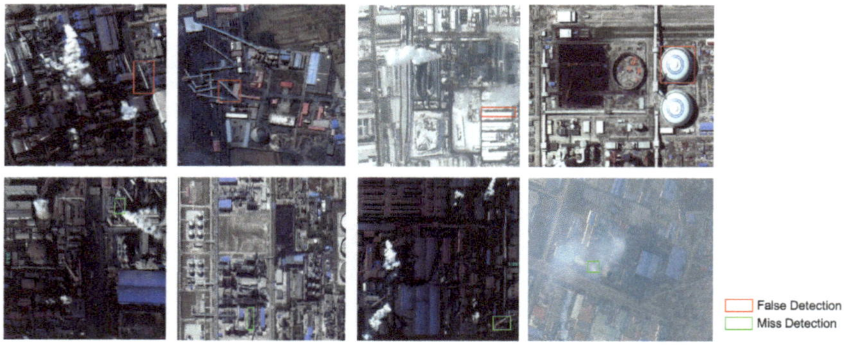

Figure 12. The visualization of false detection and misdetection of MUREN results. The red rectangles denote false detection, and the green rectangles denote the misdetection.

7. Conclusions

In this paper, we propose a MUltistage Recursive Enhanced Network (MUREN) to tackle composite CFPP detection, which consists of three procedures: the symmetrically enhanced networks embedded into a residual block, including a channel-enhanced subnetwork (CEN) and a spatial-enhanced subnetwork (SEN), recursive connections in FPN, and multistage detectors. We assess the MUREN method using the test datasets collected from the Beijing–Tianjin–Hebei region and Guangzhou–Foshan–Zhaoqing region. Compared with the other seven deep-learning-based object detection methods, MUREN achieved the best performance with an mAP of 92.46%, which improves this index by 4.57–21.38%. Moreover, we conducted a large-scale application in the Guangzhou–Foshan–Zhaoqing region, which resulted in nine complete CFPPs and forty-three incomplete CFPPs . The results indicate MUREN has great ability and potential in composite CFPP detection and counting. MUREN could bring effectiveness to large-scale composite CFPP detection and support carbon neutrality to the extent of CFPP monitoring. In the future, we will explore domain adaptation methods in remote sensing to employ MUREN in more large-scale and complex applications. We will also explore rotational object detection methods, continually boosting performance in real large-scale applications.

Author Contributions: Conceptualization, S.Y.; Methodology, S.Y.; Validation, J.Z.; Data curation, L.Z. and R.D.; Writing—original draft, S.Y.; Writing—review & editing, J.Z., L.Z., R.D., R.C.C.C. and H.F.; Visualization, S.Y. and J.Z.; Supervision, R.C.C.C. and H.F.; Project administration, R.C.C.C. and H.F.; Funding acquisition, R.C.C.C. and H.F. All authors have read and agreed to the published version of the manuscript.

Funding: This research was supported in part by CityU Research Grant No. 9239059, the National Key Research and Development Plan of China (Grant No. 2020YFB0204800), the National Natural Science Foundation of China (Grant No. T2125006), the Jiangsu Innovation Capacity Building Program (Grant No. BM2022028).

Institutional Review Board Statement: Not applicable.

Informed Consent Statement: Not applicable.

Data Availability Statement: Code and data are published on: https://github.com/yuanshuai0914/MUREN (accessed on: 14 April 2023).

Conflicts of Interest: The authors declare that they have no known competing financial interest or personal relationships that could have appeared to influence the work reported in this paper.

Appendix A

Some examples of detection results and comparisons by eight methods.

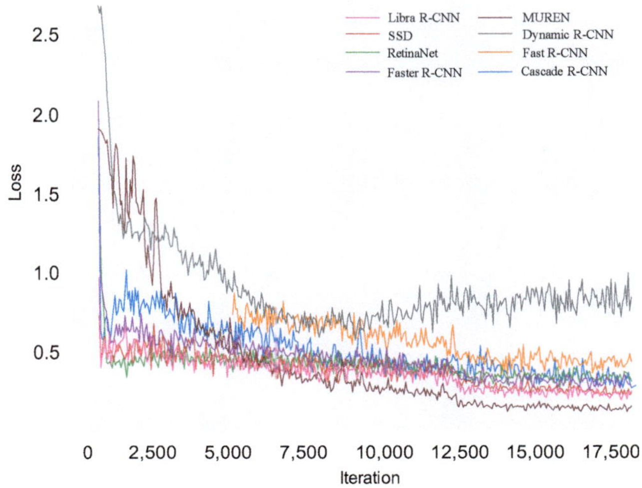

Figure A1. The training loss comparison of eight methods. Our MUREN achieves the best loss convergence of about 0.15.

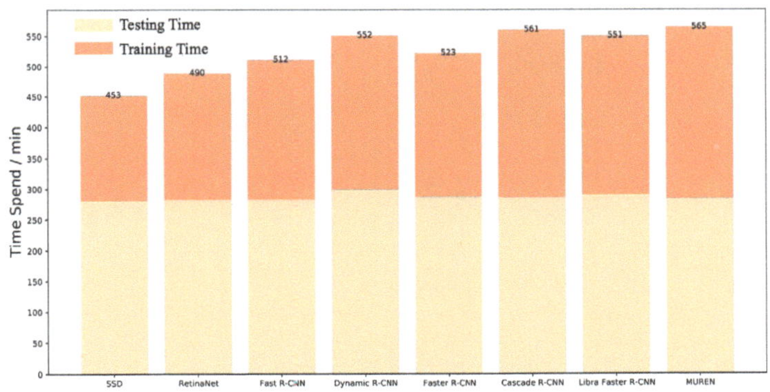

Figure A2. The time required by the eight methods in training and testing. Though MUREN has the maximum total time spending, the testing time of each method is almost the same, which is the real time consumption in large-scale applications.

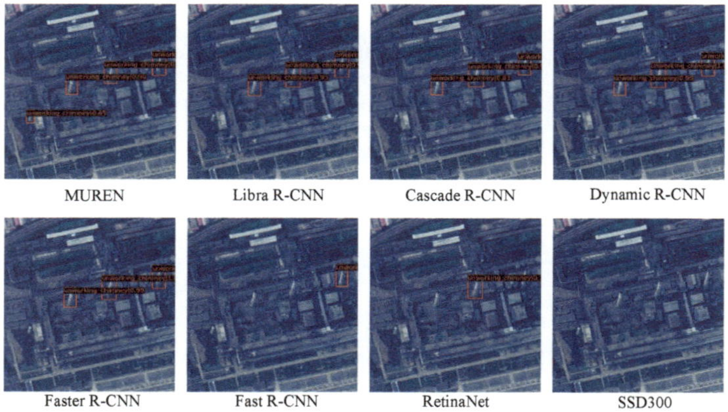

Figure A3. The detection results of the eight methods. The box denotes the location, and the text denotes the category and working status. The number denotes the confidence value.

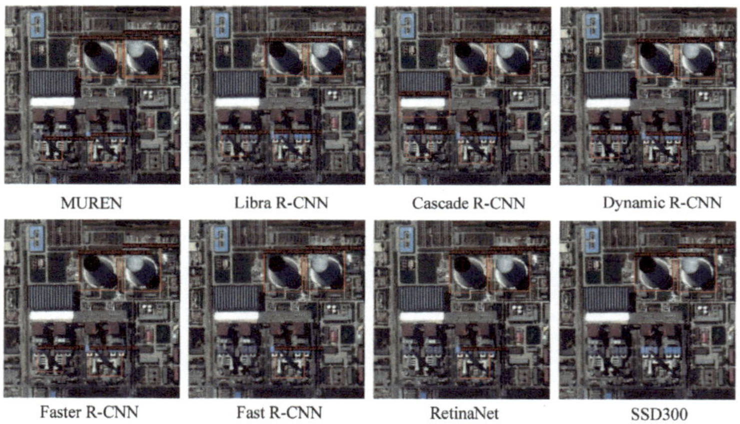

Figure A4. The detection results of the eight methods. The box denotes, the location and the text denotes the category and working status. The number denotes the confidence value.

References

1. Liu, Z.; Guan, D.; Crawford-Brown, D.; Zhang, Q.; He, K.; Liu, J. A low-carbon road map for China. *Nature* **2013**, *500*, 143–145. [CrossRef] [PubMed]
2. Li, Z.Z.; Li, R.Y.M.; Malik, M.Y.; Murshed, M.; Khan, Z.; Umar, M. Determinants of carbon emission in China: How good is green investment? *Sustain. Prod. Consum.* **2021**, *27*, 392–401. [CrossRef]
3. He, K.; Huo, H.; Zhang, Q.; He, D.; An, F.; Wang, M.; Walsh, M.P. Oil consumption and CO_2 emissions in China's road transport: Current status, future trends, and policy implications. *Energy Policy* **2005**, *33*, 1499–1507. [CrossRef]
4. Zhou, S.; Wei, W.; Chen, L.; Zhang, Z.; Liu, Z.; Wang, Y.; Kong, J.; Li, J. Impact of a coal-fired power plant shutdown campaign on heavy metal emissions in China. *Environ. Sci. Technol.* **2019**, *53*, 14063–14069. [CrossRef] [PubMed]
5. Gao, H.; Yang, W.; Yang, Y.; Yuan, G. Analysis of the air quality and the effect of governance policies in China's Pearl River Delta, 2015–2018. *Atmosphere* **2019**, *10*, 412. [CrossRef]
6. Zhong, L.; Louie, P.K.; Zheng, J.; Yuan, Z.; Yue, D.; Ho, J.W.; Lau, A.K. Science–policy interplay: Air quality management in the Pearl River Delta region and Hong Kong. *Atmos. Environ.* **2013**, *76*, 3–10. [CrossRef]
7. Zhang, H.; Deng, Q. Deep learning based fossil-fuel power plant monitoring in high resolution remote sensing images: A comparative study. *Remote. Sens.* **2019**, *11*, 1117. [CrossRef]
8. Han, C.; Li, G.; Ding, Y.; Yan, F.; Bai, L. Chimney detection based on faster R-CNN and spatial analysis methods in high resolution remote sensing images. *Sensors* **2020**, *20*, 4353. [CrossRef]
9. Deng, Q.; Zhang, H. Chimney and condensing tower detection based on FPN in high resolution remote sensing images. In Proceedings of the Image and Signal Processing for Remote Sensing XXV, International Society for Optics and Photonics, Strasbourg, France, 9–12 September 2019; Volume 11155, p. 111552B.

10. Yao, Y.; Jiang, Z.; Zhang, H.; Cai, B.; Meng, G.; Zuo, D. Chimney and condensing tower detection based on faster R-CNN in high resolution remote sensing images. In Proceedings of the 2017 IEEE International Geoscience and Remote Sensing Symposium (IGARSS), Fort Worth, TX, USA, 23–28 July 2017; pp. 3329–3332.
11. Deng, J.; Dong, W.; Socher, R.; Li, L.J.; Li, K.; Fei-Fei, L. Imagenet: A large-scale hierarchical image database. In Proceedings of the 2009 IEEE Conference on Computer Vision and Pattern Recognition, Miami, FL, USA, 20–25 June 2009; pp. 248–255.
12. Everingham, M.; Van Gool, L.; Williams, C.K.; Winn, J.; Zisserman, A. The pascal visual object classes (voc) challenge. *Int. J. Comput. Vis.* **2010**, *88*, 303–338. [CrossRef]
13. Liu, W.; Anguelov, D.; Erhan, D.; Szegedy, C.; Reed, S.; Fu, C.Y.; Berg, A.C. Ssd: Single shot multibox detector. In Proceedings of the European Conference on Computer Vision, Amsterdam, The Netherlands, 11–14 October 2016; Springer: Berlin/Heidelberg, Germany, 2016; pp. 21–37.
14. Lin, T.Y.; Goyal, P.; Girshick, R.; He, K.; Dollár, P. Focal loss for dense object detection. In Proceedings of the IEEE International Conference on Computer Vision, Venice, Italy, 22–29 October 2017; pp. 2980–2988.
15. Redmon, J.; Divvala, S.; Girshick, R.; Farhadi, A. You only look once: Unified, real-time object detection. In Proceedings of the IEEE Conference on Computer Vision and Pattern Recognition, Las Vegas, NV, USA, 27–30 June 2016; pp. 779–788.
16. Girshick, R. Fast r-cnn. In Proceedings of the IEEE International Conference on Computer Vision, Santiago, Chile, 7–13 December 2015 ; pp. 1440–1448.
17. Ren, S.; He, K.; Girshick, R.; Sun, J. Faster r-cnn: Towards real-time object detection with region proposal networks. *Adv. Neural Inf. Process. Syst.* **2015**, *28*, 91–99. [CrossRef]
18. Zhang, H.; Chang, H.; Ma, B.; Wang, N.; Chen, X. Dynamic R-CNN: Towards high quality object detection via dynamic training. In Proceedings of the European Conference on Computer Vision, Glasgow, UK, 23–28 August 2020; Springer: Berlin/Heidelberg, Germany, 2020; pp. 260–275.
19. Pang, J.; Chen, K.; Shi, J.; Feng, H.; Ouyang, W.; Lin, D. Libra r-cnn: Towards balanced learning for object detection. In Proceedings of the IEEE/CVF Conference on Computer Vision and Pattern Recognition, Long Beach, CA, USA, 15–20 June 2019; pp. 821–830.
20. Cai, Z.; Vasconcelos, N. Cascade r-cnn: Delving into high quality object detection. In Proceedings of the IEEE Conference on Computer Vision and Pattern Recognition, Salt Lake City, UT, USA, 18–23 June 2018; pp. 6154–6162.
21. Shen, L.; Margolies, L.R.; Rothstein, J.H.; Fluder, E.; McBride, R.; Sieh, W. Deep learning to improve breast cancer detection on screening mammography. *Sci. Rep.* **2019**, *9*, 12495. [CrossRef]
22. Danaee, P.; Ghaeini, R.; Hendrix, D.A. A deep learning approach for cancer detection and relevant gene identification. In Proceedings of the Pacific Symposium on Biocomputing, Kohala Coast, Hawaii, 4–8 January 2017; pp. 219–229.
23. Barbedo, J.G.A.; Koenigkan, L.V.; Santos, T.T.; Santos, P.M. A study on the detection of cattle in UAV images using deep learning. *Sensors* **2019**, *19*, 5436. [CrossRef] [PubMed]
24. Duporge, I.; Isupova, O.; Reece, S.; Macdonald, D.W.; Wang, T. Using very-high-resolution satellite imagery and deep learning to detect and count African elephants in heterogeneous landscapes. *Remote Sens. Ecol. Conserv.* **2020**, *7*, 369–381. [CrossRef]
25. Kulkarni, R.; Dhavalikar, S.; Bangar, S. Traffic light detection and recognition for self driving cars using deep learning. In Proceedings of the 2018 Fourth International Conference on Computing Communication Control and Automation (ICCUBEA), Pune, India, 16–18 August 2018; pp. 1–4.
26. Wang, J.G.; Zhou, L.B. Traffic light recognition with high dynamic range imaging and deep learning. *IEEE Trans. Intell. Transp. Syst.* **2018**, *20*, 1341–1352. [CrossRef]
27. Najibi, M.; Singh, B.; Davis, L.S. Fa-rpn: Floating region proposals for face detection. In Proceedings of the IEEE/CVF Conference on Computer Vision and Pattern Recognition, Long Beach, CA, USA, 16–20 June 2019; pp. 7723–7732.
28. Sun, X.; Wu, P.; Hoi, S.C. Face detection using deep learning: An improved faster RCNN approach. *Neurocomputing* **2018**, *299*, 42–50. [CrossRef]
29. Li, C.; Wang, R.; Li, J.; Fei, L. Face detection based on YOLOv3. In *Recent Trends in Intelligent Computing, Communication and Devices*; Springer: Singapore, 2020; pp. 277–284.
30. Zheng, J.; Fu, H.; Li, W.; Wu, W.; Zhao, Y.; Dong, R.; Yu, L. Cross-regional oil palm tree counting and detection via a multi-level attention domain adaptation network. *ISPRS J. Photogramm. Remote Sens.* **2020**, *167*, 154–177. [CrossRef]
31. Zheng, J.; Fu, H.; Li, W.; Wu, W.; Yu, L.; Yuan, S.; Tao, W.Y.W.; Pang, T.K.; Kanniah, K.D. Growing status observation for oil palm trees using Unmanned Aerial Vehicle (UAV) images. *ISPRS J. Photogramm. Remote Sens.* **2021**, *173*, 95–121. [CrossRef]
32. Li, W.; Dong, R.; Fu, H.; Yu, L. Large-scale oil palm tree detection from high-resolution satellite images using two-stage convolutional neural networks. *Remote. Sens.* **2019**, *11*, 11. [CrossRef]
33. Chen, Z.; Chen, D.; Zhang, Y.; Cheng, X.; Zhang, M.; Wu, C. Deep learning for autonomous ship-oriented small ship detection. *Saf. Sci.* **2020**, *130*, 104812. [CrossRef]
34. Cui, Z.; Li, Q.; Cao, Z.; Liu, N. Dense attention pyramid networks for multi-scale ship detection in SAR images. *IEEE Trans. Geosci. Remote Sens.* **2019**, *57*, 8983–8997. [CrossRef]
35. Ammour, N.; Alhichri, H.; Bazi, Y.; Benjdira, B.; Alajlan, N.; Zuair, M. Deep learning approach for car detection in UAV imagery. *Remote. Sens.* **2017**, *9*, 312. [CrossRef]
36. Xu, Y.; Yu, G.; Wang, Y.; Wu, X.; Ma, Y. Car detection from low-altitude UAV imagery with the faster R-CNN. *J. Adv. Transp.* **2017**, *2017*, 2823617. [CrossRef]

37. Hamaguchi, R.; Hikosaka, S. Building detection from satellite imagery using ensemble of size-specific detectors. In Proceedings of the IEEE Conference on Computer Vision and Pattern Recognition Workshops, Salt Lake City, UT, USA, 18–23 June 2018; pp. 187–191.
38. Zhang, L.; Dong, R.; Yuan, S.; Li, W.; Zheng, J.; Fu, H. Making low-resolution satellite images reborn: A deep learning approach for super-resolution building extraction. *Remote Sens.* **2021**, *13*, 2872. [CrossRef]
39. Cheng, G.; Han, J. A survey on object detection in optical remote sensing images. *ISPRS J. Photogramm. Remote Sens.* **2016**, *117*, 11–28. [CrossRef]
40. Zhu, X.X.; Tuia, D.; Mou, L.; Xia, G.S.; Zhang, L.; Xu, F.; Fraundorfer, F. Deep learning in remote sensing: A comprehensive review and list of resources. *IEEE Geosci. Remote Sens. Mag.* **2017**, *5*, 8–36. [CrossRef]
41. Li, K.; Wan, G.; Cheng, G.; Meng, L.; Han, J. Object detection in optical remote sensing images: A survey and a new benchmark. *ISPRS J. Photogramm. Remote Sens.* **2020**, *159*, 296–307. [CrossRef]
42. Peng, J.; Liu, Y. Model and context-driven building extraction in dense urban aerial images. *Int. J. Remote Sens.* **2005**, *26*, 1289–1307. [CrossRef]
43. Chaudhuri, D.; Kushwaha, N.K.; Samal, A. Semi-automated road detection from high resolution satellite images by directional morphological enhancement and segmentation techniques. *IEEE J. Sel. Top. Appl. Earth Obs. Remote Sens.* **2012**, *5*, 1538–1544. [CrossRef]
44. Eikvil, L.; Aurdal, L.; Koren, H. Classification-based vehicle detection in high-resolution satellite images. *ISPRS J. Photogramm. Remote Sens.* **2009**, *64*, 65–72. [CrossRef]
45. Yao, X.; Han, J.; Guo, L.; Bu, S.; Liu, Z. A coarse-to-fine model for airport detection from remote sensing images using target-oriented visual saliency and CRF. *Neurocomputing* **2015**, *164*, 162–172. [CrossRef]
46. Zhu, C.; Zhou, H.; Wang, R.; Guo, J. A novel hierarchical method of ship detection from spaceborne optical image based on shape and texture features. *IEEE Trans. Geosci. Remote Sens.* **2010**, *48*, 3446–3456. [CrossRef]
47. Pang, J.; Li, C.; Shi, J.; Xu, Z.; Feng, H. R^2-CNN: Fast Tiny Object Detection in Large-scale Remote Sensing Images. *IEEE Trans. Geosci. Remote Sens.* **2019**, *57*, 5512–5524. [CrossRef]
48. Li, W.; Fu, H.; Yu, L.; Cracknell, A. Deep learning based oil palm tree detection and counting for high-resolution remote sensing images. *Remote. Sens.* **2017**, *9*, 22. [CrossRef]
49. Deng, Z.; Sun, H.; Zhou, S.; Zhao, J.; Lei, L.; Zou, H. Multi-scale object detection in remote sensing imagery with convolutional neural networks. *ISPRS J. Photogramm. Remote Sens.* **2018**, *145*, 3–22. [CrossRef]
50. Fu, K.; Chang, Z.; Zhang, Y.; Xu, G.; Zhang, K.; Sun, X. Rotation-aware and multi-scale convolutional neural network for object detection in remote sensing images. *ISPRS J. Photogramm. Remote Sens.* **2020**, *161*, 294–308. [CrossRef]
51. Li, K.; Cheng, G.; Bu, S.; You, X. Rotation-insensitive and context-augmented object detection in remote sensing images. *IEEE Trans. Geosci. Remote Sens.* **2017**, *56*, 2337–2348. [CrossRef]
52. Liu, W.; Ma, L.; Chen, H. Arbitrary-oriented ship detection framework in optical remote-sensing images. *IEEE Geosci. Remote Sens. Lett.* **2018**, *15*, 937–941. [CrossRef]
53. Li, C.; Xu, C.; Cui, Z.; Wang, D.; Zhang, T.; Yang, J. Feature-attentioned object detection in remote sensing imagery. In Proceedings of the 2019 IEEE International Conference on Image Processing (ICIP), Taipei, Taiwan, China, 22–25 September 2019; pp. 3886–3890.
54. Sun, X.; Wang, P.; Wang, C.; Liu, Y.; Fu, K. PBNet: Part-based convolutional neural network for complex composite object detection in remote sensing imagery. *ISPRS J. Photogramm. Remote Sens.* **2021**, *173*, 50–65. [CrossRef]
55. Zhu, H.; Chen, X.; Dai, W.; Fu, K.; Ye, Q.; Jiao, J. Orientation robust object detection in aerial images using deep convolutional neural network. In Proceedings of the 2015 IEEE International Conference on Image Processing (ICIP), Quebec City, QC, Canada, 27–30 September 2015; pp. 3735–3739.
56. Farooq, A.; Hu, J.; Jia, X. Efficient object proposals extraction for target detection in VHR remote sensing images. In Proceedings of the 2017 IEEE International Geoscience and Remote Sensing Symposium (IGARSS), Fort Worth, TX, USA, 23–28 July 2017; pp. 3337–3340.
57. Qian, W.; Yang, X.; Peng, S.; Guo, Y.; Yan, J. Learning modulated loss for rotated object detection. *arXiv* **2019**, arXiv:1911.08299.
58. Zhao, P.; Qu, Z.; Bu, Y.; Tan, W.; Guan, Q. Polardet: A fast, more precise detector for rotated target in aerial images. *Int. J. Remote Sens.* **2021**, *42*, 5821–5851. [CrossRef]
59. Bao, S.; Zhong, X.; Zhu, R.; Zhang, X.; Li, Z.; Li, M. Single shot anchor refinement network for oriented object detection in optical remote sensing imagery. *IEEE Access* **2019**, *7*, 87150–87161. [CrossRef]
60. Tian, H.; Wang, Y.; Xue, Z.; Qu, Y.; Chai, F.; Hao, J. Atmospheric emissions estimation of Hg, As, and Se from coal-fired power plants in China, 2007. *Sci. Total Environ.* **2011**, *409*, 3078–3081. [CrossRef] [PubMed]
61. Büke, T.; Köne, A.Ç. Estimation of the health benefits of controlling air pollution from the Yatağan coal-fired power plant. *Environ. Sci. Policy* **2011**, *14*, 1113–1120. [CrossRef]
62. Mittal, M.L.; Sharma, C.; Singh, R. Estimates of emissions from coal fired thermal power plants in India. In Proceedings of the 2012 International Emission Inventory Conference, Tampa, FL, USA, 13–16 August 2012; pp. 13–16.
63. Yin, W.; Diao, W.; Wang, P.; Gao, X.; Li, Y.; Sun, X. PCAN—Part-Based Context Attention Network for Thermal Power Plant Detection in Remote Sensing Imagery. *Remote Sens.* **2021**, *13*, 1243. [CrossRef]

64. He, K.; Zhang, X.; Ren, S.; Sun, J. Deep residual learning for image recognition. In Proceedings of the IEEE Conference on Computer Vision and Pattern Recognition, Las Vegas, NV, USA, 27–30 June 2016; pp. 770–778.
65. Hu, J.; Shen, L.; Sun, G. Squeeze-and-excitation networks. In Proceedings of the IEEE Conference on Computer Vision and Pattern Recognition, Salt Lake City, UT, USA, 18–23 June 2018; pp. 7132–7141.
66. Lin, M.; Chen, Q.; Yan, S. Network in network. *arXiv* **2013**, arXiv:1312.4400.
67. Ioffe, S.; Szegedy, C. Batch normalization: Accelerating deep network training by reducing internal covariate shift. In Proceedings of the International Conference on Machine Learning (PMLR), Lille, France, 7–9 July 2015; pp. 448–456.
68. Nair, V.; Hinton, G.E. Rectified linear units improve restricted boltzmann machines. In Proceedings of the Icml, Haifa, Israel, 21–24 June 2010.
69. Woo, S.; Park, J.; Lee, J.Y.; Kweon, I.S. Cbam: Convolutional block attention module. In Proceedings of the European Conference on Computer Vision (ECCV), Munich, Germany, 8–14 September 2018; pp. 3–19.
70. Vaswani, A.; Shazeer, N.; Parmar, N.; Uszkoreit, J.; Jones, L.; Gomez, A.N.; Kaiser, Ł.; Polosukhin, I. Attention is all you need. In Proceedings of the Advances in Neural Information Processing Systems, Long Beach, CA, USA, 4–9 December 2017; pp. 5998–6008.
71. Qiao, S.; Chen, L.C.; Yuille, A. Detectors: Detecting objects with recursive feature pyramid and switchable atrous convolution. In Proceedings of the IEEE/CVF Conference on Computer Vision and Pattern Recognition, Nashville, TN, USA, 20–25 June 2021; pp. 10213–10224.
72. Lin, T.Y.; Dollár, P.; Girshick, R.; He, K.; Hariharan, B.; Belongie, S. Feature pyramid networks for object detection. In Proceedings of the IEEE Conference on Computer Vision and Pattern Recognition, Honolulu, HI, USA, 21–26 July 2017; pp. 2117–2125.
73. Chen, L.C.; Papandreou, G.; Kokkinos, I.; Murphy, K.; Yuille, A.L. Deeplab: Semantic image segmentation with deep convolutional nets, atrous convolution, and fully connected crfs. *IEEE Trans. Pattern Anal. Mach. Intell.* **2017**, *40*, 834–848. [CrossRef]
74. Alvarez, J.; Petersson, L. Decomposeme: Simplifying convnets for end-to-end learning. *arXiv* **2016**, arXiv:1606.05426.
75. Paszke, A.; Gross, S.; Massa, F.; Lerer, A.; Bradbury, J.; Chanan, G.; Killeen, T.; Lin, Z.; Gimelshein, N.; Antiga, L.; et al. Pytorch: An imperative style, high-performance deep learning library. *Adv. Neural Inf. Process. Syst.* **2019**, *32*, 8026–8037.
76. Chen, K.; Wang, J.; Pang, J.; Cao, Y.; Xiong, Y.; Li, X.; Sun, S.; Feng, W.; Liu, Z.; Xu, J.; et al. MMDetection: Open mmlab detection toolbox and benchmark. *arXiv* **2019**, arXiv:1906.07155.
77. Bottou, L. Stochastic gradient descent tricks. In *Neural Networks: Tricks of the Trade*; Springer: Berlin/Heidelberg, Germany, 2012; pp. 421–436.
78. Shrivastava, A.; Gupta, A.; Girshick, R. Training region-based object detectors with online hard example mining. In Proceedings of the IEEE Conference on Computer Vision and Pattern Recognition, Las Vegas, NV, USA, 27–30 June 2016; pp. 761–769.
79. Buslaev, A.; Iglovikov, V.I.; Khvedchenya, E.; Parinov, A.; Druzhinin, M.; Kalinin, A.A. Albumentations: Fast and flexible image augmentations. *Information* **2020**, *11*, 125. [CrossRef]
80. Bodla, N.; Singh, B.; Chellappa, R.; Davis, L.S. Soft-NMS–improving object detection with one line of code. In Proceedings of the IEEE International Conference on Computer Vision, Honolulu, HI, USA, 21–26 July 2017; pp. 5561–5569.

remote sensing

MDPI

Article

Cross-Viewpoint Template Matching Based on Heterogeneous Feature Alignment and Pixel-Wise Consensus for Air- and Space-Based Platforms

Tian Hui, Yuelei Xu *, Qing Zhou, Chaofeng Yuan and Jarhinbek Rasol

Institute of Unmanned System Research, Northwestern Polytechnical University, Xi'an 710000, China
* Correspondence: xuyuelei@nwpu.edu.cn

Abstract: Template matching is the fundamental task in remote sensing image processing of air- and space-based platforms. Due to the heterogeneous image sources, different scales and different viewpoints, the realization of a general end-to-end matching model is still a challenging task. Considering the abovementioned problems, we propose a cross-view remote sensing image matching method. Firstly, a spatial attention map was proposed to solve the problem of the domain gap. It is produced by two-dimensional Gaussian distribution and eliminates the distance between the distributed heterogeneous features. Secondly, in order to perform matching at different flight altitudes, a multi-scale matching method was proposed to perform matching on three down-sampling scales in turn and confirm the optimal result. Thirdly, to improve the adaptability of the viewpoint changes, a pixel-wise consensus method based on a correlation layer was applied. Finally, we trained the proposed model based on weakly supervised learning, which does not require extensive annotation but only labels one pair of feature points of the template image and search image. The robustness and effectiveness of the proposed methods were demonstrated by evaluation on various datasets. Our method accommodates three types of template matching with different viewpoints, including SAR to RGB, infrared to RGB, and RGB to RGB.

Keywords: cross-viewpoint; heterogeneous alignment; template matching; spatial attention; multi-scale matching; pixel-wise consensus

Citation: Hui, T.; Xu, Y.; Zhou, Q.; Yuan, C.; Rasol, J. Cross-Viewpoint Template Matching Based on Heterogeneous Feature Alignment and Pixel-Wise Consensus for Air- and Space-Based Platforms. *Remote Sens.* **2023**, *15*, 2426. https://doi.org/10.3390/rs15092426

Academic Editor: Silvia Liberata Ullo

Received: 23 February 2023
Revised: 20 April 2023
Accepted: 26 April 2023
Published: 5 May 2023

1. Introduction

Remote sensing image matching is used to identify visual correspondences between two images and is fundamental to tasks such as visual localization, topographic mapping, and environmental inspection [1]. However, during the image acquisition process, remote sensing images may include heterogeneous images from various sources, with differences in scale and viewpoint, due to the differences in air- and space-based platforms, as well as differences in sensors and observation altitudes. The image matching process based on air- and space-based platforms is shown in Figure 1, where the observation viewpoint of the space-based platform is close to vertical, while the observation viewpoint of the air-based platform is usually an oblique angle. At this point, due to differences in the observation angle and distance, there will be differences in scale and viewpoint for the same target or area. On the other hand, air-based platforms mostly carry optical sensors, while space-based platforms can carry a wider variety of sensor types, including optical and radar sensors. This leads to a problem of data heterogeneity in image matching due to differences in sensors. In this work, images acquired by satellites are referred to as template images, while images acquired by drones are referred to as search images. For the template matching task, the template image is a part of the search image, and the primary objective of our work is to match them to the same coordinates. In summary, there are three objective problems that need to be addressed in the task of matching images from air- and space-based platforms:

(1) Data heterogeneity between template and search images;
(2) Scale differences between template and search images;
(3) Observation angle differences between template and search images.

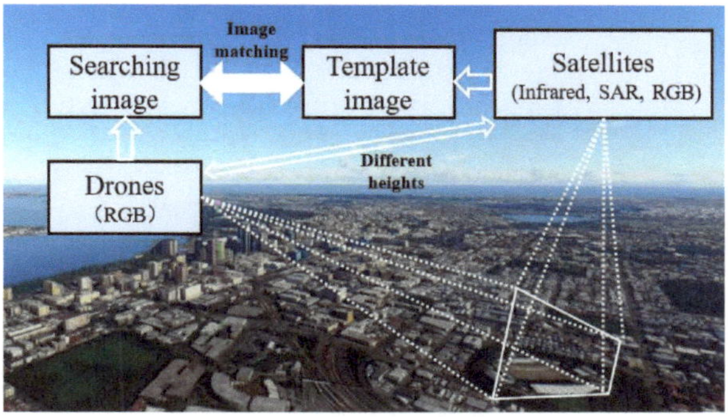

(**a**) Principle schematic of our work

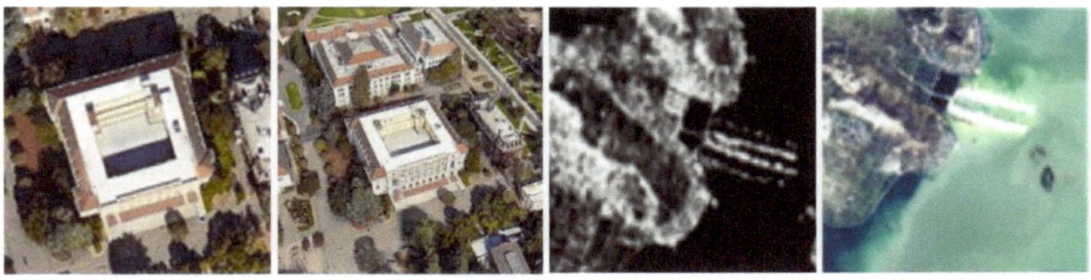

(**b**) Template images and searching images

Figure 1. Comparison of different sources and different view images.

Traditionally, visual correspondences consistent with geometric transformations, such as affine transformations, are processed using local features [2]. These features can be divided into point-based and region-based matching. Point-based matching uses feature points as local representations of image features, representing the original pixel intensities and being robust to luminance variations, affine variations, and noise. However, this method is only suitable for similar scenes with similar grayscale intensities, and performance will significantly decline when facing different image sources and views. Region-based matching [3–5] finds local features through feature descriptor similarity. Local features have good integrity and stability in cases of deformation and area change. However, this method lacks flexibility and adaptability when dealing with heterogeneous images and changes in viewpoint.

In recent years, deep neural networks (DNNs) have gradually been applied to remote sensing image processing. Due to their high flexibility and adaptiveness, DNNs have shown good performance in many tasks, such as semantic segmentation, change detection, and pixel-based classification [6–8]. In particular, the two-stream parallel network has become a common choice for many works [9]. Han et al. [10] proposed MatchNet, which segments the image into patches and then classifies them using the fully connected layer. Similarly, D2Net [11] and CMM-Net [12] segment the feature map to obtain feature descriptors and match points. The segmentation of images or feature maps, using the properties of a two-stream parallel network to calculate the similarity between two features, is a very effective method. However, there are some problems when faced with large-scale variations. Some researchers [13–15] proposed combining feature point extraction methods (SIFT [16], SURF [17]) with DCNN. These methods first use feature extraction operators to calculate possible match points and extract patches based on those points. Then, they use DCNN to calculate the similarity of the two patches, thus maintaining the correct and eliminating the incorrect points. The effectiveness and accuracy of these methods have been confirmed through numerous experiments. However, these methods are divided into two stages and cannot be trained as an end-to-end model. Rocco et al. [18] proposed a trainable model for affine transformation. They labeled the parameters and trained the model with the images in order to transform them. Furthermore, some works used DCNN to learn more reliable similarity metrics between descriptors, such as SuperPoint [19], SuperGlue [20], NCNet [21], and Patch2pix [22]. These models consider feature maps as feature descriptors and regress matching points by comparing the similarity between two feature maps. However, as in our work, remote sensing images often contain many complex and repeated textures, making it difficult to maintain accuracy against this background for pre-trained models based on point and edge datasets. On the other hand, images originate from different sensors (such as infrared, SAR, and RGB) and different platforms (such as satellite platforms and aviation platforms); hence, extracting and characterizing homologous features from heterogeneous features in multimodal images remains a challenge.

Considering the scale differences and viewpoint differences between the template image and the searching image, the object tracking methods are well suited to our work. Siamese neural network is the basic framework of object tracking methods. Broadly speaking, the two-stream parallel network is optimized and improved based on it and used for remote sensing image processing. Object trackers [23–25] usually use the first frame of a video as a template to track the next frame. Although the template has some deformation with the reference image of the next frame, this difference is slight. However, in our work, the templates were usually fixed and provided by a satellite, but the search view based on drones was constantly changing. This is a challenge for image matching, meaning that the trackers are not directly applicable to our work. Drawing lessons from the advantages of deep feature descriptors and object trackers, Zhang et al. [14] proposed a multi-modal remote sensing image matching method based on DNN. It was produced by a Siamese fully convolutional network, and a similarity score between two input images was obtained.

The abovementioned methods have been verified on different datasets. However, considering the heterogeneity of image features, scales and viewpoints in template matching tasks, an end-to-end trainable model is still required. In our work, we decomposed the template image matching of air- and space-based platforms into three main questions. The contributions of our work are as follows:

(1) We propose a heterogeneous feature alignment method, aimed at addressing the problem of data heterogeneity between template and search images caused by sensor distances. Our approach utilizes the Siamese FC [23] as the main model for image matching and addresses domain shift issues by introducing a spatial attention map based on a 2D Gaussian distribution. This method forms an adaptive spatial activation using the 2D Gaussian distribution, which dynamically adjusts the weight of positive and negative samples in the loss function. This allows for convergence during training, reducing the distribution distance between heterogeneous features and enabling the model to better learn regional features for matching template and search images.

(2) We propose a multi-scale matching method aimed at solving the problem of scale differences between search images and template images caused by observation distances. The template images provided by satellite are usually fixed, while the search images provided by aerial platforms vary with flight trajectories. Therefore, we propose a multi-scale matching method based on multi-level sub-sampling comparison. We extract feature maps of the template images at different sub-sampling levels of the model, respectively match them with the search images, obtain matching points, and determine the optimal matching points by comparing the Euclidean distance between each matching point and the ground truth.

(3) We proposed a pixel-wise consensus method aimed at addressing the problem of viewpoint differences between template and search images caused by different observation

angles. After applying the two aforementioned methods, the model could determine the position of the template image in the search image and achieve image region-level matching. To further achieve image registration at different viewpoints, we proposed a pixel-wise consensus method based on a correlation layer. This method constructs a correlation map between the template feature map and the search feature map, and regresses the matching point pairs by solving for the points with the maximum correlation value.

(4) The annotation of labels for image matching is a labor-intensive and time-consuming task. Therefore, the weakly supervised learning method was proposed to reduce the labeling workload. In this method, we only labeled one point, which was the centroid of the template image, and indicated its position in the search image. The model could learn the local features during training and predict the location of the template image in the search image. We compared the distance between the ground truth and the prediction to determine the positive and negative samples, labeled as 1 and −1, respectively. Then, we implemented the regression process on the correlation map to produce pairs of matching points.

(5) We conducted three types of matching experiments, including SAR to RGB, infrared to RGB, and RGB to RGB, each with viewpoint differences. Furthermore, to demonstrate the robustness of our method in the context of viewpoint changes, we conducted simulations at several exact angles. Additionally, we compared all of the experiments using different methods, including handcrafted and deep-learning-based methods, to verify the effectiveness of the proposed method and demonstrate its feasibility.

The remainder of this paper is structured as follows. Section 2 details the methods used for feature extraction, heterogeneous feature alignment, multi-scale matching, pixel-wise consensus, and weakly supervised learning. In Section 3, we present the experimental details, the dataset statement, and comparisons of the experimental results with several state-of-the-art methods, including ours. Finally, we provide conclusions in Section 4. An overview of our work is presented in Figure 2.

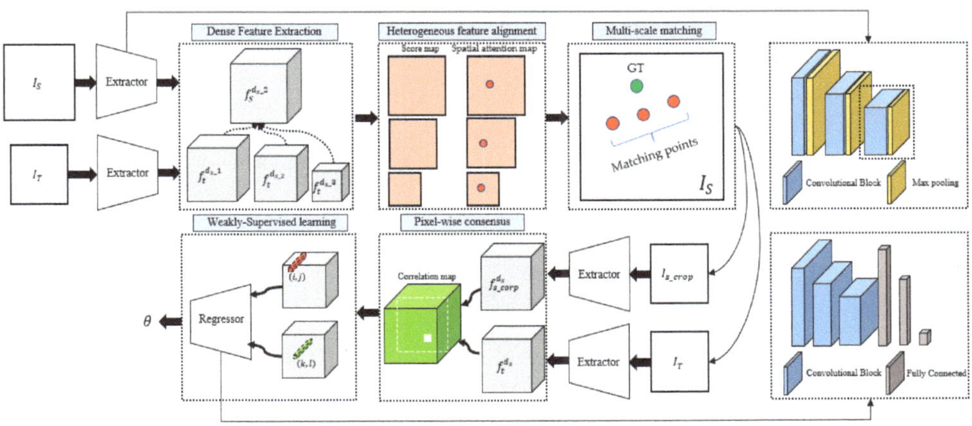

Figure 2. Our method consists of a two-stage end-to-end matching model. In the first stage, the extractor processes both the template and search images, and feature maps are generated. We then perform multi-scale matching and select the high confidence score map. Simultaneously, we complete heterogenous feature alignment based on the spatial attention map. In the second stage, we crop the search image based on the location generated by the score map and obtain the cropped search feature map and template feature map through the extractor. Finally, we use the correlation layer to obtain the correlation map and regress the matching points through the regressor.

2. Methods

2.1. Dense Feature Extraction

Here, we follow the common practice of using CNN as a feature extractor, and the backbone of our method is a pre-trained VGG16 [26]. We consider the pre-trained model based on the open-source dataset to learn some edge and corner feature information, which is helpful for this task. We remove the full connection layer and retain the former 8 layers. Given the template image I_T and the search image I_S, dense feature maps are produced by the feature extractor, and they can be formulated as:

$$f_s^d = \phi_1(I_S) \tag{1}$$

$$f_t^d = \phi_1(I_T) \tag{2}$$

where ϕ_1 denotes the feature extractor, and we consider the last layer of the feature extractor to be the activation maps. The spatial resolutions of the search image and template image are (H, W) and (M, N), respectively. Similarly, (h, w) and (m, n) denote the height and width of the search feature map f_s^d and template feature map f_t^d, respectively. For the search image I_S, the spatial dimension of its feature map f_s^d is $\frac{H}{2^d}$, where d denotes the down-sampling operation. In our method, the number of down-samplings was set to 1, 2 and 3.

2.2. Heterogeneous Feature Alignment

The search feature map f_s^d and template feature map f_t^d contain the feature descriptions of the search image and template image. Many methods process these two feature maps directly. However, in our work, the template image was part of the search image, which means that the global information of the search image was redundant for pixel-wise registration. In order to further process the match points, we must first achieve the template feature matching to select the most relevant local information. Determining the most relevant area in the search image is a significant challenge. Usually, the template image obtained by satellite is RGB or SAR. As drones are often equipped with visible and infrared cameras, the search image is RGB. When the source sensors of the template image and search image are different, the problem of a domain gap arises. The same object has various features in different domains. For example, RGB images contain information on edge and color, but these types of data cannot be displayed in long-wave infrared images. (In our work, we used an infrared sensor with a wavelength of 8 μm, which is long-wave infrared and relatively lacking in edge information).

We followed the methods of previous object tracking works [26] and proposed the local area consensus network based on the Siamese network. Thus, $f_s^d \in \mathbb{R}^{h\times w\times c}$ and $f_t^d \in \mathbb{R}^{m\times n\times c}$ are processed through the convolutional operation, and then f_t^d can be seen as a convolution kernel, which slides on f_s^d to obtain the score map $s \in \mathbb{R}^{(h-m+1)\times(w-n+1)}$. Since the model structure is clear and fixed, the coordinates of each point on the score map correspond to the location of f_t^d in f_s^d. We define the process in the following way:

$$s = Conv2d\left(f_t^d, f_s^d\right) \tag{3}$$

$$loss_{cla} = \frac{1}{n}\sum_n -[y_n \cdot \log(s_n) + (1-y_n)\cdot \log(1-s_n)] \tag{4}$$

where s denotes the score map, which consists of predicted scores $s_n = (s_1, s_2, s_3 \ldots s_n)$, the convolutional stride is 1, and y_n is the ground truth labelled by 0 and 1, as shown in Figure 3b.

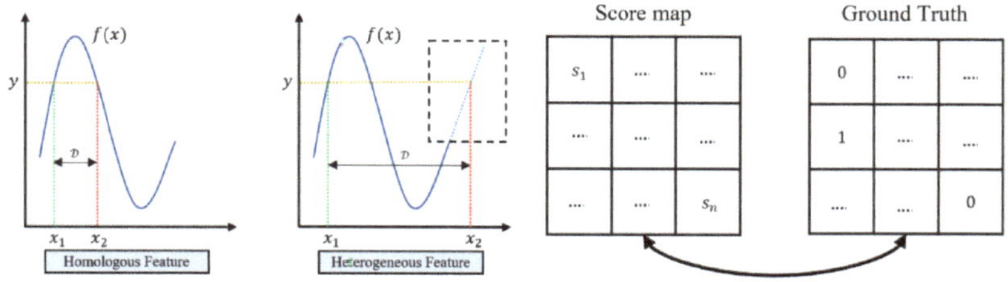

(**a**) Feature distance (**b**) Score map and ground truth

Figure 3. The processing of heterogeneous feature alignment. For the template image matching, intuitively, the feature distance of the heterogenous image is more significant than that of the homologous image, and the model fitting is difficult. Therefore, it is necessary to propose a method that can aid in model fitting when faced with heterogenous image matching.

Note that there is a problem of class imbalance, according to which we only have one positive sample but $(h-m+1)(w-n+1)-1$ as a negative sample. Ideally, the model will fit the position of the positive sample (labelled as 1). It is successful for images from the same source with the same viewpoint (as in object tracking, in which adjacent frames are used for processing). However, in our work, the template image and search image originated from different sources with different viewpoints. Clearly, the feature relevance was weaker, and the problem of class imbalance was amplified by the heterogeneous features, which prevented the model from converging during training, as shown in Figure 4.

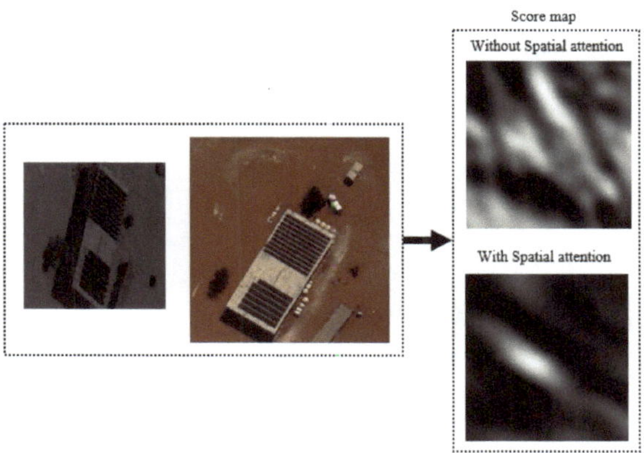

Figure 4. The high-score regions on the score map, represented by white areas, indicate the potential location of the template image. In our experiment, we chose the region with the highest score as the location of the template image. It is evident that the score map without spatial attention diverges entirely and cannot converge to the correct area.

In order to alleviate the domain gap, a feasible processing method is to diminish the high-level information distance of the template image and the search image. Let the coordinates of the score map be (p_n, q_n). The spatial attention map is introduced based on a two-dimensional Gaussian distribution. This provides a solid incentive for the positive sample and weakens the redundant part of the global information. It also alleviates the domain gap and brings the heterogeneous features closer together in high-level features, which can be formulated as:

$$g(p_n, q_n) = \frac{1}{2\pi\sigma_1\sigma_2\sqrt{1-\rho^2}} e^{[-\frac{1}{2(1-\rho^2)}(\frac{(p_n-\mu_1)^2}{\sigma_1{}^2} - \frac{2\rho(p_n-\mu_1)+(q_n-\mu_2)}{\sigma_1\sigma_2} + \frac{(q_n-\mu_2)^2}{\sigma_2{}^2})]} \quad (5)$$

where (μ_1, μ_2) is the location of the ground truth of the positive sample, the center point of the matching area (template image in search image). Here, σ_1 and σ_2 determine the magnitude of the two-dimensional Gaussian distribution, which, intuitively, points adjacent to the ground truth by one pixel and, given an incentive, can facilitate the model fitting, as seen in Figure 5. Therefore, we set σ_1, σ_2 to 0.1. ρ indicates the correlation coefficient. Since the pixels are independent, we set ρ to 0. The two-dimensional Gaussian distribution can be simplified as follows:

$$g(p_n, q_n) = \frac{1}{2\pi\sigma_1\sigma_2} e^{[-\frac{1}{2}(\frac{(p_n-\mu_1)^2}{\sigma_1} + \frac{(q_n-\mu_2)^2}{\sigma_2})]} \quad (6)$$

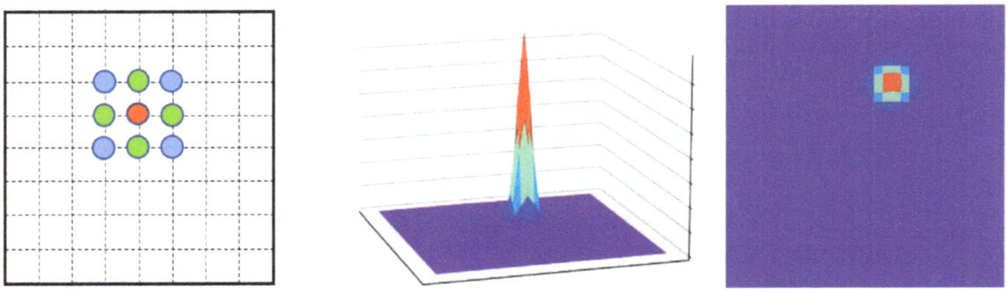

(**a**) Ground truth and adjacent points (**b**) Spatial attention map

Figure 5. We use the spatial attention map to weight the score map. The red location (ground truth) indicates a strong signal, while the green and blue locations (adjacent points) indicate weak signals. Other points are filtered out.

During training, the spatial attention map dynamically adjusts the location of the incentive according to the ground truth on each matching scale and performs the calculation with the loss function:

$$loss_{cla} = \frac{1}{n} \sum_{n} -g(p_n, q_n)[y_n \cdot \log(s_n) + (1 - y_n) \cdot \log(1 - s_n)] \quad (7)$$

2.3. Multi-Scale Matching

The spatial resolution of the template image provided by the satellite is fixed, but the search area changes with the flight altitude of drones. Detectors [27,28] and trackers [24,25] usually use RPN and Anchor to accommodate object scale changes. However, in template matching, it is necessary to consider not only the features of the object itself but also the background information. Furthermore, our label only records the location of one pair of matching points; the size of the matching area does not participate in the regression calculation. Inspired by Yolo's three-scale detection path, we introduced a multi-scale matching method to accommodate different flight altitudes. We applied the down-sampling times as different scale extraction standards. Then, three-scale template image feature maps of f_t^d were obtained. After the convolutional operation, we could also obtain three-scale score maps, which can be defined as:

$$s_d = Conv2d\left(f_t^d, f_s^2\right) \quad (8)$$

The multi-scale matching process can be seen in Figure 6. In the training stage, three-scale template feature maps can cover different ranges in the search feature map, alleviating the problem of varying search horizons with different flight altitudes of drones. We constantly obtain three match points on the three-scale score maps by match point re-localization, and the model selects one score map for the loss calculation, which is determined by the Euclidean distance. The score map selection can be defined as:

$$(p^*, q^*) = \underset{(p^d, q^d)}{\operatorname{argmin}} \sqrt{\left(p^d - p^{gt}\right)^2 + \left(q^d - q^{gt}\right)^2}, d = 1,\ 2,\ 3 \tag{9}$$

where (p^{gt}, q^{gt}) is the ground truth of the labelled points, $\left(p^d,\ q^d\right)$ denotes the inference points with the highest score of three scale, and $(p^*,\ q^*)$ denotes the optimal inference points, which indicates the best score map.

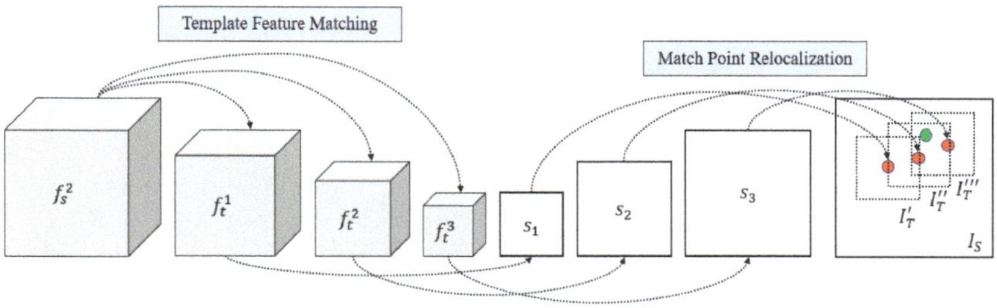

Figure 6. Three-scale score maps, each of which can predict the matching points (red points) with the highest score, the most accurate prediction among $(s_1, s_2,\ s_3)$. We compared the pixel errors with the ground truth (green point) and chose the score map corresponding to the minimum error for the loss calculation.

2.4. Pixel-Wise Consensus

After the abovementioned processing, we obtained a certain area of the search image, which was highly correlated with the template image. To further achieve pixel-level registration, we introduced a correlation layer, which was first proposed in the study on FlowNet [29]. The correlation layer measures the similarity between two feature maps and does not need to be trained. Some researchers [18,21] have also used a correlation layer to achieve affine transformation and neighborhood consensus. In our work, we used the correlation layer to predict pairs of matching points, as shown in Figure 7. Assuming that we obtained a good match point $(p^*,\ q^*)$ in s_d, we could then crop the search image I_S to obtain I_S^{crop}. Let $d = 2$, then the correlation feature map is:

$$f_{s'} = \phi_1\left(I_S^{crop}\right) \tag{10}$$

$$f_t = \phi_1(I_T) \tag{11}$$

$$f_{t\to s'} = \psi(f_{s'}, f_t) \tag{12}$$

$$f_{s'\to t} = \psi(f_t, f_{s'},) \tag{13}$$

where $f_{s'}$ denotes the feature map of the cropped search image. ψ is the correlation layer, and $f_{t\to s'} \in \mathbb{R}^{mn\times h'\times w'}$ is the template image dose for the correlation operation with the cropped search image. Correspondingly, $f_{s'\to t} \in \mathbb{R}^{h'w'\times m\times n}$ is the cropped search image dose for the correlation operation with the template image. $f_{t\to s'}$ and $f_{s'\to t}$ are transpose relations, which contain all of the pairwise points.

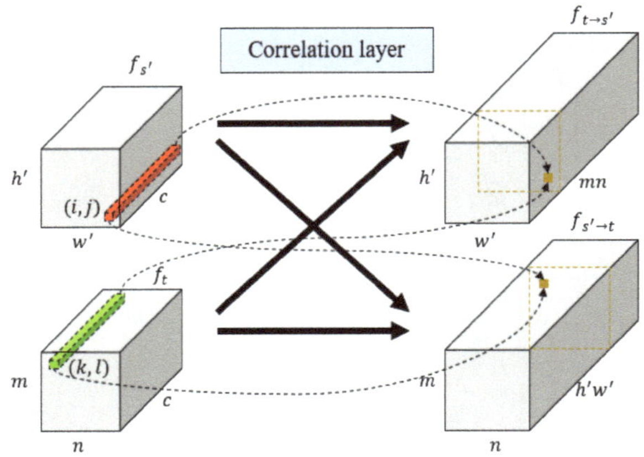

Figure 7. Supposing that $(i,\ j)$ and (k, l) are a pair of matching points, so that $(c, i,\ j)$ and $(c, k,\ l)$ can be seen as the feature description. The coordinates of the most robust activation in $f_{t\to s'}$ and the coordinates of the strongest activation in $f_{s'\to t}$ can be solved for each other.

Determining pairs of matching points, they can be defined as:

$$(i^{c_{k,l}}, j^{c_{k,l}}) = \underset{(i,j)}{\operatorname{argmax}} f_{t\to s'}(c_{k,l}, i, j), \quad 1 \le k \le m, 1 \le l \le n \tag{14}$$

$$(k^{c_{i,j}}, l^{c_{i,j}}) = \underset{(k,l)}{\operatorname{argmax}} f_{s'\to t}(c_{i,j}, k, l), \quad 1 \le i \le h', 1 \le j \le w' \tag{15}$$

$$c_{k,l} := kn + l, \quad c_{i,j} := iw' + j \tag{16}$$

$$k^* = \left\lfloor \frac{c_{k,l}}{m} \right\rfloor \tag{17}$$

$$l^* = c_{k,l}\ mod\ m \tag{18}$$

where (k^*, l^*) denotes the point solved by $f_{t\to s'}$. If $k^* = k\ \&\ l^* = l$, then (i, j) and (k, l) can be seen as pairs of matching points.

2.5. Weakly Supervised Learning

In this section, we define the weakly supervised training conducted in our work. Usually, fully supervised training is the preferred method, but in work such as ours, dense annotation consisting of pairs of points is time-consuming and complicated. Sparse annotation is one option, but it may cause imprecision. Therefore, it is desirable to propose a weakly supervised training method to train using a pair of images and to fit pairs of matching points. For this purpose, we proposed a training method that only needs one point in the annotation. Supposing the matching image (I_T, I_S) and the matching feature map $\left(f_t^d, f_s^d\right)$, the er can be obtained. The positive and negative can be defined as:

$$er = \sqrt{\left(p^d - p^{gt}\right)^2 + \left(q^d - q^{gt}\right)^2} \tag{19}$$

$$Positive: \begin{cases} er \le 5, & 0 < epoch \le 50 \\ er \le 2, & 50 < epoch \le 100 \end{cases} \tag{20}$$

$$Negative: \begin{cases} er > 5, & 0 < epoch \le 50 \\ er > 2, & 50 < epoch \le 100 \end{cases} \tag{21}$$

where *epoch* denotes the training epochs. Based on this definition, the $pos(f_{t\to s'}, f_{s'\to t})$ and $neg(f_{t\to s'}, f_{s'\to t})$ are obtained, and they are labelled as:

$$pos(f_{t\to s'}, f_{s'\to t}) = 1 \tag{22}$$

$$neg(f_{t\to s'}, f_{s'\to t}) = -1 \tag{23}$$

From Section 2.4, we can obtain correlation feature maps $f_{t\to s'}$ and $f_{s'\to t}$. Due to their reciprocal transposition, we only choose one to form the regression. Ideally, the number of correct matching points of two perfectly matched images is $h' * w'$, but in the cross-view matching task, the number of matching points is much smaller than this value. Note that the size of the correlation map is (mn, h', w'), which means that there is a significant amount of noise and a number of incorrect matching features. Therefore, we proposed the regression module ϕ_2, a cascade neural network consisting of a convolution layer and fully connected layer. First, considering the fact that the feature is available in the neighborhood of the correct point, we used the 1-kernel of the convolution layer to force the model to filter out redundant information and noise, each followed by Relu non-linearities. Through several layers, the channel dimension of the correlation map is reduced but maintains the spatial scale. Therefore, in this sparse tensor, a large amount of redundant information is filtered, while the stronger activated points are retained. The number of channel dimension reductions is set to $\frac{mn}{t}$, $t = 2, 4, 6, 8$. Then, the regression is processed based on $f_{t\to s'}$ by the regression module ϕ_2:

$$\theta = \phi_2(f_{t\to s'}) \tag{24}$$

where ϕ_2 is a cascade neural network, and θ is the prediction. Additionally, the loss function can be defined, and it consists of two parts: the first is image region classification loss, illustrated in Section 2.2, Formula (7), and the second is match point regression loss. These parts can be defined as:

$$loss_{reg} = \frac{1}{M}\sum_{i=1}^{M}(y_i - \theta_i)^2 \tag{25}$$

$$loss = loss_{cla} + loss_{reg} \tag{26}$$

2.6. Match Point Re-Localization

The area prediction was processed on the score map, and the match point regression was processed on the correlation map. Both were down-sampled during inference, and the pixel error was accumulated. In order to locate the exact location of the match points in the original resolution, we proposed a match point re-localization method.

Suppose that the cropped searching feature map $f_{s'}$, and the template feature map f_t and match points (i, j, k, l) are obtained through Formulas (12)–(17). $f_{s'}$ and f_t are up-sampled by bilinear interpolation. Note that the maximum pooling size in our model is $2 * 2$; thus, each point corresponds to four points in the corresponding area of the up-sampled feature map. Furthermore, we use the cosine distance to compare the similarity between feature points. We first compare the most similar points in the up-sampled feature map; then, we confirm the most similar points between the two feature maps and proceed to the original resolution according to the down-sampling times. The processing stage is illustrated in Figure 8 and can be formulated as follows:

$$f_{s'}{}' = \mathcal{F}(f_{s'}) \tag{27}$$

$$f_t{}' = \mathcal{F}(f_t) \tag{28}$$

$$conf(f_{s'}, f_{s'}{}') = max_{top-2}\left\{ \frac{f_{ij} \cdot f_{i_n j_n}}{||f_{ij}||_2 ||f_{i_n j_n}||_2} \right\}, \text{ n} = 1,2,3,4 \tag{29}$$

$$conf(f_t, f_t{}') = max_{top-2}\left\{ \frac{f_{kl} \cdot f_{k_n l_n}}{||f_{kl}||_2 ||f_{k_n l_n}||_2} \right\}, \text{ n} = 1,2,3,4 \tag{30}$$

$$conf(f_t{}', f_{s'}{}') = \left\{ \frac{f_{i'j'} \cdot f_{k'l'}}{||f_{i'j'}||_2 ||f_{k'l'}||_2}, \frac{f_{i''j''} \cdot f_{k'l'}}{||f_{i''j''}||_2 ||f_{k'l'}||_2}, \frac{f_{i'j'} \cdot f_{k''l''}}{||f_{i'j'}||_2 ||f_{k''l''}||_2}, \frac{f_{i''j''} \cdot f_{k''l''}}{||f_{i''j''}||_2 ||f_{k''l''}||_2} \right\} \tag{31}$$

where $\mathcal{F}$ is the bilinear interpolation, and $conf(f_{s'}, f_{s'}{}')$ is the top-2 similarity between $f_{s'}$ and the up-sampled feature map $f_{s'}{}'$, the same as $conf(f_t, f_t{}')$. Based on the confidence value, four candidate points can be obtained from each feature map, denoted as (i', j'), (i'', j''), (k', l') (k'', l''). Then, the two $conf(f_t{}', f_{s'}{}')$ feature maps are obtained, indicating the match points in the up-sampled feature maps.

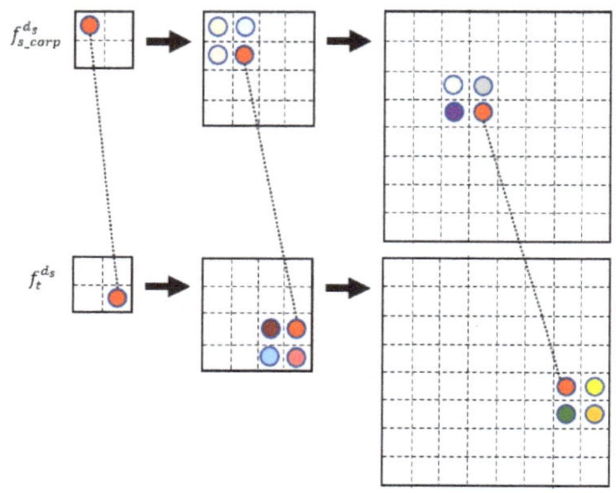

Figure 8. In the match point re-localization process, we use bilinear interpolation to restore the original resolution in a step-by-step manner and find the new match point in the corresponding area with the minimum distance.

3. Experiments

3.1. Experimental Datasets

In this section, we present the details of three types of template matching experiments that we conducted to demonstrate the effectiveness and robustness of our proposed methods. The experiments were carried out as follows:

(1) SAR to RGB: The SEN1-2 Dataset [30] was used for this part of the experiments. The dataset contained SAR images from Sentinel-1 satellites and multispectral images from Sentinel-2 satellites. The dataset included 32 cities and terrain such as mountains, forests, lakes, rivers, wastelands, buildings, etc., spanning a wide range of categories and thus being challenging for heterogeneous image matching. We cropped the SAR image as the template image and the RGB image as the search image. The total number of image pairs was 1200.

(2) Infrared to RGB: The Drone-view, which is our self-built dataset, was used in this part of the experiments. We used the drone's infrared and visible sensors to photograph the ground at 200 m and 500 m altitudes, respectively, as shown in Figure 9. The dataset contained a variety of scenarios, such as lakes, roads, mountains, etc. The images acquired at the 500 m altitude were template images, and the images acquired at the 200 m altitude were considered as search images. The total number of image pairs to conduct the experiment in this part was 3482.

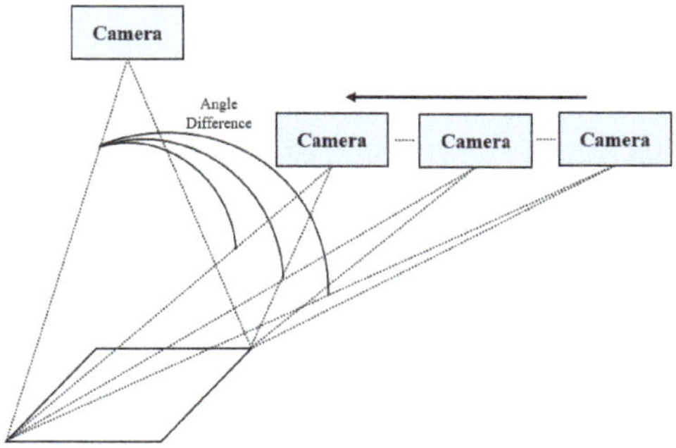

Figure 9. We collected images with various angle differences.

(3) RGB to RGB: The University-1652 [31] was used in this part of the experiments. This dataset contained maps of 1652 colleges and universities derived from three platforms: drones, satellites and ground cameras. We used the data of the drones and satellites, and the differences in viewpoints between the two was obvious. The images of the satellite were considered as direct template images, and the images of drones were regarded as search images. The total number of image pairs in this part of the evaluation was 701.

All image pairs were labelled, as shown in Figure 10. We moved the template image to the search image in order to find the matching area and label the center point coordinates. The coordinates can be imprecise due to the changing viewpoint; hence, we must specify the matching area. For the three types of template matching (SAR, infrared and RGB), the former (SAR) was the template image, and the last (RGB) was the search image. This process simulated the search view of the drones and the three types of templates provided by the satellites. The number of training sets and the test set were divided according to the ratio of 8:2, and the distribution used for each dataset is shown in Table 1. In our experiments, we used data augmentation to increase the size of the dataset, including random brightness, random grayscale, and adding Gaussian noise. During training, each augmentation strategy was applied once.

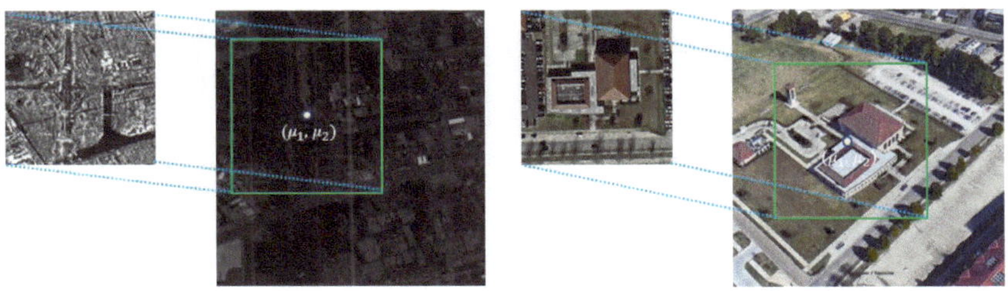

Figure 10. The process of labelling [The green box means the matching area and the "()" represents the center point coordinates of this area.].

Table 1. Dataset distribution.

Types \ Datasets	The SEN1-2 Dataset [30] (Contains Mountains, Cities, Lakes, Rivers, etc.)		The Drone-View (Contains Roads, Lakes, Rivers, etc.)		The University-1652 [31] (Contains Cities, Buildings, etc.)	
	Training	**Testing**	**Training**	**Testing**	**Training**	**Testing**
SAR to RGB	960	240	N/A	N/A	N/A	N/A
Infrared to RGB	N/A	N/A	2786	696	N/A	N/A
RGB to RGB	N/A	N/A	N/A	N/A	561	140

3.2. Experimental Details and Metrics

During training, 20% of the data was selected for testing, and 80% was selected for training with each dataset. We resized the search image to 512*512 and the template image to 128*128. The model parameters were optimized by SGD with a momentum of 0.9. The learning rate (LR) was set to 5×10^{-3} in the first 50 epochs and decayed every 50 epochs by $0.1 \times 5 \times 10^{-3}$. All of the experiments were implemented with an NVIDIA 2080Ti. In order to illustrate the effectiveness of our method, we compared both handcrafted and deep feature methods, including SIFT [16], SURF [17], D2-Net [11], NCnet [21], SuperPoint [19], SuperGlue [20], Patch2pix [22], RIFT [32], STM [33] and SAR-SIFT [34]. For the performance evaluation of the different methods, we introduced the Precision (P), Recall (R) and root-mean-square error ($RMSE$) to perform a quantitative analysis. They can be formulated as follows:

$$P = \frac{TP}{TP + FP} \tag{32}$$

$$R = \frac{TP}{TP + FN} \tag{33}$$

$$RMSE = \sqrt{\frac{1}{TP}\sum_{i=1}^{TP}\left(x_i' - x_i\right)^2 + \left(y_i' - x_i\right)^2} \tag{34}$$

where P is Precision and R is Recall. TP, FN, FP and TN can be divided by the following rules:
TP: Matching error ≤ 3 pixels.
FN: Matching pairs ignored by the model.
FP: Matching error > 3 pixels.
TN: Remaining pixel pairs.

We also compared the inference times, counting the average time taken by each method to process a single image on the test set. In our experiments, the error rate of a point being less than 3 pixels was considered as the correct matching. In our work, we used the vectors of the coordinates where the matching pairs were located as feature descriptors. For each matching result, we could obtain a list of matching pairs, which could be recorded as $pairs = \{[[i_1, j_1], [k_1, l_1]], \ldots . [[i_n, j_n], [k_n, l_n]]\}$. Furthermore, the vectors could be extracted as $pairs\_vec = \left\{\left[f_{ij}^1, f_{kl}^1\right], \ldots \left[f_{ij}^n, f_{kl}^n\right]\right\}$. Therefore, the distance between feature descriptors could be defined as:

$$D\left(f_{ij}, f_{kl}\right) = ||f_{ij} - f_{kl}||_2 \tag{35}$$

We sorted the distances and then used the 50th one as the threshold, and 50 matching pairs were obtained to calculate the image transfer parameters. Suppose that the matched pairs of feature points are $\{(x, y), (\widetilde{x}, \widetilde{y})\}$, and the transferred coordinates are (x', y'), which originate from $(\widetilde{x}, \widetilde{y})$. MA represents the proportion of correctly matched pairs, and $RMSE$ indicates the deviation of TP from the ground truth.

3.3. Experimental Results of SAR to RGB

From the results in Table 2, it can be seen that deep-learning-based methods have stronger adaptability and robustness compared to handcrafted methods in heterogeneous and complex scenarios. SIFT [16] and SURF [17] use pixel intensity for feature point matching; however, there are nonlinear intensity differences in heterogeneous images, and handcrafted algorithms are sensitive to such variations. They are not suitable for heterogeneous template matching.

Table 2. Comparison of experimental results of SAR to RGB.

Methods	P	R	RMSE	Time
SIFT [16]	32.43%	31.28%	3.795	7.63 s
SURF [17]	33.47%	31.57%	3.783	2.40 s
SAR-SIFT [34]	39.73%	38.26%	3.658	4.48 s
STM [33]	43.77%	42.16%	3.364	0.95 s
D2Net [11]	62.16%	63.39%	2.393	1.62 s
RIFT [32]	64.81%	64.14%	2.327	5.32 s
NCnet [21]	74.02%	72.39%	1.824	0.82 s
SuperPoint [19]	72.48%	71.16%	1.867	0.12 s
SuperGlue [20]	76.14%	75.26%	1.761	0.14 s
Patch2pix [22]	78.68%	77.12%	1.603	0.76 s
Ours	81.49%	80.04%	1.528	0.09 s

Compared with handcrafted methods, deep-learning-based methods can extract high-level information depending on the intensity of the pixels, which provides the model with better adaptability. D2Net [11] and NCnet [21] extract a large number of feature points in terrain types such as harbors and rivers, but when faced with landscapes such as cities and mountains, the models extract many duplicate feature points. Because of the large number of duplicate textures in these types of terrain, these models are misleading. Relatively speaking, methods (SuperPoint [19], SuperGlue [20], Patch2pix [22] and ours) for determining feature points using image patches have better performance. Image patches provide more information than pixels, achieving a higher accuracy and robustness. Our method obtained 81.49%, 80.04% and 1.528 of P, R and RMSE, respectively. Compared with traditional methods (SIFT [16], SURF [17]), the metrics were increased by 49.06%, 48.76% and 2.267 and 48.02%, 48.47% and 2.255, respectively. On the other hand, our method has better performance than other deep learning methods. For example, compared with pixel-level methods (NCnet [21], SuperPoint [19] and SuperGlue [20]), we achieved a higher P, with results of 7.47%, 9.01% and 5.35%, respectively. The main reason for this phenomenon is that the model directly extracts feature points from global information. Thus, it is difficult for the model to determine the exact position of the template image in the search image when matching using feature points. Additionally, when the location of the template image shows deviation, the generation of feature points will also show deviation.

Relative to pixel-level methods, the RMSE values of patch–pixel-level methods (Patch2pix [22] and ours) are smaller, and the prediction deviation of feature points is less significant. They have better adaptability to viewpoint changes. Compared with Patch2pix [22], we achieved a higher P, R and RMSE of 2.81%, 2.92% and 0.075, respectively. Compared with multi-modal matching methods, such as RIFT [32] and SAR-SIFT [34], the P was increased by 16.68% and 41.76%, and the R was increased by 15.90% and 41.78%. SuperPoint [19], SuperGlue [20], and our method are the fastest-running methods. Compared to the former two methods, ours is approximately 0.05 s faster for inference. Our method has an advantage in regard to the template image location due to the heterogeneous feature alignment. Therefore, our method can extract more accurate feature points from the base mountain with precise template positioning. From the analysis of the abovementioned experimental results, including those of the deep methods and traditional methods, it can be seen that the template matching capability of the proposed method for SAR to RGB is better than that of the other methods.

3.4. Experimental Results of Infrared to RGB

The template image matching experiment was implemented on our own dataset. This dataset consisted of infrared and visible light data. We used a drone to shoot vertically at the altitude of 500 m for the template image and then took pictures at the altitude of 200 m with a 30° RGB axis, which was different from the vertical direction, as the search image, as shown in Figure 8. Table 3 shows that the traditional methods obtained better performance in this experiment than SAR-to-RGB template matching. The main reason for this is that the background of our experimental site was a lake and shore, and the image background was relatively clean and straightforward.

Table 3. Comparison of experimental results for infrared to RGB.

Methods	P	R	RMSE	Time
SIFT [16]	44.79%	43.28%	3.228	7.71 s
SURF [17]	42.16%	42.03%	3.297	2.37 s
SAR-SIFT [34]	45.31%	46.21%	3.101	4.54 s
STM [33]	52.61%	51.47%	2.851	0.97 s
D2Net [11]	53.34%	52.71%	2.814	1.61 s
RIFT [32]	57.62%	55.18%	2.723	5.35 s
NCnet [21]	58.03%	57.83%	2.678	0.80 s
SuperPoint [19]	59.31%	59.12%	2.573	0.11 s
SuperGlue [20]	66.17%	65.44%	2.215	0.14 s
Patch2pix [22]	75.13%	74.67%	1.783	0.77 s
Ours	78.33%	77.14%	1.614	0.08 s

However, for the pixel-level methods, the infrared to RGB matching performance was lower than that for SAR-to-RGB matching, because although the image background was simple, there were many repeated texture features, which led to erroneous matching. The patch–pixel-level methods are more suitable for this situation. Compared with Patch2pix [22], our method obtained 78.33%, 77.14% and 1.614 for P, R and RMSE and achieved higher values of 3.20%, 2.47% and 0.169 for P, R and RMSE, respectively. According to the above experimental analysis, whether SAR to RGB or infrared to RGB, the crucial aim is to achieve heterogeneous feature alignment. All of the methods captured low-level information; however, this is not enough for heterogeneous image template matching. Due to the spatial attention map, our method maintained good performance for heterogeneous feature alignment.

3.5. Experimental Results for RGB to RGB

From the results in Table 4, it can be seen that all of the methods obtained better performance than those described in Sections 3.3 and 3.4. The main reason for this phenomenon is that the template image and the search image were homologous. There was little difference between handcrafted and deep-learning-based methods in homologous image matching because the homologous images had similar pixel intensities in the matched feature points. However, the experiments still reflected the methods' resistance capability for viewpoint change. Our method obtained 85.16%, 86.32% and 1.397 for P, R and RMSE, respectively. Compared with SuperGlue [20] and Patch2pix [22], we achieved higher values of 3.87% and 1.68% for P, respectively. On the other hand, the traditional method, SIFT [16], which does not involve training, obtained 64.16% and 65.43% for P and R, being approximately 10% lower than the values for the deep learning methods.

Compared with the previous experiments, this shows that the deep learning methods have more advantages in more complex situations, such as repeated textures, complicated image backgrounds and heterogeneous images. This requires that the model extract high-level information about feature points to avoid errors. From the above analysis, although there was no apparent gap between the methods in terms of performance, it still can be seen that our method had higher performance in RGB-to-RGB template matching due to the cross-viewpoint matching method of the pixel-wise consensus method.

Table 4. Comparison of experimental results for RGB to RGB.

Methods	P	R	RMSE	Time
SIFT [16]	64.16%	65.43%	1.924	7.36 s
SURF [17]	61.47%	60.18%	2.416	2.32 s
SAR-SIFT [34]	67.42%	68.79%	1.902	4.52 s
STM [33]	72.02%	71.68%	1.887	0.96 s
D2Net [11]	73.67%	73.28%	1.761	1.58 s
RIFT [32]	74.48%	74.16%	1.748	5.32 s
NCnet [21]	75.93%	76.17%	1.682	0.78 s
SuperPoint [19]	78.41%	77.84%	1.603	0.13 s
SuperGlue [20]	81.29%	80.16%	1.519	0.13 s
Patch2pix [22]	83.48%	83.73%	1.451	0.76 s
Ours	85.16%	86.32%	1.397	0.09 s

3.6. Experimental Results for Different Viewpoints

In the previous experiments, we evaluated the performance of our method on each of the three datasets. However, it was unclear how well the model performed at specific viewing angles. Therefore, we conducted an additional experiment on the infrared to RGB dataset to simulate a range of angular differences using both a fixed camera and a movable camera to mimic space-based observations. The process is illustrated in Figure 9.

As shown in Table 5, we conducted experiments for six angles increasing every 5°, and the maximum angle difference was 30°. In each angle experiment, our method obtained the best performance. In the experiments with a slight angle difference, such as 5° and 10°, the performance of the different methods was not clear. Our method achieved higher values of 3.89% and 2.33% for P than Patch2pix [22] and Super-Glue [20], and the performance of the traditional methods was very close to some deep learning methods. However, in the experiments with a large angle difference, our proposed method demonstrated stability and robustness. In particular, in the experiments with 25° and 30° angle differences, our method obtained 74.81% and 73.43% for P and 73.64% and 72.57% for R, respectively, while the performance of the remaining methods was severely degraded. It can be seen from Figure 11 that our method had the least performance degradation as the angle difference increased, which demonstrates the better robustness and adaptability of our proposed method. Some matching results can be seen in Figure 12, which shows the matching effects of different methods on SAR to RGB, Infrared to RGB, and RGB to RGB. We visualized the checkerboard mosaic for each of the three matching types, and the results are shown in Figure 13. From the results in the figure, we can see that the template image effectively covered the corresponding region in the search image, indicating that our method rendered a template image highly suitable for the matching of the correct position.

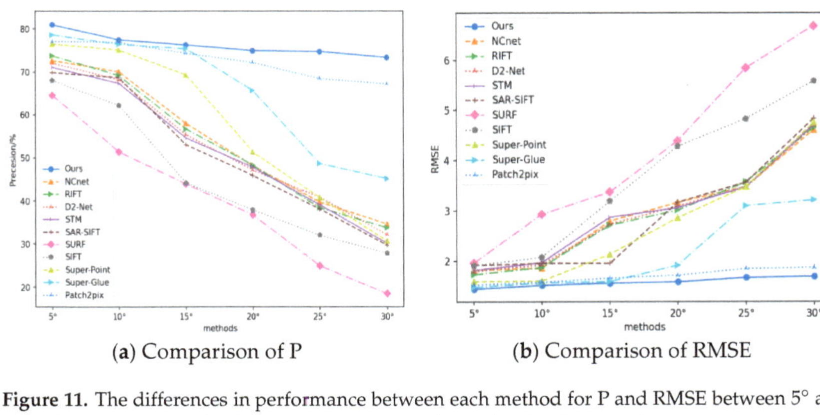

(**a**) Comparison of P (**b**) Comparison of RMSE

Figure 11. The differences in performance between each method for P and RMSE between 5° and 30°, where the smaller the difference is, the better the method's robustness is.

Table 5. Comparison of experimental results for different viewpoints.

Angle/Metrics \ Methods		SIFT [16]	SURF [17]	SAR-SIFT [34]	STM [33]	D2-Net [11]	RIFT [32]	NCnet [21]	SuperPoint [19]	SuperGlue [20]	Patch2pix [22]	Ours
	P	68.08%	64.62%	69.91%	71.13%	72.11%	73.81%	72.72%	76.53%	78.68%	77.12%	81.01%
5°	R	67.52%	63.18%	68.82%	70.16%	71.68%	72.93%	71.67%	75.83%	77.15%	76.88%	80.57%
	RMSE	1.911	1.957	1.914	1.817	1.792	1.732	1.818	1.594	1.478	1.524	1.439
	P	62.17%	51.42%	68.74%	67.43%	68.28%	69.27%	70.13%	75.16%	76.46%	76.92%	77.53%
10°	R	61.72%	50.73%	67.93%	66.81%	67.65%	68.12%	69.65%	74.83%	75.93%	75.36%	76.57%
	RMSE	2.071	2.928	1.953	1.968	1.924	1.876	1.857	1.602	1.569	1.581	1.517
	P	44.21%	43.96%	53.03%	54.62%	55.38%	56.72%	58.02%	69.41%	75.48%	74.37%	76.31%
15°	R	43.82%	43.15%	53.67%	53.73%	54.91%	56.87%	57.68%	68.16%	74.37%	73.68%	75.29%
	RMSE	3.191	3.372	2.984	2.869	2.731	2.713	2.792	2.134	1.593	1.663	1.561
	P	37.93%	36.74%	45.92%	47.83%	47.17%	48.36%	48.18%	51.37%	65.74%	72.28%	75.03%
20°	R	36.58%	35.73%	44.78%	46.37%	46.82%	49.21%	47.62%	50.19%	64.43%	71.37%	74.62%
	RMSE	4.278	4.391	3.167	3.061	3.093	3.012	3.167	2.861	1.921	1.718	1.585
	P	32.01%	24.93%	38.13%	39.14%	40.93%	38.46%	39.78%	40.81%	48.73%	68.52%	74.81%
25°	R	32.14%	23.69%	37.82%	38.76%	39.63%	37.53%	38.64%	39.73%	47.92%	67.68%	73.64%
	RMSE	4.829	5.847	3.568	3.481	3.468	3.578	3.483	3.479	3.105	1.852	1.671
	P	27.82%	18.45%	29.68%	30.15%	32.15%	33.78%	34.64%	30.67%	45.28%	67.28%	73.43%
30°	R	26.82%	17.67%	28.52%	29.27%	31.76%	32.41%	33.71%	29.68%	44.62%	66.69%	72.57%
	RMSE	5.581	6.686	4.836	4.712	4.696	4.659	4.613	4.752	3.217	1.873	1.692

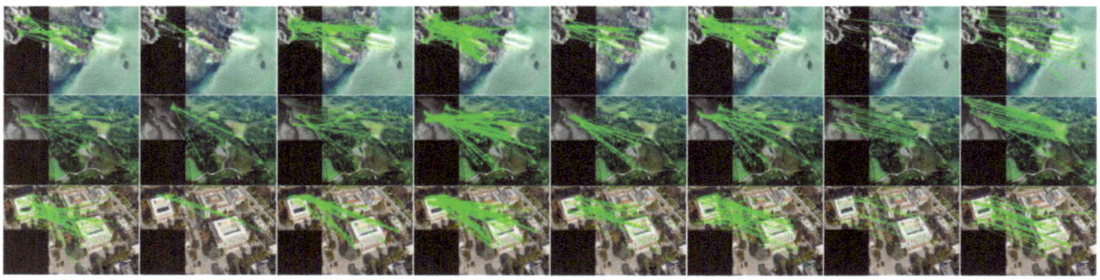

Figure 12. Comparison results. The first row is SAR to RGB, the second row is infrared to RGB, and the third row is RGB to RGB.

(**a**) SAR to RGB (**b**) RBG to RGB (**c**) Infrared to RGB

Figure 13. Comparison results for checkerboard mosaic.

4. Conclusions

In this work, we proposed a novel method for cross-view heterogeneous template matching that can be applied to both air- and space-based platforms. However, due to the inherent differences in their working environments and payloads, we faced three significant challenges that needed to be addressed. Firstly, the images produced by different sensors were heterogeneous. Secondly, we observed scale variations caused by different heights. Lastly, feature distortion was caused by different viewing angles. To tackle these issues, we proposed the following methods:

(1) To address the first challenge of the heterogeneous images produced by different sensors, we proposed a method called the heterogeneous feature alignment method based on the spatial attention map. In this approach, we added a two-dimensional Gaussian distribution to the loss function to minimize the distance between the distributed heterogeneous features. By doing so, we can effectively match the features of the template image with those of the search image, allowing them to correspond to each other.

(2) To address the second challenge of the scale variation caused by different heights, we proposed a multi-scale matching method based on multi-layer sampling point regression. With this approach, we perform regression on the matching points at different down-sampling scales to preserve the result with the smallest distance error. By doing so, we can effectively match the template and search images at different scales, overcoming the challenge of scale variation caused by different heights.

(3) To address the third challenge of feature distortion caused by different viewing angles, we proposed a pixel-wise consensus method based on the correlation layer. With this method, we use a correlation layer to extract the pixel points with the highest correlation between the feature maps of the search image and the template image, thus obtaining the matching feature points. By doing so, we can effectively overcome the feature distortion caused by different viewing angles and achieve accurate template matching.

Furthermore, considering the complexity and time-consuming nature of annotation, we proposed a weakly supervised learning method to minimize the annotation efforts required. With this method, we only need to label one point to indicate the location of the template image in the search image, thereby completing the entire training process. Four parts of the experiment demonstrated the robustness and effectiveness of our proposed methods. In the experiment on different viewpoints, our method showed good adaptability to changes in viewpoint and maintained a high level of accuracy. Specifically, when the maximum viewpoint difference was 30 degrees, our method achieved a precision of 73.43%, a recall of 72.57%, and an RMSE of 1.692. These results demonstrate that our proposed method is well-suited for cross-view heterogeneous template matching with air- and space-based platforms.

Author Contributions: Conceptualization, T.H.; methodology, T.H.; software, T.H.; validation, T.H., Q.Z., J.R. and C.Y.; formal analysis, T.H. and Q.Z.; investigation, Y.X.; resources, Y.X.; data curation, T.H. and J.R.; writing—original draft preparation, T.H.; writing—review and editing, T.H. and Y.X.; visualization, T.H. and C.Y.; supervision, Y.X. All authors have read and agreed to the published version of the manuscript.

Funding: The Nature Science Foundation of Shaanxi: 2022JQ-653.

Data Availability Statement: The SAR to RGB and RGB to RGB datasets are openly available in a public repository. They can be downloaded at https://mediatum.ub.tum.de/1436631, accessed on 10 June 2022. and https://drive.google.com/file/d/1iVnP4gjw-iHXa0KerZQ1IfIO0i1jADsR/view?usp=sharing, accessed on 1 March 2022. The infrared to RGB dataset is available on request from the authors. The data that support the findings of this study are available from the corresponding author upon reasonable request.

Conflicts of Interest: The authors declare no conflict of interest.

References

1. Sattler, T.; Maddern, W.; Toft, C.; Torii, A.; Hammarstrand, L.; Stenborg, E.; Safari, D.; Okutomi, M.; Pollefeys, M.; Sivic, J.; et al. Benchmarking 6DOF Outdoor Visual Localization in Changing Conditions. In Proceedings of the 2018 IEEE/CVF Conference on Computer Vision and Pattern Recognition, Salt Lake City, UT, USA, 18–23 June 2018; pp. 8601–8610. [CrossRef]
2. Taira, H.; Okutomi, M.; Sattler, T.; Cimpoi, M.; Pollefeys, M.; Sivic, J.; Pajdla, T.; Torii, A. InLoc: Indoor Visual Localization with Dense Matching and View Synthesis. *IEEE Trans. Pattern Anal. Mach. Intell.* **2019**, *43*, 1293–1307. [CrossRef] [PubMed]
3. Dalal, N.; Triggs, B. Histograms of oriented gradients for human detection. In Proceedings of the 2005 IEEE Computer Society Conference on Computer Vision and Pattern Recognition, San Diego, CA, USA, 20–25 June 2005; pp. 886–893. [CrossRef]
4. Ham, B.; Cho, M.; Schmid, C.; Ponce, J. Proposal Flow. In Proceedings of the 2016 IEEE Conference on Computer Vision and Pattern Recognition, Las Vegas, NV, USA, 27–30 June 2016; pp. 3475–3484. [CrossRef]
5. Lowe, D.G. Distinctive Image Features from Scale-Invariant Keypoints. *Int. J. Comput. Vis.* **2004**, *60*, 91–110. [CrossRef]
6. Zhao, Z.; Jiao, L.; Zhao, J.; Gu, J.; Zhao, J. Discriminant deep belief network for high-resolution SAR image classification. *Pattern Recognit.* **2017**, *61*, 686–701. [CrossRef]
7. Alali, F.M.; Tarakji, B.; Alqahtani, A.S.; Alqhtani, N.R.; Nabhan, A.B.; Alenzi, A.; Alrafedah, A.; Robaian, A.; Noushad, M.; Kujan, O.; et al. SAR Image segmentation based on convolutional-wavelet neural network and markov random field. *Pattern Recognit.* **2017**, *64*, 255–267. [CrossRef]
8. Zhao, W.; Jiao, L.; Ma, W.; Zhao, J.; Zhao, J.; Liu, H.; Cao, X.; Yang, S. Superpixel-Based Multiple Local CNN for Panchromatic and Multispectral Image Classification. *IEEE Trans. Geosci. Remote Sens.* **2017**, *55*, 4141–4156. [CrossRef]
9. Altwaijry, H.; Trulls, E.; Hays, J.; Fua, P.; Belongie, S. Learning to Match Aerial Images with Deep Attentive Architectures. In Proceedings of the 2016 IEEE Conference on Computer Vision and Pattern Recognition, Las Vegas, NV, USA, 27–30 June 2016; pp. 3539–3547. [CrossRef]
10. Han, X.; Leung, T.; Jia, Y.; Sukthankar, R.; Berg, A.C. Matchnet: Unifying feature and metric learning for patch-based matching. In Proceedings of the IEEE Conference on Computer Vision and Pattern Recognition, Boston, MA, USA, 7–12 June 2015; pp. 3279–3286. [CrossRef]
11. Dusmanu, M.; Rocco, I.; Pajdla, T.; Pollefeys, M.; Sivic, J.; Torii, A.; Sattler, T. D2-Net: A Trainable CNN for Joint Description and Detection of Local Features. In Proceedings of the 2019 IEEE/CVF Conference on Computer Vision and Pattern Recognition, Long Beach, CA, USA, 15–20 June 2019; pp. 8084–8093. [CrossRef]
12. Cui, S.; Ma, A.; Wan, Y.; Zhong, Y.; Luo, B.; Xu, M. Cross-Modality Image Matching Network with Modality-Invariant Feature Representation for Airborne-Ground Thermal Infrared and Visible Datasets. *IEEE Trans. Geosci. Remote Sens.* **2022**, *60*, 1–14. [CrossRef]
13. Zhu, H.; Jiao, L.; Ma, W.; Liu, F.; Zhao, W. A Novel Neural Network for Remote Sensing Image Matching. *IEEE Trans. Neural Netw. Learn. Syst.* **2019**, *30*, 2853–2865. [CrossRef] [PubMed]
14. Zhang, H.; Ni, W.; Yan, W.; Xiang, D.; Wu, J.; Yang, X.; Bian, H. Registration of Multi-modal Remote Sensing Image Based on Deep Fully Convolutional Neural Network. *IEEE J. Sel. Top. Appl. Earth Obs. Remote Sens.* **2019**, *12*, 3028–3042. [CrossRef]
15. Li, Z.; Zhang, H.; Huang, Y. A Rotation-Invariant Optical and SAR Image Registration Algorithm Based on Deep and Gaussian Features. *Remote Sens.* **2021**, *13*, 2628. [CrossRef]
16. Lowe, D.G. Object recognition from local scale-invariant features. In Proceedings of the Seventh IEEE International Conference on Computer Vision, Kerkyra, Greece, 20–27 September 1999; Volume 2, pp. 1150–1157. [CrossRef]
17. Bay, H.; Tuytelaars, T. Van Gool, LSurf: Speeded up robust features. In *Computer Vision—ECCV 2006, Proceedings of the 9th European Conference on Computer Vision, Graz, Austria, 7–13 May 2006*; Springer: Berlin/Heidelberg, Germany, 2006; pp. 404–417. [CrossRef]
18. Rocco, I.; Arandjelovic, R.; Sivic, J. Convolutional Neural Network Architecture for Geometric Matching. In Proceedings of the 2017 IEEE Conference on Computer Vision and Pattern Recognition, Honolulu, HI, USA, 21–26 July 2017; pp. 39–48. [CrossRef]
19. DeTone, D.; Malisiewicz, T.; Rabinovich, A. SuperPoint: Self-Supervised Interest Point Detection and Description. In Proceedings of the 2018 IEEE/CVF Conference on Computer Vision and Pattern Recognition Workshops, Salt Lake City, UT, USA, 18–22 June 2018; pp. 337–350. [CrossRef]
20. Sarlin, P.E.; DeTone, D.; Malisiewicz, T.; Rabinovich, A. SuperGlue: Learning Feature Matching with Graph Neural Networks. In Proceedings of the 2020 IEEE/CVF Conference on Computer Vision and Pattern Recognition, Seattle, WA, USA, 13–19 June 2020; pp. 4937–4946. [CrossRef]
21. Rocco, I.; Cimpoi, M.; Arandjelović, R.; Torii, A.; Pajdla, T.; Sivic, J. NCNet: Neighbourhood Consensus Networks for Estimating Image Correspondences. *IEEE Trans. Pattern Anal. Mach. Intell.* **2022**, *44*, 1020–1034. [CrossRef]
22. Zhou, Q.; Sattler, T.; Leal-Taixé, L. Patch2Pix: Epipolar-Guided Pixel-Level Correspondences. In Proceedings of the 2021 IEEE/CVF Conference on Computer Vision and Pattern Recognition, Nashville, TN, USA, 20–25 June 2021; pp. 4667–4676. [CrossRef]
23. Bertinetto, L.; Valmadre, J.; Henriques, J.F.; Vedaldi, A.; Torr, P.H.S. Fully-Convolutional Siamese Networks for Object Tracking. In *Computer Vision—ECCV 2016 Workshops. ECCV 2016*; Springer: Cham, Switzerland, 2016; pp. 850–865. [CrossRef]
24. Li, B.; Yan, J.; Wu, W.; Zhu, Z.; Hu, X. High Performance Visual Tracking with Siamese Region Proposal Network. In Proceedings of the 2018 IEEE/CVF Conference on Computer Vision and Pattern Recognition, Salt Lake City, UT, USA, 18–23 June 2018; pp. 8971–8980. [CrossRef]

25. Li, B.; Wu, W.; Wang, Q.; Zhang, F.; Xing, J.; Yan, J. SiamRPN++: Evolution of Siamese Visual Tracking with Very Deep Networks. In Proceedings of the 2019 IEEE/CVF Conference on Computer Vision and Pattern Recognition, Long Beach, CA, USA, 15–20 June 2019; pp. 4277–4286. [CrossRef]
26. Simonyan, K.; Zisserman, A. Very Deep Convolutional Networks for Large-Scale Image Recognition. 2014. Available online: https://arxiv.org/abs/1409.1556 (accessed on 1 October 2021).
27. Bochkovskiy, A.; Wang, C.; Liao, H. YOLOv4: Optimal Speed and Accuracy of Object Detection. *arXiv* **2020**, arXiv:2004.10934.
28. Ge, Z.; Liu, S.; Wang, F.; Li, Z.; Sun, J. YOLOX: Exceeding YOLO Series in 2021. *arXiv* **2021**, arXiv:2107.08430.
29. Dosovitskiy, A.; Fischer, P.; Ilg, E.; Hausser, P.; Hazirbas, C.; Golkov, V.; Van Der Smagt, P.; Cremers, D.; Brox, T. FlowNet: Learning Optical Flow with Convolutional Networks. In Proceedings of the 2015 IEEE International Conference on Computer Vision, Santiago, Chile, 7–13 December 2015; pp. 2758–2766. [CrossRef]
30. Schmitt, M.; Hughes, L.H.; Zhu, X. The SEN1-2 Dataset for Deep Learning in SAR-RGB Data Fusion. *arXiv* **2018**, arXiv:1807.01569.
31. Zheng, Z.; Wei, Y.; Yang, Y. University-1652: A Multi-view Multi-source Benchmark for Drone-based Geo-localization. *arXiv* **2020**, arXiv:2002.12186.
32. Li, J.; Hu, Q.; Ai, M. RIFT: Multi-Modal Image Matching Based on Radiation-Variation Insensitive Feature Transform. *IEEE Trans. Image Process.* **2020**, *29*, 3296–3310. [CrossRef] [PubMed]
33. Li, L.; Han, L.; Ding, M.; Cao, H.; Hu, H. A deep learning semantic template matching framework for remote sensing image registration. *J. Photogramm. Remote Sens.* **2021**, *181*, 205–217. [CrossRef]
34. Dellinger, F.; Delon, J.; Gousseau, Y.; Michel, J.; Tupin, F. SAR-SIFT: A SIFT-Like Algorithm for SAR Images. *IEEE Trans. Geosci. Remote Sens.* **2015**, *53*, 453–466. [CrossRef]

remote sensing

MDPI

Article

SIVED: A SAR Image Dataset for Vehicle Detection Based on Rotatable Bounding Box

Xin Lin [1,2,3], Bo Zhang [1,2,*], Fan Wu [1,2], Chao Wang [1,2,3], Yali Yang [1,4] and Huiqin Chen [1,4]

1 Key Laboratory of Digital Earth Science, Aerospace Information Research Institute, Chinese Academy of Sciences, Beijing 100094, China
2 International Research Center of Big Data for Sustainable Development Goals, Beijing 100094, China
3 College of Resources and Environment, University of Chinese Academy of Sciences, Beijing 100049, China
4 School of Computer and Information Engineering, Heilongjiang University of Science and Technology, Harbin 150022, China
* Correspondence: zhangbo@radi.ac.cn

Abstract: The research and development of deep learning methods are heavily reliant on large datasets, and there is currently a lack of scene-rich datasets for synthetic aperture radar (SAR) image vehicle detection. To address this issue and promote the development of SAR vehicle detection algorithms, we constructed the SAR Image dataset for VEhicle Detection (SIVED) using Ka, Ku, and X bands of data. Rotatable bounding box annotations were employed to improve positioning accuracy, and an algorithm for automatic annotation was proposed to improve efficiency. The dataset exhibits three crucial properties: richness, stability, and challenge. It comprises 1044 chips and 12,013 vehicle instances, most of which are situated in complex backgrounds. To construct a baseline, eight detection algorithms are evaluated on SIVED. The experimental results show that all detectors achieved high mean average precision (mAP) on the test set, highlighting the dataset's stability. However, there is still room for improvement in the accuracy with respect to the complexity of the background. In summary, SIVED fills the gap in SAR image vehicle detection datasets and demonstrates good adaptability for the development of deep learning algorithms.

Keywords: SIVED; vehicle detection; synthetic aperture radar (SAR); complex scenarios; rotatable bounding box; deep learning

Citation: Lin, X.; Zhang, B.; Wu, F.; Wang, C.; Yang, Y.; Chen, H. SIVED: A SAR Image Dataset for Vehicle Detection Based on Rotatable Bounding Box. *Remote Sens.* **2023**, *15*, 2825. https://doi.org/10.3390/rs15112825

Academic Editors: Yanni Dong, Xiaochen Yang and Qian Du

Received: 6 March 2023
Revised: 19 May 2023
Accepted: 26 May 2023
Published: 29 May 2023

1. Introduction

The synthetic aperture radar (SAR), as an active sensor, has the capability of all-day, all-weather imaging. Due to the long wavelength of emitted electromagnetic waves, SAR can effectively identify camouflage and penetrate masked objects. Hence, with increasing resolution, SAR can acquire high-quality images and is being widely used in target surveillance and reconnaissance. The appearance features of a target are varied and influenced by its near environments. How to detect, i.e., find and locate, the target in a complex background has been an important research direction for SAR image applications. Vehicles, as a common type of land transportation, have gradually become increasingly a point of focus for researchers.

The most commonly used traditional SAR target detection method is the Constant False Alarm Rate (CFAR) algorithm, a typical detection operator based on statistical characteristics. Nevertheless, this method has a drawback in that it necessitates the manual definition of the background clutter distribution and guarding area, which ultimately leads to detection outcomes being influenced by complex scenarios and human factors.

However, deep networks have a remarkable feature abstraction ability and can automatically perform target detection in complex backgrounds. Due to their powerful performance, deep networks are at present widely used for SAR object detection, and many researchers have designed different networks or improved on existing detection networks.

For SAR ship detection, Jiao et al. [1] proposed a multi-scale neural network based on Faster R-CNN for densely connected convolutional network layers to solve the multi-scale and multi-scene SAR ship detection problem. Cui et al. [2] proposed a ship detection method based on Dense Attention Pyramid Network, which also adds attention modules to the convolutional network layer and connects them for multi-scale ship detection. Zhang et al. [3] proposed a novel balance learning network (BL-Net) to solve four imbalance problems in SAR ship detection. They also carried out work on high-speed SAR ship detection [4–6]. In addition, researchers have used a rotatable bounding box-based detection network [7–10] to cope with dense scenes, thus further improving localization accuracy. For SAR airplane detection, He et al. [11] proposed a component-based multi-layer parallel network to solve the sparsity and diversity problems that arise from the SAR scattering mechanism. Wang et al. [12] proposed a new fast detection framework, named the Efficient Weighted Feature Fusion and Attention Network (EWFAN), which conducts the automatic and rapid detection of aircrafts with high accuracy. Moreover, many researchers [13–15] have carried out work related to SAR aircraft detection. Other types of SAR object detection include oil-tank detection [16,17] and video SAR shadow detection [18], to name but a few.

Since the deep network is a data-driven approach, a considerable amount of input samples are required as training data to achieve a better performance. Therefore, corresponding datasets must be created for network training purposes. Over the past few years, the reason for the more rapid development of SAR ship detection is the richness of the SAR ship dataset [19–23]. However, the only public dataset containing vehicle targets is Moving and Stationary Target Acquisition and Recognition (MSTAR) [24], which comprises a series of military vehicle chips and clutter images and was originally intended for classification studies. In recent years, some researchers have carried out detection work using MSTAR. For example, Long et al. [25] proposed a simple rotated detection model on the dataset, where the vehicles were embedded into the clutter images for detection experiments. Zhang et al. [26] selected eight types of vehicle chips and integrated them into the background images to construct the SAR_OD dataset first and then utilized data enhancement to improve detection accuracy. Sun et al. [27] constructed a small dataset called LGSVOD by manually labeling three target classes, and proposed an improved YOLOv5 network for detection. As mentioned above, the studies all embedded vehicle targets into the clutter images for detection. In this case, although there are many vehicle categories in the MSTAR dataset, the background is homogeneous, and the distribution of vehicles is ideal and not sufficiently dense. Therefore, for civilian vehicle detection in urban-area scenes, some researchers have used MiniSAR images [28] or FARAD images [28], which are publicly available at Sandia National Laboratories (USA). Wang et al. [29] took MiniSAR images as the basis, added MSTAR data as an expansion to form a detection dataset, and used transfer learning based on the Single Shot MultiBox Detector (SSD). Zou et al. [30] proposed a SCEDet network on a band of FARAD images, and the produced dataset contained 1661 vehicles for experimental validation. Tang et al. [31] employed a CFAR-guided SSD algorithm using five MiniSAR images for experimental validation. As mentioned, in the above studies, only a portion of those images were selected for experiments, and the corresponding data were not fully explored and required a unified evaluation metric.

In this paper, we construct a SAR Image dataset for VEhicle Detection based on a rotatable bounding box named SIVED, which collects data in the X, Ku, and Ka bands from MSTAR, MiniSAR, and FARAD. The imaging areas of MiniSAR and FARAD are mainly concentrated in urban areas with complex backgrounds composed of trees, buildings, roads, and clutter. SIVED contains more than 270 dense scene chips. The rotatable bounding boxes are adopted to avoid information redundancy and reduce interference from the background and adjacent targets in dense scenes. This is also convenient for orientation estimations and aspect ratio calculations. The creation of SIVED consists of three main steps. The first step involves data preprocessing, including the removal of scenes without targets. Then, an automatic annotation algorithm based on CFAR and YOLOv5 is proposed to establish semi-automatic annotation. Finally, chips and annotation files are automatically organized to build the dataset. After construction, a complete analysis is performed for the dataset characteristics, eight rotated detection algorithms are selected to verify the stability of and challenge the dataset, and the corresponding baseline is built for the reference of relevant researchers.

The main contributions of this paper are as follows.

- Using publicly available high-resolution SAR data that includes vehicle targets, we construct the first SAR image dataset in three bands for vehicle detection. The rotatable bounding box annotation is adapted to reduce redundant background clutter and accurately position targets in dense scenes. This dataset can advance vehicle detection development and facilitate vehicle monitoring in complex terrestrial environments.
- An algorithm, combined with a detection network, is proposed for the annotation of MSTAR data to increase annotation efficiency. The annotated files contain enriched information, expanding the potential for various applications.
- Experiments are conducted using eight state-of-the-art rotated detection algorithms, which establish a relevant baseline to evaluate this dataset. The experimental results confirm the dataset's stability and the adaptability of the current algorithms to vehicle targets.

The rest of this paper is organized as follows. Section 2 presents the basic information about SIVED. Section 3 describes the construction of SIVED. In Section 4 is an analysis of the characteristics of SIVED. Section 5 introduces the architectures of the selected eight rotated detection algorithms. Presented in Section 6 are experiments conducted based on SIVED toward establishing a baseline to analyze the dataset characteristics. Section 7 provides a detailed discussion of the dataset based on the experimental results from the eight algorithms. Finally, the conclusions and outlook are given in Section 8.

2. Basic Information about SIVED

The previously mentioned open data MSTAR (X-band), MiniSAR (Ku-band) [28], and FARAD (Ka and X-bands) [28] are used to construct the dataset; the basic information of the above open data is shown in Table 1. For MSTAR, 5168 vehicle chips at 17° and 15° depression angles are selected. SIVED consists of a training, test, and validation set, and the chip size is set to 512 × 512. The statistics for the number of chips and the number of targets are shown in Table 2. Chips for vehicles located in urban areas are selected for display in Figure 1a, and the scenes contain parking lots, buildings, tree coverings, both sides of roads, etc. Since the original MSTAR chip size is 128 × 128, the chips are stitched in groups of 16 (4 × 4) to form slices of corresponding settings, as shown in Figure 1b.

Table 1. Basic information of open data.

Data	Source	Location	Band	Polarization	Resolution
FARAD	Sandia National Laboratory	Albuquerque, NM, USA	Ka/X	VV/HH	0.1 m × 0.1 m
MiniSAR	Sandia National Laboratory	Albuquerque, NM, USA	Ku	-	0.1 m × 0.1 m
MSTAR	U.S. Air Force	-	X	HH	0.3 m × 0.3 m

Table 2. Statistics on the sub-set number of SIVED in two scenarios.

	Scene	Train	Valid	Test	Total	
number of chips	urban	578	72	71	721	1044
	MSTAR	259	32	32	323	
number of vehicles	urban	5417	710	718	6845	12,013
	MSTAR	4144	512	512	5168	

(a)

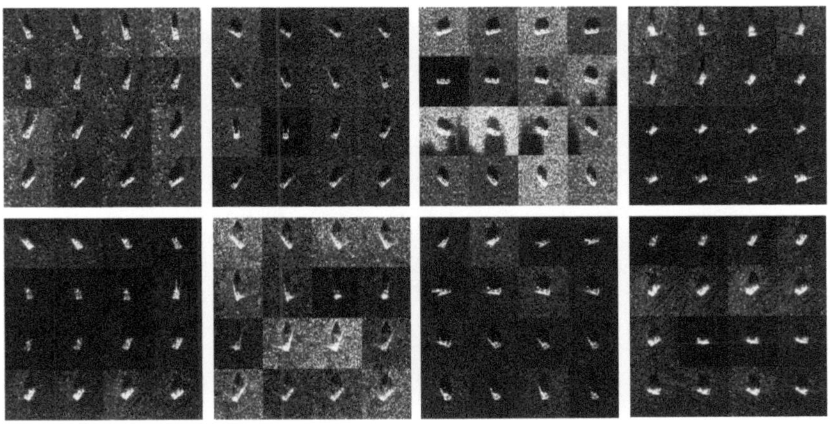

(b)

Figure 1. The chips with the size of 512 × 512 in SIVED. (**a**) Vehicles in different urban scenarios; (**b**) MSTAR chips after splicing.

SIVED is annotated with rotatable bounding boxes; one of the chips is annotated with bounding boxes as shown in Figure 2a. The row direction of the image matrix is defined as the y-axis, the column direction is defined as the x-axis, and the coordinates follow the form of (x, y) as recorded. Meanwhile, two formats of annotation files are provided. One format is derived from the DOTA [32] annotation format, using TXT files to record the annotation information of each chip, which is characterized by conciseness and clarity as well as the possibility of being directly applied to most rotated detection networks. The corresponding file format is shown in Figure 2b, where the corner coordinates are arranged clockwise. It should be mentioned that the difficulty indicates whether the labeled target is a detection difficulty instance (0 means no difficulty, and 1 means difficulty). Another from the PASCAL VOC [33] annotation format, using XML files, is employed to record the detailed information of each chip, including source data, band, resolution, polarization mode, target azimuth, etc., which is convenient for researchers to exploit these for further research; the annotation file is shown in Figure 3.

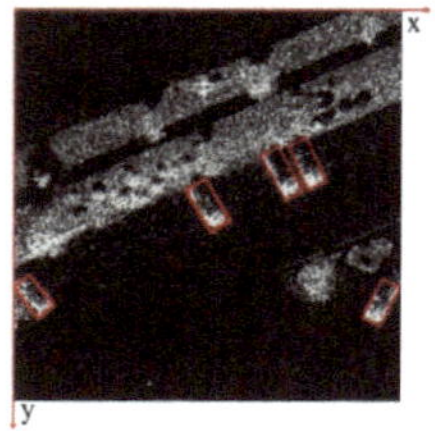

(**a**)

510 364 487 352 461 403 484 414 Vehicle 0
227 237 253 221 286 276 261 291 Vehicle 0
326 192 349 179 382 238 359 250 Vehicle 0
362 179 386 167 413 225 389 237 Vehicle 0
0 364 20 347 53 388 32 405 Vehicle 0

x1 y1 x2 y2 x3 y3 x4 y4 category difficult (format)

(**b**)

Figure 2. Presentation of annotation example. (**a**) Visualization of target annotation; (**b**) presentation of annotation TXT file.


```
<annotation>
  <folder>SIVED</folder>
  <filename>FARAD_183.xml</filename>
  <source>FARAD</source>
  <band>Ka</band>
  <resolution>0.1</resolution>
  <polarization>VV</polarization>
  <size>
    <width>512</width>
    <height>512</height>
    <depth>1</depth>
  </size>
  <segmented>0</segmented>
  <object>
    <type>rbndbox</type>
    <name>Vehicle</name>
    <difficult>0</difficult>
    <rbndbox>
      <cx>485.5</cx>                         // x-coordinate of the center point
      <cy>383.0</cy>                         // y-coordinate of the center point
      <w>57.245087125446844</w>              // width
      <h>25.942243542145693</h>              // height
      <angle>-62.52556837372287</angle>      // unit is degree(°)
    </rbndbox>
  </object>
  </object>...<object>
  </object>...<object>
  </object>...<object>
  </object>...<object>
</annotation>
```


Figure 3. Presentation of an annotation XML file.

3. Dataset Construction

This section explains the details of the dataset construction process, as shown in Figure 4. It consists of three main steps: data preprocessing and selection, semi-automatic annotation, and dataset production. The critical step is semi-automatic annotation, designed with a CFAR operator and neural network.

3.1. Data Preprocessing and Selection

The public dataset comprises 351 FARAD-Ka images, 61 FARAD-X images, and 20 MiniSAR images, all of which were converted into TIFF files using the officially provided script files. Upon visual interpretation, it was discovered that specific images did not contain vehicles, while others had reduced resolution due to sampling issues, rendering vehicle extraction ineffective. This situation is demonstrated in Figure 5. Ultimately, a total of 17 MiniSAR images, 53 FARAD-Ka images, and 13 FARAD-X images were selected as annotated images.

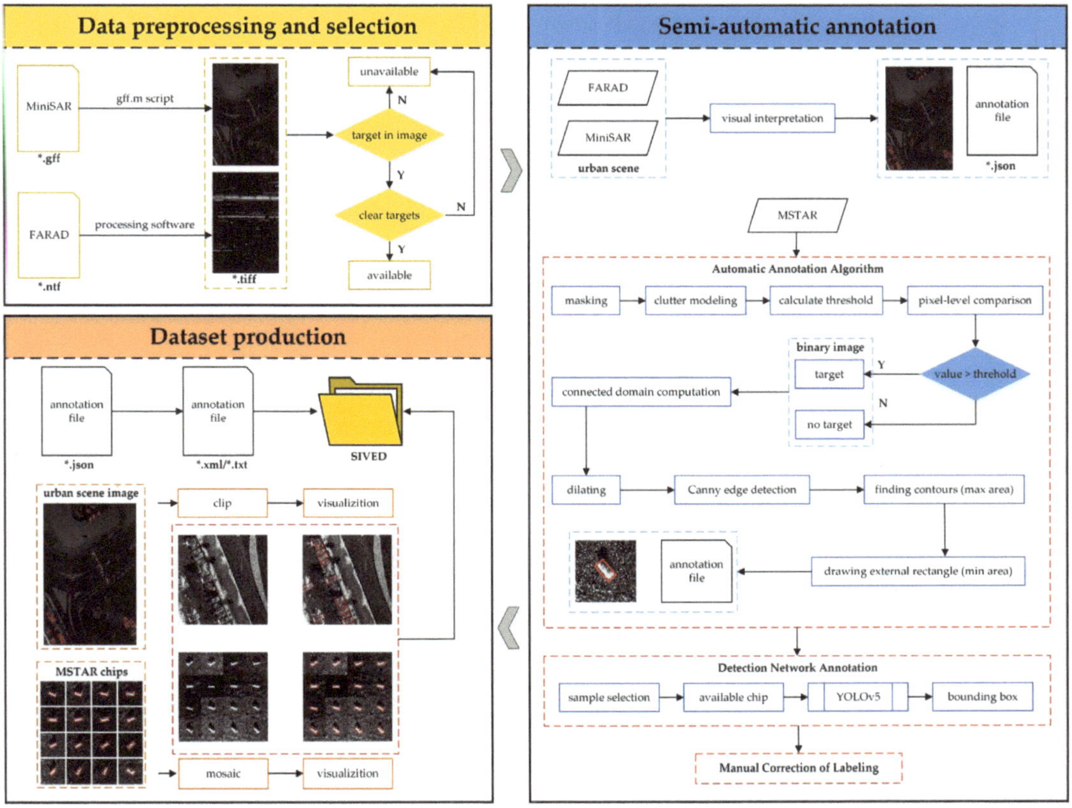

Figure 4. The construction workflow of SIVED, * represents file name.

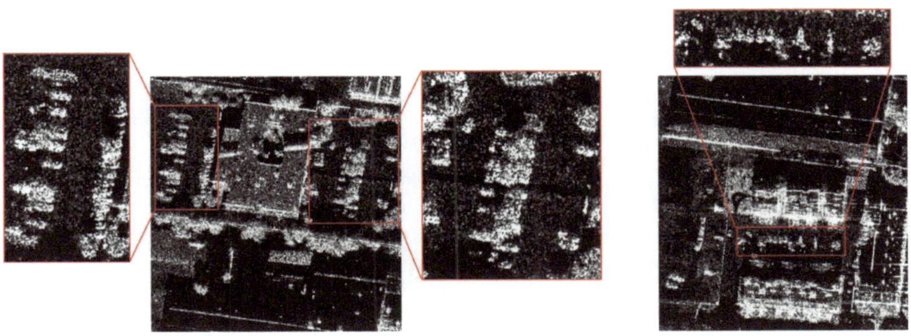

(a) (b)

Figure 5. The presentation of vehicles that cannot be extracted effectively: (**a**) area 1; (**b**) area 2.

Before the annotation process, special cases need to be described. In urban scene images, there are usually small buildings that match the size of vehicles and show strong scattering features. They are easily confused, but the geographic location does not change, so we used Google Earth to compare and identify them. The corresponding example is shown in Figure 6.

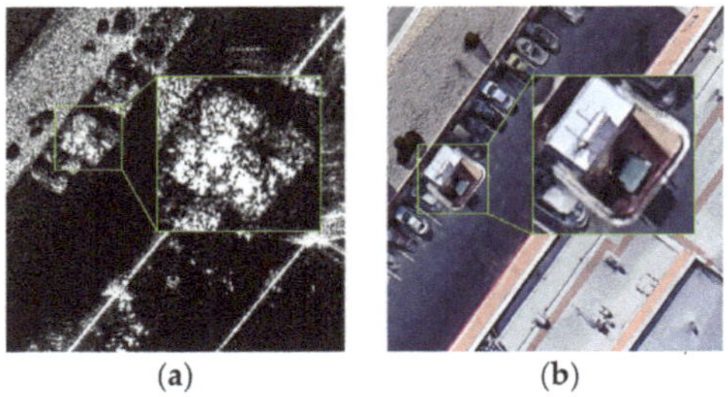

Figure 6. Example of a small building easily confused with vehicle targets in (**a**) SAR image and (**b**) Google Earth.

Meanwhile, there will be regions where trees and vehicle targets overlap each other, which can be divided into two cases: the target is obscured by trees but exhibits strong scattering properties and contains the combined scattering information of vehicles and trees, as shown in Figure 7a; the target is partially obscured by trees, and the obscured part is entirely invisible, as shown in Figure 7b. In the first case, the target is labeled according to the complete profile, and in the second case, only the visible part of the target is marked.

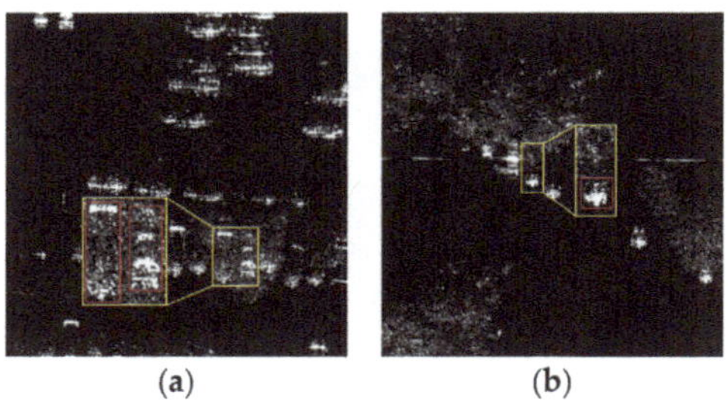

Figure 7. Example in regions where trees and vehicles overlap each other, red boxes represent labeled vehicles: (**a**) completely covered but visible situation; (**b**) partially covered and partially visible case.

3.2. Semi-Automatic Annotation of SIVED

The vehicle targets in urban scenes are manually annotated by visual interpretation. For the vehicles in MSTAR, semi-automatic annotation is used; the specific process is shown in Figure 4, including the automatic annotation algorithm, detection network annotation, and manual correction. The details of each step are as follows.

1. Algorithm Automatic Annotation: In this paper, an automatic annotation algorithm is designed for MSTAR; the input is the MSTAR chips, and the output is the visual annotation box and the coordinate values of the four corner points. After the chip is input, a 30 × 30 area in the center is first selected for masking. The main purpose is to roughly mask the target area to prevent pixel leakage, which affects the clutter distribution estimation. The second step for clutter estimation is to select the Rayleigh distribution [34], whose probability density function (PDF) is described as follows:

$$p(x) = \frac{x}{\sigma^2} e^{-\frac{x^2}{2\sigma^2}}, x > 0 \quad (1)$$

where x denotes the pixel values that the range is defined as [0, 255], and σ^2 denotes the variance of the pixel values in the estimated region.

In the third step, given the false alarm rate $P_{fa} = 0.00001$, according to PDF and P_{fa} to calculate the threshold T, the relationship between P_{fa} and T as follows:

$$P_{fa} = 1 - \int_0^T p(x)dx \quad (2)$$

In the fourth step, a pixel-by-pixel comparison is performed; those larger than the threshold are determined as target pixels for forming a binary image. In the fifth step, the connected domain is calculated for the binary image, the one with the largest area will be extracted as the target, and the target is then expanded to make the target structure more complete. In the sixth step, Canny edge detection [35] is performed; next, the edges are extracted, and the contour containing the largest area is then found, and the minimum external moments are drawn. Finally, the coordinates are output.

2. Detection Network Annotation: The previous step automatically outputs 5168 annotation boxes for all chips. Next, a total of 2893 chips, about 56% of the total number of chips, are confirmed by visual interpretation, and the position of its annotation boxes are accurate. Then, these chips form a sample set that is fed into the rotated YOLOv5 network for precise detection. Finally, with a total of 50 epochs trained and the weights retained, the remaining 2275 chips are fed into the trained network, and the coordinates of the detection boxes are output.
3. Manual Correction of Labeling: Eventually, 162 detected boxes are offset, accounting for about 7% of the network output chips. They only contain part of the target or contain more non-target pixels and are corrected by manual labeling. Finally, MSTAR data annotation is complete.

Compared with manual labeling, the proposed semi-automatic annotation algorithm improves efficiency and saves on labor cost. Ultimately, the manually modified chips account for about 3% of the total, significantly reducing the workload and labeling time.

3.3. Dataset Production

If the pixel area of the vehicle in the chips is larger than 80% of the original pixel area during the crop, it is labeled as a positive sample, which means a sample that is retained in the chip. The pixel values are converted to 8-bit integer data, and the annotation information is output to DOTA format and PASCAL VOC format as TXT files and XML files, respectively. For MSTAR chips, the stitching operation is performed, and the annotation files are integrated to add offsets to the coordinates of each target. Next is visualization of the annotation boxes. Finally, all chips are divided into the training, test, and validation sets according to the ratio of 8:1:1.

3.4. File Structure

The dataset is constructed, and its file structure is presented in Figure 8. The Annotations folder contains the annotation files in XML format, while JPEGImages holds all the image chips. VisualImages contains the annotated visual chips, and ImageSets holds the dataset that has been divided into the training, test, and validation sets. These sets mainly contain image chips and annotation files in TXT format. It is worth noting that all images and annotation files are named in the form of source with a serial number, such as FARAD_1.*, and the front sub-name indicates its source to enable researchers to freely select either urban scene data (FARAD and MiniSAR) or MSTAR simple scene data. Additionally, relevant experiments will be conducted in later chapters to provide benchmark metrics under different scenes, thereby facilitating convenience for the researchers.

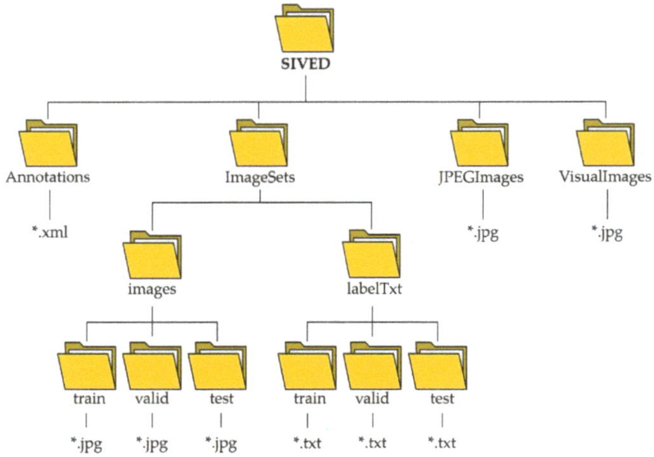

Figure 8. Structure of SIVED, * represents file name.

4. Dataset Analysis

Two primary metrics are considered to provide an overall analysis of the dataset: the scale distribution and the angle distribution of the vehicles. The scale distribution is specific to all datasets and can be used to analyze the target scale that the dataset focuses on and its level of challenge for detection algorithms. Meanwhile, the angle distribution takes the imaging mechanism of SAR into consideration. Assuming all other conditions are the same, the target features will significantly differ at different azimuth angles. The presence of targets at various angles will contribute to the network's ability to extract features and rapidly converge toward stability in the training procedure.

4.1. Scale Distribution Analysis

The scale of a target can be defined as the pixel area it occupies. In the COCO dataset [36], targets are classified as small, medium, or large based on their pixel area range, as shown in Table 3. The detection of small targets is generally considered to be the most challenging task, and this is attributed to three main characteristics of small-target detection:

1. Few features are available, meaning less visual information, and extracting features with discriminatory power is complicated. Moreover, this process is easily disturbed by environmental factors.
2. High localization accuracy is required because small targets cover a small area in the image. An offset of even a single pixel point from the bounding box can cause significant errors in the prediction process.
3. Dense scenes are common, and small targets in proximity are prone to interference from neighboring targets.

Table 3. The official definition of the COCO dataset for the size of targets.

Type	Area
Small target	$0 < a \leq 32^2$
Medium target	$32^2 < a \leq 96^2$
Large target	$a > 96^2$

The target scale distribution of our dataset in each set is shown in Figure 9. Small targets make up 54.5% of the dataset, and 99.8% of the targets have an area less than 3072 (which is three times the area of small targets). This indicates that the dataset mainly comprises small- and medium-sized targets, making it challenging for detection algorithms to accurately identify them.

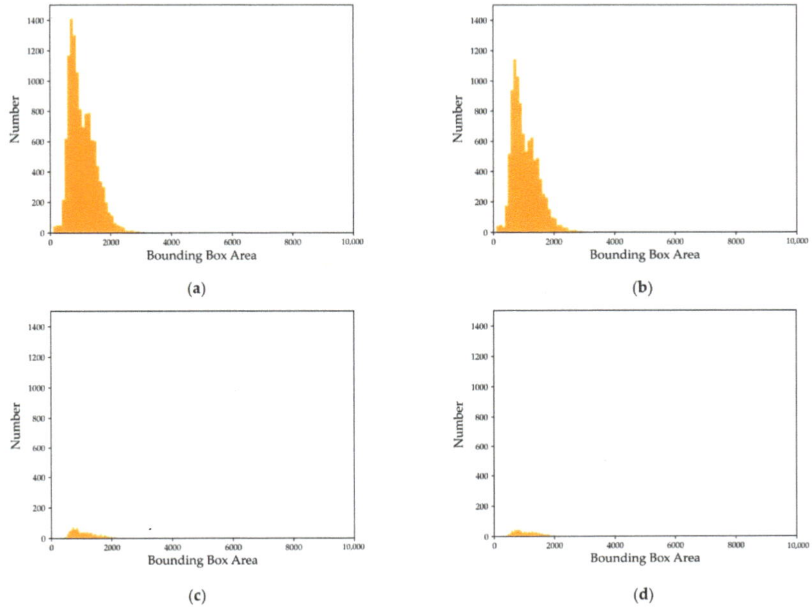

Figure 9. Statistical chart of vehicle scale size in SIVED: (**a**) entire dataset; (**b**) training set; (**c**) valid set; (**d**) test set.

4.2. Angle Distribution Analysis

Due to the difficulty in determining the direction of a vehicle's head, the azimuthal definition of 0°~360° cannot be used to evaluate the target angle. Instead, the long edge definition method is employed. As depicted in Figure 10, the range is defined as [−90°,90°), where the clockwise direction is positive. This method takes into account the relatively extensive and symmetrically distributed natural angle range, resulting in more uniform characterization of the target angle distribution. Figure 11 presents the angle distribution statistics of SIVED, which demonstrate that the targets are distributed at different angles, thereby promoting the integrity of feature extraction and dataset stability.

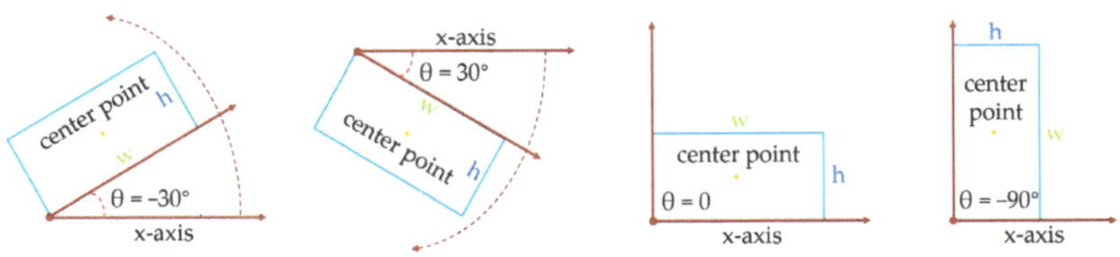

Figure 10. Rotatable bounding box long edge definition method.

4.3. Properties of SIVED

SIVED has three main properties: richness, stability, and challenge. It contains SAR data in X, Ku, and Ka bands, which enrich the target features compared to previous SAR datasets that lack multi-source image. Additionally, SIVED includes vehicle targets in multiple scenarios and with various properties, making it suitable for meeting actual mission requirements.

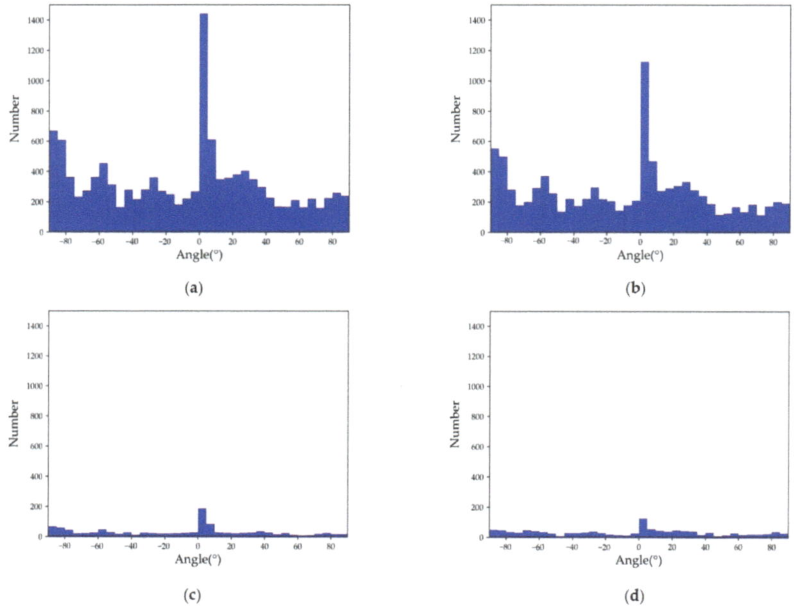

Figure 11. Statistical chart of vehicle angle in SIVED: (**a**) entire dataset; (**b**) training set; (**c**) valid set; (**d**) test set.

All the vehicles in SIVED are distributed at different azimuths, which is beneficial for maintaining the integrity of target characteristics under SAR imaging. In the later section, advanced detection network experiments will be used to verify its stability.

SIVED primarily focuses on small targets and complex backgrounds, and it has been designed to satisfy this requirement, as analyzed in Section 4.1. Detecting small targets is always challenging in object detection tasks, and urban areas, in particular, pose an added challenge with multiple features and tiny structures that can be easily confused with vehicle targets.

5. Methodology

Currently, there are two mainstream classification methods for deep detection algorithms. The first method divides algorithms into two-stage and one-stage based on their process. The second method categorizes algorithms as anchor-based or anchor-free, depending on whether they use anchors. Additionally, there are different regression strategies or designed modules for detecting rotatable bounding boxes. In this paper, we will experiment with eight different models to verify the stable reliability of the dataset. We will describe these models in the following sections.

5.1. Object Detection Framework Overview

Most object detection networks consist of four parts: input, backbone, neck, and head, as shown in Figure 12. The backbone is responsible for feature extraction and typically uses classical CNN networks such as VGG [37] or ResNet [38]. In this paper, the backbone of all networks is ResNet50 [38]. The neck plays a top-down role and is mainly used for feature fusion, which makes the features learned by the network richer and more diverse and improves the detection performance. The neck used in this paper in all cases is feature pyramid networks (FPN) [39]. The head, also known as the detection head, is responsible for generating network outputs, making predictions using the extracted features, and ultimately outputting the object's class and location.

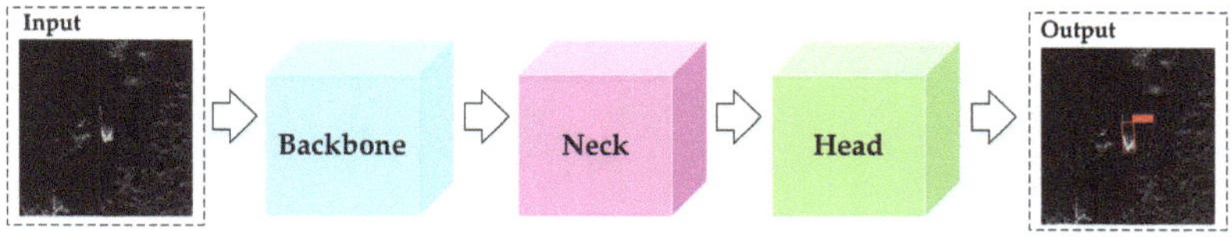

Figure 12. The architecture of the object detection framework.

5.2. Rotated Faster-RCNN

Faster-RCNN [40] is a two-stage detection algorithm with a structure shown in Figure 13. The main parts include the feature extraction network, region proposal network (RPN), and ROI pooling. As mentioned earlier, the feature extraction network is the backbone, while the RPN and ROI pooling layers form the head. The RPN generates candidate regions, and the ROI pooling layer extracts those regions from the feature map for final classification and regression. Angular parameters are added to the network for regression to form the final Rotated Faster R-CNN.

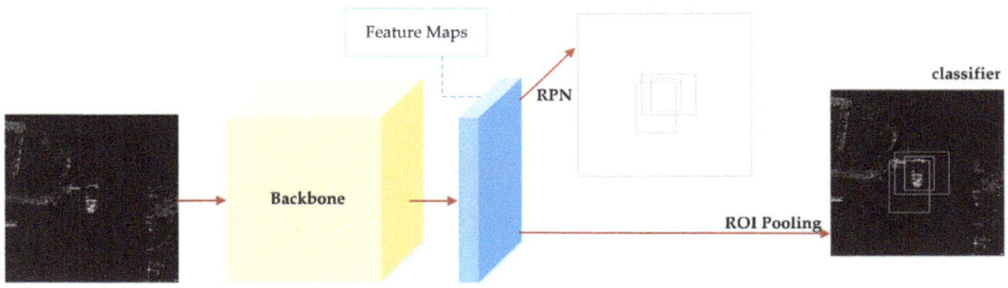

Figure 13. The architecture of Faster R-CNN.

5.3. Rotated RetinaNet

RetinaNet [41] is a one-stage network with a typical structure, as shown in Figure 14. The feature maps of different dimensions extracted by the ResNet50 backbone are input to the FPN, and the classification and box regression sub-networks complete the detection process. Focal Loss, proposed by the authors to balance positive and negative samples, is the core contribution of the paper. By introducing Focal Loss, one-stage detection performance surpasses that of two-stage detection networks for the first time. The same technique is applied for angle parameter regression, resulting in the creation of Rotated RetinaNet.

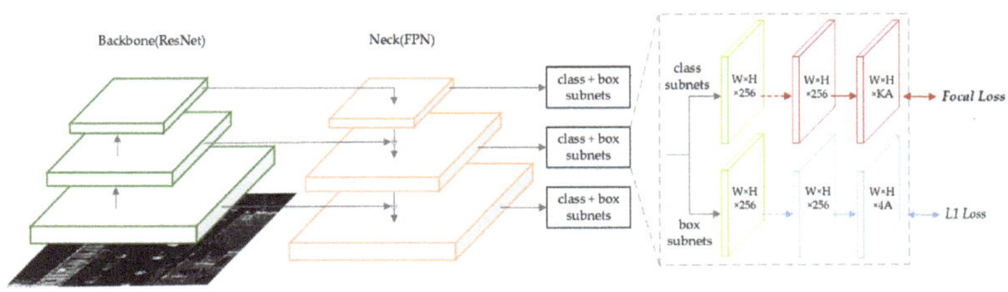

Figure 14. The architecture of RetinaNet.

5.4. Rotated FCOS

FCOS [42] is a typical one-stage anchor-free network with a structure shown in Figure 15. It outputs fused feature maps through the backbone network and FPN. Later, it outputs feature vectors at each point location on each feature map through the detection head network, which is used to represent the final bounding box results. In contrast to anchor-based networks, the detection head network in FCOS does not use anchors for regression but introduces center-ness to describe the distance from the point position in the feature map to the center of the target to achieve target localization. By including the calculation of angular loss, it becomes Rotated FCOS.

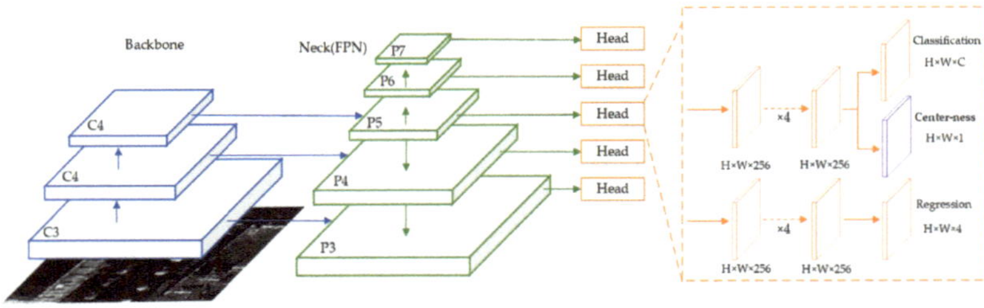

Figure 15. The architecture of FCOS.

5.5. S^2A-Net

Single-Shot Alignment Network (S^{2}A-Net) [43] is a one-stage detector for rotated objects with the structure shown in Figure 16. It consists of two modules: the Feature Alignment Module (FAM) and the Orientation Detection Module (ODM). The FAM can generate a high-quality anchor using the Anchor Refinement Network and adaptively align the convolutional features according to the anchor box using novel alignment convolution. The ODM first employs an active rotation filter for encoding, and direction-sensitive and direction-invariant features are then generated to mitigate the inconsistency between classification scores and localization accuracy.

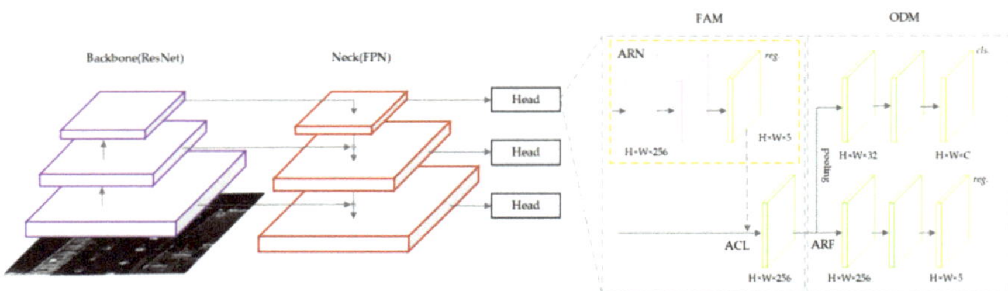

Figure 16. The architecture of S^{2}A-Net.

5.6. RoI Transformer

RoI Transformer [44] is a module used in rotated two-stage detectors. It consists of two parts: the first part is the supervised rotated RoI learner, which learns the transformation parameters from horizontal RoIs to rotated RoIs; the second part is the Rotated Position Sensitive RoI Alignment module, which extracts rotationally invariant features from rotated RoIs to enhance target classification and boundary regression.

5.7. Gliding Vertex

The Gliding Vertex [45] is also a simple yet effective strategy for rotated box regression. It replaces the direct regression of the four vertices of the rotated box with sliding of the vertices of the horizontal bounding box on each corresponding edge to accurately describe multi-directional objects. The parameter representation is shown in Figure 17. Specifically, the algorithm regresses four length ratios that characterize the relative glide offsets of each corresponding face. This approach facilitates the learning of offsets and avoids confusion in the sequential labeling points of oriented objects. To further address the confusion problem for near-horizontal objects, the authors introduce a skewness factor based on the area ratio between an object and its horizontal bounding box. This factor guides the choice of horizontal or oriented detection for each object. When using this method, the overall network architecture is referred to as the structure of Rotated Faster-RCNN, and five additional variables are simply added to the regression headers.

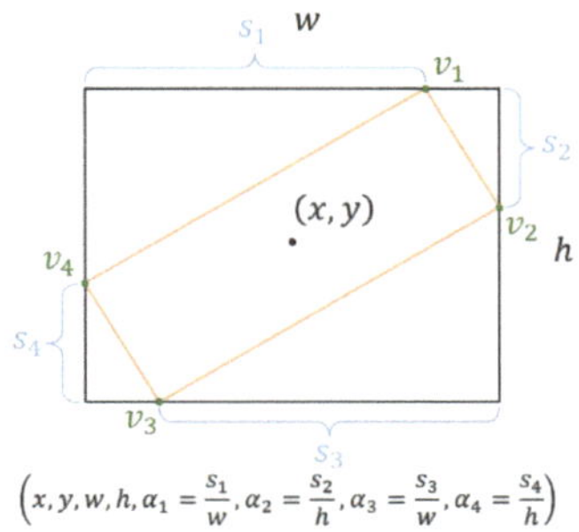

Figure 17. Parameter representation of Gliding Vertex.

5.8. Oriented RepPoints

Oriented RepPoints [46] adopts a strategy for rotated bounding box regression that generates adaptive point sets for geometric structure representation using an adaptive points learning method. It employs three directed transformation functions that aid in accurate classification and localization with proper orientation. The architecture of Oriented RepPoints is shown in Figure 18. To select a representative sample of oriented points during training, an adaptive points assessment and assignment scheme is proposed for adaptive point learning that can capture non-axis aligned features from neighboring objects or background noise. A spatial constraint is introduced to penalize outlier points to enable adaptive learning. This method uses a one-stage network architecture, which is similar to RetinaNet while retaining Focal Loss. The corresponding detection network is formed by adding the corresponding strategy to the head.

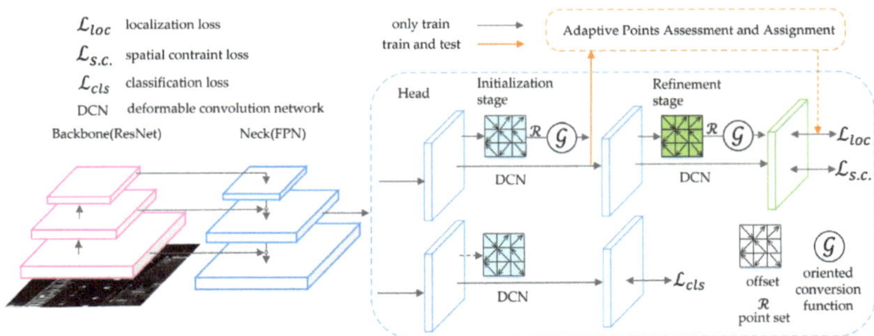

Figure 18. The architecture of Oriented RepPoints.

5.9. KLD

KLD [47] is a strategy for rotated bounding box regression. Specifically, it first converts the rotated bounding box into a two-dimensional Gaussian distribution and then calculates the Kullback–Leibler divergence (KLD) between the Gaussian distributions as the regression loss. KLD can also be understood as a new loss function and will be added to the Rotate RetinaNet framework in the experiments of this paper.

Previous studies have successfully applied the Faster-RCNN, Retinanet, and FCOS frameworks to SAR ship detection [10,48–50], demonstrating their versatility in characterizing and learning from SAR images. Meanwhile, the Oriented RepPoints, Gliding Vertex RoI Transformer, and KLD methods are novel regression strategies, and S^2A-Net is a model tailored to optical remote sensing images; these methods are awaiting evaluated for dataset compatibility. These various methods represent both stable classical and emerging algorithms. Thus, they are suitable for establishing a baseline and validating dataset performance.

6. Experimental Results

In this section, we will conduct experiments to validate the dataset using the rotated detection algorithms selected in Section 5. The purpose of these experiments is to test the dataset's stability, compare the performance of different algorithms, and analyze the algorithm that achieves the best results.

6.1. Experimental Setup and Evaluation Metrics

The experiments were conducted with Ubuntu 18.04, using an Intel(R) Xeon(R) Gold 6230R CPU @ 2.10 GHz, and an NVIDIA Geforce RTX 3090 GPU. The detection algorithms are based on the mmrotate [51] deep learning framework, with the SGD optimizer, initial learning rate set to 0.01, momentum set to 0.937, and weight decay set to 0.0005. The batch-size is set to 8, the number of training epochs is set to 150, and a warm-up of 500 iterations is performed before training, with the warm-up learning rate set as one-third of the initial learning rate. During training, the validation set was used for verification, and the optimal weights were saved for testing.

To evaluate the effectiveness of the detection models on the constructed dataset, various metrics such as recall (*R*), precision (*P*), and mean average precision (*mAP*) were introduced. These metrics evaluate the model's ability to recognize objects in each image as a whole. The definitions for these metrics are as follows:

$$P = \frac{TP}{TP+FP}; R = \frac{TP}{TP+FN}; mAP = \int_0^1 P(R)dR \tag{3}$$

the *mAP* metric is calculated using *TP* (the number of detected targets), *FN* (the number of missed targets), and *FP* (the number of false alarms). For a detailed explanation of the *mAP* metric, please refer to Table 4, where the IOU represents the overlap ratio between the predicted bounding box and the actual labeled box to the total area of both boxes.

Table 4. Definition of metrics.

Metrics	Explanation
mAP	mAP when IOU = 0.5:0.05:0.95
mAP_{50}	mAP when IOU = 0.5
mAP_{75}	mAP when IOU = 0.75

6.2. Experiments for Baselines

The evaluation experiments for the dataset were performed using the eight networks described in Section 5. The corresponding test set evaluation metric values for each network are presented in Table 5. The results show that all networks achieved an mAP_{50} above 95% except the RoI Transformer. Additionally, the recall remained above 97% for all networks except the Rotated FCOS and RoI Transformer, and precision remained above 90% for all networks except the RoI Transformer. Notably, Oriented RepPoints demonstrated the best performance, maintaining the highest mAP across different definitions.

Table 5. Metric values of different networks on SIVED.

Network	Recall *	Precision *	mAP *	mAP_{75} *	mAP_{50} *
RoI Transformer	95.61	84.40	37.45	16.91	93.47
Rotated FCOS	88.86	**96.47**	50.40	48.13	95.60
S^2A-Net	97.48	90.90	55.49	57.32	97.72
Rotated RetinaNet	97.48	92.73	53.11	50.93	97.76
KLD	98.05	93.34	57.49	64.48	97.92
Rotated Faster R-CNN	97.80	95.55	56.03	59.00	98.09
Gliding Vertex	**98.13**	95.72	56.06	59.59	98.33
Oriented RepPoints	98.05	95.11	**60.15**	**70.69**	**99.13**

* All metric values are in percentages (%).

Figure 19 illustrates the detection results of various algorithms in typical scenes. The first and second rows depict the background of strongly scattering buildings, while the third row presents a dense scene with trees. The fourth row displays easily confused small objects, and the fifth and sixth rows depict cluttered scenes. Finally, the seventh row displays MSTAR chips. In urban scenarios, all algorithms exhibited instances of missing detections and false alarms.

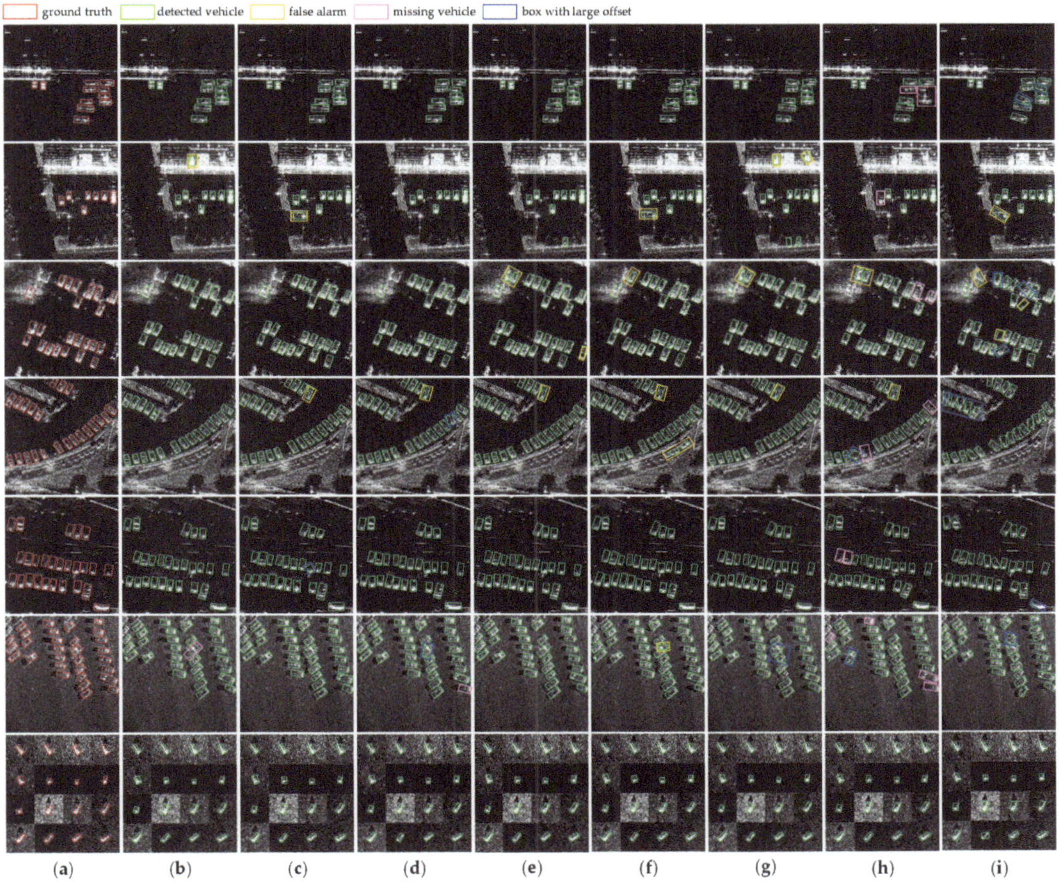

Figure 19. Ground truth and detection results of eight detection networks. The red boxes in the first column represent ground truths, while the green boxes in other columns denote detected vehicles. False alarms are shown in yellow boxes, missing vehicles in pink boxes, and bounding boxes with large offsets in blue boxes. (**a**) Ground truth; (**b**) Oriented RepPoints; (**c**) Gliding Vertex; (**d**) Rotated Faster R-CNN; (**e**) KLD; (**f**) Rotated RetinaNet; (**g**) S^2A-Net; (**h**) Rotated FCOS; (**i**) RoI Transformer.

6.3. Additional Experiments

We conducted two additional experiments, and the details are as follows:

1. An experimental test to establish baseline metrics for two different scenarios: urban and MSTAR simple. The training set comprises data from all scenarios. These metrics can serve as a point of reference for researchers when selecting challenging datasets for their own experiments. As shown in Table 6, the mAP_{50} score is higher for the MSTAR scene test than for the urban scene, reflecting the greater complexity and interference that exists in urban settings.

2. To verify the superiority of the SIVED dataset over FARAD and MSTAR, we chose a simple structured Rotated RetinaNet, trained with those three datasets, respectively. We compared the performance of the network trained with those datasets via the specific metric values shown in Table 7. The results show that the constructed SVID dataset improved the network performance.

Table 6. Values of mAP_{50} in different scenarios.

Network	mAP_{50}_all *	mAP_{50}_urban *	mAP_{50}_MSTAR *
RoI Transformer	93.47	89.23	97.91
Rotated FCOS	95.60	92.19	98.86
S^2A-Net	97.72	96.69	97.91
Rotated RetinaNet	97.76	96.45	99.32
KLD	97.92	97.46	98.83
Rotated Faster R-CNN	98.09	97.53	98.76
Gliding Vertex	98.33	96.71	**100**
Oriented RepPoints	**99.13**	**98.34**	99.73

* All metric values are expressed as percentages (%). The metric mAP50_all indicates that the test set contains data from all scenarios, while mAP50_urban and mAP50_MSTAR indicate that the test sets only contain data from the urban (FARAD and MiniSAR) and MSTAR scenarios, respectively.

Table 7. Values of mAP_{50} in different scenarios.

Training data	mAP_{50}_FARAD *	mAP_{50}_MSTAR *
FARAD	96.53	-
MSTAR	-	99.24
SIVED	**96.77**	**99.32**

* All metric values are expressed as percentages (%). The metric mAP50_FARAD indicates that the test set only contains data from FARAD, while mAP50_MSTAR indicates that the test set only contains data from MSTAR.

7. Discussion

As discussed in the previous section, currently available datasets for SAR image target detection are primarily focused on ship targets. However, as ships are much larger than vehicles in terms of physical scale, our dataset is positioned to address the detection of small targets defined based on image pixels. Thus, SIVED would serve as a valuable complement to constructing a multiple-target dataset with various scales. Compared to the typical SAR ship dataset SSDD [21], SIVED offers rich associated information in its annotation file. Although SSDD employs rotatable bounding boxes, the annotation file only includes the position and angle of the target. In contrast, SIVED is annotated with the source of the slices and the basic information of the corresponding sensors, such as band, polarization mode, and resolution. This additional information will be useful in small-target detection research based on SAR imaging mechanisms. SSDD contains 1160 chips and 2587 ship targets, with an average of two targets per chip. On the other hand, SIVED's urban area comprises 721 chips and 6845 vehicle targets, indicating an average of nine targets per chip. This suggests that SIVED's distribution of targets is denser while the target capacity is larger, which puts forward a higher challenge for the positioning accuracy of the detection algorithm. Furthermore, rich land features and clutters constitute a complex background.

The characteristics of vehicles in SAR imaging are closely related to the wavelength of the band used. Typically, longer wavelengths lead to a deeper penetration but worse characterization ability for target details, whereas shorter wavelengths result in a weaker penetration but better characterization ability for target details. In this paper, we constructed a dataset of Ka, Ku, and X bands, with sequential increases in their wavelengths. Figure 20 shows the imaging results of different bands in FARAD, and it is evident that the vehicles in Ka-band images exhibit more texture features than those in X-band images. The inclusion of images of different bands indicates that the features become rich, enabling the network to learn more knowledge and improve the generalization ability. In addition, due to the side-view imaging of SAR and the penetration of the microwave, it can form a mixed area of vehicles and trees as shown in Figure 7, where the pixel information is formed by the combination of trees and vehicles, and the vehicle targets in the above scenes are labeled, which further enriches the features of the dataset while making full use of SAR imaging characteristics.

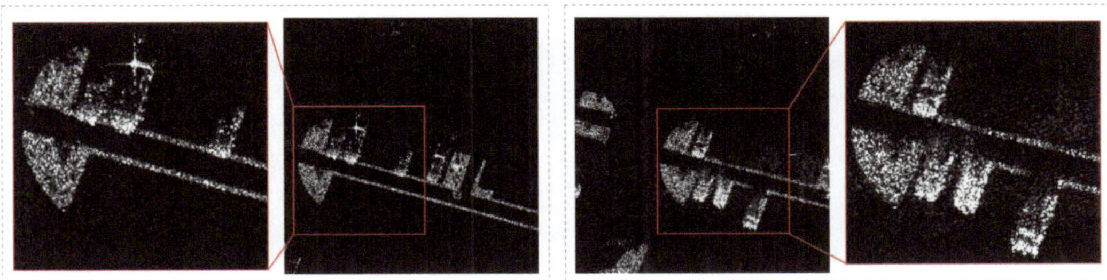

(a) (b)

Figure 20. Chips of FARAD of different bands: (**a**) Ka band; (**b**) X band.

As a high recall indicates a high rate of target detection, the results of the experiment in Table 4 suggest that the dataset is relatively stable. However, the precision does not reach the same level as the recall, resulting in more false alarms, which means that the background was wrongly identified as the target. This finding verifies the complexity of the dataset's background. It provides further evidence that the dataset is challenging but still maintains stability. Generally, the one-stage network has a lower detection accuracy compared to the two-stage network since the two-stage network distinguishes the background from the target in the RPN, whereas the one-stage network performs regression directly, resulting in an imbalance between the categories of background and target. Nonetheless, Oriented RepPoints utilizes the Focal Loss in RetinaNet to provide appropriate weight control to cross-entropy loss and focus loss calculation on target categories, which improves upon the influence caused by the imbalance between background and target. Combined with the adaptive improvements for rotatable bounding box regression, it thus obtains the highest performance.

The visual detection results presented in Figure 18 reveal that the phenomenon of missing detection and false alarm exists in the detection results of different algorithms. This finding is in line with the actual scenario, where there is greater interference and difficulty in distinguishing dense targets in urban environments, ultimately highlighting the challenge of the dataset.

When constructing SIVED, multiple sources of data are used, making full use of the SAR imaging mechanism, and sensor information is recorded during annotation, which endows the dataset with richness. Compared with the typical ship dataset, SIVED mainly focuses on small targets, and contains more dense scenes and complex backgrounds, which make the dataset challenging. The targets are distributed at different angles, and the different algorithms maintain high recall values in the experiment, which demonstrates the stability of the dataset. In summary, SIVED exhibits three properties: richness, stability, and challenge.

8. Conclusions

This paper presents the construction of a new SAR image vehicle detection dataset, named SIVED, which addresses the lack of rich scene datasets in the field. It includes 120,133 vehicle targets and more than 270 dense scene chips. Targets are annotated by rotatable bounding boxes, which improve localization accuracy and reduce interference from background and adjacent targets, and a semi-automatic annotation method is proposed to improve efficiency. Robust baselines are created through eight models in a unified framework. SIVED has three main properties: richness, reliability, and challenge. It comprises complex urban context and rich dense scenes, which are not available in other SAR datasets. Although the images currently in use are limited, we believe SIVED will be a valuable resource for advancing the development of vehicle detection technology for SAR imagery. We will also make further additions to the dataset in the future.

Author Contributions: Conceptualization, X.L. and B.Z.; methodology, X.L.; software, X.L.; validation, B.Z.; formal analysis, X.L.; investigation, B.Z.; resources, B.Z.; data curation, X.L.; data analysis, Y.Y. and H.C.; writing—original draft reparation, X.L.; writing—review and editing, B.Z. and F.W.; visualization, X.L.; supervision, B.Z.; project administration, C.W. and B.Z.; funding acquisition, C.W. All authors have read and agreed to the published version of the manuscript.

Funding: This research was funded by the National Natural Science Foundation of China (Grant No. 41930110).

Data Availability Statement: The dataset was shared on the GitHub web for researchers. The link is https://github.com/CAESAR-Radi/SIVED (accessed on 15 May 2023).

Acknowledgments: The authors would like to thank Sandia National Laboratory (USA) for providing public MiniSAR and FARAD data online and the U.S. Air Force for providing MSTAR data online.

Conflicts of Interest: The authors declare no conflict of interest.

References

1. Jiao, J.; Zhang, Y.; Sun, H.; Yang, X.; Gao, X.; Hong, W.; Fu, K.; Sun, X. A densely connected end-to-end neural network for multiscale and multiscene SAR ship detection. *IEEE Access* **2018**, *6*, 20881–20892. [CrossRef]
2. Cui, Z.; Li, Q.; Cao, Z.; Liu, N. Dense attention pyramid networks for multi-scale ship detection in SAR images. *IEEE Trans. Geosci. Remote Sens.* **2019**, *57*, 8983–8997. [CrossRef]
3. Zhang, T.; Zhang, X.; Liu, C.; Shi, J.; Wei, S.; Ahmad, I.; Zhan, X.; Zhou, Y.; Pan, D.; Li, J. Balance learning for ship detection from synthetic aperture radar remote sensing imagery. *ISPRS J. Photogramm. Remote Sens.* **2021**, *182*, 190–207. [CrossRef]
4. Zhang, T.; Zhang, X. High-speed ship detection in SAR images based on a grid convolutional neural network. *Remote Sens.* **2019**, *11*, 1206. [CrossRef]
5. Zhang, T.; Zhang, X.; Shi, J.; Wei, S. Depthwise separable convolution neural network for high-speed SAR ship detection. *Remote Sens.* **2019**, *11*, 2483. [CrossRef]
6. Zhang, T.; Zhang, X.; Shi, J.; Wei, S. HyperLi-Net: A hyper-light deep learning network for high-accurate and high-speed ship detection from synthetic aperture radar imagery. *ISPRS J. Photogramm. Remote Sens.* **2020**, *167*, 123–153. [CrossRef]
7. An, Q.; Pan, Z.; Liu, L.; You, H. DRBox-v2: An improved detector with rotatable boxes for target detection in SAR images. *IEEE Trans. Geosci. Remote Sens.* **2019**, *57*, 8333–8349. [CrossRef]
8. Chen, C.; He, C.; Hu, C.; Pei, H.; Jiao, L. MSARN: A deep neural network based on an adaptive recalibration mechanism for multiscale and arbitrary-oriented SAR ship detection. *IEEE Access* **2019**, *7*, 159262–159283. [CrossRef]
9. Chen, S.; Zhang, J.; Zhan, R. R2FA-Det: Delving into high-quality rotatable boxes for ship detection in SAR images. *Remote Sens.* **2020**, *12*, 2031. [CrossRef]
10. Xu, C.; Zhang, B.; Gao, J.; Wu, F.; Zhang, H.; Wang, C. FCOSR: An anchor-free method for arbitrary-oriented ship detection in SAR images. *J. Radars* **2022**, *11*, 1–12. [CrossRef]
11. He, C.; Tu, M.; Xiong, D.; Tu, F.; Liao, M. A Component-Based Multi-Layer Parallel Network for Airplane Detection in SAR Imagery. *Remote Sens.* **2018**, *10*, 1016. [CrossRef]
12. Wang, J.; Xiao, H.; Chen, L.; Xing, J.; Pan, Z.; Luo, R.; Cai, X. Integrating weighted feature fusion and the spatial attention module with convolutional neural networks for automatic aircraft detection from SAR images. *Remote Sens.* **2021**, *13*, 910. [CrossRef]
13. Zhao, Y.; Zhao, L.; Liu, Z.; Hu, D.; Kuang, G.; Liu, L. Attentional Feature Refinement and Alignment Network for Aircraft Detection in SAR Imagery. *IEEE Trans. Geosci. Remote Sens.* **2021**, *60*, 1–16. [CrossRef]
14. Zhang, P.; Xu, H.; Tian, T.; Gao, P.; Li, L.; Zhao, T.; Zhang, N.; Tian, J. SEFEPNet: Scale Expansion and Feature Enhancement Pyramid Network for SAR Aircraft Detection with Small Sample Dataset. *IEEE J. Sel. Top. Appl. Earth Obs. Remote Sens.* **2022**, *15*, 3365–3375. [CrossRef]
15. Bao, W.; Hu, J.; Huang, M.; Xu, Y.; Ji, N.; Xiang, X. Detecting Fine-Grained Airplanes in SAR Images With Sparse Attention-Guided Pyramid and Class-Balanced Data Augmentation. *IEEE J. Sel. Top. Appl. Earth Obs. Remote Sens.* **2022**, *15*, 8586–8599. [CrossRef]
16. Ma, C.; Zhang, Y.; Guo, J.; Hu, Y.; Geng, X.; Li, F.; Lei, B.; Ding, C. End-to-end method with transformer for 3-D detection of oil tank from single SAR image. *IEEE Trans. Geosci. Remote Sens.* **2021**, *60*, 1–19. [CrossRef]
17. Wu, Q.; Zhang, B.; Xu, C.; Zhang, H.; Wang, C. Dense Oil Tank Detection and Classification via YOLOX-TR Network in Large-Scale SAR Images. *Remote Sens.* **2022**, *14*, 3246. [CrossRef]

18. Xu, X.; Zhang, X.; Zhang, T.; Yang, Z.; Shi, J.; Zhan, X. Shadow-Background-Noise 3D Spatial Decomposition Using Sparse Low-Rank Gaussian Properties for Video-SAR Moving Target Shadow Enhancement. *IEEE Geosci. Remote Sens. Lett.* **2022**, *19*, 1–5. [CrossRef]
19. Wang, Y.; Wang, C.; Zhang, H.; Dong, Y.; Wei, S. A SAR dataset of ship detection for deep learning under complex backgrounds. *Remote Sens.* **2019**, *11*, 765. [CrossRef]
20. Wei, S.; Zeng, X.; Qu, Q.; Wang, M.; Su, H.; Shi, J. HRSID: A high-resolution SAR images dataset for ship detection and instance segmentation. *IEEE Access* **2020**, *8*, 120234–120254. [CrossRef]
21. Zhang, T.; Zhang, X.; Li, J.; Xu, X.; Wang, B.; Zhan, X.; Xu, Y.; Ke, X.; Zeng, T.; Su, H. Sar ship detection dataset (ssdd): Official release and comprehensive data analysis. *Remote Sens.* **2021**, *13*, 3690. [CrossRef]
22. Xu, C.; Su, H.; Li, J.; Liu, Y.; Yao, L.; Gao, L.; Yan, W.; Wang, T. RSDD-SAR: Rotated ship detection dataset in SAR images. *J. Radars* **2022**, *11*, 581–599. [CrossRef]
23. Zhang, T.; Zhang, X.; Ke, X.; Zhan, X.; Shi, J.; Wei, S.; Pan, D.; Li, J.; Su, H.; Zhou, Y. Ls-ssdd-v1. 0: A deep learning dataset dedicated to small ship detection from large-scale sentinel-1 sar images. *Remote Sens.* **2020**, *12*, 2997. [CrossRef]
24. Keydel, E.R.; Lee, S.W.; Moore, J.T. MSTAR extended operating conditions: A tutorial. In Proceedings of the SPIE 2757, Algorithm for Synthetic Aperture Radar Imagery III, Orlando, FL, USA, 10–12 April 1996; SPIE: Bellingham, WA, USA, 1996; pp. 228–242. [CrossRef]
25. Long, Y.; Jiang, X.; Liu, X.; Zhang, Y. Sar Atr with Rotated Region Based on Convolution Neural Network. In Proceedings of the IGARSS 2019—2019 IEEE International Geoscience and Remote Sensing Symposium, Yokohama, Japan, 28 July–2 August 2019. [CrossRef]
26. Zhang, X.; Chai, X.; Chen, Y.; Yang, Z.; Liu, G.; He, A.; Li, Y. A Novel Data Augmentation Method for SAR Image Target Detection and Recognition. In Proceedings of the 2021 IEEE International Geoscience and Remote Sensing Symposium IGARSS, Brussels, Belgium, 11–16 July 2021; pp. 3581–3584. [CrossRef]
27. Sun, Y.; Wang, W.; Zhang, Q.; Ni, H.; Zhang, X. Improved YOLOv5 with transformer for large scene military vehicle detection on SAR image. In Proceedings of the 2022 7th International Conference on Image, Vision and Computing (ICIVC), Xi'an, China, 26–28 July 2022; pp. 87–93. [CrossRef]
28. Complex SAR Data. Available online: https://www.sandia.gov/radar/pathfinder-radar-isr-and-synthetic-aperture-radar-sar-systems/complex-data/ (accessed on 11 October 2022).
29. Wang, Z.; Du, L.; Mao, J.; Liu, B.; Yang, D. SAR target detection based on SSD with data augmentation and transfer learning. *IEEE Geosci. Remote Sens. Lett.* **2018**, *16*, 150–154. [CrossRef]
30. Zou, B.; Qin, J.; Zhang, L. Vehicle detection based on semantic-context enhancement for high-resolution SAR images in complex background. *IEEE Geosci. Remote Sens. Lett.* **2021**, *19*, 1–5. [CrossRef]
31. Tang, T.; Wang, Y.; Liu, H.; Zou, S. CFAR-Guided Dual-Stream Single-Shot Multibox Detector for Vehicle Detection in SAR Images. *IEEE Geosci. Remote Sens. Lett.* **2022**, *19*, 1–5. [CrossRef]
32. Xia, G.-S.; Bai, X.; Ding, J.; Zhu, Z.; Belongie, S.; Luo, J.; Datcu, M.; Pelillo, M.; Zhang, L. DOTA: A large-scale dataset for object detection in aerial images. In Proceedings of the IEEE Conference on Computer Vision and Pattern Recognition, Salt Lake City, UT, USA, 18–23 June 2018; pp. 3974–3983. [CrossRef]
33. Everingham, M.; Van Gool, L.; Williams, C.K.; Winn, J.; Zisserman, A. The pascal visual object classes (voc) challenge. *Int. J. Comput. Vis.* **2009**, *88*, 303–308. [CrossRef]
34. Ward, K. Compound representation of high resolution sea clutter. *Electron. Lett.* **1981**, *7*, 561–565. [CrossRef]
35. Canny, J. A computational approach to edge detection. *IEEE Trans. Pattern Anal. Mach. Intell.* **1986**, *6*, 679–698. [CrossRef]
36. Lin, T.-Y.; Maire, M.; Belongie, S.; Hays, J.; Perona, P.; Ramanan, D.; Dollár, P.; Zitnick, C.L. Microsoft coco: Common objects in context. In Proceedings of the Computer Vision–ECCV 2014: 13th European Conference, Zurich, Switzerland, 6–12 September 2014; Proceedings, Part V 13, 2014. pp. 740–755. [CrossRef]
37. Simonyan, K.; Zisserman, A. Very Deep Convolutional Networks for Large-Scale Image Recognition. In Proceedings of the International Conference on Learning Representations (ICLR), San Diego, CA, USA, 7–9 May 2015. [CrossRef]
38. He, K.; Zhang, X.; Ren, S.; Sun, J. Deep residual learning for image recognition. In Proceedings of the Proceedings of the IEEE Conference on Computer Vision and Pattern Recognition, Las Vegas, NV, USA, 26 June–1 July 2016; pp. 770–778. [CrossRef]
39. Lin, T.-Y.; Dollár, P.; Girshick, R.; He, K.; Hariharan, B.; Belongie, S. Feature pyramid networks for object detection. In Proceedings of the IEEE Conference on Computer Vision and Pattern Recognition, Honolulu, HI, USA, 21–26 July 2017; pp. 2117–2125. [CrossRef]
40. Ren, S.; He, K.; Girshick, R.; Sun, J. Faster r-cnn: Towards real-time object detection with region proposal networks. *IEEE Trans. Pattern Anal. Mach. Intell.* **2017**, *39*, 1137–1149. [CrossRef]
41. Lin, T.-Y.; Goyal, P.; Girshick, R.; He, K.; Dollár, P. Focal loss for dense object detection. In Proceedings of the IEEE International Conference on Computer Vision, Venice, Italy, 22–29 October 2017; pp. 2980–2988. [CrossRef]
42. Tian, Z.; Shen, C.; Chen, H.; He, T. Fcos: Fully convolutional one-stage object detection. In Proceedings of the IEEE/CVF International Conference on Computer Vision, Seoul, Republic of Korea, 27 October–2 November 2019; pp. 9627–9636. [CrossRef]
43. Han, J.; Ding, J.; Li, J.; Xia, G.-S. Align deep features for oriented object detection. *IEEE Trans. Geosci. Remote Sens.* **2021**, *60*, 1–11. [CrossRef]

44. Ding, J.; Xue, N.; Long, Y.; Xia, G.-S.; Lu, Q. Learning roi transformer for oriented object detection in aerial images. In Proceedings of the IEEE/CVF Conference on Computer Vision and Pattern Recognition, Long Beach, CA, USA, 15–20 June 2019; pp. 2849–2858. [CrossRef]
45. Xu, Y.; Fu, M.; Wang, Q.; Wang, Y.; Chen, K.; Xia, G.-S.; Bai, X. Gliding vertex on the horizontal bounding box for multi-oriented object detection. *IEEE Trans. Pattern Anal. Mach. Intell.* **2020**, *43*, 1452–1459. [CrossRef] [PubMed]
46. Li, W.; Chen, Y.; Hu, K.; Zhu, J. Oriented reppoints for aerial object detection. In Proceedings of the IEEE/CVF Conference on Computer Vision and Pattern Recognition, New Orleans, LA, USA, 18–24 June 2022; pp. 1829–1838. [CrossRef]
47. Yang, X.; Yang, X.; Yang, J.; Ming, Q.; Wang, W.; Tian, Q.; Yan, J. Learning high-precision bounding box for rotated object detection via kullback-leibler divergence. *Adv. Neural Inf. Process. Syst.* **2021**, *34*, 18381–18394. [CrossRef]
48. Lin, Z.; Ji, K.; Leng, X.; Kuang, G. Squeeze and excitation rank faster R-CNN for ship detection in SAR images. *IEEE Geosci. Remote Sens. Lett.* **2018**, *16*, 751–755. [CrossRef]
49. Wang, Y.; Wang, C.; Zhang, H.; Dong, Y.; Wei, S. Automatic ship detection based on RetinaNet using multi-resolution Gaofen-3 imagery. *Remote Sens.* **2019**, *11*, 531. [CrossRef]
50. Sun, Z.; Dai, M.; Leng, X.; Lei, Y.; Xiong, B.; Ji, K.; Kuang, G. An anchor-free detection method for ship targets in high-resolution SAR images. *IEEE J. Sel. Top. Appl. Earth Obs. Remote Sens.* **2021**, *14*, 7799–7816. [CrossRef]
51. Zhou, Y.; Yang, X.; Zhang, G.; Wang, J.; Liu, Y.; Hou, L.; Jiang, X.; Liu, X.; Yan, J.; Lyu, C. Mmrotate: A rotated object detection benchmark using pytorch. In Proceedings of the 30th ACM International Conference on Multimedia, Lisbon, Portugal, 10 October 2022; pp. 7331–7334. [CrossRef]

remote sensing MDPI

Article

Improving YOLOv7-Tiny for Infrared and Visible Light Image Object Detection on Drones

Shuming Hu, Fei Zhao *, Huanzhang Lu, Yingjie Deng, Jinming Du and Xinglin Shen

National Key Laboratory of Science and Technology on Automatic Target Recognition, College of Electronic Science and Technology, National University of Defense Technology, Changsha 410073, China
* Correspondence: zhaofei08@nudt.edu.cn

Abstract: To address the phenomenon of many small and hard-to-detect objects in drone images, this study proposes an improved algorithm based on the YOLOv7-tiny model. The proposed algorithm assigns anchor boxes according to the aspect ratio of ground truth boxes to provide prior information on object shape for the network and uses a hard sample mining loss function (HSM Loss) to guide the network to enhance learning from hard samples. This study finds that the aspect ratio difference of vehicle objects under drone perspective is more obvious than the scale difference, so the anchor boxes assigned by aspect ratio can provide more effective prior information for the network than those assigned by size. This study evaluates the algorithm on a drone image dataset (DroneVehicle) and compares it with other state-of-the-art algorithms. The experimental results show that the proposed algorithm achieves superior average precision values on both infrared and visible light images, while maintaining a light weight.

Keywords: object detection; drone images; DroneVehicle dataset; YOLOv7-tiny; infrared and visible light

Citation: Hu, S.; Zhao, F.; Lu, H.; Deng, Y.; Du, J.; Shen, X. Improving YOLOv7-Tiny for Infrared and Visible Light Image Object Detection on Drones. *Remote Sens.* **2023**, *15*, 3214. https://doi.org/10.3390/rs15133214

Academic Editors: Yanni Dong, Xiaochen Yang and Qian Du

Received: 17 April 2023
Revised: 17 June 2023
Accepted: 18 June 2023
Published: 21 June 2023

1. Introduction

Object detection algorithms and drone technology have become increasingly mature and play important roles in many fields, such as traffic management, agricultural irrigation, forest patrol, and battlefield reconnaissance. These tasks can achieve higher accuracy and efficiency by using drones while also reducing casualties. However, many problems still need to be solved for object detection in remote sensing images because of the small object size and complex background in drone perspective. Moreover, the computation of algorithms is restricted by the limited computing power of edge platforms on drones, which poses challenges for their practical application.

Object detection algorithms aim to obtain the location and category of objects in images. In recent years, deep-learning-based object detection algorithms have developed rapidly. General object detection algorithms include two-stage algorithms based on candidate boxes, such as RCNN [1], Fast RCNN [2], Mask RCNN [3], etc. These algorithms first generate several candidate boxes that may contain objects in the image, then classify these candidate boxes, and finally detect all objects in the image: one-stage algorithms based on regression, such as YOLO [4–9] series, SSD [10] series, RetinaNet [11], etc. These algorithms divide the image into several cells and directly predict whether each cell contains an object, as well as the category and location of the object. These methods have good detection performance on general object detection datasets such as VOC [12] and COCO [13] datasets. Among them, one-stage algorithms often achieve real-time detection speed and are easier to apply to edge devices. General object detection algorithms use horizontal bounding boxes (HBB) to surround objects, which are simple and intuitive in form but not accurate enough for object localization. Oriented object detection algorithms [14–20] predict oriented bounding boxes (OBB) to surround objects, which are closer to the shape of objects and can also obtain the motion direction of objects. RoI Transformer [18] is a two-stage object detection algorithm that achieves oriented object detection by transforming a horizontal region of interest (RoI) into a rotated RoI. However, steps such as horizontal candidate box generation, RoI alignment and oriented candidate box generation require a large amount of computation; oriented R-CNN [19] proposes an oriented candidate box generation network (oriented RPN), which reduces the computational cost of generating oriented candidate boxes, but as a two-stage algorithm, it is still difficult to meet real-time detection; S^2ANet [20] aligns convolutional features and arbitrary orientation objects through a feature alignment module (FAM) and then uses an oriented detection module (ODM) to achieve one-stage rotation object detection. Oriented bounding boxes require more parameters, and post-processing such as non-maximum suppression (NMS) for oriented bounding boxes is more complex than horizontal bounding boxes, which is unacceptable for embedded devices in terms of computational cost. In addition, some algorithms explore object detection algorithms that fuse infrared and visible light images. D-ViTDet [21] and LAIIFusion [22] use illumination perception modules to perceive the illumination difference in each region of the image, providing more suitable reference for cross-modal image fusion. UA-CMDet [23], RISNet [24], and ECISNet [25] optimize the cross-modal mutual information utilization to improve the detection performance. TSFADet [26] aligns cross-modal objects from translation, scaling, and rotation aspects through a network.

There are some datasets for object detection from a drone perspective, such as [23,27–31]. However, only a small portion of the dataset contains both infrared and visible light images. The VEDAI [29] dataset comprises infrared and visible light dual-band images that include 1200 pairs of images and 9 different ground vehicle categories. The DroneVehicle [23] dataset is the first large-scale cross-modal dataset that covers all time periods under a drone perspective. It contains infrared and visible light dual-modal data that include 28,439 pairs of images and 5 different vehicle categories: car, truck, bus, van, and freight car. The objects are annotated with polygons and there are 953,087 annotations in total. The visible light images have 389,779 cars, 22,123 trucks, 15,333 buses, 11,935 vans, and 13,400 freight cars. The infrared images have 428,086 cars, 25,960 trucks, 16,590 buses, 12,708 vans, and 17,173 freight cars. The images cover scenes during day, evening, and night and show drones at different heights (80 m, 100 m, and 120 m) and angles (15°, 30°, and 45°). Each pair of images undergoes affine transformation and region cropping to ensure that most cross-modal image pairs are aligned. The DroneVehicle dataset is divided into a training set with 17,990 image pairs, a validation set with 1469 image pairs, and a test set with 8980 image pairs. All experiments are conducted in a fair comparison according to the given division method. Because oriented bounding boxes can provide additional direction information compared to horizontal bounding boxes, but they also introduce more complex post-processing operations and larger computational overheads, this paper mainly focuses on object location and algorithm speed; therefore, this paper uses horizontal bounding boxes.

Due to the large distance between the ground objects and the drone, the objects occupy a small number of pixels in the image, resulting in unsatisfactory detection performance. Moreover, since the object size span is small under the drone perspective, some methods such as feature pyramid network (FPN) [32] are not as effective as in natural scenes with multi-scale objects. Most of the existing object detection algorithms are developed on visible light image datasets, which are greatly affected by illumination. Visible light images have richer texture information than infrared images when the illumination condition is good, but they have poor detection performance when the illumination condition is bad. As shown in Figure 1, the objects are clearer in visible light images during daytime, but many objects are invisible in visible light images at night, while they are more obvious in infrared images. Combining different sensor information can improve the adaptability of drones to complex environments. However, due to the limited payload of drones, only some low-power embedded devices can be mounted. To achieve real-time object detection on drone images, the algorithm needs to be simple and fast.

(a) (b)

Figure 1. Images of ground vehicles from a drone perspective, where (**a**) is a visible light image and (**b**) is an infrared image.

The above-mentioned problems limit the application of object detection algorithms on drones. To enable drones to adapt to complex environments, this paper conducts research on both infrared and visible light modalities of data and improves the lightweight model YOLOv7-tiny to make it run in real time on edge devices. To address the small size span of objects in drone images and the class imbalance problem, this paper proposes a new anchor box assignment method to balance the imbalanced class samples. This method can provide more prior information that matches the size characteristics of objects under drone perspective and improve the accuracy of the network. This paper also improves the loss function to deal with the hard–easy sample imbalance problem and proposes HSM Loss to dynamically adjust the weights of hard-to-detect samples and enhance the network's learning of hard-to-detect samples.

In summary, the main contributions of this paper are as follows:

1. We propose a novel anchor assignment method for drone images, AR-anchor (aspect ratio-anchor), which alleviates the sample imbalance problem, provides more effective prior information for the network, and improves the object detection performance.
2. We propose a hard-sample mining loss function (HSM Loss), which alleviates the hard–easy sample imbalance problem and enhances the detection ability of the network without increasing computational cost.
3. We demonstrate that our algorithm can be effective on both infrared and visible light modalities of data, enhancing the robustness and generalization of drone image object detection.
4. We present a lightweight algorithm that can be more efficiently applied in practice.

The structure of this paper is as follows: Section 2 introduces the network structure and method in detail. Section 3 presents our work and experimental results and compares them with related methods to verify the effectiveness of our approach. Finally, Section 4 discusses the research content, and Section 5 concludes the paper.

2. Materials and Methods

This section first introduces the framework and idea of deep-learning-based object detection algorithms, as well as their advantages and disadvantages, then introduces the structure and principle of the YOLOv7-tiny algorithm, and finally introduces the idea and details of our proposed anchor assignment method and improved loss function.

2.1. Overall Network Framework

Deep-learning-based object detection algorithms usually consist of three components: a backbone network for extracting image features, a neck for fusing deep semantic information and shallow detail information, and a detection head for predicting object classes and locations. The backbone network can be a network for image classification with the last classification layer removed, such as ResNet [33], or a backbone network designed for object detection tasks such as DarkNet [6]. These networks are often pre-trained on large-scale datasets to learn the ability to extract image features, which can significantly improve the performance for small-scale and similar-scene object detection datasets. Images are features extracted by the network to obtain feature maps. To reduce the computational complexity, the resolution of the feature maps gradually decreases in the network. Most backbone networks are designed to perform downsampling 5 times. The last layer of feature maps has a width and height of 1/32 of the original image. An object occupying 32 $\times$ 32 pixels in the image only occupies one point in the last layer of feature maps. As the network depth increases, the feature maps have stronger semantic information, but lose detail information. Therefore, object detection algorithms do not only detect objects from the last layer of feature maps but also do multi-scale detection from the last 3 layers of feature maps with different resolutions. By using the feature pyramid structure of the neck, the characteristics of different layers of the network are fully utilized to improve the adaptability of the network to multi-scale objects. The detection head realizes the classification and localization of objects. Different algorithms have different detection mechanisms. Some detection heads divide the image into several cells; each cell shares the extracted features and predicts the confidence, category, location, and width and height of the object, such as YOLOv7. Some algorithms use decoupled detection heads, which separate classification and regression tasks, such as Efficientdet [34].

The network structure is shown in Figure 2. Its backbone network uses a multi-branch stacked structure for feature extraction, which can extract more effective features while reducing computation. After the input image goes through the backbone network for feature extraction, the feature layers after downsampling 3, 4, and 5 times are used to predict objects; at the last feature layer of the backbone network, there is an SPP [35] module with a CSP [36] structure, which has max pooling layers with different sizes and can enable the network to learn more effective features from different receptive fields, allowing deep convolutional networks to extract richer semantic information; the neck part uses a PANet [37] structure to fuse deep-feature map upsampling to shallow-feature maps to enhance the semantic information of shallow networks, while fusing shallow-feature map downsampling to deep-feature maps to supplement the detail information of deep-feature maps; a detection head with shared weights is used for final object classification and localization. Each layer uses three anchor boxes with different shapes, which are responsible for detecting objects with different shapes. Finally, non-maximum suppression is used to filter out multiple anchor boxes' repeated predictions for the same object; the reparameterization technique is used, and there are different network branches on the same layer with different sizes of convolution kernels. During inference, convolution kernels of different branches are equivalently replaced by convolution kernels with the same size and then fused into a single branch. This improves the parallelism of computation and reduces storage and inference speed on the GPU when inferring. During training, YOLOv7-tiny uses the simOTA strategy to assign positive and negative samples. This strategy calculates the cost matrix between each ground truth box and anchor box then assigns one ground truth box to k anchor boxes. These anchor boxes are regarded as positive samples. If an anchor box is assigned multiple ground truth boxes for prediction, then this anchor box only predicts the ground truth box with the lowest cost. This strategy can greatly increase the number of positive samples.

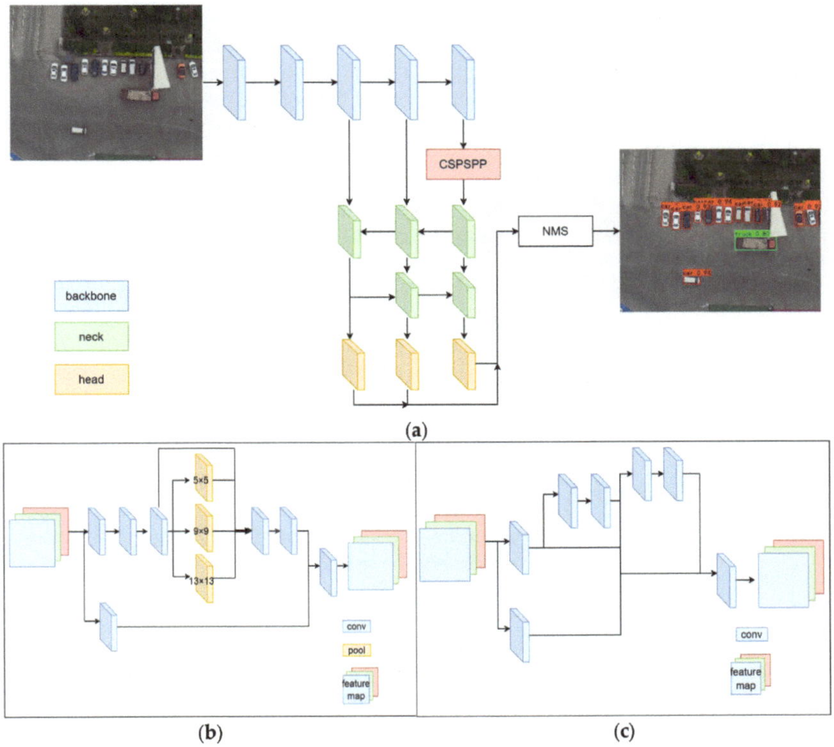

Figure 2. YOLOv7-tiny structure, where (**a**) is the overall algorithm structure, (**b**) is the CSPSPP structure, and (**c**) is the multi-branch stacked structure. The backbone extracts the features of the input image and then neck fuses the features of different scales. The detection head predicts the position and type of the object and finally obtains the final detection result through NMS. The CSPSPP structure greatly increases the receptive field and separates the most significant contextual features, while the multi-branch stack module enables the network to learn more features and has stronger robustness by controlling the shortest and longest gradient paths.

2.2. Anchor Allocation Method Based on Aspect Ratio

The network predicts the width and height of the object by predicting the offset of the object size relative to the anchor box. Therefore, objects that are close to the size of the anchor box are easier to detect, while objects that have a large difference in size from the anchor box are easily ignored. Usually, the design of anchor boxes includes objects of various scales and shapes. Different feature layers with different receptive fields of the network are assigned different sizes of anchor boxes. Multiple anchor boxes on the same layer cover different shapes of objects according to the aspect ratio. Carefully designed anchor boxes provide effective prior information for the network and greatly improve its detection ability. However, under the drone perspective, the object size difference is not obvious. The algorithm in this paper groups the objects from the perspective of aspect ratio and assigns anchor boxes that are closer to the object shape distribution, providing more effective prior information for the network.

YOLOv7-tiny uses 9 different sizes of anchor boxes: 3 small anchor boxes on the feature layer with 3 times downsampling to detect small objects, 3 large anchor boxes on the feature layer with 5 times downsampling to detect large objects, and each layer has 3 anchor boxes with aspect ratios greater than 1, less than 1, and close to 1, respectively, to detect objects with different scales and shapes. The design method of anchor boxes is mainly based on clustering algorithms, such as K-means, on the bounding boxes to obtain the object sizes as prior information for the network. This design method can reflect the sizes of most objects in the data, but most datasets have a long-tail distribution situation where some object samples are far beyond the rest of the samples. The obtained anchor boxes mainly reflect the sizes of objects with more samples, resulting in better detection performance for categories with more samples and worse detection performance for other categories. In addition, in natural scenes, objects can be very close or far away from the camera; plus, size differences between objects exist from point objects to objects that occupy almost all the pixels of an image. Using different sizes of anchor boxes to guide different layers of the network to detect different sizes of objects is a good strategy, but when flying a drone, objects are far away from the camera, so they are mostly small objects without obvious size differences like natural scenes. At this time, what differs greatly among bounding boxes is the aspect ratio, so the original allocation method is not that effective anymore.

Based on the above problems, this paper proposes a new anchor box allocation method for object detection under drone perspective, abandoning the original strategy of allocating anchor boxes to different layers according to object size and using different aspect ratios of anchor boxes on different layers, making the network possess stronger detection ability for objects under drone perspective and allocating prior boxes separately for different types of objects to alleviate the impact of long-tail distribution of sample categories. Specifically, this paper first statistically analyzes the object size and anchor box size in the DroneVehicle dataset. As shown in Figure 3, the horizontal axis represents width and the vertical axis represents height in units of pixel number. Since the objects are obtained from drone perspective, their size does not exceed 350×350. The default anchor boxes of YOLOv7-tiny differ greatly from object size distribution. Using the k-means algorithm to cluster on the dataset obtains 9 anchor boxes and 15 anchor boxes, which are more consistent with object size than the original default anchor boxes, but they will be affected by sample number difference among different categories. The sample categories are counted as shown in Figure 4. Car has 428,086 samples in the infrared images and 389,779 samples in the visible light images, which are far more than other categories. Most anchor box sizes obtained directly by k-means are close to a car, so they are seriously affected by long-tail distribution of samples. Therefore, this paper's algorithm first clusters each category's object size using the k-means algorithm, separately obtaining 3 anchor boxes for each category. The DroneVehicle dataset has 5 categories, so it obtains a total of 15 anchor boxes. At this time, obtained anchor boxes can take into account each category's size. Then, it explores allocation methods for the anchor boxes. From Figure 3, it can be seen that most samples gather near the angle bisector line of the first quadrant of the coordinate system and approximate axial symmetry distribution. If it classifies objects according to the original allocation method by size as shown in Figure 3a–c, it is equivalent to dividing objects into three parts according to the area of the rectangle formed by the coordinates with the origin point, which introduces prior information that does not match the characteristics of vehicle object size distribution under drone perspective into the network very well, so according to this paper's method, it allocates according to the object aspect ratio, which is equivalent to using two straight lines with different slopes dividing object shape into three areas with different colors, as shown in Figure 3d. At this time, the rule dividing objects becomes the aspect ratio, which is more suitable for object size distribution under drone perspective. Since all objects are vehicles, their own sizes have aspect ratios less than one, so when the drone direction is the same or perpendicular to the object, the bounding box shape is close to the object shape, but when the drone direction is at a 45° angle with respect to the object, the bounding box shape differs greatly from the object shape, so at this time, the bounding box aspect ratio is close to 1, so the network needs stronger semantic information for learning the difference between object and bounding box, so it puts each category's aspect ratio close-to-one prior box on the last layer predicting network. That is to say, the green area samples in the sample distribution are allocated predicting the network with a larger receptive field and stronger semantic information deeper layer network, while because there are more samples with the same direction between the drone and vehicle than the perpendicular direction, it allocates a less-than-one aspect ratio prior box, that is, the red area samples predicting a network with less downsampling times, higher resolution, and shallower layer network.

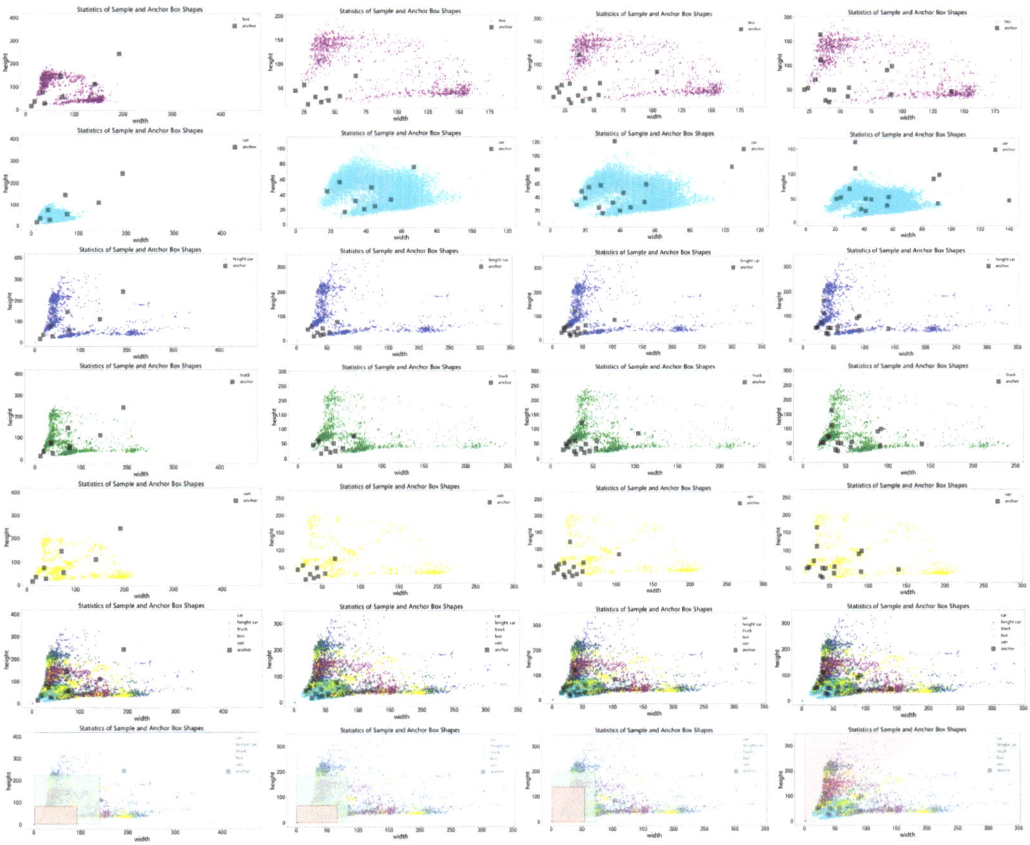

(a) (b) (c) (d)

Figure 3. The object sizes of different classes were statistically analyzed: (**a**) shows the distribution of the original anchor boxes of YOLOv7-tiny on objects; (**b**) shows nine anchor boxes obtained by k-means clustering; (**c**) shows 15 anchor boxes obtained by k-means clustering; (**d**) shows the proposed method, where the first five rows are the object distributions of each class, the sixth row is the overall distribution, and the last row is the sample interval for anchor box division. Different color regions represent different groups of anchor boxes that are mainly responsible for predicting objects with different shapes.

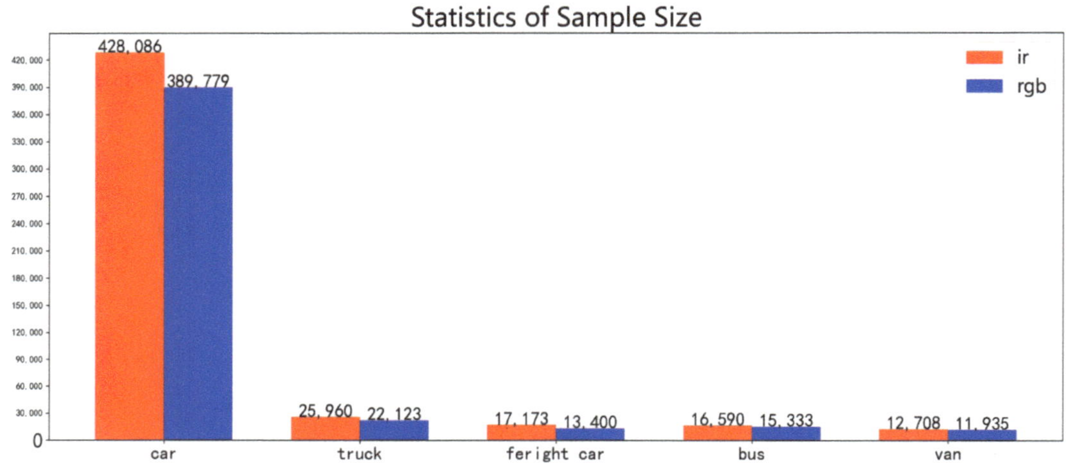

Figure 4. Statistical analysis of sample numbers for different classes. Red represents the sample count in infrared images, while blue represents the sample count in visible light images.

2.3. HSM Loss (Hard-Sample Mining Loss)

Object detection from a drone's perspective faces many challenges, mainly because objects in drone-captured images usually have small sizes, resulting in insufficient features, the contrast of visible light images is low under low illumination, and the contrast of infrared images is low when the ambient temperature and the object are close, making it difficult to distinguish between the object and the background and the motion blur of moving objects. These factors result in hard-to-detect objects that are difficult to accurately locate and identify in images, thus affecting the performance of drones in various application scenarios. For hard-sample mining, there are methods such as focal loss [11] that adjust the weights of easy and hard samples according to confidence scores, the gradient harmonizing mechanism (GHM) [38] that adjusts weights according to gradient contributions, online hard example mining (OHEM) [39] that increases the number of hard samples in training samples, etc. The core of these methods is to increase the weight of hard samples during training. Based on focal loss, this paper conducts research on hard sample mining.

The loss function of object detection algorithms consists of two parts: classification and regression. YOLOv7 matches positive and negative samples based on the aspect ratio of ground truth boxes and anchor boxes, instead of using IoU as in previous versions. Classification uses cross-entropy loss.

$$\text{Cross} - \text{Entropy Loss} = -\log(p_t) \tag{1}$$

$$p_t = \begin{cases} \hat{p} & \text{if } y = 1 \\ (1 - \hat{p}) & \text{otherwise} \end{cases} \tag{2}$$

For positive samples, the larger the prediction value, the smaller the loss; for negative samples, the larger the prediction value, the larger the loss. Focal loss is proposed to mainly solve two problems: class imbalance and easy–hard sample imbalance. Since one-stage object detection algorithms divide images into several grids, each grid has multiple anchor boxes to predict the different shapes of objects. Each prediction generates a large number of bounding boxes, among which only a small part are assigned as positive samples. The problem of class imbalance is serious.

$$\text{Focal Loss}\,(y, \hat{p}) = -\alpha_t (1 - p_t)^{\gamma} \log(p_t) \tag{3}$$

$$\alpha_t = \begin{cases} \alpha & \text{if } y = 1 \\ (1 - \alpha) & \text{otherwise} \end{cases} \tag{4}$$

In focal loss, parameter γ is used to control the weight of different samples. p_t is a number between 0 and 1. For negative samples, most of them are background pixels. Since there is a large number of detection boxes that are negative samples in the network, a modulation factor $(1 - p_t)^{\gamma}$ is multiplied by the cross-entropy loss to suppress the weight of negative samples and achieve the purpose of balancing positive and negative samples. For positive samples, focal loss also has the ability to dynamically adjust the network learning difficulty of samples. When p_t is close to 1, the sample belongs to an easy-to-detect sample, and the network reduces its weight. Smaller p_t values indicate positive samples that are difficult to detect, and the network increases their loss. Due to excessive suppression of a large number of negative samples by the modulation factor, α is used to adjust the weight of positive and negative samples, so that the network achieves a balance between positive and negative samples. The typical values for α and γ are 0.25 and 2, respectively [11].

Focal loss has been proven to balance positive and negative samples and hard and easy samples in many experiments, but it performs poorly in the YOLO series, sometimes even leading to network performance degradation. This paper analyzes this problem and finds that YOLOv7-tiny itself has the ability to adjust positive and negative samples, using the simOTA strategy to assign positive and negative samples. Under this strategy, a ground truth box can match multiple anchor boxes as positive samples, and there are different modulation factors for positive and negative samples. Therefore, the imbalance of positive and negative samples in YOLOv7-tiny is not obvious, and directly introducing focal loss will cause negative samples in the network to be suppressed. The most intuitive phenomenon of this is that after introducing focal loss, the network performance does not increase but decreases, and when the confidence threshold is set low, the network will have a large amount of background predicted as objects, and negative samples are suppressed too much. At the same time, because focal loss dynamically allocates the weights of hard and easy samples by multiplying a modulation factor greater than 0 and less than 1 on the basis of cross-entropy loss, it always shows a suppressive effect on classification loss. For objects with a confidence of 0.9, the loss is only one percent of the cross-entropy loss. Although this can make the network learn more fully from hard samples, it limits the learning of easy samples. This paper improves this problem. First, this paper believes that balancing hard and easy samples should not always suppress them but also amplify them. We define hard samples as those with confidence lower than 0.5, which is consistent with the threshold used for prediction. When applying the algorithm to real images, we use 0.5 as the confidence threshold and take the objects with network prediction confidence higher than 0.5 as the prediction results. Therefore, this paper adds a modulation factor to make the weights of easy detection samples with confidence greater than 0.5 decay and the weights of hard detection samples with confidence less than 0.5 amplify; therefore, the loss of hard samples is amplified, while easy samples are suppressed. At this point, there is still a problem. For simple samples, as the confidence increases, the decay factor decreases exponentially, which will cause the network to stop learning after learning to a certain extent for easy samples. Therefore, a lower limit is set for the modulation coefficient to reduce the impact on the hard and easy sample boundaries as much as possible. A 0.1 is added to the modulation coefficient, and then, the lower limit of decay for easy samples is 10 times, and the impact on the hard and easy boundaries is relatively small; when $\gamma = 2$, the boundary is about 0.53. This bias will cause greater changes in the boundary of hard and easy samples, while a bias that is too small will cause the lower limit of the modulation coefficient decay to be too low. Therefore, 0.1 is a suitable bias. At this time, the modulation factor ranges from 0.1 to 4.1. Finally, in focal loss, after adjusting the weights of positive and negative samples by focal loss, the weight of negative samples is less than that of positive samples, which is used to reduce the weight of positive samples. This paper's algorithm does not need this parameter to adjust the weight and removes it to obtain HSM Loss. The images of three loss functions are shown in Figure 5. The focal loss value is always smaller than the cross-entropy loss, while the HSM Loss proposed in this paper is smaller than cross-entropy loss when the confidence is greater than 0.5 and larger than cross-entropy loss when the confidence is less than 0.5, which can better mine hard samples.

$$\text{HSM loss } = -(2^{\gamma}(1-\mathrm{p_t})^{\gamma}+0.1)\log(\mathrm{p_t}) \tag{5}$$

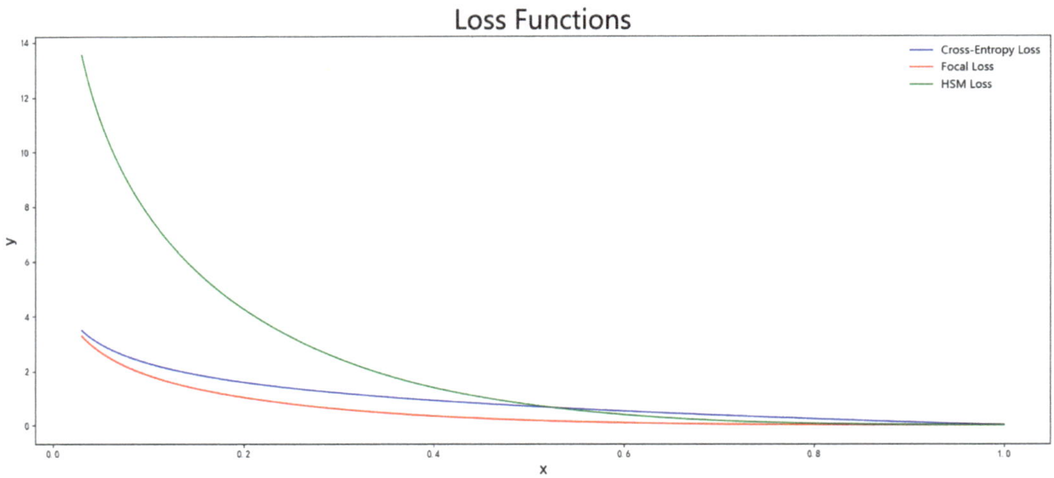

Figure 5. Loss function of cross-entropy, focal loss, and HSM Loss. When the confidence level of HSM Loss is less than 0.5, compared with the cross-entropy loss, it will be amplified, which is conducive to the learning of hard samples. In addition, when the confidence level is high, there is an upper limit for the attenuation, and the samples with a high confidence level still have the ability to continue learning.

3. Results

This paper conducted a large number of experiments on the public dataset DroneVehicle. This section first introduces the DroneVehicle dataset, then designs a series of ablation experiments to prove the effectiveness of the method proposed in this paper and finally compares it with some advanced object detection algorithms to demonstrate its superiority.

3.1. Implementation Details

The experimental environment setup of this paper includes CPU: i7-12700, GPU: 1080TI, and pytorch1.10.2. We use mAP as the evaluation metric and consider a predicted box as a positive sample when its IoU with the ground truth box is larger than 0.5. The mAP obtained by testing is the highest value on the test set. The data annotation is oriented bounding boxes, and we obtain horizontal bounding boxes by taking the minimum horizontal circumscribed rectangle of each box. All data splits are based on the original splits of each dataset. Parameter settings during training: the image resolution for training models is 640 × 640. We pad the original images with a resolution of 840 × 712 to have equal width and height and, then, resize them to 640 × 640 to ensure that the image content maintains its original aspect ratio; we use mosaic data augmentation and mixup data augmentation; when using pre-trained weights, we freeze the backbone part for training for the first 30 epochs and, then, train the whole network; the batch size is 64 when freezing the backbone and 32 when training the whole network; the pre-trained weights are YOLOv7-tiny trained on the COCO dataset; the learning rate is set to 0.001 and adjusted by cosine annealing strategy; the optimizer is Adam with a momentum of 0.937; when using focal loss, we set $\alpha = 0.25$ and $\gamma = 2$.

3.2. Ablation Study

This paper designed a series of ablation experiments on the proposed anchor box design method and loss function improvement, which fully demonstrated the effectiveness of our method.

3.2.1. AR-Anchor

First, we conducted an ablation experiment on the anchor box assignment method and verified it on both infrared and visible light images. This experiment aimed to validate that our method of assigning prior boxes according to aspect ratio was more effective than assigning them according to size under the drone perspective and that it could alleviate the problem of poor detection performance for classes with few samples because of class imbalance. The default anchor boxes of YOLOv7-tiny are shown in Table 1, and Table 2 shows the mAP of using different anchor boxes on infrared and visible light images. We first tested with the default anchor boxes of YOLOv7-tiny and obtained a mAP of 73.79% on visible light images and 76.89% on infrared images. Then, we clustered the ground truth boxes of the training set of the DroneVehicle dataset and obtained nine different shapes of anchor boxes, which we grouped by size. Since the DroneVehicle dataset contains night scenes where the objects are hard to see in visible light images and the infrared and visible light images are not strictly aligned, there are differences in the annotations. Therefore, we clustered separately on infrared and visible light images and obtained three groups of anchor boxes as shown in Table 1. The size differences among different groups of anchor boxes are not large. The results of training with these anchor boxes were a mAP of 74.88% on visible light images and 78.29% on infrared images. To prove that the performance improvement was brought by our proposed algorithm, we used this result as the baseline for subsequent comparison, rather than making an unfair comparison with the result using default anchor boxes. This also reflected the impact of prior information of anchor boxes on network performance. Appropriate prior information could greatly improve network performance and accelerate convergence speed.

Table 1. Loss function of cross-entropy, focal loss, and HSM Loss.

Methods	Anchor Boxes
YOLO-anchor	(12, 16), (19, 36), (40, 28); (36, 75), (76, 55), (72, 146); (142, 110), (192, 243), (459, 401)
Kmeans (3 × 3, RGB)	(28, 17), (39, 21), (18, 45); (34, 32), (45, 25), (25, 57); (54, 34), (67, 76), (43, 50)
Kmeans (3 × 5, RGB)	(29, 15), (16, 29), (35, 21), (18, 44), (43, 22), (30, 33), (20, 52), (47, 28), (25, 58), (44, 40), (35, 52), (64, 34), (51, 58), (35, 120), (98, 84)
Ours (RGB)	(21, 49), (27, 69), (28, 70), (34, 111), (37, 64); (42, 24), (49, 26), (58, 37), (91, 39), (52, 35); (43, 49), (62, 48), (85, 93), (91, 93), (30, 26)
YOLO-anchor	(12, 16), (19, 36), (40, 28); (36, 75), (76, 55), (72, 146); (142, 110), (192, 243), (459, 401)
Kmeans (3 × 3, IR)	(17, 31), (33, 19), (20, 50), (40, 48), (44, 25), (27, 58), (52, 34), (55, 58), (85, 110)
Kmeans (3 × 5, IR)	(15, 29), (30, 16), (28, 24), (20, 39), (40, 20), (18, 49), (46, 25), (36, 32), (22, 55), (29, 58), (54, 33), (42, 47), (55, 59), (37, 121), (104, 84)
Ours (IR)	(21, 50), (34, 164), (30, 71), (34, 111), (24, 53); (41, 24), (140, 46), (56, 36), (91, 40), (38, 28); (45, 49), (41, 50), (88, 90), (92, 98), (57, 54)

Table 2. Detection results of different anchor boxes.

Anchor	Modality	Car	Truck	Freight Car	Bus	Van	mAP
YOLO-anchor		96.06	73.66	53.52	94.35	51.34	73.79
Kmeans (3 × 3)	RGB	**96.28**	74.77	53.99	94.44	54.94	74.88
Kmeans (3 × 5)		96.17	**75.89**	55.97	**94.81**	55.03	75.57

Table 2. *Cont.*

Anchor	Modality	Car	Truck	Freight Car	Bus	Van	mAP
Ours	RGB	96.19	75.75	**57.34**	94.52	**56.68**	**76.10**
YOLO-anchor	IR	97.89	76.11	62.88	94.16	53.42	76.89
Kmeans (3 × 3)		98.03	76.79	65.27	94.62	56.74	78.29
Kmeans (3 × 5)		**98.08**	77.89	66.48	94.54	58.34	79.07
Ours		98.02	**79.47**	**67.22**	**94.82**	**58.59**	**79.62**

This paper proposes an anchor box assignment method that clusters the bounding boxes of five vehicle categories on the DroneVehicle dataset separately, obtaining 15 anchor boxes. Then, according to the aspect ratio, the anchor boxes are grouped into different network layers: the ones with an aspect ratio less than 1 are assigned to the shallow layer, the ones with an aspect ratio close to 1 are assigned to the deep layer, and the ones with an aspect ratio greater than 1 are assigned to the middle layer. The grouping results based on aspect ratio are shown in Table 1. The experimental results show that the proposed method achieves a mAP of 76.10% on visible light images and 79.62% on infrared images. To verify the effectiveness of the proposed anchor box assignment method, we compare it with a baseline method that clusters 15 anchor boxes based on their size and assigns them to different network layers. The anchor boxes obtained by the baseline method for infrared and visible light images are also shown in Table 1. The experimental results show that the proposed method achieves a mAP of 75.57% on visible light images and 79.07% on infrared images. Compared with the baseline method, the proposed method improves the mAP by 1.33% on infrared images and 1.22% on visible light images. Moreover, using the proposed assignment method has an additional improvement of 0.55% on infrared images and 0.53% on visible light images compared with using more anchor boxes. This demonstrates that the proposed algorithm based on aspect ratio assignment is more suitable for object detection from drone perspective. In addition, from the AP of each category, it can be seen that except for bus, which has a small improvement, the detection performance of categories with fewer samples has a more significant improvement. From the perspective of the samples themselves, the anchor boxes obtained by the direct clustering method tend to have the size of a car, which has the most samples. The convolutional neural network extracts features by using shared convolutional kernels, and the network is more inclined to extract features of categories with more samples. Since bus and car are close in size, although there are few bus samples in the data, they benefit from the network's preference for car, and the AP of bus is much higher than the other categories with few samples. The anchor box allocation method proposed in this paper reduces the impact of sample imbalance on the network and anchor boxes. The AP of car, which has the largest sample size, even decreases slightly, but the AP of other categories increases. Since bus is closer to car, the network and anchor boxes themselves have a certain preference for it, so its AP baseline is high, resulting in a less obvious improvement by the proposed method.

3.2.2. HSM Loss

We designed ablation experiments to improve the loss function. This section still uses the baseline obtained in the previous section. This subsection experimentally verifies that focal loss balances positive and negative samples by suppressing negative samples and dynamically adjusts the weights of positive samples to mine hard learning samples. At the same time, combining focal loss and simOTA strategy, we propose HSM Loss for YOLOv7-tiny to achieve a balance between positive and negative samples and easy and hard samples. Table 3 shows the mAP using different loss functions on infrared and visible light images. First, we use the original focal loss for experiments. The mAP on visible light data is 71.90%, and the mAP on infrared images is 76.15%. Compared with the baseline, it can be seen that the introduction of focal loss has a negative effect. At the same time, when the network confidence threshold is set low, there will be a large amount of background being regarded as objects, indicating that the network's learning of background is insufficient and that negative samples are suppressed too much. HSM Loss reduces the suppression degree of easy samples, resulting in a mAP of 75.72% on visible light data and a mAP of 78.76% on infrared images. It can be seen that compared with the original focal loss, performance has been greatly improved. Moreover, since visible light images are greatly affected by illumination, there are many images taken in the evening or at night in the data, where visible light images have more hard objects to detect. HSM Loss outperforms the baseline by 0.84% on visible light images and by 0.47% on infrared images. The larger improvement on visible light images demonstrates the effectiveness of HSM Loss in mining hard samples.

Table 3. Detection results of different loss functions.

Loss Function	Modality	Car	Truck	Freight Car	Bus	Van	mAP
Cross Entropy	RGB	96.28	74.77	53.99	94.44	54.94	74.88
Focal Loss		96.15	71.04	47.79	93.71	50.82	71.90
HSM Loss		**96.43**	**74.96**	**55.86**	**94.96**	**56.38**	**75.72**
Cross Entropy	IR	98.03	76.79	65.27	94.62	56.74	78.29
Focal Loss		98.22	74.98	61.88	93.84	51.82	76.15
HSM Loss		**98.34**	**77.29**	**66.44**	**94.91**	**56.81**	**78.76**

We conducted experiments combining the anchor box allocation method and loss function improvement method proposed in this paper. The results are shown in Table 4. The mAP of our algorithm on visible light data is 76.31%, and the mAP on infrared images is 79.82%. Compared with the baseline, our algorithm has an increase of 1.53% on visible light images mAP and an increase of 1.43% on infrared images mAP, which fully proves the effectiveness of our method.

Table 4. Detection results of different methods.

Methods	Modality	Car	Truck	Freight Car	Bus	Van	mAP
Baseline	RGB	96.28	74.77	53.99	94.44	54.94	74.88
Baseline + HSM Loss		**96.43**	74.96	55.86	94.96	56.38	75.72
Baseline + AR-anchor		96.19	75.75	57.34	94.52	56.68	76.10
Ours		96.09	**75.89**	**57.35**	**95.19**	**57.00**	**76.31**
Baseline	IR	98.03	76.79	65.27	94.62	56.74	78.29
Baseline + HSM Loss		**98.34**	77.29	66.44	94.91	56.81	78.76
Baseline + AR-anchor		98.02	**79.47**	67.22	94.82	58.59	79.62
Ours		98.10	79.31	**67.56**	**95.31**	**58.81**	**79.82**

Some image detection results are shown in Figure 6. For the car object with the largest sample size, both YOLOv7-tiny and our algorithm can effectively detect it, and our algorithm has a higher confidence level. For the hard-to-detect objects, the original YOLOv7-tiny missed some of them, while our proposed method can detect them more effectively.

3.3. Comparison with Other Object Detection Algorithms

In this paper, we compare our method with some state-of-the-art real-time object detection algorithms on the DroneVehicle dataset. Since our method is based on horizontal bounding boxes, we first select some advanced real-time object detection methods for experiments, including YOLOv5-s and YOLOX-tiny [40]. YOLOv5-s is an anchor-based algorithm, while YOLOX-tiny is an anchor-free algorithm. Then, we convert our method into an oriented object detection algorithm and compare it with some mainstream oriented object detection algorithms. We trained YOLOv5-s with the same strategy as YOLOv7-tiny, using the same anchors, learning rate, Adam optimizer, resolution, and pre-trained weights on the COCO dataset. We froze the backbone for the first 30 epochs and trained the whole network for the remaining epochs. YOLOX-tiny did not have anchor hyperparameters and showed gradient instability when using Adam optimizer, resulting in difficulty in convergence. Therefore, we used SGD optimizer with a learning rate of 0.01 and ignored samples with large loss values during training. The final results are shown in Table 5.

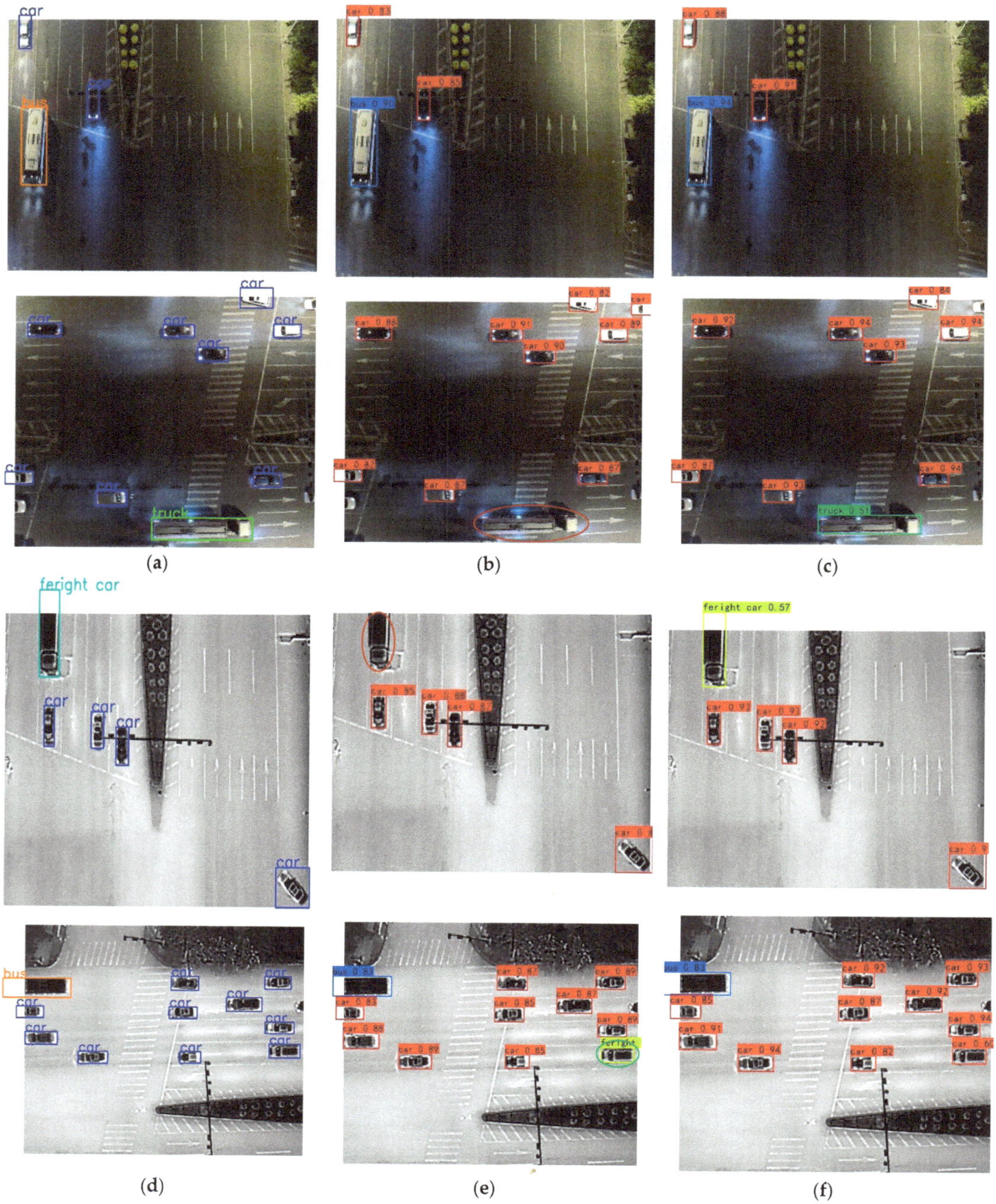

Figure 6. Detection results of YOLOv7-tiny and our proposed algorithm: (**a**,**d**) are ground truth for these images, (**b**,**e**) are for YOLOv7-tiny, while (**c**,**f**) are for our proposed algorithm; (**a**–**c**) are visible light images, while (**d**–**f**) are infrared images. Red circles indicate missed detection, while green circles indicate false detection.

Table 5. Detection results of different algorithms.

Algorithm	Modality	Car	Trunk	Freight Car	Bus	Van	mAP
YOLOv5-s	RGB	95.89	74.07	53.45	94.85	49.31	73.52
YOLOX		96.11	74.18	55.35	94.74	54.34	74.95
YOLOv7-tiny		**96.28**	74.77	53.99	94.44	54.94	74.88
Ours		96.09	**75.89**	**57.35**	**95.19**	**57.00**	**76.31**
YOLOv5-s	IR	97.72	74.71	62.24	94.64	49.23	75.71
YOLOX		97.99	76.60	65.73	94.46	56.98	78.35
YOLOv7-tiny		98.03	76.79	65.27	94.62	56.74	78.29
Ours		**98.10**	**79.31**	**67.56**	**95.31**	**58.81**	**79.82**

The results in the table show that the anchor-free algorithm YOLOX can achieve good detection performance without introducing prior information artificially, and it surpasses YOLOv5-s and YOLOv7-tiny on both infrared and visible light images, which indirectly indicates the effectiveness of the prior information provided by our method. We add angle prediction to the network output and KLD loss [41] to this part, then obtain the network anchors according to the ground truth of oriented bounding boxes and retrain the network. The detection results are shown in Figure 7, which demonstrate that our method can accurately detect the objects in both infrared and visible light images. We compare our method with some oriented object detection algorithms as shown in Table 6, and it can be seen that our method has good detection performance on both infrared and visible light images. Moreover, by comparing the detection performance of horizontal boxes and oriented boxes, we find that the IoU calculation of oriented boxes is stricter, and it requires more accurate bounding boxes to have IoU greater than 0.5 with the ground truth. Therefore, the mAP of oriented object detection is lower than that of horizontal object detection. In addition, oriented object detection can obtain the orientation information of the objects.

Table 6. Detection results of different oriented object detection algorithms.

Algorithm	Modality	Car	Trunk	Freight Car	Bus	Van	mAP
ROI Transformer	RGB	61.55	55.05	42.26	85.48	44.84	61.55
S^2ANet		79.86	50.02	36.21	82.77	37.52	57.28
Oriented R-CNN		80.26	55.39	42.12	86.84	46.92	62.30
Ours		**92.42**	**68.12**	**50.85**	**93.11**	**53.79**	**71.66**
ROI Transformer	IR	89.64	50.98	53.42	88.86	44.47	65.47
S^2ANet		89.71	51.03	50.27	88.97	44.03	64.80
Oriented R-CNN		89.63	53.92	53.86	89.15	40.95	65.50
Ours		**97.03**	**69.49**	**58.32**	**93.09**	**52.64**	**74.12**

We also compare the parameter size, computation amount, and FPS of several algorithms as shown in Table 7, where FPS is measured on 1080TI. Due to the convolutional operation with shared parameters used by CNN, the parameter size and computation amount are different, and the network inference speed is more closely related to the computation amount. Therefore, YOLOv7-tiny is faster than YOLOX, and YOLOv5-s is the slowest. The overall network speed also includes the post-processing computation, in which NMS takes the longest time. The HBB algorithm proposed in this paper has a slightly higher computation amount than YOLOv7-tiny, but it enhances the network detection performance, reduces redundant detection boxes, and lowers the post-processing time. Therefore, it has the fastest speed among all algorithms. The OBB object detection algorithm proposed in this paper has the slowest speed because its NMS needs to calculate the IoU of rotated boxes, which requires a lot of time.

Figure 7. *Cont.*

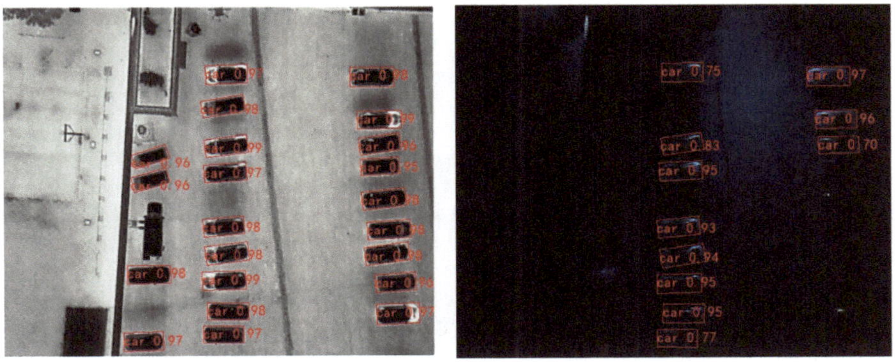

(a) (b)

Figure 7. Detection results of our proposed algorithm with oriented object detection: (**a**) shows detection results on infrared images; (**b**) shows detection results on visible light images.

Table 7. The parameter size, computational cost, and speed of different algorithms.

Algorithm	Parameter (M)	FLOPs (G)	FPS
YOLOv5-s (HBB)	7.033	15.980	96.3
YOLOX (HBB)	**5.034**	15.238	102.6
YOLOv7-tiny (HBB)	6.025	**13.215**	113.3
Ours (HBB)	6.043	13.273	**118.0**
Ours (OBB)	6.047	13.287	88.7

4. Discussion

4.1. AR-Anchor

Object detection algorithms based on anchor box mechanism rely on the prior information provided by anchor boxes. For datasets with known ground truth object sizes, using anchor boxes can make the network more targeted to detect objects. Therefore, object detection algorithms based on an anchor box mechanism often perform better and train more stably than anchor-free algorithms. The prior information provided by anchor boxes guides different layers of the network to pay more attention to objects with different characteristics. In natural scenes, object sizes span a large range, and different layers of the network have different resolutions and receptive fields. Therefore, grouping anchor boxes according to size and matching them with the network's receptive field can bring effective prior information of object size to the network. When increasing the number of anchor boxes, different anchor boxes are responsible for predicting more clear object shapes, and the detection effect is better. At the same time, more anchor boxes also bring greater computational cost. However, under drone perspective, object size differences are not that large. Grouping anchor boxes clustered from ground truth boxes in the dataset according to size results in small size differences between adjacent groups of anchor boxes, which is not ideal for detection performance. In this paper, we use a more suitable vehicle object distribution characteristic under drone perspective and group anchor boxes according to aspect ratio. Different layers predict objects with different orientations, and three anchor boxes are assigned for each class of objects, which alleviates the influence of long-tail effect of sample quantity and achieves better detection performance. The reasons for the AR-anchor's effectiveness include (1) using more anchor boxes, (2) using a distribution method that is more in line with the object characteristics of drone perspective, and (3) balancing sample quantity imbalance.

4.2. HSM Loss

Since the number of objects in the image is always much smaller than anchor boxes in the one-stage object detection algorithm, and some objects have less obvious features, there are problems of positive and negative sample imbalance and hard and easy sample imbalance. Many methods have been proposed to solve these problems, such as focal loss. However, when focal loss is directly introduced into YOLOv7-tiny, the detection effect will become worse. The main reason is that focal loss is aimed at the phenomenon that the number of negative samples in the single-stage algorithm is much larger than that of positive samples, and the negative samples are almost easy samples. It balances positive and negative samples by suppressing negative samples. The lower the confidence of negative samples, the more severe the loss decay. The background with a confidence of 0.01 will be decayed to one-ten-thousandth of the cross-entropy loss. YOLOv7-tiny itself has the ability to adjust positive and negative samples. After adding focal loss, it leads to too much suppression of negative samples during training, which destroys the original balance. This paper reduces the suppression of focal loss on negative samples by controlling the lower limit of sample modulation coefficient and scales the modulation coefficient. The final value range is from 0.1 to 4.1, which makes the loss of hard samples with confidence less than 0.5 amplified and obtains HSM Loss that can mine hard samples better. This loss function can balance hard and easy samples on the basis of maintaining the original YOLOv7-tiny positive and negative sample balance and improve network detection effect. HSM Loss mainly reduces the suppression of negative samples by controlling the lower limit of modulation coefficient and makes it have stronger hard sample mining ability by scaling modulation coefficient.

5. Conclusions

This paper aims to improve the object detection performance under drone perspective, which is a challenging problem because of the complex background and small object size. A novel algorithm based on YOLOv7-tiny is proposed, which uses an aspect-ratio-based anchor box assignment method and a hard sample mining loss function. The designed anchor boxes provide more prior information that matches the object size characteristics under drone perspective and also consider the class imbalance problem among samples. The anchor boxes are designed separately for each class, reducing the influence of classes with more samples on classes with less samples and enhancing the network's ability to detect objects. The designed loss function strengthens the learning of hard samples and improves the network's detection performance on hard samples. Compared with the original model, our algorithm achieves 1.53% improvement in mAP on infrared images and 1.43% improvement on visible light images, with only a 0.3% increase in parameters and 0.4% increase in computational cost. It retains the lightweight feature of the original model, making it easier to apply to drones and more practical.

Infrared and visible light images each have their advantages and strong complementarity. By fusing the two types of images, one can obtain both the rich detail information of visible light images and the interference resistance of infrared images, which can lead to better applications. This paper will explore the algorithms of image fusion and detection and will validate the effectiveness of the proposed algorithms on more datasets.

Author Contributions: Conceptualization, S.H.; methodology, S.H. and J.D.; software, S.H. and Y.D.; validation, S.H., Y.D. and F.Z.; formal analysis, S.H., F.Z. and H.L.; investigation, S.H. and F.Z.; resources, S.H.; data curation, S.H. and Y.D.; writing—original draft preparation, S.H.; writing—review and editing, S.H., F.Z., X.S. and J.D.; visualization, S.H. and F.Z.; supervision, F.Z. and H.L.; project administration, F.Z.; funding acquisition, X.S. All authors have read and agreed to the published version of the manuscript.

Funding: This work was supported by the National Natural Science Foundation of China (61901489).

Data Availability Statement: The DroneVehicle remote sensing dataset was obtained from https://github.com/VisDrone/DroneVehicle, accessed on 29 December 2021.

Conflicts of Interest: The authors declare no conflict of interest.

References

1. Girshick, R.; Donahue, J.; Darrell, T. Rich feature hierarchies for accurate object detection and semantic segmentation. In Proceedings of the IEEE Conference on Computer Vision and Pattern Recognition, Columbus, OH, USA, 23–28 June 2014; pp. 580–587.
2. Girshick, R. Fast r-cnn. In Proceedings of the IEEE International Conference on Computer Vision, Santiago, Chile, 7–13 December 2015; pp. 1440–1448.
3. He, K.; Gkioxari, G.; Dollár, P. Mask r-cnn. In Proceedings of the IEEE International Conference on Computer Vision, Venice, Italy, 22–29 October 2017; pp. 2961–2969.
4. Redmon, J.; Divvala, S.; Girshick, R. You only look once: Unified, real-time object detection. In Proceedings of the IEEE Conference on Computer Vision and Pattern Recognition, Las Vegas, NV, USA, 27–30 June 2016; pp. 779–788.
5. Redmon, J.; Farhadi, A. YOLO9000: Better.; faster.; stronger. In Proceedings of the IEEE Conference on Computer Vision and Pattern Recognition, Honolulu, HI, USA, 21–26 July 2017; pp. 7263–7271.
6. Redmon, J.; Farhadi, A. Yolov3: An incremental improvement. *arXiv* **2018**, arXiv:1804.02767.
7. Bochkovskiy, A.; Wang, C.Y.; Liao, H.Y.M. Yolov4: Optimal speed and accuracy of object detection. *arXiv* **2020**, arXiv:2004.10934.
8. Li, C.; Li, L.; Jiang, H. YOLOv6: A single-stage object detection framework for industrial applications. *arXiv* **2022**, arXiv:2209.02976.
9. Wang, C.Y.; Bochkovskiy, A.; Liao, H.Y.M. YOLOv7: Trainable bag-of-freebies sets new state-of-the-art for real-time object detectors. *arXiv* **2022**, arXiv:2207.02696.
10. Liu, W.; Anguelov, D.; Erhan, D. Ssd: Single shot multibox detector. In Proceedings of the European conference on computer vision, Amsterdam, The Netherlands, 11–14 October 2016; Springer: Cham, Switzerland; pp. 21–37.
11. Lin, T.Y.; Goyal, P.; Girshick, R. Focal loss for dense object detection. In Proceedings of the IEEE International Conference on Computer Vision, Venice, Italy, 22–29 October 2017; pp. 2980–2988.
12. Everingham, M.; Eslami, S.M.A.; Van Gool, L. The pascal visual object classes challenge: A retrospective. *Int. J. Comput. Vis.* **2015**, *111*, 98–136. [CrossRef]
13. Lin, T.Y.; Maire, M.; Belongie, S. Microsoft coco: Common objects in context. In Proceedings of the European Conference on Computer Vision, Zurich, Switzerland, 6–12 September 2014; Springer: Cham, Switzerland; pp. 740–755.
14. Li, C.; Xu, C.; Cui, Z. Feature-attentioned object detection in remote sensing imagery. In Proceedings of the 2019 IEEE International Conference on Image Processing (ICIP), Taipei, Taiwan, 22–25 September 2019; IEEE: Piscataway, NJ, USA; pp. 3886–3890.
15. Liu, Z.; Wang, H.; Weng, L. Ship rotated bounding box space for ship extraction from high-resolution optical satellite images with complex backgrounds. *IEEE Geosci. Remote Sens. Lett.* **2016**, *13*, 1074–1078. [CrossRef]
16. Ma, J.; Shao, W.; Ye, H. Arbitrary-oriented scene text detection via rotation proposals. *IEEE Trans. Multimed.* **2018**, *20*, 3111–3122. [CrossRef]
17. Xu, Y.; Fu, M.; Wang, Q. Gliding vertex on the horizontal bounding box for multi-oriented object detection. *IEEE Trans. Pattern Anal. Mach. Intell.* **2020**, *43*, 1452–1459. [CrossRef] [PubMed]
18. Ding, J.; Xue, N.; Long, Y. Learning roi transformer for oriented object detection in aerial images. In Proceedings of the IEEE/CVF Conference on Computer Vision and Pattern Recognition, Long Beach, CA, USA, 15–20 June 2019; pp. 2849–2858.
19. Xie, X.; Cheng, G.; Wang, J. Oriented R-CNN for object detection. In Proceedings of the IEEE/CVF International Conference on Computer Vision, Montreal, BC, Canada, 11–17 October 2021; pp. 3520–3529.
20. Han, J.; Ding, J.; Li, J. Align deep features for oriented object detection. *IEEE Trans. Geosci. Remote Sens.* **2021**, *60*, 1–11. [CrossRef]
21. Fang, Z.; Zhang, T.; Fan, X.H. A ViTDet based dual-source fusion object detection method of UAV. In Proceedings of the 2022 International Conference on Image Processing, Computer Vision and Machine Learning (ICICML), Xi'an, China, 28–30 October 2022; IEEE: Piscataway, NJ, USA; pp. 628–633.
22. Wu, J.; Shen, T.; Wang, Q. Local Adaptive Illumination-Driven Input-Level Fusion for Infrared and Visible Object Detection. *Remote Sens.* **2023**, *15*, 660. [CrossRef]
23. Sun, Y.; Cao, B.; Zhu, P. Drone-based RGB-infrared cross-modality vehicle detection via uncertainty-aware learning. *IEEE Trans. Circuits Syst. Video Technol.* **2022**, *32*, 6700–6713. [CrossRef]
24. Wang, Q.; Chi, Y.; Shen, T. Improving RGB-infrared object detection by reducing cross-modality redundancy. *Remote Sens.* **2022**, *14*, 2020. [CrossRef]
25. An, Z.; Liu, C.; Han, Y. Effectiveness Guided Cross-Modal Information Sharing for Aligned RGB-T Object Detection. *IEEE Signal Process. Lett.* **2022**, *29*, 2562–2566. [CrossRef]
26. Yuan, M.; Wang, Y.; Wei, X. Translation, Scale and Rotation: Cross-Modal Alignment Meets RGB-Infrared Vehicle Detection. In *Proceedings of the Computer Vision–ECCV 2022: 17th European Conference, Tel Aviv, Israel, 23–27 October 2022*; Proceedings, Part IX; Springer Nature: Cham, Switzerland, 2022; pp. 509–525.
27. Zhu, P.; Wen, L.; Bian, X. Vision meets drones: A challenge. *arXiv* **2018**, arXiv:1804.07437.
28. Xia, G.S.; Bai, X.; Ding, J. DOTA: A large-scale dataset for object detection in aerial images. In Proceedings of the IEEE Conference on Computer Vision and Pattern Recognition, Salt Lake City, UT, USA, 18–23 June 2018; pp. 3974–3983.
29. Razakarivony, S.; Jurie, F. Vehicle detection in aerial imagery: A small target detection benchmark. *J. Vis. Commun. Image Represent.* **2016**, *34*, 187–203. [CrossRef]

30. Pang, D.; Ma, P.; Feng, Y. Tensor Spectral k-support Norm Minimization for Detecting Infrared Dim and Small Target against Urban Backgrounds. *IEEE Trans. Geosci. Remote Sens.* **2023**. [CrossRef]
31. Fu, R.; Fan, H.; Zhu, Y. A dataset for infrared time-sensitive target detection and tracking for air-ground application. *China Sci. Data* **2022**, *7*, 206–221.
32. Lin, T.Y.; Dollár, P.; Girshick, R. Feature pyramid networks for object detection. In Proceedings of the IEEE Conference on Computer Vision and Pattern Recognition, Honolulu, HI, USA, 21–26 July 2017; pp. 2117–2125.
33. He, K.; Zhang, X.; Ren, S. Deep residual learning for image recognition. In Proceedings of the IEEE Conference on Computer Vision and Pattern Recognition, Las Vegas, NV, USA, 27–30 June 2016; pp. 770–778.
34. Tan, M.; Pang, R.; Le, Q.V. Efficientdet: Scalable and efficient object detection. In Proceedings of the IEEE/CVF Conference on Computer Vision and Pattern Recognition, Seattle, WA, USA, 13–19 June 2020; pp. 10781–10790.
35. He, K.; Zhang, X.; Ren, S. Spatial pyramid pooling in deep convolutional networks for visual recognition. *IEEE Trans. Pattern Anal. Mach. Intell.* **2015**, *37*, 1904–1916. [CrossRef] [PubMed]
36. Wang, C.Y.; Liao, H.Y.M.; Wu, Y.H. CSPNet: A new backbone that can enhance learning capability of CNN. In Proceedings of the IEEE/CVF Conference on Computer Vision and Pattern Recognition Workshops, Seattle, WA, USA, 14–19 June 2020; pp. 390–391.
37. Liu, S.; Qi, L.; Qin, H. Path aggregation network for instance segmentation. In Proceedings of the IEEE Conference on Computer Vision and Pattern Recognition, Salt Lake City, UT, USA, 18–23 June 2018; pp. 8759–8768.
38. Li, B.; Liu, Y.; Wang, X. Gradient harmonized single-stage detector. In Proceedings of the AAAI Conference on Artificial Intelligence, Honolulu, HI, USA, 29–31 January 2019; Volume 33, pp. 8577–8584.
39. Shrivastava, A.; Gupta, A.; Girshick, R. Training region-based object detectors with online hard example mining. In Proceedings of the IEEE Conference on Computer Vision and Pattern Recognition, Las Vegas, NV, USA, 27–30 June 2016; pp. 761–769.
40. Ge, Z.; Liu, S.; Wang, F. Yolox: Exceeding yolo series in 2021. *arXiv* **2021**, arXiv:2107.08430.
41. Yang, X.; Yang, X.; Yang, J. Learning high-precision bounding box for rotated object detection via kullback-leibler divergence. *Adv. Neural Inf. Process. Syst.* **2021**, *34*, 18381–18394.

remote sensing

Article

Unmanned Aerial Vehicle Perspective Small Target Recognition Algorithm Based on Improved YOLOv5

He Xu [1,2,3], Wenlong Zheng [1,2], Fengxuan Liu [1,2], Peng Li [1,2,3,*] and Ruchuan Wang [1,2,3]

1 School of Computer Science, Nanjing University of Posts and Telecommunications, Nanjing 210023, China; xuhe@njupt.edu.cn (H.X.); 1022040809@njupt.edu.cn (W.Z.); 1221045635@njupt.edu.cn (F.L.); wangrc@njupt.edu.cn (R.W.)
2 Jiangsu High Technology Research Key Laboratory for Wireless Sensor Networks, Nanjing 210023, China
3 Jiangsu HPC and Intelligent Processing Engineer Research Center, Nanjing 210023, China
* Correspondence: lipeng@njupt.edu.cn

Abstract: Small target detection has been widely used in applications that are relevant to everyday life and have many real-time requirements, such as road patrols and security surveillance. Although object detection methods based on deep learning have achieved great success in recent years, they are not effective in small target detection. In order to solve the problem of low recognition rate caused by factors such as low resolution of UAV viewpoint images and little valid information, this paper proposes an improved algorithm based on the YOLOv5s model, called YOLOv5s-pp. First, to better suppress interference from complex backgrounds and negative samples in images, we add a CA attention module, which can better focus on task-specific important channels while weakening the influence of irrelevant channels. Secondly, we improve the forward propagation and generalisation of the network using the Meta-ACON activation function, which adaptively learns to adjust the degree of linearity or nonlinearity of the activation function based on the input data. Again, the SPD Conv module is incorporated into the network model to address the problems of reduced learning efficiency and loss of fine-grained information due to cross-layer convolution in the model. Finally, the detection head is improved by using smaller, smaller-target detection heads to reduce missed detections. We evaluated the algorithm on the VisDrone2019-DET and UAVDT datasets and compared it with other state-of-the-art algorithms. Compared to YOLOv5s, mAP@.5 improved by 7.4% and 6.5% on the VisDrone2019-DET and UAVDT datasets, respectively, and compared to YOLOv8s, mAP@.5 improved by 0.8% and 2.1%, respectively. For improving the performance of the UAV-side small target detection algorithm, it will help to enhance the reliability and safety of UAVs in critical missions such as military reconnaissance, road patrol and security surveillance.

Keywords: small target detection head; improving YOLOv5s; activation function; attention mechanism; spd conv module

Citation: Xu, H.; Zheng, W.; Liu, F.; Li, P.; Wang, R. Unmanned Aerial Vehicle Perspective Small Target Recognition Algorithm Based on Improved YOLOv5. *Remote Sens.* **2023**, *15*, 3583. https://doi.org/10.3390/rs15143583

Academic Editor: Qian Du, Yanni Dong, Xiaochen Yang

Received: 19 June 2023
Revised: 11 July 2023
Accepted: 16 July 2023
Published: 17 July 2023

1. Introduction

In recent years, the unmanned aerial vehicle (UAV) industry has developed rapidly, and UAVs have been widely used in various industries. Low-cost and easy-to-use consumer-grade UAVs have been developed and used in aerial photography, highway traffic, military reconnaissance, agriculture, construction and other fields, bringing great convenience to people. However, when using target recognition algorithms such as YOLO for small target detection tasks from the perspective of UAVs, there are still many challenges that need to be overcome which pose a significant challenge to target recognition tasks on the UAV side. According to the definition of the international organization SPIE, targets with sizes less than 0.12% of the original image are considered as small targets in this paper.

The YOLO (You Only Look Once) algorithm is widely used in image recognition tasks, and has also received a lot of attention in drone or remote sensing image recognition. The first version of YOLO, YOLOv1 [1], was first proposed by Redmon et al. in 2015, followed by YOLOv2 [2] and YOLOv3 [3], both also by Redmon et al. In 2020, Bochkovskiy et al. proposed YOLOv4 [4], and since then YOLOv5, YOLOv6, YOLOv7, YOLOv8 and other models have been developed, all of which play an important role in computer vision.

This article cites several previous studies. In Reference [5], this paper investigates how replacing certain structural elements of a model (as well as their connections and other parameters) affects performance and inference time. In this process, a series of models at different scales are proposed and named YOLO-Z. In Reference [6], the paper proposes to add a prediction head to detect objects of different scales based on YOLOv5. Then, the original prediction head is replaced with a transformer prediction head to explore the potential of self-attention mechanism. The Convolutional Block Attention Module (CBAM) is also integrated to find attention regions in scenes with dense objects. Attention regions are searched for in scenes with dense objects. Reference [7] selects YOLOv5s as the benchmark method of the article. By introducing the feature extraction structure of shufflenetv2, the backbone of YOLOv5s is improved to reduce the amount of calculation of the network, and then the attention mechanism CBMA module is added to the network. Reference [8] addressed the issue of imbalanced quantities of car model pictures in the dataset by using image processing techniques such as image flipping, adding Gaussian noise, and color changes to balance the number of car models. They also added a SimAM attention mechanism without parameters to the bottleneck to improve the network's feature extraction ability. Reference [9] proposed a strategy combining the Faster-RCNN object detection model and three convolutional neural networks, ZF, VGG-16 and ResNet-101, which demonstrated good generalization ability on the dataset. Reference [10] presents an improved U-Net (SU-Net) model for water body detection by fusing deep neural networks for the interference of mixed and dark features on water body detection. Reference [11] proposes the use of a UAV to acquire aerial imagery of the study area through a predetermined mission. Then, the image data were processed using Structure in Motion (SfM) software to obtain a digital surface model (DSM), digital orthophoto map (DOM), point cloud and 3D model. Finally, the generated spatial data are analysed and studied by spatial feature extraction. Reference [12] proposes to segment the image using multi-resolution image segmentation technique with appropriate scale, compactness and smoothing to form homogeneous image objects. Various parameters (spectral, texture, background and elevation) are calculated for VHR UAV images. Urban feature extraction and classification rules are developed especially for roads and buildings. Segmented roads are classified based on their width and connectivity. Extracted buildings are classified based on their elevation and size. Finally, in Reference [13], a lightweight remote sensing rotating object detection model based on YOLOv5 is proposed in this study to improve the performance of remote sensing image object detection. To solve the problem of detecting rotating objects at any angle, the angle regression problem is transformed into a classification problem, enabling the model to predict an additional angle information that is used to rotate the horizontal bounding box to enclose the rotating objects at any angle. In response to the issue of dense distribution of small objects in remote sensing images, the detection layer structure of the model is improved and an object detection scale is added to pay more attention to small objects.

In summary, based on the application characteristics of the YOLOv5 object detection algorithm in small target recognition from the perspective of unmanned aerial vehicles (UAVs) and the advantages and disadvantages of previous research literature, this article designs a small target detection algorithm for the UAV perspective based on the YOLOv5s network structure, named YOLOv5s-pp. The following improvements were made: (1) The CA [14] attention mechanism was introduced, which can consider lightweight attention in both channel and spatial dimensions and can solve problems such as long-distance dependence. (2) The Meta-ACON [15] adaptive activation function was used, which can adaptively adjust the linear or nonlinear degree of the activation function based on the input data. (3) The SPD Conv [16] module was integrated into the network to reduce the loss of fine-grained information caused by cross-layer convolution and feature representation with low learning efficiency, improving the small target feature extraction ability. (4) The detection head of the network was optimized, using a smaller detection head which can reduce the overall loss and the occurrence of missed and false detections, improving the recognition performance of small target information features. Experimental results show that the improved YOLOv5s-pp algorithm has an mAP@.5 of 33.8% on the VisDrone2019-DET dataset, which is 7.4 percentage points higher than the original YOLOv5s network.

The remainder of this article is structured as follows. In Section 2, we briefly review the related work of this article. Section 3 briefly introduces the network structure of the Yolov5 algorithm. Section 4 provides detailed information on the modules of the proposed Yolov5s-pp algorithm. Section 5 presents the experimental results. Finally, Section 6 presents the conclusions.

2. Related Work

In this section, we provide a summary of work related to the research presented in this paper. The summary is divided into two parts: general object detection strategies and methods for detecting small objects.

2.1. General Object Detection

Currently, deep convolutional neural networks dominate the field of object detection. Object detection algorithms can be divided into two categories: single-stage and two-stage (region-free and region-based). Single-stage detectors (region-free detectors) directly predict the location of objects without extracting suggestions, such as SSD [17], YOLO and RetinaNet [18]. Two-stage detectors (region-based detectors) generate a set of region proposals and then predict the object categories within the proposed regions of interest (RoIs) and refine the proposals based on the defined position and size. This approach has been adopted in methods such as Fast R-CNN [19], Faster R-CNN [9] and Cascade R-CNN [20]. Typically, single-stage methods are faster than two-stage methods, while two-stage methods have higher accuracy than single-stage methods. With improvements in recent years, single-stage methods (such as YOLO, RetinaNet, etc.) have also made progress, and the overall performance of the algorithms is superior to two-stage methods. However, these algorithms are designed for general object detection and are not good at detecting small and dense objects.

2.2. Small Object Detection

Compared to general image detection, object detection in aerial drone imagery is more challenging. This is because the objects to be detected in aerial imagery are relatively small, their distribution is uncertain and they can be particularly sparse or concentrated, making the distribution uneven and resulting in a large number of overlapping objects to be detected. As a result, many studies have focused specifically on the detection of small objects. For example, in [21], VariFocal is used to replace the binary cross-entropy loss function in the original classification to overcome the problem of uneven sample distribution and improve detection recall. Meanwhile, the Coordinating Attention (CA) mechanism is added to improve the detection accuracy. In [22], the Cross-Layer Context Fusion Module (CCFM) is proposed, which learns different scales of context information in parallel through multiple branches to improve the representational ability of feature information and the recognition ability of the network. To address the problem of spatial information loss of small objects, the Spatial Information Enhancement Module (SIEM) is designed to adaptively learn the weak spatial information to be protected for small objects. In [23], anchor boxes are assigned according to the aspect ratio of the ground truth box to provide the network with prior information about the shape of the object. The network uses the Hard Sample Mining Loss (HSM Loss) function to guide learning and provide prior information about the shape of the object. In [24], appropriate spatial information is adaptively obtained through multi-scale receptive fields to improve the feature extraction capability. In addition, a Segmentation Fusion (SF) submodule is proposed to effectively reduce the time complexity of ASPConv, and a Fast Multi-Scale Fusion (FMF) module is proposed to alleviate the shortcomings of semantic and spatial information fusion.

The aim of this paper is to address the challenges of detecting small and unevenly distributed objects in drone aerial imagery. We choose the relatively lightweight Yolov5s network model as the base algorithm and adopt the Meta-ACON adaptive activation function. This function adaptively learns and adjusts the linear or nonlinear degree of the activation function based on the input data in order to fully learn the feature information in the input data. To reduce the fine-grained information loss caused by cross-layer convolution and the decrease in the efficiency of feature representation, we integrate the SPD Conv module into the integrated network. To address the problem of small objects, we optimise the detection head by using a smaller detection head, which can not only reduce the overall loss, but also reduce the number of missed detections and false detections. To address the problem of information loss due to long-range dependencies, we introduce the CA attention mechanism, which considers lightweight attention in both the channel and spatial dimensions in parallel to improve the feature extraction capability.

3. YOLOv5s Algorithm

The YOLOv5 algorithm is a single-stage target detection algorithm that can complete target detection in a single step. YOLOv5 offers four models in different sizes: YOLOv5s, YOLOv5m, YOLOv5l and YOLOv5x. To improve the detection accuracy while maintaining the detection speed, we chose the YOLOv5s lightweight model. This model follows the detection idea of the YOLO series, which is to divide the input image into a grid and detect the object by detecting the center point of the object within a certain grid.

The YOLOv5s model consists of four parts: input, Backbone, Neck, and output (also called head/predict).This paper briefly introduces the YOLOv5s network model using the YOLOv5s-5.0 version as the baseline network. The overall network structure is shown in Figure 1.

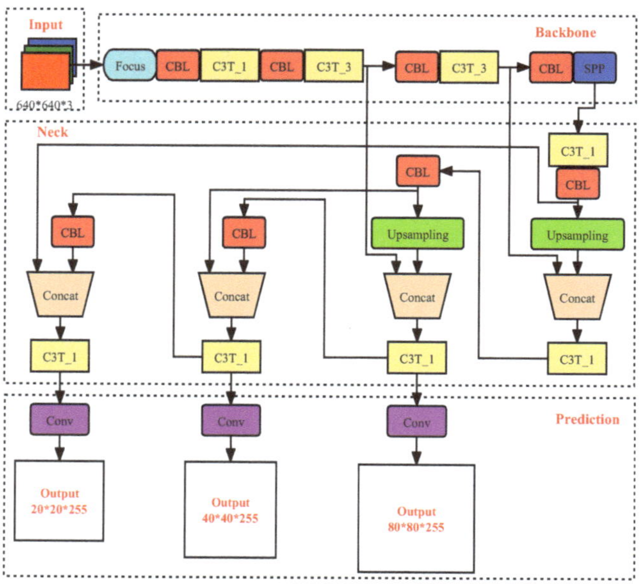

Figure 1. Overall structure diagram of YOLOv5s.

The Focus module mainly performs slicing and reorganization of the input data, and then passes the output data to the next module through CBL operation. This module can better capture the details of objects while reducing computation and memory consumption. The structure diagram of the Focus module is shown in Figure 2.

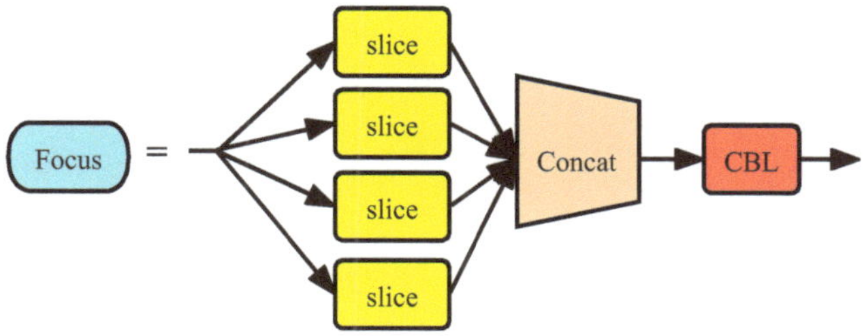

Figure 2. Structure diagram of the Focus module.

The CBL module consists mainly of a convolution operation, followed by batch normalization and the SiLU [25] activation function, and then outputs the data to the next module. The Res-unit (residual module) consists mainly of a 1×1 convolution and a 3×3 convolution. By introducing skip connections, the residual module allows information to flow more smoothly through the network, reducing the problems of gradient vanishing and gradient explosion. Additionally, the residual module can improve the network's expressive power, enabling it to better fit complex nonlinear functions. The structure diagrams of the CBL module and the Res-unit module are shown in Figure 3.

Figure 3. Structure diagram of the CBL and Res-unit module.

The YOLOv5s model extensively uses the C3 module structure, which is an improvement over the bottleneckCSP structure. In addition, residual modules can be added to the C3 module to make parameter passing more stable, ensuring the overall learning stability of the network. The structure diagram of the C3 module is shown in Figure 4.

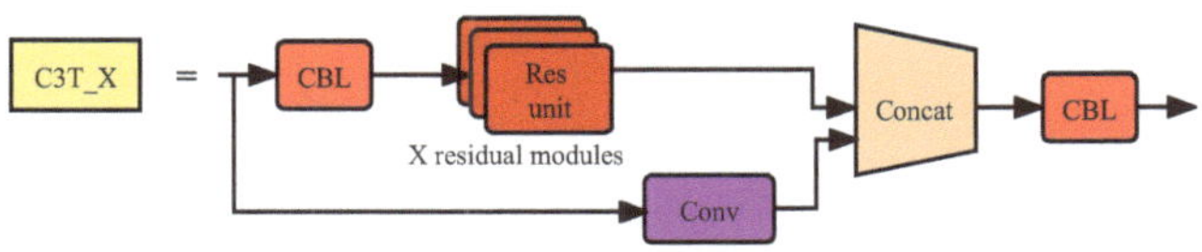

Figure 4. Structure diagram of the C3 module.

The SPP (Spatial Pyramid Pooling) module is a convolutional neural network module used to enhance feature extraction capabilities. This module can accept input feature maps of any size and output fixed-size feature vectors. The SPP module mainly consists of two convolutional layers, a pooling layer and a concatenation layer. The module uses max pooling to reduce the computational load of network parameters, while also increasing the receptive field. The structure diagram of the SPP module is shown in Figure 5.

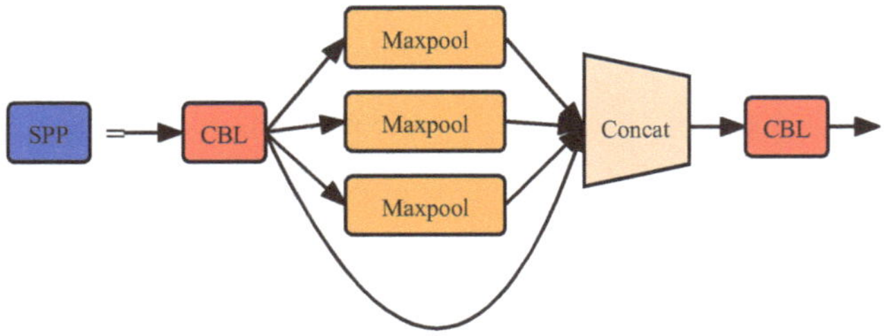

Figure 5. Structure diagram of the SPP module.

4. YOLOv5s-pp Algorithm

The optimization of the YOLOv5s-pp algorithm compared to the YOLOv5s algorithm mainly involves four aspects: using the Meta-ACON activation function, introducing the CA attention mechanism, optimizing the small target detection head and integrating the SPD Conv module. These optimization measures can effectively improve the recognition performance of the model, making the YOLOv5s-pp algorithm perform better in small target detection tasks. The Meta-ACON activation function can improve the model's generalization ability and robustness, the CA attention mechanism can improve the model's focus on important features, the small target detection head can improve the model's small target detection ability and the SPD Conv module can further improve the model's feature representation ability. The overall network structure of YOLOv5s-pp is shown in Figure 6.

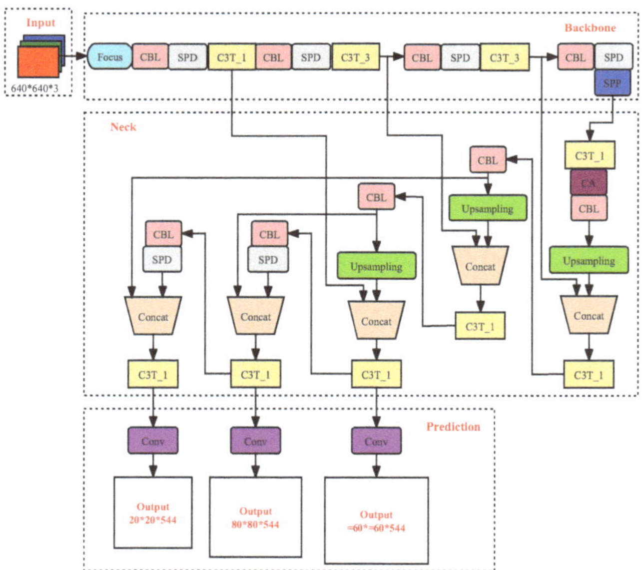

Figure 6. Overall structure diagram of YOLOv5s-pp.

The improved YOLOv5s-pp network model has a certain increase in depth compared to the YOLOv5s model, and the number of parameters in the whole model increases by about 3.3M compared to YOLOv5s. As the width and depth of the network model increases, the amount of computation required by the model increases and the inference speed of the model decreases, so we hope to make the small target detection algorithm achieve better results with as little increase in the complexity of the network model as possible.

4.1. Meta-ACON Activation Function

To improve the forward propagation and generalization capabilities of the YOLOv5s model, we used the Meta-ACON activation function instead of the original SiLU activation function. The Meta-ACON activation function is a function that combines the ReLU [26] and Swish [27] activation functions, and it can not only better solve the problem of neuron death, but also adaptively choose whether to activate neurons and adaptively choose the degree of linearity or nonlinearity of the activation function. The introduction of this activation function can effectively improve the performance and generalization ability of the model, making it more robust and adaptive.

The formula for $max(x_1, \ldots, x_n)$ can be obtained from the definition of smooth maximum [28], as shown in equation:

$$S_\beta(x_1, \ldots, x_n) = \frac{\sum_{i=1}^{n} x_i e^{\beta x_i}}{\sum_{i=1}^{n} e^{\beta x_i}} \tag{1}$$

Assuming that the function $f = max(\eta_a(x), \eta_b(x))$, and then using the above Smooth maximum formula to obtain its smooth approximation function, the following formula can be obtained, as shown in this equation:

$$\begin{aligned} & S_\beta(\eta_a(x), \eta_b(x)) \\ & = \eta_a(x) \cdot \frac{e^{\beta \eta_a(x)}}{e^{\beta \eta_a(x)} + e^{\beta \eta_b(x)}} + \eta_b(x) \cdot \frac{e^{\beta \eta_b(x)}}{e^{\beta \eta_a(x)} + e^{\beta \eta_b(x)}} \\ & = \eta_a(x) \cdot \frac{1}{1 + e^{-\beta(\eta_a(x) - \eta_b(x))}} + \eta_b(x) \cdot \frac{1}{1 + e^{-\beta(\eta_b(x) - \eta_a(x))}} \\ & = \eta_a(x) \cdot \sigma[\beta(\eta_a(x) - \eta_b(x))] + \eta_b(x) \cdot \sigma[\beta(\eta_b(x) - \eta_a(x))] \\ & = (\eta_a(x) - \eta_b(x)) \cdot \sigma[\beta(\eta_a(x) - \eta_b(x))] + \eta_b(x) \end{aligned} \tag{2}$$

Let $\eta_a(x) = P_1 x$, $\eta_b(x) = P_2 x$, and the expression of Meta-ACON can be obtained, as shown in this equation:

$$Meta_{ACON}(x) = (P_1 - P_2)x \cdot \sigma[\beta(P_1 - P_2)x] + P_2 x \tag{3}$$

The parameters P_1 and P_2 of Meta-ACON are two learnable parameters that control the upper and lower limits of the function. The σ parameter represents the Sigmoid function. The β parameter is responsible for dynamically controlling the linearity or nonlinearity of the activation function and adapting the degree of nonlinearity for each layer. In order to make the linear and nonlinear capabilities of the Meta-ACON activation function adaptive, a module $G(x)$ is proposed to dynamically learn β from the input feature $x(C \times W \times H)$, in order to dynamically adaptively control the linear or nonlinear capabilities of the function. This customised behaviour of the activation function helps to improve generalization and propagation performance. The formula is shown in this equation:

$$\beta_c = \sigma W_1 W_2 \sum_{h=1}^{H} \sum_{w=1}^{W} x_{c,h,w} \tag{4}$$

The design space of the adaptive function includes channels, pixels and layers, and the shared parameter β of the channel is used here. First, the mean values of the H and W dimensions are calculated, respectively, and then two 1×1 convolution layers are used to make all pixels in each channel share a weight. Finally, the Sigmoid activation function is used to obtain β. It should be noted that a scaling parameter r is added between $W_1 \in R^{C \times C/r}$ and $W_2 \in R^{C/r \times C}$ to reduce the parameters, which is set to 16 by default.

4.2. SPD Conv Module

Convolutional Neural Networks (CNNs) have achieved tremendous success in computer vision for object detection and image classification tasks. However, for small object detection tasks, especially those involving small objects with overlapping occlusions, their performance can drop dramatically. One of the reasons for this is that the use of cross-convolutional layers can lead to the loss of fine-grained information and low learning efficiency of the feature representation. To solve this problem, we introduce the SPD Conv module to eliminate the effect of each cross-convolution layer. This approach can perform better in small object detection tasks. The SPD Conv module can effectively preserve the detailed information in the image while improving the learning efficiency and feature representation ability of the model.

Consider any intermediate feature map X of size $S \times S \times C$, and slice, merge and convolve a series of subfeature maps as shown in Figure 7.

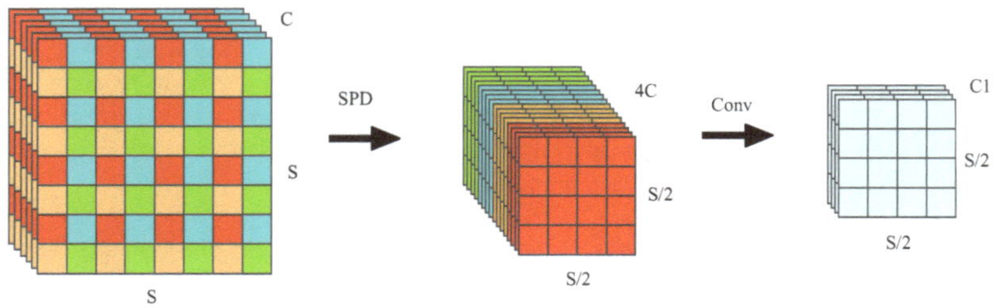

Figure 7. Structure diagram of the SPD Conv module.

The SPD component integrates the input feature map transformation technique into the entire YOLOv5 network, downsampling the feature maps throughout the network. The specific splitting formula is shown in equation:

$$\begin{aligned}
&f_{0,0} = X[0:S:scale, 0:S:scale], \\
&f_{1,0} = X[1:S:scale, 0:S:scale], \ldots, \\
&f_{scae-1,0} = X[scae-1:S:scale, 0:S:scale]; \\
&\ldots \\
&f_{0,scae-1} = X[0:S:scale, scae-1:S:scale], \\
&f_{1,scae-1} = X[1:S:scale, scae-1:S:scale], \ldots, \\
&f_{scae-1,scae-1} = X[scae-1:S:scale, scae-1:S:scale]
\end{aligned} \tag{5}$$

Given any feature map X, the submaps $f_{x,y}$ are formed by all entries $X(i,j)$ that can be divided proportionally by $i+x$ and $j+y$. Therefore, each submap downsamples X by a proportional factor, resulting in four submaps $f_{0,0}$, $f_{1,0}$, $f_{0,1}$, and $f_{1,1}$, each downsampling X by a factor of 2x. These submaps are then concatenated along the channel dimension to obtain a feature map X', which has reduced spatial dimensions by a proportional factor and increased channel dimensions by a proportional factor two.

After the SPD operation, a non-strided convolution with a stride of one is used to process the output feature map. This is done to preserve as much feature map information as possible. Using a strided convolution could result in the loss of some feature map information. In addition, channel compression on the feature map after the SPD operation does not result in a significant loss of feature information, as each pixel along the channel direction is a sparse matrix and most of the compressed information is not significant. Therefore, we use non-strided convolution to preserve the feature map information and channel compression to reduce the computational complexity of the model while maintaining its performance stability.

4.3. Small Target Detection Head

When detecting objects, YOLOv5s algorithm may encounter problems detecting small targets from the drone's perspective, such as missed detection or poor detection performance. Although YOLOv5's default detection heads can complete multi-scale object detection tasks, they may not perform well on detecting small targets. Therefore, we use smaller detection heads to address this issue. Directly adding a small target detection head would increase the number of model parameters and decrease the inference speed, and the overall detection performance may suffer. Therefore, we decided to change the size of the three detection heads in YOLOv5 and set the corresponding prior box sizes. This approach can effectively improve the model's detection capability and accuracy without significantly affecting the model's computational speed and inference efficiency.

The structure of the output section of the YOLOv5s-pp head is shown in Figure 8.

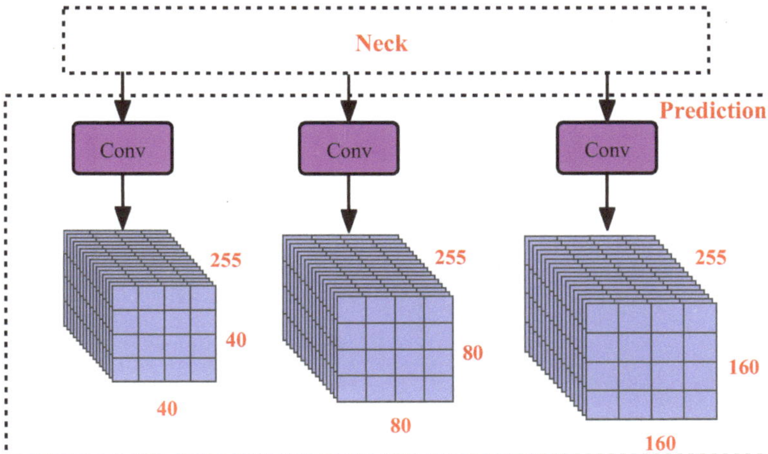

Figure 8. Structure diagram of the small target detection head module.

To match the changed size of the detection heads, the anchor size of the prior boxes must be changed to improve detection performance on small targets. The anchor sizes of the prior boxes are set as shown in Table 1.

Table 1. Improved anchors parameters.

Anchors	Anchor Size
Anchor-1	[4, 5 5, 9 9, 7]
Anchor-2	[8, 14 15, 12 10, 23]
Anchor-3	[30, 18 24, 28 58, 46]

On the YOLOv5s-pp model, we introduced a new detection head through the first feature of the C3 layer, shown in Figure 6. The detection head has a resolution of 160×160 pixels, which corresponds to only two downsampling operations in the backbone network and contains richer low-level feature information. A further upsampling operation is performed in the Neck network to obtain features at the same scale as the input feature map from the backbone network, followed by feature fusion by concatenation. The output feature is the fusion result of multiple input features, which enables the detection head to detect small targets more effectively. The modified small target detection head can better focus on the small target features in the feature map and is more sensitive to small targets. Adding this detection head adds almost no computational and memory overhead to the model, but significantly improves the model's ability to detect small targets.

4.4. Coordinate Attention Mechanism

In the YOLOv5s-pp network model, we incorporated the Coordinate Attention (CA) mechanism, which allows the model to focus on key information from a large amount of information and better attend to the detailed features of the input feature map, thus improving the recognition performance of small targets. In convolutional neural networks, attention mechanisms such as SENet [29–31] and CBAM [32,33] are widely used, but the SENet attention mechanism only focuses on channel dimension information and does not consider spatial dimension information, while the CBAM attention mechanism, although it integrates both channel and spatial dimension information, cannot solve the problem of long-distance dependencies in spatial dimensions. The CA attention mechanism is a lightweight attention mechanism that concurrently considers both channel and spatial dimensions and can solve problems such as long-distance dependencies. Therefore, we chose to incorporate the CA attention mechanism into YOLOv5s-pp to improve the model's recognition capability and accuracy for small targets.

The flowchart of the Coordinate Attention (CA) mechanism is shown in Figure 9.

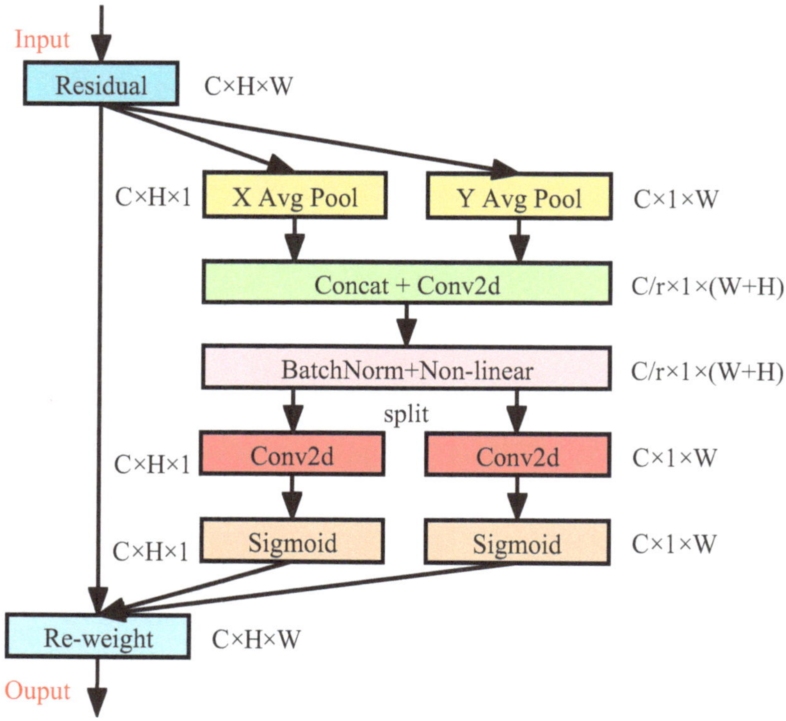

Figure 9. Structure diagram of the CA module.

For the input feature map of size $H \times W \times C$, average pooling operations are performed along the horizontal and vertical directions of the feature map. A $1 \times W$ pooling kernel is used in the horizontal direction (X direction) to obtain an $H \times 1 \times C$ information feature map through global average pooling, as shown in this equation:

$$Z_c^h(h) = \frac{1}{W} \sum_{0 \le i < W} x_c(h, i), Z_c^h \in R^{C \times H \times 1} \tag{6}$$

Similarly, an $H \times 1$ pooling kernel is used in the vertical direction (Y direction) to obtain a $1 \times W \times C$ information feature map through global average pooling, as shown in this equation:

$$Z_c^w(w) = \frac{1}{H} \sum_{0 \le j < H} x_c(j, w), Z_c^w \in R^{C \times 1 \times W} \tag{7}$$

The two information feature maps Z_c^h and Z_c^w, obtained by pooling operations, are concatenated along the spatial dimension. The concatenated feature map is then passed through a 1×1 convolutional layer with an activation function. The resulting feature map is then sliced along the spatial dimension to obtain two separate feature maps. Each of these feature maps is passed through a convolutional layer with an activation function, resulting in two attention vectors g^h and g^w. The specific equations for these operations are shown in this equation:

$$f = \delta(F_1([Z^h, Z^w])), f \in R^{C/r \times 1 \times (H+W)} \tag{8}$$

where: $f^h \in R^{C/r \times H \times 1}, f^w \in R^{C/r \times 1 \times W}$.

$$g^h = \sigma(F_h(f^h)) \tag{9}$$

$$g^w = \sigma(F_w(f^w)) \tag{10}$$

The two attention vectors $g^h \in C \times H \times 1$ and $g^w \in C \times 1 \times W$ obtained in the previous step are broadcast to the dimensions of $C \times H \times W$ and multiplied element-wise with the input feature map X_c after residual operations to obtain the final attention features. The equation is shown here:

$$y_c(i,j) = x_c(i,j) \times g_c^h(i) \times g_c^w(j) \tag{11}$$

The CA attention mechanism considers not only the importance of different channel information, but also the importance of spatial information. It concatenates the global features in the horizontal and vertical directions to form the whole global feature, which helps to solve the problem of long-range dependencies to some extent. This allows coordinate attention to more accurately locate the position of the object of interest, thereby improving the recognition performance of the whole model.

5. Experiments

5.1. Experimental Environment Introduction

In the experimental part of this paper, we have chosen the UAV view small target detection datasets VisDrone2019-DET and UAVDT as the experimental datasets. Where the VisDrone2019-DET dataset was used as the main test dataset for the experiments, the UAVDT dataset was used to further test the robustness and applicability of the model. All the images in these two datasets are from the aerial view of UAVs, which are typical UAV view small target detection datasets. Among them, VisDrone2019-DET is a dataset collected and published by the AISKYEYE team [34], which ensures the rigour of the dataset. The UAVDT dataset is a dataset proposed by Dawei Du et al. [35]. at ICCV2018. Both datasets contain images taken with different models of UAV equipment in different scenarios, weather and lighting conditions, respectively, and are highly representative.

The VisDrone2019-DET dataset is a dataset mainly used for image classification and object detection experiments, consisting of 10,209 images divided into four subsets as shown in Table 2. The training subset is used to train the algorithm and contains 6471 images. The validation subset is used to verify the performance of the algorithm during the training process and contains 548 images. The test-challenge subset is used for competition and contains 1580 images. The test-dev subset is the default test set for public evaluation and contains 1610 images. The VisDrone2019-DET dataset contains 10 categories, including motor, bus, awning-tricycle, tricycle, truck, van, car, bicycle, person and pedestrian.

Table 2. Number of VisDrone2019-DET dataset images.

Train	Val	Test-Challenge	Test-Dev
6471 images	584 images	1580 images	1610 images

Dividing the dataset is one of the most important steps in the training of a network model. Firstly, dividing the dataset avoids the problem of over-fitting the model to the training data, which can lead to degradation in generalisation performance. Secondly, dividing the dataset helps us to evaluate the performance of the model more accurately. Finally, partitioning the dataset also ensures the independence of the training, validation and testing datasets, which is important for evaluating the performance of the model and avoiding data leakage problems. We will use the following three subsets in the next experiments: the training set, the validation set and the test-dev set. The train set is used to train the model, the validation set is used to evaluate the performance of the model during training and to perform hyperparameter tuning, and the test-dev set is used to finally evaluate the performance of the model.

The UAVDT dataset is specifically designed for vehicle detection tasks from an unmanned aerial vehicle (UAV) perspective. It contains images taken by UAVs at different heights and angles on urban roads, along with corresponding annotation information such as vehicle position, bounding box and category in each frame. Vehicle categories in the dataset include private cars, buses and trucks. The dataset has features such as different perspectives, rich vehicle categories, large data volume and high annotation accuracy.

The UAVDT dataset contains a total of 40,735 images, including 15 scenes and over 20,000 vehicle instances. The dataset is divided into three subsets as shown in Table 3. The training subset is used to train the algorithm and contains 31,515 images; the validation subset is used to verify the performance of the algorithm during training and contains 2396 images; and the test subset is used to evaluate the generalization ability of the overall model and contains 6824 images.

Table 3. Number of UAVDT dataset images.

Train	Val	Test
31,515 images	2396 images	6824 images

This experimental project is based on the PyTorch open source framework and uses the GPU to train and test the YOLOv5 algorithm. The data is accelerated by GPU computing through the CUDA and CUDNN environments, which speeds up the training process. The batch size of training samples is set to 16. The input image size is 640 × 640. The specific configuration of the training and testing environments is shown in Table 4.

Table 4. Experimental environment configuration.

Projects	Environment
Operating System	Windows 10(x64)
CPU	i5-9300HF CPU@2.40GHz
Memory size	32GB
GPU	NVIDIA GeForce GTX 2080 Ti
CUDA Version	10.2
Python Version	3.8

5.2. Evaluation Indicators

The most commonly used metrics in object detection algorithms include *Precision* (P), *Recall* (R), F1 Score, Average Precision (AP), and mean Average Precision (mAP). Here are brief introductions to their calculation:

The *Precision* (P), which represents the proportion of true positive predictions among all samples predicted as positive. Its formula is shown in this equation:

$$Precision = \frac{TP}{TP + FP} \tag{12}$$

where *TP* represents the number of true positive samples that are predicted as positive, and *FP* represents the number of false positive samples that are predicted as positive but are actually negative.

The *Recall* (R) is used to determine how many true positive samples are successfully detected among all actual positive samples. Its formula is shown in this equation:

$$Recall = \frac{TP}{TP + FN} \tag{13}$$

where *TP* represents the number of true positive samples that are predicted as positive, and *FN* represents the number of false negative samples that are predicted as negative but are actually positive.

The*F*1 score is the weighted harmonic mean of precision and recall, and is used to balance the trade-off between precision and recall. The formula for calculating the *F*1 score is shown in equation:

$$F1 = \frac{2TP}{2TP + FN + FP} \tag{14}$$

Mean Average Precision (*mAP*) is an important evaluation metric for object detection that is usually involved in research that involves detecting multiple classes. For each class, the model calculates its corresponding average precision, and the *mAP* is obtained by summing the average precision of all classes in the dataset and dividing it by the number of classes. The formula for calculating *mAP* is shown in this equation:

$$mAP = \frac{\sum AveragePrecision(c)}{N(classes)} \tag{15}$$

where *Average Precision(c)* represents the average precision of a class *c*, and *N(classes)* represents the total number of classes in the dataset.

5.3. Experimental Results

Through experiments, we can obtain the specific performance comparison between the original YOLOv5s model and the optimized YOLOv5s-pp model on the VisDrone2019-DET dataset, as shown in Table 5. It can be seen that the overall performance of the YOLOv5s-pp model is better than that of the original YOLOv5s model.

Table 5. Improved before and after model in VisDrone2019-DET dataset val subset of experimental data.

Models	F1 (%)	R (%)	mAP@.5 (%)
YOLOv5s	38.0	52.0	31.2
YOLOv5s-pp	47.0 (+9.0)	63.0 (+11.0)	41.7 (+10.5)

The validation results of the YOLOv5s model on the val subset are shown in Table 6.

Table 6. Experimental data of YOLOv5s on the val subset.

Category	P (%)	R (%)	mAP@.5 (%)
pedestrian	43.8	41.6	37.4
people	44.1	36.1	29.5
bicycle	27.2	14.0	9.8
car	57.7	75.1	71.5
van	44.5	35.5	33.4
truck	40.7	30.7	25.4
tricycle	38.3	24.1	17.8
awning-tricycle	24.1	14.2	9.2
bus	64.6	38.2	41.8
motor	47.2	44.7	36.4

The validation results of the YOLOv5s-pp model on the val subset are shown in Table 7.

Table 7. Experimental data of YOLOv5s-pp on the val subset.

Category	P (%)	R (%)	mAP@.5 (%)
pedestrain	61.9 (+18.1)	49.9 (+8.3)	51.7 (+14.3)
people	56.4 (+12.3)	38.2 (+2.1)	39.6 (+10.1)
bicycle	40.4 (+13.2)	19.3 (+5.3)	19.0 (+9.2)
car	72.5 (+14.8)	81.8 (+6.7)	82.1 (+10.6)
van	57.0 (+12.5)	42.4 (+6.9)	44.1 (+10.7)
truck	58.7 (+18.0)	33.6 (+2.9)	36.0 (+10.6)
tricycle	48.6 (+10.3)	29.6 (+5.5)	26.3 (+8.5)
awning-tricycle	38.3 (+14.2)	17.9 (+3.7)	14.7 (+5.5)
bus	68.3 (+3.7)	51.4 (+13.2)	55.3 (+13.5)
motor	63.0 (+15.8)	47.2 (+2.5)	48.2 (+11.8)

The overall performance of the trained YOLOv5s and YOLOv5s-pp models on the VisDrone2019-DET dataset test-dev subset is shown in Table 8.

Table 8. Improved before and after model in VisDrone2019-DET dataset test-dev subset of experimental data.

Models	F1 (%)	R (%)	mAP@.5 (%)
YOLOv5s	33.0	52.0	26.4
YOLOv5s-pp	40.0 (+7.0)	60.0 (+8.0)	33.8 (+7.4)

The validation results of the YOLOv5s model on the test-dev subset are shown in Table 9.

Table 9. Experimental data of YOLOv5s on the test-dev subset.

Category	P (%)	R (%)	mAP@.5 (%)
pedestrain	39.9	27.1	23.6
people	40.6	16.5	13.3
bicycle	26.8	7.2	6.4
car	56.7	72.5	67.3
van	31.0	37.9	29.5
truck	32.4	36.1	26.1
tricycle	23.4	20.6	12.8
awning-tricycle	34.8	16.7	13.1
bus	62.9	49.1	50.2
motor	38.8	30.0	22.0

The validation results of the YOLOv5s-pp model on the test-dev subset are shown in Table 10.

Table 10. Experimental data of YOLOv5s-pp on the test-dev subset.

Category	P (%)	R (%)	mAP@.5 (%)
pedestrain	49.9 (+10.0)	35.5 (+8.4)	33.9 (+10.3)
people	46.4 (+5.8)	21.0 (+4.5)	19.5 (+6.2)
bicycle	31.4 (+4.6)	12.7 (+5.5)	10.7 (+4.3)
car	65.8 (+9.1)	77.5 (+5.0)	75.0 (+7.7)
van	44.9 (+13.9)	39.6 (+1.7)	37.8 (+8.3)
truck	44.5 (+12.1)	40.0 (+3.9)	34.9 (+8.8)
tricycle	29.3 (+5.9)	31.1 (+10.5)	20.2 (+7.4)
awning-tricycle	40.1 (+5.3)	22.5 (+5.8)	18.6 (+5.5)
bus	67.1 (+4.2)	52.6 (+3.5)	56.1 (+5.9)
motor	47.4 (+8.6)	36.5 (+6.5)	31.1 (+9.1)

From Tables 5–10, it can be observed that the performance of the YOLOv5s-pp model on the VisDrone2019-DET dataset is superior to the original YOLOv5s model for both the val and test subsets. The overall optimization and generalization ability of the YOLOv5s-pp model is better.

The performance parameter results of YOLOv5s trained on the VisDrone2019-DET dataset are shown in Figure 10.

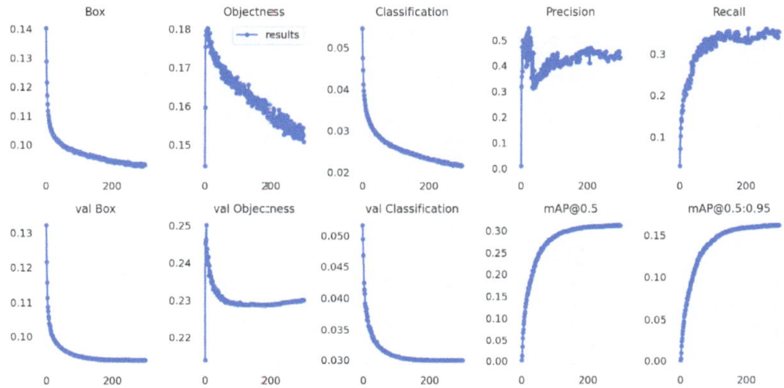

Figure 10. YOLOv5s performance parameter results diagram.

The performance parameter results of YOLOv5s-pp trained on the VisDrone2019-DET dataset are shown in Figure 11.

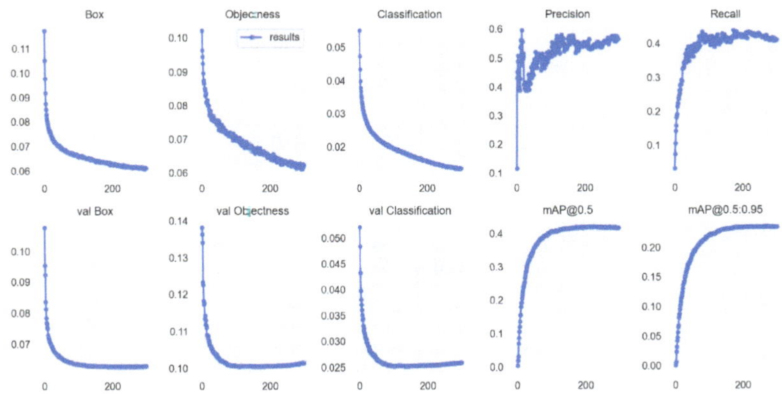

Figure 11. YOLOv5s-pp performance parameter results diagram.

From the experimental results shown in Figures 10 and 11, we can observe the improvement curves of the overall optimization performance of the model before and after the improvements. The figures not only display performance curves such as Precision, Recall and mAP, but also demonstrate the loss curves of model parameter optimization. Among them, the Box curve is the loss of bounding box using GIOU Loss in YOLOv5s. The smaller the value, the more accurate the selected target is. The Objectness curve is the mean loss inferred as the target, and the smaller the value, the more accurate the target detection is. The Classification curve is the mean loss inferred as classification, and the smaller the value, the more accurate the target classification is. From the before-and-after optimization result figures, it can be seen that the optimized YOLOv5s-pp model has better overall loss and detection performance.

Through experiments, we evaluated the effects of the CA attention mechanism, Meta-ACON activation function, SPD Conv module, and small object detection head on the YOLOv5s-pp network model. Specifically, we refer to the YOLOv5s-pp model with the CA attention mechanism removed as the YOLOv5s-A model; the YOLOv5s-pp model with the Meta-ACON activation function removed as the YOLOv5s-B model; the YOLOv5s-pp model with the SPD Conv module removed as the YOLOv5s-C model; and the YOLOv5s-pp model with the small object detection head modified as the YOLOv5s-D model.

We trained each of the above models on the VisDrone2019-DET dataset and tested them on the validation subset. The comparison results of the test data are shown in Table 11.

Table 11. Validating the performance impact of each module on a val subset of the VisDrone2019-DET dataset.

Models	F1 (%)	R (%)	mAP@.5 (%)
YOLOv5s	38.0	52.0	31.2
YOLOv5s-A	45.0	60.0	38.5
YOLOv5s-B	46.0	61.0	40.6
YOLOv5s-C	45.0	62.0	39.3
YOLOv5s-D	40.0	55.0	33.2
YOLOv5s-pp	47.0	63.0	41.7

After training, each model was tested on the test-dev subset, and the comparison results of the test data are shown in Table 12.

Table 12. Validating the performance impact of each module on a test-dev subset of the VisDrone2019-DET dataset.

Models	F1 (%)	R (%)	mAP@.5 (%)
YOLOv5s	33.0	52.0	26.4
YOLOv5s-A	38.0	57.0	31.3
YOLOv5s-B	39.0	56.0	31.6
YOLOv5s-C	37.0	58.0	32.5
YOLOv5s-D	36.0	54.0	28.8
YOLOv5s-pp	40.0	60.0	33.8

Through Tables 11 and 12, we can see that when any of the CA attention mechanism, Meta-ACON activation function, SPD Conv module, or any module in the small object detection head is removed from the YOLOv5s-pp model, it will have a negative impact on the overall performance of the model. It is worth noting that the change in the small object detection head has a greater impact on the recognition performance of the entire model, while the impact of the other modules on the model is relatively small. However, overall, the recognition performance of these modified models is better than that of the YOLOv5s model.

In order to further verify the advantages of the YOLOv5s-pp algorithm, we compared it with other object detection models and conducted tests on the test set and validation set of the VisDrone2019-DET dataset. The model training information is shown in Table 13, and the model performance comparison results are shown in Tables 14 and 15.

The algorithms used for performance comparison in Table 14 and Table 15 are all based on the VisDrone2019-DET dataset, where the FocSDet, ClusDet, HRDNet, YOLOv5sm+, AF-O-YOLOv5, QueryDet, and RRNet algorithms are cited from the training results of other researchers, and since the research results are not explicitly listed in the results of some researchers' papers, we do not fill in the corresponding F1 scores and Rs in Tables 14 and 15.The rest of the YOLOv5 versions and the YOLOv8 versions are trained by ourselves.

From Tables 13–15, we can see that the overall performance of the YOLOv5s-pp model is superior to the four different sizes of YOLOv5 models. From Table 15, we can see that the mAP@.5 value of the YOLOv5s-pp algorithm on the test-dev subset is 7.4% higher than that of the YOLOv5s, 4.1% higher than that of the YOLOv5m, 2.6% higher than that of the YOLOv5l and 1.3% higher than that of the YOLOv5x. From Table 13, we can see that the YOLOv5s-pp algorithm has 10.5 M parameters, which is 3.3 M more than the YOLOv5s model, but far less than the parameters of the other three versions of YOLOv5. The main reason for the different numbers of parameters between the different versions of YOLOv5 is the difference in the depth of the network model, which to some extent reflects the width and depth of the network model. As the width and depth of the network model increases, the spatial and temporal complexity of the model also increases, which means that the GPU resources occupied during model training also increase, and the training time and forward inference time of the model will also increase to some extent. This can be reflected from the model training time and model parameters in Table 13. From Table 15, we can also see that the overall performance of the YOLOv5s-pp model is better than that of the YOLOv8s model. The mAP@.5 value of the YOLOv5s-pp algorithm on the test-dev subset is 0.8% higher than that of the YOLOv8s, and its overall performance is comparable to that of the YOLOv8m model, but lower than that of the YOLOv8l and YOLOv8x models. Although the mAP@.5 values of YOLOv8l and YOLOv8x models on the test-dev subset are 3.1% and 4.2% higher than that of the YOLOv5s-pp model, respectively, their parameter amounts are 4.1 and 6.5 times that of the YOLOv5s-pp model. Their network model depth and parameter amounts far exceed those of the YOLOv5s-pp model, which requires more GPU resources and time to train and infer the model.

Table 13. Details of each model trained on the VisDrone2019-DET dataset.

Models	Size (Pixels)	Batch-Size	Epoch	Training Hours	Params (M)
YOLOv8s	640	16	300	8.8	11.2
YOLOv8m	640	16	300	11.1	25.9
YOLOv8l	640	16	300	13.3	43.7
YOLOv8x	640	16	300	16.9	68.2
YOLOv5s	640	16	300	9.4	7.2
YOLOv5m	640	16	300	12.2	21.2
YOLOv5l	640	16	300	13.7	46.5
YOLOv5x	640	16	300	20.5	86.7
YOLOv5s-pp	640	16	300	11.2	10.5

Table 14. Experimental results of various models on the val subset of the VisDrone2019-DET dataset.

Models	F1 (%)	R (%)	mAP@.5 (%)
FocSDet [36]	-	41.1	30.8
ClusDet [37]	-	-	32.4
HRDNet [38]	-	46.6	35.5
YOLOv5sm+ [39]	-	-	36.7
AF-O-YOLOv5 [40]	45.0	61.0	41.5
YOLOv8s	44.0	61.0	40.9
YOLOv8m	46.0	62.0	43.3
YOLOv8l	49.0	62.0	45.6
YOLOv8x	50.0	63.0	46.9
YOLOv5s	38.0	52.0	31.2
YOLOv5m	42.0	57.0	35.1
YOLOv5l	44.0	59.0	36.9
YOLOv5x	45.0	60.0	38.8
YOLOv5s-pp	47.0	63.0	41.7

Table 15. Experimental results of various models on the test-dev subset of the VisDrone2019-DET dataset.

Models	F1 (%)	R (%)	mAP@.5 (%)
QueryDet [41]	-	-	28.4
RRNet [42]	-	-	29.1
AF-O-YOLOv5 [40]	39.0	58.0	32.6
YOLOv8s	38.0	56.0	33.0
YOLOv8m	40.0	58.0	34.4
YOLOv8l	43.0	58.0	36.9
YOLOv8x	43.0	59.0	38.0
YOLOv5s	33.0	52.0	26.4
YOLOv5m	37.0	55.0	29.7
YOLOv5l	38.0	56.0	31.2
YOLOv5x	40.0	57.0	32.5
YOLOv5s-pp	40.0	60.0	33.8

The images in Figures 12 and 13 are selected from real scenes taken from the UAV's perspective. To better test the model's performance, we have selected images from both daytime and nighttime scenes, respectively. As we can see in Figure 12, the YOLOv5s model did not recognize the two motorbikes in the lower left corner of the image in the night environment; however, the YOLOv5s-pp model detected the motorbikes successfully. As we can see in Figure 13, in the above set of test comparisons, the YOLOv5s-pp model identified more small targets than the YOLOv5s model, and for the relatively more distant and smaller targets, the YOLOv5s model showed a large number of missed detections. Also in the upper left and right corners of the basketball court, the YOLOv5s model failed to fully detect the pedestrians in the image, while the YOLOv5s-pp model successfully detected the pedestrians in it. In contrast, the YOLOv5s-pp model is more accurate in detecting small targets in the image.

(a) (b)

Figure 12. Plot of test results for the example in a daytime environment (Based on VisDrone2019-DET dataset). (**a**) shows the test results for the yolov5s model and (**b**) shows the test results for the yolov5s-pp model.

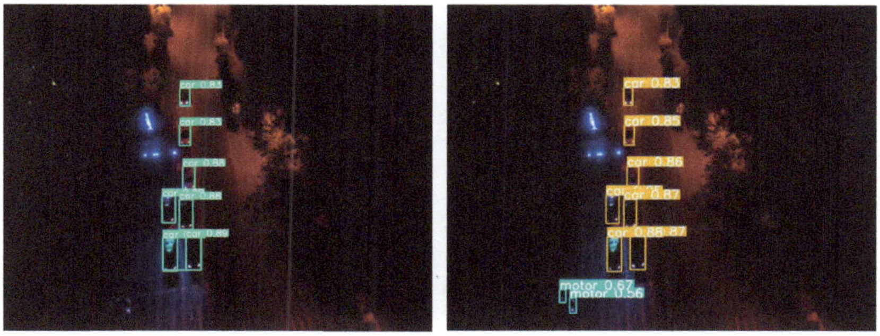

(a) (b)

Figure 13. Plot of test results for the example in the nighttime environment (Based on VisDrone2019-DET dataset). (**a**) shows the test results for the yolov5s model and (**b**) shows the test results for the yolov5s-pp model.

To further test the applicability of the YOLOv5s-pp model, we evaluated the model again on the UAVDT dataset. In this evaluation, we only compared the relatively lightweight YOLOv5s, YOLOv5m, YOLOv8s and YOLOv8m models, as the UAVDT dataset has a larger amount of data and does not train well on the same training environment as heavier models such as YOLOv5l, YOLOv5x, YOLOv8l and YOLOv8x. Detailed information related to the training of the models is presented in Table 16, while Tables 17 and 18 present the final performance comparison results.

Table 16. Details of each model trained on the UAVDT dataset.

Models	Size (Pixels)	Batch-Size	Epoch	Training Hours	Params (M)
YOLOv8s	640	16	200	18.5	11.2
YOLOv8m	640	16	200	31.3	25.9
YOLOv5s	640	16	200	22.7	7.2
YOLOv5m	640	16	200	33.6	21.2
YOLOv5s-pp	640	16	200	24.1	10.5

Table 17. Experimental results of various models on the val subset of the UAVDT dataset.

Models	F1 (%)	R (%)	mAP@.5 (%)
YOLOv8s	45.0	61.0	41.6
YOLOv8m	47.0	62.0	42.8
YOLOv5s	41.0	55.0	35.4
YOLOv5m	44.0	58.0	37.7
YOLOv5s-pp	49.0	64.0	42.1

Table 18. Experimental results of various models on the test subset of the UAVDT dataset.

Models	F1 (%)	R (%)	mAP@.5 (%)
YOLOv8s	40.0	58.0	33.6
YOLOv8m	44.0	60.0	35.9
YOLOv5s	37.0	54.0	29.2
YOLOv5m	41.0	57.0	31.5
YOLOv5s-pp	44.0	61.0	35.7

Combining Tables 16–18, we can see that the YOLOv5s-pp model has better overall performance than the three different-sized models of YOLOv5s, YOLOv5m and YOLOv8s. In Table 18, it can be observed that the YOLOv5s-pp algorithm's mAP@.5 value on the test subset is 6.5% higher than that of YOLOv5s, 4.2% higher than that of YOLOv5m and 2.1% higher than that of YOLOv8m. Compared to the YOLOv8m model, the mAP@.5 value of the YOLOv5s-pp model on the test subset is slightly lower by 0.2%. However, it should be noted that the parameter size of the YOLOv8m model is 2.5 times that of the YOLOv5s-pp model. The overall network depth and parameter size of YOLOv8m are much larger than those of YOLOv5s-pp, which will have a negative impact on the training and inference speed of the model.

The images in Figures 14 and 15 are also selected from real scene images taken from the UAV viewpoint, and again we selected images from both daytime and nighttime scenes. From Figure 14, we can see that in the night environment, the YOLOv5s model does not recognise the vehicle further away in the upper left corner of the picture; however, the YOLOv5s-pp model detects all the vehicles in the picture smoothly. As we can see from Figure 15, in the above set of test comparisons, the YOLOv5s-pp model identifies more small targets on the motorway than the YOLOv5s model, and there is a significant amount of under-detection by the YOLOv5s model of vehicles that are relatively farther away and smaller in the picture. In addition, near the toll booths in the following set of images, the YOLOv5s model fails to fully detect vehicles further away in the images, while the YOLOv5s-pp model detects more targets. In contrast, the YOLOv5s-pp model has better detection performance for small targets in the pictures.

(a) (b)

Figure 14. Plot of test results for the example in a daytime environment (Based on UAVDT dataset). (**a**) shows the test results for the yolov5s model and (**b**) shows the test results for the yolov5s-pp model.

(a) (b)

Figure 15. Plot of test results for the example in the nighttime environment (Based on UAVDT dataset). (**a**) shows the test results for the yolov5s model and (**b**) shows the test results for the yolov5s-pp model.

6. Conclusions

In order to improve the performance of small target recognition algorithms for drone's perspective images, this paper selected YOLOv5s model as the basic model and made improvements on it. Firstly, we replaced the original SiLU activation function with the Meta-ACON activation function, so that the activation function of the entire model can adaptively select the degree of activation between linear and nonlinear. Secondly, we used the SPD Conv module to enhance the feature extraction capability of small targets and optimized the fine-grained information loss and learning rate decrease caused by cross-layer convolution. In addition, we incorporated the CA attention mechanism into the network model, considering both channel dimensions and spatial dimensions, to solve problems such as long-range dependencies. Finally, we improved the size of the detection head, which not only reduced the overall loss but also improved the accuracy of target detection. The experimental results show that compared with the benchmark model YOLOv5s, the improved YOLOv5s-pp model improves the performance by 7.4% on the test-dev subset of the VisDrone2019-DET dataset and by 6.5% on the test subset of the UAVDT dataset. Compared with the YOLOv8s model, the improved YOLOv5s-pp model improves the performance by 0.8% on the test-dev subset of the VisDrone2019-DET dataset and 2.1% on the test subset of the UAVDT dataset, and the overall performance of the small-target recognition outperforms that of the original YOLOv5s and YOLOv8s models. In further research, we will continue to study the situation where the target cannot be accurately recognised due to occlusion or overlap in the UAV view, and we will also incorporate deep learning-based super-resolution techniques into our research, hoping to further improve the performance of the small target recognition algorithms through this work, and hopefully make some contribution to this field.

Author Contributions: Conceptualization, H.X., W.Z., P.L. and R.W.; methodology, H.X. and W.Z.; software, P.L., F.L. and R.W.; validation, H.X., W.Z. and P.L.; formal analysis, R.W.; investigation, H.X., W.Z., F.L., P.L. and R.W.; resources, H.X. and P.L.; data curation, W.Z. and F.L.; writing—original draft preparation, H.X. and W.Z.; writing—review and editing, H.X., W.Z. and F.L.; visualization, P.L. and R.W.; supervision, P.L. and R.W.; project administration, H.X. and P.L.; funding acquisition, H.X. and P.L. All authors have read and agreed to the published version of the manuscript.

Funding: This work was supported in part by the National Key R&D Program of China under Grant 2019YFB2103003, in part by the Scientific and Technological Support Project of Jiangsu Province under Grant BE2019740, Six Talent Peaks Project of Jiangsu Province (RJFW-111), and in part by the Postgraduate Research and Practice Innovation Program of Jiangsu Province under Grant SJCX22_0267, SJCX22_0275 and SJCX23_0274.

Data Availability Statement: Not applicable.

Conflicts of Interest: The authors declare no conflict of interest.

References

1. Redmon, J.; Divvala, S.; Girshick, R.; Farhadi, A. You Only Look Once: Unified, Real-Time Object Detection. In Proceedings of the 2016 IEEE Conference on Computer Vision and Pattern Recognition (CVPR), Las Vegas, NV, USA, 27–30 June 2016; pp. 779–788.
2. Redmon, J.; Farhadi, A. YOLO9000: Better, Faster, Stronger. In Proceedings of the 2017 IEEE Conference on Computer Vision and Pattern Recognition (CVPR), Honolulu, HI, USA, 21–26 July 2017; pp. 6517–6525.
3. Redmon, J.; Farhadi, A. YOLOv3: An Incremental Improvement. *arXiv* **2018**, arXiv:1804.02767. [CrossRef]
4. Bochkovskiy, A.; Wang, C.; Liao, H. M. YOLOv4: Optimal Speed and Accuracy of Object Detection. *arXiv* **2020**, arXiv:2004.10934. [CrossRef]
5. Benjumea, A.; Teeti, I.; Cuzzolin, F. YOLO-Z: Improving small object detection in YOLOv5 for autonomous vehicles. *arXiv* **2023**, arXiv:2112.11798. [CrossRef]
6. Zhu, X.K.; Lyu, S.C.; Wang, X. TPH-YOLOv5: Improved YOLOv5 Based on Transformer Prediction Head for Object Detection on Drone-captured Scenarios. In Proceedings of the 2021 IEEE/CVF International Conference on Computer Vision Workshops (ICCVW), Montreal, BC, Canada, 11–17 October 2021; pp. 2778–2788.
7. Huang, Y.; Cui, H.; Ma, J.; Hao, Y. Research on an aerial object detection algorithm based on improved YOLOv5. In Proceedings of the 2022 International Conference on Computer Engineering and Applications (ICCEA), Changchun, China, 20–22 May 2022; pp. 396–400.

8. Shao, L.; Wu, H.; Li, C.; Li, J. A Vehicle Recognition Model Based on Improved YOLOv5. *Electronics* **2023**, *12*, 1323. [CrossRef]
9. Ren, S.Q.; He, K.; Sun, J.; Girshick, R. Faster R-CNN: Towards Real-Time Object Detection with Region Proposal Networks. *IEEE Trans. Pattern Anal. Mach. Intell.* **2017**, *39*, 1137–1149. [CrossRef] [PubMed]
10. Li, W.; Li, Y.; Gong, J.; Feng, Q.; Zhou, J.; Sun, J.; Shi, C.; Hu, W. Urban Water Extraction with UAV High-Resolution Remote Sensing Data Based on an Improved U-Net Model. *Remote Sens.* **2021**, *13*, 3165. [CrossRef]
11. Teng, Z.; Li, C. T.; Zhao, W. J.; Wang, Z. X.; Li, R.; Zhang, X.; Song, Y. Extraction and Analysis of Spatial Feature Data of Traditional Villages Based on the Unmanned Aerial Vehicle (UAV) Image. *Mob. Inf. Syst.* **2022**, *2022*, 4663740. [CrossRef]
12. Sharma, S. K.; Kumar, M.; Maithani, S.; Kumar, P. Feature Extraction in Urban Areas Using UAV Data. In Proceedings of the UASG 2021: Wings 4 Sustainability, Roorkee, India, 2–4 April 2021; Springer: Cham, Switzerland, 2023; pp. 87–98. [CrossRef]
13. Li, Y.; Li, M.; Li, S.; Li, Y. Improved YOLOv5 for Remote Sensing Rotating Object Detection. In Proceedings of the 2021 6th International Conference on Communication, Image and Signal Processing (CCISP), Chengdu, China, 19–21 November 2021; pp. 64–68.
14. Hou, Q.B.; Zhou, D.Q.; Feng, J.S. Coordinate Attention for Efficient Mobile Network Design. In Proceedings of the 2021 IEEE/CVF Conference on Computer Vision and Pattern Recognition (CVPR), Nashville, TN, USA, 20–25 June 2021; pp. 13708–13717.
15. Ma, N.; Zhang, X.; Sun, J.; Liu, M. Activate or Not: Learning Customized Activation. In Proceedings of the 2021 IEEE/CVF Conference on Computer Vision and Pattern Recognition (CVPR), Nashville, TN, USA, 20–25 June 2021; pp. 8028–8038.
16. Sunkara, R.; Luo, T. No More Strided Convolutions or Pooling: A New CNN Building Block for Low-Resolution Images and Small Objects. *arXiv* **2022**, arXiv:2208.03641. [CrossRef]
17. Liu, W.; Anguelov, D.; Erhan, D.; Szegedy, C.; Reed, S.; Fu, C.-Y; Berg, A.C. SSD: Single Shot MultiBox Detector. *arXiv* **2015**, arXiv:1512.02325. [CrossRef]
18. Lin, T.Y.; Goyal, P.; Girshick, R.B.; He, K.; Dollár, P. Focal Loss for Dense Object Detection. In Proceedings of the 2017 IEEE International Conference on Computer Vision (ICCV), Venice, Italy, 22–29 October 2017; pp. 2999–3007. [CrossRef]
19. Girshick, R.B. Fast R-CNN. In Proceedings of the 2015 IEEE International Conference on Computer Vision (ICCV), Washington, DC, USA, 7–13 December 2015; pp. 1440–1448.
20. Cai, Z. W.; Vasconcelos, N. Cascade R-CNN: Delving Into High Quality Object Detection. In Proceedings of the 2018 IEEE/CVF Conference on Computer Vision and Pattern Recognition, Salt Lake City, UT, USA, 18–23 June 2018; pp. 6154–6162. [CrossRef]
21. Hou, H.; Chen, M.; Tie, Y.; Li, W. A Universal Landslide Detection Method in Optical Remote Sensing Images Based on Improved YOLOX. *Remote Sens.* **2022**, *14*, 4939. [CrossRef]
22. Yuan, Y.; Bai, H.; Wu, P.; Guo, H.; Deng, T.; Qin, W. An Intelligent Detection Method for Small and Weak Objects in Space. *Remote Sens.* **2023**, *15*, 3169. [CrossRef]
23. Hu, S.; Zhao, F.; Lu, H.; Deng, Y.; Du, J.; Shen, X. Improving YOLOv7-Tiny for Infrared and Visible Light Image Object Detection on Drones. *Remote Sens.* **2023**, *15*, 3214. [CrossRef]
24. Qi, G.; Zhang, Y.; Wang, K.; Mazur, N.; Liu, Y.; Malaviya, D. Small Object Detection Method Based on Adaptive Spatial Parallel Convolution and Fast Multi-Scale Fusion. *Remote Sens.* **2022**, *14*, 420. [CrossRef]
25. Elfwing, S.; Uchibe, E.; Doya, K. Sigmoid-Weighted Linear Units for Neural Network Function Approximation in Reinforcement Learning. *Neural Netw. Off. J. Int. Neural Netw. Soc.* **2017**, *107*, 3–11. [CrossRef] [PubMed]
26. Glorot, X.; Bordes, A.; Bengio, Y. Deep Sparse Rectifier Neural Networks. *J. Mach. Learn. Res.* **2011**, *15*, 315–323.
27. Ramachandran, P.; Zoph, B.; Le, Q. V. Swish: A Self-Gated Activation Function. *arXiv* **2017**, arXiv:1710.05941. [CrossRef]
28. Lange, M.; Holz, O.; Villmann, T. Applications of lp-Norms and their Smooth Approximations for Gradient Based Learning Vector Quantization. In Proceedings of the European Symposium on Artificial Neural Networks, Bruges, Belgium, 23–25 April 2014.
29. Hu, J.; Shen, L.; Sun, G. Squeeze-and-Excitation Networks. In Proceedings of the 2018 IEEE/CVF Conference on Computer Vision and Pattern Recognition, Salt Lake City, UT, USA, 18–23 June 2018; pp. 7132–7141.
30. Tian, T.; Li, L.L.; Chen, W.T.; Zhou, H. SEMSDNet: A Multiscale Dense Network With Attention for Remote Sensing Scene Classification. *IEEE J. Sel. Top. Appl. Earth Obs. Remote. Sens.* **2021**, *14*, 5501–5514. [CrossRef]
31. Guo, P.Y.; Song, C. Facial Expression Recognition with Squeeze-and-Excitation Network. In Proceedings of the 2022 7th International Conference on Intelligent Computing and Signal Processing (ICSP), Xi'an, China, 15–17 April 2022; pp. 962–967.
32. Woo, S.; Park, J.; Lee, J.; Kweon, I. CBAM: Convolutional Block Attention Module. In Proceedings of the European Conference on Computer Vision, Munich, Germany, 8–14 September 2018; Springer International Publishing: Cham, Switzerland, 2018; pp. 3–19. [CrossRef]
33. Zhu, Y.H.; Liu, C.L.; Jiang, S.Q. Multi-Attention Meta Learning for Few-Shot Fine-Grained Image Recognition. In Proceedings of the Twenty-Ninth International Joint Conference on Artificial Intelligence, Yokohama, Japan, 7–15 January 2021; Volume 152, pp. 1090–1096.
34. Zhu, P.; Du, D.; Wen, L.; Bian, X.; Ling, H.; Hu, Q.; Peng, T.; Zheng, J.; Wang, X.; Zhang, Y.; et al. VisDrone-VID2019: The Vision Meets Drone Object Detection in Video Challenge Results. In Proceedings of the 2019 IEEE/CVF International Conference on Computer Vision Workshop (ICCVW), Seoul, Republic of Korea, 27–28 October 2019; pp. 227–235.
35. Du, D.; Qi, Y.; Yu, H.; Yang, Y.F.; Duan, K.; Li, G.; Zhang, W.; Huang, Q.; Tian, Q. The Unmanned Aerial Vehicle Benchmark: Object Detection and Tracking. In Proceedings of the European Conference on Computer Vision, Munich, Germany, 8–14 September 2018. [CrossRef]
36. Zhang, Z.; Yi, H. H.; Zheng, J. Focusing on Small Objects Detector in Aerial Images. *Acta Electronica Sinica* **2023**, *51*, 944.

37. Yang, F.; Fan, H.; Chu, P.; Blasch, E.; Ling, H. Clustered Object Detection in Aerial Images. In Proceedings of the 2019 IEEE/CVF International Conference on Computer Vision (ICCV), Seoul, Republic of Korea, 27 October–2 November 2019; pp. 8310–8319.
38. Liu, Z.L.; Gao, G.Y.; Sun, L. HRDNet: High-Resolution Detection Network for Small Objects. In Proceedings of the 2021 IEEE International Conference on Multimedia and Expo (ICME), Shenzhen, China, 5–9 July 2021; pp. 1–6.
39. Ren, S.Q.; He, K.; Sun, J.; Girshick, R. Real-time Object Detection in UAV Images Based on Improved YOLOv5s. *Opto-Electron. Eng.* **2022**, *49*, 69–81.
40. Li, D.N. Research on Small Object Detection Model Based on Optimized YOLOv5. Master's Thesis, Xinjiang Normal University, Xinjiang, China, 2022.
41. Yang, C.H.; Huang, Z.H.; Wang, N.Y. QueryDet: Cascaded Sparse Query for Accelerating High-Resolution Small Object Detection. In Proceedings of the 2022 IEEE/CVF Conference on Computer Vision and Pattern Recognition (CVPR), New Orleans, LA, USA, 18–24 June 2022; pp. 13658–13667.
42. Chen, C.R.; Zhang, Y.; Lv, Q. .; Wei, S.; Wang, X. . RRNet: A Hybrid Detector for Object Detection in Drone-Captured Images. In Proceedings of the 2019 IEEE/CVF International Conference on Computer Vision Workshop (ICCVW), Seoul, Republic of Korea, 27–28 October 2019; pp. 100–108.

remote sensing

Article

UCDnet: Double U-Shaped Segmentation Network Cascade Centroid Map Prediction for Infrared Weak Small Target Detection

Xiangdong Xu [1,2], Jiarong Wang [1,*], Ming Zhu [1], Haijiang Sun [1], Zhenyuan Wu [1,2], Yao Wang [1,2], Shenyi Cao [3] and Sanzai Liu [1,2]

1 Changchun Institute of Optics, Fine Mechanics and Physics, Chinese Academy of Sciences, Changchun 130033, China; xuxiangdong21@mails.ucas.ac.cn (X.X.); zhuming@ciomp.ac.cn (M.Z.); sunhj@ciomp.ac.cn (H.S.); wuzhenyuan22@mails.ucas.ac.cn (Z.W.); wangyao222@mails.ucas.ac.cn (Y.W.); liusanzai22@mails.ucas.ac.cn (S.L.)

2 University of Chinese Academy of Sciences, Beijing 100049, China

3 Faculty of Electronic and Information Engineering, Xi'an Jiaotong University, Xi'an 710000, China; caoshenyi@stu.xjtu.edu.cn

* Correspondence: wangjiarong@cust.edu.cn; Tel.: +86-138-4406-4774

Abstract: In recent years, the development of deep learning has brought great convenience to the work of target detection, semantic segmentation, and object recognition. In the field of infrared weak small target detection (e.g., surveillance and reconnaissance), it is not only necessary to accurately detect targets but also to perform precise segmentation and sub-pixel-level centroid localization for infrared small targets with low signal-to-noise ratio and weak texture information. To address these issues, we propose UCDnet (Double U-shaped Segmentation Network Cascade Centroid Map Prediction for Infrared Weak Small Target Detection) in this paper, which completes "end-to-end" training and prediction by cascading the centroid localization subnet with the semantic segmentation subnet. We propose the novel double U-shaped feature extraction network for point target fine segmentation. We propose the concept and method of centroid map prediction for point target localization and design the corresponding Com loss function, together with a new centroid localization evaluation metrics. The experiments show that ours achieves target detection, semantic segmentation, and sub-pixel-level centroid localization. When the target signal-to-noise ratio is greater than 0.4, the IoU of our semantic segmentation results can reach 0.9186, and the average centroid localization precision can reach 0.3371 pixels. On our simulated dataset of infrared weak small targets, the algorithm we proposed performs better than existing state-of-the-art networks in terms of semantic segmentation and centroid localization.

Keywords: infrared weak small target; target detection; semantic segmentation; centroid localization; sub-pixel-level localization

Citation: Xu, X.; Wang, J.; Zhu, M.; Sun, H.; Wu, Z.; Wang, Y.; Cao, S.; Liu, S. UCDnet: Double U-Shaped Segmentation Network Cascade Centroid Map Prediction for Infrared Weak Small Target Detection. *Remote Sens.* **2023**, *15*, 3736. https://doi.org/10.3390/rs15153736

Academic Editors: Yanni Dong, Xiaochen Yang and Qian Du

Received: 5 July 2023
Revised: 23 July 2023
Accepted: 25 July 2023
Published: 27 July 2023

1. Introduction

Infrared small target detection technology, as the main technical support for surveillance and reconnaissance [1,2], precise localization [3–5], and attitude estimation [6], has been widely applied in various fields. In the field of surveillance and reconnaissance, such as drone tracking and search and rescue operations, not only the rough detection of infrared targets is required but also precise segmentation and centroid localization of the targets. This is essential for effective prediction of target motion trajectories, thereby enabling early warning and appropriate measures. In the field of the Internet of Things (IoT), such as smart transportation and smart agriculture, precise detection is a prerequisite for achieving perception and decision-making. In the aforementioned application scenarios, targets often occupy a small number of pixels in the image and suffer from issues such as texture loss and low signal-to-noise ratio.

Challenges:

On one hand, as the distance between the imaging device and the target increases, the size of the target in the image becomes smaller, and the target appears in a faint and weak state. Existing semantic segmentation algorithms perform well in segmenting large-sized or clearly bounded targets, but they show limited effectiveness in segmenting infrared small targets with low signal-to-noise ratio and weak texture information.

On the other hand, existing infrared weak small target detection methods focus on bounding box detection and pay less attention to the problem of accurate centroid localization of these targets. When the infrared small targets have irregular shapes and undergo continuous changes in posture, the center of the bounding box cannot effectively represent the centroid of the target. Existing deep learning-based methods for predicting target centroids are still limited to pixel-level predictions, only able to predict which pixel the centroid belongs to in the image. In certain specific scenarios, a deviation of one pixel in predicting the target centroid can result in several meters of actual distance, and this deviation may be further amplified as the process continues.

Existing Methods:

To date, there are various traditional methods for infrared small target detection. Common approaches include: (1) filter-based methods [7–9], (2) methods based on the human visual system [10–14], and (3) methods based on low-rank sparse recovery [15,16]. However, filter-based detection algorithms can only suppress uniform background to a certain extent and fail to address complex background problems, resulting in low detection performance and poor robustness. Methods based on the human visual system rely on contrasting the differences between small target regions and their surrounding areas, making them unsuitable for detecting weak small targets. Low-rank sparse recovery methods can improve detection precision by applying certain constraints. However, in infrared images with weak small targets and complex backgrounds, some strong clutter signals may be as sparse as the target signal, leading to increased false alarm. Overall, traditional algorithms heavily rely on manual craftsmanship, lack adaptability, and are difficult to be widely applied in infrared target detection in various application scenarios.

Recent approaches have turned to deep learning methods to address these issues. For example, Liu et al. [17] developed a CNN-based method for infrared small target detection. Y. Dai et al. [18] encoded spatial information and proposed an Asymmetric Context Module (ACM). They also introduced the SIRST dataset, which has been widely used in the field of infrared small target detection. Other researchers, such as Q. Hou et al. [19], proposed a robust infrared small target detection network (RISTDnet) based on deep learning. They constructed a feature extraction framework that combines handcrafted features and convolutional neural networks, established a mapping network between the feature map and the likelihood of small targets in the image, and applied threshold segmentation to identify real targets. X. Zhou et al. [20] unfolded the sparse low-rank regularization model into a deep neural network. They used the infrared patch-image (IPI) model to convert the original infrared image into patch images through local patch construction. Zhao et al. [21] proposed a detection pattern based on generative adversarial networks (GAN) to automatically learn the features of the target and directly predict the intensity of the target.

Our Contribution:

In this paper, to address the sub-pixel-level accurate centroid localization of infrared weak small targets, we propose UCDnet (Double U-shaped Segmentation Network Cascade Centroid Map Prediction for Infrared Weak Small Target Detection), and its prediction results are shown in Figure 1. The algorithm takes the input infrared image and passes it through a semantic segmentation subnet for feature extraction, feature fusion, and dense prediction, outputting pixel-level predictions (i.e., semantic segmentation results) of the targets (Figure 1a). Meanwhile, the input infrared image is also fed into the centroid localization subnet, which generates a centroid map with target centroid position information. The pixel-level predictions from the semantic segmentation subnet are mapped onto the centroid map, allowing the centroid map to retain only the results corresponding to the target's pixel locations. Finally, through decoding operations, accurate centroid results of the targets can be obtained (Figure 1b). Additionally, we have created an infrared simulation dataset specifically tailored for weak small targets.

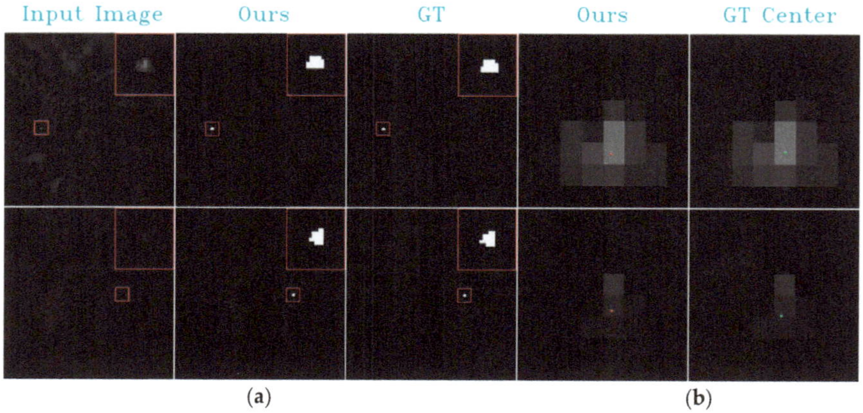

Figure 1. The performance of our algorithm on our simulated datasets for semantic segmentation and centroid localization. (**a**) The **left column** shows our UCDnet semantic segmentation results, and the **right column** shows the corresponding ground truth. (**b**) The **left column** shows our UCDnet centroid localization results, and the **right column** shows the corresponding ground truth.

Our contributions can be summarized as follows:

We have designed a novel sub-pixel-level centroid localization method for infrared weak small targets. We innovatively propose the concept and method of centroid map prediction, addressing the challenge of accurately locating the centroid of point targets. Our method utilizes a lightweight fully convolutional structure and incorporates a novel Com loss function. We also introduce a new evaluation metric for centroid localization. Experimental results demonstrate that our average centroid localization precision achieves 0.3371 pixels when the signal-to-noise ratio of the targets is above 0.4.

We have proposed a feature extraction network for semantic segmentation of infrared weak small targets. Our method employs a novel double U-shaped structure, which effectively combines the feature extraction module and attention module to finer predict the easily overlooked edge pixels, which is crucial for precise centroid localization. Experimental results show that our semantic segmentation IoU achieves 0.9186 when the signal-to-noise ratio of the targets is above 0.4.

We have simulated a series of infrared weak small target datasets for training, validation, and testing purposes. These datasets include various forms of infrared small targets. To replicate real-world scenarios, we have introduced multiple types of vertical line noise and Gaussian noise in the images. We have also specifically simulated the dim and weak states of targets when they pass through cloud, which pose challenges for detection algorithms.

2. Materials and Methods

The main idea of UCDnet is to construct an infrared small target detection network that achieves dense prediction and sub-pixel-level centroid localization. Its network structure is shown in Figure 2 and consists of two parts: the semantic segmentation subnet and the centroid localization subnet. Section 2.1 introduces the simulation content of the dataset. Section 2.2 presents the proposed semantic segmentation method, while Section 2.3 describes the proposed sub-pixel-level centroid localization method. Finally, Section 2.4 discusses the loss function and training method used during network training.

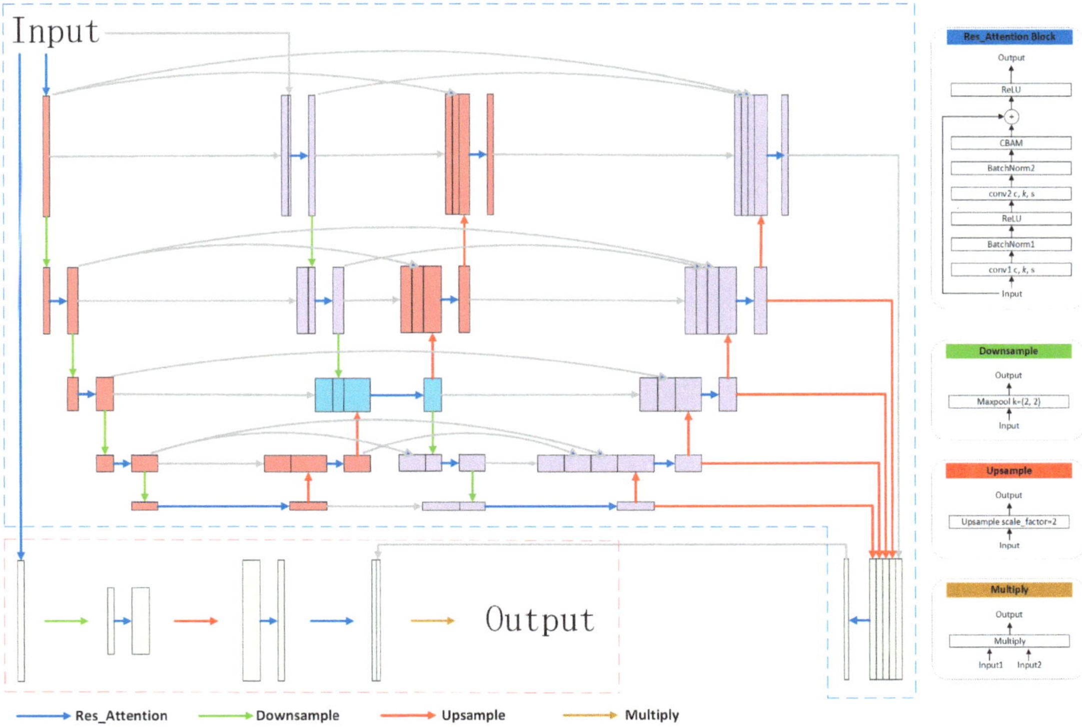

Figure 2. The network structure diagram of our proposed algorithm, UCDnet, consists of a semantic segmentation subnet and a centroid localization subnet. In the semantic segmentation subnet, a double U-shaped backbone network, which embeds attention modules, is used to extract features from the input infrared images. A feature fusion prediction branch is then used to perform pixel-wise binary classification based on the fused features at different levels, generating a semantic segmentation mask. In the centroid localization subnet, a "floating-point" centroid map is predicted based on the input infrared images. The semantic segmentation mask generated by the semantic segmentation subnet is mapped into the centroid map to narrow down the predicted range of the centroid, enabling accurate calculation of the target's centroid. The details of the Res_Attention block, Downsample block, Upsample_n block, and Multiply block are shown in Table A1 in Appendix A.

2.1. Dataset Simulation

The dataset used in this paper is simulated based on satellite imagery (with a size of around 10,000 × 10,000 pixels) collected by meteorological satellites. The simulated targets are generated based on the grayscale distribution of real targets, and a total of 125 targets were simulated. The extraction of target features from infrared images is closely related to background information and noise [22]. The image data from different backgrounds exhibit significant differences. For meteorological satellite images, cloud cover is the main source of interference, and the presence of high-intensity noise further complicates target detection. We simulated 8 to 14 frames of motion trajectories for each target individually. In order to make the dataset more representative of real-world scenarios, slight variations in the shape and pixel values of the targets were introduced within each trajectory, resulting in a total of 1306 images. Finally, random cropping was performed on these images to obtain sub-images of size 256 × 256 pixels containing the targets (the simulation process of the infrared weak small target dataset is illustrated in Figure A1 in Appendix A). The size of the targets ranged from 6 × 4 pixels to 10 × 13 pixels, and the signal-to-noise ratio (SNR) of the images varied from 0.4 to 6.4. It can also be observed from Figure 3 that the SNR of the image complies with the central limit theorem. Figure 4 illustrates the range of target sizes, with blue dots representing the number of targets for each size. Figure 5 showcases the approximate shapes of the targets.

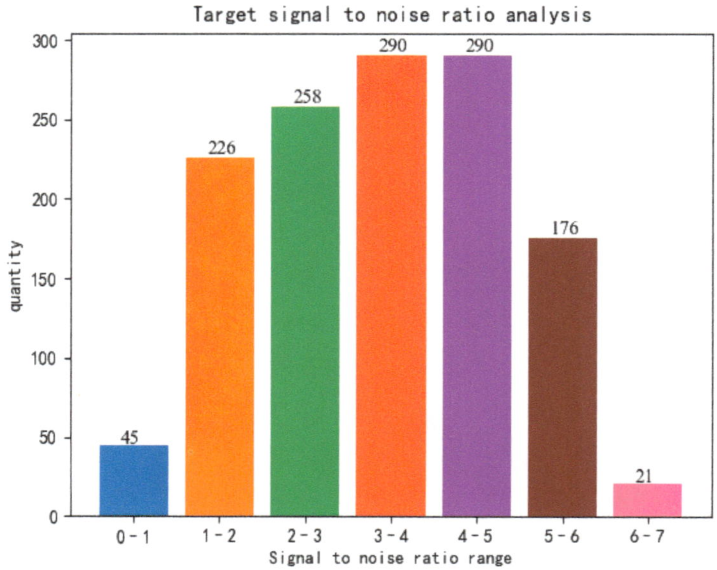

Figure 3. Histogram of the target signal-to-noise ratio distribution.

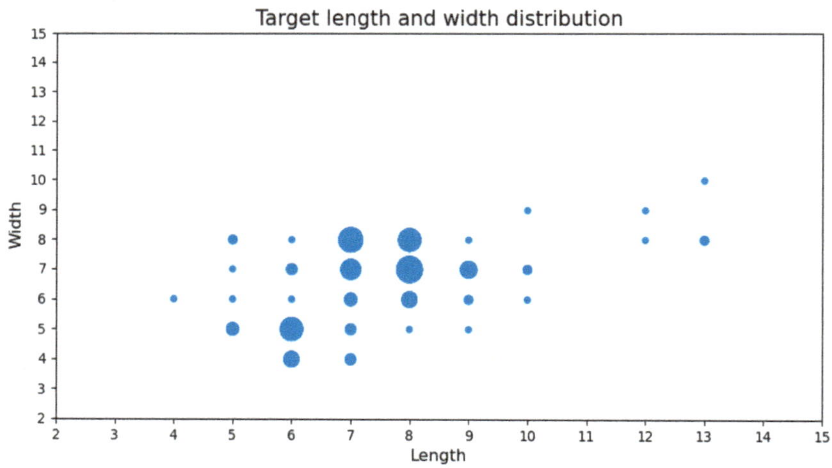

Figure 4. Distribution of target's length and width, along with the quantity distribution for each size.

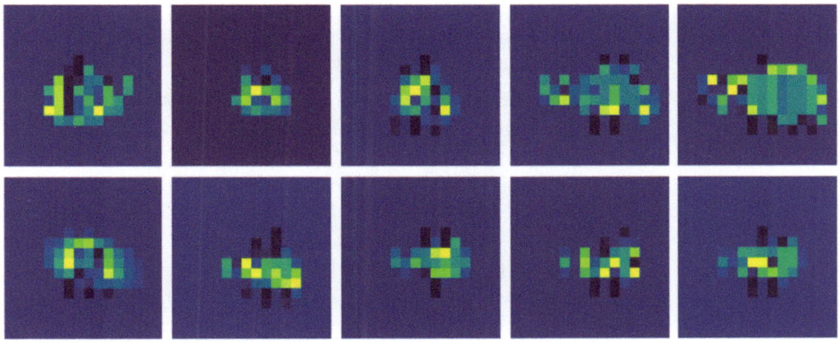

Figure 5. The schematic representation of the shapes of certain targets in the dataset used in this paper.

2.2. Semantic Segmentation Subnet

The semantic segmentation subnet consists of a feature extraction backbone network and a feature fusion prediction branch. The feature extraction backbone network adopts a double U-shaped structure, focusing on extracting edge features of the targets to handle low signal-to-noise ratio situations. It incorporates embedded attention modules to enhance segmentation precision. The feature fusion prediction branch performs feature fusion at different hierarchical levels and conducts per-pixel binary classification based on the fused results to generate semantic segmentation masks.

The double U-shaped network structure is shown in Figure 6. Inspired by the U-Net [23] and DNAnet [24] network structures, we have made improvements to the main feature extraction network. It follows the overall Encoder–Decoder structure, which is similar to the combination of two large U-shaped networks and one small U-shaped network. The two large U-shaped networks perform four downsampling operations, where the shallow network extracts more target texture information, and the deep network allows for a larger field of view and captures deeper semantic information [25]. The small U-shaped network just performs two downsampling operations, reserving more target positional information than bigger ones. Compared to the U-Net network, our approach uses more feature concatenation to achieve precise extraction of edge features.

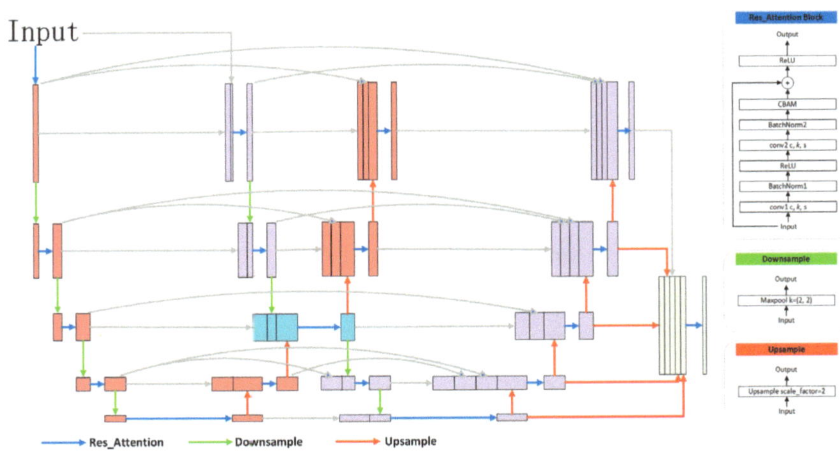

Figure 6. The structure of semantic segmentation subnet. Firstly, the input infrared image is processed by a double U-shaped backbone network with embedded attention modules to extract features. Then, the feature fusion prediction branch utilizes the fused features at different hierarchical levels to perform per-pixel binary classification and generate semantic segmentation masks.

In contrast to most existing infrared small target semantic segmentation methods, inspired by the DNAnet network structure, each layer of our main feature extraction network incorporates a specific attention module [26]. Each attention module in a layer includes channel attention and spatial attention components, enabling adaptive learning of the importance of different channels and spatial positions in the image. The channel attention dynamically adjusts the impact of each channel in the output feature map, while the spatial attention dynamically searches for the precise location of the targets on the feature map. The combination of two attentions helps the network pay more attention to the positional information of small targets and some detailed edge information in each feature map. The feature extraction network determines the overall position of the infrared small targets based on their grayscale and contrast information, while the attention modules focus more on the edge information of the targets.

The basic and core component of the entire network is the Res_Attention Block. In each layer, the input feature map undergoes a series of operations, including a 3 × 3 convolution, normalization, activation function, another 3 × 3 convolution, and normalization, followed by a hybrid attention mechanism. The output of this series of operations is then added to the original input to obtain the output of the Res_Attention Block. The hybrid attention mechanism consists of a channel attention mechanism followed by a spatial attention mechanism. For an input F, the equations for the channel attention mechanism (Mc(F)) and the spatial attention mechanism (Ms(F)) can be expressed as:

$$M_c(F) = \sigma(MLP(AvgPool(F)) + MLP(MaxPool(F))) \tag{1}$$

$$M_s(F) = \sigma\left(f^{7\times 7}([AvgPool(F); MaxPool(F)])\right) \tag{2}$$

where σ represents the sigmoid operation, MLP represents a convolution operation with a kernel size of 1 × 1, AvgPool represents average pooling, MaxPool represents maximum pooling, [;] represents channel concatenation, and $f^{7\times 7}$ represents a convolution with a kernel size of 7 × 7.

2.3. Centroid Localization Subnet

The centroid localization subnet (shown in Figure 7) consists of the centroid feature extraction network and the mapping module. The feature extraction network includes a hybrid attention mechanism, downsampling, and upsampling structures. The hybrid attention mechanism focuses on learning the relationship between the image's grayscale, contrast, and target centroids. The mapping module performs pixel-wise multiplication, multiplying the binary segmentation result with the centroid map. This operation filters out the parts that do not belong to the target, resulting in more accurate centroid localization.

The centroid localization subnet predicts the "floating-point" centroid map (as shown in Figure 8) based on the semantic segmentation results to calculate the target centroid. The centroid map, which is different from the heatmap proposed in CenterNet [27], is first introduced and applied by us. In the centroid map, the value of each pixel represents the proximity to the target centroid. To achieve sub-pixel-level centroid localization, we do not designate a specific pixel as the centroid in both the ground truth label map and the predicted map. Instead, we employ an indirect constraint method to obtain the floating-point centroid position (sub-pixel-level). Specifically, during the training phase, the labels for the centroid localization subnet are calculated using a two-dimensional Gaussian function [28]. When the target centroid position is (x_0, y_0), the label value at any position in the centroid map can be obtained using the two-dimensional Gaussian function as follows:

$$V_{ij} = e^{-\frac{(x_0-j)^2+(y_0-i)^2}{2}} \tag{3}$$

where j represents the current pixel's column, i represents the current pixel's row, and V_{ij} represents the corresponding label value at that position. As shown in Figure 8, when the centroid position is $x_0 = 3.3$, $y_0 = 4.6$, we can calculate the values of V_{43}, V_{53}, and V_{54} as follows:

$$V_{43} = e^{-\frac{(3.3-3)^2+(4.6-4)^2}{2}} \approx 0.79852 \text{ (yellow pixel)}$$

$$V_{53} = e^{-\frac{(3.3-3)^2+(4.6-5)^2}{2}} \approx 0.88250 \text{ (red pixel)}$$

$$V_{54} = e^{-\frac{(3.3-4)^2+(4.6-5)^2}{2}} \approx 0.72253 \text{ (blue pixel)}$$

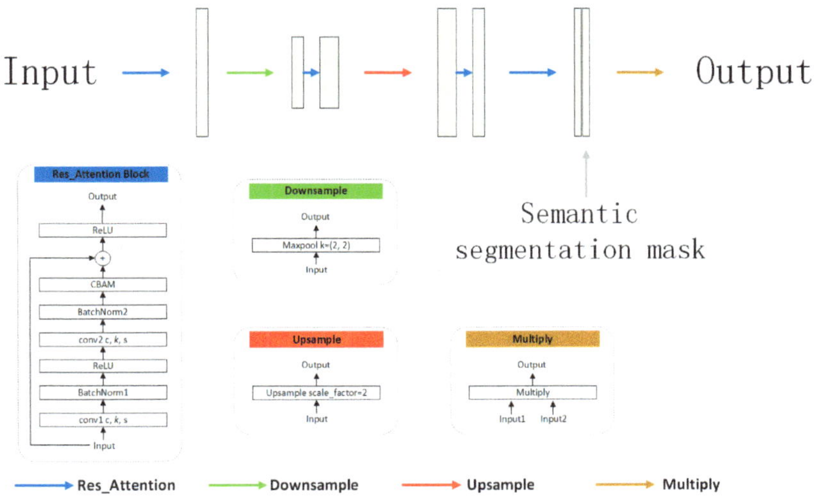

Figure 7. The structure of centroid localization subnet. Firstly, the feature extraction module with embedded attention modules is used to extract the regions where all suspected target centroids are located in the image. Then, the more accurate pixel-level prediction results of the targets generated from semantic segmentation subnet are mapped to the centroid map. This process eliminates false target centroid regions and preserves the true target centroid regions (i.e., centroid map).

0. 00000	0. 00000	0. 00001	0. 00002	0. 00002	0. 00001	0. 00000	0. 00000
0. 00001	0. 00011	0. 00066	0. 00147	0. 00120	0. 00036	0. 00004	0. 00000
0. 00015	0. 00242	0. 01463	0. 03255	0. 02665	0. 00803	0. 00089	0. 00004
0. 00120	0. 01974	0. 11943	0. 26580	0. 21762	0. 06555	0. 00726	0. 00030
0. 00361	0. 05931	0. 35880	0. 79852	0. 65377	0. 19691	0. 02182	0. 00089
0. 00399	0. 06555	0. 39653	0. 88250	0. 72253	0. 21762	0. 02411	0. 00098
0. 00162	0. 02665	0. 16122	0. 35880	0. 29376	0. 08848	0. 00980	0. 00040
0. 00024	0. 00399	0. 02411	0. 05366	0. 04394	0. 01323	0. 00147	0. 00006

Figure 8. The label for training centroid localization subnet and also the predicted centroid map.

During the prediction phase, the actual centroid positions are calculated based on the predicted centroid map, following these steps:

1. Firstly, determine the number of targets N in our predicted centroid map, which corresponds to N clusters of pixels.
2. For each cluster of pixels mentioned above, select the maximum value $V_{i_1 j_1}$ from our predicted results as the centroid reference point $A_{i,i\in[1,\,N]}$ (represented by the red position in Figure 8).
3. Take the centroid reference point A_i as the center and find the maximum value $V_{i_2 j_2}$ within the vertical neighborhood $N_4(A_i)$ to serve as the centroid vertical correction point $B_{i,i\in[1,\,N]}$ (represented by the yellow position in Figure 8).
4. Similarly, take the centroid reference point A_i as the center and find the maximum value $V_{i_3 j_3}$ within the horizontal neighborhood $N_4(A_i)$ to serve as the centroid horizontal correction point $C_{i,i\in[1,\,N]}$ (represented by the blue position in Figure 8).
5. Combining the centroid reference point A_i, the vertical correction point B_i, and the horizontal correction point C_i, calculate the sub-pixel-level precise positions x_0 and y_0 of each centroid using the following equations:

$$V_{i_1 j_1} = e^{-\frac{(x_0 - j_1)^2 + (y_0 - i_1)^2}{2}} \tag{4}$$

$$V_{i_2 j_2} = e^{-\frac{(x_0 - j_2)^2 + (y_0 - i_2)^2}{2}} \tag{5}$$

$$V_{i_3 j_3} = e^{-\frac{(x_0 - j_3)^2 + (y_0 - i_3)^2}{2}} \tag{6}$$

where $i_1 = i_3$ and $j_1 = j_2$, the above three equations can be used to obtain the following:

$$x_0 = \frac{2\ln\frac{V_{i_3 j_3}}{V_{i_1 j_1}} - (j_1^2 - j_3^2)}{2(j_3 - j_1)} \tag{7}$$

$$y_0 = \frac{2\ln\frac{V_{i_2 j_2}}{V_{i_1 j_1}} - (i_1^2 - i_2^2)}{2(i_2 - i_1)} \tag{8}$$

6. Repeat the above steps 2 to 5 until all N predicted targets have been traversed.

In summary, compared to directly predicting the centroid positions (x_0, y_0) using the network, predicting the centroid map makes the network training more easily convergent (due to the presence of more constraints). Furthermore, compared to predicting bounding boxes and then post-processing to obtain the centroids, the proposed method of predicting the centroid map and then extracting the centroids is more efficient and accurate.

2.4. Loss Function

The loss function used in our method consists of two parts. For the semantic segmentation part, as shown in Equation (9), we employ Dice Loss as the loss function. For the centroid localization part, as shown in Equation (10), we have developed our own loss function based on the MSE Loss, which we refer to as the Center of Mass Loss (Com Loss).

$$\text{Dice Loss} = 1 - 2\frac{|A \cap B|}{|A| + |B|} \tag{9}$$

$$\text{Com Loss} = \begin{cases} (y-p)^2 & if\ y > 0.1 \\ (y-p)^2(1-y)^4 & other \end{cases} \tag{10}$$

In Equation (9), A and B represent the semantic segmentation label (per-pixel) and the predicted semantic segmentation results (per-pixel), respectively. In Equation (10), y represents the centroid map label (per-position), and p represents the predicted centroid map (per-position).

To improve the algorithm's performance, we first optimize the semantic segmentation results. During the first 10 epochs, we only calculate the loss value for semantic segmentation. After the 10th epoch, we add to calculate the loss value for the centroid localization module. The weights assigned to the loss values of the semantic segmentation and centroid localization parts follow specific function distributions: $\frac{1}{1+e^{-0.1(60-epoch)+0.5}} + 1$ for the semantic segmentation part and $\left(1 - \frac{1}{1+e^{-0.1(60-epoch)+0.5}}\right) \times 0.8 + 0.2$ for the centroid localization part (shown in Figure 9). This approach improves the overall efficiency of the network by filtering out unnecessary computations and further enhances the performance of the centroid localization module.

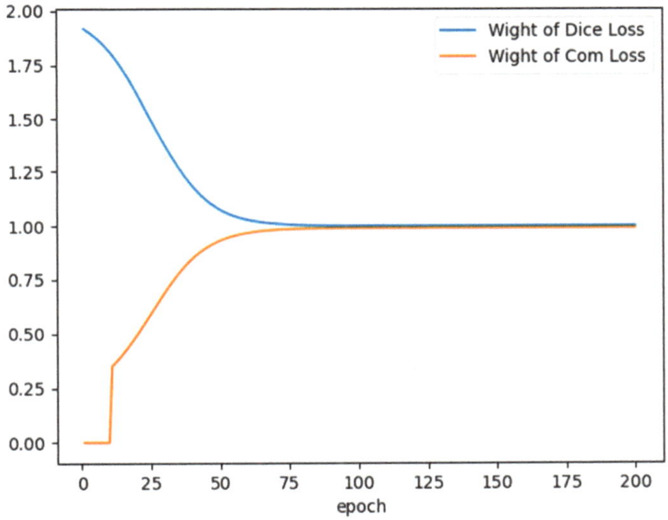

Figure 9. Illustration of the weights of Dice Loss and Com Loss in the overall loss.

3. Results

This section presents the experimental results of the proposed method in this paper, along with some comparative results, which demonstrate that our method outperforms other approaches. Section 3.1 provides a detailed introduction to the evaluation metrics used in this study. Section 3.2 presents all the experimental procedures and analyses.

3.1. Evaluation Metrics

3.1.1. Semantic Segmentation Evaluation Metrics

Since there is only one semantic segmentation class in this paper, the main evaluation metrics used are IoU, Precision, and Recall. In semantic segmentation, the definition of IoU is the ratio of the intersection to the union of the ground truth and predicted sets. This ratio can be expressed as the ratio of True Positive (TP) to the sum of TP, False Positive (FP), and False Negative (FN), which represents the overall segmentation precision. Precision is defined as the ratio of the intersection to the predicted segmentation set. This ratio can be expressed as the ratio of TP to the sum of TP and FP, reflecting the precision of the predicted results. Recall is defined as the ratio of the intersection to the ground truth segmentation set. This ratio can be expressed as the ratio of TP to the sum of TP and FN, reflecting the recall or sensitivity of the predicted results. TP, FP, FN, and TN are shown in Figure 10.

$$\text{IoU} = \frac{\text{TP}}{\text{FP} + \text{FN} + \text{TP}} \quad (11)$$

$$\text{Precision} = \frac{\text{TP}}{\text{FP} + \text{TP}} \quad (12)$$

$$\text{Recall} = \frac{\text{TP}}{\text{FN} + \text{TP}} \quad (13)$$

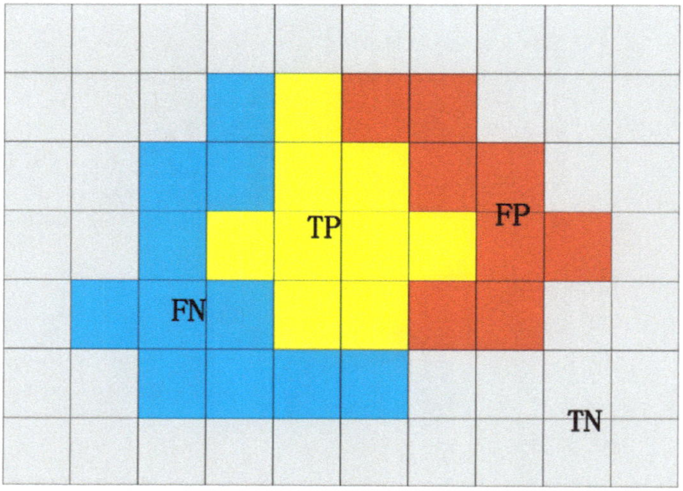

Figure 10. Illustration of IoU evaluation metric. TP (True Positive): the predicted result is positive, and it is actually positive. FP (False Positive): the predicted result is positive, but it is actually negative. FN (False Negative): the predicted result is negative, but it is actually positive. TN (True Negative): the predicted result is negative, and it is actually negative.

3.1.2. Centroid Localization Evaluation Metrics

If there is only one target in each image, we can directly use the Euclidean distance between the predicted centroid and the actual centroid of the target as the evaluation metric for centroid localization. However, considering the cases of multiple targets or false alarms, this paper introduces a novel centroid localization evaluation metric. Specifically, clustering is performed with the real centroids as centers, and all detected centroids are assigned to the nearest real centroid. Then, the sum of distances between all predicted centroids and their closest real centroids is calculated. This distance sum is divided by the smaller value between the number of real centroids and the number of detected centroids. The calculation equation is as follows:

$$\text{mDis} = \frac{\sum_{i=0}^{m} \min\limits_{0 \ll j \ll n} \left(\sqrt{\left(x_i - x_j\right)^2 + \left(y_i - y_j\right)^2} \right)}{\min(m, n)} \quad (14)$$

where n represents the number of actual target centroids, m represents the number of detected target centroids, x_i represents the horizontal coordinate of the detected target centroid, x_j represents the horizontal coordinate of the real target centroid, y_i represents the vertical coordinate of the detected target centroid, and y_j represents the vertical coordinate of the real target centroid.

3.2. Experimental Results and Analysis

3.2.1. Experimental Setup

In this experiment, the input image size of the network is 256 × 256 pixels, and the batch size is set to 16. The training and testing dataset are divided approximately in a ratio of 5:1. The semantic segmentation subnet in the experiment utilizes Dice Loss as the loss function, while the centroid localization subnet employs Com Loss as the loss function. The optimization method for the loss function is Adaptive Gradient (AdaGrad), which is a first-order optimization method [29]. The advantage of using AdaGrad is that it allows the manual adjustment of learning rate to be replaced by setting hyperparameters. The hardware device mainly used in the experiment is an NVIDIA RTX 3090 graphics card with 24 GB of VRAM. The software programs primarily used include the deep learning framework PyTorch, the plotting tool Matplotlib, the image processing library OpenCV, and the scientific computing library NumPy.

3.2.2. Comparison with Other State-of-the-Art Methods

In this section, we evaluate our implementation by comparing it with traditional image processing method MLCM [30], classic semantic segmentation network U-Net [23], and the latest semantic segmentation networks DNAnet [24], HRNet [31], and MTU-Net [32]. We perform a quantitative and qualitative analysis of the experimental results.

(1) Quantitative Analysis

Table 1 compares the performance of our algorithm and the other five state-of-the-art algorithms in terms of evaluation metrics such as IoU, Precision, Recall, F1-score, and GFLOPs. The experimental results demonstrate that our method ranks first in all evaluation metrics.

Table 1. Comparison of our semantic segmentation algorithm with five other algorithms.

Semantic Segmentation Algorithm	IoU	Precision	Recall	F1-Score	GFLOPs
MLCM [30]	0.1810	0.1810	1.0000	0.3065	-
U-Net [23]	0.4905	0.6435	0.5774	0.6087	46.02
HRNetv2-w18 [31]	0.7216	0.8307	0.8443	0.8374	4.64
MTU-Net [32]	0.8712	0.9403	0.9229	0.9315	5.48
DNAnet [24]	0.8862	0.9566	0.9236	0.9398	14.05
UCDnet (ours)	0.9186	0.9673	0.9480	0.9576	15.56

(2) Qualitative Analysis

We visualized the semantic segmentation results of three classical algorithms and our UCDnet. As shown in Figure 11, yellow annotations represent false alarms, red annotations represent missed detections, green annotations represent correctly predicted targets, and blue annotations indicate zoomed-in views of the targets. Due to the influence of factors such as complex background in infrared images and noise resembling target shapes, the existing algorithms and network detection performance are unsatisfactory. From the prediction results, it can be observed that the traditional MLCM algorithm has significant limitations, with excessive false alarms and almost complete loss of target edge information. The U-Net misses some targets in particular situations but retains some edge information compared to traditional algorithms. The DNAnet's prediction results show a good segmentation of the edges, but still have some errors compared to the ground truth. However, it is evident that our algorithm's predicted target edges closely match the ground truth, demonstrating high robustness without false alarms or missed targets. Our algorithm outperforms other comparative approaches in terms of detection and segmentation performance.

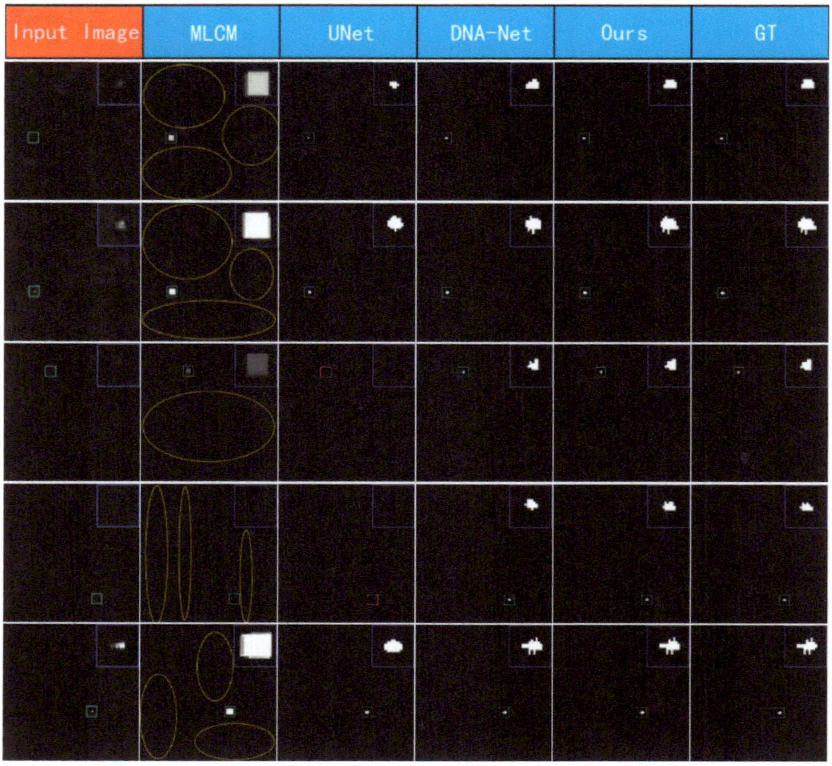

Figure 11. Comparison of our semantic segmentation results with the semantic segmentation results of three classical algorithms.

3.2.3. Analysis of UCDnet Feature Extraction Network

(1) Quantitative Analysis

To validate the superior performance of the proposed network structures in feature extraction compared to other similar network structures, we conducted the following ablation experiments on the main feature extraction network of the semantic segmentation subnet. Throughout the ablation experiments, we kept the structure of the centroid localization subnet unchanged and investigated the impact of variations in the structure of the semantic segmentation subnet on the network performance. As illustrated in Figure 12, "ablation study1" represents the network structure of a single U-shaped structure shown in Figure 12a, "ablation study2" represents a network structure similar to double U-shaped structure shown in Figure 12b, "ablation study3" represents a network structure based on Figure 12b with an added attention module shown in Figure 12c, and "ablation study4" represents a network structure with the addition of a feature fusion strategy based on Figure 12c. Training each of the four networks for 100 epochs yielded the results shown in Table 2 and Figure 13.

From Table 2, it can be observed that the double U-shaped network structure achieves higher IoU, Precision, and Recall compared to the single U-shaped network structure. By incorporating attention modules into the network, the IoU of the network in the test results improves by two percentage points, while Recall increases by three percentage points. Additionally, the introduction of a feature fusion strategy leads to an IoU improvement of over one percentage point. Considering the high IoU requirement for the output results in this study, the network structure depicted in Figure 12d is adopted.

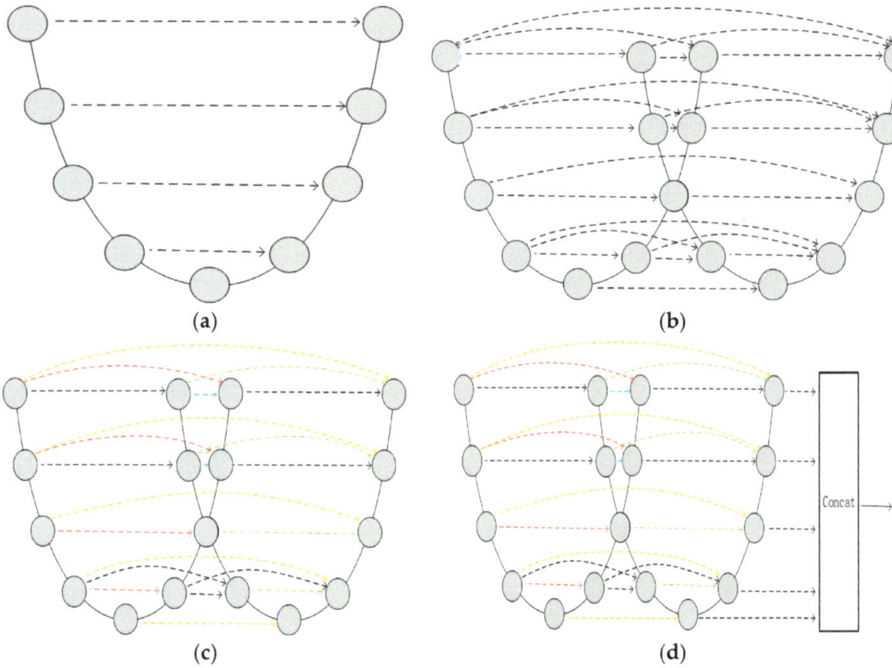

Figure 12. Architecture diagrams of the four networks in the ablation experiments, where (**d**) represents the network structure proposed in this paper. The figure (**a**) represents a single U-shaped structure without an attention module, the figure (**b**) represents a double U-shaped structure without an attention module, and the figure (**c**) represents a double U-shaped structure with an added attention module. The colored dotted lines in the figures indicate the presence of the attention module in the corresponding operation step.

Table 2. Comparison of results from four ablation studies.

Semantic Segmentation Algorithm	IoU	Precision	Recall
single U-shaped structure (ablation study1)	0.8665	0.9419	0.9155
double U-shaped structure (ablation study2)	0.8728	0.9600	0.9057
double U-shaped structure + attention module (ablation study3)	0.9070	0.9599	0.9428
double U-shaped structure + attention module + feature fusion (ours)	0.9186	0.9673	0.9480

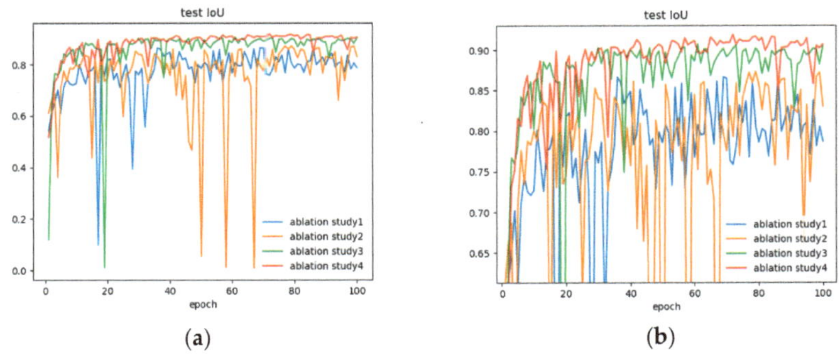

Figure 13. The curve depicting the variation of IoU for each ablation experiment across epochs. (**b**) presents a zoomed-in view of (**a**).

According to the curve shown in Figure 13, it can be observed that the network structure proposed in this paper consistently outperforms the other three ablation study network structures in terms of the evaluation metric IoU. Furthermore, the IoU values exhibit minimal fluctuations with increasing epochs. Considering the comprehensive set of ablation experiments, the proposed network structure in this paper demonstrates stronger feature extraction capabilities and better stability, while exhibiting superior learning ability for capturing detailed target features.

(2) Qualitative Analysis

From Figure 14, it can be observed that in the encoding process of the first U-shaped structure, as the downsampling iterations increase, the network pays more attention to the brighter regions in the image. In the decoding process, with the increase in upsampling iterations, the network can further distinguish between the target and the background. In the encoding process of the second U-shaped structure, FM (2,1) serves both as an encoding and decoding component, enabling more accurate discrimination between the background and the target. With an increase in downsampling iterations during the encoding process, FM (4,1) is able to fully differentiate the background regions surrounding the target. Comparing FM (0,2) and FM (0,3), although the rough shape of the target can already be observed at the output of the first U-shaped structure, the output of the second U-shaped structure is closer to the actual target.

3.2.4. Analysis of Centroid Localization Network

(1) Quantitative Analysis

As shown in Table 3, we compared the performance between three classical centroid localization methods and our method. Our approach demonstrated the best localization precision, outperforming even the suboptimal squared weighted centroid method by 0.2009 pixels.

Table 3. Comparison of four centroid localization methods.

Centroid Localization Methods	Localization Precision (Pixels)
center of bounding box	0.6096
gray value centroid method	0.5775
squared weighted centroid method	0.5380
UCDnet (Ours)	0.3371

The solution method for the gray value centroid method is as follows: For image block B, determine the center of the minimum bounding rectangle enclosing the target. Establish a Cartesian coordinate system with this center as the origin. Then, calculate the centroid using the following equations:

$$\mathrm{m00} = \sum_{x,y \in B} I(x,y) \tag{15}$$

$$\mathrm{m10} = \sum_{x,y \in B} x \times I(x,y) \tag{16}$$

$$\mathrm{m01} = \sum_{x,y \in B} y \times I(x,y) \tag{17}$$

$$\mathrm{C} = \left(\frac{m10}{m00}, \frac{m01}{m00}\right) \tag{18}$$

where $I(x,y)$ represents the pixel value at position (x, y), and C represents the coordinates of the centroid obtained on the Cartesian coordinate system established with the center of the image block.

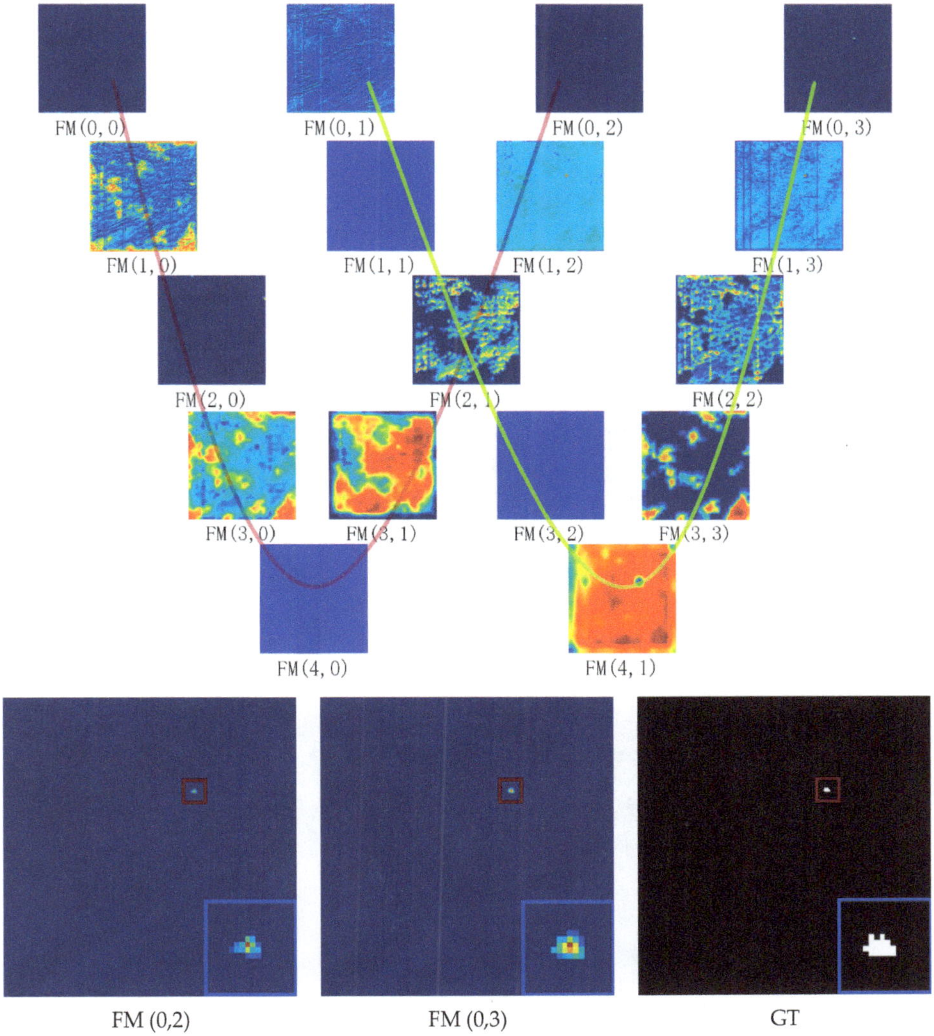

Figure 14. This is a heatmap generated using Grad-CAM. It can represent the contribution of each part of the network to the output.

The solution method for the squared weighted centroid method is as follows: For image block B, determine the center of the minimum bounding rectangle enclosing the target. Establish a Cartesian coordinate system with this center as the origin. Then, calculate the centroid using the following equations:

$$\text{m00} = \sum_{x,y \in B} I^2_{(x,y)} \tag{19}$$

$$\text{m10} = \sum_{x,y \in B} x \times I^2_{(x,y)} \tag{20}$$

$$\text{m01} = \sum_{x,y \in B} y \times I^2_{(x,y)} \tag{21}$$

$$\text{C} = \left(\frac{m10}{m00}, \frac{m01}{m00} \right) \tag{22}$$

where $I(x, y)$ represents the pixel value at position (x, y), and C represents the coordinates of the centroid obtained in the Cartesian coordinate system established with the center of the image block.

(2) Qualitative Analysis

We selected two images from the dataset and fed them into our UCDnet. We saved the final centroid localization results. Then, we extracted the target pixel blocks from the images, enlarged them eight times, and mapped our centroid localization results onto the magnified images (as shown in Figure 15). The centroid localization precision achieved sub-pixel-level precision.

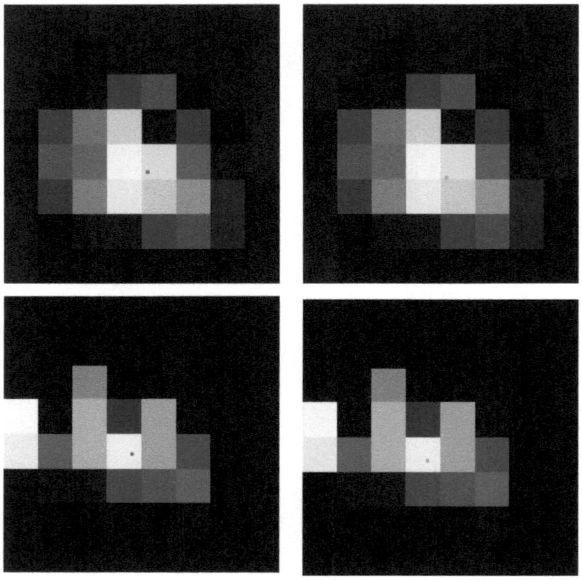

Figure 15. Centroid localization result image, with our detected results **on the left** and ground truth **on the right**. The red dots indicate the centroids of the detected targets, and the green dots represent the true positions of the centroids.

3.2.5. Discussion

Currently, numerous classical target detection algorithms have been applied in the field of infrared weak small target detection. We selected five representative algorithms, namely, MLCM, U-Net, HRNet, MTU-Net, and DNANet, for comparative experiments with our proposed method and obtained outstanding results. The IoU of our semantic segmentation results reached a leading level because of the proposed double U-shaped structure with attention mechanism. Furthermore, we innovatively introduced a deep learning-based approach for sub-pixel-level centroid localization, significantly improving the centroid localization precision compared to existing methods. Extensive experiments validated that our proposed algorithm exhibits favorable performance and robustness.

Additionally, to verify the applicability and generalization ability of our proposed algorithm beyond the field of surveillance and reconnaissance, we evaluated our algorithm on three different publicly available datasets from diverse fields. On an industrial field dataset for motor magnetic tile defect detection, our algorithm demonstrated excellent segmentation results even when facing weak texture information of motor magnetic defects (see experimental results in the first row of Figure A2 in Appendix A). On a medical field dataset for cell semantic segmentation [33,34], our algorithm accurately extracted edges of irregularly shaped cells (see experimental results in the second row of Figure A2 in Appendix A). On an agricultural field dataset for olive fruit semantic segmentation, our algorithm achieved precise segmentation for small-sized olive fruits, remaining unaffected by factors like crowding and occlusion (see experimental results in the third row of Figure A2 in Appendix A). These supplementary experiments further demonstrate the high robustness of our proposed algorithm. Our algorithm proves to be effective not only on our simulated dataset but also on various datasets from other fields, showcasing its excellent performance.

The proposed algorithm in this paper also has some limitations: (1) Point targets in infrared images are prone to confusion with high-frequency noise and similar objects. The method proposed in this paper focuses on fine segmentation and localization but requires further improvement in false alarm removal. (2) Infrared weak small targets have low signal-to-noise ratio and can easily be overwhelmed by background clutter. The algorithm in this paper may experience missed detections when the target signal-to-noise ratio is below 1. This is unacceptable for applications such as security monitoring and autonomous driving, which demand high safety requirements. These limitations will be the focus of our future work.

4. Conclusions

This paper presents a novel method called UCDnet for infrared weak small target detection and centroid localization. The innovation of our proposed semantic segmentation subnet lies in the improved U-Net and DNAnet structures, with the integration of attention modules. The double U-shaped backbone feature extraction network in our approach enables more accurate segmentation of target edges, while the addition of attention modules improves the capturing of target positional information. The innovation of our centroid detection subnet lies in the ability to overcome the constraint of unit pixel size in the original image, achieving sub-pixel-level centroid localization and minimizing the difference between predicted and ground truth centroid positions. Extensive comparative experiments demonstrate the superiority of our proposed semantic segmentation and centroid localization methods in terms of detection precision and robustness compared to existing mainstream methods.

In the future, we plan to expand our research in three directions: (1) How to use the motion information of targets in sequential images to reduce false alarms. The algorithm proposed in this paper is based on single-frame image for target detection, which may encounter challenges in removing false alarms in certain low-quality images or complex scenes. We will explore the temporal characteristics of targets in sequential images, combining spatial and temporal features to improve target discriminability. (2) How to construct an integrated network for enhancement and detection to address the issue of missed detections when the target is extremely weak. In our future work, we will embed more efficient enhancement modules or super-resolution reconstruction modules into the detection network to enhance target features and capture target details, aiming to achieve or even surpass the human eye's detection limit. (3) How to utilize multi-source image information fusion for handling weak small targets [35]. The infrared images used in this paper are obtained from thermal radiation imaging. Due to the technological limitations of sensors, issues such as loss of target texture information, high noise, and low signal-to-noise ratio may exist. In the future, we will develop a weak small target detection network based on multi-source data, incorporating data from multiple platforms and payloads to achieve stable and continuous detection and localization capabilities.

Author Contributions: Methodology, X.X., J.W., M.Z., H.S. and S.C.; Software, X.X., Z.W., Y.W. and S.L.; Writing—original draft, X.X., Z.W., Y.W., S.C. and S.L.; Writing—review & editing, J.W., M.Z. and H.S. All authors have read and agreed to the published version of the manuscript.

Funding: This research was funded by the Science and Technology Department of Jilin Province, China, under grant number 20210201137GX.

Conflicts of Interest: The authors declare no conflict of interest.

Appendix A

In Appendix A, we provide additional technical details and more experimental results.

Appendix A.1. Simulation Details of Infrared Weak Small Target Dataset

The production process of the proposed infrared weak small target dataset is shown in Figure A1. The overall steps include simulating background images, adding noise, de-signing target motion trajectories, adding targets to the background, randomly cropping image patches with targets, and generating semantic segmentation labels.

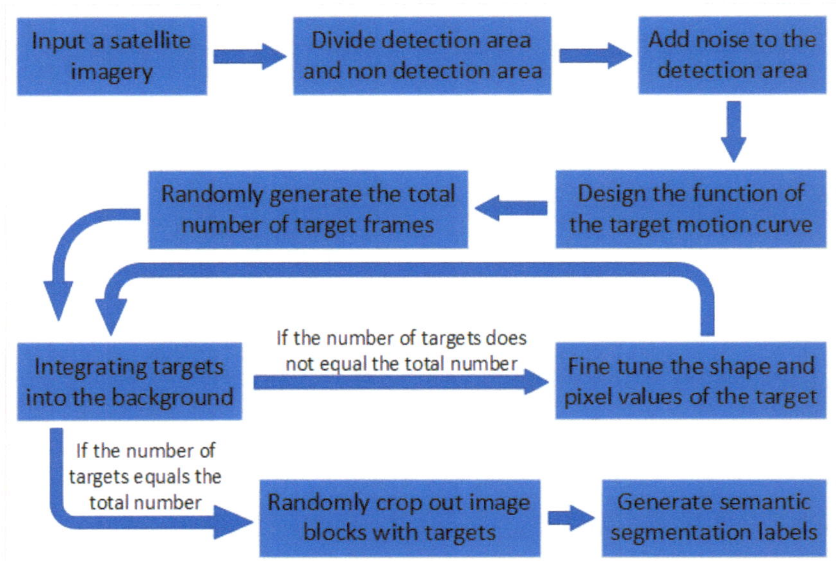

Figure A1. The schematic diagram of the simulation process for the proposed dataset of infrared weak small targets in this paper.

Appendix A.2. Details of Network Architectures

In Table A1, we list the details of the Res_Attention block, Downsample block, Up-sample_n block, and Multiply block used in our proposed network. These details include kernel size, stride, padding, upsampling and downsampling ratios, among others.

Table A1. Constituent elements of each block in the proposed network architecture.

Block	Layer	Input Channels	Output Channels	Kernel Size	Stride	Padding
Res_Attention	Conv2d	in_channels	out_channels	3	1	1
	BatchNorm2d	out_channels	out_channels	-	-	-
	ReLU	-	-	-	-	-
	Conv2d	out_channels	out_channels	3	1	1
	BatchNorm2d	out_channels	out_channels	-	-	-
	CBAM	out_channels	out_channels	-	-	-
	ReLU	-	-	-	-	-
Downsample	MaxPool2d	in_channels	in_channels	2	2	-
Upsample_n	Upsample	in_channels	in_channels	-	n	-
Multiply	-	1, 1	1	-	-	-

Appendix A.3. Experimental Results in Other Field

In Figure A2, we demonstrate the semantic segmentation results of our UCDnet in the industrial field (top row), medical field (middle row), and agricultural field (bottom row). The results reveal the good performance of our algorithm in these fields.

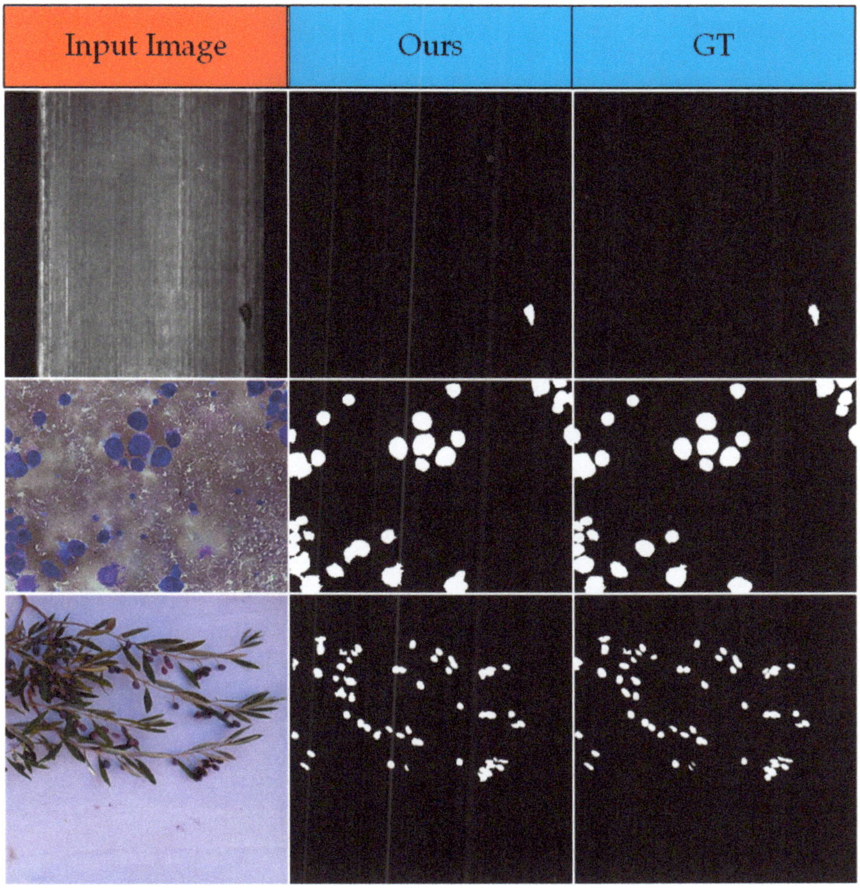

Figure A2. The first row in the figure shows the semantic segmentation results of motor magnetic tile defects, **the second row** presents the semantic segmentation results of cells, and **the third row** exhibits the semantic segmentation results of olive fruits.

References

1. Chen, G.; Wang, H.T.; Chen, K.; Li, Z.J.; Song, Z.D.; Liu, Y.L.; Chen, W.K.; Knoll, A. A Survey of the Four Pillars for Small Object Detection: Multiscale Representation, Contextual Information, Super-Resolution, and Region Proposal. *IEEE Trans. Syst. Man Cybern.-Syst.* **2022**, *52*, 936–953. [CrossRef]
2. Sun, X.L.; Guo, L.C.; Zhang, W.L.; Wang, Z.; Yu, Q.F. Small Aerial Target Detection for Airborne Infrared Detection Systems Using LightGBM and Trajectory Constraints. *IEEE J. Sel. Top. Appl. Earth Observ. Remote Sens.* **2021**, *14*, 9959–9973. [CrossRef]
3. Sun, Y.; Yang, J.G.; An, W. Infrared Dim and Small Target Detection via Multiple Subspace Learning and Spatial-Temporal Patch-Tensor Model. *IEEE Trans. Geosci. Remote Sens.* **2021**, *59*, 3737–3752. [CrossRef]
4. Liu, C.; Xie, F.Y.; Dong, X.M.; Gao, H.X.; Zhang, H.P. Small Target Detection From Infrared Remote Sensing Images Using Local Adaptive Thresholding. *IEEE J. Sel. Top. Appl. Earth Observ. Remote Sens.* **2022**, *15*, 1941–1952. [CrossRef]
5. Yu, W.B.; Liu, C.; Yang, H.T.; Wang, G.X. A method for improving the detection accuracy of subpixel edge. In Proceedings of the Chinese Automation Congress (CAC), Hangzhou, China, 22–24 November 2019; pp. 158–162.
6. Kosari, A.; Sharifi, A.; Ahmadi, A.; Khoshsima, M. Remote sensing satellite's attitude control system: Rapid performance sizing for passive scan imaging mode. *Aircr. Eng. Aerosp. Technol.* **2020**, *92*, 1073–1083. [CrossRef]

7. Li, Y.S.; Li, Z.Z.; Zhang, C.; Luo, Z.F.; Zhu, Y.; Ding, Z.Q.; Qin, T.Q. Infrared maritime dim small target detection based on spatiotemporal cues and directional morphological filtering. *Infrared Phys. Technol.* **2021**, *115*, 19. [CrossRef]
8. Li, Y.S.; Li, Z.Z.; Shen, Y.; Li, J. Infrared Small Target Detection Based on 1-D Difference of Guided Filtering. *IEEE Geosci. Remote Sens. Lett.* **2023**, *20*, 5. [CrossRef]
9. Zhang, L.; Lin, Z. Infrared Small Target Detection Based on Anisotropic Contrast Filter. In Proceedings of the 2020 IEEE 5th International Conference on Signal and Image Processing (ICSIP), Nanjing, China, 23–25 October 2020.
10. Han, J.H.; Ma, Y.; Huang, J.; Mei, X.G.; Ma, J.Y. An Infrared Small Target Detecting Algorithm Based on Human Visual System. *IEEE Geosci. Remote Sens. Lett.* **2016**, *13*, 452–456. [CrossRef]
11. Yang, L.N.; Jia, B.; Liu, Q. Adaptive Small Target Detection Based on Least Squares and Human Visual System. In Proceedings of the IEEE 17th International Conference on Communication Technology (ICCT), Chengdu, China, 27–30 October 2017; pp. 1565–1568.
12. Qiang, W.; Liu, H.K. An Infrared Small Target Fast Detection Algorithm in the Sky Based on Human Visual System. In Proceedings of the 4th Annual International Conference on Network and Information Systems for Computers (ICNISC), Wuhan, China, 20–22 April 2018; pp. 176–181.
13. Faramarzi, I.; Han, J.H.; Chen, Y.Y. Infrared Dim and Small Targets Detection Based on Multi-scale Local Contrast Measure Utilizing Efficient Spatial Filters. In Proceedings of the 6th International Conference on Signal Processing and Intelligent Systems (ICSPIS), Sadjad Univ, Mashhad, Iran, 23–24 December 2020.
14. Han, J.H.; Moradi, S.; Faramarzi, I.; Liu, C.Y.; Zhang, H.H.; Zhao, Q. A Local Contrast Method for Infrared Small-Target Detection Utilizing a Tri-Layer Window. *IEEE Geosci. Remote Sens. Lett.* **2020**, *17*, 1822–1826. [CrossRef]
15. He, Y.J.; Li, M.; Zhang, J.L.; An, Q. Small infrared target detection based on low-rank and sparse representation. *Infrared Phys. Technol.* **2015**, *68*, 98–109. [CrossRef]
16. Wei, H.Y.; Tan, Y.H.; Lin, J. Robust Infrared Small Target Detection Via Temporal Low-rank and Sparse Representation. In Proceedings of the 3rd International Conference on Information Science and Control Engineering (ICISCE), Beijing, China, 8–10 July 2016; pp. 583–587.
17. Liu, M.; Du, H.Y.; Zhao, Y.J.; Dong, L.Q.; Hui, M. Image Small Target Detection based on Deep Learning with SNR Controlled Sample Generation. In Proceedings of the 2nd International Conference on Computer Science and Mechanical Automation (CSMA), Wuhan, China, 10–12 November 2016; pp. 211–220.
18. Dai, Y.M.; Wu, Y.Q.; Zhou, F.; Barnard, K. Asymmetric Contextual Modulation for Infrared Small Target Detection. In Proceedings of the IEEE Winter Conference on Applications of Computer Vision (WACV), Electr Network, Waikoloa, HI, USA, 5–9 January 2021; pp. 949–958.
19. Hou, Q.Y.; Wang, Z.P.; Tan, F.J.; Zhao, Y.; Zheng, H.L.; Zhang, W. RISTDnet: Robust Infrared Small Target Detection Network. *IEEE Geosci. Remote Sens. Lett.* **2022**, *19*, 5. [CrossRef]
20. Zhou, X.Y.; Zhang, Y.; Hu, Y. Infrared Small Target Detection Via Learned Infrared Patch-Image Convolutional Network. In Proceedings of the IEEE International Geoscience and Remote Sensing Symposium (IGARSS), Kuala Lumpur, Malaysia, 17–22 July 2022; pp. 867–870.
21. Zhao, B.; Wang, C.P.; Fu, Q.; Han, Z.S. A Novel Pattern for Infrared Small Target Detection With Generative Adversarial Network. *IEEE Trans. Geosci. Remote Sens.* **2021**, *59*, 4481–4492. [CrossRef]
22. Cao, S.Y.; Lian, R.Y.; Zhang, Y.W.; Wu, F.Y.; Peng, Z.M. Infrared dim target detection via hand-crafted features and deep information combination. In Proceedings of the 17th Conference on Industrial Electronics and Applications (ICIEA), Chengdu, China, 16–19 December 2022; pp. 632–635.
23. Ronneberger, O.; Fischer, P.; Brox, T. U-Net: Convolutional Networks for Biomedical Image Segmentation. In Proceedings of the 18th International Conference on Medical Image Computing and Computer-Assisted Intervention (MICCAI), Munich, Germany, 5–9 October 2015; pp. 234–241.
24. Li, B.Y.; Xiao, C.; Wang, L.G.; Wang, Y.Q.; Lin, Z.P.; Li, M.; An, W.; Guo, Y.L. Dense Nested Attention Network for Infrared Small Target Detection. *IEEE Trans. Image Process.* **2023**, *32*, 1745–1758. [CrossRef] [PubMed]
25. Hu, H.G.; Zheng, Y.X.; Zhou, Q.W.; Xiao, J.; Chen, S.Y.; Guan, Q. MC-Unet: Multi-scale Convolution Unet for Bladder Cancer Cell Segmentation in Phase-Contrast Microscopy Images. In Proceedings of the IEEE International Conference on Bioinformatics and Biomedicine (BIBM), San Diego, CA, USA, 18–21 November 2019; pp. 1197–1199.
26. Zhao, Z.; Chen, K.; Yamane, S. CBAM-Unet++:easier to find the target with the attention module "CBAM". In Proceedings of the 2021 IEEE 10th Global Conference on Consumer Electronics (GCCE), Las Vegas, NY, USA, 12 October 2021.
27. Zhou, X.; Wang, D.; Krähenbühl, P. Objects as Points. *arXiv* **2019**, arXiv:1904.07850.
28. Oakley, J.P. Statistical properties of local extrema in two-dimensional Gaussian random fields. *IEEE Trans. Signal Process.* **1998**, *46*, 130–140. [CrossRef]
29. Fang, J.K.; Fong, C.M.; Yang, P.; Hung, C.K.; Lu, W.L.; Chang, C.W. AdaGrad Gradient Descent Method for AI Image Management. In Proceedings of the 7th IEEE International Conference on Consumer Electronics—Taiwan (ICCE-Taiwan), Taoyuan, Taiwan, 28–30 September 2020.
30. Heydarian, M.; Doyle, T.E.; Samavi, R. MLCM: Multi-Label Confusion Matrix. *IEEE Access* **2022**, *10*, 19083–19095. [CrossRef]

31. Sun, K.; Xiao, B.; Liu, D.; Wang, J.D.; Soc, I.C. Deep High-Resolution Representation Learning for Human Pose Estimation. In Proceedings of the 32nd IEEE/CVF Conference on Computer Vision and Pattern Recognition (CVPR), Long Beach, CA, USA, 16–20 June 2019; pp. 5686–5696.
32. Wu, T.H.; Li, B.Y.; Luo, Y.H.; Wang, Y.Q.; Xiao, C.; Liu, T.; Yang, J.G.; An, W.; Guo, Y.L. MTU-Net: Multilevel TransUNet for Space-Based Infrared Tiny Ship Detection. *IEEE Trans. Geosci. Remote Sens.* **2023**, *61*, 15. [CrossRef]
33. Cui, Z.M.; Li, C.J.; Chen, N.L.; Wei, G.D.; Chen, R.N.; Zhou, Y.F.; Shen, D.G.; Wang, W.P. TSegNet: An efficient and accurate tooth segmentation network on 3D dental model. *Med. Image Anal.* **2021**, *69*, 12. [CrossRef] [PubMed]
34. Hossain, M.S. Microc alcification Segmentation Using Modified U-net Segmentation Network from Mammogram Images. *J. King Saud Univ.-Comput. Inf. Sci.* **2020**, *32*, 1225–1226. [CrossRef]
35. Ghaderizadeh, S.; Abbasi-Moghadam, D.; Sharifi, A.; Tariq, A.; Qin, S.J. Multiscale Dual-Branch Residual Spectral-Spatial Network With Attention for Hyperspectral Image Classification. *IEEE J. Sel. Top. Appl. Earth Observ. Remote Sens.* **2022**, *15*, 5455–5467. [CrossRef]

remote sensing

MDPI

Article

Discarding–Recovering and Co-Evolution Mechanisms Based Evolutionary Algorithm for Hyperspectral Feature Selection

Bowen Liao [1], Yangxincan Li [1], Wei Liu [2], Xianjun Gao [3] and Mingwei Wang [1,*]

[1] School of Computer Science, Hubei University of Technology, Wuhan 430068, China; 2010301013@hbut.edu.cn (B.L.); 2010412235@hbut.edu.cn (Y.L.)
[2] Institute of Geological Survey, China University of Geosciences, Wuhan 430074, China; wliu@cug.edu.cn
[3] School of Geosciences, Yangtze University, Wuhan 430100, China; junxgao@yangtzeu.edu.cn
* Correspondence: wmwscola@hbut.edu.cn

Abstract: With the improvement of spectral resolution, the redundant information in the hyperspectral imaging (HSI) datasets brings computational, analytical, and storage complexities. Feature selection is a combinatorial optimization problem, which selects a subset of feasible features to reduce the dimensionality of data and decrease the noise information. In recent years, the evolutionary algorithm (EA) has been widely used in feature selection, but the diversity of agents is lacking in the population, which leads to premature convergence. In this paper, a feature selection method based on discarding–recovering and co-evolution mechanisms is proposed with the aim of obtaining an effective feature combination in HSI datasets. The feature discarding mechanism is introduced to remove redundant information by roughly filtering the feature space. To further enhance the agents' diversity, the reliable information interaction is also designed into the co-evolution mechanism, and if detects the event of stagnation, a subset of discarded features will be recovered using adaptive weights. Experimental results demonstrate that the proposed method performs well on three public datasets, achieving an overall accuracy of 92.07%, 92.36%, and 98.01%, respectively, and obtaining the number of selected features between 15% and 25% of the total.

Keywords: hyperspectral imaging; feature selection; data dimensionality reduction; evolutionary algorithm; discarding–recovering mechanism; co-evolution mechanism

Citation: Liao, B.; Li, Y.; Liu, W.; Gao, X.; Wang, M. Discarding–Recovering and Co-Evolution Mechanisms Based Evolutionary Algorithm for Hyperspectral Feature Selection. *Remote Sens.* **2023**, *15*, 3788. https://doi.org/10.3390/rs15153788

Academic Editors: Qian Du, Yanni Dong and Xiaochen Yang

Received: 9 June 2023
Revised: 17 July 2023
Accepted: 27 July 2023
Published: 30 July 2023

1. Introduction

The advancement of hyperspectral remote sensing leads to its widespread use in scanning continuous, narrow spectral bands, as it enables the acquisition of information on the reflection or radiation spectrum of objects at various wavelengths [1–3]. Digital number (DN) or reflectance value is considered as the feature value for each band and represented as a feature vector. However, there is a large amount of redundant information collected through hyperspectral imaging (HSI) with the electromagnetic spectrum and visible light infrared technology, resulting in high dimensionality [4]. In essence, data dimensionality reduction helps to trim the redundancy and noise [5] and improves classification accuracy, which has become an important topic in the processing of HSI datasets.

Generally, there are two methods for removing the redundancy of the dataset: feature extraction and feature selection. Feature extraction involves the linear or nonlinear transformation of the original high-dimensional features, such as combining different features into a new feature set [6], where the features lose their original physical meaning. Feature selection involves selecting the most representative feature combination from the dataset; it detects representative features and decreases redundant information or noise from data, which improves classification accuracy and enhances comprehensibility [7]. Due to the difficulty in interpreting selected features from feature extraction, feature selection is widely used in the processing of HSI datasets.

There are three feature selection strategies based on the search rule, namely, filter, wrapper, and embedded [8]. The filtering strategy analyzes each feature using a proxy measure [9] and selects a combination with a specified number of features based on the score ranking. However, the score only reflects the correlation with labels, ignoring the feature interactivity that some feature with a low correlation to labels provides greater performance improvement than those with a high correlation to labels. The wrapper strategy combines the process of feature selection with the agent to identify an appropriate combination of features. However, this strategy requires a continuous measure of the feature combination, resulting in high computational complexity and inadequate generalization ability [10]. The embedded strategy selects features in the learning process [11,12], and incorporates the feature selection in training, avoiding the overfitting that may occur in other strategies by adjusting the weights of features. The embedded strategy is usually combined with some iterative searching; the weight is used to guide the next iteration.

Evolutionary algorithm (EA), which mimics the adaptation and survival of the fittest observed in living organisms in nature, uses the searched heuristic information as the guidance; the genetic material of these combinations is then assembled to create new offspring, and the process is repeated over many generations to allow the population evolves towards better solutions [13]. Traditional EAs update the genetic material through mutation and crossover operations, which are then passed to the next generation. However, those algorithms only consider the stochasticity between agents, but not the similarity between them, which will lead to the occurrence of premature convergence and overfitting phenomena [14,15]. To address these limitations, distance-based EAs have been proposed; these algorithms are designed to calculate the distance between agents to determine their similarity and select some of them for crossover and mutation based on the similarity [16–20], thereby helping to maintain diversity in the population and prevent premature convergence. Nonetheless, due to the absence of competition or collaboration, the information interaction between agents is insufficient, making it difficult for the EA to overcome local optima and leading to stagnation in the iterative process.

The co-evolution mechanism is a means of enhancing information interaction ability. Due to its robustness, this mechanism has received extensive attention and has been widely used in various fields, including natural language processing and image retrieval [21,22]. By combining EA, the co-evolution mechanism improves the search efficiency of the EA in feature selection to some extent [23–25]. It divides the original feature set into many subsets; subpopulations are formed based on the agents generated by these subsets, then this mechanism enhances the diversity by information interacting between the agents in different subpopulations. However, the current information interaction only takes into account exchanging the solution encoding with weak representation, leading to the low diversity of agents and the subpopulation imbalance where some agents obtain better combinations after searching than others most of the time. Therefore, a co-evolution mechanism with prominent reliability is necessary to be further searched to fully realize its potential.

In this paper, a feature selection method based on discarding–recovering and co-evolution mechanisms is proposed to obtain a reduced feature combination of the HSI datasets with adequate accuracy. The feature discarding mechanism is introduced to filter out redundant features from the original dataset. Moreover, the co-evolution mechanism is combined with EA to enhance the diversity of agents, and a reliable information interaction is used to enable collaborative search between agents and help EA to jump out of the local optima. To avoid the erroneous discarding of the interactive features that have a low correlation with labels and improve the generalization ability, feature recovery is introduced to raise the probability of discarded features. The purpose of this work is that propose a feature selection method to select an effective feature combination and decrease the redundant information in HSI datasets. The co-evolution mechanism is utilized to promote the subpopulations of EA consistently. Moreover, the feature discarding and recovering mechanisms are used to avoid meaningless searching and enhance the generalization ability. The main contributions of this work are listed as follows:

(1) The discarding–recovering mechanism is designed to enhance the generalization ability and decrease the computational load, which filters the original feature space and recovers some features into the population.
(2) The co-evolution mechanism is combined with EA, which divides two subpopulations to co-evolve and utilizes reliable information interaction to enhance the diversity of agents in subpopulations.
(3) A feature selection method based on discarding–recovering and co-evolution mechanisms is proposed to obtain an effective feature combination, which has a prominent performance in HSI datasets.

The rest of this paper is structured as follows: Section 2 provides the background information; Section 3 details the proposed feature selection method; Section 4 presents the experimental results from different perspectives; Section 5 exhibits the discussion of the proposed method and Section 6 outlines the conclusions.

2. Related Work

2.1. The Feature Selection Method Based on Distance-Based EA

The feature selection method based on distance-based EA has received much attention for its effectiveness in data dimensionality reduction, as it iteratively uses heuristic information to guide the next iteration. Wu et al. [26] developed the particle swarm optimizer (PSO) to reduce the dimensionality of the HSI dataset, where the chaotic sequence was used to initialize the feature space, helping PSO jump out of local optima. Su et al. [27] proposed a novel feature selection method based on the improved firefly algorithm (FA), which largely outperformed the conventional covariance method. Xie et al. [28] proposed a comprehensive feature selection method based on the artificial bee colony algorithm (ABC) and subspace division, achieving prominent overall classification accuracy (OA) while reducing a small amount of redundant information. Wang et al. [29] presented an optimized feature selection method based on the grey wolf optimizer (GWO) in the HSI dataset, which uses the adaptive weight to regulate the balance between optimal individuals and chaos operation to set correlative parameters. Tschannerl et al. [30] proposed an unsupervised feature selection method based on information theory and a modified discrete gravitational search algorithm (GSA), obtaining a more informative subset of features. However, with the increase of the data dimensionality, the ability of EA for further dimensionality reduction gradually decreases since the monotonous agents, leading to the selected feature combination, are redundant to some extent, and distinguishing between the approximate labels is difficult.

2.2. The Co-Evolution Mechanism of Feature Selection

The co-evolution mechanism uses the "divide and conquer" approach to divide the population, identify the current optimal subsets in the feature space, and eventually join them together into a global subset. Song et al. [31] proposed an adaptive subpopulation size adjustment mechanism based on co-evolution and a feature importance-oriented spatial partition strategy, decreasing the calculating time of particle evaluation and providing a competitive solution for the feature selection of high-dimensional data. Zhao et al. [32] proposed a multiple populations co-evolution mechanism and multi-stage interaction learning (OL) mechanism to fully search the prospective features in the stagnant state and increase the possibility of jumping out of local optima. Zhou et al. [33] proposed a feature selection method based on a cooperative co-evolution mechanism (CC-DFS). This method used a heterogeneous model to search for feature combinations with cut-off points and feature combinations without cut-off points, resulting in improved performance and generalization ability. Rashid et al. [34] proposed a feature selection method based on a cooperative co-evolution mechanism and random feature grouping (CC-RFG). Three ways were introduced to decompose the feature set dynamically to ensure the interactive features were divided into the same subpopulation. However, the above co-evolution mechanisms for feature selection only exchange the feature combination with weak representation, leading to the difficulty of regulating those features.

2.3. Motivation

To tackle the problem of the EA in data dimensionality reduction caused by the large feature space and redundant information, the preliminary filtering of the original feature set is required, which helps to decrease the redundancy information of the dataset. To further enhance the performance and effectiveness of EA, it is important to speed up the search process, increase the diversity of agents, and facilitate effective information interaction to improve the quality of the selected features.

Regarding the co-evolution mechanism, when agents from different subpopulations interact, they exchange information that is likely to improve the OA or decrease the number of selected features searched by agents. However, if weak features are not considered, it will lead to an imbalance problem. To overcome these limitations, increasing the probability of selecting the weak features and promoting diverse information interaction between agents is necessary. By this, the co-evolution mechanism achieves a balanced and effective optimization process, leading to a prominent result in HSI datasets.

In all, to improve search efficiency, it is necessary to remove the redundant features in the original feature set while recovering some of these features when detecting update stagnation. Additionally, the co-evolution mechanism is introduced to enhance the diversity of agents in corresponding subpopulations, given that interaction with diverse information is required to maintain the balance between subpopulations. All these measures help improve the performance and stability of agents in feature selection, making them more effective for real-world applications.

3. The Proposed Method

There is a certain of redundant information in the HSI dataset, and the performance in the reduction of data dimensionality has room to improve for EA. As a result, the feature discarding mechanism is implemented that uses some measure criteria to roughly filter the feature space, and the co-evolution mechanism is utilized to divide the population and take the reliable information interaction between agents to enhance the generalization ability. During the iteration process, if a stagnation phenomenon is detected, it is likely caused by the previous erroneous discarding of the interactive features, so the feature recovering mechanism is detonated to increase the selected probability of weak features by adaptive weights, and some of them will be recovered into the subpopulations.

3.1. The Feature Discarding Mechanism

Given the high degree of redundant features in the original dataset, removing it on a large scale is necessary. This eliminates the need for a thorough analysis of each feature and allows for a fast return of selected features. The evaluation measure for each feature is defined as follows:

$$S^n = \sum_{K=1}^{n} X_n^K, n = 1,2,3\ldots\ldots m \tag{1}$$

$$X_n^K = \frac{2}{\left|\left(\theta^k\right)^T H\theta^k - \left(\theta^k\right)^T H^{(-t)}\theta^k\right|} \tag{2}$$

where $H \in \mathbb{R}^{n\times n}$, $H_{i,j} = kernel(x_i, x_j)$, and $H^{(-t)} = kernel\left(x_i^{-t}, x_j^{-t}\right)$ the $(-t)$ indicates that t^{th} feature is discarded. Note that $kernel(x_i, x_j)$ is the kernel function mapped to a high-dimensional space, and θ represents the optimized parameters obtained from the SVM-based classifier [35].

The feature discarding mechanism, which is based on forward filtering and reverse learning, is implemented to obtain the ranking of feature scores using Equations (1) and (2), drops the specified number of features, and recovers groups of features with low score ranking through reverse learning [36]. In addition, to improve the generalization ability, the feature discarding mechanism calculates the compromise value of recovery groups [37]. The mathematical model is defined as follows:

$$U_i = \max_{1<k<m}\left[w_k \frac{o_k^* - x_{ik}}{o_k^* - o_k^-}\right], R_i = \sum_{k=1}^{m} w_k \frac{o_k^* - x_{ik}}{o_k^* - o_k^-} \tag{3}$$

$$Q_i = \frac{v(R_i - R^*)}{R^- - R^*} + \frac{(1-v)(U_i - U^*)}{U^- - U^*} \tag{4}$$

where R_i and U_i denote the utility measure and the regret measure between m features, respectively, W is the weight vector, O_k^* is the maximum value of kth feature of the decision matrix, O_k^- is the minimum value of kth feature of the decision matrix. R^* and U^* are the maximum value of R *and* U, respectively. S^-, U^- are the minimum value of R *and* U, respectively. Q_i represents the compromise value for each sample. The feature discarding mechanism obtains the compromise value of feature groups using Equations (3) and (4); the smallest one is selected as the original feature set.

3.2. The EA-Based Co-Evolution Mechanism

After feature discarding, the original feature set still has a high degree of redundant features, necessitating further decrease. The EA-based co-evolution mechanism can effectively search for the remaining features. Specifically, it divides the population into many subpopulations and uses information interaction to achieve a balance between them.

3.2.1. The Population Division Based on Feature Correlation

Generally, the population division involves partitioning the original feature set into multiple clusters (i.e., feature subsets) and initializing the agents generated in subpopulations based on these clusters. In addition, agents only search for features within their corresponding subsets while obtaining the rest via information interaction. Ideally, the population division considers the correlation between features or between features and labels commonly to minimize the correlation between features and maximize the correlation between features and labels [38]. However, when interactive features are partitioned into different subsets, subpopulations may fall into local traps that are not the local optimum of the original feature set and rather the local optimum resulting from the incorrect division. Therefore, the population division should ensure the feature subsets corresponding to subpopulations are sufficiently different, and the interactive features are partitioned together as much as possible, with the correlation between features considered.

Furthermore, generating many subsets requires an equal number of subpopulations to match them, leading to a large computation load. Additionally, interactive features may be divided into different subsets, resulting in mature convergence. To minimize the redundant features of the entire dataset, the population division decomposes the original feature set into two subsets, generating agents to form subpopulations within them. Figure 1 shows an example of the population division. The original feature set is partitioned according to the correlation between features, assuming that it has m features waiting for selection, two subsets are formed after the population division, and the number of features is q. To maintain subpopulations' balance, q is equal to $\lfloor m/2 \rfloor$, the $\lfloor\rfloor$ is the integer-value function.

3.2.2. The Reliable Information Interaction

The agents in different subpopulations are designed to exchange the information in parallel to facilitate interaction. If a feature does not belong to the current feature subset, it is searched with a probability of 0. Moreover, subpopulations should be provided with representative information to keep balance. Features with unsatisfactory scores may be the result of not finding other interactive features [39]. With the representative information, the features' performance will be boosted. The representative information is defined as the best and worst combinations searched by agents in the corresponding subpopulation, and one of them is selected as the interaction object to enhance the reliability of the co-evolution mechanism. Figure 2 illustrates the reliable information interaction between subpopulations. It can be seen that during the interaction, each subpopulation receives representative information from the other. The agent then combines this information to make an overall evaluation after conducting a search. By following this process, features will be fully searched to obtain a prominent classification accuracy through the support vector machine (SVM)-based classifier on the testing set.

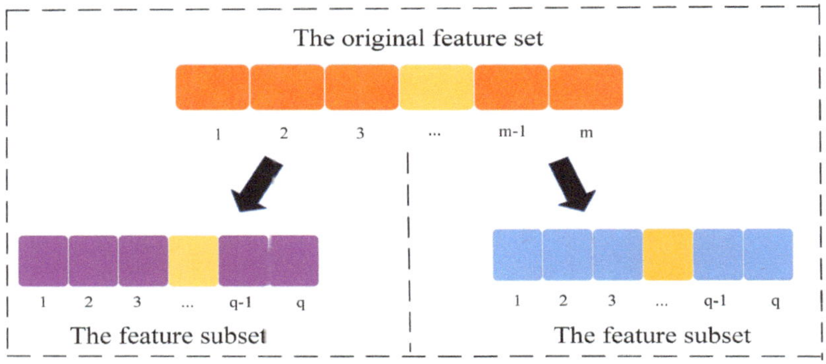

Figure 1. The process of population division.

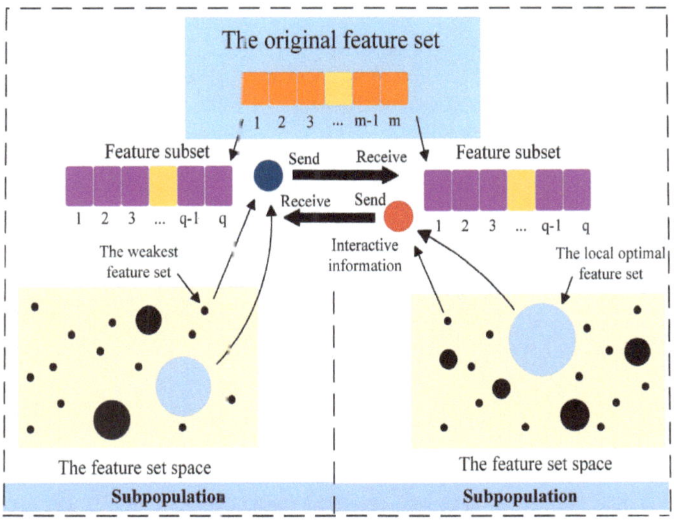

Figure 2. The process of reliable information interaction between subpopulations.

In subpopulations, the position of the next iteration of agents ($\vec{X}^{(iter+1)}$) is updated based on the distance ($\vec{d}_{IS}$) between the current agent's position ($\vec{X}^{iter}$) and the optimal position. Here, the positions of the current agents are updated based on the optimal agent obtained using Equation (5).

$$\vec{X}^{(iter+1)} = \vec{X}^{iter} - \vec{L} \cdot \vec{d}_{IS} \tag{5}$$

where $\vec{d_{IS}}$ refers to the distance between the current agent and the global optimum, while $\vec{L}$ denotes the social status of the optimum with a random value selected from the range of $[-1, 1]$.

3.3. The Feature Recovering Mechanism

After the feature discarding, only the features with high ranking are selected, but the interactive features are not considered, which may lead to stagnation [40]. Consequently, the performance of feature subsets searched by agents may fall into local optima. Moreover, EA generally operates within the original feature set, and it is difficult to recycle the discarded features. To address these limitations, the feature recovering mechanism is applied to incorporate the recycled discarded features into the recovery subset, thereby increasing the probability of selection. The feature recovery mechanism has two stages. The first stage is reverse learning, which increases the probability of selecting features with low score ranking. Moreover, if training stagnation is detected, indicating that the subpopulation is not improved in successive iterations, some of the discarded features should be recycled. This will allow agents to fully search features later.

More attention should be paid to the features with low scores in the evaluative measures when recovering features. However, the low score does not necessarily mean that corresponding features should be simply discarded. Therefore, the lower-ranked features will receive higher weights. Assuming the dimension of input data is m, the calculation for weight is described below:

$$W^m = 1 - \frac{S^m}{\sum_{i=1}^{m} S_i^m} \tag{6}$$

where W^m denotes the feature weight, and S^m represents the feature score set obtained from feature discarding, S_i^m represents the ith feature score. More weights obtained through Equation (6) are assigned to weak features, thus increasing their chances of being selected. As illustrated in Figure 3, after the features recovered through weighted screening are added to the corresponding subpopulation's feature subset, a new feature space is generated for agents.

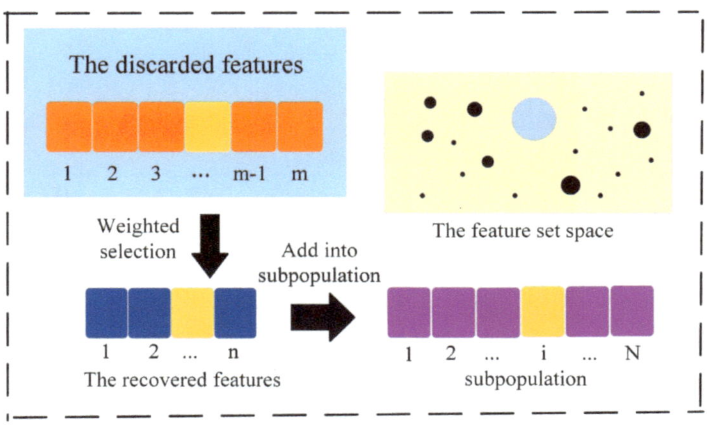

Figure 3. The process of feature recovery.

3.4. The Objective Function

The main target of feature selection is obtaining a representative feature combination from the original feature set to maximize the OA [41], which is an important evaluation criterion, but how to decrease the number of selected features is also a crucial target in feature selection. In this paper, the objective function is used to evaluate the feature combination searched by agents [42]; it is described in Equation (7).

$$fitness = \alpha \cdot OA + (1 - \alpha) \cdot \lg\left(\frac{n_c}{n_s}\right) \quad (7)$$

where $fitness$ represents the fitness value of the feature combination searched by agents, OA represents the overall classification accuracy obtained by SVM. Note that n_c and n_s are the number of total features in the dataset and the number of selected features. α is a weight factor of OA and the number of selected features; it takes $\alpha = 0.9$ in this paper.

3.5. Implementation of the Proposed Method

The proposed feature selection method updates the agent based on distance, and its key process involves the information interaction between agents. Moreover, in the event of stalling, it recycles some of the discarded features, thereby improving the probability of features with low score ranking. The proposed feature selection method is described as follows (Algorithm 1):

Algorithm 1: Discarding–recovering and co-evolution mechanisms for HSI feature selection

Input: the n × m dataset D, the agent size Agesize, the number of feature groups M by reverse learning, and the maximum number of iterations Maxiter
Output: the effective feature combination selected by agents

Undergo the feature discarding process through the feature discarding mechanism and obtain the SS and DS using the Equations (1)–(4) by reverse learning M feature groups
for i **in** SS:
do
$W_i^m \leftarrow 1 - \frac{S_i^m}{\sum_{k=1}^{m} S_k^m}$
end for
The SS is divided into two subsets SS_1 and SS_2 based on the correlation
$Selectgroup_1 \leftarrow SS_1$, $Selectgroup_2 \leftarrow SS_2$
Two subpopulations are generated in $Selectgroup_1$ and $Selectgroup_2$ and Agesize agents are obtained
$t \leftarrow 0$
while t < Maxiter:
do
Update the location of each agent by Equation (5)
Update the fitness value of each agent by Equation (7)
if the optimal solution has been updated **then**
Exchange the information through interaction
end if
if one of the subpopulations has stalled **then**
h ← Recover(subset)
SS ← Add(SS, h), DS ← Sub(DS, h), $Selectgroup_{1,2} \leftarrow$ Add ($Selectgroup_{1,2}$, h),
S ← Add(s, h)
end if
$t \leftarrow t + 1$
end while
return The OA of effective feature combination

In the beginning, the feature evaluation is performed on all agents to discard the features with low score ranking, resulting in a selected set (SS) of features and a discarded set (DS) of features. The SS is then divided into two subsets: $selectgroup_1$ and $selectgroup_2$. Representative information is exchanged between these subsets when the optimal agent is updated. Moreover, if the stagnation phenomenon is detected, the feature recovering mechanism is triggered to recycle a recovery feature subset with a certain number based on the adaptive weight W. These features will be added to the SS and removed from the DS. The iteration process continues until the maximum iteration is reached.

4. Experimental Results

The proposed feature selection method is implemented using Python 3.8 on a personal computer that has a 2.30 GHz CPU, 8.00 GB RAM, and the Windows 8 operating system. To evaluate the performance of the proposed method, three HSI datasets, namely KSC (176 bands), Salinas (204 bands), and Longkou (270 bands), are used in the study. The experimental results are compared with some feature selection methods with EA-based, co-evolution mechanism-based, and others, and each independent experiment is performed in 30 operations with 50 iterations of each operation.

4.1. Dataset Description

The first dataset was acquired by NASA at the Kennedy Space Center (KSC) in Florida. It was obtained from a distance of approximately 20 km and contained 224 bands with a spatial resolution of 18 m. After removing bands with water absorbance and low signal-to-noise ratio, 176 bands were used for verification. The image consists of 512 × 614 pixels.

The second HSI dataset, named Salinas, was obtained by an AVIRIS sensor in the Salinas Valley of California, USA. It consists of 204 bands with a spatial resolution of 3.7 m and a pixel size of 512 × 217. The spectral range of the dataset spans from 0.4 to 2.5 μm, and the spectral resolution is 10 nanometers.

The third dataset was obtained in Longkou Town, Jingzhou City, Hubei Province, China, and includes six classes in an agrarian context. The UAV flew at an altitude of 500 m, and the spatial resolution of the airborne hyperspectral image is approximately 0.463 m. The image size is 550 × 400, with 270 bands ranging from 400 to 1000 nm.

The class names and corresponding sample numbers of three HSI datasets are described in Table 1. The image scene and ground truth of them are shown in Figure 4.

Table 1. The land-cover classes of three HSI datasets.

Class Number	Class Name	Sample Number	Class Name	Sample Number	Class Name	Sample Number
1	Scrub	761	Brocoli_green weeds_1	2009	Corn	34,511
2	Willow swamp	243	Brocoli_green weeds_2	3726	Cotton	8374
3	Cabbage palm Hammock	256	Fallow	1976	Sesame	3031
4	Cabbage palm/Oak hammock	252	Fallow_rough plow	1394	Broad-leaf soybean	63,212
5	Slash pine	161	Fallow_smooth	2678	Narrow-leaf soybean	4151
6	Oak/Broadleaaf hammock	229	Stubble	3959	Rice	11,854
7	Hardwood swamp	105	Celery	3579	Water	67,056
8	Graminoid marsh	431	Grapes_Untrained	11,271	Roads and houses	7124
9	Spartina marsh	520	Soil vineyard develop	6203	Mixed weed	5229
10	Cattaial marsh	404	Corn_senesced greed_weeds	3278	-	-
11	Salt marsh	419	Lettuce_romaine 4 wk	1068	-	-
12	Mud flats	503	Lettuce_romaine 5 wk	1927	-	-
13	Water	927	Lettuce_romaine 6 wk	916	-	-
14	-	-	Lettuce_romaine 7 wk	1070	-	-
15	-	-	Vinyard_untrained	7268	-	-
16	-	-	Vinyard_vertical_trellis	1087	-	-
-	Total	5211	Total	54,129	Total	204,542

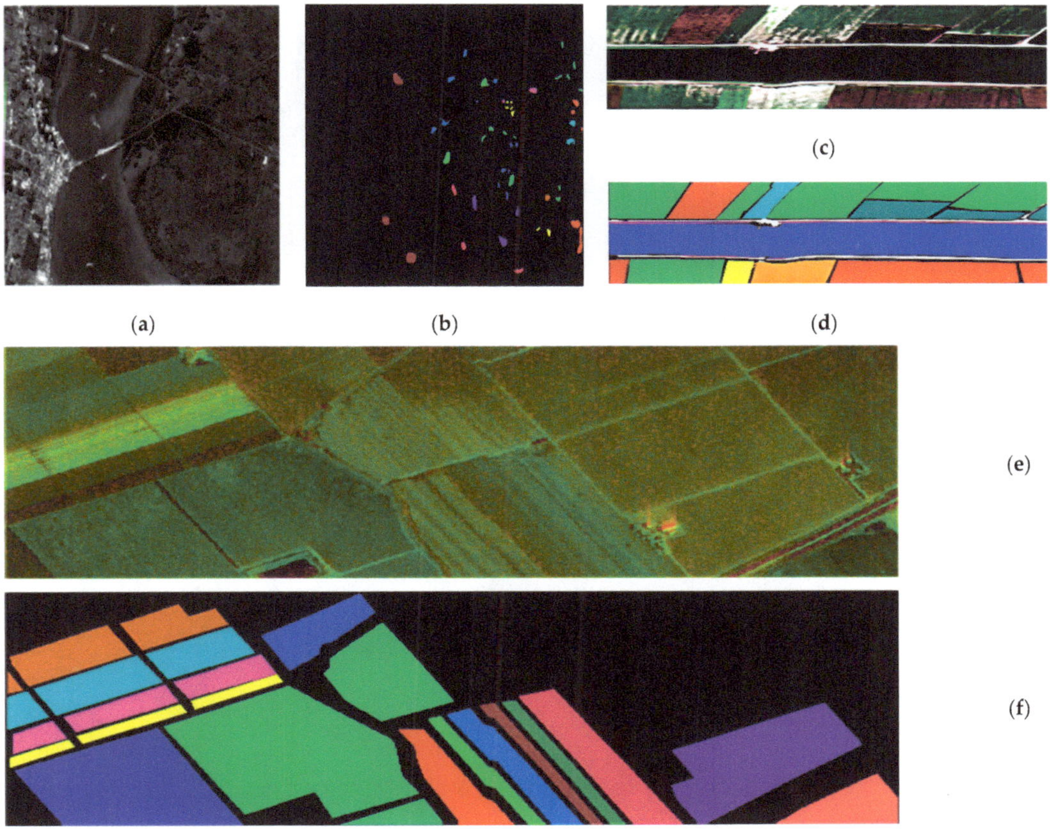

Figure 4. Image scene and ground truth of three HSI datasets: (a) KSC image (b) KSC ground truth (c) Salinas image (d) Salinas ground truth (e) Longkou image (f) Longkou ground truth.

4.2. Parameters Setting of EAs

Before running, some parameters of EAs should be set for the heuristic search. The performance of the effective feature combination is dependent on the setting of parameters to some extent. In this paper, several EA-based feature selection methods, including PSO [43], FA [44], GWO [45], and GSA [46], are adopted to provide an intuitive performance comparison with the proposed method. Table 2 shows the parameters setting by these EAs.

Table 2. Parameters setting of each algorithm.

Parameters	Values
Size of agents	15
Dimension	Number of features
The number of iterations per algorithm	50
The acceleration constant c_1, c_2 in PSO	2
Min-max inertia weight ω_{min}, ω_{max} in PSO	0.2, 0.9
The light intensity absorption coefficient I in FA	1
The step factor α in FA	0.97
Min-max attraction β_{min}, β_{max} in FA	0.2, 1
α Correlation coefficient in GWO	[2, 0]
The initial universal gravitational constant in GSA	100
Number of the subpopulation in co-evolution mechanism	2

4.3. Experiments for the Search Ability

Table 3 presents the OA and Kappa coefficient after 30 independent operations, while WTL is the win/tie/loss indicator of the fitness value. Table 4 shows the number of features (Num) and CPU time (Time) selected by the 30 independent operations. To demonstrate the prominent OA of the effective feature combination achieved in each iteration, the average number of features and OA are obtained for each iteration, as shown in Figure 5, and the fitness value is shown in Figure 6.

Table 3. The OA and Kappa coefficient of EA-based feature selection methods.

Datasets	Metrics	PSO	FA	GWO	GSA	Proposed
KSC	OA (%)	90.97 ± 0.33	90.26 ± 0.24	90.92 ± 0.31	90.71 ± 0.38	92.07 ± 0.12
	Kappa	0.904 ± 0.002	0.887 ± 0.0016	0.897 ± 0.001	0.892 ± 0.002	0.917 ± 0.001
	WTL	+	−	+	+	28/0/2
Salinas	OA (%)	91.77 ± 0.15	91.63 ± 0.23	91.65 ± 0.17	91.82 ± 0.26	92.36 ± 0.21
	Kappa	0.907 ± 0.001	0.908 ± 0.0014	0.905 ± 0.001	0.909 ± 0.0018	0.915 ± 0.0013
	WTL	+	−	+	−	27/1/2
Longkou	OA (%)	97.36 ±0.11	96.89 ± 0.27	97.12 ± 0.41	97.74 ± 0.33	98.01 ± 0.14
	Kappa	0.964 ±0.0002	0.956 ± 0.001	0.960 ± 0.0024	0.969 ± 0.006	0.979 ± 0.002
	WTL	+	−	+	+	30/0/0

Table 4. Number of selected features and CPU time of EA-based feature selection methods.

Datasets	Metrics	PSO	FA	GWO	GSA	Proposed
KSC	Num	49.6	50.3	50.7	55.3	42.2
	Time	190.1176	152.4127	204.1769	195.3568	150.6149
Salinas	Num	67.5	55.3	56.7	55.6	43.1
	Time	3600.5043	2714.3658	3536.9347	3742.5928	2427.6280
Longkou	Num	95.7	68.7	76	95.4	62.5
	Time	1564.3620	1269.3795	1514.6489	1892.3481	1204.7301

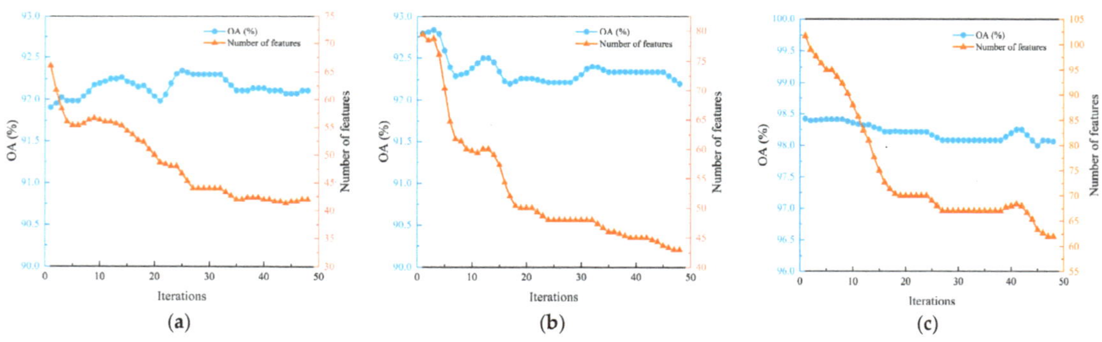

Figure 5. The trajectory of the OA and number of features on three HSI datasets: (**a**) KSC, (**b**) Salinas, (**c**) Longkou.

According to Table 3, the proposed method outperforms PSO, FA, GWO, and GSA in search capability, which achieves a prominent OA, surpassing PSO, FA, GWO, and GSA by 1.1%, 1.81%, 1.15%, and 1.36%, respectively. Those experimental results exhibit the superior searchability of the proposed method. Moreover, it enhances the development potential of local search by using a feature recovery mechanism. Its winning frequency is higher than 27, especially in Longkou, where it reached 30. These demonstrate the prominent stability of the proposed method and the superior exploration ability.

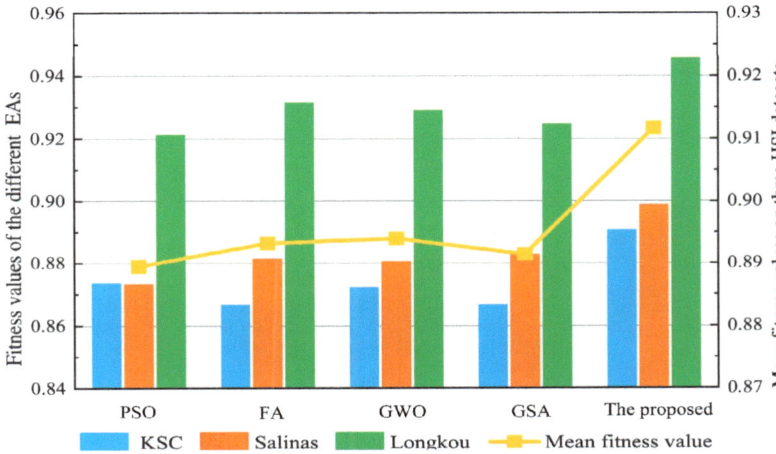

Figure 6. The fitness values of the EA-based and the proposed feature selection method on three HSI datasets. The bar chart shows the fitness values on three HSI datasets, and the curve shows the mean fitness value.

According to Table 4, the proposed method exhibits significantly higher reductive efficiency than other EA-based feature selection methods. Specifically, it selects less than 20% of the features from the HSI dataset, resulting in the selection of only 42 features out of a total of 176 bands in the KSC dataset while achieving a prominent OA. The Salinas dataset, it selects approximately half number of features compared with GSA yet achieves a better OA. On average, other methods select 58.7 features, whereas the proposed method selects only 43.1 features, indicating superior performance. In addition, the feature discarding mechanism substantially reduces redundant features, thereby shrinking the feature space and improving the computation time, especially in the Longkou dataset.

As shown in Figure 5, after feature discarding, the number of selected features searched by agents is still high, and the number of features is decreased after the heuristic search, while the OA is little to no fluctuation, demonstrating the prominent stability of the proposed method. Furthermore, the feature recovering mechanism effectively updates agents before the iteration ends, indicating that the proposed feature selection method possesses a prominent ability to escape from local optima. According to Figure 6, the fitness value is visualized to comprehensively evaluate the searching ability of each algorithm, and the proposed feature selection method achieves promising results on three HSI datasets and ranks 1st in terms of average fitness value, followed by GWO, FA, PSO, and GSA. Moreover, the proposed method achieves the optimal fitness value on three HSI datasets compared with other EA-based methods, proving it has a prominent search ability for feature selection.

4.4. Comparison with Other Feature Selection Methods

To assess the impact on each class, some feature selection methods in HSI datasets are compared in the experiment: maximum information minimum redundancy (MRMR) [47], joint mutual information with class correlation (JOMIC) [48], joint mutual information maximization (JMIM) [49], conditional mutual information maximization (CMIM) [50] and shallow-to-deep feature enhancement (SDFE) [51]. The experiments are performed on 10% to 25% of the total features. The accuracy for each class and Kappa coefficient are shown in Tables 5–16.

Table 5. The results for the KSC dataset using 10% of the total number of features.

Class Number	MRMR	JOMIC	JMIM	CMIM	SDFE	Proposed
1	92.5714	86.6579	88.4363	81.4724	92.3717	92.4264
2	86.2559	86.2222	93.2990	86.7841	86.1751	89.3519
3	63.4675	69.9670	74.7170	83.8542	85.4077	77.4074
4	52.9070	60.1990	52.1886	51.9231	55.5556	57.7670
5	62.5000	73.1183	75.0000	70.3704	68.3761	65.0407
6	60.8434	60.9929	70.4918	64.8936	83.4711	71.0059
7	67.9487	78.4091	71.4286	83.3333	74.7826	74.0741
8	76.5661	79.4304	76.7442	68.6981	87.6238	93.9058
9	82.6430	86.4341	84.4530	90.9692	90.2390	89.9606
10	87.6081	100.0000	100.0000	99.1562	99.8727	99.9618
11	95.5959	95.4054	98.8950	99.1892	98.5836	93.6869
12	91.3551	79.9213	76.8642	74.0876	92.7039	97.9499
13	99.5175	99.9482	99.2894	99.8801	99.6247	99.9915
OA (%)	85.0320	86.0128	85.7143	84.5203	90.2132	90.3624
Kappa	0.8406	0.8543	0.8418	0.8370	0.8908	0.8917

Table 6. The results for the KSC dataset using 15% of the total number of features.

Class Number	MRMR	JOMIC	JMIM	CMIM	SDFE	Proposed
1	91.8733	92.5104	88.2108	88.9655	92.8870	92.2006
2	89.6226	90.4762	72.0755	82.0961	88.7324	90.6977
3	70.7071	69.2557	77.7328	84.6154	80.2469	83.4711
4	48.4733	56.6845	56.7742	58.2996	57.9336	60.2996
5	82.0513	79.4872	73.3333	52.9032	75.2381	86.6667
6	78.4314	66.6567	71.5789	66.6667	65.1007	73.6842
7	70.1923	73.4513	81.5789	72.4771	74.2268	70.5357
8	80.9645	87.7612	85.0785	72.2343	89.5408	88.5856
9	82.7778	78.1570	87.1401	89.1304	91.6667	89.9408
10	87.8873	91.9162	100.0000	100.0000	99.7183	96.8927
11	98.3562	96.3824	98.4000	99.4350	97.6378	95.1157
12	92.9440	93.7811	94.4056	93.1925	98.1352	97.6526
13	98.4597	98.1065	99.9919	66.6667	100.0000	100.0000
OA (%)	86.5885	87.0362	87.9104	87.0362	90.5970	90.6397
Kappa	0.8514	0.8560	0.8644	0.8581	0.8925	0.8936

Table 7. The results for the KSC dataset using 20% of the total number of features.

Class Number	MRMR	JOMIC	JMIM	CMIM	SDFE	Proposed
1	92.9178	90.7609	87.2679	86.2924	92.0833	93.1642
2	91.7476	86.1472	88.8889	88.4793	95.5801	88.5416
3	73.6462	72.5632	85.8537	90.7407	85.7759	79.4165
4	60.5634	48.8189	56.2500	59.3156	58.0000	68.2564
5	65.9091	79.6875	72.7273	58.2677	78.5047	78.2196
6	65.1613	77.0833	69.3548	67.7419	78.8618	72.2519
7	77.1429	80.6818	74.5763	72.0339	92.0930	85.6429
8	82.9268	88.8298	81.7043	82.4818	86.8106	93.4579
9	88.5177	88.2828	88.3268	88.7574	92.6295	90.1245
10	91.3793	92.6686	99.9948	99.4444	97.7077	99.9094
11	97.6501	95.3728	99.9859	98.1283	98.9333	99.9654
12	92.2902	92.4242	95.7143	97.2772	97.3872	95.6413
13	99.6407	99.7599	100.0000	99.9618	100.0000	100.0000
OA (%)	88.3156	88.2942	88.7846	88.8699	90.9595	91.7057
Kappa	0.8712	0.8628	0.8750	0.8735	0.8992	0.9047

Table 8. The results for the KSC dataset using 25% of the total number of features.

Class Number	MRMR	JOMIC	JMIM	CMIM	SDFE	Proposed
1	94.5428	93.5897	90.2878	89.9015	91.5395	94.2413
2	89.6714	91.9431	86.5169	92.3729	88.5572	98.5618
3	69.4444	73.6301	77.7778	88.0000	87.1111	87.1594
4	56.3636	54.1176	60.3960	60.5442	61.9910	69.5432
5	83.3333	85.5072	75.0000	72.7273	75.8929	84.5621
6	62.4309	73.4463	71.4286	68.2353	68.0203	77.8654
7	69.5238	78.2178	91.0448	79.3103	75.2137	72.5613
8	86.9674	91.1111	82.1306	83.3333	90.0990	93.6421
9	88.1764	88.5375	88.0299	95.1807	92.0892	92.6578
10	98.8338	94.9861	100.0000	99.9185	99.7175	99.9153
11	97.9058	98.5836	99.9103	97.1963	97.7901	97.2541
12	97.5717	94.0552	90.5983	92.6923	97.5501	98.4623
13	99.9826	99.5910	99.8459	100.0000	100.0000	99.8917
OA (%)	89.4883	89.8507	89.1721	90.3300	91.1514	92.8144
Kappa	0.8742	0.8819	0.8893	0.8991	0.9014	0.9164

Table 9. The results for the Salinas dataset using 10% of the total number of features.

Class Number	MRMR	JOMIC	JMIM	CMIM	SDFE	Proposed
1	99.5343	98.8636	100.0000	100.0000	99.8862	99.9654
2	97.2044	98.1765	97.9009	98.0724	98.5891	98.1319
3	92.7253	94.6309	89.8361	92.1729	94.4625	91.2998
4	98.0614	97.9032	99.1948	97.9592	97.1919	98.8606
5	97.8351	98.0897	95.3859	92.2832	98.4127	99.1625
6	99.4410	99.9438	99.3292	99.6081	99.7199	99.9438
7	97.7411	97.6190	99.9373	98.9480	99.0081	99.4410
8	74.1664	73.8863	70.0681	71.0198	74.8498	74.2189
9	97.9937	98.0613	96.5974	93.7715	99.1081	98.1501
10	94.0709	90.5249	86.0244	87.9890	94.5692	92.3497
11	94.6352	97.6852	76.1431	73.6961	95.5032	87.8981
12	91.0464	93.7778	93.0586	93.3839	95.7589	96.8362
13	93.1193	96.4200	90.2004	93.3180	93.7500	94.8598
14	98.6957	98.7179	95.7356	98.4305	98.0088	98.4749
15	77.0670	78.4153	77.4679	75.3187	80.4813	91.4327
16	98.5255	98.8032	99.6269	99.5019	99.6193	99.7484
OA (%)	89.3674	89.5685	87.3106	87.2573	90.4799	90.1081
Kappa	0.8797	0.8816	0.8657	0.8659	0.8936	0.8905

Table 10. The results for the Salinas dataset using 15% of the total number of features.

Class Number	MRMR	JOMIC	JMIM	CMIM	SDFE	Proposed
1	99.1982	99.9989	100.0000	100.0000	99.7517	100.0000
2	97.8947	98.2456	98.5832	98.3510	99.3765	99.5247
3	93.9427	91.5418	90.2646	94.9324	93.1071	94.8578
4	94.4870	97.7707	98.7302	97.5000	97.9528	99.0476
5	98.1387	97.6840	97.7215	97.1193	98.8215	98.1148
6	99.9987	100.0000	99.9438	99.9846	99.9438	100.0000
7	98.3446	99.0087	99.8118	99.0093	99.3808	99.2560
8	74.3518	74.5124	75.0246	75.4254	76.8367	78.0508
9	99.1783	99.0046	96.3605	97.5506	99.3585	99.2479
10	91.3971	93.1818	86.8185	86.8385	93.8272	95.9028
11	92.8421	97.1111	87.4157	87.6518	96.9365	96.0526
12	92.8414	94.9381	96.9629	96.4206	95.7589	96.2963
13	94.6262	96.1814	90.8072	92.8899	98.0583	94.6009

Table 10. *Cont.*

Class Number	MRMR	JOMIC	JMIM	CMIM	SDFE	Proposed
14	98.9035	98.2684	97.3154	96.3907	96.9199	94.6030
15	77.8520	78.5059	82.9680	82.0026	78.9930	81.5315
16	99.0826	98.9717	98.5258	99.5031	99.4987	99.0111
OA (%)	89.6342	90.0324	89.7656	90.0283	91.0142	91.7074
Kappa	0.8825	0.8876	0.8815	0.8936	0.9013	0.9105

Table 11. The results for the Salinas dataset using 20% of the total number of features.

Class Number	MRMR	JOMIC	JMIM	CMIM	SDFE	Proposed
1	99.4420	99.7738	100.0000	99.9843	99.4562	99.9913
2	99.4069	98.8797	98.9399	99.5258	98.9399	99.5848
3	95.0166	94.3844	92.3409	91.9181	94.1748	94.8408
4	97.3101	95.8333	97.8056	98.4227	99.3620	98.1132
5	98.5062	98.8235	98.3165	97.3177	97.9525	99.2437
6	100.0000	99.7205	99.9437	99.9439	99.8426	99.9439
7	99.4420	99.2551	99.9958	100.0000	99.6278	99.4406
8	75.6456	74.8615	75.5213	75.4221	78.1847	78.7730
9	98.6116	98.9015	97.1339	98.2715	99.2855	99.4298
10	93.5461	94.0072	88.2393	91.5014	96.0114	97.5300
11	91.0537	94.7253	91.1700	85.3061	95.2174	93.8819
12	95.3933	95.2009	97.0688	95.9866	96.8433	95.2486
13	95.2381	94.1452	95.0588	95.0588	96.6746	97.8208
14	97.2574	98.4881	98.0728	97.0402	96.5092	98.3368
15	77.3140	79.7858	83.1199	83.0685	78.8020	83.4154
16	98.3668	98.0964	99.3797	99.2583	97.4421	99.3750
OA (%)	90.3198	90.3321	90.3937	90.4347	91.3294	92.2411
Kappa	0.8907	0.8925	0.8901	0.8968	0.9047	0.9200

Table 12. The results for the Salinas dataset using 25% of the total number of features.

Class Number	MRMR	JOMIC	JMIM	CMIM	SDFE	Proposed
1	99.7743	99.4388	100.0000	99.9846	100.0000	100.0000
2	98.9971	99.1124	98.6382	98.4174	98.9393	99.8322
3	96.5324	96.8750	91.7559	92.9638	93.9297	96.1354
4	97.4724	97.9528	99.0461	98.4202	98.8906	99.3782
5	97.9305	99.0033	97.0906	98.5012	99.6619	99.5342
6	100.0000	99.9439	99.9438	99.8315	99.9438	99.9473
7	99.3180	99.4389	98.9467	99.7492	99.5644	99.9544
8	77.0343	75.3308	76.4241	75.4392	77.2064	80.7261
9	98.9362	99.0747	98.0524	97.1100	99.4294	99.4388
10	92.7785	93.4540	91.5855	93.6324	98.3138	97.7492
11	93.9759	96.4835	92.4406	92.7602	95.5789	98.9252
12	95.8567	95.8520	94.9283	96.0894	96.4126	98.7631
13	93.1034	95.6627	94.7743	96.8750	97.1429	99.3150
14	98.6813	95.8763	95.3586	97.3029	97.2973	98.5960
15	76.6654	80.1782	84.4185	84.4747	82.6695	84.2496
16	98.6076	98.2478	99.6264	99.5037	98.4029	99.5761
OA (%)	90.6893	90.7098	90.7960	90.7591	91.7607	93.2614
Kappa	0.8845	0.8914	0.8946	0.8935	0.9079	0.9251

Table 13. The results for the Longkou dataset using 10% of the total number of features.

Class Number	MRMR	JOMIC	JMIM	CMIM	SDFE	Proposed
1	96.3905	98.3974	73.6477	72.1819	98.6880	98.0647
2	88.8722	80.9339	38.6952	56.6265	86.0244	89.4598

Table 13. *Cont.*

Class Number	MRMR	JOMIC	JMIM	CMIM	SDFE	Proposed
3	95.2000	96.5035	0.0000	0.0000	96.8468	90.7834
4	94.9534	95.8103	72.4373	73.0363	97.4190	96.8578
5	74.8555	79.7403	0.0000	16.6667	75.0000	83.3333
6	95.7386	93.5909	0.0000	0.0000	93.8721	95.0276
7	99.9834	99.9834	84.6953	85.1456	100.0000	100.0000
8	87.3580	92.5262	0.0000	48.6330	91.6667	94.0458
9	70.6941	74.0634	38.3442	36.0714	86.9010	90.5724
OA (%)	95.4965	95.9963	72.9248	74.4785	96.9090	97.1208
Kappa	0.9731	0.9733	0.6845	0.7055	0.9793	0.9815

Table 14. The results for the Longkou dataset using 15% of the total number of features.

Class Number	MRMR	JOMIC	JMIM	CMIM	SDFE	Proposed
1	97.7496	97.6906	73.8964	74.6577	99.2295	99.3867
2	90.7609	91.1444	38.6128	58.0488	82.5495	89.7638
3	95.4802	98.8571	0.0000	0.0000	96.5665	95.7219
4	95.8955	96.1853	73.4588	75.8225	97.8005	96.4383
5	82.0442	81.6216	0.0000	25.0000	86.4407	85.7520
6	94.1784	93.5574	0.0000	0.0000	95.3447	96.2829
7	99.9834	99.9834	85.3745	87.0789	99.9503	99.9834
8	90.0000	90.3226	100.0000	50.1294	92.2272	94.4109
9	84.7458	83.3333	41.9118	38.4615	94.6844	93.0818
OA (%)	96.5721	96.6319	73.5984	76.5754	97.4413	97.4522
Kappa	0.9552	0.9560	0.6456	0.7110	0.9863	0.9867

Table 15. The results for the Longkou dataset using 20% of the total number of features.

Class Number	MRMR	JOMIC	JMIM	CMIM	SDFE	Proposed
1	96.3905	98.4955	86.5508	89.1432	98.8821	99.2585
2	88.8722	89.0357	56.8080	55.0672	85.9626	88.7850
3	95.2000	96.9231	0.0000	0.0000	99.9917	94.2731
4	94.9534	96.1932	83.9454	86.5783	97.4028	98.1774
5	74.8555	80.6283	3.8462	0.0000	82.4289	84.6753
6	95.7386	96.1825	0.0000	0.0000	96.3470	95.0673
7	99.9834	99.9834	92.4528	90.3115	99.9503	100.0000
8	87.3580	90.1306	60.3004	55.6107	93.2635	93.8160
9	70.6941	90.9091	73.4375	96.5217	95.3642	96.2865
OA (%)	95.4965	96.9035	84.5719	85.2890	97.4957	97.8596
Kappa	0.9731	0.9767	0.8060	0.8114	0.9849	0.9894

Table 16. The results for the Longkou dataset using 25% of the total number of features.

Class Number	MRMR	JOMIC	JMIM	CMIM	SDFE	Proposed
1	99.4819	99.3237	91.1993	90.2739	99.6143	99.1987
2	92.2865	87.6963	60.2791	60.7280	87.4667	87.3533
3	97.8723	99.4382	99.2701	97.7778	99.0868	97.7376
4	96.0814	96.7295	94.9780	95.6297	97.3448	98.3615
5	81.3602	79.3367	31.2500	14.0351	83.3333	86.5169
6	96.4674	96.3504	85.2273	69.6970	96.0253	97.7860
7	99.9669	99.9668	87.4238	87.3244	99.9669	99.9503
8	92.5150	92.3767	83.2187	83.4019	92.5595	91.8999
9	92.1311	86.7069	89.6970	78.4504	92.9012	97.5430
OA (%)	97.3055	97.1480	88.8309	88.3909	97.5771	98.0389
Kappa	0.9715	0.9852	0.8583	0.8535	0.9880	0.9910

4.4.1. The Result of the KSC Dataset

Tables 5–8 show the OA and Kappa coefficients for the KSC dataset using 10–25% of the total number of features.

Based on Tables 5–8, it is concluded that the proposed feature selection method outperforms MRMR, JOMIC, JMIM, CMIM, and SDFE in terms of the OA for different numbers of features, with an improvement of over 0.7%. Furthermore, when using 20% of the total number of features, the Kappa coefficient reaches 0.9, demonstrating that its OA is basically anastomotic with the labels. For 25% of the total number of features, other feature selection methods have an OA of below 91.2%, while the proposed method achieves the OA and Kappa coefficients exceeding 92.8% and 0.916, respectively. Moreover, the proposed method takes the OA of over 97% for five classes, with Willow swamp, Cattaial marsh, and Mudflats even reaching 98%. In summary, it is a practical feature selection method for the KSC dataset.

4.4.2. The Result of the Salinas dataset

Tables 9–12 present the OA and Kappa coefficients for the Salinas datasets using a fixed number of features.

The experimental results demonstrate that the proposed method outperforms other commonly used feature selection methods, achieving an OA of over 92% while obtaining the total number of features by less than 20%. Moreover, the Kappa coefficient for 25% of the total number of features is 0.2 higher than that of other methods, and the OA is higher for each class, with an OA of over 96% for all 14 classes. Notably, the samples of Brocoli_green_beads_1 are all correctly identified. These indicate that it achieves a prominent OA and Kappa coefficient for each class of the Salinas datasets, demonstrating the superiority of the proposed method.

4.4.3. The Result of the Longkou Dataset

Tables 13–16 present the OA and Kappa coefficients for the Longkou datasets using a fixed number of features.

Tables 13–16 present the OA and Kappa coefficients in the Longkou dataset. It is evident that the proposed method obtains prominent OA and Kappa coefficients, and it maintains a clear advantage in the classification of a small number of features. In the experimental comparison using 10% of the total number of features, MRMR, JOMIC, and SDFE achieve an OA of below 97%, while the proposed method achieves an OA of as high as 97.1%, which is 1.6%, 1.1% and 0.2% higher than MRMR, JOMIC, and SDFE, respectively. The OA of JMIM and CMIM is lower than 89%. The Kappa coefficient also demonstrates an overall advantage for the proposed method. Those results indicate that it is a robust and feasible feature selection method for the Longkou dataset.

5. Discussion

5.1. Design Analysis of the Proposed Method

EA is an effective strategy to obtain a feature combination of HSI datasets with a preferable OA in a limited time, the OA obtained on three HSI datasets exceeds 90%, and some even reach 98%. However, it is prone to stagnation during iteration due to the insufficient interactivity of agents. Co-evolution is a prominent mechanism to improve the agents' diversity, the original feature set is divided into some subsets, and agents are generated by those to form subpopulations. Moreover, information interaction exchanges the optimal feature combination searched by agents to maintain the balance of subpopulations, but solely exchanging the optimal feature combination reduces the selected probability of interactive features. The proposed method incorporates reliable information interaction and a series of mechanisms focusing on features to address this. The trajectory of the OA for each iteration indicates that the stability of the proposed method, is decreased by less than 0.5% as the feature space condenses, and the computational time is also reduced by an average of 15%.

The proposed method has a prominent OA in most of the classes and even reaches 100% for Brocoli_green_beads_1 in the KSC dataset and Water in the Longkou dataset. Although it is lower than other feature selection methods in a few classes, the difference is not apparent in the class with small samples. Although other methods based on measure criteria stand out in terms of efficiency, it is difficult to distinguish interactive features as the number of instances increases. Feature discarding is an effective mechanism for eliminating redundant information and improving the computational load. Similar to other feature selection methods, the OA is negatively impacted due to improper discarding. To counterbalance this effect, the feature recovering mechanism is employed to improve the generalization ability while maintaining the OA at a high level. Experimental results indicate that the OA of the proposed method surpasses other feature selection methods by an average of 3%, and important features are adequately restored by the feature recovery mechanism, thereby improving the performance and reliability of the proposed method.

5.2. Discussion for the Training Size

In Section 4.1, three HSI datasets, namely, KSC, Salinas, and Longkou, are introduced to validate the performance of the proposed method. The OA of effective feature combination and computation time is influenced by the size of the training set. Several tests are conducted on the proportion of the training set, ranging from 5% to 25%, because of the small-sample learning properties to determine the appropriate size of the training set. The change curves for the number of features and the OA of different training sets are shown in Figure 7.

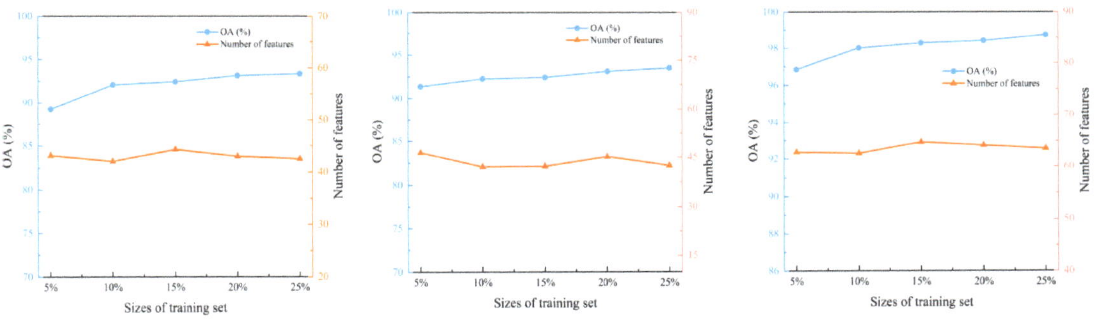

(a) (b) (c)

Figure 7. The average number of selected features and the OA under the different training sizes of three HSI datasets: (**a**) KSC, (**b**) Salinas, (**c**) Longkou.

The experimental results indicate that the increasing size of the training set from 5% to 10% leads to a significant improvement in the OA. However, further increasing the proportion from 10% to 25% only results in a minimal increase, while the computation time also decreases to some extent. Additionally, the number of selected features does not show significant fluctuations, so the size of the training set is designated as 10%. This size strikes a balance between the OA and computational load, making it a practical and effective choice for feature selection in HSI datasets.

5.3. Comparison with Other Co-Evolution Mechanisms

To verify the search efficiency of the co-evolution mechanism in the proposed method, it is compared with other co-evolution mechanisms named CC-DFS [33] and CC-RFG [34]; the average fitness value of each iteration on three HSI datasets is shown in Figure 8.

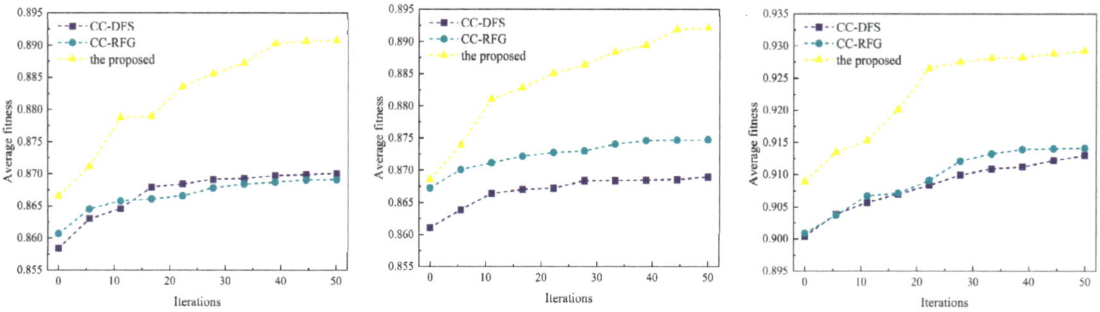

(**a**) (**b**) (**c**)

Figure 8. The average fitness value of different co-evolution mechanisms on three HSI datasets: (**a**) KSC, (**b**) Salinas, (**c**) Longkou.

In the beginning, the fitness value of the proposed method is higher than that of CC-DFS and CC-RFG in three HSI datasets, which demonstrates that the feature discard mechanism effectively removes the redundant features. With further iterations, the fitness trajectory of CC-DFS and CC-RFG gradually stabilizes while that of the proposed method remains upward. This indicates that the co-evolution mechanism enhances the search efficiency of agents and suggests a prominent ability to escape from local optima. As a result, the reliable co-evolution mechanism effectively interacts with more representative information, largely avoiding the occurrence of stagnation.

6. Conclusions

A feature selection method based on discarding–recovering and co-evolution mechanisms is proposed in this study with the aim of obtaining effective feature combinations in HSI datasets. According to the experimental results, the proposed method outperforms other EA-based feature selection methods, including PSO, FA, GWO, and GSA, in terms of optimization ability and search speed in the feature space. It achieves a prominent OA with a small number of selected features, outperforming other feature selection methods in this regard, and exhibits satisfied stability. In addition, through comparing with the other co-evolution mechanism, the fitness trajectory exhibits that the reliable co-evolution mechanism could interact with more representative information between agents, making them continuously improve. The performance limitations caused by feature discarding are improved through the recovery of dropped features, which guarantees the generalization ability and decreases computational load.

Furthermore, the proposed method outperforms MRMR, JOMIC, JMIM, CMIM, and SDFE in terms of the OA with varying numbers of features, and the reliable information interaction ensures a more balanced learning process, which maintains a positive balance between classification accuracy and the number of selected features, making it become a suitable choice for feature selection. In future studies, more representative criteria will be synthesized in the information interaction to further improve the diversity of agents. Moreover, it is interesting to use feature clustering to take the population division and further avoid population imbalance.

Author Contributions: Conceptualization, B.L. and M.W.; methodology, B.L. and Y.L.; software, M.W. and Y.L.; validation, B.L. and W.L.; formal analysis, Y.L.; investigation, B.L. and M.W.; resources, B.L. and X.G.; writing—original draft preparation, B.L. writing—review and editing, B.L. and M.W.; visualization, X.G. All authors have read and agreed to the published version of the manuscript.

Funding: This work was funded by the National Natural Science Foundation of China under Grant No. 41901296 and the Key Laboratory for National Geographic Census and Monitoring, National Administration of Surveying, Mapping and Geoinformation under Grant No. 2018NGCM06.

Data Availability Statement: The datasets presented in this paper can be obtained through https://www.ehu.eus/ccwintco/index.php/Hyperspectral_Remote_Sensing_Scenes.

Conflicts of Interest: The authors declare no conflict of interest.

References

1. Gwon, Y.; Kim, D.; You, H.; Nam, S.-H.; Kim, Y.D. A Standardized Procedure to Build a Spectral Library for Hazardous Chemicals Mixed in River Flow Using Hyperspectral Image. *Remote Sens.* **2023**, *15*, 477. [CrossRef]
2. Liu, J.; Li, Y.; Zhao, F.; Liu, Y. Hyperspectral Remote Sensing Images Feature Extraction Based on Spectral Fractional Differentiation. *Remote Sens.* **2023**, *15*, 2879. [CrossRef]
3. Wei, X.; Xiao, J.; Gong, Y. Blind Hyperspectral Image Denoising with Degradation Information Learning. *Remote Sens.* **2023**, *15*, 490. [CrossRef]
4. Wang, J.; Mao, X.; Wang, Y.; Tao, X.; Chu, J.; Li, Q. Automatic generation of pathological benchmark dataset from hyperspectral images of double stained tissues. *Opt. Laser Technol.* **2023**, *163*, 109331. [CrossRef]
5. Tang, C.; Liu, X.; Li, M.; Wang, P.; Chen, J.; Wang, L.; Li, W. Robust unsupervised feature selection via dual self-representation and manifold regularization. *Knowl. Based Syst.* **2018**, *145*, 109–120. [CrossRef]
6. Ruan, W.; Sun, L. Robust latent discriminative adaptive graph preserving learning for image feature extraction. *Knowl. Based Syst.* **2023**, *268*, 110487. [CrossRef]
7. Ba, J.; Wang, P.; Yang, X.; Yu, H.; Yu, D. Glee: A granularity filter for feature selection. *Eng. Appl. Artif. Intell.* **2023**, *122*, 106080. [CrossRef]
8. Ma, W.; Zhou, X.; Zhu, H.; Li, L.; Jiao, L. A two-stage hybrid ant colony optimization for high-dimensional feature selection. *Pattern Recognit.* **2021**, *116*, 107933. [CrossRef]
9. Cekik, R.; Uysal, A.K. A novel filter feature selection method using rough set for short text data. *Expert Syst. Appl.* **2020**, *160*, 113691. [CrossRef]
10. Cilia, N.D.; D'alessandro, T.; De Stefano, C.; Fontanella, F.; di Freca, A.S. Comparing filter and wrapper approaches for feature selection in handwritten character recognition. *Pattern Recognit. Lett.* **2023**, *168*, 39–46. [CrossRef]
11. Deng, T.; Huang, Y.; Yang, G.; Wang, C. Pointwise mutual information sparsely embedded feature selection. *Int. J. Approx. Reason.* **2022**, *151*, 251–270. [CrossRef]
12. Paja, W. Generational Feature Elimination to Find All Relevant Feature Subset. *Syst. Technol.* **2017**, *72*, 140–148. [CrossRef]
13. Aranha, C.; Villalón, C.L.C.; Campelo, F.; Dorigo, M.; Ruiz, R.; Sevaux, M.; Sörensen, K.; Stützle, T. Metaphor-based metaheuristics, a call for action: The elephant in the room. *Swarm Intell.* **2022**, *16*, 1–6. [CrossRef]
14. Qin, Y.; Li, Z.; Ding, J.; Zhao, F.; Meng, M. Automatic optimization model of transmission line based on GIS and genetic algo-rithm. *Array* **2023**, *17*, 100266. [CrossRef]
15. Zheng, K.; Zhang, Q.; Peng, L.; Zeng, S. Adaptive memetic differential evolution-back propagation-fuzzy neural network algo-rithm for robot control. *Inf. Sci.* **2023**, *637*, 118940. [CrossRef]
16. Ong, P.; Zainuddin, Z. An optimized wavelet neural networks using cuckoo search algorithm for function approximation and chaotic time series prediction. *Decis. Anal. J.* **2023**, *6*, 100188. [CrossRef]
17. Qu, L.; He, W.; Li, J.; Zhang, H.; Yang, C.; Xie, B. Explicit and size-adaptive PSO-based feature selection for classification. *Swarm Evol. Comput.* **2023**, *77*, 101249. [CrossRef]
18. Su, F.; Duan, C.; Wang, R. Analysis and improvement of GSA's optimization process. *Appl. Soft Comput.* **2021**, *107*, 107367. [CrossRef]
19. Al-Tashi, Q.; Md Rais, H.; Abdulkadir, S.-J.; Mirjalili, S.; Alhussian, H. A Review of Grey Wolf Optimizer-Based Feature Se-lection Methods for Classification. *Evol. Mach. Learn. Tech. Algorithms Intell. Syst.* **2017**, *12*, 273–286.
20. Chary, V.; Rosalina, K. Analysis of transmission line modeling routines by using offsets measured least squares regression ant lion optimizer. ORPD and ELD problems. *Heliyon* **2023**, *9*, 13387. [CrossRef]
21. Bezginov, A.; Clark, G.W.; Charlebois, R.L.; Dar, V.-U.; Tillier, E.R. Coevolution Reveals a Network of Human Proteins Originating with Multicellularity. *Mol. Biol. Evol.* **2013**, *30*, 332–346. [CrossRef]
22. Qi, S.; Wang, R.; Zhang, T.; Dong, N. Cooperative coevolutionary competition swarm optimizer with perturbation for high-dimensional multi-objective optimization. *Inf. Sci.* **2023**, *644*, 119253. [CrossRef]
23. Zhong, R.; Zhang, E.; Munetomo, M. Cooperative coevolutionary differential evolution with linkage measurement minimization for large-scale optimization problems in noisy environments. *Complex Intell. Syst.* **2023**, *573*, 1–18. [CrossRef]
24. Tian, J.; Li, M.; Chen, F. Dual-population based coevolutionary algorithm for designing RBFNN with feature selection. *Expert Syst. Appl.* **2010**, *37*, 6904–6918. [CrossRef]
25. Too, J.; Abdullah, A.R.; Saad, N.M. A New Co-Evolution Binary Particle Swarm Optimization with Multiple Inertia Weight Strategy for Feature Selection. *Informatics* **2019**, *6*, 21. [CrossRef]
26. Wu, Y.; Xue, W.; Xu, L.; Guo, X.; Xue, D.; Yao, Y.; Zhao, S.; Li, N. Optimized least-squares support vector machine for predicting aero-optic imaging deviation based on chaotic particle swarm optimization. *Optik* **2020**, *206*, 163215. [CrossRef]
27. Su, H.; Li, Q.; Du, P. Hyperspectral band selection using firefly algorithm. In Proceedings of the 2014 6th Workshop on Hyperspectral Image and Signal Processing: Evolution in Remote Sensing, Lausanne, Switzerland, 24-27 June 2014; pp. 1–4.

28. Xie, F.; Li, F.; Lei, C.; Yang, J.; Zhang, Y. Unsupervised band selection based on artificial bee colony algorithm for hyperspectral image classification. *Appl. Soft Comput.* **2019**, *75*, 428–440. [CrossRef]
29. Wang, M.; Liu, W.; Chen, M.; Huang, X.; Han, W. A band selection approach based on a modified gray wolf optimizer and weight updating of bands for hyperspectral image. *Appl. Soft Comput.* **2021**, *112*, 107805. [CrossRef]
30. Tschannerl, J.; Ren, J.; Yuen, P.; Sun, G.; Zhao, H.; Yang, Z.; Wang, Z.; Marshall, S. MIMR-DGSA: Unsupervised hyperspectral band selection based on information theory and a modified discrete gravitational search algorithm. *Inf. Fusion* **2019**, *51*, 189–200. [CrossRef]
31. Song, X.-F.; Zhang, Y.; Guo, Y.-N.; Sun, X.-Y.; Wang, Y.-L. Variable-Size Cooperative Coevolutionary Particle Swarm Optimization for Feature Selection on High-Dimensional Data. *IEEE Trans. Evol. Comput.* **2020**, *24*, 882–895. [CrossRef]
32. Zhao, F.; Bao, H.; Wang, L.; Cao, J.; Tang, J. A multipopulation cooperative coevolutionary whale optimization algorithm with a two-stage orthogonal learning mechanism. *Knowl. Based Syst.* **2022**, *246*, 108664. [CrossRef]
33. Zhou, Y.; Kang, J.; Zhang, X. A Cooperative Coevolutionary Approach to Discretization-Based Feature Selection for High-Dimensional Data. *Entropy* **2020**, *22*, 613. [CrossRef] [PubMed]
34. Rashid, A.N.M.B.; Ahmed, M.; Sikos, L.F.; Haskell-Dowland, P. Cooperative co-evolution for feature selection in Big Data with random feature grouping. *J. Big Data* **2020**, *7*, 107. [CrossRef]
35. Fernández, D.; Adermann, E.; Pizzolato, M.; Pechenkin, R.; Rodríguez, C.G.; Taravat, A. Comparative Analysis of Machine Learning Algorithms for Soil Erosion Modelling Based on Remotely Sensed Data. *Remote Sens.* **2023**, *15*, 482. [CrossRef]
36. Guo, Y.; Zhang, Z.; Tang, F. Feature selection with kernelized multi-class support vector machine. *Pattern Recognit.* **2021**, *117*, 107988. [CrossRef]
37. Chen, T.-Y. An evolved VIKOR method for multiple-criteria compromise ranking modeling under T-spherical fuzzy uncertainty. *Adv. Eng. Inform.* **2022**, *54*, 101802. [CrossRef]
38. Yan, X.; Jia, M. Intelligent fault diagnosis of rotating machinery using improved multiscale dispersion entropy and mRMR feature selection. *Knowl. Based Syst.* **2019**, *163*, 450–471. [CrossRef]
39. Maldonado, J.; Riff, M.C.; Neveu, B. A review of recent approaches on wrapper feature selection for intrusion detection. *Expert Syst. Appl.* **2022**, *198*, 116822. [CrossRef]
40. Liu, Z.; Yang, J.; Wang, L.; Chang, Y. A novel relation aware wrapper method for feature selection. *Pattern Recognit.* **2023**, *140*, 109566. [CrossRef]
41. Maha, N.; Ghaith, M.; Ouajdi, K. Advances in nature-inspired metaheuristic optimization for feature selection problem: A comprehensive survey. *Comput. Sci. Rev.* **2023**, *49*, 100559.
42. Zhuang, Z.; Pan, J.-S.; Li, J.; Chu, S.-C. Parallel binary arithmetic optimization algorithm and its application for feature selection. *Knowl. Based Syst.* **2023**, *275*, 110640. [CrossRef]
43. Du, W.; Ma, J.; Yin, W. Orderly charging strategy of electric vehicle based on improved PSO algorithm. *Energy* **2023**, *271*, 127088. [CrossRef]
44. Kumar, V.; Kumar, D. A Systematic Review on Firefly Algorithm: Past, Present, and Future. *Arch. Comput. Methods Eng.* **2021**, *28*, 3269–3291. [CrossRef]
45. Achom, A.; Das, R.; Pakray, P. An improved Fuzzy based GWO algorithm for predicting the potential host receptor of COVID-19 infection. *Comput. Biol. Med.* **2022**, *151*, 106050. [CrossRef] [PubMed]
46. Biabani, F.; Shojaee, S.; Hamzehei-Javaran, S. A new insight into metaheuristic optimization method using a hybrid of PSO, GSA, and GWO. *Structures* **2022**, *44*, 1168–1189. [CrossRef]
47. Esmaeili, A.; Hamidi, J.K.; Mousavi, A. Determination of sublevel stoping layout using a network flow algorithm and the MRMR classification system. *Resour. Policy* **2023**, *80*, 103265. [CrossRef]
48. Robindro, K.; Clinton, U.B.; Hoque, N.; Bhattacharyya, D.K. JoMIC: A joint MI-based filter feature selection method. *J. Comput. Math. Data Sci.* **2023**, *6*, 100075. [CrossRef]
49. Kumar, C.; Chatterjee, S.; Oommen, T.; Guha, A. Automated lithological mapping by integrating spectral enhancement techniques and machine learning algorithms using AVIRIS-NG hyperspectral data in Gold-bearing granite-greenstone rocks in Hutti, India. *Int. J. Appl. Earth Obs. Geoinf.* **2020**, *86*, 102006.
50. Souza, F.; Premebida, C.; Araújo, R. High-order conditional mutual information maximization for dealing with high-order dependencies in feature selection. *Pattern Recognit.* **2022**, *131*, 108895. [CrossRef]
51. Zhou, L.; Ma, X.; Wang, X.; Hao, S.; Ye, Y.; Zhao, K. Shallow-to-Deep Spatial–Spectral Feature Enhancement for Hyperspectral Image Classification. *Remote Sens.* **2023**, *15*, 261. [CrossRef]

remote sensing

MDPI

Article

National-Standards- and Deep-Learning-Oriented Raster and Vector Benchmark Dataset (RVBD) for Land-Use/Land-Cover Mapping in the Yangtze River Basin

Pengfei Zhang [1], Yijin Wu [1,*], Chang Li [1,*], Renhua Li [2], He Yao [2], Yong Zhang [2], Genlin Zhang [1] and Dehua Li [1]

[1] Key Laboratory for Geographical Process Analysis & Simulation of Hubei Province, College of Urban and Environmental Science, Central China Normal University, Wuhan 430079, China; zhangpf@mails.ccnu.edu.cn or pengf_zhang@126.com (P.Z.); glz@mails.ccnu.edu.cn (G.Z.); lidehua@mails.ccnu.edu.cn (D.L.)

[2] Yangtze River Basin Monitoring Center Station for Soil and Water Conservation, Changjiang Water Resources Commission, Wuhan 430010, China; cwjczxz@163.com (R.L.); yaoxiaohe@163.com (H.Y.); yongzchn@163.com (Y.Z.)

* Correspondence: wuyijin@ccnu.edu.cn (Y.W.); lcshaka@126.com or lichang@ccnu.edu.cn (C.L.)

Abstract: A high-quality remote sensing interpretation dataset has become crucial for driving an intelligent model, i.e., deep learning (DL), to produce land-use/land-cover (LULC) products. The existing remote sensing datasets face the following issues: the current studies (1) lack object-oriented fine-grained information; (2) they cannot meet national standards; (3) they lack field surveys for labeling samples; and (4) they cannot serve for geographic engineering application directly. To address these gaps, the national-standards- and DL-oriented raster and vector benchmark dataset (RVBD) is the first to be established to map LULC for conducting soil water erosion assessment (SWEA). RVBD has the following significant innovation and contributions: (1) it is the first second-level object- and DL-oriented dataset with raster and vector data for LULC mapping; (2) its classification system conforms to the national industry standards of the Ministry of Water Resources of the People's Republic of China; (3) it has high-quality LULC interpretation accuracy assisted by field surveys rather than indoor visual interpretation; and (4) it could be applied to serve for SWEA. Our dataset is constructed as follows: (1) spatio-temporal-spectrum information is utilized to perform automatic vectorization and label LULC attributes conforming to the national standards; and (2) several remarkable DL networks (DenseNet161, HorNet, EfficientNetB7, Vision Transformer, and Swin Transformer) are chosen as the baselines to train our dataset, and five evaluation metrics are chosen to perform quantitative evaluation. Experimental results verify the reliability and effectiveness of RVBD. Each chosen network achieves a minimum overall accuracy of 0.81 and a minimum Kappa of 0.80, and Vision Transformer achieves the best classification performance with overall accuracy of 0.87 and Kappa of 0.86. It indicates that RVBD is a significant benchmark, which could lay a foundation for intelligent interpretation of relevant geographic research about SWEA in the Yangtze River Basin and promote artificial intelligence technology to enrich geographical theories and methods.

Keywords: remote sensing dataset; deep learning; soil water erosion assessment; object-oriented image classification; land-use/land-cover mapping

Citation: Zhang, P.; Wu, Y.; Li, C.; Li, R.; Yao, H.; Zhang, Y.; Zhang, G.; Li, D. National-Standards- and Deep-Learning-Oriented Raster and Vector Benchmark Dataset (RVBD) for Land-Use/Land-Cover Mapping in the Yangtze River Basin. *Remote Sens.* **2023**, *15*, 3907. https://doi.org/10.3390/rs15153907

Academic Editors: Qian Du, Yanni Dong and Xiaochen Yang

Received: 8 June 2023
Revised: 3 August 2023
Accepted: 5 August 2023
Published: 7 August 2023

1. Introduction

Soil water erosion has become a serious environmental hazard around the global world, which impacts climate change, agricultural production, and socio-economic-ecological sustainable development [1–3]. Land-use/land-cover (LULC) information reflects the interaction between human activities and natural ecosystems, which has been identified as a decisive factor to accelerate global land degradation and soil water erosion [4]. The People's Republic of China faces a critical challenge from soil water erosion [5,6]. Based on deep learning (DL) and remote sensing technology, high-accuracy LULC information could be extracted to perform large-scale and extensive intelligent ground monitoring, which is beneficial to conducting soil water erosion assessment (SWEA) in a cost-efficient manner [7]. Compared with traditional models that rely on statistics or physical knowledge, DL networks are trained with massive samples that could automatically learn remote sensing parameter characteristics of ground objects [8–11]. This means that the performance of DL-based approaches strongly depends on the quality and quantity of the provided dataset [12]. Therefore, the construction of a high-quality remote sensing image interpretation dataset contributes to enhancing the generalization ability of DL networks and further improving the accuracy of LULC mapping.

At present, many research institutions and scholars are devoted to remote sensing dataset research for mapping LULC. There are two main categories of datasets: scene classification datasets and object detection datasets. Object detection datasets focus on recognizing ground objects with bounding boxes to predict the location and LULC categories [13], such as FAIR1M [12], TAS [14], ImageNet [15], PASCCAL VOC [16], SZTAKI-INRIA [17], MSCOCO [18], UCAS-AOD [19], DLR 3K [20], NWPU VHR-10 [21], VEDAI [22], HRSC2016 [23], COWC [24], RSOD [25], LEVIR [26], ITCVD [27], DOTA [28], DIOR [14], and RSSOD [29]. Scene classification is for categorizing remote sensing images into a series of LULC categories with the image patches [29], such as UC Merced Land-Use [30], WHU-RS19 [31], RSSCN7 [32], Brazilian Coffee Scene [33], RSC11 [34], SIRI-WHU [35], NWPU-RESISC45 [29], RSD46-WHU [25], AID [36], AID++ [37], OPTIMAL-31 [38], PatternNet [39], OSAR [40], RSI-CB [41], DIOR [42], Eurosat [43], Bigearthnet [44], MLRSNet [45], BigEarthNet-MM [46], MRSB [47], AIFS-DATASET [48], MRSID [49], and LuoJiaSET [50]. The available datasets could offer great potential in mitigating the highly nonlinear and overparameterized restrictions of DL networks [51].

It is crucial and urgently necessary to construct thematic remote sensing datasets for conducting SWEA. There still exists an application gap between geographic research and specific geo-engineering applications. It is worth noting that different scholars or institutions have constructed diverse LULC classification systems to meet their respective research needs. However, this diversity of current datasets is not conducive to their universality across various geographical researches. To be specific, there is still a lack of thematic remote sensing datasets for conducting SWEA. Hence, it is crucial and urgently needed to construct datasets in accordance with a standard and authoritative LULC classification system. This could improve the application and universality of remote sensing datasets for SWEA research.

It is crucial and urgently necessary to construct refined fine-grained LULC datasets, which can significantly enhance the monitoring and assessing capacities of soil water erosion. LULC data are a significant factor for calculating the Chinese soil loss equation (CSLE) model [5,52]. It is also recognized as the authoritative model to quantitatively evaluate the magnitude and distribution of soil erosion in the People's Republic of China, and it adopts the LULC classification system conforming to the national industry standards of the Standards for Classification and Gradation of Soil Erosion (SL 190-2007) published by the Ministry of Water Resources of the People's Republic of China. It is generally regarded as the vegetation cover and biological practice factor, which reflects the impact of vegetation cover and biological practice on erosion rate under fallow conditions [47]. Therefore, improving the accuracy of LULC mapping directly affects the monitoring level of SWEA. The development of the fine-grained (i.e., second-level LULC) dataset is beneficial to implementing high-accuracy LULC mapping. It can further enhance the intra-class similarity and improve the inter-class variability, which promotes the development of the LULC classification level and, in turn, further improves the accuracy of SWEA.

It is crucial and urgently necessary to improve the quality of LULC labeling for the remote sensing dataset. The correction of sample labeling is very important in remote-sensing application research [53]. DL models rely on training using numerous labeled data to yield the high classification accuracy [54]. Due to the phenomenon of the same object with different spectrum and the foreign body with the same spectrum, different ground objects in remote sensing images have similarities in respect of color, texture, size, shape, shadow, and distribution position. That generally results in errors of LULC labeling. To alleviate the issue, some researchers have delved into various aspects, as follows: on one hand, much research adopted quality control methods to reduce human errors [55–65]. For example, Qi et al. [45] rely on many technicians to train data labeling several times until reaching a predetermined and reliable confidence score. On the other hand, remote sensing interpretation keys have a great effect on image labeling [66–75]. Remote sensing interpretation keys are sampled through field survey, which can provide accurate interpretation reference and real LULC category information rather than relying on human visual interpretation. However, there are fewer studies that rely on field surveys to improve the quality of data labeling.

In conclusion, fine-grained remote sensing classification dataset research still faces the following scarcities:

(1) Lacking object-oriented fine-grained datasets for DL-based LULC mapping. Current remote sensing scene classification datasets and object detection datasets primarily emphasize recognizing the LULC category and spatial position information. The fixed image patches generally contain heterogeneous objects, which means that the detailed geometrical information of ground objects, such as object boundaries, is still missing. That limits the progression of high-accuracy LULC mapping.

(2) Lacking datasets conforming to national LULC classification standards. The classification systems of current remote sensing datasets are diverse and are formulated according to different research needs. This diversity in LULC classification systems hinders the broader applicability of these datasets in other related fields, such as agricultural production and socio-economic–ecologically sustainable development [1–3]. Thus, developing datasets based on universal and authoritative standards, such as national industry standards, is essential to enhance their universality and application value.

(3) Lacking field surveys for LULC dataset labeling. Current remote sensing datasets generally depend on professional technicians to label the LULC category, which means that there is no process for field surveys to verify the correctness of the labeled LULC. The subjectivity of professional technicians and the complexity of remote sensing images contribute to the degradation of data labeling quality. Incorrect data labeling greatly influences the training of DL networks and reduces classification accuracy. Thus, improving the quality of image labeling will significantly enhance the quality of the dataset.

(4) Lacking datasets meeting the engineering application requirement to conduct SWEA in the Yangtze River Basin. The application gap between current published datasets and SWEA application is still not addressed. It is worth noting that there is no thematic dataset available for conducting SWEA. In addition, some samples of representative LULC categories (e.g., sloping cropland) are not sampled in current datasets, which plays an important role in soil and water conservation.

To solve the above issues, a second-level standards-of-the-Ministry-of-Water-Resources- and DL-oriented raster and vector benchmark dataset (RVBD) is the first to be established to perform LULC mapping for supporting SWEA in the Yangtze River Basin, which adheres to the notional standards published by the Ministry of Water Resources of the People's Republic of China and meets the requirements of DL. The main remarkable innovation and contributions are shown as follows:

(1) A second-level object- and DL-oriented dataset with raster and vector data is first to be established for large-scale LULC mapping to the best of our knowledge. Different from the current datasets only containing remote sensing image patches, RVBD also includes vector data. In addition, image patches from open-source Google images are homogeneous objects with geometric boundary information, which can be directly applied for mapping LULC.

(2) An LULC dataset conforming to the national industry standards is the first to be established to the best of our knowledge. The classification system of RVBD is

constructed following the water resources industry standard of the People's Republic of China, i.e., the Current Land Use Classification (GB/T 21010-2017). It is significant for improving the universality of RVBD and the application value.

(3) A high-quality LULC labeling dataset with the assist of remote sensing interpretation keys is the first to be established to the best of our knowledge. Remote sensing interpretation keys are sampled through field surveys to facilitate the interpretation of LULC categories by indoor technicians. It is equally important that the correctness of sample labeling is verified through a field survey, which significantly improves the quality of sample labeling.

(4) RVBD is the first to lay an intelligent foundation for high-accuracy LULC mapping to support SWEA to the best of our knowledge. It greatly improves the application value of RVBD. Particularly, geographical theories and methods are further enriched based on artificial intelligence (AI) technology.

2. Raster and Vector Benchmark Dataset (RVBD)

2.1. Description of RVBD

RVBD is the first to be constructed to serve for SWEA in the Yangtze River Basin of the People's Republic of China. RVBD contains 23,300 pairs of labeled samples with corresponding remote sensing image patches and vector data. Vector data are utilized as masks to generate image patches by cropping remote sensing images with geometric shapes. The remote sensing images were downloaded from Google Earth with three spectral bands (i.e., red, green, and blue bands), which were obtained in the year 2020. The image resolution is 2 m. It is worth noting that the cloud cover area of the images does not exceed 10% of the overall regions. It includes 15 LULC categories: paddy land, dry land, sloping cropland, garden land, forest, shrub land, grassland, urban construction land, rural construction land, mining land, other construction land, rural road, other transportation land, water, and barren land. Some examples of each LULC category of RVBD are shown in Figure 1.

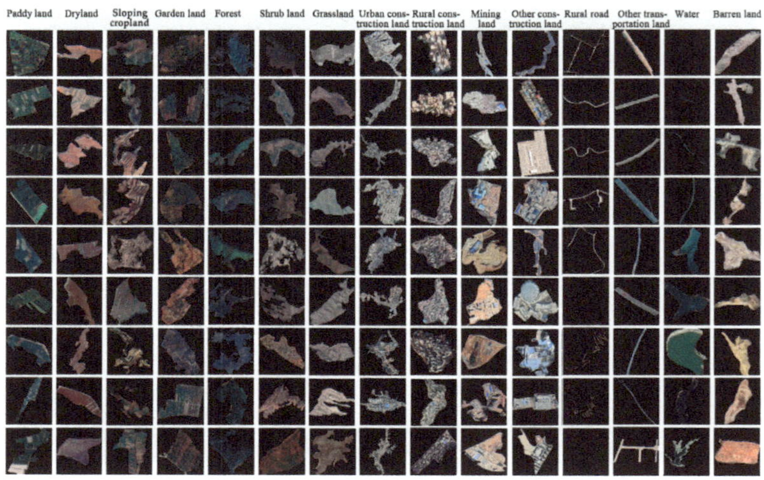

(a)

Figure 1. *Cont.*

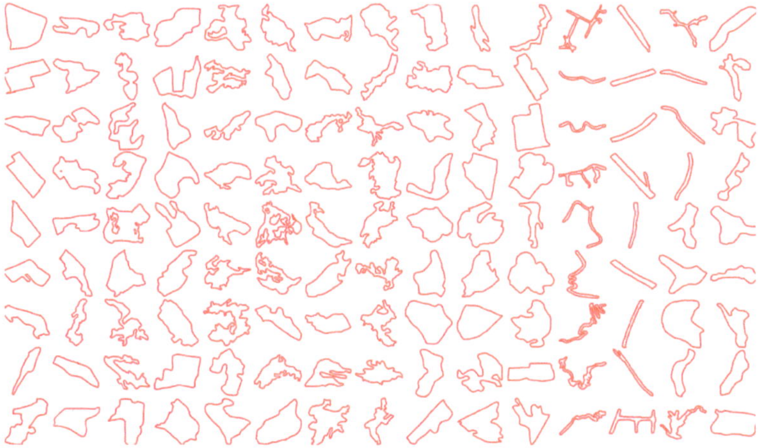

(b)

Figure 1. Some examples of each LULC category in RVBD are shown: (**a**) represents the remote sensing image patches of samples and (**b**) represents the vector data of samples.

2.2. Classification System

In order to facilitate its application in remote sensing monitoring of soil and water loss in the Yangtze River Basin, as well as to provide high-accuracy LULC mapping information for relevant geographical research in the area, the classification system of RVBD could be developed in strict compliance with the industry standards set by the Ministry of Water Resources of China. The classification system of the dataset is implemented in accordance with the national standard (i.e., the Technical Specification of Soil and Water Conservation Monitoring by Remote Sensing (SL 592-2012)). It could be applicable for soil and water conservation monitoring at various scales within China, including national, watershed, and region scales. The standard reference, the Current Land Use Classification (GB/T 21010-2017), combined with the characteristics of the soil and water conservation industry, has formed a suitable LULC classification system for SWEA. This classification system could be applied to calculate soil and water loss and provide data support for ecological environment monitoring. Table 1 shows the classification system of each category in detail with some samples of remote sensing interpretation keys from field surveys.

Table 1. The LULC classification system of the RVBD is described in detail.

Level I	Level II	Remote Sensing Interpretation Key	
Name	**Name**	**Remote Sensing Image**	**Photo**
Cultivated land	Paddy land	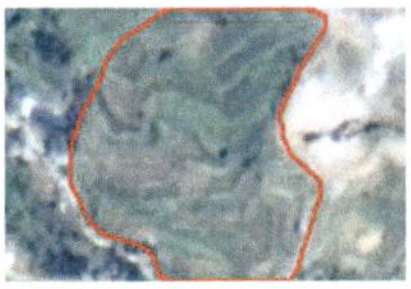	
	Dry land		

Table 1. *Cont.*

Level I	Level II	Remote Sensing Interpretation Key	
Name	**Name**	**Remote Sensing Image**	**Photo**
Cultivated land	Sloping cropland		
Garden land	Garden land		
Forest	Forest		
	Shrub land		
Grassland	Grassland	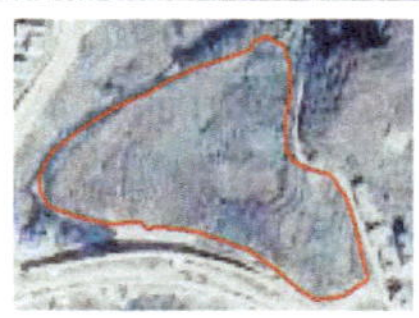	
Construction land	Urban construction land		
	Rural construction land	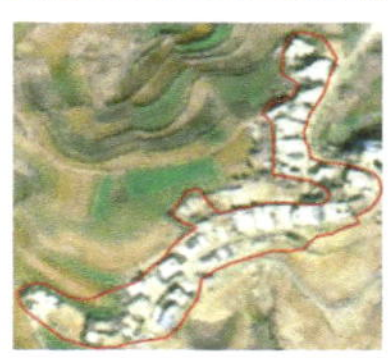	

Table 1. *Cont.*

Level I	Level II	Remote Sensing Interpretation Key	
Name	**Name**	**Remote Sensing Image**	**Photo**
Construction land	Mining land	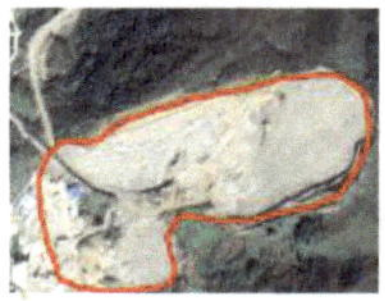	
	Other construction land		
Transportation land	Other transportation land	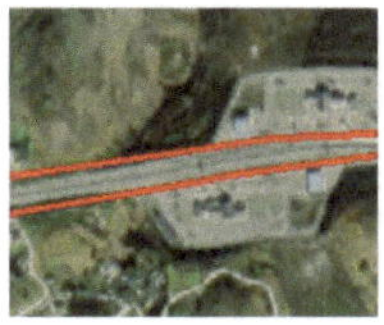	
	Rural road		
Water	Water	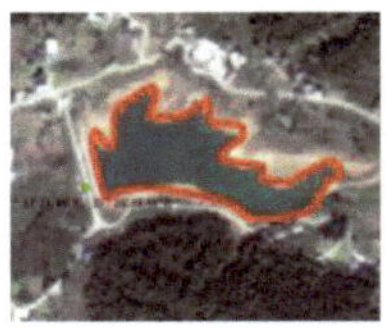	
Other land	Barren land		

2.3. Dataset Splits

The RVBD contains 23,300 pairs of samples with image patches and a vector dataset. The number of different LULC categories of RVBD varies greatly from 800 to 2000, and the sample number statistics for each LULC category is shown in Figure 2. The RVBD is divided into a training set with 13,980 pairs of objects, a validation set with 4660 pairs of objects, and a test set with 4660 pairs of objects, according to the ratio of 6:2:2, and remote sensing image patches are utilized for model training, model optimization, and model evaluation of DL baseline networks, respectively. The detailed splits are reported in Table 2.

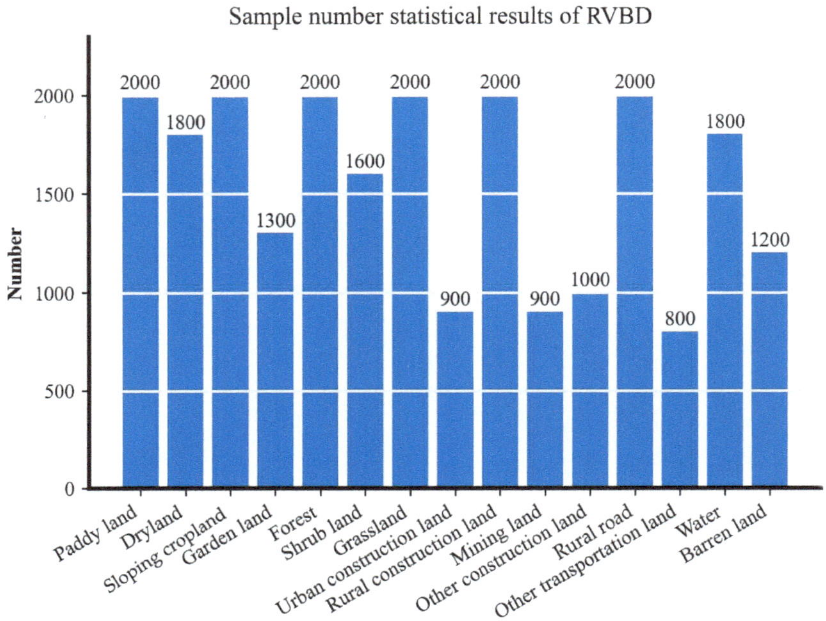

Figure 2. The sample number of each LULC category in RVBD.

Table 2. The number of objects of RVBD for each category and dataset split.

LULC Category	Training Set	Validation Set	Test Set
Paddy land	1200	400	400
Dry land	1080	360	360
Sloping cropland	1200	400	400
Garden	780	260	260
Forest	1200	400	400
Shrub land	960	320	320
Grassland	1200	400	400
Urban construction land	540	180	180
Rural construction land	1200	400	400
Mining land	540	180	180
Other construction land	600	200	200
Other transportation land	1200	400	400
Rural road	480	160	160
Water	1080	360	360
Barren land	720	240	240
Total number	13,980	4660	4660

2.4. Study Area

In this research, the study area is located in the middle and lower reaches of the Jinsha River Basin. It is one of the areas with serious soil water erosion in the Yangtze River Basin, and 13 counties are selected that run through Yunnan, Guizhou, and Sichuan provinces. Soil water erosion has become a critical factor hindering the development of the regional economy and society. The complex geological and geomorphic environment of the region, coupled with its various climatic conditions, have fostered a wide range of natural landscapes. In addition, the region is prosperous in both economy and culture, which is a key area of the Yangtze River Economic Belt. Thus, the complex interaction between natural and human activities has resulted in the formation of rich and diverse LULC patterns in this region, which is beneficial for selecting samples conforming to the national industry standards and constructing a diverse and representative dataset. The study area is shown in Figure 3.

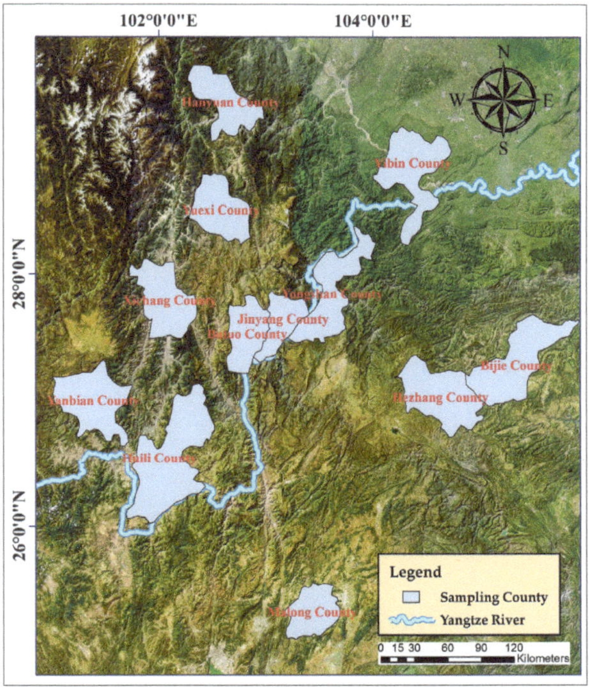

Figure 3. Study area.

2.5. Field Surveys

Field surveys are carried out to establish a representative, practical, and stable dataset in accordance with the industry standards, the Technical Specification of Soil and Water Conservation Monitoring by Remote Sensing (SL 592-2012) of the Ministry of Water Resources of the People's Republic of China. It aims to (1) sample remote sensing interpretation keys for indoor professional visual interpretation and (2) verify the correctness of labeled samples. Field surveys are sampled as follows: (1) global positioning system (GPS) points are set up according to the principle of uniform distribution in space. It is worth noting that GPS points are selected in the areas with wide vision and covering a rich LULC category. (2) Field photos and remote sensing image samples are sampled or used to verify the LULC category information. Remote sensing interpretation keys are beneficial to verify the image characteristics (i.e., color, shadow, texture, size, shape, location, etc.) and could provide exact LULC category information to assist indoor technicians in improving the correctness of labeling LULC categories of images. Field verification is very meaningful for reducing the errors caused by the phenomenon of having the same objects with different spectra and foreign bodies with the same spectra. The GPS-sampled points of the field survey are shown in Figure 4.

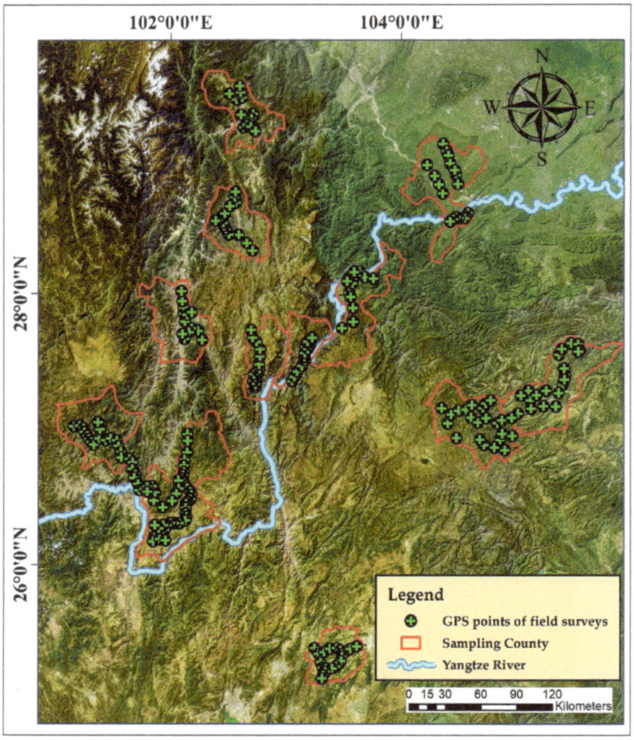

Figure 4. Sampling points of field survey.

3. Methodology

The overall workflow of the dataset construction method is shown in Figure 5, which mainly includes the following two contents:

(1) Remote sensing dataset construction driven by spatio-temporal spectrum information

Various pieces of geographic information driven by spatio-temporal spectrum big data are utilized to construct the RVBD. Radiation knowledge is provided from remote sensing images to implement large-scale monitoring for SWEA. Thematic geometry knowledge is provided from volunteered geographic information, which is regarded as the mask data to yield the high-accuracy geometry vector data, such as road, river, and construction data. Then, automatic vectorization is performed for unmasked areas based on the multi-resolution segmentation approach. Finally, LULC attribute information is labeled by professional technicians with the assist of expert knowledge from remote sensing interpretation keys.

(2) Dataset evaluation based on DL

Five outstanding DL networks obtained from two different architectures, i.e., convolutional neural network (CNN) and Transformer, are chosen to evaluate the effectiveness of the established RVBD. This dataset is divided into a training set, a validation set, and a test set for DL network training and parameter optimization. Then, the manual labeling errors and machine errors generated by the DL networks are correcting based on field surveys. Finally, the accuracy evaluation is performed to verify the effectiveness of the RVBD.

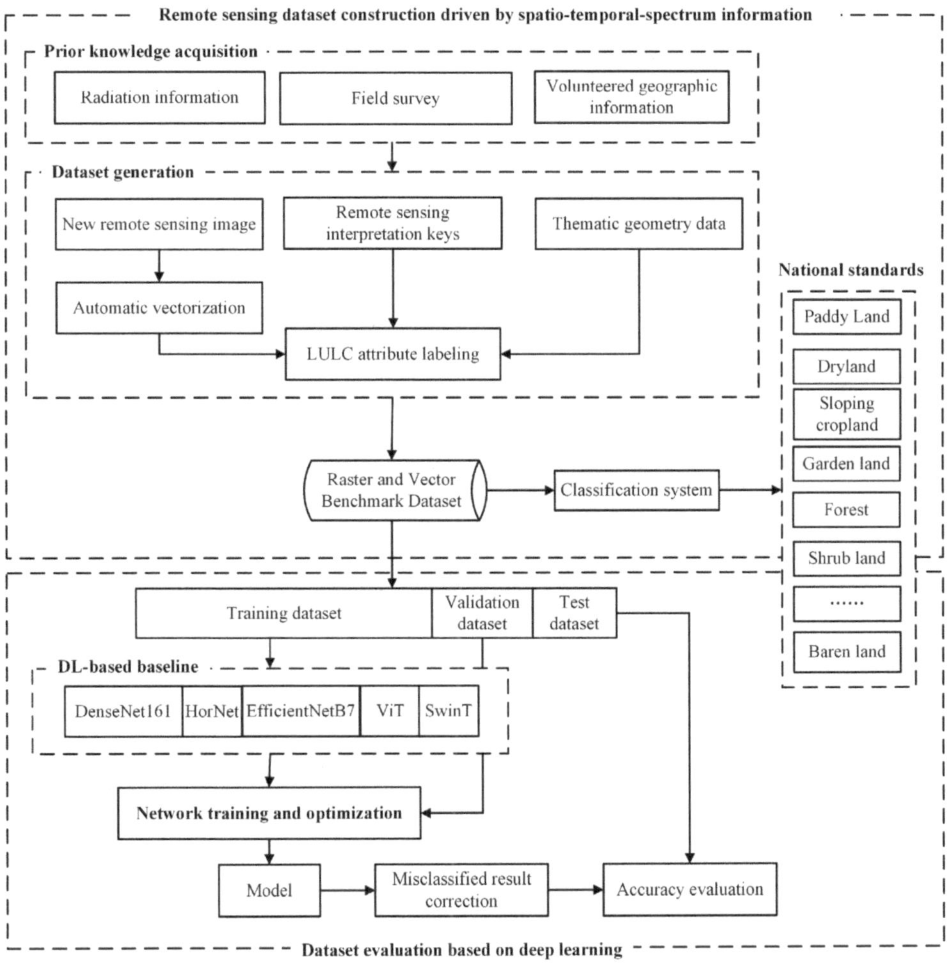

Figure 5. The overall workflow.

3.1. Remote Sensing Dataset Construction Driven by Spatio-Temporal Spectrum Information

(1) Prior knowledge acquisition from spatio-temporal spectral big data

Prior knowledge is conducive to alleviate the problem that the acquisition of large-scale labeled training data is laborious and expensive [76]. In the era of geographic big data, prior knowledge is obtained from spatio-temporal spectral big data to serve the construction of a remote sensing dataset, which includes various pieces of strictly calibrated geospatial information (e.g., radiation knowledge, thematic geometric knowledge, and expert knowledge). The spatio-temporal spectral big data provide the following data: remote sensing images are downloaded from open-source Google images and processed by strict geometric calibration and radiation calibration, high-accuracy volunteered geographic information from OpenStreetMap (OSM), and historical remote sensing interpretation keys sampled by field surveys. In addition, there are still much remote sensing data available that can be further explored, such as hyperspectral images. It is worth noting that they contain high-dimensional and multispectral information, which includes redundant data. This could increase the computational demands of DL networks and impede the convergence of DL training. Therefore, it is highly important to perform the dimensionality reduction methods, such as principal component analysis (PCA) [77].

(2) Thematic geometry masking by volunteered geographic information

Volunteered geographic information (VGI) can provide high-accuracy geographic thematic data, which are widely recognized for the reliability, availability, and time efficiency of the data acquisition [78]. OpenStreetMap data are generally regarded as the most active and widely applied VGI data [79], which can be collected by both professional and amateur volunteers. Leveraging OSM data could save time and money on large-scale sample labeling. The accurate thematic vector data (i.e., road vector, river vector, and construction vector) could be yielded from OSM, and geographical registration is performed combining the aforementioned vector data with remote sensing images to minimize geometric errors. Then, remote sensing images can be masked with the above vector to obtain prior thematic vector regions.

(3) Automatic vectorization based on multi-resolution segmentation

The multi-resolution segmentation approach is adopted to implement automatic vectorization for the remained unmasked regions [80]. It is a bottom-up region-merging approach to merge local homogeneity pixels to generate heterogenous objects, and three crucial parameters are generally utilized to optimize the segmentation results, as follows: the scale parameter is used to optimize the segmented objects and solve the phenomenon of over-segmentation and under-segmentation; the shape parameter takes the relationship between spectral and spatial uniformity into account; the compactness parameter considers spatial heterogeneity from two geometric attributes of different objects such as the perimeter and their area, the perimeter, and bounding boxes.

(4) Attribute labeling and dataset constructing

The LULC classification system is constructed for computing the model of CSLE, conforming to the national classification standard adopted by the Ministry of Water Resources of the People's Republic of China (i.e., GB/T 21010-2017). LULC samples are selected on the geometric vector data of ground objects combining the automatic vectorization result with OSM thematic vector data. Based on visual interpretation, the sample attribute is labeled by professional technicians with the assist of remote sensing interpretation keys from field surveys, which are available by field surveys and are beneficial to identify ground objects on remote sensing images. Quality control is implemented to check the correctness of the labeled LULC category based on the field survey, which is over 92%. Finally, the remote sensing images are cropped with the geometric vector data of selected samples to generate the established RVBD.

3.2. Dataset Evaluation Based on DL

3.2.1. DL-Based Baseline

Current mainstream DL architectures are traditional CNN architectures and the more recent widely recognized Transformer architectures. The CNN and Transformer architectures have discrepant architecture designs and feature extraction capabilities. CNN architecture has a stronger local perceptual ability because of built-in inductive biases that apply local convolutional filters to enhance the spatial invariance [81]. Transformer architecture generally splits the input image into a sequence of patches to model sequence-to-sequence (long-range) relations for yielding the stronger global modeling ability. Several superior DL networks of CNN and Transformer architectures are selected to evaluate the established RVBD, such as HorNet [82], DenseNet161 [83], EfficientNetB7 [84], Vision Transformer (ViT) [85], and Swin Transformer (SwinT) [86], which are briefly introduced as follows.

EfficientNetB7 is one variant of EfficientNet and has been widely recognized for effectively evaluating the performance of remote sensing datasets [84], tackling the issue that traditional CNNs are scaled only in following individual means of increasing the numbers of layers, increasing numbers of channels, or adjusting the input image size. EfficientNet designs a new scaling-up CNN approach, which develops an innovative compound coefficient to efficiently balance the scale relationships among depth, width, and resolution dimensions for yielding excellent whole-classification performance.

DenseNet161 is one variant of the Dense Convolutional Network and achieves a noteworthy improvement based on ResNet [87], which has the several following convincing benefits: alleviating the vanishing gradient issue, enhancing information flow, improving feature reuse, and compressing the number of parameters [83]. The dense block is the dominating feature extraction module in DenseNet, which regards any one of all the preceding layers as the input feature to directly access the subsequent layers. DenseNet enhances the reuse efficiency of the feature maps and yields good performance in remote sensing classification.

HorNet is inspired by the dot-product self-attention operation in Transformers and explicitly explores the spatial interaction relationship between local space and its circumjacent region [82]. Recursive gated convolution is designed to implement high-order and long-term spatial interactions with recursive gating convolution and large kernel convolutions. The network achieves efficient, extendable, and translation-equivariant performance for yielding remarkable power in image classification.

The design inspiration of ViT stems from the natural language processing (NPL) architecture. ViT is a pure and standard Transformer that splits an image into sequences of patches like sequences of word tokens in NPL, which leverages multi-head self-attention to capture the global dependency relationship for the patches [85]. In addition, position information remains based on the position-embedding module. It is a simple and scalable architecture that has shown superior performance in various visual tasks.

SwinT is different from pure visual Transformer architecture ViT, which introduces the additional visual inductive biases (i.e., locality, translation invariance, and hierarchy) with shifting windows to enhance the local modeling power [86]. Instead of using a fixed window to generate image patches, the shifted windows can compute self-attention crossing the boundaries of the previous windows for yielding stronger performance. It has lower latency and enables efficient processing of high-resolution images while maintaining a good balance between performance and computation.

3.2.2. Network Training Strategy

Transfer learning is an excellent training strategy and could achieve higher levels of generalization power with fewer training iterations, which could leverage pretrained weight parameters and can be applied to downstream tasks directly, such as remote sensing image classification [78]. For the aforementioned five DL networks for the established RVBD, the pretrained models stemming from ImageNet [15], which is a popular deep learning image classification dataset, are transferred to train with the RVBD. It is worth noting that network weights of all layers are fine tuned to yield better classification performance.

3.2.3. Misclassified Result Correction

Manual labeling can easily lead to misclassification. To address this problem, field surveys are carried out to correct the manual labeling errors and machine errors generated by DL networks.

(1) Define unreliable classification results. Utilizing the softmax function, the DL-based classification probabilities are generated as output. If the top two highest classification probabilities are approximately equal (i.e., the difference is less than 0.1), they could be considered as unreliable classification results.
(2) Correct the unreliable results by visual interpretation. The manual visual interpretation is employed to update the classification result with the assist of remote sensing interpretation keys.
(3) Verify the results by field surveys. The aforementioned, easily misclassified objects are further verified by field surveys to correct the machine errors generated by DL-based classification and human errors generated by visual interpretation. Especially

in regions with terrain or potential hazards, we employ unmanned aerial vehicles (UAVs) to facilitate the manual validation [88].

3.2.4. Evaluation Metrics

To quantitatively evaluate the quality of the established RVBD, different evaluation metrics are adopted, such as overall accuracy (OA), Kappa coefficient (Kappa), precision, recall, and F_1 score. These metrics focus on the different facets of the classification capabilities of the selected DL network. In addition, confusion matrices of predicted results of all networks are also provided to represent more performance details, which is conducive to the analysis classification results for each LULC class.

The Kappa coefficient can be calculated by the following Equation (1):

$$\text{Kappa} = \frac{l_o - l_e}{1 - l_e}, l_o = \frac{m}{n}, l_e = \frac{\sum_{i=1}^{c} p_i q_i}{n^2} \tag{1}$$

where Kappa is the Kappa coefficient, m and n are the number of correctly classified samples and total samples, respectively, and p_i and q_i are the number of real samples and the predicted samples of the ith LULC category, respectively.

The OA can be calculated by the following Equation (2):

$$\text{OA} = \frac{TP + TN}{TP + TN + FP + FN} \tag{2}$$

The precision can be calculated by the following Equation (3):

$$\text{Precision} = \frac{TP}{TP + FP} \tag{3}$$

The recall can be calculated by the following Equation (4):

$$\text{Recall} = \frac{TP}{TP + FN} \tag{4}$$

The F_1 score can be calculated by the following Equation (4):

$$F_1 = 2 \times \frac{\text{Precision} * \text{Recall}}{\text{Precision} + \text{Recall}} \tag{5}$$

In Equations (2) and (3), true positive (TP) means that the true LULC category and predicted LULC category are both positive, true negative (TN) means that the true LULC category and the predicted LULC category are both negative, false positive (FP) means that the true LULC category is negative but the predicted LULC category is positive, and false negative (FN) means that the true LULC category is positive but the predicted LULC category is negative.

4. Experiments and Results

4.1. Experimental Settings

Five popular DL networks, i.e., HorNet [82], DenseNet161 [83], EfficientNetB7 [84], ViT [85], and SwinT [86], are chosen as the baseline networks to evaluate the classification performance for the constructed benchmark RVBD.

A fine-tuned training strategy is implemented to improve the generalization capability based on the pretrained weights of ImageNet [15]. All remote sensing image patches of samples are resized to 256 × 256 pixels as the input for each network. Random horizontal and vertical flip operations are carried out for data augmentation. AdamW [89] is chosen as the optimizer with 100 epochs. The cosine annealing strategy is utilized as a learning scheduler, and the initial learning rate is set to 0.001. The parameter of the batch size of all networks is set to 32.

In addition, all aforementioned networks are implemented in the study on a work station equipped with, i.e., an Intel Core i7-8700 central processing unit (CPU) and four NVIDIA GeForce GTX 3090Ti Central Processing Unit graphics processing units (GPUs).

4.2. Results and Analysis

In this section, five fine-tuned DL networks are performed as the baseline classification networks to evaluate the effectiveness of the constructed RVBD from the following two aspects: overall classification accuracy analysis and class-wise classification accuracy analysis.

(1) Overall classification accuracy analysis

The classification performance of the aforementioned five fine-tuned networks using RVBD is analyzed in detail from the overall classification view. As reported in Table 3, it is even more evident that all five networks consistently perform well, with OA and Kappa metric values both exceeding 0.80. In particular, the values with bold font indicate the highest value in the comparative networks. It is obvious that the ViT network achieves the best performance, and its OA and Kappa metric values are 0.87 and 0.86, respectively. In addition, the relatively worse classification network is HorNet, which also yields reliable accuracy with OA of 0.81 and Kappa of 0.80. The above-excellent OA and Kappa metric values of the five networks indicate that DL networks achieve significant classification ability and the constructed dataset is effective.

Table 3. The overall classification accuracy results for RVBD.

Network Name	Overall Accuracy	Kappa
DenseNet161 [83]	0.86	0.85
EfficientNetB7 [84]	0.84	0.83
HorNet [82]	0.81	0.80
SwinT [86]	0.83	0.82
ViT [85]	0.87	0.86

(2) Class-wise accuracy analysis

To represent the classification capacity of all the aforementioned networks for each LULC category of RVBD, class-wise accuracy analysis is carried out. The confusion matrixes of all networks are shown in Figure 6, which represent the classification details for each LULC category. We find that all networks have yielded outstanding classification performance for most LULC categories. Precision, recall, and F_1-score metrics are chosen to quantitatively evaluate the classification performance for each LULC category of RVBD, which are shown in Table 4, Table 5, and Table 6, respectively. It is worth noting that the highest classification values among the five aforementioned DL networks for each LULC category are annotated with bold font.

The results can be obviously observed as the following: (1) the mean values of all metrics (i.e., precision, recall, and F_1 score) are basically higher than 0.80 for all chosen networks, which means that all networks represent the excellent and stable classification capacity for each LULC category; (2) the ViT network achieves the best performance on account of yielding the highest values of precision, recall, and F_1-score metrics among most LULC categories; (3) the HorNet network has the relatively worst classification capacity compared with other chosen networks; (4) the F_1 score is the harmonic mean of the precision and recall metrics. From the view of the F_1-score metric, water, sloping cropland, and paddy land have the best classification effect for all chosen DL networks. Rural road, mining land, and shrub land are more easily misclassified.

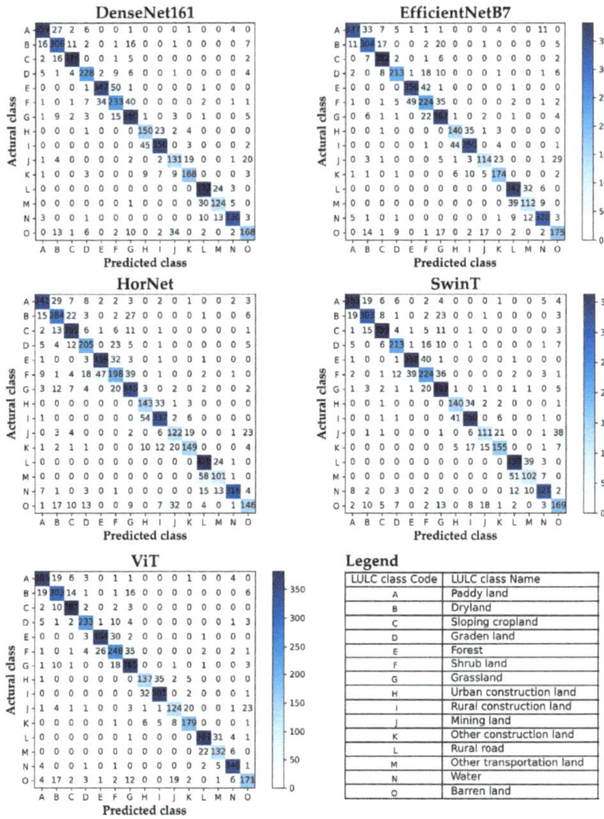

Figure 6. The confusion matrixes of classification results of all chosen DL networks are shown.

Table 4. The precision metric results of all networks for RVBD.

LULC Class \ Network Name	DenseNet161 [83]	EfficientNetB7 [84]	HorNet [82]	SwinT [86]	ViT [85]
Paddy land	0.92	0.94	0.88	0.90	0.91
Dry land	0.81	0.82	0.78	0.86	0.83
Sloping cropland	0.95	0.91	0.84	0.92	0.93
Garden	0.88	0.90	0.78	0.86	0.93
Forest	0.91	0.87	0.88	0.89	0.93
Shrub land	0.75	0.72	0.70	0.72	0.79
Grassland	0.81	0.79	0.78	0.79	0.83
Urban construction land	0.73	0.73	0.68	0.75	0.78
Rural construction land	0.91	0.87	0.84	0.84	0.90
Mining land	0.72	0.81	0.68	0.75	0.81
Other construction land	0.86	0.83	0.84	0.83	0.86
Other transportation land	0.89	0.87	0.82	0.84	0.93
Rural road	0.77	0.72	0.73	0.67	0.78
Water	0.95	0.91	0.98	0.93	0.94
Barren land	0.78	0.76	0.75	0.70	0.81
Mean Values	0.84	0.83	0.80	0.82	0.86

Table 5. The recall accuracy results of all networks for RVBD.

Network Name / LULC Class	DenseNet161 [83]	EfficientNetB7 [84]	HorNet [82]	SwinT [86]	ViT [85]
Paddy land	0.90	0.84	0.85	0.88	0.91
Dry land	0.85	0.84	0.79	0.84	0.84
Sloping cropland	0.93	0.95	0.90	0.90	0.95
Garden	0.88	0.82	0.79	0.82	0.90
Forest	0.87	0.89	0.90	0.89	0.91
Shrub land	0.73	0.70	0.62	0.70	0.77
Grassland	0.90	0.90	0.86	0.91	0.91
Urban construction land	0.83	0.78	0.79	0.78	0.76
Rural construction land	0.87	0.87	0.84	0.87	0.91
Mining land	0.73	0.63	0.68	0.62	0.69
Other construction land	0.84	0.87	0.74	0.77	0.89
Other transportation land	0.93	0.90	0.94	0.89	0.91
Rural road	0.77	0.70	0.63	0.64	0.82
Water	0.92	0.91	0.88	0.89	0.96
Barren land	0.70	0.73	0.61	0.70	0.71
Mean Values	0.84	0.82	0.79	0.81	0.86

Table 6. The F_1 score accuracy results of all networks for RVBD.

Network Name / LULC Class	DenseNet161 [83]	EfficientNetB7 [84]	HorNet [82]	SwinT [86]	ViT [85]
Paddy land	0.91	0.89	0.87	0.89	0.91
Dry land	0.83	0.83	0.78	0.85	0.84
Sloping cropland	0.94	0.93	0.87	0.91	0.94
Garden	0.88	0.86	0.78	0.84	0.91
Forest	0.89	0.88	0.89	0.89	0.92
Shrub land	0.74	0.71	0.66	0.71	0.78
Grassland	0.86	0.84	0.82	0.84	0.87
Urban construction land	0.78	0.75	0.73	0.76	0.77
Rural construction land	0.89	0.87	0.84	0.86	0.91
Mining land	0.73	0.71	0.68	0.68	0.74
Other construction land	0.85	0.85	0.79	0.80	0.88
Other transportation land	0.91	0.89	0.87	0.87	0.92
Rural road	0.77	0.71	0.68	0.65	0.80
Water	0.93	0.91	0.93	0.91	0.95
Barren land	0.74	0.75	0.67	0.70	0.76
Mean Values	0.84	0.83	0.79	0.81	0.86

4.3. Discussion

(1) The effectiveness and superiority of RVBD

This paper performs comprehensive experiments based on several outstanding DL baselines to further demonstrate the effectiveness and superiority of the established RVBD. (1) Some reliable accuracy evaluation metrics are selected to evaluate the effectiveness of RVBD. Regardless of the following two aspects of analysis, overall accuracy or class-wise accuracy, the classification results also achieve stable and outstanding performance. (2) The RVBD represents an excellent classification ability with the assist of geometric information. Some LULC categories with representative geometric shapes generally obtain better classification accuracy. For example, water is banded and planar, construction land generally has regular shapes, and other rural transportation land is generally striped. The above-mentioned LULC categories all achieve better F_1 scores in all baselines. This indicates that the geometric information provided by vector data is beneficial to improving the classification accuracy. Meanwhile, our method has some substantive and reproducible practices for establishing relative datasets: (1) multiple-source vector data are beneficial for providing references for manual visual interpretation and reducing a certain amount of the workload of sample labeling; (2) field surveys are conducive to sampling remote sensing interpretation keys as references for indoor manual visual interpretation. In addition, it is also conducive for verifying the classification results for correcting the machine errors generated by DL networks and human errors generated by visual interpretation.

(2) The applicability of RVBD

The following advantages reflect the stronger applicability of the dataset: (1) the dataset is constructed based on the object-oriented approach. This means that it has complete geometric information, which makes it beneficial for mapping LULC [73,90,91]. (2) The classification system conforms to the national industry standard adopted by the Ministry of Water Resources of the People's Republic of China. This means that it has higher adaptability to be applied to other relative research. (3) The RVBD is a thematic dataset. This means that it can be applied to conduct SWEA in the Yangtze River Basin. (4) It is conducive to updating the ground-truth data comprehensively. The established dataset integrates raster and vector data, which makes it easy to update the classification results of ground-truth data based on the DL models to the vector data for achieving high-accuracy LULC mapping. (5) It is an excellent solution for a data application flexibility plan. On one hand, our dataset conforms to the national industry standards to construct a land-use/land-cover classification system, which makes it easily and directly applied in other research [92–95]; on the other hand, the transfer learning method [96] could be adopted to train effective DL models for achieving high-accuracy land-use/land-cover mapping in other geographical areas, only depending on a limited number of samples without huge cost.

(3) The classification capacity of DL networks

DL networks are further verified to be effective in dealing with a remote sensing classification task. The aforementioned five DL networks are only selected to evaluate the established RVBD. However, the challenge still exists that samples of the established RVBD are easily misclassified, especially for similar LULC categories, because of the phenomenon of the same object with a different spectrum and a foreign body with the same spectrum. Hence, there is ample space for further optimizing DL networks to obtain better classification accuracy.

5. Conclusions

To the best of our knowledge, this paper is the first to construct an RVBD for conducting SWEA in the Yangtze River Basin based on the support of spatio-temporal spectral big data, which conforms the standards of the Ministry of Water Resources of the People's Republic of China and is based on DL. The RVBD includes 15 LULC categories and 23,300 pairs of object-based samples with corresponding image patches and vector data. It is sampled from Google images with 2 m resolution, which are spread across many countries of the Yangtze River Basin. It is worth noting that five DL networks are introduced to verify the effectiveness of the RVBD by evaluating the classification accuracy from two aspects: overall accuracy analysis and class-wise accuracy analysis. Experimental results verify the effectiveness of the RVBD. Every chosen network achieved remarkable performance, and the ViT network achieves the best classification performance with overall accuracy of 0.87 and Kappa of 0.86.

It is worth noting that the RVBD has broad and flexible applicability significance: (1) it could be utilized to provide high resolution LULC data, which contribute to the research in the Yangtze River basin, such as geographical conditions monitoring [95], prediction/simulation LULC change [94], ecosystem service [92], climate change [93], and so on; (2) it could be easily applied in other areas because of its inclusion of abundant and diverse LULC categories. On the one hand, the Yangtze River Basin basically covers all the terrain around the world. It has abundant topography with a multi-level and terraced distribution that spans from plateaus to plains; on the other hand, the Yangtze River Basin exhibits diverse climate characteristics, which vary from plateau climate to subtropical monsoon climate to subtropical maritime climate. The above characteristics of natural conditions result in a rich variety of LULC categories except for deserts and glaciers, which allows our dataset to be easily transferred to any geographical area, such as the Yellow River Basin, and even to the world.

In addition to this, our research still contains some limitations: there are no LULC classes on the snow-covered plateau. Especially in the classification data verification of super-large national land, there still are some objective difficulties, as follows: (1) remote sensing images with limited spectral and spatial resolution pose difficulties for visual interpretation. The visual interpretation capacities of different technicians are different, which could inevitably result in misclassification results. (2) Sampling remote sensing interpretation keys by field surveys is difficult. Conducting field surveys is a challenging task for sampling remote sensing interpretation keys and verifying the accuracy of classification results because of the substantial human and financial resources required. In future, we will further improve this dataset in the following promising aspects: (1) the classification ability needs to be further explored and strengthened by incorporating or designing more advanced deep learning models; (2) the generalization ability should be further investigated and enhanced, especially under conditions of few-shot samples, to broaden its application potential. In particular, it lays an intelligent data foundation for SWEA in the Yangtze River Basin, which is beneficial for the promotion and development of the intelligent application of remote sensing.

Author Contributions: Conceptualization, P.Z. and C.L.; Methodology, P.Z. and C.L.; Software, G.Z. and D.L.; Formal analysis, P.Z.; Resources, Y.W., R.L., H.Y. and Y.Z.; Data curation, Y.W., H.Y. and Y.Z.; Writing—original draft, P.Z.; Writing—review & editing, P.Z. and C.L.; Visualization, G.Z. and D.L.; Supervision, Y.W., C.L. and R.L.; Project administration, Y.W.; Funding acquisition, C.L. All authors have read and agreed to the published version of the manuscript.

Funding: This research was funded by the National Natural Science Foundation of China (NSFC) (Grant No. 41771493 and 41101407) and the Fundamental Research Funds for the Central Universities (Grant No. CCNU22QN019).

Data Availability Statement: The national-standards- and deep-learning-oriented raster and vector benchmark dataset (RVBD) could be publicly available at https://doi.org/10.5281/zenodo.8002595.

Acknowledgments: The authors are grateful for the comments and contributions of the editors, anonymous reviewers, and members of the editorial team.

Conflicts of Interest: The authors declare no conflict of interest.

References

1. Jin, F.; Yang, W.; Fu, J.; Li, Z. Effects of vegetation and climate on the changes of soil erosion in the Loess Plateau of China. *Sci. Total Environ.* **2021**, *773*, 145514. [CrossRef] [PubMed]
2. Vrieling, A. Satellite remote sensing for water erosion assessment: A review. *CATENA* **2006**, *65*, 2–18. [CrossRef]
3. Lamane, H.; Moussadek, R.; Baghdad, B.; Mouhir, L.; Briak, H.; Laghlimi, M.; Zouahri, A. Soil water erosion assessment in Morocco through modeling and fingerprinting applications: A review. *Heliyon* **2022**, *8*, e10209. [CrossRef]
4. Borrelli, P.; Robinson, D.A.; Fleischer, L.R.; Lugato, E.; Ballabio, C.; Alewell, C.; Meusburger, K.; Modugno, S.; Schütt, B.; Ferro, V. An assessment of the global impact of 21st century land use change on soil erosion. *Nat. Commun.* **2017**, *8*, 2013. [CrossRef]
5. Liu, B.; Xie, Y.; Li, Z.; Liang, Y.; Zhang, W.; Fu, S.; Yin, S.; Wei, X.; Zhang, K.; Wang, Z. The assessment of soil loss by water erosion in China. *Int. Soil Water Conserv. Res.* **2020**, *8*, 430–439. [CrossRef]
6. Wuepper, D.; Borrelli, P.; Finger, R. Countries and the global rate of soil erosion. *Nat. Sustain.* **2020**, *3*, 51–55. [CrossRef]
7. Khatami, R.; Mountrakis, G.; Stehman, S.V. A meta-analysis of remote sensing research on supervised pixel-based land-cover image classification processes: General guidelines for practitioners and future research. *Remote Sens. Environ.* **2016**, *177*, 89–100. [CrossRef]
8. Long, C.; Li, X.; Jing, Y.; Shen, H. Bishift Networks for Thick Cloud Removal with Multitemporal Remote Sensing Images. *Int. J. Intell. Syst.* **2023**, *2023*, e9953198. [CrossRef]
9. Dimitrovski, I.; Kitanovski, I.; Kocev, D.; Simidjievski, N. Current trends in deep learning for Earth Observation: An open-source benchmark arena for image classification. *ISPRS J. Photogramm. Remote Sens.* **2023**, *197*, 18–35. [CrossRef]

10. Shen, H.; Zhou, W.; Li, X. A Fast Globally Optimal Seamline Detection Method for High-Resolution Remote Sensing Images. *IEEE Geosci. Remote Sens. Lett.* **2023**, *20*, 6003305. [CrossRef]
11. Tan, Z.; Gao, M.; Li, X.; Jiang, L. A Flexible Reference-Insensitive Spatiotemporal Fusion Model for Remote Sensing Images Using Conditional Generative Adversarial Network. *IEEE Trans. Geosci. Remote Sens.* **2022**, *60*, 5601413. [CrossRef]
12. Sun, X.; Wang, P.; Yan, Z.; Xu, F.; Wang, R.; Diao, W.; Chen, J.; Li, J.; Feng, Y.; Xu, T.; et al. FAIR1M: A benchmark dataset for fine-grained object recognition in high-resolution remote sensing imagery. *ISPRS J. Photogramm. Remote Sens.* **2022**, *184*, 116–130. [CrossRef]
13. Yang, X.; Dong, M.; Wang, Z.; Gao, L.; Zhang, L.; Xue, J.-H. Data-augmented matched subspace detector for hyperspectral subpixel target detection. *Pattern Recognit.* **2020**, *106*, 107464. [CrossRef]
14. Li, K.; Wan, G.; Cheng, G.; Meng, L.; Han, J. Object detection in optical remote sensing images: A survey and a new benchmark. *ISPRS J. Photogramm. Remote Sens.* **2020**, *159*, 296–307. [CrossRef]
15. Deng, J.; Dong, W.; Socher, R.; Li, L.; Li, K.; Li, F. Imagenet: A large-scale hierarchical image database. In Proceedings of the 2009 IEEE Conference on Computer Vision and Pattern Recognition, Miami, FL, USA, 20–25 June 2009; pp. 248–255.
16. Everingham, M.; Van Gool, L.; Williams, C.K.; Winn, J.; Zisserman, A. The pascal visual object classes (voc) challenge. *Int. J. Comput. Vis.* **2010**, *88*, 303–338. [CrossRef]
17. Benedek, C.; Descombes, X.; Zerubia, J. Building Development Monitoring in Multitemporal Remotely Sensed Image Pairs with Stochastic Birth-Death Dynamics. *IEEE Trans. Pattern Anal. Mach. Intell.* **2012**, *34*, 33–50. [CrossRef]
18. Lin, T.-Y.; Maire, M.; Belongie, S.; Hays, J.; Perona, P.; Ramanan, D.; Dollár, P.; Zitnick, C.L. Microsoft coco: Common objects in context. In Proceedings of the European Conference on Computer Vision, Zurich, Switzerland, 6–12 September 2014; pp. 740–755.
19. Zhu, H.; Chen, X.; Dai, W.; Fu, K.; Ye, Q.; Jiao, J. Orientation robust object detection in aerial images using deep convolutional neural network. In Proceedings of the 2015 IEEE International Conference on Image Processing (ICIP), Quebec City, QC, Canada, 27–30 September 2015; pp. 3735–3739.
20. Liu, K.; Mattyus, G. Fast multiclass vehicle detection on aerial images. *IEEE Geosci. Remote Sens. Lett.* **2015**, *12*, 1938–1942.
21. Cheng, G.; Zhou, P.; Han, J. Learning rotation-invariant convolutional neural networks for object detection in VHR optical remote sensing images. *IEEE Trans. Geosci. Remote Sens.* **2016**, *54*, 7405–7415. [CrossRef]
22. Razakarivony, S.; Jurie, F. Vehicle detection in aerial imagery: A small target detection benchmark. *J. Vis. Commun. Image Represent.* **2016**, *34*, 187–203. [CrossRef]
23. Liu, Z.; Wang, H.; Weng, L.; Yang, Y. Ship rotated bounding box space for ship extraction from high-resolution optical satellite images with complex backgrounds. *IEEE Geosci. Remote Sens. Lett.* **2016**, *13*, 1074–1078. [CrossRef]
24. Mundhenk, T.N.; Konjevod, G.; Sakla, W.A.; Boakye, K. A large contextual dataset for classification, detection and counting of cars with deep learning. In Proceedings of the European Conference on Computer Vision, Amsterdam, The Netherlands, 11–14 October 2016; pp. 785–800.
25. Long, Y.; Gong, Y.; Xiao, Z.; Liu, Q. Accurate object localization in remote sensing images based on convolutional neural networks. *IEEE Trans. Geosci. Remote Sens.* **2017**, *55*, 2486–2498. [CrossRef]
26. Zou, Z.; Shi, Z. Random access memories: A new paradigm for target detection in high resolution aerial remote sensing images. *IEEE Trans. Image Process.* **2017**, *27*, 1100–1111. [CrossRef]
27. Yang, M.Y.; Liao, W.; Li, X.; Rosenhahn, B. Deep learning for vehicle detection in aerial images. In Proceedings of the 2018 25th IEEE International Conference on Image Processing (ICIP), Athens, Greece, 7–10 October 2018; pp. 3079–3083.
28. Xia, G.; Bai, X.; Ding, J.; Zhu, Z.; Belongie, S.; Luo, J.; Datcu, M.; Pelillo, M.; Zhang, L. DOTA: A large-scale dataset for object detection in aerial images. In Proceedings of the IEEE Conference on Computer Vision and Pattern Recognition, Salt Lake City, UT, USA, 18–23 June 2018; pp. 3974–3983.
29. Wang, Y.; Bashir, S.M.A.; Khan, M.; Ullah, Q.; Wang, R.; Song, Y.; Guo, Z.; Niu, Y. Remote sensing image super-resolution and object detection: Benchmark and state of the art. *Expert Syst. Appl.* **2022**, *197*, 116793. [CrossRef]
30. Yang, Y.; Newsam, S. Bag-of-visual-words and spatial extensions for land-use classification. In Proceedings of the 18th SIGSPATIAL International Conference on Advances in Geographic Information Systems, San Jose, CA, USA, 2–5 November 2010; pp. 270–279.
31. Hu, J.; Jiang, T.; Tong, X.; Xia, G.-S.; Zhang, L. A benchmark for scene classification of high spatial resolution remote sensing imagery. In Proceedings of the 2015 IEEE International Geoscience and Remote Sensing Symposium (IGARSS), Milan, Italy, 26–31 July 2015; pp. 5003–5006.
32. Zou, Q.; Ni, L.; Zhang, T.; Wang, Q. Deep learning based feature selection for remote sensing scene classification. *IEEE Geosci. Remote Sens. Lett.* **2015**, *12*, 2321–2325. [CrossRef]
33. Penatti, O.A.B.; Nogueira, K.; dos Santos, J.A. Do Deep Features Generalize From Everyday Objects to Remote Sensing and Aerial Scenes Domains. In Proceedings of the IEEE Conference on Computer Vision and Pattern Recognition Workshops, Boston, MA, USA, 7–12 June 2015; pp. 44–51.
34. Zhao, L.; Tang, P.; Huo, L. Feature significance-based multibag-of-visual-words model for remote sensing image scene classification. *J. Appl. Remote Sens.* **2016**, *10*, 035004. [CrossRef]
35. Zhao, B.; Zhong, Y.; Xia, G.-S.; Zhang, L. Dirichlet-derived multiple topic scene classification model for high spatial resolution remote sensing imagery. *IEEE Trans. Geosci. Remote Sens.* **2015**, *54*, 2108–2123. [CrossRef]

36. Xia, G.-S.; Hu, J.; Hu, F.; Shi, B.; Bai, X.; Zhong, Y.; Zhang, L.; Lu, X. AID: A benchmark data set for performance evaluation of aerial scene classification. *IEEE Trans. Geosci. Remote Sens.* **2017**, *55*, 3965–3981. [CrossRef]
37. Jin, P.; Xia, G.-S.; Hu, F.; Lu, Q.; Zhang, L. AID++: An Updated Version of AID on Scene Classification. In Proceedings of the IGARSS 2018—2018 IEEE International Geoscience and Remote Sensing Symposium, Valencia, Spain, 22–27 July 2018; pp. 4721–4724.
38. Wang, Q.; Liu, S.; Chanussot, J.; Li, X. Scene classification with recurrent attention of VHR remote sensing images. *IEEE Trans. Geosci. Remote Sens.* **2018**, *57*, 1155–1167. [CrossRef]
39. Zhou, W.; Newsam, S.; Li, C.; Shao, Z. PatternNet: A benchmark dataset for performance evaluation of remote sensing image retrieval. *ISPRS J. Photogramm. Remote Sens.* **2018**, *145*, 197–209. [CrossRef]
40. Zhong, Y.; Su, Y.; Wu, S.; Zheng, Z.; Zhao, J.; Ma, A.; Zhu, Q.; Ye, R.; Li, X.; Pellikka, P. Open-source data-driven urban land-use mapping integrating point-line-polygon semantic objects: A case study of Chinese cities. *Remote Sens. Environ.* **2020**, *247*, 111838. [CrossRef]
41. Li, H.; Dou, X.; Tao, C.; Wu, Z.; Chen, J.; Peng, J.; Deng, M.; Zhao, L. RSI-CB: A Large-Scale Remote Sensing Image Classification Benchmark Using Crowdsourced Data. *Sensors* **2020**, *20*, 1594. [CrossRef]
42. Barrena-González, J.; Rodrigo-Comino, J.; Gyasi-Agyei, Y.; Pulido Fernández, M.; Cerdà, A. Applying the RUSLE and ISUM in the Tierra de Barros Vineyards (Extremadura, Spain) to Estimate Soil Mobilisation Rates. *Land* **2020**, *9*, 93. [CrossRef]
43. Helber, P.; Bischke, B.; Dengel, A.; Borth, D. EuroSAT: A Novel Dataset and Deep Learning Benchmark for Land Use and Land Cover Classification. *IEEE J. Sel. Top. Appl. Earth Obs. Remote Sens.* **2019**, *12*, 2217–2226. [CrossRef]
44. Sumbul, G.; Charfuelan, M.; Demir, B.; Markl, V. Bigearthnet: A large-scale benchmark archive for remote sensing image understanding. In Proceedings of the IGARSS 2019-2019 IEEE International Geoscience and Remote Sensing Symposium, Yokohama, Japan, 28 July–2 August 2019; pp. 5901–5904.
45. Qi, X.; Zhu, P.; Wang, Y.; Zhang, L.; Peng, J.; Wu, M.; Chen, J.; Zhao, X.; Zang, N.; Mathiopoulos, P.T. MLRSNet: A multi-label high spatial resolution remote sensing dataset for semantic scene understanding. *ISPRS J. Photogramm. Remote Sens.* **2020**, *169*, 337–350. [CrossRef]
46. Sumbul, G.; De Wall, A.; Kreuziger, T.; Marcelino, F.; Costa, H.; Benevides, P.; Caetano, M.; Demir, B.; Markl, V. BigEarthNet-MM: A Large-Scale, Multimodal, Multilabel Benchmark Archive for Remote Sensing Image Classification and Retrieval. *IEEE Geosci. Remote Sens. Mag.* **2021**, *9*, 174–180. [CrossRef]
47. Hong, D.; Hu, J.; Yao, J.; Chanussot, J.; Zhu, X.X. Multimodal remote sensing benchmark datasets for land cover classification with a shared and specific feature learning model. *ISPRS J. Photogramm. Remote Sens.* **2021**, *178*, 68–80. [CrossRef]
48. Li, L.; Yao, X.; Cheng, G.; Han, J. AIFS-DATASET for Few-Shot Aerial Image Scene Classification. *IEEE Trans. Geosci. Remote Sens.* **2022**, *60*, 5618211. [CrossRef]
49. Li, Y.; Zhou, Y.; Zhang, Y.; Zhong, L.; Wang, J.; Chen, J. DKDFN: Domain Knowledge-Guided deep collaborative fusion network for multimodal unitemporal remote sensing land cover classification. *ISPRS J. Photogramm. Remote Sens.* **2022**, *186*, 170–189. [CrossRef]
50. Zhang, Z.; Zhang, M.; Gong, J.; Hu, X.; Xiong, H.; Zhou, H.; Cao, Z. LuoJiaAI: A cloud-based artificial intelligence platform for remote sensing image interpretation. *Geo-Spat. Inf. Sci.* **2023**, 1–24. [CrossRef]
51. Papoutsis, I.; Bountos, N.I.; Zavras, A.; Michail, D.; Tryfonopoulos, C. Benchmarking and scaling of deep learning models for land cover image classification. *ISPRS J. Photogramm. Remote Sens.* **2023**, *195*, 250–268. [CrossRef]
52. Zhang, H.; Zhang, R.; Qi, F.; Liu, X.; Niu, Y.; Fan, Z.; Zhang, Q.; Li, J.; Yuan, L.; Song, Y. The CSLE model based soil erosion prediction: Comparisons of sampling density and extrapolation method at the county level. *CATENA* **2018**, *165*, 465–472. [CrossRef]
53. Dong, Y.; Liang, T.; Yang, C.; Luo, H.; Zhang, Y. Joint Distance Transfer Metric Learning for Remote-Sensing Image Classification. *IEEE Geosci. Remote Sens. Lett.* **2022**, *19*, 6506205. [CrossRef]
54. Yang, C.; Dong, Y.; Du, B.; Zhang, L. Attention-Based Dynamic Alignment and Dynamic Distribution Adaptation for Remote Sensing Cross-Domain Scene Classification. *IEEE Trans. Geosci. Remote Sens.* **2022**, *60*, 5634713. [CrossRef]
55. Hosseiny, B.; Rastiveis, H.; Homayouni, S. An Automated Framework for Plant Detection Based on Deep Simulated Learning from Drone Imagery. *Remote Sens.* **2020**, *12*, 3521. [CrossRef]
56. Northcutt, C.; Jiang, L.; Chuang, I. Confident Learning: Estimating Uncertainty in Dataset Labels. *J. Artif. Intell. Res.* **2021**, *70*, 1373–1411. [CrossRef]
57. Abdollahi, A.; Pradhan, B.; Shukla, N.; Chakraborty, S.; Alamri, A. Deep learning approaches applied to remote sensing datasets for road extraction: A state-of-the-art review. *Remote Sens.* **2020**, *12*, 1444. [CrossRef]
58. Lekki, J.; Deutsch, E.; Sayers, M.; Bosse, K.; Anderson, R.; Tokars, R.; Sawtell, R. Determining remote sensing spatial resolution requirements for the monitoring of harmful algal blooms in the Great Lakes. *J. Great Lakes Res.* **2019**, *45*, 434–443. [CrossRef]
59. Velmurugan, K.; Saravanasankar, S.; Venkumar, P.; Sudhakarapandian, R.; Bona, G.D. Hybrid fuzzy AHP-TOPSIS framework on human error factor analysis: Implications to developing optimal maintenance management system in the SMEs. *Sustain. Futur.* **2022**, *4*, 100087. [CrossRef]
60. Di Bona, G.; Falcone, D.; Forcina, A.; De Carlo, F.; Silvestri, L. Quality Checks Logit Human Reliability (LHR): A New Model to Evaluate Human Error Probability (HEP). *Math. Probl. Eng.* **2021**, *2021*, e6653811. [CrossRef]
61. Liang, X.; Liu, X.; Yao, L. Review–a survey of learning from noisy labels. *ECS Sens. Plus* **2022**, *1*, 021401. [CrossRef]

62. González-Rivero, M.; Beijbom, O.; Rodriguez-Ramirez, A.; Holtrop, T.; González-Marrero, Y.; Ganase, A.; Roelfsema, C.; Phinn, S.; Hoegh-Guldberg, O. Scaling up ecological measurements of coral reefs using semi-automated field image collection and analysis. *Remote Sens.* **2016**, *8*, 30. [CrossRef]
63. Chang, C.-M.; Lee, C.-H.; Igarashi, T. Spatial labeling: Leveraging spatial layout for improving label quality in non-expert image annotation. In Proceedings of the 2021 CHI Conference on Human Factors in Computing Systems, Yokohama, Japan, 8–13 May 2021; pp. 1–12.
64. Bona, G.D.; Falcone, D.; Forcina, A.; Silvestri, L. Systematic human reliability analysis (SHRA): A new approach to evaluate human error probability (HEP) in a nuclear plant. *Int. J. Math. Eng. Manag. Sci.* **2021**, *6*, 345–362.
65. Gotovac, S.; Zelenika, D.; Marušić, Ž.; Božić-Štulić, D. Visual-based person detection for search-and-rescue with uas: Humans vs. machine learning algorithm. *Remote Sens.* **2020**, *12*, 3295. [CrossRef]
66. Gupta, E.; Das, S.; Rajani, M.B. Archaeological exploration in Srirangapatna and its environ through remote sensing analysis. *J. Indian Soc. Remote Sens.* **2017**, *45*, 1057–1063. [CrossRef]
67. Zhao, Y.; Feng, D.; Yu, L.; See, L.; Fritz, S.; Perger, C.; Gong, P. Assessing and improving the reliability of volunteered land cover reference data. *Remote Sens.* **2017**, *9*, 1034. [CrossRef]
68. Sajjad, H.; Kumar, P. Future challenges and perspective of remote sensing technology. In *Applications and Challenges of Geospatial Technology: Potential and Future Trends*; Springer: Berlin/Heidelberg, Germany, 2019; pp. 275–277.
69. Fritz, S.; See, L.; Perger, C.; McCallum, I.; Schill, C.; Schepaschenko, D.; Duerauer, M.; Karner, M.; Dresel, C.; Laso-Bayas, J.-C.; et al. A global dataset of crowdsourced land cover and land use reference data. *Sci Data* **2017**, *4*, 170075. [CrossRef]
70. Wang, X. Information extraction of tourist geological resources based on 3D visualization remote sensing image. *Int. Arch. Photogramm. Remote Sens. Spat. Inf. Sci.* **2018**, *XLII-3*, 1815–1820. [CrossRef]
71. Weiers, S.; Bock, M.; Wissen, M.; Rossner, G. Mapping and indicator approaches for the assessment of habitats at different scales using remote sensing and GIS methods. *Landsc. Urban Plan.* **2004**, *67*, 43–65. [CrossRef]
72. Duan, Y.; Li, X.; Zhang, L.; Chen, D.; Ji, H. Mapping national-scale aquaculture ponds based on the Google Earth Engine in the Chinese coastal zone. *Aquaculture* **2020**, *520*, 734666. [CrossRef]
73. Wu, Y.; Zhang, P.; Wu, J.; Li, C. Object-oriented and deep-learning-based high-resolution mapping from large remote sensing imagery. *Can. J. Remote Sens.* **2021**, *47*, 396–412. [CrossRef]
74. Zhao, Y.; Gong, P.; Yu, L.; Hu, L.; Li, X.; Li, C.; Zhang, H.; Zheng, Y.; Wang, J.; Zhao, Y.; et al. Towards a common validation sample set for global land-cover mapping. *Int. J. Remote Sens.* **2014**, *35*, 4795–4814. [CrossRef]
75. Behera, M.D.; Gupta, A.K.; Barik, S.K.; Das, P.; Panda, R.M. Use of satellite remote sensing as a monitoring tool for land and water resources development activities in an Indian tropical site. *Environ. Monit. Assess.* **2018**, *190*, 401. [CrossRef]
76. Zhang, M.; Zhao, X.; Li, W.; Zhang, Y.; Tao, R.; Du, Q. Cross-Scene Joint Classification of Multisource Data With Multilevel Domain Adaption Network. *IEEE Trans. Neural Netw. Learn. Syst.* **2023**, 1–13. [CrossRef]
77. Wold, S.; Esbensen, K.; Geladi, P. Principal component analysis. *Chemom. Intell. Lab. Syst.* **1987**, *2*, 37–52. [CrossRef]
78. Su, Y.; Zhong, Y.; Zhu, Q.; Zhao, J. Urban scene understanding based on semantic and socioeconomic features: From high-resolution remote sensing imagery to multi-source geographic datasets. *ISPRS J. Photogramm. Remote Sens.* **2021**, *179*, 50–65. [CrossRef]
79. Johnson, B.A.; Iizuka, K. Integrating OpenStreetMap crowdsourced data and Landsat time-series imagery for rapid land use/land cover (LULC) mapping: Case study of the Laguna de Bay area of the Philippines. *Appl. Geogr.* **2016**, *67*, 140–149. [CrossRef]
80. Happ, P.N.; Ferreira, R.S.; Bentes, C.; Costa, G.; Feitosa, R.Q. Multiresolution segmentation: A parallel approach for high resolution image segmentation in multicore architectures. *Int. Arch. Photogramm. Remote Sens. Spat. Inf. Sci.* **2010**, *38*, C7.
81. Liu, Z.; Mao, H.; Wu, C.-Y.; Feichtenhofer, C.; Darrell, T.; Xie, S. A ConvNet for the 2020s. *arXiv* **2022**, arXiv:2201.03545. [CrossRef]
82. Rao, Y.; Zhao, W.; Tang, Y.; Zhou, J.; Lim, S.N.; Lu, J. Hornet: Efficient high-order spatial interactions with recursive gated convolutions. *Adv. Neural Inf. Process. Syst.* **2022**, *35*, 10353–10366.
83. Huang, G.; Liu, Z.; Van Der Maaten, L.; Weinberger, K.Q. Densely connected convolutional networks. In Proceedings of the IEEE Conference on Computer Vision and Pattern Recognition, Honolulu, HI, USA, 21–26 July 2017; pp. 4700–4708.
84. Tan, M.; Le, Q. Efficientnet: Rethinking model scaling for convolutional neural networks. In Proceedings of the 36th International Conference on Machine Learning, Long Beach, CA, USA, 9–15 June 2019; pp. 6105–6114.
85. Dosovitskiy, A.; Beyer, L.; Kolesnikov, A.; Weissenborn, D.; Zhai, X.; Unterthiner, T.; Dehghani, M.; Minderer, M.; Heigold, G.; Gelly, S. An image is worth 16x16 words: Transformers for image recognition at scale. *arXiv* **2020**, arXiv:2010.11929.
86. Liu, Z.; Lin, Y.; Cao, Y.; Hu, H.; Wei, Y.; Zhang, Z.; Lin, S.; Guo, B. Swin transformer: Hierarchical vision transformer using shifted windows. In Proceedings of the IEEE/CVF International Conference on Computer Vision, Montreal, BC, Canada, 11–17 October 2021; pp. 10012–10022.
87. He, K.; Zhang, X.; Ren, S.; Sun, J. Deep Residual Learning for Image Recognition. In Proceedings of the IEEE Conference on Computer Vision and Pattern Recognition, Las Vegas, NV, USA, 26 June–1 July 2016; pp. 770–778.
88. Li, C.; Yao, J.; Li, R.; Zhu, Y.; Yao, H.; Zhang, P.; Wei, D.; Zhao, S.; Li, Y.; Wu, Y. "3S" technologies and application for dynamic monitoring soil and water loss in the Yangtze river bain, China. *Int. Arch. Photogramm. Remote Sens. Spat. Inf. Sci.* **2020**, *43*, 1563–1567. [CrossRef]
89. Loshchilov, I.; Hutter, F. Decoupled weight decay regularization. *arXiv* **2017**, arXiv:1711.05101.

90. Martins, V.S.; Kaleita, A.L.; Gelder, B.K.; da Silveira, H.L.F.; Abe, C.A. Exploring multiscale object-based convolutional neural network (multi-OCNN) for remote sensing image classification at high spatial resolution. *ISPRS J. Photogramm. Remote Sens.* **2020**, *168*, 56–73. [CrossRef]
91. Ma, L.; Liu, Y.; Zhang, X.; Ye, Y.; Yin, G.; Johnson, B.A. Deep learning in remote sensing applications: A meta-analysis and review. *ISPRS J. Photogramm. Remote Sens.* **2019**, *152*, 166–177. [CrossRef]
92. Li, C.; Fang, S.; Geng, X.; Yuan, Y.; Zheng, X.; Zhang, D.; Li, R.; Sun, W.; Wang, X. Coastal ecosystem service in response to past and future land use and land cover change dynamics in the Yangtze river estuary. *J. Clean. Prod.* **2023**, *385*, 135601. [CrossRef]
93. Xiao, R.; Cao, W.; Liu, Y.; Lu, B. The impacts of landscape patterns spatio-temporal changes on land surface temperature from a multi-scale perspective: A case study of the Yangtze River Delta. *Sci. Total Environ.* **2022**, *821*, 153381. [CrossRef]
94. Zhang, S.; Yang, P.; Xia, J.; Wang, W.; Cai, W.; Chen, N.; Hu, S ; Luo, X.; Li, J.; Zhan, C. Land use/land cover prediction and analysis of the middle reaches of the Yangtze River under different scenarios. *Sci. Total Environ.* **2022**, *833*, 155238. [CrossRef]
95. Zhang, J.; Li, W.; Zhai, L. Understanding geographical conditions monitoring: A perspective from China. *Int. J. Digit. Earth* **2015**, *8*, 38–57. [CrossRef]
96. Liu, Y.; Ding, L.; Chen, C.; Liu, Y. Similarity-Based Unsupervised Deep Transfer Learning for Remote Sensing Image Retrieval. *IEEE Trans. Geosci. Remote Sens.* **2020**, *58*, 7872–7889. [CrossRef]

remote sensing

Article

Radar Active Jamming Recognition under Open World Setting

Yupei Zhang [1], Zhijin Zhao [2,*] and Yi Bu [3]

[1] School of Electronic and Information, Hangzhou Dianzi University, Hangzhou 310018, China; 2143040036@hdu.edu.cn
[2] School of Communication Engineering, Hangzhou Dianzi University, Hangzhou 310018, China
[3] School of Electrical and Computer Engineering, Royal Melbourne Institute of Technology (RMIT University), Melbourne 3000, Australia; prototyperimac@gmail.com
* Correspondence: zhaozj03@hdu.edu.cn

Abstract: To address the issue that conventional methods cannot recognize unknown patterns of radar jamming, this study adopts the idea of zero-shot learning (ZSL) and proposes an open world recognition method, RCAE-OWR, based on residual convolutional autoencoders, which can implement the classification of known and unknown patterns. In the supervised training phase, a residual convolutional autoencoder network structure is first constructed to extract the semantic information from a training set consisting solely of known jamming patterns. By incorporating center loss and reconstruction loss into the softmax loss function, a joint loss function is constructed to minimize the intra-class distance and maximize the inter-class distance in the jamming features. Moving to the unsupervised classification phase, a test set containing both known and unknown patterns is fed into the trained encoder, and a distance-based recognition method is utilized to classify the jamming signals. The results demonstrate that the proposed model not only achieves sufficient learning and representation of known jamming patterns but also effectively identifies and classifies unknown jamming signals. When the jamming-to-noise ratio (JNR) exceeds 10 dB, the recognition rate for seven known jamming patterns and two unknown jamming patterns is more than 92%.

Keywords: radar signals; jamming recognition; deep learning; zero-shot learning; convolutional neural network

Citation: Zhang, Y.; Zhao, Z.; Bu, Y. Radar Active Jamming Recognition under Open World Setting. *Remote Sens.* **2023**, *15*, 4107. https://doi.org/10.3390/rs15164107

Academic Editors: Qian Du, Yanni Dong and Xiaochen Yang

Received: 30 June 2023
Revised: 17 August 2023
Accepted: 18 August 2023
Published: 21 August 2023

1. Introduction

Radar faces increasingly complex electronic countermeasures, with various new types of radar jamming patterns continuously emerging as challenges [1,2]. The accurate recognition of radar jamming is a precondition and key to implementing anti-jamming measures, and automatic recognition of radar jamming patterns can effectively improve the target detection and tracking performance of radar. Therefore, jamming pattern recognition has always been a research hotspot in anti-jamming technology [3,4]. Suppression jamming by emitting high-power noise signals is not only effective for linear frequency modulated (LFM) radar systems but also for other modulated radar systems, making it the most widely used [5]. Therefore, this paper focuses on the recognition of suppression jamming. Its main patterns include amplitude modulation jamming (AMJ), frequency modulation jamming (FMJ), comb spectrum jamming (CJ), phase modulation jamming (PMJ), swept jamming (SJ), etc. Conventional jamming pattern detection and classification methods are generally based on feature engineering. Firstly, multi-dimensional features of signal in time domain, frequency domain, and transform domain are extracted, including features such as moment kurtosis, moment skewness, envelope fluctuation, noise factor [6–9], singular spectrum features [10], bispectrum features [11] and other signal features. Then, with the help of machine learning-based classifiers, such as support vector machine (SVM) [12], decision trees [13], and back propagation (BP) neural networks [12], classifications are accomplished. However, feature engineering-based classification methods are time-consuming and require expert experience, especially when the transform domain features are large, the classification performance has limited room for improvement, and the recognition rate is low in strong noise environment.

With the development of deep learning (DL) technology, the feature extraction capability of neural networks has been improving, and classification methods based on DL are emerging. In [14], A novel hybrid framework of optimized deep learning models combined with multi-sensor fusion is developed for condition diagnosis of concrete arch beam, and the results demonstrate that the method can achieve the classification of structural damage with limited sensors and high levels of uncertainties.

Convolutional neural networks (CNNs), due to their network architecture which incorporates weight sharing and small local receptive fields, have significantly reduced the number of node connections compared to conventional neural networks. This simplification of network connections has made CNNs widely applied in deep learning models [15–17]. CNNs can train their parameters using jamming signals, eliminating the need for manual feature extraction and the design of decision trees for classification criteria. As a result, CNNs have been extensively used in the research of classifying and recognizing radar jamming signals [18].

In [19], a jamming recognition algorithm based on improved LeNet·CNN network was designed which extracted one-dimensional radar received signals and adjusted the network structure parameters to achieve optimal performance for the recognition of jamming signals. Ref. [20] obtained the time-frequency spectrogram of jamming signals by short-time Fourier transform, combined with the improved VGGNet·16 network model for feature learning and training, and the simulation verified that the algorithm is still effective for the identification of six kinds of mixed jamming. Ref. [21] adopted an adaptive cropping algorithm to crop most of the redundant information of the time-frequency image and kept the complete information of the jamming in the CNN for training, and finally achieved the recognition of nine kinds of jamming signals with high accuracy and fast iteration. In [22], a 1D CNN-based radar jamming signal classification model was proposed to achieve the classification of 12 typical jamming signals by putting the real and imaginary parts of jamming signals into the parallel network for training. In [23], a CNN was constructed using the real and imaginary parts of the signal as inputs. With sufficient training samples, this method demonstrated excellent recognition capabilities. The mentioned papers primarily address the issue of recognizing jamming when there are sufficient labeled samples. However, [6,24,25] considered the case of insufficient labeled samples. Ref. [6] proposed a method based on a time-frequency self-attentive mechanism. The recognition rate for most of the patterns of jamming reaches 90% when the samples with labels account for 3%. In [24], Tian et al. inputed features obtained through empirical mode decomposition and wavelet decomposition into the network. Simulations conducted on a dataset consisting of only 8400 samples showed that the recognition rates for four types of jamming were all above 90% when the JNR exceeded 6 dB. In [25], a large number of unlabeled samples were first used to train an jamming recognition network to extract valuable features. Then, a small number of labeled samples were used to improve the classification accuracy.

Although the application of deep learning technology in radar jamming recognition is rapidly developing, the current methods still suffer from the closed-set assumption, i.e., the existing methods assume that the jamming patterns are included in the training set. However, in the actual battlefield environment, the enemy may invent new jamming patterns, making it challenging to collect data for all patterns in the training set, so the actual radar jamming environment is an open set scenario, i.e., the test environment is likely to have jamming patterns that do not exist in the training jamming library. In the actual open set jamming scenario, when a jamming pattern that does not exist in the training jamming library appears in the test environment, the existing radar jamming identification methods will incorrectly identify this unknown jamming as one of the known jamming patterns. In [26], Zhou et al. first investigated the open set recognition problem for radar jamming; however, the method can only detect or reject the unknown patterns, but cannot effectively identify the unknown patterns. How to further classify these unknown pattern signals remains a challenging task, and it falls under the research domain of open world recognition (OWR).

Currently, OWR techniques have been applied in target recognition of synthetic aperture radar (SAR) images. In [27], a hierarchical embedding and incremental evolutionary network (HEIEN) was designed for when there are fewer unknown target training sets in open scenarios, which requires only a small number of unknown target samples for effective model training. A more stable feature space was built in [28], which has better interpretability. In testing, experiments on a dataset containing seven known targets and one unknown target show that the method improves the reliability of recognizing unknown targets. In [29], Song et al. used physical EM simulation images of targets at different azimuths as training data in order to learn the features of unknown targets. An accuracy of 91.93% can be achieved in a recognition task with a dataset containing nine known targets and one unknown target.

Nevertheless, OWR is just starting in radar jamming pattern recognition. Zero-shot learning (ZSL) [30,31] is an effective approach to address the challenge of open world recognition. The most typical implementation of ZSL is based on feature mapping. The goal of this approach is to learn the mapping relationship between the original signal space and the semantic feature space using the known jamming patterns. This learned relationship is then generalized to the unknown pattern dataset, enabling recognition and classification of unknown patterns using the semantic features. ZSL can be classified into two types: traditional zero-shot learning (TZSL) and generalized zero-shot learning (GZSL) [32]. TZSL assumes that the training patterns are known, while the testing patterns are unknown, and there is mutual exclusion between the training and testing patterns. GZSL assumes that the training patterns include both known and unknown patterns in the testing phase. GZSL has a more relaxed experimental setting, which better reflects real-world scenarios. Therefore, in this paper, we consider the GZSL scenario.

To address the challenge of existing methods being unable to classify unknown jamming patterns specifically, this paper adopts the idea of ZSL and conducts research on jamming patterns recognition in an open world scenario. We propose a residual convolutional autoencoder-based radar jamming open world recognition algorithm, abbreviated as RCAE-OWR.

In summary, the main contributions of this paper are as follows:

- In order to address the limitations of existing radar jamming pattern recognition methods, which are mostly closed-set recognition or simply rejecting unknown patterns, we propose a zero-shot learning approach based on residual conventional autoencoders. This method does not require prior information about the patterns of jamming and can classify both known and unknown patterns using distance-based recognition methods.
- A hybrid loss function consisting of cross-entropy loss, center loss and reconstruction loss is introduced to recognize different patterns of jamming signals. Where the cross-entropy loss makes the features obtained from the mapping network divisible, and to some extent widens the distance among different patterns, the center loss makes it easier to delineate the boundaries of the various patterns, and the reconstruction loss ensures that the most essential characterization of the pattern features is learned from the known patterns dataset.
- Through extensive experimental simulations, we evaluate the open world recognition performance of the proposed algorithm and investigate the influence of JNR and the number of unknown patterns on the algorithm's performance. The simulation results demonstrate that the proposed algorithm achieves effective recognition of both known and unknown patterns, especially in high-JNR environments.

2. Jamming Signal Modeling

In radar jamming and anti-jamming training, the signals received by the radar receiver include radar echo signals, jamming signals and noise signals, which can be expressed as follows:

$$S(t) = s_o(t) + J(t) + v(t), \tag{1}$$

where $S(t)$ denotes the total signal received by the radar, $s_o(t)$ represents the echo signal, $J(t)$ means the jamming signal, and $v(t)$ is the noise signal.

2.1. Echo Signal

Typical modulation types of radar echo signals include continuous wave (CW), linear frequency modulation (LFM), phase shift keying (PSK), frequency shift keying (FSK), etc. LFM, with its frequency linearly changing over time, is the most widely used. Therefore, in this paper, we focus on studying the typical echo signal of LFM, which can be expressed as follows [33]:

$$s_o(t) = \sum_{i=1}^{N} \mathrm{rect}\left(\frac{t - iT_r}{T_p}\right) \mathrm{e}^{\left(j2\pi\left(\left(f_o - \frac{B}{2}\right)(t-iT_r) + \frac{K_{LF}}{2}(t-iT_r)^2\right)\right)} \tag{2}$$

where i denotes the pulse sequence; $\mathrm{rect}\left(\frac{t-iT_r}{T_p}\right)$ represents the i-th rectangular pulse with a width of T_p; T_r is the pulse repetition period; T_p is the pulse width; f_o is the radar center frequency; B means the bandwidth; and $K_{LF} = \frac{B}{T_p}$ is the LFM coefficient.

2.2. Jamming Signal

In this paper, we select nine typical radar active jamming patterns as the research objects, including: radio noise jamming (RNJ) [26], amplitude modulation jamming (AMJ) [34], frequency modulation jamming (FMJ) [34], comb spectrum jamming (CJ) [13], phase modulation jamming (PMJ) [34], linear sweep frequency jamming (LSFJ) [34], non-linear sweep frequency jamming (NLSFJ) [35], hopping frequency jamming (HFJ) [35] and periodic Gaussian pulse jamming (PGJ) [36].

2.2.1. Radio Noise Jamming

RNJ is a narrowband Gaussian process generated by filtering and then amplifying the noise signal. Its mathematical model is expressed as:

$$J(t) = U_r(t)e^{j(\omega t + \varphi)} \tag{3}$$

where $U_r(t)$ obeys a Gaussian distribution, ω is the carrier frequency, and φ is the initial phase, following a uniform distribution on $[0, 2\pi]$.

2.2.2. Amplitude Modulation Jamming

The model of AMJ is represented as:

$$J(t) = [U_0 + K_{AM}U_n(t)]e^{j(\omega t + \varphi)} \tag{4}$$

where $U_n(t)$ is zero-mean Gaussian white noise, U_0 is the carrier voltage, and K_{AM} is the amplitude modulation index.

2.2.3. Frequency Modulation Jamming

FMJ is a type of barrage jamming, and its model is expressed as:

$$J(t) = U_0 \cdot e^{j\left(\omega t + 2\pi K_{FM} \int_0^t u(t')dt' + \varphi\right)}. \tag{5}$$

where K_{FM} is the frequency modulation index and $u(t)$ is a zero-mean stationary random process.

2.2.4. Comb Spectrum Jamming

CJ consists of multiple narrowband noise frequency modulation signals, and expresses as:

$$J(t) = \sum_{i=1}^{m} U_i e^{j\left[2\pi f_i t + 2\pi K_{FM} \int_0^t u(t)dt + \varphi\right]}. \quad (6)$$

where U_i is the amplitude, f_i represents the frequency where the comb teeth appear, and m is the number of frequencies.

2.2.5. Phase Modulation Jamming

The model of PMJ is:

$$J(t) = U_0 \cdot e^{j(\omega t + K_{PMU} u(t) + \varphi)} \quad (7)$$

where K_{PM} is the phase modulation index.

2.2.6. Linear Sweep Frequency Jamming

LSFJ varies linearly with time in a frequency band, and expresses as:

$$J(t) = U_0 e^{j\left(2\pi f_0 t + 2\pi K_{FM} t^2 + \varphi\right)} \quad (8)$$

where f_0 is the initial frequency.

2.2.7. Non-Linear Sweep Frequency Jamming

NLSFJ is similar to LSFJ, except that the instantaneous frequency magnitude of the jamming signal varies continuously with the square of time, and expresses as:

$$J(t) = U_0 e^{j\left(2\pi f_0 t + 2\pi K_{FM} t^3 + \varphi\right)} \quad (9)$$

2.2.8. Hopping Frequency Jamming

HFJ is a wideband non-stationary signal in which the frequency changes over time. The model of HFJ can be written as:

$$J(t) = \sum_{n=1}^{N} A_n e^{j(2\pi f_n t + \varphi_n)} \cdot p(t - nT_H). \quad (10)$$

where $\{A_n\}$ represents the amplitude sequence, $\{f_n\}$ is the pseudo-random frequency sequence, φ_n is the random phase sequence, T_H is the hop duration, and $p(t)$ is the base pulse signal with a pulse width of T_H.

2.2.9. Periodic Gaussian Pulse Jamming

PGP is a widely used active suppression jamming. The PGP is expressed as:

$$J(t) = \begin{cases} \Phi(t), 0 < t < \tau \\ 0, \tau < t < T \end{cases} \quad (11)$$

where $\Phi(t)$ represents a Gaussian function with a mean of 0 and a variance of 1, τ denotes the pulse duration, and T is the pulse period of the jamming.

The time-domain waveforms of the nine patterns are shown in Figure 1. As can be seen from the figure, there is little difference in the time domain between the various patterns of suppressed jamming signals, which are more difficult to distinguish manually and are easily disturbed by noise.

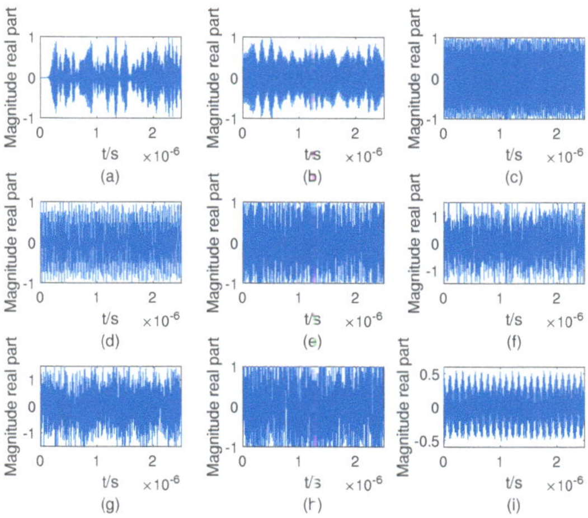

Figure 1. Time-domain waveforms of nine patterns of radar jamming signals. From (**a**) to (**i**): (**a**) RNJ, (**b**) AMJ, (**c**) FMJ, (**d**) CJ, (**e**) PMJ, (**f**) LSFJ, (**g**) NLSFJ, (**h**) HFJ, (**i**) PGJ.

3. Proposed Method

Assuming that the dataset of jamming signals received by the radar is $\mathcal{T}$, it consists of $N_c + N_u$ kinds of patterns of jamming. The subset $\mathcal{T}_c$ is composed of samples of N_c kinds of known patterns, while the subset $\mathcal{T}_u$ is composed of samples of N_u kinds of unknown patterns. The two subsets are complementary, meaning $\mathcal{T}_c \cup \mathcal{T}_u = \mathcal{T}$ and $\mathcal{T}_c \cap \mathcal{T}_u = \varnothing$. For the known patterns set $\mathcal{T}_c = \left\{ \left(\mathbf{x}_i^c, y_i^c, \mathbf{z}_{y_i^c} \right) \mid \mathbf{x}_i^c \in \mathcal{X}_c^{n_s \times N_d}, y_i^c \in \mathcal{Y}_c = \{1, 2, \cdots, N_c\}, \mathbf{z}_{y_i^c} \in \mathbf{Z}_c^{N_c \times N_k} \right\}$ containing n_c samples of N_c kinds of patterns, where $\mathbf{x}_i^c$ represents a N_d-dimensional vector of the i-th sample, y_i^c is the label of the sample, and $\mathbf{z}_{y_i}$ denotes the N_k-dimensional semantic vector describing the features of the sample's corresponding pattern to be obtained by supervised learning. Similarly, for the unknown patterns dataset consisting of n_u samples of N_c kinds of patterns, $\mathcal{T}_u = \left\{ \mathbf{x}_i^u, y_i^u, \mathbf{z}_{y_i^u}^u \right\}$, where $\mathbf{x}_i^u \in \mathcal{X}_u^{n_u \times N_d}$ is the N_d-dimensional feature vector of the i-th sample, $y_i^u \in \mathcal{Y}_u = \{N_c + 1, N_c + 2, \ldots, N_c + N_u\}$ represents the label of the sample, and $\mathbf{z}_{y_i^u} \in \mathbf{Z}_u^{N_u \times N_k}$ denotes the N_k-dimensional semantic information of the sample's corresponding pattern.

For the generalized zero-shot learning classification task considered in this paper, the supervised training phase only allows the utilization of dataset of known patterns $\mathcal{T}_c$. However, the objective is to ensure that the model trained in the supervised training phase can accurately classify $\mathcal{X}_c \cup \mathcal{X}_u$ into the $N_c + N_u$-dimensional space $\mathcal{Y}_c \cup \mathcal{Y}_u$ during the unsupervised classification phase.

Figure 2 illustrates the network framework of the RCAE-OWR algorithm, which consists of a supervised training phase and an unsupervised classification phase. In the supervised training phase, the residual convolutional autoencoder (RCAE) network is used to extract semantic features of known jamming patterns. Meanwhile, the network is trained using center loss, cross-entropy loss, and reconstruction loss. In [22,37], time-domain signals are directly used as inputs to the network to extract the deep features of different signals, and the simulation results prove that the direct time-domain signal-based recognition methods obtain good recognition performance in terms of accuracy and speed, demonstrating a huge potential for radar signal processing. Motivated by [22,37], in this paper, we also directly feed the IQ data of the jamming signals into the model. Then, after the supervised training, the unsupervised classification phase was entered. At this phase, the parameters of encoder of the RCAE are kept fixed, and both known and unknown jamming samples are input to the encoder to obtain their semantic features. Then, a distance-based discriminative method is employed to achieve open world recognition of radar jamming signals.

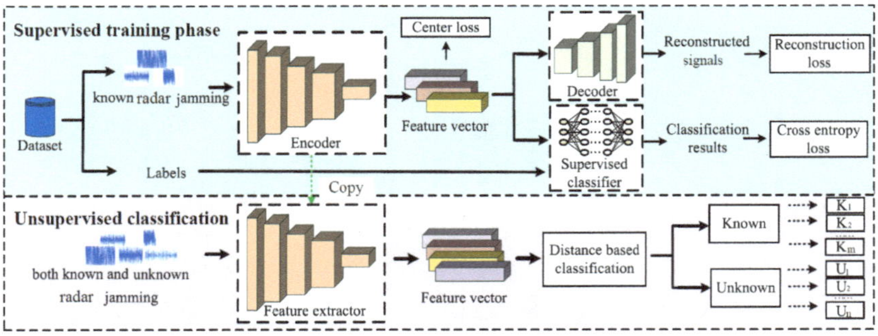

Figure 2. Radar jamming recognition framework of RCAE-OWR.

3.1. Supervised Training

The supervised training phase mainly consists of an autoencoder network (AE) and a supervised classification network classifier. The autoencoder is divided into two parts: the encoder and the decoder. In the supervised training phase, the focus is on constructing the mapping relationship between the time-domain signal and the semantic features.

3.1.1. Residual Convolutional Autoencoder

Due to the simplicity of the traditional AE structure, this study primarily considers the Residual Convolutional Autoencoder (RCAE). It replaces the fully connected layers in AE with convolutional layers and pooling layers, inheriting the characteristics of an autoencoder. This enables better feature learning and improves the efficiency of feature learning in AE. To prevent degradation in recognition performance, a residual network structure is employed. The input signals are IQ dual-channel data with a length of 512, resulting in a dimension of 2×512. Additionally, to maintain the vector dimensions after convolution, we have set the convolution kernel size to 3×3, padding = 1 and stride = 1. RCAE is based on the semantic autoencoder (SAE) architecture [31], and SAE enables mapping functions learned from known patterns to be better generalized to unknown patterns, which can effectively resist the domain shift problem [38]. In the encoding process, convolutional operations are used to extract features from input samples and obtain semantic vectors. Then, the decoder utilizes transpose convolution to reconstruct the semantic vectors and restore them to the original inputs.

The basic components of RCAE include the input layer, convolutional layers, semantic layer, deconvolutional layers, and output layer.

The RCAE designed in this paper is illustrated in Figure 3. It replaces the fully connected layers (FC) in the AE with convolutional layers and pooling layers, inheriting the characteristics of the AE for better feature learning. In addition, to prevent degradation of recognition performance, a residual structure is employed. The encoder uses convolutional operations to extract features from the input samples to obtain the semantic vector; the decoder utilizes transposed convolution to reconstruct the semantic vector and reduce it to the original signal.

The basic components of RCAE include: input layer, convolutional layer, semantic layer, deconvolutional layer and output layer.

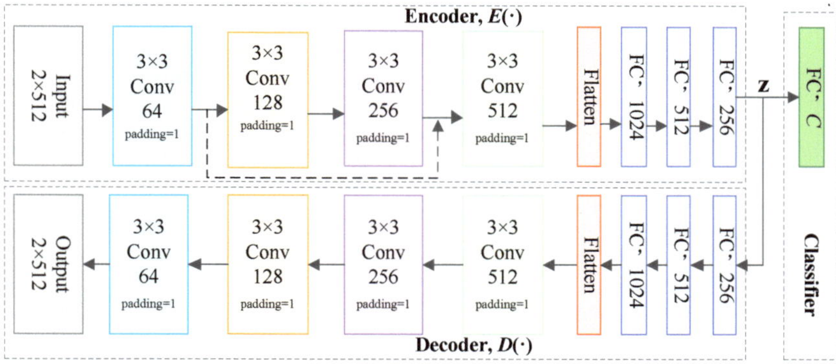

Figure 3. Residual Convolutional Autoencoder Network Structure.

In the encoding part, the input layer receives the input data $\mathbf{x}_i^c$ and passes it to the encoder. The encoder gradually extracts the semantic features of the input data through multiple convolutional layers and their residual structures, denoted as $\mathbf{z} = E\left(\mathbf{x}_i^c\right)$, where $E(\cdot)$ denotes the mapping function of the encoder.

The convolutional layers extract features from the input data $\mathbf{x}_i^c$ using convolutional operations. The data processing can be described as follows:

$$\mathbf{g}_i = f[\text{conv}(\mathbf{x}, \mathbf{k}_i^{\text{conv}}) + b_i^{\text{conv}}]. \tag{12}$$

where $\mathbf{k}_i^{\text{conv}}$ is the i-th convolutional kernel matrix, $\text{conv}(\cdot)$ denotes the convolution operation, b_i^{conv} represents the i-th bias term, $f(\cdot)$ represents the activation function. The feature layer integrates the diverse features extracted by the convolutional layers and outputs the semantic vector $\mathbf{z}$.

In the decoding part, the semantic feature $\mathbf{z}$ is up-sampled through transpose convolutional operations, aiming to reconstruct the original signal $\tilde{\mathbf{x}}_i^c$ based on the semantic features. This process is denoted as $\tilde{\mathbf{x}}_i^c = D(\mathbf{z})$, where $D(\cdot)$ denotes the mapping function of the decoder. Finally, the reconstructed results are outputted through the output layer. The data processing can be described as follows:

$$\mathbf{q}_i = f\left[\text{deconv}\left(\mathbf{z}, \mathbf{k}_i^{\text{dec}}\right) + b_i^{\text{dec}}\right] \tag{13}$$

where $\mathbf{k}_i^{\text{dec}}$ represents the i-th convolutional kernel matrix, deconv $(\cdot)$ denotes the transpose convolution operation, b_i^{dec} represents the i-th bias term.

The training process of RCAE aims to minimize the reconstruction error, ensuring the effectiveness of feature extraction. It can be expressed as:

$$L_{re} = \frac{1}{2M} \sum_{i=1}^{M} ||\tilde{\mathbf{x}}_i^c - \mathbf{x}_i^c||_2^2 \tag{14}$$

where M represents the number of samples in a batch, $\tilde{\mathbf{x}}_i^c$ is the signal reconstructed by $\mathbf{x}_i^c$ through the RCAE network.

To encourage the feature vectors of the same patterns to be close to their corresponding pattern centers and far from centers of different patterns, a center loss is introduced. During model training, the center loss assigns a center for each jamming pattern. Assuming the input sample is $\mathbf{x}_i^c$ with label y_i^c, and the center for pattern y_i is denoted as $\overline{\mathbf{z}}_{y_i^c}^c$. The center loss can be defined as:

$$L_{cl} = \frac{1}{M} \sum_{i=1}^{M} \left\| E(\mathbf{x}_i^c) - \overline{\mathbf{z}}_{y_i^c}^c \right\|_2^2 \tag{15}$$

During the model iteration process, the selection of pattern center $\overline{\mathbf{z}}^c_{y_i^c}$ is an important issue. Theoretically, the optimal center for pattern y_i^c would be the mean of all the feature vectors in that pattern. However, calculating the mean for all samples in each iteration would impose additional computational cost and reduce the efficiency of the model. To address this, the pattern center are initialized randomly, and then updated separately for each batch. The update process is as follows:

$$\Delta\overline{\mathbf{z}}^c_{y_i^c} = \frac{\sum_{m=1}^{M} \delta(y_i^c = y_m^c) \cdot \left(\overline{\mathbf{z}}^c_{y_i^c} - E(\mathbf{x}_m^c)\right)}{0.1 + \sum_{m=1}^{M} \delta(y_i^c = y_m^c)} \tag{16}$$

$$\overline{\mathbf{Z}}^{t+1}_{y_i^c} = \overline{\mathbf{Z}}^{t}_{y_i^c} - \alpha \cdot \Delta\overline{\mathbf{Z}}^{c}_{y_i^c} \tag{17}$$

where $\delta(y_i = k) = 1$ when $y_i = k$; otherwise, it is 0. $\overline{\mathbf{z}}^t_{y_i^c}$ is the semantic center of y_i^c at the t-th epoch, α is the learning rate.

3.1.2. Supervised Classifier

The encoder, acting as a feature extractor, takes the encoded features and feeds them into a fully connected layer followed by a softmax classifier, which outputs the predicted label. The loss function for this process utilizes cross-entropy between the predicted label and the true label.

$$L_{ce} = -\frac{1}{M}\sum_{i=1}^{M} \mathbf{y}_i \log(\hat{\mathbf{y}}_i) \tag{18}$$

where $\mathbf{y}_i$ represents the true label $\mathbf{x}_i^c$ in one-hot format, and $\hat{\mathbf{y}}_i$ represents the predicted probability vector.

The reconstruction loss L_{re} ensures that the reduced-dimensional semantic features are representative, the central loss L_{cl} promotes cohesion among feature vectors of the same jamming pattern, and the cross-entropy loss L_{ce} enhances the discriminative ability of feature vectors among different jamming patterns. To achieve both increased inter-class distance and reduced intra-class distance, a joint loss function is proposed:

$$L = L_{ce} + \lambda_{cl} L_{cl} + \lambda_{re} L_{re} \tag{19}$$

where λ_{cl} and λ_{re} are constants used to balance the weights of the three loss functions. The detailed gradient $\nabla_\theta L$ of L is shown in Appendix A.

The network parameters during the supervised training phase are updated as shown in Algorithm 1.

Algorithm 1 Pseudocode for supervised training of RCAE-OWR

Input: known jamming patterns dataset $\{(\mathbf{x}_1^c, y_1^c), \cdots, (\mathbf{x}_n^c, y_n^c)\}$ and hyperparameter.
Output: Optimal parameters θ.

while $epoch = 1, \ldots, N$ **do**
  for each batch with size M **do**
    Feed in a batch $\{(\mathbf{x}_1^c, y_1^c), \cdots, (\mathbf{x}_M^c, y_M^c)\}$.
    Calculate $\Delta\overline{\mathbf{Z}}^c_{y_i^c}$ via Equation (16).
    Calculate $\overline{\mathbf{Z}}^t_{y_i^c}$ via Equation (17).
    Calculate L_{ce} via Equation (18).
    Calculate L_{cl} via Equation (15).
    Calculate L_{re} via Equation (14).
    Calculate L via Equation (19).
    Update $\theta : \theta \leftarrow \theta - \eta\nabla_\theta L$.
  end for
end while

3.2. Unsupervised Classification

After learning the mapping relationship between the original time-domain signals and the semantic feature space in the supervised training phase, this learned relationship can be generalized to the unknown jamming patterns dataset. Finally, the unknown patterns can be recognized and classified using the semantic features. Inspired by [39], a distance-based classification rule is proposed.

3.2.1. Known Jamming Patterns Classification

Sequentially, the n_c known jamming samples from the training set are inputted into the trained RCAE-OWR model to extract semantic features. The semantic center vector $\overline{\mathbf{z}}_k^c$ corresponding to the k-th known jamming pattern is defined as:

$$\overline{\mathbf{z}}_k^c = \frac{\sum_{i=1}^{n_c} \delta(y_i = k)\mathbf{z}_{y_i^c}}{\sum_{i=1}^{n_c} \delta(y_i = k)}, k = 1, 2, \ldots, N_c \tag{20}$$

where $\mathbf{z}_{y_i^c}$ denotes the semantic feature extracted from the i-th input sample $\mathbf{x}_i^c$. When a test sample $\mathbf{t}$ is input, its semantic features $E(\mathbf{t})$ are extracted by the encoder, and its Mahalanobis distance to the center vectors of each known jamming pattern is calculated as:

$$d(E(\mathbf{t}), \overline{\mathbf{z}}_k^c) = \sqrt{(E(\mathbf{t}) - \overline{\mathbf{z}}_k^c)^T \mathbf{\Omega}_k^{-1} (E(\mathbf{t}) - \overline{\mathbf{z}}_k^c)} \tag{21}$$

where $\mathbf{\Omega}_k$ is the diagonal covariance matrix corresponding to $\overline{\mathbf{z}}_k$, and $\mathbf{\Omega}_k^{-1}$ is its inverse matrix.

Let $d_1 = \min_k d(E(\mathbf{t}), \overline{\mathbf{z}}_k^c)$. If $d_1 \leq \Theta_c$, then $\mathbf{t}$ belongs to the known jamming patterns. Here, $\Theta_c = \lambda_c \times 3\sqrt{N_\Omega}$ is a given threshold determined by the 3σ criterion [34], where λ_c is a coefficient and N_Ω is the dimension of $\mathbf{\Omega}_k$. In this case, the label $y_\mathbf{t}$ of $\mathbf{t}$ can be determined as follows:

$$y_\mathbf{t} = \arg\min_k (d(E(\mathbf{t}), \overline{\mathbf{z}}_k^c)) \tag{22}$$

when $d_1 > \Theta_c$, $\mathbf{t}$ belongs to the unknown jamming patterns, and its label $y_\mathbf{t}$ in this case is described in Section 3.2.2.

3.2.2. Unknown Jamming Patterns Classification

Let $\mathcal{T}_U$ represent the dataset of unknown jamming patterns. If $\mathcal{T}_U = \varnothing$, then $\mathbf{t}$ belongs to the first sample of a new unknown jamming pattern, and its label is $y_\mathbf{t} = \mathcal{U}_1$. The sample $\mathbf{t}$ is then recorded in the $\mathcal{T}_u$. If $\mathcal{T}_U \neq \varnothing$, the following rules are applied to determine whether $\mathbf{t}$ belongs to a previously recorded unknown jamming pattern or a new unknown jamming pattern. First, let $\overline{\mathbf{z}}_i^u$ represent the semantic center vector of pattern $\mathcal{U}_i$ in the $\mathcal{T}_U$:

$$\overline{\mathbf{z}}_i^u = \frac{\sum_{k \in \mathcal{T}_u} \delta(y_k = \mathcal{U}_i)\mathbf{z}_{y_k^u}}{\sum_{k \in \mathcal{T}_u} \delta(y_k = \mathcal{U}_i)}, i = 1, 2, \cdots, N \tag{23}$$

where N is the number of patterns that already exist in $\mathcal{T}_U$. $\mathbf{z}_{y_j^u}$ denotes the semantics extracted from the j-th unknown pattern sample $\mathbf{x}_j^u$. Let $d_2 = \min_{\mathbf{z}_i} d(E(\mathbf{t}), \overline{\mathbf{z}}_i^u)$. If $d_2 \leq \Theta_u$, then $\mathbf{t}$ belongs to an existing unknown jamming pattern. Here, $\Theta_u = \lambda_u \max(d_c, d_u)$ is a given threshold, where λ_u is a coefficient, d_c and d_u are defined as follows:

$$d_c = \max_{\overline{\mathbf{z}}_k^c} d(E(\mathbf{t}), \overline{\mathbf{z}}_k^c) \tag{24}$$

$$d_u = \max_{\overline{\mathbf{z}}_i^u} d(E(\mathbf{t}), \overline{\mathbf{z}}_i^u) \tag{25}$$

In this case, the label $y_\mathbf{t}$ of $\mathbf{t}$ can be determined as follows:

$$y_\mathbf{t} = \arg\min_{u_i} (d(E(\mathbf{t}), \overline{\mathbf{z}}_i^u)) \tag{26}$$

Otherwise, **t** belongs to a new unknown jamming pattern, and its label is $y_{\mathbf{t}} = \mathcal{U}_{N+1}$. The test sample **t** is then recorded in the $\mathcal{T}_u$.

In summary, the recognition rules for the unsupervised classification phase are presented in Algorithm 2.

Algorithm 2 Pseudocode for unsupervised classification of RCAE-OWR

Input: test sample $\mathbf{t}_i$ and well-trained weight of Encoder, θ_c.
Output: Jamming pattern y_i.

for $i = 1, \ldots, N_{\text{test}}$ **do**
  Calculate semantic vector $\mathbf{z}_i$ of $\mathbf{t}_i$
  $d_1 = \min_k d(E(\mathbf{t}), \bar{\mathbf{z}}_k^c)$
  $d_2 = \min_i d(E(\mathbf{t}), \bar{\mathbf{z}}_i^u)$
  $d_c = \max_{\bar{\mathbf{z}}_k^c} d(E(\mathbf{t}), \bar{\mathbf{z}}_k^c)$
  $d_u = \max_{\bar{\mathbf{z}}_i^u} d(E(\mathbf{t}), \bar{\mathbf{z}}_i^u)$
  $\Theta_u = \lambda_u \max(d_c, d_u)$
  if $d_1 \leq \Theta_c$ **then**
    $y_i = \arg\min_k (d(E(\mathbf{t}), \bar{\mathbf{z}}_k^c))$
  else if $d_1 > \Theta_c$ and $\mathcal{T}_U = \varnothing$ **then**
    add **t** to $\mathcal{T}_u$
    $y_i = \mathcal{U}_1$
  else if $d_1 > \Theta_c, \mathcal{T}_U \neq \varnothing$ and $d_2 \leq \Theta_u$ **then**
    $y_i = \arg\min_{\mathcal{U}_i} (d(E(\mathbf{t}), \bar{\mathbf{z}}_i^u))$
  else
    $y_i = \mathcal{U}_{N+1}$
    add **t to** $\mathcal{T}_u$**.**
  end if
end for

4. Performance Evaluation

4.1. Simulation Parameter Settings

The simulations are performed on a PC with an Intel(R) Core(TM) i7-9700 CPU and a GeForce RTX2070s GPU. The deep learning framework PyTorch is used for training and testing the neural network. The RCAE-OWR network is initialized with random weights, and the learning rate is set to 0.001 . The batch size is set to 256, and the number of epochs is set to 250. The values of $\lambda_{cl}, \lambda_{re}, \lambda_c$, and λ_u are set to $0.02, 10, 0.5$, and 1.5, respectively. Furthermore, grid search is applied to ascertain the optimal hyperparameters. Specifically, we train the model for each possible combination among all the candidate parameters by loop traversal and pick the parameter combination that minimizes the validation set error as the optimal hyperparameters.

4.2. Original Dataset

The jamming signals are generated using MATLAB. The dataset includes the nine kinds of radar active jamming signals mentioned in Section 2.2, with detailed parameters shown in Table 1. The data description is shown in Table 2. Gaussian white noise is added during the simulation, and the JNR ranges from -4 dB to 18 dB with a step size of 2 dB. For each JNR, 1000 samples are generated for each type of radar active jamming signal. In the following experiments, N_c kinds of the jamming patterns will be treated as known jamming, while the remaining $9 - N_c$ kinds of patterns will be treated as unknown jamming.

Table 1. Nine kinds of Radar Active Jamming Parameter Settings.

Jamming Pattern	Parameter Description
RNJ	Carrier frequency: 60∼110 MHz.
AMJ	Carrier frequency: 60∼110 MHz, bandwidth: 5∼10 MHz, K_{AM}: 0.1∼0.9
FMJ	Carrier frequency: 60∼110 MHz, bandwidth: 5∼10 MHz, K_{FM}: 0.1∼0.9
CJ	Carrier frequency: 60∼110 MHz, bandwidth: 5∼10 MHz, number of bands: 2∼4
PMJ	Carrier frequency: 60∼110 MHz, bandwidth: 5∼10 MHz, K_{PM}: 0.1∼0.9
LSFJ	Starting frequency: 1∼10 MHz, ending frequency: 50∼100 MHz.
NLSFJ	Starting frequency: 1∼10 MHz, ending frequency: 50∼100 MHz.
HFJ	$\mathcal{N}$ = 20, $\{f_c\}$: [10, 100] MHz, T_H: 32∼64 µs
PGJ	Pulse period T: [2.5, 10] µs . Duty cycle: [1/8, 1/2]

Table 2. The original dataset.

Total Samples	Samples of Each Pattern	Samples Each JNR	Feature Dimension	Patterns
108,000	12,000	1000	2 × 512	9
patterns				
RNJ, AMJ, FMJ, CJ, PMJ, LSFJ, NLSFJ, HFJ, PGJ				
number of JNR values		JNR values		
12		−4, −2, 0, 2, 4, 6, 8, 10, 12, 14, 16, 18		

4.3. Training Process

Firstly, N_c patterns are selected from the original dataset as known patterns, and the remaining $9 - N_c$ patterns are considered unknown patterns. The samples from known patterns are used not only for training the network but also for testing. However, the samples from unknown patterns are exclusively used for testing purposes. To reduce the impact of noise on the training results, during the training phase, only the known pattern samples with a JNR higher than 12dB are used to train the RCAE-OWR network. For each JNR, 700 samples are randomly selected from the known pattern samples to form the training set, and another 100 samples are randomly chosen to construct the validation set. In the testing phase, a wider range of JNR values is used, namely, JNR = −4–18 dB. For each of the nine jamming patterns, 200 samples are randomly selected at each JNR value to form the test set. It is important to note that the training set, validation set and test set for the known patterns samples are mutually exclusive and do not overlap with each other.

Figures 4–6 show the curves of cross-entropy loss L_{ce}, central loss L_{cl} and reconstruction loss L_{re} of the proposed RACE-OWR algorithm with the number of iterations. It can be seen from the figures that the three losses are gradually decreasing and converging after the epoch exceeds 50, which illustrates the effectiveness and the fast convergence and low complexity of the algorithm. In the convergence stage, the reconstruction loss approaches approximately 0.274, which indicates that the features extracted by the network designed in this paper can effectively represent the original signal and further provide a guarantee for unsupervised classification. The cross-entropy loss converges to 0.82, which contributes to the separation of different patterns, while the central loss converges to 0.12, which contributes to the aggregation of samples from the same pattern.

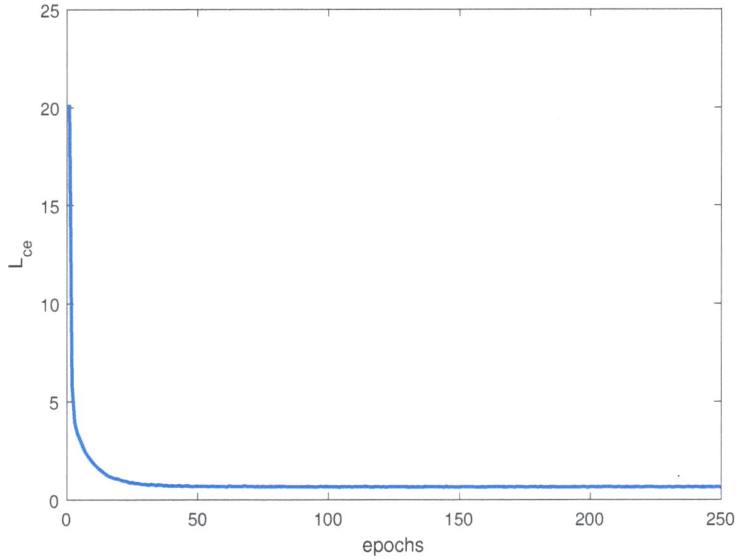

Figure 4. Cross-entropy loss.

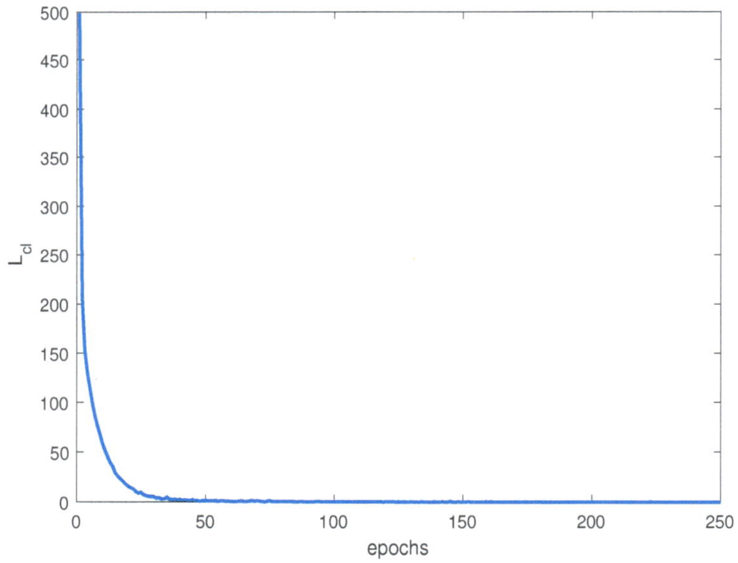

Figure 5. Center loss.

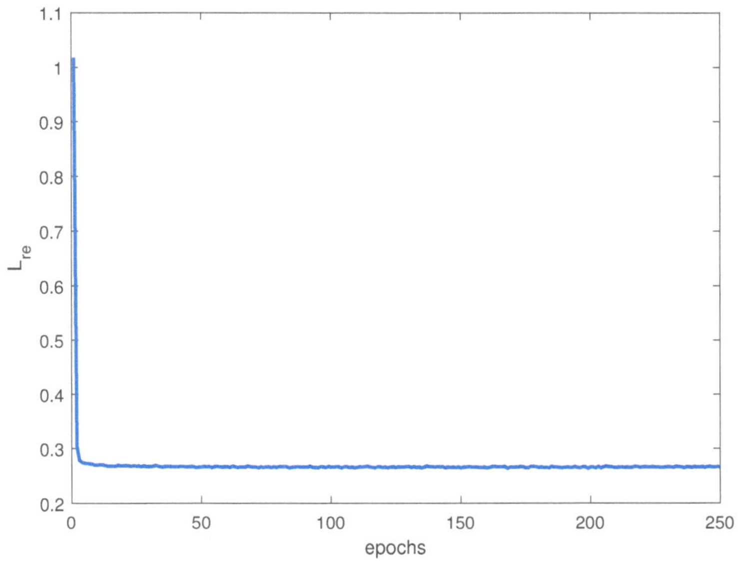

Figure 6. Reconstruction loss.

4.4. Impact of Different Combinations of Unknown Patterns on Classification Performance

In order to investigate the influence of different combinations of unknown patterns on the algorithm's classification performance, we conducted experiments by randomly selecting two jamming patterns from the dataset as unknown patterns. We define two metrics for evaluation: true known rate (TKR) and true unknown rate (TUR). TKR is calculated as TK/K, and TUR is calculated as TU/U, where TK and TU represent the number of correctly identified samples from known and unknown patterns, respectively, while K and U represent the total number of samples from known and unknown patterns. For ease of comparison, we ensure that the test set's JNR is consistent with the training set, ranging from 12 to 18 dB. The experimental results are summarized in Table 3. From the table, it can be observed that at higher JNR values, the algorithm effectively distinguishes between known and unknown patterns, validating the efficacy of our proposed approach. However, under the three combinations tested, neither TKR nor TUR could reach 1. This indicates that some samples from the known patterns are misclassified as unknown, and vice versa.

Table 3. Performance with different unknown pattern combinations.

Indicator	Unknown Patterns		
	RNJ, HFJ	**AMJ, PGJ**	**RNJ, NLSFJ**
Average accuracy	0.957	0.965	0.978
TKR	0.966	0.982	0.994
TUR	0.999	0.999	0.995

4.5. Impact of λ_c on Algorithm Performance

The value of Θ_c, which determines the proportion of jamming signal samples classified as known or unknown patterns, is influenced by the value of λ_c. In Section 3.2.1, Θ_c is defined as $\Theta_c = \lambda_c \times 3\sqrt{N_\Omega}$, where N_Ω is a constant representing the dimension of the matrix. Therefore, this section investigates the impact of λ_c on the algorithm's performance. Since TKR and TUR cannot both increase with the increase in λ_c, a weighted true rate (WTR) [39] is defined to balance these two metrics, i.e., $WTR = \xi \times TKR + (1 - \xi) \times TUR$, where ξ is a balancing factor set to 0.5 , indicating equal importance given to TKR and

TUR. Figure 7 shows the curves of TKR, TUR, and WTR as λ_c varies. It can be observed that as λ_c increases, TKR increases while TUR decreases. This is because as λ_c increases, the value of Θ_c increases, so more samples are discriminated as known jamming patterns and fewer samples are discriminated as unknown jamming patterns. The value of WTR initially increases and then decreases with the increase in λ_c, reaching its maximum at a value of approximately 0.5∼0.7.

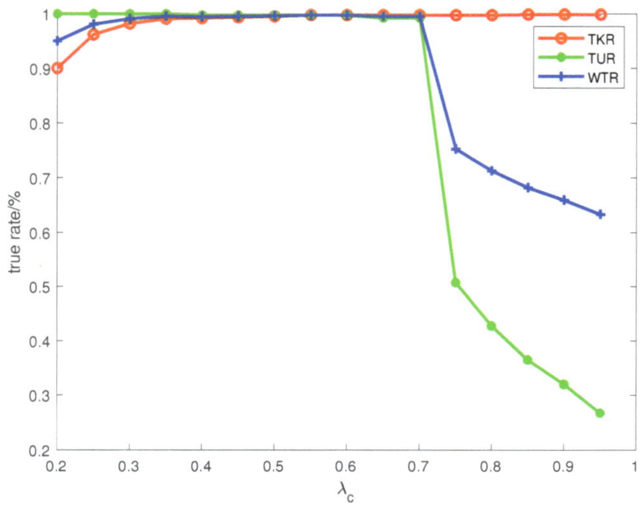

Figure 7. The impact of λ_c on recognition rate.

4.6. Ablation Study

In this research, we conduct an ablation study to comprehend the individual contributions of component loss functions in the context of the RCAE-OWR learning model. By systematically eliminating specific component loss functions, we investigate their impact on the overall performance. The experimental findings, as summarized in Table 4, reveal crucial insights into the significance of each loss function. Notably, the removal of L_{ce} (Cross-Entropy Loss) leads to the most substantial performance degradation, resulting in an accuracy of 95.1%. These results underscore the paramount role played by the cross-entropy loss in the learning process, surpassing the influence of the other two component loss functions, namely, L_{re} and L_{ct}. Furthermore, we observe that the incorporation of Lre enhances the accuracy by 3.3%. This improvement can be attributed to the ability of L_{re} to preserve essential semantic features, thereby facilitating the differentiation between known and unknown patterns. Additionally, the utilization of the center loss function, L_{ct} yields a minor enhancement in our learning model's performance.

Table 4. Ablation study on RCAE-OWR. L_{ce}: cross entropy loss. L_{cl}: center loss. L_{re}: reconstruction loss.

Variant	True Rate
RCAE-OWR(L_{ce},L_{cl} and L_{re})	0.995
without L_{ce}	0.951
without L_{cl}	0.982
without L_{re}	0.962

4.7. Open set Recognition Performance

When N_c is set to 7 as well as PGJ and NLSFJ as unknown patterns, the accuracy of the RCAE-OWR algorithm for the seven known jamming patterns and the detection rate for the two unknown jamming patterns are shown in Figure 8. It can be observed that as the JNR increases, the recognition rates for various patterns of jamming also increase.This is because with a higher JNR, the signal becomes cleaner, and the extracted features become more discriminative. When the JNR is below −2 dB, the various jamming signals are submerged in the noise, and the extracted semantic features are not sufficient to separate them. As a result, samples of known jamming patterns are often recognized as samples of unknown jamming patterns, leading to a higher detection rate for unknown jamming patterns compared to known jamming patterns.

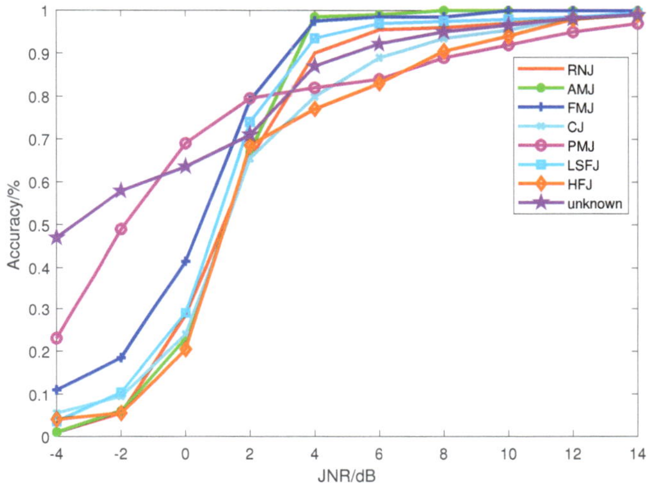

Figure 8. Open set recognition of RCAE-OWR.

Table 5 presents the accuracy of the RCAE-OWR algorithm and the supervised-based algorithm based on seven known jamming patterns when the JNR is set to 10dB. For a fair comparison, the network structure of the supervised-based algorithm is consistent with the encoder of the RCAE-OWR network in this paper. The difference lies in the fact that the supervised-based algorithm directly outputs the probabilities of the seven jamming patterns as the recognition results, while RCAE-OWR does not have prior knowledge of the specific number of patterns. Therefore, Algorithm 2 is used for jamming pattern recognition in RCAE-OWR. As the supervised-based algorithm is designed for closed-set recognition, it achieves higher detection rates. On the other hand, the RCAE-OWR algorithm is designed for open-set recognition, which means that known jamming patterns can be wrongly recognized as unknown jamming patterns, and unknown jamming patterns can be misclassified as known jamming patterns. As a result, the recognition rate of RCAE-OWR is lower than that of the supervised-based algorithm.

Table 5. Comparison of detection performance between two algorithms.

Method	RNJ	AMJ	FMJ	CJ	PMJ	LSFJ	HFJ
supervised	98%	100%	99%	95%	92%	98%	95%
RCAE-OWR	96%	100%	98.5%	93.5%	89%	97.5%	90.5%

The confusion matrix of RCAE-OWR for known jamming patterns and unknown jamming patterns is shown in Figure 9 when JNR is set to 18 dB. It can be observed from the figure that almost all samples are correctly classified into their respective patterns. This is because with higher levels of jamming, the extracted signal features better represent the original signal, resulting in increased intra-class cohesion and expanded inter-class separation. However, it is also evident that there are cases where known jamming patterns are recognized as unknown jamming patterns, and vice versa. Therefore, our algorithm needs to ensure a high detection rate for unknown jamming patterns while minimizing the impact on the recognition of known jamming patterns.

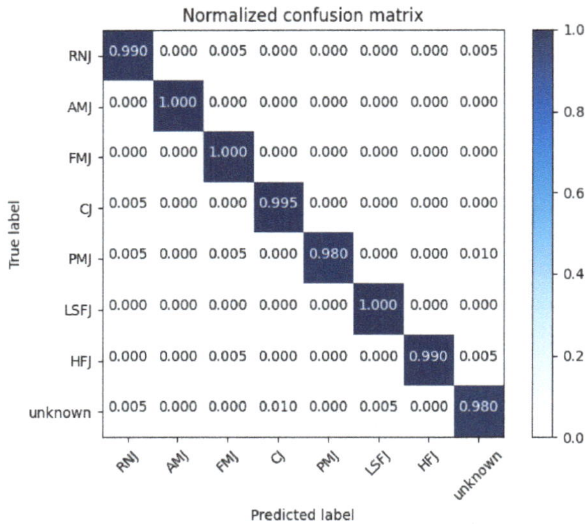

Figure 9. RCAE-OWR open set recognition results when JNR = 18 dB.

4.8. *Analysis of Unknown Jamming Patterns Recognition Performance in RCAE-OWR*

To evaluate the recognition performance of RCAE-OWR for unknown jamming patterns, *Precision*, *Recall*, and *F*1 are introduced as performance metrics [6]. *Precision* represents the ratio of the number of correctly classified samples to the sum of correctly classified samples and incorrectly classified samples. *Recall* represents the ratio of the number of correctly classified samples to the total number of samples. *F*1 is defined as the harmonic mean of *Precision* and *Recall*, given by $F1 = 2 \times Precision \times Recall/(Precision + Recall)$ [40].

4.8.1. Performance Comparison of Different Methods

In order to validate the effectiveness of our proposed RCAE-OWR, two other OWR-based methods, IOmSVM [41] and SNN-OWR [35] are implemented for comparison. The features extracted in the IOmSVM method are referenced from [12]. We utilize Openness [42] to measure the complexity of the open world task to describe the weight of the number of unknown patterns. Openness is defined as :

$$\text{Openness} = 1 - \sqrt{\frac{N_{\text{train}}}{N_{\text{test}}}} \tag{27}$$

where N_{train} denotes the number of patterns contained in the training set and N_{test} denotes the number of patterns contained in the test set. When the Openness is 0 , the task will move to closed-set recognition. In this paper, we set N_{test} to 9 and keep changing the value of N_{train} for experiments; *F*1 score is used to measure the performance of the algorithm and the results are shown in Figure 10. From the figure, it can be seen that as the Openness increases, the classification accuracy of the three methods decreases, which is attributed to the fact that more information about unknown patterns is fed into the model, posing a serious challenge to the classification of the model. Meanwhile, RCAE-OWR has the highest recognition accuracy at the same Openness, e.g., when the Openness is 0.25, the average *F*1 score of RCAE-OWR is 4.29% and 10.26% higher than that of SNN-OWR and IOmSVM, respectively.

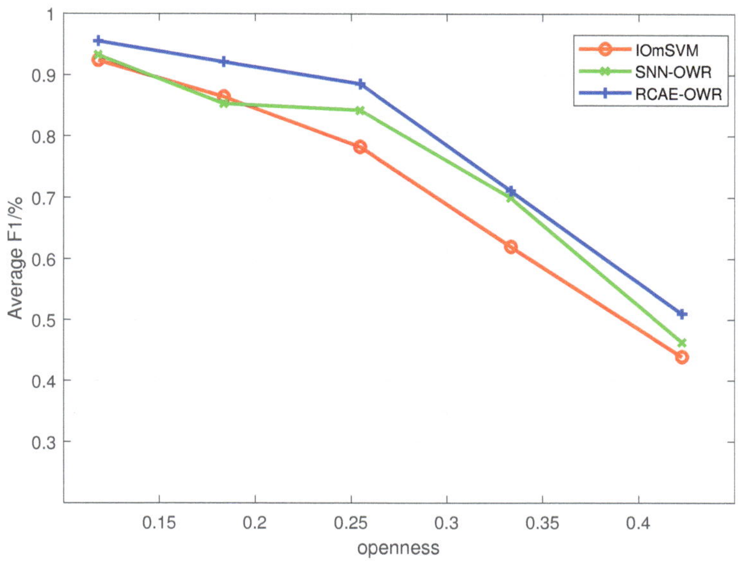

Figure 10. $F1$ scores against varying Openness under different methods.

Figure 11 illustrates the performance of the three methods as the JNR varies, with N_u set to 3. Notably, the detection performance of all three methods exhibits improvement as JNR increases, and the differences in $F1$ scores diminish with higher JNR values. Among the three OWR-based methods, RCAE-OWR stands out with the highest classification accuracy, particularly at low JNR. For instance, when the JNR is 8 dB, the average $F1$ score of RCAE-OWR surpasses that of SNN-OWR and IOmSVM by 6.85% and 12.23%, respectively.

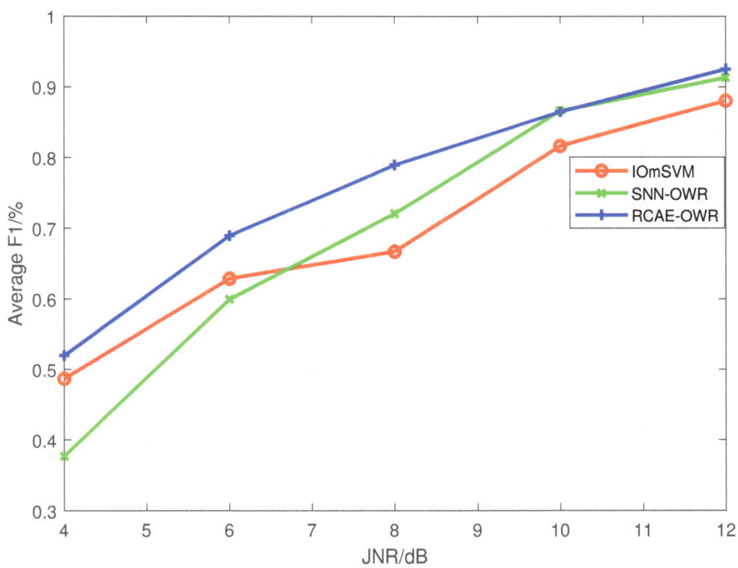

Figure 11. $F1$ scores against JNR under different methods.

Table 6 presents the offline training time and online recognition time of the three methods. It can be observed from the table that the IOmSVM method has the shortest training time, and likewise the recognition accuracy is the worst within the three methods.

Both SNN-OWR and RCAE-OWR require longer training times; however, once the models are trained effectively, they still achieve real-time jamming recognition.

Table 6. Runtime (in seconds) of different methods.

Method	Offline Training	Online Recognition
IOmSVM	753	0.0075
SNN-OWR	1506	0.0071
RCAE-OWR	1427	0.0062

4.8.2. Impact of Unknown Patterns on RCAE-OWR Performance

To investigate the influence of the number of unknown jamming patterns on the algorithm's recognition performance, we randomly selected two, three, four, five and six kinds of patterns out of the nine kinds of jamming patterns as unknown jamming patterns, while the remaining patterns were considered known jamming patterns. The experimental results are presented in Table 7, with JNR set to 12 dB. From the table, it can be observed that as the number of unknown jamming patterns decreases, the algorithm's detection performance improves, as indicated by an increase in the $F1$ score. For example, when the number of unknown jamming patterns is reduced from four to two, the average $F1$ score increases by 0.070. However, when the number of unknown patterns surpasses the number of known patterns, it poses a challenge to the algorithm's detection performance. For example, when increasing the number of unknown patterns from four to five, the average $F1$ score decreases by 0.174.

Table 7. Effect of Unknown jamming patterns on RCAE-OWR Performance.

Number of Patterns	Pattern	*Precision*	*Recall*	*F1*
2	Pattern 1	0.941	0.964	0.952
	Pattern 2	0.979	0.940	0.959
3	Pattern 1	0.916	0.940	0.927
	Pattern 2	0.929	0.928	0.928
	Pattern 3	0.922	0.900	0.910
4	Pattern 1	0.861	0.930	0.894
	Pattern 2	0.907	0.920	0.913
	Pattern 3	0.912	0.820	0.863
	Pattern 4	0.909	0.841	0.873
5	Pattern 1	0.841	0.724	0.778
	Pattern 2	0.781	0.696	0736
	Pattern 3	0.701	0.680	0.690
	Pattern 4	0.623	0.713	0.663
	Pattern 5	0.542	0.692	0.692
6	Pattern 1	0.779	0.598	0.677
	Pattern 2	0.712	0.545	0.613
	Pattern 3	0.603	0.536	0.567
	Pattern 4	0.481	0.533	0.505
	Pattern 5	0.365	0.452	0.402
	Pattern 6	0.243	0.405	0.301

4.8.3. Recognition Performance of RCAE-OWR for Unknown Jamming Patterns

The recognition rates of RCAE-OWR for unknown jamming patterns with JNR of 4∼14 dB are shown in Figure 12. It can be observed from the graph that when the number of unknown jamming patterns is fixed, the average $F1$ score increases with the increase in JNR. This is because at lower JNR levels, a large number of known jamming patterns are incorrectly classified as unknown, resulting in poor classification performance. As the JNR increases, the characteristics of different jamming signals become more distinct, and the samples of known patterns wrongly classified as unknown patterns decrease, leading to improved performance in classifying unknown patterns. Additionally, at lower JNR levels, there is a significant difference in the average $F1$ score among the three different jamming patterns. However, as the JNR increases, this difference decreases. For example, at JNR = 4 dB, the average $F1$ score for two unknown patterns is 0.3 and 0.65 higher than that for three and four unknown patterns, respectively. At JNR = 10 dB, the average $F1$ score for two unknown patterns is 0.08 and 0.15 higher than that for three and four unknown patterns, respectively. This also highlights the importance of improving the JNR of the dataset as a prerequisite and guarantee for open world recognition of radar jamming signals.

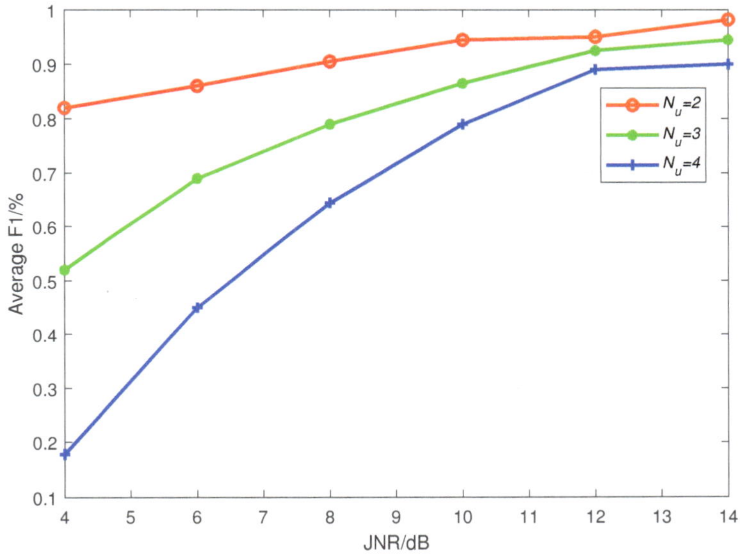

Figure 12. Recognition performance of RCAE-OWR for different unknown jamming patterns.

4.8.4. Performance of RCAE-OWR for Unknown Patterns under Different Combinations

To investigate the influence of JNR on the algorithm's performance with varying combinations of unknown patterns, we randomly extract three or four kinds of signals from the dataset to form the unknown patterns set, while the remaining are considered as known patterns for experimental evaluation. The results of these experiments are presented in Figure 13. From the figure, it can be seen that as the JNR increases, the overall classification accuracy also improves; this is because as the JNR increases, the signal is less affected by noise and is more discriminative. Notably, when the number of unknown patterns is fixed, the classification accuracy consistently increases with higher JNR values. Additionally, regardless of the specific combination of unknown patterns data sets, the classification accuracy exhibits minimal variation under the same JNR condition, highlighting the robustness of the RCAE-OWR algorithm.

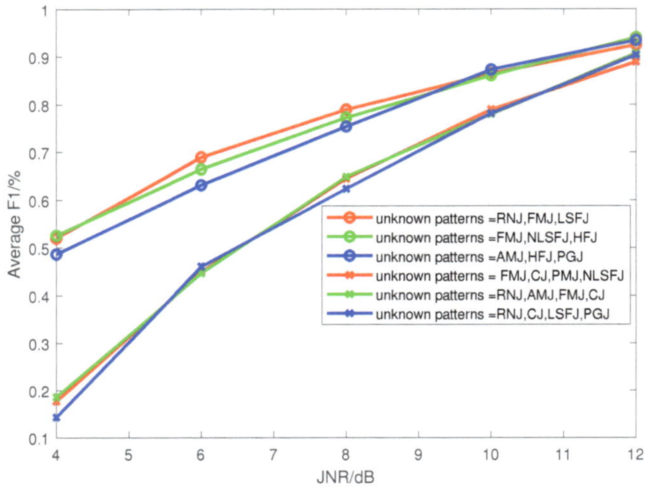

Figure 13. Performance under different unknown patterns.

5. Conclusions

In this study, we addressed the problem of open world recognition for radar jamming signals. Leveraging the idea of zero-shot learning, we proposed an open world recognition method called RCAE-OWR based on residual autoencoders. This method utilizes the features extracted by the encoder network to form the semantic centers of known jamming patterns. Then, a distance-based approach is proposed to classify known and unknown jamming patterns. Experimental results demonstrate that the RCAE-OWR algorithm can effectively recognize unknown jamming patterns, especially under high JNR. Although the proposed RCAE-OWR algorithm has demonstrated excellent classification performance in suppression jamming signals, there are still some limitations and drawbacks to address. For instance, this paper focused only on common jamming signals, while there are other jamming patterns, such as agile noise jamming and deceptive jamming. In the future, these patterns should be incorporated into our method. Additionally, deep learning has been widely applied to intelligent radar jamming signal recognition with a balanced training set. However, in most cases, an imbalanced training set is inevitable. Therefore, considering open world recognition of radar jamming under class-imbalanced conditions will be a challenging yet valuable task in future research.

Author Contributions: Conceptualization Z.Z. and Y.Z.; methodology, Y.Z. and Z.Z.; software, Y.B. and Y.Z.; validation, Z.Z.; formal analysis, Z.Z.; investigation, Z.Z. and Y.Z.; resources, Z.Z.; data curation, Y.B.; writing—original draft preparation, Y.Z.; writing—review and editing, Y.B. and Y.Z.; visualization, Z.Z.; supervision, Z.Z.; project administration, Y.Z.; funding acquisition, Z.Z. All authors have read and agreed to the published version of the manuscript.

Funding: This work was supported by the State Key Program of National Natural Science of China (Grant No. U19B2016) and Zhejiang Provincial Key Lab of Data Storage and Transmission Technology, Hangzhou Dianzi University.

Acknowledgments: The authors would like to thank an Editor who handled the manuscript and the anonymous reviewers for their valuable comments and suggestions.

Conflicts of Interest: The authors declare no conflict of interest.

Appendix A

Appendix A.1

Lemma A1. *Assuming the* $\nabla_\theta L_{ce}$, $\nabla_\theta L_{cl}$ *and* $\nabla_\theta L_{le}$ *are gradients of* L_{ce}, L_{cl} *and* $L_{e\varepsilon}$, *respectively, then there is:*

$$\begin{aligned}\nabla_\theta L &= \nabla_\theta L_{ce} + \lambda_{cl}\nabla_\theta L_{cl} + \lambda_{re}\nabla_\theta L_{re} \\ &= \frac{1}{M}\left(\sum_i (S(E(\mathbf{x}_i^c)) - y_i^c) + \sum_i E(\mathbf{x}_i^c) - \bar{\mathbf{z}}_{y_i^c}^c + \sum_i \tilde{\mathbf{x}}_i^c - \mathbf{x}_i^c\right)\end{aligned} \tag{A1}$$

Proof. because $L = L_{ce} + \lambda_{cl}L_{cl} + \lambda_{re}L_{re}, \therefore \nabla_e L$ can be notated as:

$$\nabla_\theta L_t = \nabla_\theta L_{ce} + \lambda_{rl}\nabla_\theta L_{kl} + \lambda_{ct}\nabla_\theta L_{ct} \tag{A2}$$

□

Notice that $L_{ce} = -\frac{1}{M}\sum_{i=1}^{M}\mathcal{L}_{ce}(\mathbf{y}_i, \hat{\mathbf{y}}_i)$, where $\mathcal{L}_{ce}(\mathbf{y}_i, \hat{\mathbf{y}}_i) = \sum_{j=1}^{N_c} y_{i,j}^c \log(\hat{y}_{i,j}^c)$, Let $\mathbf{u} = [u_1, u_2, \ldots, u_{N_c}]$ denote the output vector of the last layer of the network. After softmax operation, $\mathbf{u}$ yields $\hat{\mathbf{y}}_i$. That is, $\hat{\mathbf{y}}_i^j = S(E(\mathbf{x}_i^c))_j = \frac{e^{u_j}}{\sum_k e^{u_k}}$. where $S(\cdot)$ represents the softmax operation, and $S(E(\mathbf{x}_i))_j$ and u_j are the j-th elements of $S(E(\mathbf{x}_i^c))$ and $\mathbf{u}$, respectively. According to the chain derivative rule:

$$\frac{\partial \mathcal{L}_{ce}}{\partial \mathbf{u}} = \frac{\partial \mathcal{L}_{ce}}{\partial p}\frac{\partial p}{\partial \mathbf{u}} \tag{A3}$$

We have:

$$\begin{aligned}\frac{\partial \mathcal{L}_{ce}}{\partial u_j} &= \sum_m \frac{\partial \mathcal{L}_{cem}}{\partial S(E(\mathbf{x}_i^c))_m}\frac{\partial S(E(\mathbf{x}_i^c))_m}{\partial u_j} \\ &= \sum_m \frac{\partial - y_{i,m}^c \log S(E(\mathbf{x}_i^c))_m}{\partial E(\mathbf{x}_i^c)_m}\frac{\partial S(E(\mathbf{x}_i^c))_m}{\partial u_j} \\ &= \sum_m -\frac{y_{i,m}^c}{S(E(\mathbf{x}_i^c))_m}\frac{\partial S(E(\mathbf{x}_i^c))_m}{\partial u_j}\end{aligned}$$

Simplify to get:

$$\frac{\partial S(E(\mathbf{x}_i^c))_m}{\partial u_j} = \begin{cases} S(E(\mathbf{x}_i^c))_j\left(1 - S(E(\mathbf{x}_i^c))_j\right), m = j \\ -S(E(\mathbf{x}_i^c))_j S(E(\mathbf{x}_i^c))_m, m \neq j \end{cases} \tag{A4}$$

Hence, $\frac{\partial \mathcal{L}_{ce}}{\partial \mathbf{u}}$ can be rewritten as:

$$\begin{aligned}\frac{\partial \mathcal{L}_{ce}}{\partial u_j} =& \sum_m \frac{\partial \mathcal{L}_{cem}}{\partial S(E(\mathbf{x}_i^c))_m}\frac{\partial S(E(\mathbf{x}_i^c))_m}{\partial u_j} \\ =& \sum_m \frac{\partial - y_{i,m}^c \log S(E(\mathbf{x}_i^c))_m}{\partial E(\mathbf{x}_i^c)_m}\frac{\partial S(E(\mathbf{x}_i^c))_m}{\partial u_j} \\ =& S(E(\mathbf{x}_i^c))_j\left(1 - S(E(\mathbf{x}_i^c))_j\right) + \\ & \sum_{m \neq j} -\frac{y_{i,m}^c}{S(E(\mathbf{x}_i^c))_m}\left(-S(E(\mathbf{x}_i^c))_j S(E(\mathbf{x}_i^c))_m\right) \\ =& S(E(\mathbf{x}_i^c))_j - y_{i,j}^c\end{aligned}$$

therefore, $\frac{\partial \mathcal{L}_{ce}}{\partial \mathbf{u}} = S\left(E\left(\mathbf{x}_i^c\right)\right) - y_i^c$, according to $L_{ce} = -\frac{1}{M}\sum_{i=1}^{M}\mathcal{L}_{ce}(\mathbf{y}_i, \hat{\mathbf{y}}_i)$, we have:

$$\frac{\partial L_{ce}}{\partial \mathbf{u}} = \frac{1}{M}\sum_i S(E(\mathbf{x}_i^c)) - y_i^c \tag{A5}$$

Similarly, $\frac{\partial L_{cl}}{\partial \mathbf{u}} = \frac{\partial L_{cl}}{\partial E\left(\mathbf{x}_i^c\right)}$, and $L_{cl} = \frac{1}{M}\sum_{i=1}^{M}\left\|E\left(\mathbf{x}_i^c\right) - \overline{\mathbf{z}}_{v_i^c}^c\right\|_2^2$, we have:

$$\frac{\partial L_{cl}}{\partial \mathbf{u}} = \frac{\partial L_{cl}}{\partial E(\mathbf{x}_i)} = \frac{1}{M}\sum_i E(\mathbf{x}_i^c) - \overline{\mathbf{z}}_{y_i^c}^c \tag{A6}$$

Similarly,

$$\frac{\partial L_{re}}{\partial \mathbf{u}} == \frac{\partial L_{re}}{\partial E\left(\mathbf{x}_i^c\right)} = \frac{1}{M}\sum_i \tilde{\mathbf{x}}_i^c - \mathbf{x}_i^c \tag{A7}$$

In summary, it can be concluded that:

$$\begin{aligned}\nabla_\theta L &= \nabla_\theta L_{ce} + \lambda_{cl}\nabla_\theta L_{cl} + \lambda_{re}\nabla_\theta L_{re} \\ &= \frac{1}{M}\left(\sum_i (S(E(\mathbf{x}_i^c)) - y_i^c) + \sum_i E(\mathbf{x}_i^c) - \overline{\mathbf{z}}_{y_i^c}^c + \sum_i \tilde{\mathbf{x}}_i^c - \mathbf{x}_i^c\right)\end{aligned} \tag{A8}$$

References

1. Du, C.; Cong, Y.; Zhang, L.; Guo, D.; Wei, S. A Practical Deceptive Jamming Method Based on Vulnerable Location Awareness Adversarial Attack for Radar HRRP Target Recognition. *IEEE Trans. Inf. Forensics Secur.* **2022**, *17*, 2410–2424. [CrossRef]
2. Xu, H.; Quan, Y.; Zhou, X.; Chen, H.; Cui, T.J. A Novel Approach for Radar Passive Jamming Based on Multiphase Coding Rapid Modulation. *IEEE Trans. Geosci. Remote Sens.* **2023**, *61*, 1–14. [CrossRef]
3. Liu, Z.; Zhao, S. Unsupervised Clustering Method to Discriminate Dense Deception Jamming for Surveillance Radar. *IEEE Sensors Lett.* **2021**, *5*, 1–4. [CrossRef]
4. Lv, Q.; Quan, Y.; Feng, W.; Sha, M.; Dong, S.; Xing, M. Radar Deception Jamming Recognition Based on Weighted Ensemble CNN With Transfer Learning. *IEEE Trans. Geosci. Remote Sens.* **2022**, *17*, 1–11. [CrossRef]
5. Hua, Q.; Zhang, Y.; Wei, C.; Ji, Z.; Jiang, Y.; Wang, Y.; Xu, D. A Self-Supervised Method Based on CV-MUNet++ for Active Jamming Suppression in SAR Images. *IEEE Trans. Geosci. Remote Sens.* **2023**, *61*, 1–16. [CrossRef]
6. Luo, Z.; Cao, Y.; Yeo, T.S.; Wang, Y.; Wang, F. Few-Shot Radar Jamming Recognition Network via Time-Frequency Self-Attention and Global Knowledge Distillation. *IEEE Trans. Geosci. Remote Sens.* **2023**, *61*, 1–12. [CrossRef]
7. Huang, S.; Feng, Z.; Zhang, Y.; Zhang, K.; Li, W. Feature based modulation classification using multiple cumulants and antenna array. In Proceedings of the 2016 IEEE Wireless Communications and Networking Conference, Doha, Qatar, 3–6 April 2016; pp. 1–5.
8. Chen, T.; Gao, S.; Zheng, S.; Yu, S.; Xuan, Q.; Lou, C.; Yang, X. EMD and VMD Empowered Deep Learning for Radio Modulation Recognition. *IEEE Trans. Cogn. Commun. Netw.* **2023**, *9*, 43–57. [CrossRef]
9. Zhu, F.; Jiang, Q.Q.; Lin, C.; Xiao, Y. Typical wide band EMI identification based on support vector machine. *Syst. Eng. Electron.* **2021**, *43*, 2400–2406.
10. Lu, Y.; Li, S. CFAR detection of DRFM deception jamming based on singular spectrum analysis. In Proceedings of the 2017 IEEE International Conference on Signal Processing, Communications and Computing (ICSPCC), Xiamen, China , 22–25 October 2018; pp. 1–6.
11. Yang, S.; Tian, B.; Zhou, R. A jamming identification method against radar deception based on bispectrum analysis and fractal dimension. *J. Xian Jiaotong Univ.* **2016**, *50*, 128–135.
12. Xu, C.; Yu, L.; Wei, Y.; Tong, P. Research on Active Jamming Recognition in Complex Electromagnetic Environment. In Proceedings of the 2019 IEEE International Conference on Signal, Information and Data Processing (ICSIDP), Chongqing, China, 11–13 December 2019; pp. 1–5.
13. Wei, Y.; Li, Y.; Zhang, J.; Tong, P. Radar Jamming Recognition Method Based on Fuzzy Clustering Decision Tree. In Proceedings of the 2019 IEEE International Conference on Signal, Information and Data Processing (ICSIDP), Chongqing, China, 11–13 December 2019; pp. 1–5.
14. Yu, Y.; Li, J.; Li, J; Xia, Y.; Ding, Z.; Samali, B. Automated damage diagnosis of concrete jack arch beam using optimized deep stacked autoencoders and multi-sensor fusion. *Dev. Built Environ.* **2023**, *14*, 100128. [CrossRef]
15. Zhang, Y.; Zhao, Z. Limited Data Spectrum Sensing Based on Semi-Supervised Deep Neural Network. *IEEE Access* **2021**, *9*, 1813–1817. [CrossRef]
16. Zhang, Y.P.; Zhao, Z.J. Semi-supervised deep learning using pseudo labels for spectrum sensing. *J. Nonlinear Convex Anal.* **2022**, *23*, 1913–1929.

17. Yu, Y.; Hoshyar, A.N.; Samali, B.; Zhang, G.; Rashidi, M.; Mohammadi, M. Corrosion and coating defect assessment of coal handling and preparation plants (CHPP) using an ensemble of deep convolutional neural networks and decision-level data fusion. *Neural Comput. Appl.* **2023**, *35*, 18697–18718. [CrossRef]
18. Lv, Q.; Quan, Y.; Sha, M.; Feng, W.; Xing, M. Deep Neural Network-Based Interrupted Sampling Deceptive Jamming Countermeasure Method. *IEEE J. Sel. Top. Appl. Earth Obs. Remote Sens.* **2022**, *15*, 9073–9085. [CrossRef]
19. Zhao, Q.; Liu, Y.; Cai, L.; Lu, Y. Research on electronic jamming identification based on CNN. In Proceedings of the 2019 IEEE International Conference on Signal, Information and Data Processing (ICSIDP), Chongqing, China, 11–13 December 2019.
20. Li, M.; Ren, Q.; Wu, J. Interference classification and identification of TDCS based on improved convolutional neural network. *J. Phys. Conf. Ser.* **2020**, *16*, 121–155. [CrossRef]
21. Liu, Q.; Zhang, W. Deep learning and recognition of radar jamming based on CNN. In Proceedings of the 2019 12th International Symposium on Computational Intelligence and Design (ISCID), Hangzhou, China, 14–15 December 2019; pp. 208–212.
22. Shao, G.; Chen, Y.; Wei, Y. Convolutional neural network-based radar jamming signal classification with sufficient and limited samples. *IEEE Access* **2020**, *8*, 80588–80598. [CrossRef]
23. Shao, G.; Chen, Y.; Wei, Y. Deep fusion for radar jamming signal classification based on CNN. *IEEE Access* **2020**, *8*, 117236–117244. [CrossRef]
24. Tian, X.; Chen, B.; Zhang, Z. Multiresolution Jamming Recognition with Few-shot Learning. In Proceedings of the 2021 CIE International Conference on Radar (Radar), Haikou, Hainan, China, 15–19 December 2021; pp. 2267–2271.
25. Luo, H.; Liu, J.; Wu, S.; Nie, Z. A Semi-Supervised Deception Jamming Discrimination Method. In Proceedings of the 2021 IEEE 7th International Conference on Cloud Computing and Intelligent Systems (CCIS), Xi'an, China, 7–8 November 2021; pp. 428–432.
26. Zhou, Y.; Shang, S.; Song, X.; Zhang, S.; You, T.; Zhang, L. Intelligent Radar Jamming Recognition in Open Set Environment Based on Deep Learning Networks. *Remote Sens.* **2022**, *14*, 2410–2424. [CrossRef]
27. Wang, L.; Yang, X.; Tan, H.; Bai, X.; Zhou, F. Few-Shot Class-Incremental SAR Target Recognition Based on Hierarchical Embedding and Incremental Evolutionary Network. *IEEE Trans. Geosci. Remote Sens.* **2023**, *61*, 1–11. [CrossRef]
28. Wei, Q. R.; He, H.; Zhao, Y.; Li, J.A. Learn to Recognize Unknown SAR Targets From Reflection Similarity. *IEEE Geosci. Remote Sens. Lett.* **2022**, *19*, 1–5. [CrossRef]
29. Song, Q.; Chen, H.; Xu, F.; Cui, T.J. EM Simulation-Aided Zero-Shot Learning for SAR Automatic Target Recognition. *IEEE Geosci. Remote Sens. Lett.* **2020**, *17*, 1092–1096. [CrossRef]
30. Bernardino, R.P.; Torr, P.H.S. An embarrassingly simple approach to zero-shot learning. In Proceedings of the 32nd international conference on Machine learning (ICML'15), Lille, France, 6–11 July 2015.
31. Kodirov, E.; Xiang, T.; Gong, S. Semantic Autoencoder for Zero-Shot Learning. In Proceeding of the 2017 IEEE Conference on Computer Vision and Pattern Recognition (CVPR), Honolulu, HI, USA, 21–26 July 2017; pp. 4447–4456.
32. Zhang, L.; Xiang, T.; Gong, S. Learning a Deep Embedding Model for Zero-Shot Learning. In Proceeding of the 2017 IEEE Conference on Computer Vision and Pattern Recognition (CVPR), Honolulu, HI, USA, 21–26 July 2017; pp. 3010–3019.
33. Ma, B.J.; Qi, J.; Su, H.Y.; Zhang, X.F. One-dimensional Radar Active Jamming Signal Recognition Method Based on Bayesian Deep Learning. *Signal Process.* **2023**, *39*, 235–243.
34. Qu, Q.; Wei, S.; Liu, S.; Liang, J.; Shi, J. JRNet: Jamming Recognition Networks for Radar Compound Suppression Jamming Signals. *IEEE Trans. Veh. Technol.* **2020**, *69*, 15035–15045. [CrossRef]
35. Tang, Y.; Zhao, Z.; Chen, J.; Zheng, S.; Ye, X.; Lou, C.; Yang, X. Open world recognition of communication jamming signals. *China Commun.* **2023**, *20*, 199–214. [CrossRef]
36. Wang, J.Q. Study and Implementation of Radar Active Jamming Type Discrimination. Master's Thesis, Xidian University, Xi'an, China, 2014.
37. Zhang, H.; Yu, L.; Chen, Y.; Wei, Y. Fast Complex-Valued CNN for Radar Jamming Signal Recognition. *Remote Sens.* **2021**, *13*, 2867. [CrossRef]
38. Fu, Y.; Hospedales, T.M.; Xiang, T.; Gong, S. Trans-ductive multi-view zero-shot learning. *IEEE Tran. PAMI* **2015**, *37*, 2332–2345. [CrossRef]
39. Dong, Y.; Jiang, X.; Zhou, H.; Lin, Y.; Shi, Q. SR2CNN: Zero-Shot Learning for Signal Recognition. *IEEE Trans. Signal Process.* **2021**, *69*, 2316–2329. [CrossRef]
40. Yang, L.; Li, Q.; Shao, H.Z. An Open Set Recognition Algorithm of Electromagnetic Target Based on Metric Learning and Feature Subspace Projection. *Acta Electron. Sin.* **2022**, *17*, 1310–1318.
41. Jleed, H.; Martin, B. Incremental multiclass open-set audio recognition. *Int. J. Adv. Intell. Inform.* **2022**, *8*, 251–270. [CrossRef]
42. Geng, X.; Dong, G.; Xia, Z.; Liu, H. SAR Target Recognition via Random Sampling Combination in Open-World Environments. *IEEE J. Sel. Top. Appl. Earth Obs. Remote Sens.* **2023**, *16*, 331–343. [CrossRef]

remote sensing

MDPI

Article

Infrared Small Target Detection Based on a Temporally-Aware Fully Convolutional Neural Network

Lei Zhang [1,2], Peng Han [3], Jiahua Xi [3] and Zhengrong Zuo [3,*]

1 Beijing Institute of Tracking and Telecommunication Technology, Beijing 100094, China; rosa1982@163.com
2 College of Astronautics, Nanjing University of Aeronautics and Astronautics, Nanjing 211106, China
3 School of Artificial Intelligence and Automation, Huazhong University of Science and Technology, Wuhan 430074, China; hanpeng@hust.edu.cn (P.H.); m202072937@hust.edu.cn (J.X.)
* Correspondence: zhrzuo@hust.edu.cn

Abstract: In the field of computer vision, the detection of infrared small targets (IRSTD) is a crucial research area that plays an important role in space exploration, infrared warning systems, and other applications. However, the existing IRSTD methods are prone to generating a higher number of false alarms and an inability to accurately locate the target, especially in scenarios with low signal-to-noise ratio or high noise interference. To address this issue, we proposes a fully convolutional-based small target detection algorithm (FCST). The algorithm builds on the anchor-free detection method FCOS and adds a focus structure and a single aggregation approach to design a lightweight feature extraction network that efficiently extracts features for small targets. Furthermore, we propose a feature refinement mechanism to emphasize the target and suppress conflicting information at multiple scales, enhancing the detection of infrared small targets. Experimental results demonstrate that the proposed algorithm achieves a detection rate of 95% and a false alarm rate of 2.32% for IRSTD tasks. To tackle even more complex scenarios, we propose a temporally-aware fully convolutional infrared small target detection (TFCST) algorithm that leverages both spatial and temporal information from sequence images. Building on a single-frame detection network, the algorithm incorporates ConvLSTM units to extract spatiotemporal contextual information from the sequence images, boosting the detection of infrared small targets. The proposed algorithm shows fast detection speed and achieves a 2.73% improvement in detection rate and an 8.13% reduction in false alarm rate relative to the baseline single-frame detection networks.

Keywords: infrared small target detection; deep learning; time-aware network; fully convolutional neural network

Citation: Zhang, L.; Han, P.; Xi, J.; Zuo, Z. Infrared Small Target Detection Based on a Temporally-Aware Fully Convolutional Neural Network. *Remote Sens.* **2023**, *15*, 4198. https://doi.org/10.3390/rs15174198

Academic Editors: Yanni Dong, Xiaochen Yang and Qian Du

Received: 17 July 2023
Revised: 13 August 2023
Accepted: 21 August 2023
Published: 26 August 2023

1. Introduction

Infrared small target detection (IRSTD) has been an important subject and research focus in the field of computer vision. Infrared detection has a series of advantages, such as good environmental adaptability, high concealment, and strong anti-interference ability. With the continuous improvement of the performance of infrared detection systems, infrared detection technology has been widely used in civilian fields such as medical diagnosis of diseased cells and industrial testing, as well as in military fields such as infrared warning and infrared guidance. In these fields, real-time detection of small infrared targets with high detection rates and low false alarm rates is an essential requirement. However, in most infrared imaging systems, the distance between the target to be detected and the detector is far, the signal is severely attenuated in the atmosphere, and the target has characteristics such as small imaging area, low signal-to-noise ratio, and unclear shape and texture features. Furthermore, the complex and changing background in practical scenes poses a great challenge to the detection of infrared small targets.

Currently, the mainstream infrared small target detection algorithms rely on handcrafted features for detection, which suffer from high false alarm rates and poor adaptability in complex scenes. In recent years, deep learning-based object detection techniques have rapidly developed, achieving high levels of both speed and accuracy. Deep learning-based approaches have made great progress in many areas of computer vision, such as image segmentation [1,2], action recognition [3,4], and object detection [5,6]. However, research on applying deep learning techniques to infrared small target detection tasks is relatively rare. The main challenge lies in the small size and lack of contour information of the targets, which makes it difficult for the detection network to train and converge on them. Considering the powerful feature extraction and generalization capabilities of deep learning methods, adjusting the network structure and designing detection strategies specifically for small targets could undoubtedly lead to better performance in meeting the requirements of infrared small target detection tasks.

This study addresses the issue of detecting small infrared objects, where existing deep learning-based algorithms may lose feature information due to the complex background. To overcome this, we adjust the structure of the feature extraction network and adopt feature refinement mechanisms to enhance feature representation and suppress background interference, resulting in precise detection results generated from both classification and regression subnets. For more complex scenarios, we introduce a convolutional long short-term memory network to effectively combine spatio-temporal and feature information for higher detection performance.

The key contributions of our study are:

(1) We propose a temporally-aware fully convolutional infrared small target detection (TFCST) algorithm that leverages both spatial and temporal information from sequence images to address the problem of high false alarm rate in infrared small target detection.

(2) We introduce a fully convolutional-based small target detection algorithm (FCST) to address the issue of the reduced robustness of existing IRSTD methods in complex backgrounds. Furthermore, we propose a feature refinement mechanism to emphasize the target and suppress conflicting information at multiple scales, enhancing the IRSTD.

(3) Based on the single-frame FCST, we incorporate ConvLSTM units to extract spatiotemporal contextual information from the sequence images, boosting the detection of infrared small targets.

2. Related Work

To overcome these challenges, researchers have conducted research on the detection of infrared small targets and have achieved some research results. In the process of detecting small infrared targets, detection algorithms can be divided into two categories based on the differences in their spatial and temporal handling: algorithms based on single-frame images and those based on sequence images.

2.1. Single-Frame Image-Based Infrared Small Target Detection

Based on the current state of technology, there exist four categories of infrared small target detection algorithms for a single frame image. These categories include those based on filtering, visual contrast mechanisms, image data structures, and deep learning.

Filtering-based methods can be classified into two categories: spatial-domain filtering and frequency-domain filtering. Spatial-domain filtering methods estimate the background signal of the infrared image first, then perform differential operations on the original image and the estimated background image, and finally use threshold segmentation in the differential image to achieve small infrared target detection. Typical spatial-domain filtering methods include spatial domain high-pass filtering [7], maximum mean/median filtering [8], two-dimensional least mean square (TDLMS) filtering [9], bilateral filtering [10], and top-hat transformation [11], among others.

Frequency-domain filtering methods mainly design appropriate filters in the transform domain to remove the background and noise. The earliest frequency-domain filtering methods included ideal high-pass filtering, Gaussian high-pass filtering, and Butterworth filtering.

Recently, researchers have leveraged the powerful capabilities of the human visual system (HSV) in detecting small targets in infrared images. One approach that has been successfully used is the local contrast measure (LCM) proposed by Chen et al. [12], which is based on the contrast sensitivity mechanism of human vision. Subsequently, Han et al. [13] proposed an improved version of LCM, called the improved local contrast measure (ILCM), which employs adaptive processes and attention transfer mechanisms to effectively address the over-enhancement of noise points in LCM.

In recent years, deep learning-based target detection technology has rapidly developed, and relevant research has emerged in the field of infrared small target detection. In 2019, Wang et al. used adversarial learning to balance missed detection (MD) and false alarm (FA), using two flow models processed by adversarial training for sub-tasks, with each flow focusing on reducing missed detection or false alarm, achieving good results in infrared small target detection [14]. Inspired by the unique distribution features of infrared small targets, Shi et al. [15] considered infrared small targets as "noise" and transformed the task of small target detection into a denoising problem. In 2020, Zhao et al. [16] constructed a generative adversarial network (GAN) model to automatically learn target features and directly predict target intensity. Overall, deep learning methods can more effectively extract the features of targets and background and distinguish them, which has significant advantages over traditional detection algorithms and represents a new breakthrough direction in IRSTD technology. Deep-IRTarget [17] combines frequency domain and spatial domain features to form dual-domain features, aiming to improve the accuracy of target detection. Zhang et al. [18] proposed transforming the features extracted by a pre-trained self-supervised feature extractor into a Gaussian-like distribution to reduce feature distribution mismatch.

2.2. Multi-Frame Image-Based Infrared Small Target Detection

In the detection of small infrared moving targets, the shape, grayscale variations, and motion trajectory of the targets exhibit continuity over time, a crucial characteristic for effectively distinguishing between targets and noise. Therefore, detection methods based on sequential images need to consider both spatial and temporal information.

One such method is the pipeline filtering method, which relies on the continuity of the target's motion trajectory. The basic idea is to first detect some suspected targets in the first frame of the image and then establish a temporal pipeline along the time axis, t, in the image sequence, using these suspected targets as centers. The length of the pipeline represents the number of consecutive frames for detection, and the radius of the pipeline corresponds to the target's neighborhood size. When the suspected target appears in the pipeline a predetermined number of times, it is judged as a real target; otherwise, it is considered interference.

When the signal strength of a target is low, direct detection methods can be challenging, while energy accumulation can effectively enhance the target's energy and achieve the final detection. Zhang et al. proposed a representative energy-accumulation-based method for small target detection. This method first projects consecutive multiple infrared images into a two-dimensional space and then accumulates the energy of small targets along the four possible motion directions in that space. Subsequently, there have been some modified energy-accumulation-based methods [19] which achieved good detection results but with limited real-time performance.

Furthermore, it is worth noting that most existing infrared small target detection methods based on deep learning are based on single-frame images. To further improve detection performance, it is necessary to consider incorporating temporal information of the target in sequential images into detection networks.

Compared to static image object detection, the focus of research in video object detection is on how to fully utilize temporal information to reduce inter-frame redundant computation and improve detection quality. Inspired by the recurrent neural network (RNN) [20], some researchers have proposed utilizing the memory-based learning of the

RNN structure to learn the spatiotemporal information in videos. Xiao and Lee developed the spatial-temporal memory network model (STMN) [21], which effectively leverages the temporal information in videos by preserving long-term appearance and motion information. Chen et al. [22] combined ConvLSTM [23] with SSD [24] to construct a correlated loss function between multiple frames of video images, strengthening the consistency of detection results across adjacent video frames and effectively improving the robustness of video object detection. Duran-Vega et al. used a quasi-recurrent neural network (QRNN) [25] to extract the temporal information in videos on the basis of Yolov5, which significantly improved the detection performance while maintaining fast detection speed.

3. Method

In this section, we first introduce the overall architecture of our TFCST. Then, we present the details of the FCST block (Section 3.2) and the temporally-aware block (Section 3.3), followed by the loss function in Section 3.4.

3.1. Overall Architecture

As shown in Figure 1, the temporally-aware fully convolutional neural network (TFCST) consists of four main components:

(1) Feature Extraction Network: Firstly, the Focus structure is used to integrate the width and height information of the single-frame image into the channel space. Then, the OSA-Block convolutional unit, based on the single aggregation idea, is used to extract multi-scale features, with a total of four downsampling operations. This produces feature maps $\{C_1, C_2, C_3, C_4\}$.

(2) Feature Fusion Network: Firstly, FPN is used to add deep semantic information to the shallow feature maps with higher resolution, which yields $\{F_1, F_2, F_3, F_4\}$. Then, the feature refinement module (FRM) is used on the multi-scale features with semantic differences to filter out information that is favorable for small target detection, and the fused features are represented as $\{P_2, P_3\}$.

(3) Recurrent Neural Network: The ConvLSTM units are used to model the appearance and motion information of the targets on two feature layers, P_2 and P_3, respectively. The target information in the sequence images is then propagated and fused to produce $\{R_2, R_3\}$.

(4) Target Prediction Network: The main idea of the FCOS anchor-free detection is used to predict target class, regression boxes, and centrality for the temporally-enhanced features R_2 and R_3.

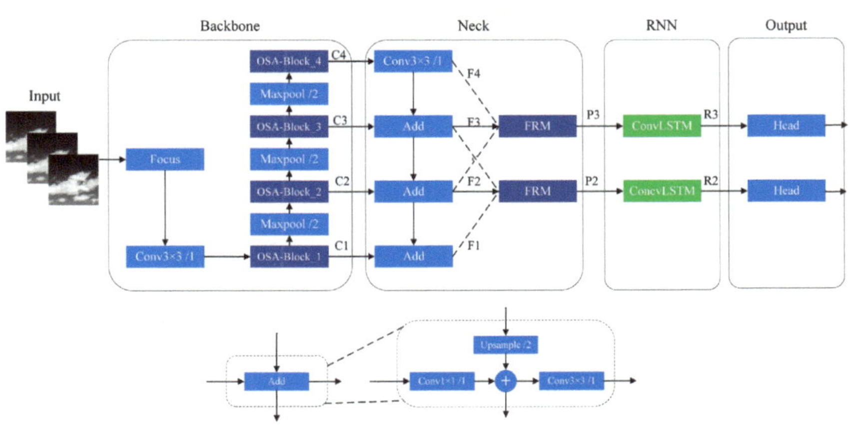

Figure 1. The overall architecture of temporally-aware fully convolutional neural network. TFCST consists of four main components: feature extraction network, feature fusion network, recurrent neural network, and target prediction network.

3.2. Fully Convolutional Neural Network for Infrared Small Target Detection

The model-driven IRSTD algorithm does not require learning and is easy to compute. It has achieved satisfactory detection results in some scenarios, but it can be prone to a large number of false alarms and poor robustness in complex and diverse environments. In contrast, deep learning methods possess powerful feature extraction and generalization capabilities and have demonstrated outstanding detection performance in various test environments. Therefore, deep learning is a new breakthrough direction in the field of infrared small target detection.

Based on the analysis above, we primarily investigate the method of detecting infrared small targets using deep learning. We propose a fully convolutional neural network for infrared small target detection (FCST), which is based on the anchor-free mechanism. The algorithm adopts the idea of pixel-level prediction proposed by fully convolutional one-stage object detection (FCOS) [26], which directly predicts the target's category and regression detection box at each position of the feature map, thus avoiding the complex calculations and poor generalization caused by anchor box design. In the feature extraction stage, FCST reduces the number of downsampling operations and combines the idea of feature aggregation to ensure that the features of small targets are not lost. In the feature fusion stage, we use the feature refining mechanism to highlight the target's characteristics and suppress conflicting information, thereby improving the detection accuracy of infrared small targets.

Modern object detection methods can be categorized into anchor-based and anchor-free methods based on whether prior boxes are used. While anchor boxes have long been considered an essential part of high-precision object detection, their mechanism is not conducive to detecting small objects. This is mainly due to the following reasons: Firstly, fixed-size, aspect ratio, and number of anchor boxes make it difficult to cover all types of objects comprehensively, which can affect the detector's generalization ability and detection accuracy. Secondly, to achieve higher recall rates, detectors densely distribute anchor boxes on the input image, but most of them are on the background region, resulting in a severe imbalance between positive and negative samples during the training process. Lastly, the detection accuracy is sensitive to the IOU threshold when assigning training labels to anchor boxes, particularly for small objects. Therefore, recent research has focused on developing anchor-free methods that overcome these limitations and offer better performance. These methods do not rely on predefined anchor boxes and can handle objects of various sizes and aspect ratios more effectively. Additionally, they can alleviate the issue of imbalance in positive and negative samples resulting from anchor boxes' use.

Target detection algorithms without anchor boxes are mainly based on center-based and keypoint-based methods. These approaches eliminate the need for generating anchor boxes, which leads to faster detection speeds and better generalization abilities. In recent years, one-stage fully convolutional object detection (FCOS) has become a popular anchor-box-free detection algorithm. Similar to semantic segmentation, FCOS employs pixel-level prediction to directly predict the class and regression of an object at each position on the feature map, without the need for predefined anchor boxes or candidate regions. Compared to anchor-based detection methods, FCOS is simpler and more computationally efficient, which consumes fewer computational resources.

In this paper, we applied the anchor-box-free detection mechanism of FCOS to design a fully convolutional neural network (FCST) for detecting small infrared targets. The structure of FCST is shown in Figure 2, and the model is designed based on two aspects: feature extraction and feature fusion networks. The target prediction network is based on the three-branch structure of FCOS.

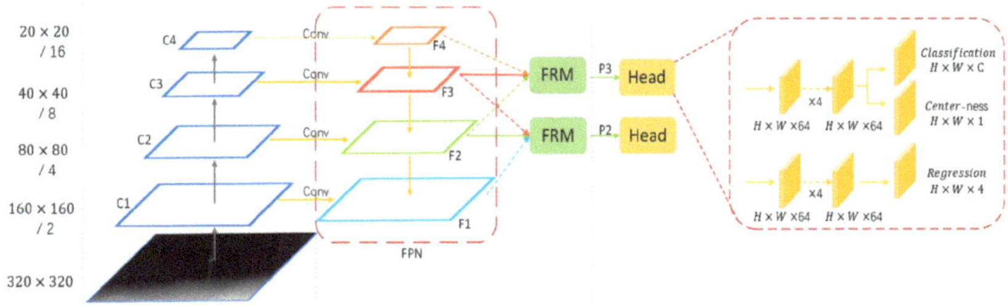

Figure 2. The structure of fully convolutional neural network for infrared small target detection (FCST). The model is designed based on two aspects: feature extraction and feature fusion networks. The target prediction network is based on the three-branch structure of FCOS.

During the feature extraction stage, in order to prevent the loss of small target feature information caused by excessive downsampling, the network reduces the downsampling frequency and enhances the feature expression by aggregating them in a single pass, resulting in a feature map $\{C_1, C_2, C_3, C_4\}$ after downsampling. In the feature fusion stage, the deep semantic information is added from top to bottom into the low-level feature maps with high resolution, generating the fused feature maps $\{F_1, F_2, F_3, F_4\}$. Furthermore, to address the issue of semantic conflicts caused by multi-scale feature fusion, a feature refining mechanism is employed to suppress conflicting information and reduce false alarms. Additionally, considering the differences in size between infrared small targets and general targets, FCST focuses on detecting on feature maps $\{P_2, P_3\}$ with downsampling rates of $4\times$ and $8\times$, which correspond to small point targets and large patch targets with high pixel occupancy, respectively.

In the context of developing feature extraction networks, the issue of losing important target information due to downsampling has been identified. As a solution, the Focus structure has been introduced as the first layer of the feature extraction network. As shown in Figure 3, this structure slices the input data and samples it at intervals to obtain 4 feature maps, thus expanding the input channel by a factor of 4. The obtained feature maps are then subject to convolutional operations to fuse the channel information, resulting in a downsampling feature map that retains all information, without loss. This approach not only reduces the model's parameters and computational complexity but also integrates the width and height information of the image into the channel space, thereby preserving more complete image information for subsequent feature extraction. Ultimately, this helps improve the detection performance of the network model.

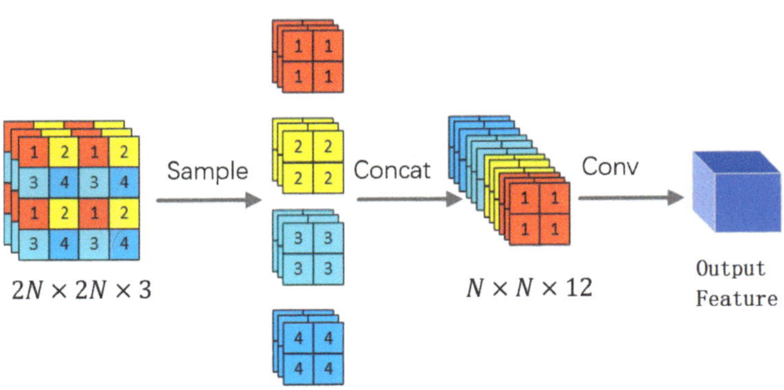

Figure 3. The structure of the Focus block. It not only reduces the model's parameters and computational complexity but also integrates the width and height information of the image into the channel space.

ResNet [27] is currently the most commonly used feature extractor in deep learning object detection models. Its design of residual structures enhances the information exchange between the front and back layers, which mitigates training difficulties and convergence issues that occur when the network is deepened. The ResNet structure is shown in Figure 4.

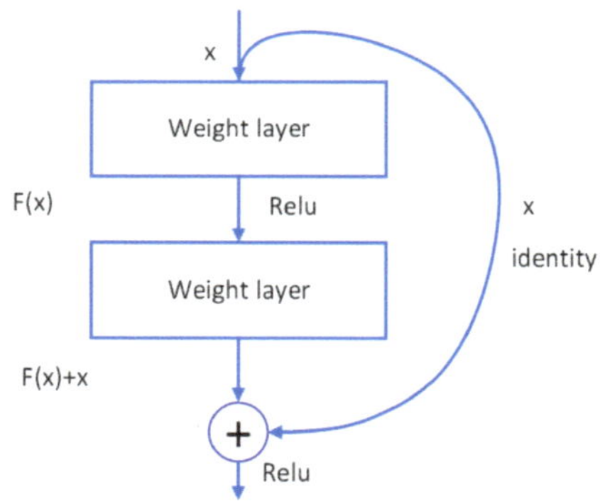

Figure 4. The structure of the Residual block. Its design of residual structures enhances the information exchange between the front and back layers, which mitigates training difficulties and convergence issues that occur when the network is deepened.

On the other hand, DenseNet [28] establishes dense connections between all the previous layers and subsequent layers to maximize the information exchange between them in the channel dimension. This achieves better performance than ResNet, with fewer parameters and computations. Although DenseNet is effective in object detection, its speed is slow due to the high memory access cost and energy consumption caused by its dense agglomeration method. Furthermore, DenseNet causes feature redundancy because the subsequent features already learn the crucial information from earlier layers.

To overcome this information redundancy, VoVNet [29] proposed One-Shot Aggregation (OSA), which aggregates only the features from all the previous layers at the final aggregation step. In brief, OSA enhances the network's ability to express features while still achieving higher computational efficiency than dense aggregation. Inspired by the residual connection and single-stage aggregation, the OSA-Block, which is a convolutional block, is constructed as depicted in Figure 5.

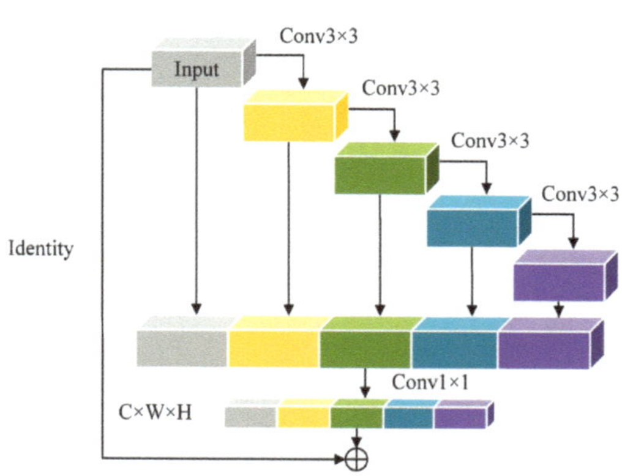

Figure 5. The structure of the OSA-Block. OSA enhances the network's ability to express features while still achieving higher computational efficiency than dense aggregation.

In the context of detecting small infrared targets, it is important to consider different factors when designing the feature extraction network. One major concern is the limited amount of pixel coverage that these targets occupy in the image, combined with the subtle features they possess. This means that using overly deep networks can lead to a loss of important information and increased false negatives. Additionally, large model parameters can hinder inference speed. Therefore, it is more suitable to use relatively shallow networks that can be optimized for this specific task.

The shallow feature space contains detailed spatial information, but it lacks sufficient semantic information and a comprehensive understanding of the overall scene. Therefore, using only shallow features for detecting small objects results in poor performance. On the other hand, deep features extract rich semantic information efficiently, but they sacrifice too much fine-grained information about the target. To make the most of the semantic information in deep network features, while preserving the spatial information of the shallow feature map, feature fusion between shallow and deep layers is necessary. Based on this idea, FPN [30] employs a top-down information propagation approach to fuse the semantic features in deep layers with the spatial details of shallow features through addition or concatenation. This method enhances feature representation capability without significant computational cost, and to some extent improves the detection performance of small objects. However, since different scales of features contain semantic differences, direct fusion can lead to redundancy and conflicts and weaken the expression ability of multi-scale features.

To address the aforementioned issues, a Feature Refinement Module (FRM) is proposed, as illustrated in Figure 6. The FRM generates adaptive weights on both spatial and channel dimensions in two parallel branches, which filter multiple scale features and preserve useful information. Before extracting adaptive weights, the feature dimensions of different scales need to be consistent. Let X^m denote the feature of the m-th layer, and X^{m-1} and X^{m+1} denote the adjacent layer features. For high-level features with smaller resolutions X^{m+1}, a 1×1 convolution is first used to adjust the channels to be consistent with X^m and then upsampled to the same size as X^m by nearest neighbor interpolation. For shallow features with larger resolutions X^{m-1}, a 3×3 convolution with a stride of 2 is applied to simultaneously adjust the channels and resolutions.

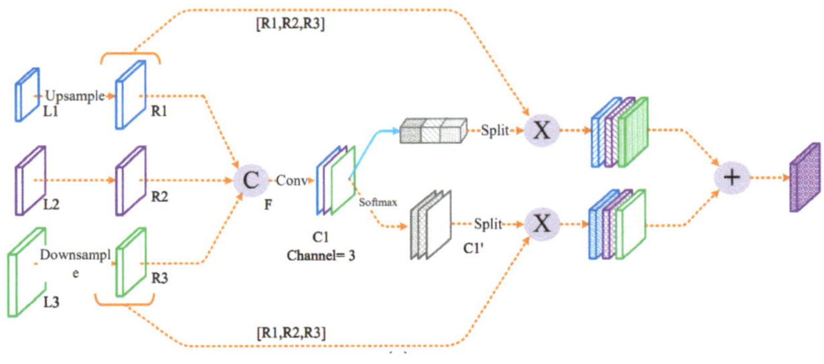

Figure 6. Illustration of feature refinement module (FRM). The FRM generates adaptive weights on both spatial and channel dimensions in two parallel branches, which filter multiple scale features and preserve useful information.

Let ${X_{x,y}}^{n \to m}$ be the result of adjusting the feature dimension of layer n to layer m. Then the process of channel-wise feature refinement is defined as follows:

$$K^m_{x,y} = a^m \cdot X^{m-1\to m}_{x,y} + b^m \cdot X^m_{x,y} + c^m \cdot X^{m+1\to m}_{x,y} \quad (1)$$

In the above equation, $K^m_{x,y}$ is the feature vector located at (x, y) of layer m, and a, b, c represent adaptive channel fusion weights with a dimension of $1 \times 1 \times 1$. The idea of obtaining these weights is to combine adaptive average pooling and maximum pooling on the spatial dimension to compress the input features and aggregate spatial information that can represent global characteristics of the image. The computation can be expressed as follows:

$$[a^m, b^m, c^m] = \sigma[AP(F) + MP(F)] \tag{2}$$

where F denotes the output of multi-scale features concatenation, as shown in Figure 7, and σ is the sigmoid operation, while AP and MP represent adaptive average pooling and maximum pooling, respectively.

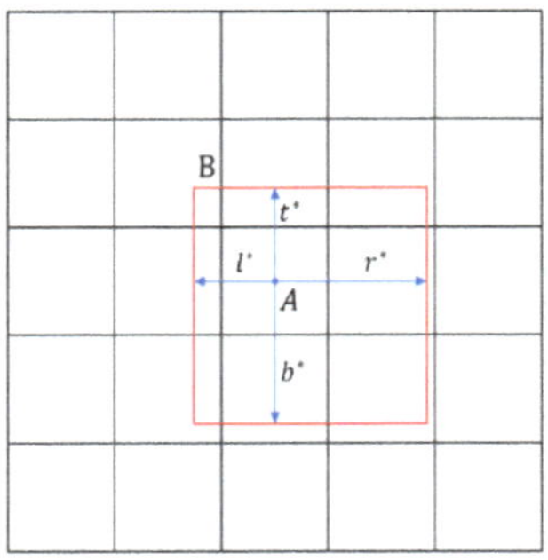

Figure 7. Regression diagram of FCOS, where A is the ground-truth target point, B represents the predicted target box and (l^*, t^*, r^*, b^*) are used for regression calculation. The target prediction network also outputs a four-dimensional vector, which represent the distances to the four edges of the detection box.

The process of spatial-wise feature refinement for multi-scale features can be defined as follows:

$$\Phi^m_{x,y} = \sum_{c=1}^{3} \sum_{k,x,y} (\alpha^m_{c,x,y} \cdot X^{m-1\to m}_{k,x,y} + \beta^m_{c,x,y} \cdot X^m_{k,x,y} + \gamma^m_{c,x,y} \cdot X^{m+1\to m}_{k,x,y}) \tag{3}$$

In the above equation, x and y represent spatial positions in the feature map, and k represents the channels. $\Phi^m_{x,y}$ is the output of spatial refinement features at (x, y). The attention weights for the spatial dimension are denoted as $\alpha^m_{c,x,y}$, $\beta^m_{c,x,y}$ and $\gamma^m_{c,x,y}$, which can be computed as follows:

$$[\alpha^m, \beta^m, \gamma^m] = Softmax(F) \tag{4}$$

where $softmax$ is used to normalize the features along the channel direction to obtain the relative weights of different channel directions at the same spatial position.

The overall output of FRM is a combination of refined features along the channel and spatial dimensions, which is expressed as follows:

$$p^m = \Phi^m + K^m \tag{5}$$

By using this method, multi-scale features are fused under the guidance of adaptive weights, enhancing the feature representation of small targets and suppressing background information. Finally, P_2 is obtained by fusing F_1, F_2, and F_3, and F_2, F_3, and F_4 are used as inputs for the target detection network.

In Figure 2, we can see that the structure of the object detection network, which is based on the anchor box, is similar to that of RetinaNet [30]. The target prediction network has two branches on each detection layer (which consist of four consecutive convolutional layers), one for category prediction and the other for detection box regression. However, unlike traditional anchor-based approaches, the idea of the anchor-free mechanism is to treat each pixel on the feature map as a sample and directly predict both the class and detection box regression without relying on fix-sized and fixed-number anchor boxes.

During the training phase, each position (x, y) on the feature map with a downsampling rate of s is mapped back to the original image through calculations. Only if the predicted class label is consistent with the true class label and the detection box regression falls within the range of the true annotation box is it regarded as a positive sample. Otherwise, it is a negative sample. Compared to the traditional method of IOU threshold-based division, this positive and negative sample selection method is more conducive to improving the recall rate for small object detection, which is also one of the significant reasons why we constructed the network based on FCOS.

In addition to class prediction, the target prediction network also outputs a four-dimensional vector, $v^* = (l^*, t^*, r^*, b^*)$, for the regression calculation of the detection box at that position, where it represent the distances to the four edges of the detection box, as shown in Figure 7. If the sample at that position is associated with detection box $B^i(B^i = (x_0^{(i)}, y_0^{(i)}, x_1^{(i)}, y_1^{(i)} \in R^4)$, then the regression target trained at that position is defined as follows:

$$\begin{cases} l^* = x - x_0^{(i)} \\ t^* = y - y_0^{(i)} \\ r^* = x_1^{(i)} - x \\ b^* = y_1^{(i)} - y \end{cases} \tag{6}$$

Furthermore, the approach of predicting on multiple feature maps can increase the recall rate of object detection, but it can also introduce many low-quality prediction boxes that deviate from the true boxes. To address this issue, FCOS proposes a simple yet effective strategy without introducing any hyperparameters. Specifically, the network adds an additional "center-ness" branch in parallel with the classification branch, which measures the offset of the predicted location relative to the center of the object and adds a new loss term based on center-ness. The formula for center-ness is as follows:

$$center\text{-}ness^* = \sqrt{\frac{min(l^*, r^*)}{max(l^*, r^*)} \times \frac{min(t^*, b^*)}{max(t^*, b^*)}} \tag{7}$$

As can be seen from the formula, the center-ness value ranges from 0 to 1, with a higher value indicating a prediction closer to the true box center, and vice versa. During the testing phase, the final score of the detection box is obtained by multiplying the object class score and the center-ness value. Therefore, center-ness can reduce the score of prediction boxes that deviate from the true box center and improve the detection accuracy by filtering them out through a non-maximum suppression (NMS) process.

3.3. Temporally-Aware Fully Convolutional Neural Network

In typical detection systems, video sequences rather than static images are acquired, and relying solely on spatial information from a single frame may not be sufficient for detecting infrared small targets in scenes with complex backgrounds and weak target signals. To achieve more robust detection capabilities, it is necessary to consider incorporating both temporal and spatial information from sequence images. While existing infrared small target detection methods based on sequence images have demonstrated high detection accuracy by processing multiple frames simultaneously, they often sacrifice detection speed and require significant computational resources, making it challenging to apply them in real-time settings.

We presents a novel approach for detecting infrared small targets by combining a fully convolutional neural network with a recurrent neural network. Inspired by RNN-based video object detection methods, we extended a single-stage anchor-free detection network to create the temporally-aware fully convolutional neural network for infrared small target (TFCST), which effectively utilizes the ConvLSTM [23] architecture to propagate and refine features across frames. As a result, TFCST exhibits enhanced detection accuracy in more complex scenarios, while also remaining capable of real-time online detection tasks. This was demonstrated through various experiments conducted on the proposed method.

Sequential image data consists of both temporal and spatial information. To learn spatiotemporal features in near-term precipitation forecasting, the convolutional long short-term memory (ConvLSTM) [23] model was proposed based on the long short-term memory (LSTM) [21] architecture. The basic unit of LSTM is referred to as a memory block, which consists of three gate units and a memory cell. The three gate units are the input gate, output gate, and forget gate, which are responsible for controlling the flow of information. The memory cell is used to store the current state of the network. During forward propagation, the input gate controls the flow of information into the memory cell, while the output gate controls the flow of information from the memory cell to other memory blocks. During backpropagation, the output gate regulates the flow of error and gradients into the memory cell, whereas the input gate controls the flow of error and gradients out of the memory cell. The forget gate determines the extent of information retention and governs the internal recurrent state of the memory cell. The ConvLSTM takes the input at time t and the memory block output, concatenates them, and performs a convolution operation to extract spatiotemporal features.

As shown in Figure 8, ConvLSTM has the same structure as the LSTM unit, which includes input gates, forget gates, output gates, and memory cells. However, what sets ConvLSTM apart from LSTM is that the input at time t and the output from the memory block at time t-1 are concatenated and then convolved, which is crucial for extracting spatiotemporal information. ConvLSTM can be described as follows:

$$i_t = \sigma(W_i * [X_t, H_{t-1}] + b_i) \tag{8}$$

$$f_t = \sigma(W_f * [X_t, H_{t-1}] + b_f) \tag{9}$$

$$o_t = \sigma(W_o * [X_t, H_{t-1}] + b_o) \tag{10}$$

$$\hat{C}_t = tanh(W_c * [X_t, H_{t-1}] + b_c) \tag{11}$$

$$C_t = f_t \circ C_{t-1} + i_t \circ \bar{C}_t \tag{12}$$

$$H_t = o_t \circ tanh(C_t) \tag{13}$$

The aim of this study is to detect infrared small targets in sequence images quickly and robustly. Building upon the single-frame detector FCST, we introduce a novel infrared small target detection model, TFCST. In the detection process, the main idea is to use a recurrent neural network structure to propagate and refine features across frames, enabling the network to have a temporal aware ability. Given the extremely fast detection speed of the basic network FCST, in order to fully utilize the temporal and spatial information of sequence images, TFCST performs detection on each frame and uses ConvLSTM to save the temporal contextual status information of historical frames. Then, based on this temporal contextual information, the features of the current frame are enhanced to improve the small target detection performance. It is worth noting that the idea of TFCST is not to post-process the results of single-frame detection, but to directly propagate temporal contextual information between feature layers, which can achieve online detection, as illustrated in Figure 9.

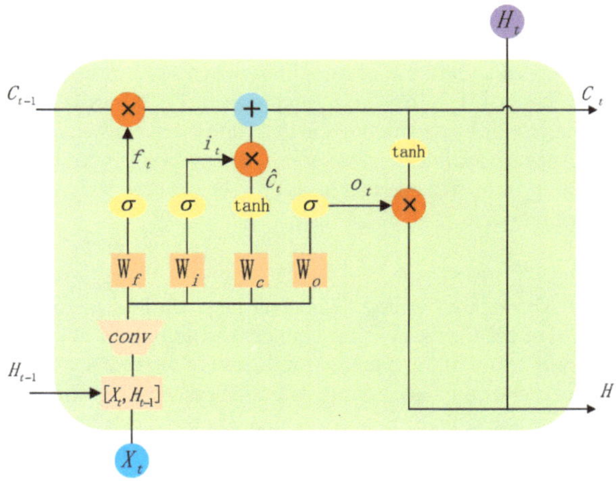

Figure 8. The architecture of ConvLSTM. The block is utilized to save the temporal contextual status information of historical frames. ConvLSTM extends LSTM by incorporating convolutional operations into the LSTM structure. It introduces convolutional layers within the LSTM cells, enabling the network to process both spatial and temporal dependencies in sequential data simultaneously.

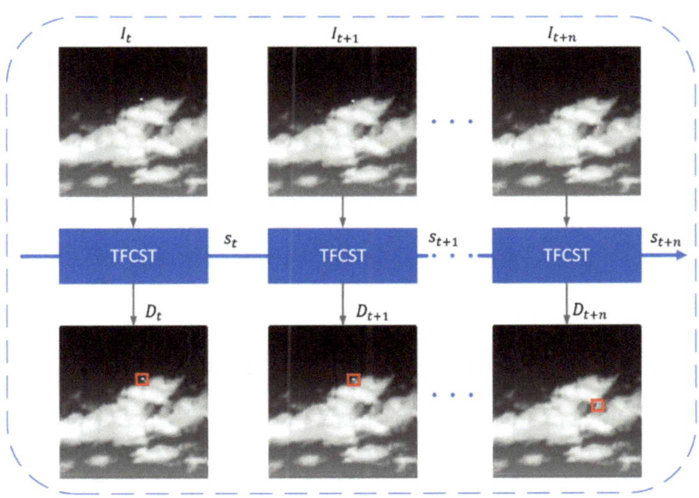

Figure 9. The detection idea of TFCST. TFCST directly propagates temporal contextual information between feature layers, which can achieve online detection.

The goal of the network model for infrared sequence $\mathcal{V} = \{I_1, I_2, \ldots I_n\}$ with n frames is to output the correct detection results $\{D_1, D_2, \ldots D_n\}$ for each frame, where D_k is a set of multiple detection boxes and target class prediction results are obtained on a single frame I_k, relying only on the image data between I_k frames. The small target detection model TFCST based on the sequence images can be represented as function $\mathcal{F}(I_t, s_{t-1}) = (D_t, s_t)$, where $s_k = \{s_k^1, s_k^2, \ldots, s_k^m\}$ represents the state information extracted by the network in the first k frames. Based on the structure of single-frame image detection model FCST, TFCST approximates the equation above by adding m ConvLSTM units with memory characteristics. The ConvLSTM unit takes the previous state vector extracted from the historical frames as input and outputs the current state vector at the current detection moment. During the detection process of the infrared sequence images, running the

TFCST detector on each frame in chronological order will yield the detection results for the entire sequence.

In consideration of the different sizes of infrared small targets with various shapes, such as point sources and patchy ones, TFCST still employs a multi-scale detection approach, mainly conducting the detection on two down-sampled feature layers at 4× and 8× scales, respectively. Furthermore, to propagate temporal and spatial contextual information among frames at each scale, TFCST incorporates ConvLSTM on each feature layer after feature fusion.

3.4. Loss Function

During the training stage of the network, loss functions need to be calculated separately for the categories, detection boxes, and centrality of the targets to optimize the network parameters. For the category prediction of infrared small targets, the Focal Loss [31] is used to calculate the loss, mainly due to the small proportion of target pixels. Focal Loss was first proposed in RetinaNet to alleviate the problem of imbalanced positive and negative samples during training. The formula is as follows:

$$FL(p_t) = -\alpha_t(1 - p_t)^{\gamma} log(p_t) \tag{14}$$

In the above formula, α_t is used to adjust the weight of the positive and negative sample loss values, γ is used to adjust the proportion of easy-to-classify samples in the total samples, and p_t is the predicted target confidence score.

Regarding regression loss, higher requirements are imposed on the location prediction in small target detection. Therefore, this study adopts the *GIOU Loss* [32] to perform positional regression on the targets, which can reflect the proximity relationship between the predicted box and the target box and effectively avoid the problem of zero gradients in the loss function due to the predicted box not intersecting with the actual box. The calculation formula for *GIOU* is as follows:

$$GIOU_{B_{GT},B_P} = \frac{|B_{GT} \cap B_P|}{|B_{GT} \cup B_P|} - \frac{B/(B_{GT} \cup B_P)}{B} \tag{15}$$

In the above formula, B_{GT} represents the ground truth box of the target, B_P represents the predicted box, and B represents the minimum bounding rectangle that can surround B_{GT} and B_P. The definition of GIOU Loss is:

$$Loss_{GIOU} = 1 - GIOU_{B_{GT},B_P} \tag{16}$$

For the calculation of centrality loss, the centrality loss can be calculated using binary cross-entropy loss. Therefore, the loss function of the target detection model can be expressed as a joint loss of classification loss L_{cls}, regression loss L_{reg}, and centrality loss L_{cen}:

$$L(c_{x,y}, t_{x,y}, e_{x,y}) = \frac{1}{N_{pos}} \sum_{x,y} L_{cls}(c_{x,y}, c^*_{x,y}) \tag{17}$$

$$+ \frac{1}{N_{pos}} \sum_{x,y} \mathbb{1}_{\{c^*_{x,y}>0\}} L_{reg}(t_{x,y}, t^*_{x,y}) \tag{18}$$

$$+ \frac{1}{N_{pos}} \sum_{x,y} L_{cen}(e_{x,y}, e^*_{x,y}) \tag{19}$$

4. Experiments

To validate the effectiveness of the proposed methods, we conducted comprehensive experiments on both FCST and TFCST. In this section, we first introduce the experimental data and training parameters. Subsequently, we performed ablation experiments and comparative experiments. The experiments were conducted on a computer equipped with a 2.90-GHz Intel Core i7-10 700 CPU, an RTX 3090 GPU, and 32-GB RAM. Python 3.7 was utilized as the programming language for all experiments.

4.1. Data Introduction

In order to verify the effectiveness of TFCST, we present a sequential image infrared small target dataset. The training set consists of 144 infrared sequences, comprising a total of 11,551 frames, while the test set consists of 42 sequences, comprising a total of 3268 frames. The length of each sequence varies between 50 and 260 frames. In these datasets, some targets are real, some are simulated, and the simulated targets were generated by our own target simulation system, in which the three-dimensional motion trajectory of the targets are projected to obtain its 2D positions in the images. In the target simulation system, the camera frame rate, the optical system point-spread-function, and the target signal-to-noise ratio can be set. The different motion rate of the target on the image can be simulated by setting different frame rates, and the intensity of the target can be calculated by the set target signal-to-noise ratio.

Additionally, considering the diversity of infrared small targets in practical scenarios, random rotation and scale adjustments were added during the construction of target images, making the targets exhibit various shapes such as points, spots, and stripes. The background samples mainly come from the single-frame infrared small target dataset proposed before. When constructing infrared sequential data, background images with more clutter interference were specifically selected. Some of the images are shown in Figure 10.

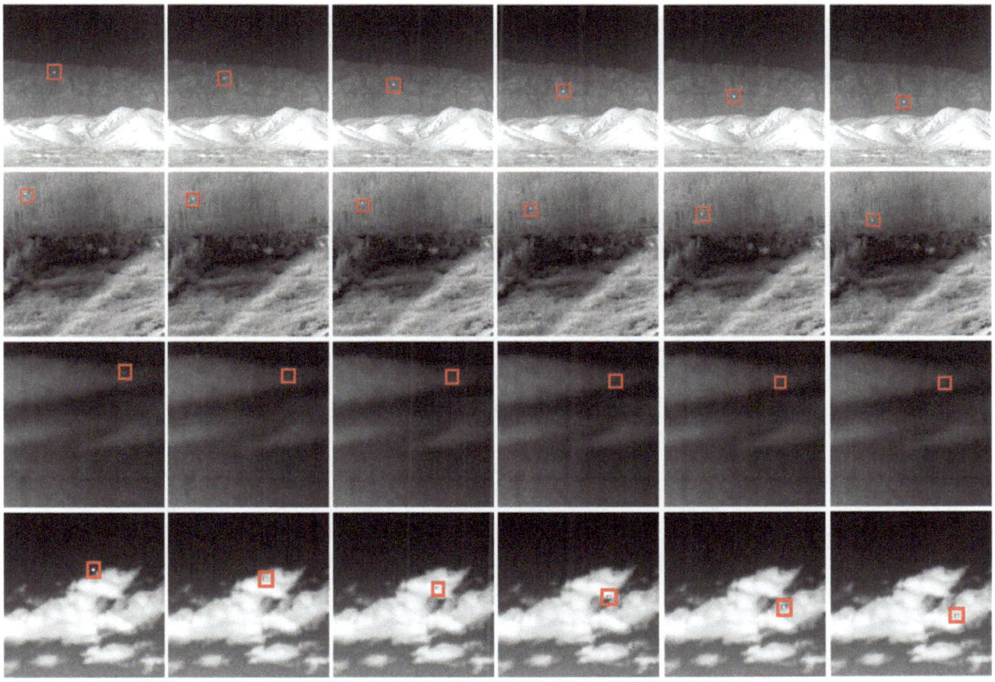

Figure 10. Illustration of the dataset. When constructing infrared sequential data, background images with more clutter interference were specifically selected.

4.2. Training Parameter

Considering the poor diversity of sequence image datasets, the TFCST network first pre-trains the FCST single-frame detector model and then further trains it on the constructed infrared sequence image dataset. During training, the weights of the feature extraction network and feature fusion network are frozen, and the ConvLSTM and the remaining parts of the network are trained using the Adam optimizer. The learning rate is adaptively adjusted based on the size of the loss function value, with an initial value of 1×10^{-3}. The tolerance for the network performance to remain unchanged is three times, and a total of 100 rounds of training are performed with a batch size of 4. After training is completed, the best evaluation result is selected as the final model.

4.3. Ablation Experiment

First, the impact of extracting continuous frame image information of length seq_len on detection performance during network training was studied, and the experimental results are shown in Figures 11 and 12.

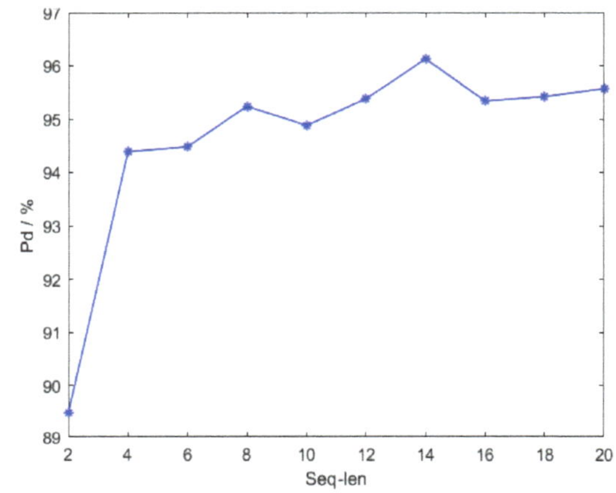

Figure 11. Illustration of the positive detection (P_d) performance. P_d is the ratio of the number of positive targets detected to the total number of all positive targets.

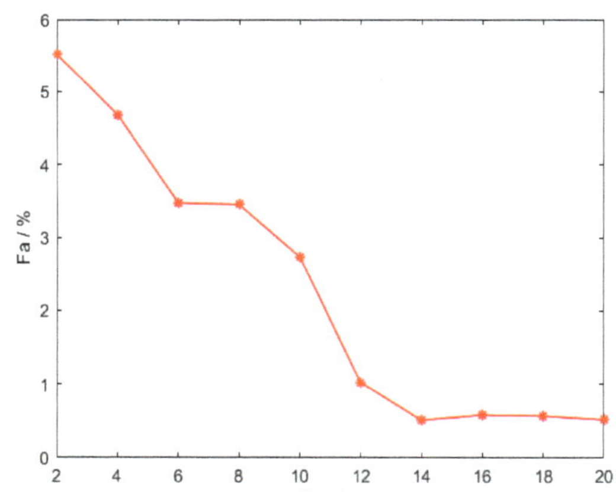

Figure 12. Illustration of the false alarm (F_a) performance. F_a is the ratio of the number of false targets detected to the total number of targets detected.

From Figures 11 and 12, it can be seen that during network training, the more video frames that the network can utilize, the more sufficient is the spatiotemporal information extracted by the ConvLSTM, which is more conducive to the detection of infrared small targets. However, when the selected length of continuous frames reaches a certain value, the network's performance also reaches saturation. In the experiment, when seq_len = 14, the detection rate of the TFCST network in the test set of sequence image infrared small targets is 96.13%, and the false alarm rate is 0.51%, which achieves the best detection effect. When increasing the length of seq_len, the detection performance of the network is difficult to further improve.

It is worth noting that as the length of seq_len increases, the overall trend of false alarm rate changes is constantly decreasing, indicating that the combination of the ConvLSTM based on the single-frame detector can improve the problem of high false alarm rate of the infrared small target detection algorithm in complex backgrounds. The trend of the detection rate, however, does not always increase, and fluctuates with the increase of seq_len. The reason may be related to the random jumping sampling strategy adopted during network training. In some cases, the samples extracted from the infrared sequence are more conducive to the ConvLSTM learning, and the feature information extracted from the historical frames can be more conducive to the detection of targets in the current frame.

Next, the impact of different temporal models on network detection performance was studied. In the experiment, two RNN variants, ConvGRU [33] and Bottleneck-LSTM [34], were selected. ConvGRU is a recurrent network model built on the basis of gated recurrent units (GRU) [35]. Compared with LSTM, GRU optimizes the structure of the memory cells, reduces parameters, and accelerates the training speed. Bottleneck-LSTM is a more efficient RNN structure proposed on the basis of ConvLSTM, which uses depth-separable convolution and reduces the number of related convolution operations in memory cells.

In the experiment, the length of the continuous frame in the sequence image during the training stage was uniformly set to 14, and multiple sets of experiments were repeated to obtain more stable results, and the statistical results were averaged. In addition, we also compared the detection speed of different RNNs with frames per second and model parameter count. The detection results of the TFCST network on the test set when different RNN models are used are shown in Table 1, where positive detection (Pd), false alarm (Fa), frames per second (FPS), and model parameter count are included.

Table 1. The detection performance of TFCST with different RNNs.

RNN	Pd	Fa	FPS	#Param(M)
ConvGRU	95.14%	3.56%	63.5	2.50
Bottleneck-LSTM	93.84%	2.52%	65.3	**2.37**
ConvLSTM	**95.25%**	**0.58%**	**66.7**	2.65

It can be observed that when using ConvGRU and Bottleneck-LSTM, the number of model parameters is smaller. However, the detection rate, false alarm rate, and even the detection speed were inferior to the performance when using ConvLSTM. Since TFCST is built on the basis of the lightweight single-frame detector FCST, if we do not consider the constraints of computing resources and model weight size, using ConvLSTM is a better choice as it can more fully extract spatiotemporal information in sequence images compared to other optimized RNN models.

4.4. Contrast Test

To thoroughly compare the performance of the FCST network with other detection methods, we compared it with a total of nine different types of detection algorithms. Among them, in the model-driven methods, we selected the morphological filtering algorithms TopHat [11], MPCM [36], based on visual contrast mechanism, IPI [37], based on image data structure, NRAM [38], based on matrix optimization, and PSTNN [39], based on tensor optimization. For data-driven methods, considering detection speed, we primarily selected one-stage object detection algorithms. Among them, for anchor-based methods, we chose Retinanet [30] and YOLOv3 [40], while for anchor-free detection methods, we selected YOLOX [41].

We selected four typical infrared scenes, and the detection results of each algorithm are shown in Figures 13–16 . In the figures, the red boxes represent correct detections, the yellow boxes represent missed detections, and the blue boxes represent false alarms. From the figure, it can be observed that traditional infrared small object detection algorithms exhibit significant variations in detection performance across different scenes. Specifically, the model-driven methods showed excellent detection performance in Scene 2 but exhibited a higher number of missed detections in Scenes 3 and 4. The TopHat algorithm had a high number of false alarms in Scene 1 but showed higher detection rates compared to other model-driven methods in Scenes 3 and 4. The reason for these issues is that model-driven methods rely on fixed hyperparameters and segmentation thresholds, leading to poor adaptability across different scenes. On the other hand, data-driven object detection algorithms, as compared to model-driven methods, generally demonstrated better detection performance. Among them, Retinanet had false alarms and missed detections in Scenes 1 and 4, YOLOv3 only predicted one false object in Scene 2, while the anchor-free detection algorithm YOLOX and the proposed FCST achieved accurate detection of small objects across different scenes.

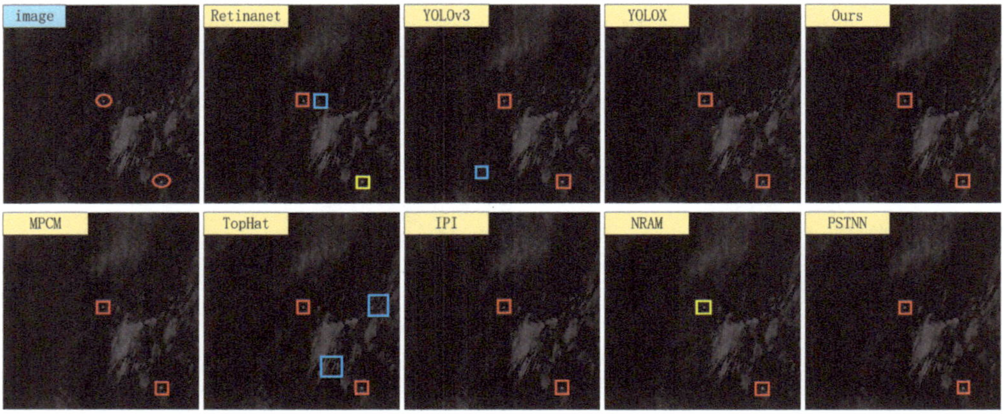

Figure 13. Illustration of detection results on Scene 1. Boxes in red, yellow, and blue respectively signify true positive targets with corresponding confidence scores, miss detected targets and, false alarms.

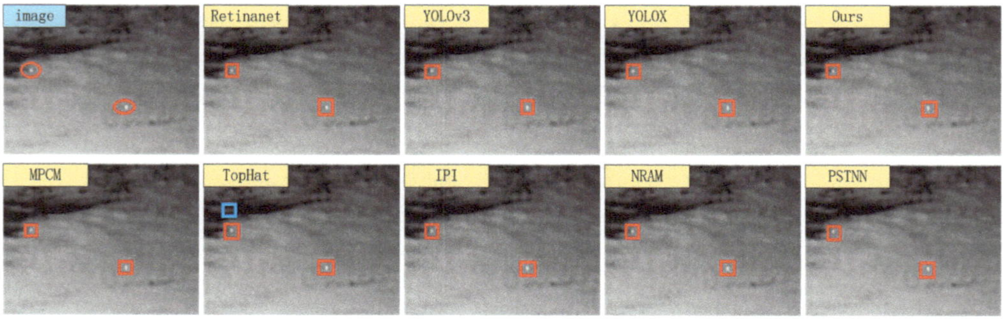

Figure 14. Illustration of detection results on Scene 2. Boxes in red, yellow, and blue respectively signify true positive targets with corresponding confidence scores, miss detected targets, and false alarms.

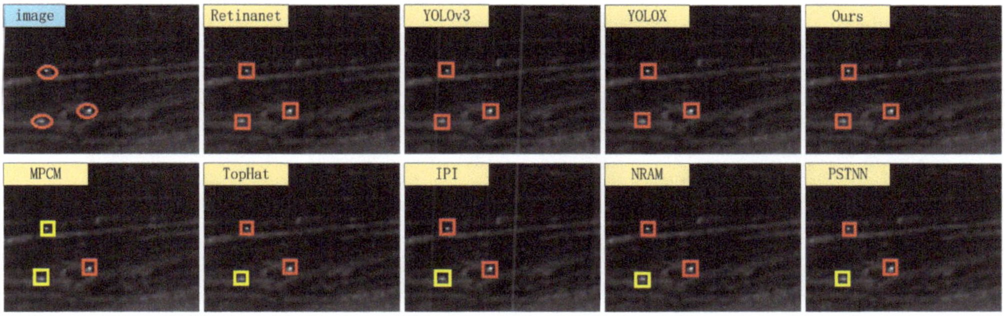

Figure 15. Illustration of detection results on Scene 3. Boxes in red, yellow, and blue respectively signify true positive targets with corresponding confidence scores, miss detected targets, and false alarms.

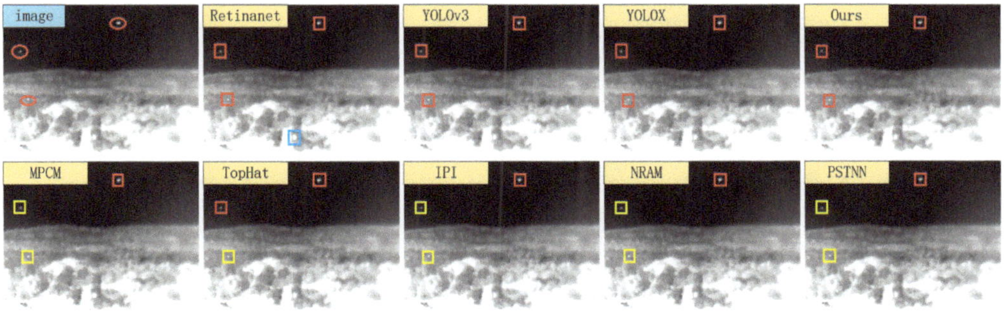

Figure 16. Illustration of detection results on Scene 4. Boxes in red, yellow, and blue respectively signify true positive targets with corresponding confidence scores, miss detected targets, and false alarms.

We conducted experiments on a test set consisting of 600 infrared images to evaluate the performance of each detection algorithm. Specifically, we compared these methods based on three metrics, P_d, F_a, and FPS, which are defined as follows:

(1) Detection rate (P_d) and false alarm rate (F_a) are defined as follows:

$$P_d = \frac{N_p}{N_t}; F_a = \frac{N_f}{N_d} \tag{20}$$

where N_p represents the number of detected true targets, N_t represents the total number of true targets, N_p represents the number of false detections, and N_d represents the total number of detected targets.

(2) Frames per second (FPS) denotes the processing speed of the detection algorithm, measured in frames processed per second.

We obtained the experimental results, which are summarized in Table 2. These results provide insights into the performance of each algorithm in terms of detection accuracy (Pd), false alarm rate (Fa), and processing speed (FPS). It is evident that the FCST algorithm achieved the highest Pd and lowest Fa among the tested methods, demonstrating its effectiveness in accurately detecting targets in the infrared images. Moreover, it maintained a competitive processing speed of 66.4 FPS, ensuring real-time performance.

Table 2. The detection performance of FCST with different methods.

RNN	Pd	Fa	FPS
MPCM	79.23%	26.74%	16.2
TopHat	87.36%	57.45%	82.9
IPI	84.81%	18.57%	0.21
NRAM	80.62%	24.27%	0.47
PSTNN	82.60%	26.31%	5.3
Retinanet	87.85%	9.1%	38.9
YOLOv3	92.35%	7.65%	54.2
YOLOX	91.99%	5.43%	48.6
FCOS	91.62%	4.80%	**68.3**
FCST	**94.85%**	**2.32%**	66.4

To analyze and demonstrate the advantages of TFCST over FCST, we selected four complex infrared image sequences from the test set for experimental comparison, and their descriptions are shown in Table 3.

Table 3. Description of image sequences. We selected a change of images to conduct the experiment.

Input	Frame	Image Size	Target Size	Background Description
Sequence_1	80	202 × 202	11 × 6	Less cluster of Building
Sequence_2	60	256 × 256	8 × 7	Heavy cluster of Ground and Building
Sequence_3	260	288 × 288	3 × 3	Heavy cluster of Cloud
Sequence_4	125	256 × 256	11 × 5	Heavy cluster of Ground

In the experiment, FCST was also trained on the constructed sequence infrared image dataset based on pre-trained weights. The visual detection results of the two detection networks on four sequences are shown in Figures 17–20, where the red box represents correct detection, the yellow box represents missed detection, and the blue box represents false alarm.

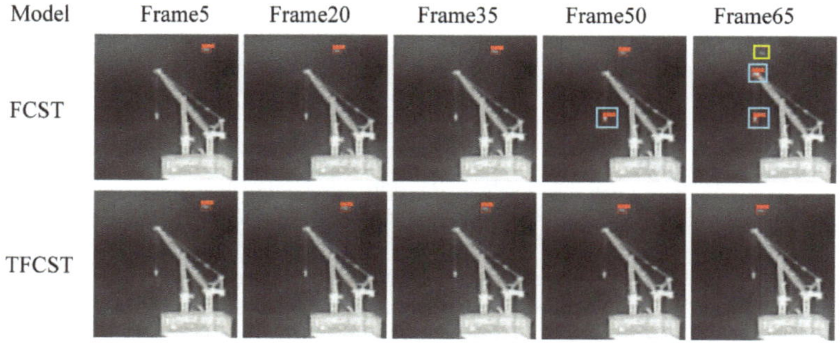

Figure 17. Illustration of detection results on Sequence 1. Boxes in red, yellow, and blue respectively signify true positive targets with corresponding confidence scores, miss detected targets, and false alarms.

It is observed that in Infrared Image Sequence 1, the artificial structure (crane) in the background has a high-intensity interference section, which is very similar to the small infrared target, making it difficult for the detection algorithm to distinguish effectively. The single-frame image-based detection algorithm FCST showed false alarms in frames 50 and 65 of sequence 1. However, the TFCST based on the sequence images utilized the state information extracted from historical frames by the detector, thus enhancing the robustness of the detection during the current frame, and hence no occasional false detections occurred during the detection process. In the Infrared Image Sequence 2, the target passes through a region with severe clutter interference during its motion. The detection method FCST, which only uses spatial information in the infrared image, cannot detect the target correctly in this situation. A missed detection occurred in frame 38 of sequence 2, and the clutter interference in the image was misjudged as a target, indicating comparatively poor robustness. In contrast, TFCST correctly detected the targets in all consecutive frames, with stable output of detection results, exhibiting better detection performance.

Furthermore, it is worth noting that even with the utilization of temporal information in the detection algorithm, TFCST, when the target is heavily mixed in the background clutter region, there may still be missed detections due to low target confidence, as shown in frame 25 of Infrared Image Sequence 3 and frame 61 of Infrared Image Sequence 4. To comprehensively compare the performance of the two detection methods tested, we further calculated the detection results for 60 infrared sequences (3050 frames in total) in the test dataset, as presented in Table 4.

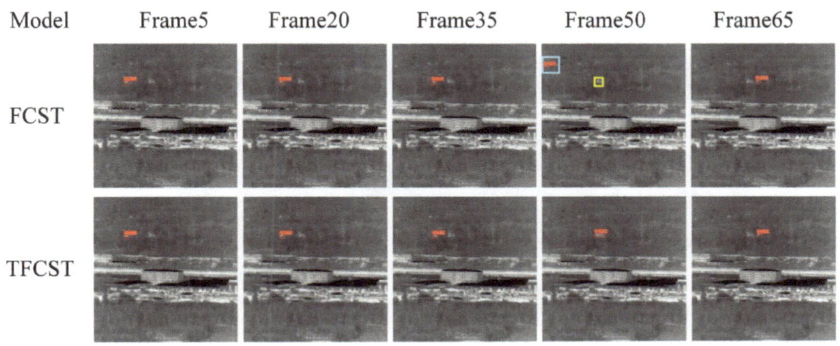

Figure 18. Illustration of detection results on Sequence 2. Boxes in red, yellow, and blue respectively signify true positive targets with corresponding confidence scores, miss detected targets, and false alarms.

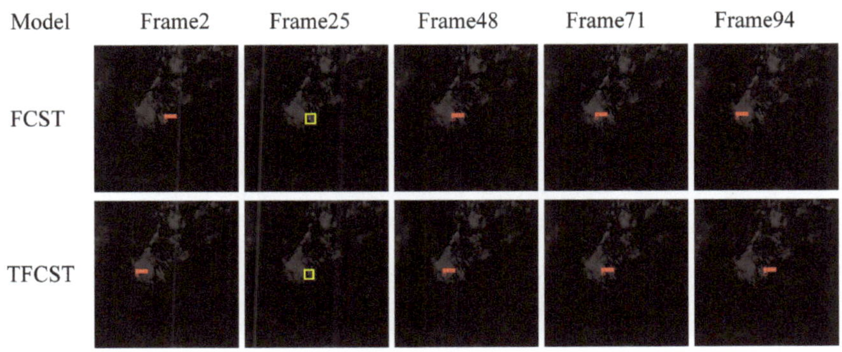

Figure 19. Illustration of detection results on Sequence 3. Boxes in red, yellow, and blue respectively signify true positive targets with corresponding confidence scores, miss detected targets, and false alarms.

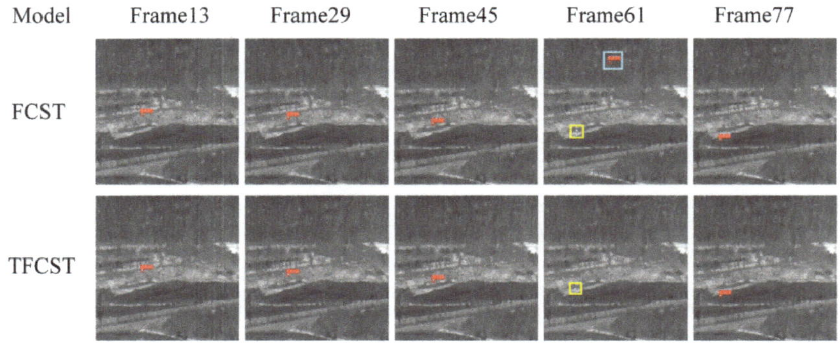

Figure 20. Illustration of detection results on Sequence 4. Boxes in red, yellow, and blue respectively signify true positive targets with corresponding confidence scores, miss detected targets, and false alarms.

Table 4. Detection results of FCST compared with TFCST. The table compares positive detection (Pd), false alarm (Fa), and frames per second (FPS).

Model	Pd	Fa	FPS
FCST	93.40%	8.64%	**66.4**
TFCST	**96.13%**	**0.51%**	65.5

From Table 4, it is evident that TFCST, with a slightly lower inference speed, achieved a 2.73% increase in detection rate compared to the FCST network, significantly reducing the number of false target predictions, with a false alarm rate of only 0.51%. This result confirms that by combining the extraction of spatiotemporal domain information using ConvLSTM, the performance of a single-frame detector can be effectively enhanced. Moreover, TFCST also has extremely fast detection speed and hence meets the requirements for real-time detection.

5. Conclusions

To address the problems of high false alarm rates and poor robustness of existing infrared small target detection algorithms, a fully convolutional-based approach is proposed. Built upon a mechanism without anchor boxes, the algorithm avoids issues related to anchor box design and unfavorable effects on small target detection by pixel-level prediction. To address the issue of losing small target feature information with a deep network, a lightweight feature extraction network is designed, incorporating reduced downsampling and single-stage aggregation that efficiently extract small target features. Additionally, a feature refinement mechanism is proposed to deal with semantic conflicts when fusing features of different scales. By learning adaptive fusion weights in both spatial and channel dimensions, this mechanism enhances small target feature representation while suppressing background information. To cope with the infrared small target detection task in more complicated backgrounds with weaker target signals, a time-series-aware fully convolutional-based infrared small target detection approach is proposed. Built upon the mechanism without anchor boxes, the algorithm combines ConvLSTM to model the target's appearance and movement information in sequential images, thus propagating and fusing temporal and spatial contextual information. This mechanism enhances the detection accuracy of infrared small targets in more complex scenarios.

Author Contributions: Conceptualization, Z.Z. and L.Z.; Methodology, Z.Z., J.X. and P.H.; Software, J.X. and P.H.; Validation, L.Z., P.H. and J.X.; Formal analysis, L.Z. and J.X.; Investigation, Z.Z. and P.H.; Resources, Z.Z.; Data curation, P.H. and J.X.; Writing – original draft, P.H. and L.Z.; Writing—review and editing, P.H., Z.Z. and L.Z.; Visualization, J.X. and P.H.; Supervision, Z.Z.; Project administration, L.Z.; Funding acquisition, Z.Z. and L.Z. All authors have read and agreed to the published version of the manuscript.

Funding: This research received no external funding.

Data Availability Statement: The data presented in this study are available on request from the corresponding author.

Conflicts of Interest: The authors declare no conflict of interest.

References

1. Chen, Y.; Lin, Y.; Yang, M.; Huang, J. Show, Match and Segment: Joint Weakly Supervised Learning of Semantic Matching and Object Co-Segmentation. *IEEE Trans. Pattern Anal. Mach. Intell.* **2021**, *43*, 3632–3647. [CrossRef] [PubMed]
2. Izadinia, H.; Sadeghi, F.; Divvala, S.; Hajishirzi, H.; Choi, Y.; Farhadi, A. Segment-Phrase Table for Semantic Segmentation, Visual Entailment and Paraphrasing. In Proceedings of the 2015 IEEE International Conference on Computer Vision (ICCV), Santiago, Chile, 7–13 December 2015; p. 10.
3. Huang, J.; Li, N.; Li, T.; Liu, S.; Li, G. Spatial-Temporal Context-Aware Online Action Detection and Prediction. *IEEE Trans. Circuits Syst. Video Technol.* **2020**, *30*, 2650–2662. [CrossRef]
4. Chen, Y.; Chen, M.; Gu, Q. Class-wise boundary regression by uncertainty in temporal action detection. *IET Image Process.* **2022**, *16*, 3854–3862. [CrossRef]
5. Chen, C.; Wei, J.; Peng, C.; Qin, H. Depth-Quality-Aware Salient Object Detection. *IEEE Trans. Image Process.* **2021**, *30*, 2350–2363. [CrossRef] [PubMed]
6. Shivappriya, S.; Priyadarsini, M.; Stateczny, A.; Puttamadappa, C.; Parameshachari, B. Cascade Object Detection and Remote Sensing Object Detection Method Based on Trainable Activation Function. *Remote Sens.* **2021**, *13*, 200. [CrossRef]
7. Zhao, K.; Zhao, K.; Kong, X. Discussion on the suppression and method of background noise of small target infrared image. *Opt. Optoelectron. Technol.* **2004**, *2*, 9–12.
8. Deshpande, S.; Deshpande, S.; Er, M.; Venkateswarlu, R.; Chan, P. Max-mean and max-median filters for detection of small targets. *Proc. SPIE* **1999**, *3809*, 74–83.
9. Hadhoud, M.; Hadhoud, M.; Thomas, D. The Two-Dimensional Adaptive LMS (TDLMS) Algorithm. *IEEE Trans. Circuits Syst.* **1988**, *35*, 485–494. [CrossRef]
10. Bae, T.; Bae, T.; Sohng, K. Small Target Detection Using Bilateral Filter Based on Edge Component. *J. Infrared Millim. Terahertz Waves* **2010**, *31*, 735–743. [CrossRef]
11. Zeng, M. The design of top-hat morphological filter and application to infrared target detection. *Infrared Phys. Technol.* **2006**, *48*, 67–76. [CrossRef]
12. Chen, C.; Chen, C.; Li, H.; Wei, Y.; Xia, T.; Tang, Y. A Local Contrast Method for Small Infrared Target Detection. *IEEE Trans. Geosci. Remote Sens.* **2013**, *52*, 574–581. [CrossRef]
13. Han, J.; Han, J.; Yong, M.; Bo, Z. A Robust Infrared Small Target Detection Algorithm Based on Human Visual System. *IEEE Geosci. Remote Sens. Lett.* **2014**, *11*, 2168–2172.
14. Wang, H.; Wang, H.; Zhou, L.; Wang, L. Miss Detection vs. False Alarm: Adversarial Learning for Small Object Segmentation in Infrared Images. In Proceedings of the International Conference on Computer Vision 2019, Seoul, Republic of Korea, 27 October–2 November 2019; pp. 8508–8517.
15. Shi, M.; Shi, M.; Wang, H. Infrared Dim and Small Target Detection Based on Denoising Autoencoder Network. *Mob. Netw. Appl.* **2020**, *25*, 1469–1483. [CrossRef]
16. Zhao, B.; Zhao, B.; Wang, C.; Fu, Q.; Han, Z. A Novel Pattern for Infrared Small Target Detection With Generative Adversarial Network. *IEEE Trans. Geosci. Remote Sens.* **2020**, *99*, 1–12 [CrossRef]
17. Zhang, R.; Zhang, R.; Xu, L.; Yu, Z.; Shi, Y.; Xu, M. Deep-IRTarget: An automatic target detector in infrared imagery using dual-domain feature extraction and allocation. *IEEE Trans. Multimed.* **2022**, *24*, 1735–1749. [CrossRef]
18. Zhang, R.; Zhang, R.; Yang, S.; Zhang, Q.; Xu, L.; He, Y.; Zhang, F. Graph-based few-shot learning with transformed feature propagation and optimal class allocation. *Neurocomputing* **2022**, *470*, 247–256. [CrossRef]
19. Ren, X.; Ren, X.; Wang, J.; Ma, T.; Bai, K.; Wang, Y. Infrared dim and small target detection based on three-dimensional collaborative filtering and spatial inversion modeling. *Infrared Phys. Technol.* **2019**, *101*, 13–24. [CrossRef]
20. Irsoy, O.; Cardie, C. Opinion mining with deep recurrent neural networks. In Proceedings of the Conference on Empirical Methods in Natural Language Processing 2014, Doha, Qatar, 26–28 October 2014; pp. 720–728.
21. Xiao, F.; Xiao, F.; Lee, Y. Video Object Detection with an Aligned Spatial-Temporal Memory. In Proceedings of the European Conference on Computer Vision (ECCV), Munich, Germany, 8–14 September 2018; pp. 485–501.

22. Lu, Y.; Lu, Y.; Lu, C.; Tang, C. Online video object detection using association LSTM. In Proceedings of the IEEE International Conference on Computer Vision, Venice, Italy, 22–29 October 2017; pp. 2344–2352.
23. Shi, X.; Shi, X.; Chen, Z.; Wang, H.; Yeung, D.; Wong, W.; Woo, W. Convolutional LSTM network: A machine learning apporach for precipitation nowcasting. In Proceedings of the 28th International Conference on Neural Infromation Processing Systems, Montreal, QC, Canada, 7–12 December 2015; pp. 802–810.
24. Liu, W.; Liu, W.; Anguelov, D.; Erhan, D. *SSD: Single Shot MultiBox Detector*; Springer International Publishing: Berlin/Heidelberg, Germany, 2016; pp. 21–37.
25. Wei, K.; Wei, K.; Fu, Y. ; Huang, H. 3-d quasi-recurrent neural network for hyperspectral image denoising. *IEEE Trans. Neural Netw. Learn. Syst.* **2020**, *32*, 363–375. [CrossRef]
26. Tian, Z.; Tian, Z.; Shen, C.; Chen, H.; He, T. FCOS: Fully convolutional one-stage object detection. In Proceedings of the IEEE International Conference on Computer Vision (ICCV), Seoul, Republic of Korea, 27 October–2 November 2019; pp. 9626–9635.
27. He, K.; He, K.; Zhang, X.; Ren, S.; Sun, J. Deep residual learning for image recognition. In Proceedings of the IEEE Conference on Computer Vision and Pattern Recognition (CVPR), Las Vegas, NV, USA, 27–30 June 2016; pp. 770–778.
28. Huang, G.; Huang, G.; Liu, Z.; Laurens, V. Densely Connected Convolutional Networks. In Proceedings of the 2017 IEEE Conference on Computer Vision and Pattern Recognition (CVPR), Honolulu, HI, USA, 21–26 July 2017.
29. Lee, Y.; Lee, Y.; Hwang, J.; Lee, S.; Bae, Y.; Park, J. An Energy and GPU-Computation Efficient Backbone Network for Real-Time Object Detection. In Proceedings of the 2019 IEEE/CVF Conference on Computer Vision and Pattern Recognition Workshops (CVPRW), Long Beach, CA, USA, 16–17 June 2019.
30. Lin, T.; Lin, T.; Dollar, P.; Girshick, R.; He, K.; Hariharan, B.; Belongie, S. Feature Pyramid Networks for Object Detection. In Proceedings of the IEEE/CVF Conference on Computer Vision & Pattern Recognition (CVPR), Honolulu, HI, USA, 21–26 July 2017; pp. 936–944.
31. Lin, T.; Lin, T.; Goyal, P.; Girshick, R.; He, K.; Dollar, P. Focal Loss for Dense Object Detection. *IEEE Trans. Pattern Anal. Mach. Intell.* **2017**, *99*, 2999–3007.
32. Rezatofighi, H.; Rezatofighi, H.; Tsoi, N.; Gwak, J.; Sadeghian, A.; Reid, I.; Savarese, S. Generalized Intersection over Union: A Metric and A Loss for Bounding Box Regression. In Proceedings of the IEEE/CVF Conference on Computer Vision and Pattern Recognition, Long Beach, CA, USA, 15–20 June 2019; pp. 658–666.
33. Ballas, N.; Ballas, N.; Yao, L.; Pal, C.; Courville, A. Delving Deeper into Convolutional Networks for Learning Video Representations. *Comput. Sci.* **2015** . [CrossRef]
34. Liu, M.; Zhu, M. Mobile Video Object Detection with Temporally-Aware Feature Maps.In Proceedings of the 2018 IEEE/CVF Conference on Computer Vision and Pattern Recognition, Salt Lake City, UT, USA, 18–23 June 2018.
35. Cho, K.; Cho, K.; Van Merrienboer, B., Gulcehre, C. Learning Phrase Representations using RNN Encoder-Decoder for Statistical Machine Translation. *Comput. Sci.* **2014**. [CrossRef]
36. Wei, Y.; Wei, Y.; You, X.; Li, H. Multiscale patch-based contrast measure for small infrared target detection. *Pattern Recognit.* **2016**, *58*, 216–226. [CrossRef]
37. Gao, C.; Gao, C.; Meng, D.; Yang, Y.; Wang, Y.; Zhou, X. Infrared patch-image model for small target detection in a single image. *IEEE Trans. Image Process.* **2013**, *22*, 4996–5009. [CrossRef] [PubMed]
38. Zhang, L.; Zhang, L.; Peng, L.; Zhang, T.; Cao, S.; Peng, Z. Infrared Small Target Detection via Non-Convex Rank Approximation Minimization Joint l2,1 Norm. *Remote Sens.* **2018**, *10*, 1821. [CrossRef]
39. Zhang, L.; Zhang, L.; Peng, Z. Infrared small target detection based on partial sum of the tensor nuclear norm. *Remote Sens.* **2019**, *11*, 382. [CrossRef]
40. Redmon, J.; Redmon, J.; Farhadi, A. YOLOv3: An Incremental Improvement. *arXiv* **2018**, arXiv:1804.02767.
41. Ge, Z.; Ge, Z.; Liu, S.; Wang, F.; Sun, J.. YOLOX: Exceeding YOLO Series in 2021. *arXiv* **2021**, arXiv:2107.08430.

remote sensing

MDPI

Article

Long-Tailed Object Detection for Multimodal Remote Sensing Images

Jiaxin Yang †, Miaomiao Yu †, Shuohao Li, Jun Zhang * and Shengze Hu

Laboratory for Big Data and Decision, National University of Defense Technology, Changsha 410073, China; yangjiaxin21@nudt.edu.cn (J.Y.); yumiaomiaonudt@nudt.edu.cn (M.Y.); lishuohao@nudt.edu.cn (S.L.); springsun@nudt.edu.cn (S.H.)
* Correspondence: zhangjun1975@nudt.edu.cn
† These authors contributed equally to this work.

Abstract: With the rapid development of remote sensing technology, the application of convolutional neural networks in remote sensing object detection has become very widespread, and some multimodal feature fusion networks have also been proposed in recent years. However, these methods generally do not consider the long-tailed problem that is widely present in remote sensing images, which limits the further improvement of model detection performance. To solve this problem, we propose a novel long-tailed object detection method for multimodal remote sensing images, which can effectively fuse the complementary information of visible light and infrared images and adapt to the imbalance between positive and negative samples of different categories. Firstly, the dynamic feature fusion module (DFF) based on image entropy can dynamically adjust the fusion coefficient according to the information content of different source images, retaining more key feature information for subsequent object detection. Secondly, the instance-balanced mosaic (IBM) data augmentation method balances instance sampling during data augmentation, providing more sample features for the model and alleviating the negative impact of data distribution imbalance. Finally, class-balanced BCE loss (CBB) can not only consider the learning difficulty of specific instances but also balances the learning difficulty between categories, thereby improving the model's detection accuracy for tail instances. Experimental results on three public benchmark datasets show that our proposed method achieves state-of-the-art performance; in particular, the optimization of the long-tailed problem enables the model to meet various application scenarios of remote sensing image detection.

Keywords: remote sensing image; object detection; long-tailed distribution; multimodality image fusion

Citation: Yang, J.; Yu, M.; Li, S.; Zhang, J.; Hu, S. Long-Tailed Object Detection for Multimodal Remote Sensing Images. *Remote Sens.* **2023**, *15*, 4539. https://doi.org/10.3390/rs15184539

Academic Editors: Qian Du, Yanni Dong and Xiaochen Yang

Received: 12 August 2023
Revised: 7 September 2023
Accepted: 12 September 2023
Published: 15 September 2023

1. Introduction

Remote sensing [1] is a technology that obtains information about the earth by using sensors mounted on aircraft or spacecraft to receive electromagnetic signals from the earth. Among them, remote sensing images are an important product of remote sensing technology, which can provide rich geographic information and help people better understand and manage the earth. In recent years, with the rapid development of aerospace technology, remote sensing technology has also reached a new stage [2]. It is mainly manifested in two aspects: One is that the use scenarios are more extensive, such as meteorology, agriculture, environment, disasters, etc. The second is that the data obtained are more abundant, such as multi-time and multi-band images. Faced with these massive data, how to analyze and extract semantic information from them has become an important research topic, which includes how to perform efficient and accurate object detection. Object detection is an important task in the field of computer vision, which aims to locate and identify the category of objects in an image. At present, a large number of studies [3–7] have been proposed. Object detection has a wide range of applications in remote sensing images, such as military reconnaissance [8], urban planning [9], traffic monitoring [10], etc. However, object detection in remote sensing images also faces some challenges.

The first challenge is how to efficiently fuse multimodal information. Remote sensing images acquired from a long distance are affected by multiple factors such as clouds, fog and lighting conditions. Using only a single band of sensor data will result in low detection accuracy, which will greatly limit the use scenarios of object detection algorithms. Most of the current [3–5] object detection techniques are mainly oriented to single modalities, such as visible light images or infrared images. If multiple wavelengths of remote sensing images can be fused, the detection accuracy can be greatly improved by complementing the advantages of each modality. As shown in Figure 1, in good lighting conditions during the day, visible light images can provide more color and texture information, while infrared images can show clearer object contours at night. Unfortunately, image data obtained from different sensors have problems such as different resolutions, different amounts of information carried and image misalignment [11], so simply combining remote sensing images cannot improve detection performance.

(a)
LLVIP

(b)
VEDAI

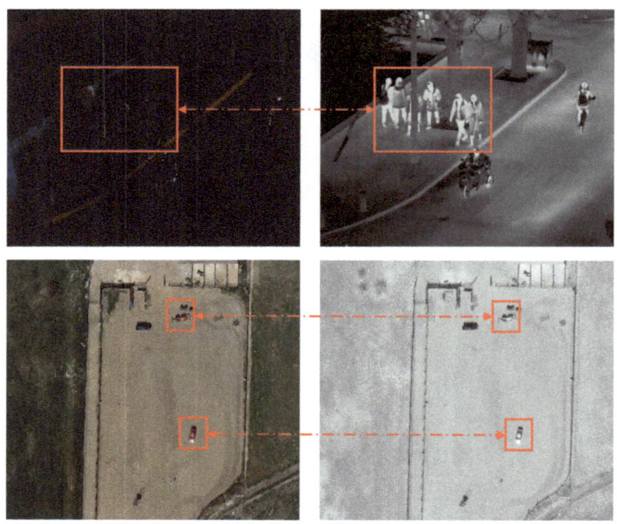

Figure 1. Comparison of visible light and infrared images. (**a**) Image examples from the LLVIP dataset. In low-light environments, infrared images can provide more information than visible light images. (**b**) Image examples from the VEDAI dataset. In good lighting conditions, visible light images can provide more color, texture and other detail information.

The second challenge is how to deal with long-tailed data with natural state distribution. Real-world data often show an open long-tailed distribution [12], as shown in Figure 2, where a small number of categories in the dataset have far more instances than other categories, resulting in a serious imbalance in data distribution. As more data are acquired, new category instances will appear, further exacerbating the long-tailed characteristics of the dataset. Although existing deep learning-based network models have begun to pay attention to the long-tailed problem [13–15], most of the long-tailed research in the remote sensing field focuses on long-tailed recognition [16,17], and there is less research on the more complex long-tailed object detection task. We can see that the red line representing detection accuracy in Figure 2 is positively correlated with instance number, which shows that the key to improving model detection accuracy is to improve the detection ability for tail classes.

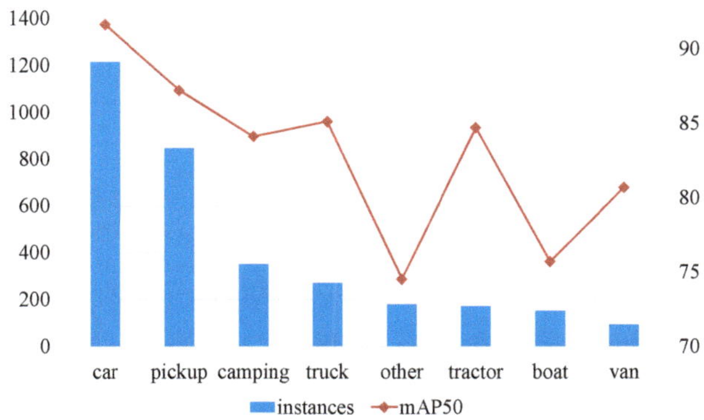

Figure 2. A combination plot of instance number and detection accuracy on the VEDAI dataset.

The third challenge is how to reduce algorithm complexity for real-time detection. In scenarios such as disaster relief where timely detection and emergency handling are required, real-time performance will be a very important consideration factor. However, due to space limitations, detection models deployed on satellites or aircraft cannot increase performance by increasing complexity. Therefore, model complexity needs to be strictly controlled when fusing multimodal information and dealing with long-tailed problems.

In order to solve these challenges, based on the above analysis, we propose a multi-source fusion object detection method for natural long-tailed datasets, which uses infrared and visible light images to improve the detection accuracy of the model on natural long-tailed datasets. Considering that multimodal image fusion needs to be based on image sample features, we propose a dynamic feature fusion module based on image information entropy [18]. By calculating the amount of information in the image, the model can dynamically allocate fusion coefficients during the training and inference stages, which enables the network to retain more key feature information and provide more accurate image features for subsequent object detection. In order to ensure that the model can meet the requirements of real-time detection and lightweight, we choose the small-sized YOLOv8s [19] as our baseline model. Considering the characteristics of natural data long-tailed distribution, we propose an instance-balanced mosaic method by improving Mosaic [20] to achieve balanced sampling, which provides more sample features for the model and alleviates the negative impact of data distribution imbalance. At the same time, in order to balance the gradient information during the training stage and reduce the impact of long-tailed distribution, we propose class-balanced BCE loss for object detection. This loss can dynamically adjust the loss weight according to the number of samples of different categories, thereby reducing the model's overfitting on head categories and underfitting on tail categories.

Finally, in order to verify the performance of our proposed method, we conducted experimental verification on three public benchmark datasets and conducted ablation analysis. The experimental results proved that our proposed method achieves state-of-the-art performance; in particular, the optimization of the long-tailed problem enables the model to meet various application scenarios of remote sensing image detection.

In summary, our contributions can be summarized as follows:

- We propose a dynamic feature fusion module based on image information entropy, which dynamically adjusts the fusion coefficient according to the different information entropy of images, enabling the model to capture more features. Compared with other similar methods, this module helps the model significantly improve detection accuracy without significantly increasing computational complexity, and this method is simple and can be easily inserted into other object detection networks to achieve feature fusion for remote sensing images.

- We propose an instance-balanced mosaic data augmentation method based on instance number, which solves the long-tailed problem by providing rich tail-class features for the model through resampling during data augmentation.
- We propose class-balanced BCE loss for long-tailed object detection. This loss provides more balanced loss information for the model according to sample number, improving the detection accuracy of tail instances.
- Based on three public benchmark datasets, we constructed a large number of experiments to verify the performance of our method. Compared with baseline methods, our method can greatly improve the performance of baseline models. The experimental results and ablation analysis prove the effectiveness of our proposed method.

The rest of this paper is organized as follows: Section 2 reviews related work. Section 3 introduces the details of our proposed method. In Section 4, we conduct experiments to verify the effectiveness of our method. Section 5 deploys ablation experiments and gives corresponding performance analysis. In the final Section 6, we summarize this paper and give possible future work.

2. Related Work

We first summarize the latest research achievements in the fields of remote sensing image object detection and deep long-tailed object detection, and analyze the shortcomings of the state-of-the-art methods in the corresponding fields. Based on this, we also point out the importance of long-tailed object detection for remote sensing image object detection.

2.1. Long-Tailed Object Detection

Long-tailed object detection is an important and challenging problem in the field of object detection, which refers to the uneven distribution of sample numbers of different categories in the training data, resulting in much lower detection performance of minority categories (tail categories) than majority categories (head categories). In order to solve this problem, many long-tailed object detection studies have emerged in recent years, which can be divided into three categories: based on sample resampling, based on loss reweighting and based on knowledge transfer [21]. Methods based on sample resampling balance the training data by changing the sampling probability of different category samples, such as instance-aware repeat factor sampling [22], class-balanced grouping and sampling [23], etc.; methods based on loss reweighting balance the training data by adjusting the loss weight of different category samples, such as focal loss [24], equalization loss [25], etc.; methods based on knowledge transfer enhance the representation of tail categories by using the knowledge of head categories, such as semi-supervised and long-tailed object detection with CascadeMatch [14], etc. However, most of these methods are designed for long-tailed object detection in natural images. In remote sensing images, there are other challenges besides the long-tailed problem. First, remote sensing images are generally multimodal images obtained by different sensors, such as visible light, infrared, radar, etc., which contain different physical characteristics and complementary information. If only one modality of information is used, a lot of useful information will be lost, resulting in a decline in model performance; secondly, the objects in remote sensing images generally have small size and low resolution, which makes object detection more difficult. For this purpose, we propose a long-tailed object detection method for multimodal remote sensing images, which can effectively fuse multimodal information and adapt to the imbalance problem between positive and negative samples of different categories.

2.2. Multimodal Feature Fusion

With the rapid development of remote sensing technology, multimodal feature fusion has become an important means to improve the performance of remote sensing image object detection. Multimodal feature fusion refers to using different types of images obtained by different sensors, such as visible light, infrared, radar, etc., to obtain more rich and robust object representation through feature extraction and fusion. In the past, there have been some studies on feature fusion. For example, Liu et al. [26] proposed a deep learning-based feature fusion method that uses convolutional neural networks (CNN) and bilinear interpolation (Bilinear Interpolation) to achieve feature extraction and fusion for visible light and infrared images; the authors of [27] proposed SuperYOLO, an accurate and fast small-object detection method for remote sensing images, which uses auxiliary super-resolution learning to fuse multimodal data and perform high-resolution (HR) object detection for multi-scale objects. In recent years, some researchers have also tried to use attention mechanism (Attention Mechanism) or multi-scale feature alignment (Multi-scale Feature Alignment) to improve the effect of multimodal feature fusion, such as [28–31], etc. However, these methods often require multiple stages, cannot detect in real time or do not consider the long-tailed problem, resulting in low detection accuracy for tail instances. We found through research that combining long-tailed object detection with multi-source object detection can significantly improve the detection ability of remote sensing images. For this purpose, we propose a long-tailed object detection method for multimodal remote sensing images, which achieves the efficient detection of multimodal remote sensing images.

3. Methodology

In this section, we first introduce the basic architecture of the method, and then describe the proposed modules in detail. The overall framework of our model is shown in Figure 3, which mainly consists of three parts: the dynamic feature fusion module, instance-balanced mosaic and class-balanced BCE loss. Through the collaboration of these three modules, we achieve the detection of multimodal long-tailed datasets.

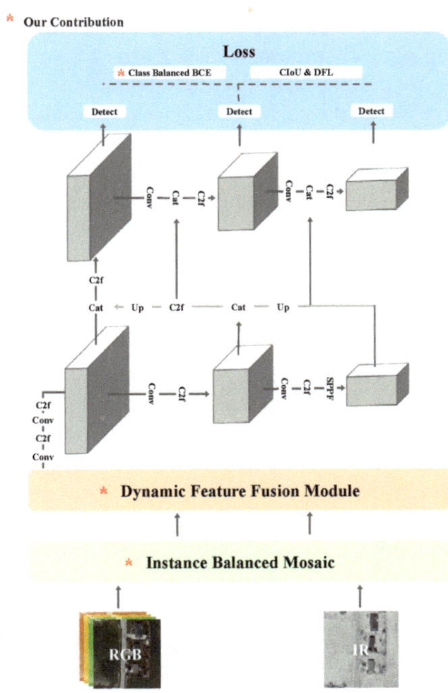

Figure 3. An overview of the model's overall framework. Our model is based on YOLOv8 architecture, and the main contributions are as follows: (1) we propose the dynamic feature fusion module to achieve efficient and accurate fusion of multi-source images; (2) the proposed instance-balanced mosaic reduces the underfitting of the model for tail instances by balancing the sampling of instances; (3) class-balanced BCE loss not only considers the learning difficulty of instances, but also balances the learning difficulty between classes.

3.1. Basic Architecture

As shown in Figure 3, the baseline architecture YOLOv8 [19] network used by our method consists of two main components: a backbone network and a head module. The backbone network is used to extract multi-scale feature maps, including low-level texture features and high-level semantic features. Then, these feature maps are fed into the head for object detection and classification. Compared with YOLOv5 [32], YOLOv8 replaces C3 with a C2f module for lighter weight, and uses split instead of conv to layer the features. It also uses a decoupled head in the head module, which aligns the classification and regression tasks by introducing task-alignment learning, making the model design more reasonable and efficient.

YOLOv8, as the latest SOTA model, has strong performance in object detection, but it still has some limitations.

Limitation 1: YOLOv8 cannot perform multimodal fusion object detection, resulting in low detection accuracy. Because there are differences between different modal images, simply fusing them cannot fully utilize the complementary information between them. For example, in good lighting conditions, visible light images contain more information such as texture, color, etc., than infrared images. Conversely, in low-visibility environments, especially when there is slight occlusion, infrared images can capture the outline of the target and other information significantly. These problems affect the accuracy and robustness of object detection, so it is necessary to design a more effective multimodal fusion method to transform the YOLO network, to extract the complementary information between different modalities and eliminate the differences and noise between different modalities.

Limitation 2: The mosaic data augmentation method used by YOLOv8 is not suitable for remote sensing object detection; because remote sensing target samples are small and scattered, cropping and stitching will lead to loss of tail-class samples, resulting in a decrease in the overall detection accuracy of the model. The mosaic data augmentation method can increase the diversity of data without increasing the size of the dataset, thereby improving the generalization ability of deep network models. The specific method is to randomly crop four images and then combine and stitch them into a new image input network. However, this method is not suitable for remote sensing object detection; because the targets in remote sensing images are usually small and sparse, random cropping will cause many targets to be cut off or only retain a part, thus affecting the model's detection and localization of targets. Especially for tail class targets, since they are rare in themselves, if they are cropped off, it will cause the model to lack enough training samples, thereby reducing the model's detection accuracy for tail classes.

Limitation 3: The loss function of YOLOv8 only considers hard samples, not tail class samples learning, resulting in low detection accuracy for tail-class instances. YOLOv8 network uses the CIoU loss function [33] to optimize the regression of target boxes. It is a loss function based on IoU (intersection over union), which can consider factors such as overlap degree, center distance, aspect ratio, etc. between target boxes, and improves the localization accuracy of target boxes. However, this loss function only focuses on hard samples, i.e., prediction boxes with low overlap with ground truth boxes, and ignores the detection difficulty difference between classes. For tail-class targets, since they are rare and difficult to detect in themselves, if only hard samples are considered and tail-class samples are ignored, this will cause the model to calculate inaccurate losses, thereby reducing the model's detection accuracy for tail classes. Therefore, it is necessary to design a loss function that can consider both hard samples and tail-class samples at the same time to improve the model's detection performance for tail classes.

3.2. Dynamic Feature Fusion Module

To efficiently utilize multi-source information for object detection, we propose the dynamic feature fusion module. As shown in Figure 4, the module consists of two parts, namely the calculation of image entropy module and the multimodal feature fusion module. The image information quantity calculation module can accurately perceive the target scene by calculating the image entropy and dynamically assigning fusion coefficients for the images. The feature fusion module then fuses the images according to the information quantity, providing more salient features for subsequent object detection.

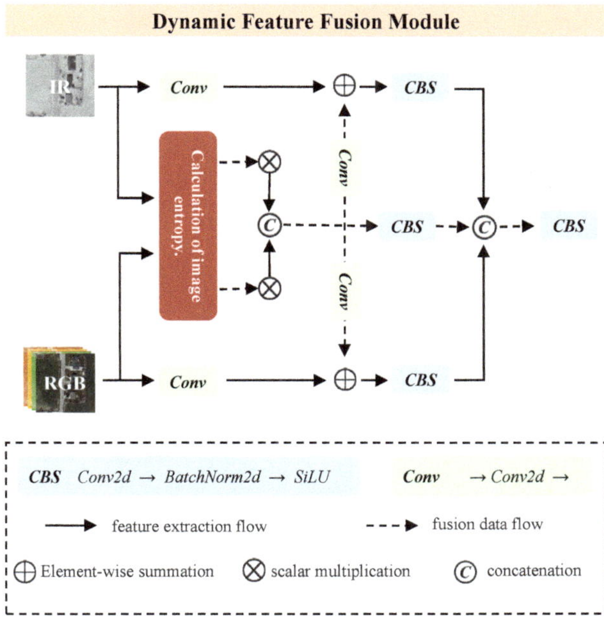

Figure 4. Dynamic feature fusion module. It consists of two parts: a calculation of image entropy module and a multimodal feature fusion module composed of multiple Conv layers. Among them, IR represents infrared image and RGB represents visible light image.

3.2.1. Calculation of Image Entropy

Preliminaries. The model input in the image multimodal object detection task is multimodal images, where we use $I_{\text{RGB}} \in \mathbb{R}^{3\times H\times W}$, $I_{\text{IR}} \in \mathbb{R}^{1\times H\times W}$ to represent the corresponding visible light image and infrared image. For convenience of calculating the image information quantity, we separate the visible light image into $I_{\text{RGB}} = \{I_{\text{R}}, I_{\text{G}}, I_{\text{B}}\}$, where $I_{\text{R}}, I_{\text{G}}, I_{\text{B}} \in \mathbb{R}^{1\times H\times W}$.

First, calculate the gradient magnitude and direction of the input images. The gradient magnitude is the change amplitude in the direction of maximum gray level change, and the gradient direction is the direction of maximum gray level change. Previous researchers have proposed many operators to calculate the gradient magnitude and direction, such as Sobel operator [34], Prewitt operator [35], Roberts operator [36], etc., which are all based on the difference between adjacent pixels in the image to estimate the gradient. In order to maintain the stability of model detection, we choose Sobel operator to calculate the image gradient.

$$G_h = \begin{bmatrix} -1 & 0 & 1 \\ -2 & 0 & 2 \\ -1 & 0 & 1 \end{bmatrix} Conv\ I_s \tag{1}$$

$$G_v = \begin{bmatrix} -1 & -2 & -1 \\ 0 & 0 & 0 \\ 1 & 2 & 1 \end{bmatrix} Conv\ I_s \tag{2}$$

where I is the image matrix, $s \in \{R, G, B, IR\}$, $Conv$ is convolution operation, G_h and G_v are gradients in horizontal and vertical directions, respectively. Then, the image gradient magnitude and direction can be obtained by the following formulas:

$$G = \sqrt{G_h^2 + G_v^2} \tag{3}$$

$$\theta = \arctan \frac{G_h}{G_v} \tag{4}$$

where G is gradient magnitude, θ is gradient direction. After obtaining the gradient magnitude and direction of each pixel, calculate the two-dimensional joint histogram n of gradient magnitude and direction. Each element n_{ij} can be calculated by

$$n_{ij} = \sum_{x=0}^{W-1} \sum_{y=0}^{H-1} F(x, y, i, j) \tag{5}$$

where W and H are the width and height of the image, $0 \leq i, j \leq 255$, $0 \leq x \leq W$, $0 \leq y \leq H$, $F(x, y, i, j)$ is an indicator function that obtains the gradient distribution map by judging each pixel one by one, i.e.,

$$F(x, y, i, j) = \begin{cases} 1 & \text{if } G(x, y) = i \text{ and } \theta(x, y) = j \\ 0 & \text{otherwise} \end{cases} \tag{6}$$

Further, the pixel probability r_{ij} can be obtained. Then, the probability of gradient magnitude being i and direction value being j is

$$r_{ij} = \frac{n_{ij}}{W * H} \tag{7}$$

Based on the probability map r, the final two-dimensional gradient entropy H_s [37] of the image can be calculated.

$$H = -\sum_{i=0}^{n-1} \sum_{j=0}^{m-1} r_{ij} \log r_{ij} \tag{8}$$

Using the above method, we can obtain the image entropy $H_{\mathrm{R}}, H_{\mathrm{G}}, H_{\mathrm{B}}, H_{\mathrm{IR}}$ of the images $I_{\mathrm{R}}, I_{\mathrm{G}}, I_{\mathrm{B}}, I_{\mathrm{IR}}$, respectively.

3.2.2. Multimodal Feature Fusion

To fuse multimodal information for object detection, we propose multimodal feature fusion. As shown in Figure 4, we extract the shallow features of the visible light and infrared images $I_{\mathrm{RGB}}, I_{\mathrm{IR}}$ input to the model using 1×1 convolution module while sending them to the image entropy calculation module.

$$f_{RGB} = Conv(I_{\mathrm{RGB}}),\ f_{IR} = Conv(I_{\mathrm{IR}}) \tag{9}$$

Then, we fuse each channel of the image according to the calculated information quantity, and the obtained features can be expressed as

$$f_{full} = Concat(I_R * H_R,\ I_G * H_G,\ I_B * H_B,\ I_{IR} * H_{IR}) \tag{10}$$

In the formula, $Concat()$ represents the concatenation operation along the channel axis; $*$ is the multiplication of image entropy and image. Further, we use 1×1 convolution module to extract global features and fuse different modal features,

$$f_{RGB} = f_{RGB} \oplus Conv\left(f_{full}\right),\ f_{IR} = f_{IR} \oplus Conv\left(f_{full}\right) \tag{11}$$

On the basis of preliminary fusion of shallow features, in order to extract more spatial information, we use *CBS* module to extract global features and fusion features of different modalities.

$$f_{RGB} = CBS(f_{RGB}),\ f_{full} = CBS\left(f_{full}\right),\ f_{IR} = CBS(f_{IR}) \quad (12)$$

Finally, we use a CBS module to fuse all the features calculated by the above formula and pass them to the subsequent backbone network for high-level semantic information extraction.

$$f = CBS\left(Concat\left(f_{full},\ f_{RGB},\ f_{IR}\right)\right) \quad (13)$$

The final output of this module is f that fuses visible light and infrared images.

3.3. Instance-Balanced Mosaic

As mentioned in Section 3.1, the default data augmentation module of YOLOv8 network is not suitable for remote sensing object detection. Because the targets in remote sensing images are small and sparse, especially when the instances present a long-tailed distribution, the tail-class instances with fewer numbers will be discarded, resulting in low detection accuracy of the model for these classes.

In order to improve the model's detection accuracy for tail classes, we propose the instance-balanced mosaic module, which considers the long-tailed problem of data and tries to preserve tail-class instances as much as possible during data augmentation. By stitching part of the content and annotation information of four original images onto a gray image to increase the diversity of the training set, this method enhances the model's detection performance for remote sensing images. As shown in Figure 5, the process can be divided into three steps.

Step 1: We generate a gray image $I(x,y)$, where W and H represent the width and height of the target image, respectively. We randomly select four images from the training set, denoted as I_1, I_2, I_3, I_4.

Step 2: In the rectangular area on the gray image that satisfies $x_c \in [W/4, 3W/4]$, $y_c \in [H/4, 3H/4]$, we randomly select a point as the stitching center point (x_c, y_c), and split the gray image from this point along horizontal and vertical directions to form stitching areas.

For each selected image I_i, we traverse each instance's class C_{ij} and count the number of that class in the training set N_j. Then, we randomly select an instance S_{ij} of the class with the least number $C_{min} = \arg\min N_j$, and calculate the Euclidean distance d_1, d_2, d_3, d_4 between its center point (x_{ij}, y_{ij}) and the four vertices $(0,0), (0,H), (W,0), (W,H)$ of the image, and select the closest vertex as the stitching point P_k. This distance can be calculated by

$$d_k = \sqrt{(x_{ij} - x_k)^2 + (y_{ij} - y_k)^2} \quad (14)$$

where (x_k, y_k) are the coordinates of the k-th vertex, $k \in \{1,2,3,4\}$.

Step 3: For each image I_i, according to the stitching point and direction, we stitch it onto the gray image so that the stitching point P_k is aligned with the stitching center point (x_c, y_c). If the two points are not aligned, we need to rotate, flip or perform other transformations on the original image; the coordinates of the aligned stitching point P_k can be expressed as (x_i', y_i'). The stitching transformation can be expressed by

$$I'(x,y) = \begin{cases} I_i(x - x_c + x_i', y - y_c + y_i'), & \text{if } (x,y) \in R_i \\ I(x,y), & \text{otherwise} \end{cases} \quad (15)$$

where $I'(x,y)$ is the stitched image, R_i is the i-th quadrant of stitching area on the gray image. For each image I_i, we also copy its target instance annotation information (such as bounding box, class etc.) to the gray image and adjust it according to stitching position. This way, we obtain a new data-augmented image that contains part of content and annotation information from four original images.

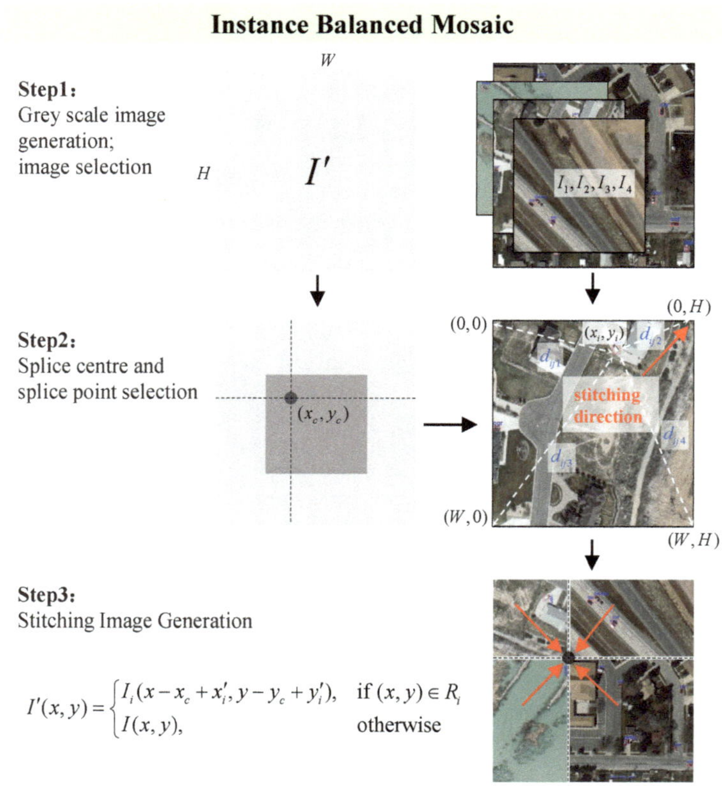

Figure 5. A schematic diagram of the instance-balanced mosaic method. It consists of three steps: first, generating a gray image as the image to be stitched and randomly selecting four images; second, randomly selecting a stitching center point in the gray image, and calculating the stitching point and direction of the selected images in turn; third, combining the four images into the gray image according to the stitching point and direction to generate a new image.

3.4. Class-Balanced BCE Loss

The overall loss function of our network is consistent with the baseline network, which consists of three parts: box loss box, classification loss cls, and regression loss dfl [38], which can be expressed as

$$L_{\text{total}} = c_1 L_{box} + c_2 L_{cls} + c_3 L_{dfl} \tag{16}$$

where c_1, c_2, c_3 are the balance coefficients set at the beginning of training. In the calculation of classification loss, compared with the original YOLOv5 network, YOLOv8 learns from TOOD [39] and introduces dynamic positive and negative sample allocation tasks, which aligns the classification and regression tasks. Unfortunately, the calculation of classification loss still does not consider the impact of long-tailed distribution of datasets. The datasets with long-tailed distribution have extremely unbalanced sample numbers among different classes, with a few minority classes having many samples, while most classes have few samples. This will cause the model to pay more attention to majority-class samples during training, and ignore minority-class samples. In order to better recognize tail class samples, and thus improve the model's detection accuracy on long-tailed datasets, we adjust the loss to class-balanced BCE loss based on the original BCE loss.

In order to assign reasonable weights to classes based on the number of instances per class, we first need to establish a formula that assigns higher weights to classes with fewer samples. Suppose we have N samples, the weight of class c can be

$$w_c = \frac{N - n_c}{n_c}, \tag{17}$$

where n_c is the number of samples belonging to class c. Based on the calculated weight w_c, we calculate the target score. The calculation of positive and negative samples is consistent with the original method [39], which first calculates the alignment degree between candidate anchor points and objects

$$M_{i,k,j} = S^{\alpha}_{i,j,c} \times IoU(B_{i,j}, G_{i,c})^{\beta}, \tag{18}$$

where α and β are two hyperparameters that control the weights of classification and localization. $S_{i,j,k}$ represents the score of class c for anchor point j of sample i. $B_{i,j}$ represents the predicted box corresponding to anchor point j of sample i. $G_{i,k}$ represents the ground truth box corresponding to anchor point j of sample i. IoU is the intersection over union calculated by CIoU method [33] between two boxes.

Then, we obtain the target score related to alignment degree $T = T(M, l)$, where l is the corresponding sample class label. By weighting, we obtain the final target score

$$y_{ic} = \left(1 + \gamma * w_c^{'}\right) * T_{ic}, \tag{19}$$

where γ is a hyperparameter that controls the weighted strength of a class, and $w_c^{'}$ is the normalized weight of w_c for more stable balancing of importance between classes.

The weighted classification loss assigns larger weights to minority classes and smaller weights to majority classes according to the distribution of different classes in the dataset, which makes the model pay more attention to minority-class samples. The final classification loss continues to use cross-entropy loss function

$$L_{cls} = -\frac{1}{N}\sum_{i=1}^{N}\sum_{c=1}^{C} y_{ic}\log(\hat{y}_{ic}). \tag{20}$$

where $\hat{y}_{ic}$ is the predicted probability that sample i belongs to class c.

4. Experiment and Results

4.1. Dataset

We conducted relevant experiments on three common datasets to test the effectiveness of our method.

(1) VEDAI dataset. VEDAI (vehicle detection in aerial imagery) dataset is a dataset designed specifically for vehicle detection in aerial images, released by the University of Caen in 2015. The images in this dataset contain various types of vehicles, which have differences in illumination, occlusion and orientation. In addition, each image contains two bands of visible light and infrared, and two resolutions of 512 × 512 and 1024 × 1024. In order to better reproduce the experimental results and compare with other experiments, it provides a unified division of training set and test set data, which makes this dataset have the ability to verify whether the model has complex multimodal object detection ability. The VEDAI dataset contains a total of 1246 images, including eight different types of vehicles, such as cars, pickups, trucks, etc., and each image has corresponding annotation information such as location.

(2) LLVIP dataset. LLVIP dataset is a dataset created specifically for low-light vision tasks, collecting a large number of visible light and infrared images taken under low-

illumination conditions. These images are captured by a dual-spectral camera with a bird's-eye-view monitoring angle, containing a lot of pedestrians and cyclists on the street, and each pair of infrared and visible light images is spatially and temporally aligned. Therefore, this dataset can be used to study visual tasks such as image fusion, pedestrian detection and image-to-image translation. Since most of the images are taken at night with poor lighting conditions, it can well reflect the complementarity between visible light and infrared images, thus testing the model's fusion detection ability for multimodal images. All pedestrians in this dataset have accurate annotation files, recording their categories and locations. According to statistics, it contains 33,672 images, that is, 16,836 pairs of images; each image has a pixel size of 640 $\times$ 512.

(3) FLIR dataset. FLIR dataset is a visible light and infrared dataset for training neural network object detection, mainly containing daytime and nighttime scenes on the highway, aiming to test the model's fusion ability of complementary information from multispectral images. However, with regard to the original version released by Teledyne FLIR company with 26,442 fully annotated images, although annotated with various types such as people, cars, traffic lights, etc., there are a lot of unaligned images in time and space, which makes the model training too complicated and unable to reasonably test the model's performance. Therefore, ref. [40] released an "aligned" version, which contains 5142 pairs of strictly aligned infrared and visible light images, excluding other categories besides people, cars and bicycles. In our experiments, we also choose to use this "aligned" version of the dataset to test and verify our model.

Table 1 summarizes the details of the datasets used in our experiments. Figure 6 shows the class imbalance distribution of FLIR and VEDAI datasets.

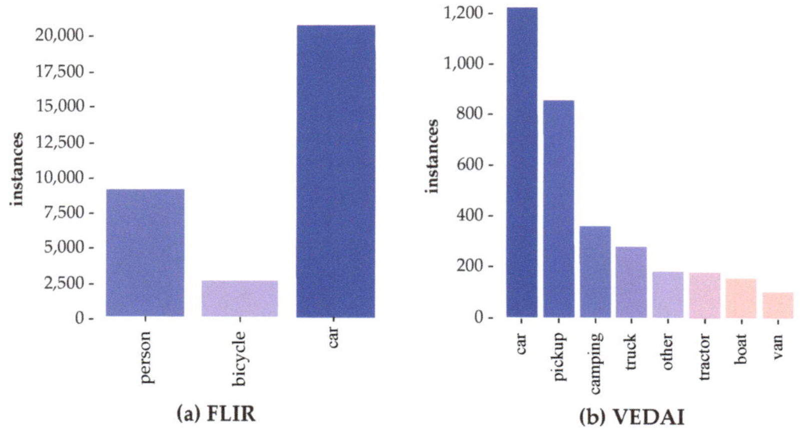

Figure 6. (**a**) Class distribution diagram of FLIR dataset. (**b**) Class distribution diagram of VEDAI dataset. (Because the LLVIP dataset has only one category, this figure does not show it).

Table 1. Summary of the dataset.

Dataset	#Classes	Image Size	#Train	#Test
FLIR	3	640 × 512	4129	1013
LLVIP	1	1280 × 1024	12,025	3463
VEDAI	8	1024 × 1024	1089	121

4.2. Implementation Details

In this experiment, we deployed our model using the PyTorch framework based on Python 3.8. The model training used a server with two NVIDIA GeForce 3090 GPUs, with a CPU of inter 13,900 k and a memory of 128 G. During the training process, we set momentum to 0.9 and weight_decay to 5×10^{-3}. We chose the optimizer corresponding to the dataset according to the dataset, and chose AdamW for FLIR and VEDAI datasets, and SGD for LLVIP dataset. Because the LLVIP dataset has large image size and many samples in the training set, the model has more iterations during training, and SGD can better escape local optimal solutions. On the contrary, although the AdamW optimizer may fall into local optimal solutions, it can dynamically adjust the learning rate according to the gradient size of parameters, and converge faster on small datasets. The initial learning rate was set to 1×10^{-2} and the final learning rate was 1×10^{-4}. The batch size was fixed at 16 and the total epoch was 100.

4.3. Accuracy Metrics

In the model evaluation, we chose to use the evaluation metric mean average precision (mAP) created by the MS-COCO dataset to evaluate the performance of the model, which represents the average value of average precision (AP), and is used to measure the performance of detection models. The higher the mAP, the stronger the detection ability, that is, the higher the better. It can be calculated by

$$\text{mAP} = \frac{1}{n}\sum_{i=0}^{n}\text{AP}_i = \frac{1}{n}\sum_{i=0}^{n}\int_0^1 P(r)\text{d}r. \tag{21}$$

where $P(r)$ is precision and r is recall, and they are calculated as follows:

$$\begin{aligned} P(r) &= \frac{\text{TP}}{\text{TP} + \text{FP}} \\ r &= \frac{\text{TP}}{\text{TP} + \text{FN}} \end{aligned} \tag{22}$$

where TP stands for true positive, which is the number of positive-class samples that are correctly predicted by the model under a certain IoU threshold; FP stands for false positive, which is the number of negative-class samples that are misreported by the model under this IoU threshold; FN stands for false negative, which is the number of positive-class samples that are missed by the model. As the most widely used evaluation metric in object detection field, mAP has various different IoU threshold standards such as mAP50 and mAP50:95. mAP50 means the average value of AP values of all classes at IoU = 0.50, and mAP50:95 means the average value when IoU = 0.50:0.05:0.95. We choose these metrics to evaluate our proposed model.

4.4. Comparisons with Previous Methods

4.4.1. Experiment Results of the VEDAI Dataset

The VEDAI dataset is a long-tailed remote sensing image dataset composed of eight classes; as shown in Figure 6, the car class has a large number of samples, while tail classes such as vans have very few samples. In this paper, we conducted extensive comparative experiments on this dataset to verify the effectiveness of our method for remote sensing object detection data with long-tailed instance distribution.

The results in Table 2 show that our proposed method achieves the highest detection accuracy on VEDAI dataset. This fully verifies the effectiveness of our proposed method for long-tailed remote sensing object detection data. Specifically, our method can achieve superior performance over other methods for two main reasons. One is that the DFF module fuses complementary information from infrared and visible light images, and obtains an mAP50 of 81% after fusion, which is much higher than 76.6% using only infrared images and 72.7% using only visible light images. These data further illustrate the performance of our proposed DFF module. The other is that by balancing classes between instances, we reduce the model's overconfidence for head instances. As shown in Figure 7, our method has the highest performance compared to other methods on tail instances such as boats, thereby improving the overall performance of the model.

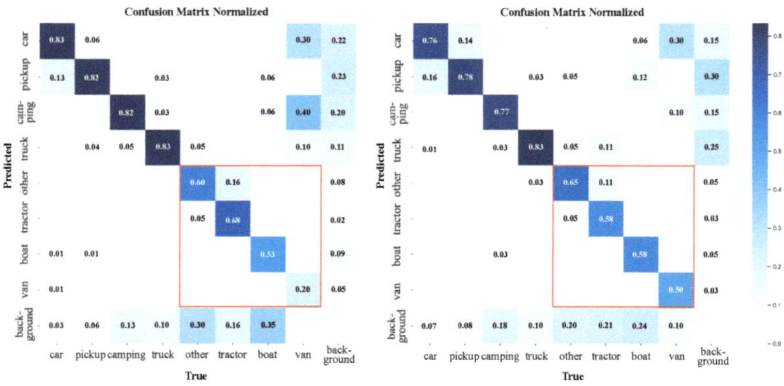

(a) YOLOv8s **(b) ours**

Figure 7. (**a**) Normalized confusion matrix of the baseline methods. (**b**) Our method.

Table 2. Comparison of our method with other state-of-the-art methods on the VEDAI dataset. The values in the table are mAP50(%).

Category		Car	Pickup	Camping	Truck	Other	Tractor	Boat	Van	Total
Instances		134	95	39	30	20	19	17	10	364
YOLOv3 [41]	IR	80.2	67.0	65.6	47.8	25.9	40.1	32.7	53.3	51.5
	RGB	83.1	71.5	69.1	59.3	48.9	67.3	33.5	55.7	61.1
	Multi	84.6	72.7	67.1	62.0	43.0	65.2	37.1	58.3	61.3
YOLOv4 [20]	IR	80.5	67.9	68.8	53.7	30.0	44.2	25.4	51.4	52.8
	RGB	83.7	73.4	71.2	59.1	51.7	65.9	34.3	60.3	62.4
	Multi	85.5	72.8	72.4	62.8	48.9	69.0	34.3	54.7	62.6
YOLOv5s [32]	IR	77.3	65.3	66.5	51.6	25.9	42.4	21.9	48.9	49.9
	RGB	80.1	58.0	66.1	51.5	45.8	64.4	21.6	40.9	54.8
	Multi	80.8	68.5	69.1	54.7	46.8	64.3	24.3	46.0	56.8
YOLOrs [42]	IR	82.0	73.9	63.8	54.2	44.0	54.4	22.0	43.4	54.7
	RGB	85.3	72.9	70.3	50.7	42.7	76.8	18.7	38.9	57.0
	Multi	84.2	78.3	68.8	52.6	46.8	67.9	21.5	57.9	59.7
YOLO-Fine [43]	IR	76.8	74.4	64.7	63.5	45.0	78.1	70.0	77.9	68.2
	RGB	79.7	74.5	77.1	81.0	37.3	70.7	60.8	63.6	68.8
YOLOFusion [44]	IR	86.7	75.9	66.6	77.1	43.0	62.3	70.7	**84.3**	70.8
	RGB	91.1	82.3	75.1	78.3	33.3	81.2	71.8	62.2	71.9
	Multi	**91.7**	85.9	78.9	78.1	54.7	71.9	71.7	75.2	75.9
SuperYOLO [27]	IR	87.9	81.4	76.9	61.6	39.4	60.6	46.1	71.0	65.6
	RGB	90.3	82.7	76.7	68.6	53.9	79.5	58.1	70.3	72.5
	Multi	91.1	85.7	79.3	70.2	57.3	80.4	60.2	76.5	75.1
Ours	IR	88.9	**87.4**	81.3	78.4	60.2	80.7	70.1	75.7	76.6
	RGB	88.4	80.1	77.7	78.8	51.3	69.8	64.0	71.4	72.7
	Multi	91.6	87.2	**84.1**	**85.1**	**72.0**	**84.7**	**75.7**	80.7	**81.0**

Note: The best results are shown in bold.

4.4.2. Experiment Results of the FLIR Dataset

As shown in Figure 6a, FLIR is a remote sensing object detection dataset that contains three categories. We conducted extensive comparative experiments on this dataset with the current state-of-the-art methods to verify the effectiveness of our proposed method for long-tailed remote sensing object detection datasets. The experimental results are shown in Figure 8 and Table 3.

Table 3. Comparison of our method with other state-of-the-art methods on the FLIR dataset.

Model	Data	Backbone	mAP50	mAP50:95
FasterR-CNN [5]	RGB	ResNet50	64.9	28.9
FasterR-CNN [5]	IR	ResNet50	74.4	37.6
SSD [4]	RGB	VGG16	52.2	21.8
SSD [4]	IR	VGG16	65.5	29.6
YOLOv3 [41]	RGB	Darknet53	58.3	25.7
YOLOv3 [41]	IR	Darknet53	73.6	36.8
YOLOv5 [32]	RGB	CSPD53	67.8	31.8
YOLOv5 [32]	IR	CSPD53	73.9	39.5
YOLOv8 [19]	RGB	-	60.7	28.8
YOLOv8 [19]	IR	-	73.7	38.3
CFR3 [40]	Multi	VGG16	72.4	-
GAFF [45]	Multi	ResNet18	72.9	37.5
YOLOFusion [44]	Multi	CFB	**78.7**	40.2
SuperYOLO [27]	Multi	YOLO	74.6	39.4
Ours	Multi	YOLO	78.5	**42.6**

Note: The best results are shown in bold.

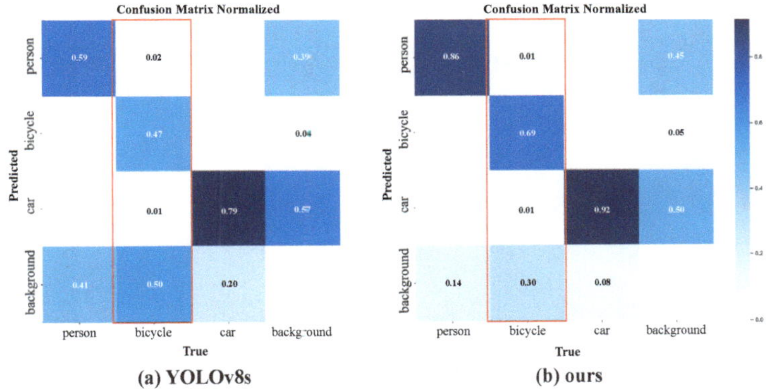

Figure 8. (**a**) Normalized confusion matrix of the baseline method. (**b**) Normalized confusion matrix of our method.

The results in Table 3 show that our proposed method achieves the highest average detection accuracy on the FLIR dataset. It can be observed that compared to single-modality methods, multi-source fusion methods generally have higher performance. Our method can surpass other multi-source fusion methods, mainly due to the solution of the long-tailed problem. From Figure 8, it can also be seen that our proposed method can solve the long-tailed distribution problem in remote sensing object detection by significantly improving the detection accuracy of tail-category instances, while maintaining or slightly improving the detection accuracy of head-category instances. Since the data obtained by remote sensing usually show a long-tailed distribution, our proposed method is more consistent with the real-world situation and has broad application value. The experimental data on the FLIR dataset fully demonstrate that our proposed method has very good detection performance on long-tailed remote sensing detection data, especially for better detection of tail-category instances.

The visual detection results of the baseline method YOLOv8s and our method in five randomly selected scenes are shown in Figure 9, where the yellow ellipses represent the targets missed by the model in detection, and the purple ellipses indicate the targets falsely detected by the model. We can observe that our method is significantly better than the baseline method in tail-class samples and complex scenes. In Figure 9a, our model detects two tail-class samples of bicycles that are missed by the baseline model, and moreover, the confidence of the only bicycle detected by the baseline model is also lower than that of our model. As marked by the purple ellipses in scenes Figure 9c,e, the baseline model also erroneously detects cars and bicycles, further demonstrating the advantage of our method.

Figure 9. (**a**–**e**) Detection visualization on FLIR dataset. Yellow ellipses represent the targets missed by the model in detection, and purple ellipses indicate the targets falsely detected by the model.

4.4.3. Experiment Results of the LLVIP Dataset

The LLVIP dataset is a single-target remote sensing dataset that contains multiple scene categories. Although this dataset only has one target of pedestrians and does not have a long-tailed distribution, the scenes in this dataset are mostly under dark light conditions, which can well test the performance of the DFF module in our method. Therefore, we conducted extensive comparative experiments on this dataset with current advanced similar methods to verify the effectiveness of our proposed method for multi-source data fusion. The experimental results are shown in Table 4.

The results in Table 4 show that our proposed method achieves the highest average detection accuracy on this dataset. At the same time, other similar algorithms such as YOLOFusion [44] also have very good performance, which indicates that multimodal fusion methods are very effective remote sensing object detection methods. In contrast, the detection accuracy of methods based on single modality is generally not high. This fully illustrates the powerful performance of the DDF module in our proposed method.

Table 4. Comparison of our method with other state-of-the-art methods on the LLVIP dataset.

Model	Data	Backbone	mAP50	mAP50:95
FasterR-CNN [5]	RGB	ResNet50	91.4	49.2
FasterR-CNN [5]	IR	ResNet50	96.1	61.1
SSD [4]	RGB	VGG16	82.6	39.8
SSD [4]	IR	VGG16	90.2	53.5
YOLOv3 [41]	RGB	Darknet53	87.1	-
YOLOv3 [41]	IR	Darknet53	94.0	-
YOLOv5 [32]	RGB	CSPD53	90.8	-
YOLOv5 [32]	IR	CSPD53	96.5	-
YOLOv8 [19]	RGB	-	92.5	54.1
YOLOv8 [19]	IR	-	96.6	63.2
YOLOFusion [44]	Multi	CFB	97.5	63.6
SuperYOLO [27]	Multi	YOLO	96.7	65.2
Ours	Multi	YOLO	**97.7**	**67.6**

Note: The best results are shown in bold.

5. Discussion

This study aims to solve the long-tailed object detection problem in remote sensing images, and the experimental results show that our method has better performance than other similar methods. In order to verify the effectiveness of each module in our method and find the optimal parameter settings, we conducted ablation experiments and parameter analysis experiments in this section.

5.1. Ablation Study

To verify the effectiveness of the modules in our proposed method, we conducted extensive ablation experiments on three datasets. The ablation experiments were consistent with the settings in Section 4.4, having the same backbone network and training strategy. The results of the ablation experiments are shown in Tables 5 and 6, which fully demonstrate the effectiveness of each module in our proposed method.

Table 5. Ablation experiments on VEDAI, FLIR, LLVIP datasets.

Dataset	DFF	IBM	CBB	mAP50	mAP50:95
VEDAI				72.7	44.3
	✓			75.4	50.1
	✓	✓		77.7	51.4
	✓	✓	✓	**81.0**	**54.5**
FLIR				60.7	28.8
	✓			75.2	40.4
	✓	✓		76.2	41.3
	✓	✓	✓	**78.5**	**42.6**
LLVIP				92.5	54.1
	✓			**97.7**	**67.6**

Note: The best results are shown in bold.

Table 6. The impact of data augmentation. The evaluation metric is mAP50 (%).

Dataset	VEDAI	FLIR	LLVIP
Mosaic	79.7	76.2	97.2
Shear	76.2	75.1	96.6
Perspective	74.4	76.0	96.9
IBM (ours)	**81.0**	**78.5**	**97.7**

Note: The best results are shown in bold.

5.1.1. Ablation Studies for Each Module

Specifically, on the VEDAI dataset, the IBM and CBB modules contribute the most to the performance improvement, with a cumulative increase of 5.6% in mAP50, due to the typical long-tailed distribution of instance categories on this dataset. Meanwhile, this dataset has good illumination conditions for data acquisition, and the visible light images already contain a lot of useful information, which makes the DFF module that fuses visible light and infrared images have limited performance improvement, with only a 2.7% increase in mAP50. On the FLIR dataset, DFF contributes more to the performance improvement than the other two modules, because this dataset has large variations in illumination conditions for data acquisition, and visible light and infrared images can provide complementary information, but there are only three instance categories, and the data is not very long-tailed. At the same time, we can see that the IBM and CBB modules have limited improvement on mAP50:95, because these two modules mainly balance the categories of samples, and mAP50:95 has higher requirements for localization accuracy. The LLVIP dataset is a remote sensing dataset collected under low-light conditions, with only one category. The baseline model achieves more than 5% improvement in mAP50 accuracy after adding our DFF module.

5.1.2. Impact of Instance-Balanced Mosaic

As described in Section 3.1, the baseline model YOLOv8 uses mosaic data augmentation by default, which crops out tail-class samples, resulting in lower recognition accuracy for tail-class samples in long-tailed distribution datasets. To address this issue, we propose an instance-balanced mosaic, which balances the distribution of head and tail classes by restricting the image stitching direction. To verify the effectiveness of our proposed IBM method, we compare three commonly used data augmentation methods on three datasets. The first method is mosaic data augmentation [20], which stitches four different images into a large image as the input for training, which can improve the model's detection ability for objects with different scales and backgrounds. The second method is shear augmentation, which shears the image along an axis, thereby improving the model's detection ability for objects with different shapes and angles. The third method is perspective augmentation [46], which performs perspective transformation on the image to improve the model's detection ability for objects with different viewpoints and distances. The experimental results are shown in Table 6, which demonstrate that our proposed IBM data augmentation method can achieve the highest detection performance on all three datasets.

5.2. Parameter Analysis

A suitable hyperparameter can make the model have more powerful performance, so in order to obtain a suitable parameter, we conducted a parameter sensitivity experiment on the VEDAI and FLIR datasets, and the experimental results are shown in Figure 10. The hyperparameter γ is used to weight the classes. The larger the value, the higher the positive weight given to the tail classes. As can be seen from the figure, as the hyperparameter γ value gradually increases, the detection accuracy of our method experiences a process of gradually increasing and then decreasing. This is because at the beginning, by weighting the tail class samples, the model achieves balanced attention for each class. However, as the value further increases, it causes the model to over-focus on the tail classes, thereby reducing the detection accuracy of the model. When $\gamma = 0.5$, the model achieves the highest performance on the two datasets, achieving 81% and 78.9% mAP50, respectively.

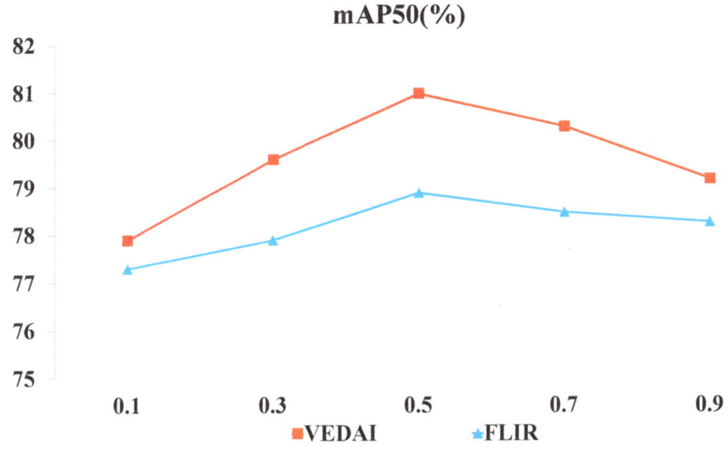

Figure 10. Parameter analysis of hyperparameter γ on VEDAI and FLIR datasets.

5.3. Effect of Operators

In Section 3.2, we use the Sobel operator to calculate the image entropy, which assigns reasonable weights for image fusion. In previous studies, many operators have been proposed, such as Sobel operator [34], Prewitt operator [35], Roberts operator [36], etc. To select a suitable operator for our model, we evaluate the performance of each operator on three datasets. As shown in Table 7, the experimental results show that our method is robust to different operators, and using different operators has no significant impact on the detection accuracy of the model.

Table 7. The impact of using different operators in the DFF module. The evaluation metric is mAP50 (%).

Dataset	SobSobel	Prewitt	Robert
VEDAI	81.0	80.1	80.6
FLIR	78.5	78.3	78.6
LLVIP	97.7	97.7	97.6

6. Conclusions

In this paper, we propose a new method for long-tailed object detection in multimodal remote sensing images, which can effectively fuse the complementary information of visible light and infrared images, and adapt to the imbalance between positive and negative samples of different classes. The main contributions of this paper are as follows: (1) Based on image entropy, the dynamic feature fusion module can dynamically adjust the fusion coefficient according to the information content of different source images, and retain more key feature information for subsequent object detection. (2) The instance-balanced mosaic data augmentation method can balance the sampling of instances in the data augmentation stage, provide more sample features for the model, and reduce the negative impact of data distribution imbalance. (3) Class-balanced BCE loss can consider both instance learning difficulty and class learning difficulty, and improve the model's detection accuracy for tail instances. The experimental results on three public benchmark datasets show that proposed method achieves state-of-the-art performance.

In summary, we propose a new method for long-tailed object detection in multimodal remote sensing images, which can effectively fuse the complementary information of visible light and infrared images, and adapt to the imbalance between positive and negative samples of different classes. Our method can improve detection accuracy and robustness of remote sensing objects, especially for tail classes. However, our method is not robust to multi-source images that are not geometrically aligned. In future, we plan to extend our method to other modalities such as radar and hyperspectral images, and explore more effective methods to deal with long-tailed data distribution.

Author Contributions: Conceptualization, J.Z. and S.H.; methodology, J.Y.; software, J.Y.; validation, J.Y., M.Y. and S.L.; formal analysis, J.Y.; writing—original draft preparation, J.Y.; funding acquisition, S.L. All authors have read and agreed to the published version of the manuscript.

Funding: This research was funded by the National Natural Science Foundation of China (grant numbers 62101571 and 61806215).

Data Availability Statement: All the datasets presented in this paper can be found through the referenced papers.

Conflicts of Interest: The authors declare no conflict of interest.

References

1. Earthdata. What Is Remote Sensing? | Earthdata. 2021. Available online: https://www.earthdata.nasa.gov/learn/backgrounders/remote-sensing (accessed on 11 September 2023).
2. Chi, M.; Plaza, A.; Benediktsson, J.A.; Sun, Z.; Shen, J.; Zhu, Y. Big Data for Remote Sensing: Challenges and Opportunities. *Proc. IEEE* **2016**, *104*, 2207–2219. [CrossRef]
3. Redmon, J.; Divvala, S.; Girshick, R.; Farhadi, A. You only look once: Unified, real-time object detection. In Proceedings of the IEEE Conference on Computer Vision and Pattern Recognition, Las Vegas, NV, USA, 26 June–1 July 2016; pp. 779–788.
4. Liu, W.; Anguelov, D.; Erhan, D.; Szegedy, C.; Reed, S.; Fu, C.Y.; Berg, A.C. Ssd: Single shot multibox detector. In Proceedings of the European Conference on Computer Vision (ECCV), Munich, Germany, 8–14 September 2016; Springer: Berlin/Heidelberg, Germany, 2016; pp. 21–37.
5. Ren, S.; He, K.; Girshick, R.; Sun, J. Faster r-cnn: Towards real-time object detection with region proposal networks. In Proceedings of the Advances in Neural Information Processing Systems, Lake Tahoe, NV, USA, 6–14 December 2015; pp. 91–99.
6. He, K.; Gkioxari, G.; Dollár, P.; Girshick, R. Mask R-CNN. In Proceedings of the IEEE International Conference on Computer Vision, Venice, Italy, 22–29 October 2017; pp. 2961–2969.
7. Girshick, R.; Donahue, J.; Darrell, T.; Malik, J. Rich feature hierarchies for accurate object detection and semantic segmentation. In Proceedings of the IEEE Conference on Computer Vision and Pattern Recognition, Columbus, OH, USA, 23–28 June 2014; pp. 580–587.
8. Janakiramaiah, B.; Kalyani, G.; Karuna, A.; Prasad, L.N.; Krishna, M. Military object detection in defense using multi-level capsule networks. *Soft Comput.* **2021**, *27*, 1045–1059. [CrossRef]
9. Ennouri, N.; Mourshed, M.; Bray, M. Advances in remote sensing applications for urban sustainability. *Euro-Mediterr. J. Environ. Integr.* **2016**, *1*, 1–15.
10. Rezaei, M.; Azarmi, M.; Pour Mir, F.M. Traffic-Net: 3D Traffic Monitoring Using a Single Camera. *arXiv* **2021**, arXiv:2109.09165.
11. Ma, T.J. Remote sensing detection enhancement. *J. Big Data* **2021**, *8*, 1–13. [CrossRef]
12. Liu, Z.; Miao, Z.; Zhan, X.; Wang, J.; Gong, B.; Yu, S.X. Large-scale long-tailed recognition in an open world. In Proceedings of the IEEE Conference on Computer Vision and Pattern Recognition, Long Beach, CA, USA, 15–20 June 2019; pp. 2537–2546.
13. Li, B.; Yao, Y.; Tan, J.; Zhang, G.; Yu, F.; Lu, J.; Luo, Y. Equalized focal loss for dense long-tailed object detection. In Proceedings of the IEEE/CVF Conference on Computer Vision and Pattern Recognition, New Orleans, LA, USA, 18–24 June 2022; pp. 1–10.
14. Zang, Y.; Zhou, K.; Huang, C.; Loy, C.C. Semi-Supervised and Long-Tailed Object Detection with CascadeMatch. *Int. J. Comput. Vis.* **2023**, *131*, 987–1001. [CrossRef]
15. Wang, T.; Zhu, Y.; Zhao, C.; Zeng, W.; Wang, J.; Tang, M. Adaptive class suppression loss for long-tail object detection. In Proceedings of the 2021 IEEE/CVF Conference on Computer Vision and Pattern Recognition (CVPR), Nashville, TN, USA, 20–25 June 2021; pp. 3103–3112.
16. Zhao, W.; Liu, J.; Liu, Y.; Zhao, F.; He, Y.; Lu, H. Teaching teachers first and then student: Hierarchical distillation to improve long-tailed object recognition in aerial images. *IEEE Trans. Geosci. Remote Sens.* **2022**, *60*, 1–12. [CrossRef]
17. Duan, Y.; Liu, X.; Jatowt, A.; Yu, H.T.; Lynden, S.; Kim, K.S.; Matono, A. Long-Tailed Graph Representation Learning via Dual Cost-Sensitive Graph Convolutional Network. *Remote Sens.* **2022**, *14*, 3295. [CrossRef]
18. Shannon, C.E. A Mathematical Theory of Communication. *Bell Syst. Tech. J.* **1948**, *27*, 379–423. [CrossRef]
19. YOLOv8. 2021. Available online: https://github.com/ultralytics/ultralytics (accessed on 11 September 2023).
20. Bochkovskiy, A.; Wang, C.Y.; Liao, H.Y.M. YOLOv4: Optimal Speed and Accuracy of Object Detection. *arXiv* **2020**, arXiv:2004.10934.

21. Liu, X.; Li, H.; Wang, J.; Li, L.; Ouyang, W. Deep learning for generic object detection: A survey. *Int. J. Comput. Vis.* **2020**, *128*, 261–318. [CrossRef]
22. Yaman, B.; Mahmud, T.; Liu, C.H. Instance-Aware Repeat Factor Sampling for Long-Tailed Object Detection. *arXiv* **2023**, arXiv:2305.08069.
23. Li, Y.; Wang, T.; Kang, B.; Tang, S.; Wang, C.; Li, J.; Feng, J. Overcoming classifier imbalance for long-tail object detection with balanced group softmax. In Proceedings of the IEEE/CVF Conference on Computer Vision and Pattern Recognition, Seattle, WA, USA, 13–19 June 2020; pp. 10991–11000.
24. Lin, T.Y.; Goyal, P.; Girshick, R.; He, K.; Dollár, P. Focal loss for dense object detection. In Proceedings of the IEEE International Conference on Computer Vision, Venice, Italy, 22–29 October 2017; pp. 2980–2988.
25. Tan, J.; Wang, C.; Li, B.; Li, Q.; Ouyang, W.; Yin, C.; Yan, J. Equalization loss for long-tailed object recognition. In Proceedings of the IEEE/CVF Conference on Computer Vision and Pattern Recognition, Seattle, WA, USA, 13–19 June 2020; pp. 11662–11671.
26. Liu, J.; Fan, X.; Jiang, J.; Liu, R.; Luo, Z. Learning a Deep Multi-Scale Feature Ensemble and an Edge-Attention Guidance for Image Fusion. *IEEE Trans. Circuits Syst. Video Technol.* **2021**, *32*. 105–119. [CrossRef]
27. Zhang, J.; Lei, J.; Xie, W.; Fang, Z.; Li, Y.; Du, Q. SuperYOLO: Super Resolution Assisted Object Detection in Multimodal Remote Sensing Imagery. *IEEE Trans. Geosci. Remote Sens.* **2023**, *61*, 1–15. [CrossRef]
28. Chen, C.; Zhao, X.; Wang, J.; Li, D.; Guan, Y.; Hong, J. Dynamic graph convolutional network for assembly behavior recognition based on attention mechanism and multi-scale feature fusion. *Sci. Rep.* **2022**, *12*, 7394. [CrossRef]
29. Li, J.; Li, B.; Jiang, Y.; Cai, W. MSAt-GAN: A generative adversarial network based on multi-scale and deep attention mechanism for infrared and visible light image fusion. *Complex Intell. Syst.* **2022**, *8*, 4753–4781. [CrossRef]
30. Ai, Y.; Liu, X.; Zhai, H.; Li, J.; Liu, S.; An, H.; Zhang, W. Multi-Scale Feature Fusion with Attention Mechanism Based on CGAN Network for Infrared Image Colorization. *Appl. Sci.* **2023**, *13*, 4686. [CrossRef]
31. Ahmed, M.R.; Ashrafi, A.F.; Ahmed, R.U.; Shatabda, S.; Islam, A.M.; Islam, S. DoubleU-NetPlus: A novel attention and context-guided dual U-Net with multi-scale residual feature fusion network for semantic segmentation of medical images. *Neural Comput. Appl.* **2023**, *35*, 14379–14401. [CrossRef]
32. YOLOv5. 2021. Available online: https://github.com/ultralytics/yolov5 (accessed on 14 September 2023).
33. Zheng, Z.; Wang, P.; Liu, W.; Li, J.; Ye, R.; Ren, D. Distance-IoU Loss: Faster and Better Learning for Bounding Box Regression. *Proc. Aaai Conf. Artif. Intell.* **2020**, *34*, 12993–13000. [CrossRef]
34. Sobel, I. *Camera Models and Machine Perception*; Technical Report; Stanford University: Standford, CA, USA, 1968.
35. Prewitt, J.M. Object enhancement and extraction. *Pict. Process. Psychopictorics* **1970**, *10*, 15–19.
36. Roberts, L.G. Machine perception of three-dimensional solids. In *Optical and Electro-Optical Information Processing*; Massachusetts Institute of Technology: Cambridge, MA, USA, 1963; pp. 159–197.
37. Kapur, J.; Sahoo, P.; Wong, A. A new method for gray-level picture thresholding using the entropy of the histogram. *Comput. Vision Graph. Image Process.* **1986**, *29*, 273–285. [CrossRef]
38. Li, X.; Wang, W.; Wu, L.; Chen, S.; Hu, X.; Li, J.; Tang, J.; Yang, J. Generalized focal loss: Learning qualified and distributed bounding boxes for dense object detection. *Adva. Neural Inf. Process. Syst.* **2020**, *33*, 21002–21012.
39. Feng, C.; Zhong, Y.; Gao, Y.; Scott, M.R.; Huang, W. TOOD: Task-aligned one-stage object detection. In Proceedings of the 2021 IEEE/CVF Conference on Computer Vision and Pattern Recognition (CVPR), Nashville, TN, USA, 20–25 June 2021; pp. 12893–12902.
40. Zhang, H.; Fromont, E.; Lefevre, S.; Avignon, B. Multispectral fusion for object detection with cyclic fuse-and-refine blocks. In Proceedings of the 2020 IEEE International Conference on Image Processing (ICIP), Virtual, 25–28 October 2020; pp. 1016–1020.
41. Redmon, J.; Farhadi, A. YOLOv3: An Incremental Improvement. *arXiv* **2018**, arXiv:1804.02767.
42. Sharma, M.; Dhanaraj, M.; Karnam, S.; Chachlakis, D.G.; Ptucha, R.; Markopoulos, P.P.; Saber, E. YOLOrs: Object Detection in Multimodal Remote Sensing Imagery. *IEEE J. Sel. Top. Appl. Earth Obs. Remote Sens.* **2021**, *14*, 1497–1508. [CrossRef]
43. Pham, M.T.; Courtrai, L.; Friguet, C.; Lefèvre, S.; Baussard, A. YOLO-Fine: One-Stage Detector of Small Objects under Various Backgrounds in Remote Sensing Images. *Remote Sens.* **2020**, *12*, 2501. [CrossRef]
44. Fang, Q.; Wang, Z. Cross-Modality Attentive Feature Fusion for Object Detection in Multispectral Remote Sensing Imagery. *Pattern Recognit.* **2022**, *130*, 108786.
45. Zhang, H.; Fromont, E.; Lefevre, S.; Avignon, B. Guided attentive feature fusion for multispectral pedestrian detection. In Proceedings of the 2021 IEEE Winter Conference on Applications of Computer Vision (WACV), Waikoloa, HI, USA, 3–8 January 2021. [CrossRef]
46. Wang, K.; Fang, B.; Qian, J.; Yang, S.; Zhou, X.; Zhou, J. Perspective transformation data augmentation for object detection. *IEEE Access* **2019**, *8*, 4935–4943. [CrossRef]

remote sensing

Article

Mask R-CNN–Based Landslide Hazard Identification for 22.6 Extreme Rainfall Induced Landslides in the Beijiang River Basin, China

Zhibo Wu [1,2], Hao Li [2], Shaoxiong Yuan [2], Qinghua Gong [2,*], Jun Wang [2] and Bing Zhang [1]

1 Faculty of Land Resource Engineering, Kunming University of Science and Technology, Kunming 650031, China; 20212101078@stu.kust.edu.cn (Z.W.); zhangbing@kmust.edu.cn (B.Z.)

2 Guangzhou Institute of Geography, Guangdong Open Laboratory of Geospatial Information Technology and Application, Guangdong Academy of Sciences, Guangzhou 510070, China; lihao@gdas.ac.cn (H.L.); yuanshx@gdas.ac.cn (S.Y.); wangjun@gdas.ac.cn (J.W.)

* Correspondence: gongqh@gdas.ac.cn

Abstract: Landslides triggered by extreme precipitation events pose a significant threat to human life and property in mountainous regions. Therefore, accurate identification of landslide locations is crucial for effective prevention and mitigation strategies. During the prolonged heavy rainfall events in Guangdong Province between 21 May and 21 June 2022, shallow and clustered landslides occurred in the mountainous regions of the Beijiang River Basin. This research used high-resolution satellite imagery and integrated the Mask R-CNN algorithm model with spectral, textural, morphological and physical characteristics of landslides in remote sensing imagery, in addition to landslide-influencing factors and other constraints, to interpret the landslides induced by the event through remote sensing techniques. The detection results show that the proposed methodology achieved a high level of accuracy in landslide identification, with a precision rate of 81.91%, a recall rate of 84.07% and an overall accuracy of 87.28%. A total of 3782 shallow landslides were detected, showing a distinct clustered distribution pattern. The performance of Mask R-CNN, Faster-CNN, U-Net and YOLOv3 models in landslide identification was further compared, and the effects of setting the rotation angle and constraints on the identification results of the Mask R-CNN algorithm model were investigated. The results show that each model improves the evaluation indices, but the Mask R-CNN model has the best detection performance; the rotation angle can effectively improve the generalization ability and robustness of the model, and the landslide-inducing factor data and texture feature sample data are the best for landslide identification. The research results provide valuable references and technical support for deepening our understanding of the distribution patterns of rainfall-triggered shallow and cluster landslides in the Beijiang River Basin.

Keywords: landslide identification; mass landslides; Mask R-CNN; Beijiang River Basin

Citation: Wu, Z.; Li, H.; Yuan, S.; Gong, Q.; Wang, J.; Zhang, B. Mask R-CNN–Based Landslide Hazard Identification for 22.6 Extreme Rainfall Induced Landslides in the Beijiang River Basin, China. *Remote Sens.* **2023**, *15*, 4898. https://doi.org/10.3390/rs15204898

Academic Editors: Qian Du, Yanni Dong and Xiaochen Yang

Received: 13 August 2023
Revised: 15 September 2023
Accepted: 18 September 2023
Published: 10 October 2023

1. Introduction

The southern region of China experiences a high frequency of rainfall-induced landslides, particularly under extreme precipitation conditions. This can lead to the development of a distinct form of shallow, mass landslide hazards within specific basins. These hazards exhibit unique characteristics, including small individual scales, prolonged durations, multiple hazard locations and distinct watershed subordination features. Consequently, they generate a cumulative and amplifying impact throughout the hazard event, resulting in substantial damage [1–3]. In addition, the complex geological conditions in the area are influenced by a variety of natural factors, resulting in a large number of hazards from mass landslides, including casualties, property damage, infrastructure damage, economic loss, ecological damage and many other aspects [4–6]. Such landslides induced by extreme rainfall are not uncommon around the world, such as in the Andean region [7], Hong

Kong [8], America [9] and Malaysia [10], where rainfall-induced landslide disasters have caused severe human and economic losses.

Traditional methods of landslide identification are time-consuming, labor-intensive and relatively difficult to identify for large-scale areas [11]. However, remote sensing technology is a very effective means of obtaining topographic and geomorphological information to identify potential landslides and landslide histories, which in turn can lead to landslide hazard assessment and landslide risk analysis [12]. Although predicting rainfall can effectively prevent geological disasters from occurring in advance, and some scholars are now conducting research in this field [13], the identification of regional landslides after extreme heavy rainfall weather events not only helps to reveal the causes and mechanisms of landslides and provide a scientific basis for landslide prevention and control but also helps to predict and allows for warning of landslides, thus reducing disaster losses. Therefore, the accurate identification and monitoring of regional landslides is of great academic and practical significance and can make a positive contribution to safeguarding people's lives and property, maintaining social stability and promoting economic development.

Currently, four methods are widely used in traditional large-scale landslide identification studies of remote sensing imagery: visual interpretation, image-based methods, object-oriented methods and deep learning. The visual interpretation method is more accurate, but it relies on expert knowledge and experience, which is costly in terms of time and manpower, and has a relatively small scope of application and cannot be interpreted quickly [14–17]. The image element–based method has the advantages of fast recognition and high accuracy, but it requires analysis under the same resolution of the geographical environment, which is prone to information waste and the "salt and pepper" phenomenon [18,19]. The object-oriented method can classify objects with similar spectral texture features into one category, which effectively improves recognition accuracy and avoids the "salt and pepper" phenomenon, but many small noises will be generated in the process of image object segmentation, and these units need to be ignored during classification, which presents certain limitations [20–23].

With the rapid development of computer image recognition technology, the cross-fertilization between the field of deep learning and the field of landslide detection has become one of the current research hotspots. Many scholars have carried out research work in this area, and some of them have proposed new algorithmic models using convolutional neural networks to develop the principle, such as U-Net [24], DeepLab V3+ [25], Faster R-CNN [26], Mask R-CNN [27], Yolov [28] and so on. Other scholars have improved the algorithmic models based on their predecessors by optimizing the model structure or combining them with other algorithms to improve the model feature extraction capability, thus increasing the recognition efficiency [29–33]. Some other scholars believe that deep learning network models are still inadequate in extracting features at multiple scales and need to be improved by adding constraints to their design [34–38]. Deep learning methods have the advantages of being adaptable and having good recognition effects, which can better solve the problems in traditional methods and thus present a new direction for landslide recognition.

Rainfall-induced landslides in South China are small in monolithic scale and not well recognizable from other features such as bare land, roads and mining areas. Located in the Beijiang basin area of South China, the area is heavily covered by vegetation and is often affected by cloudy and rainy weather, which leads to the inability of visible light and microwaves to penetrate effectively through the surface vegetation; the areas with a high degree of vegetation cover in this region are also usually steeply sloped and densely vegetated, which makes it obviously difficult to carry out field investigations [39]. Therefore, accurate remote sensing interpretation of the landslide hazard remains a significant challenge. The Mask R-CNN algorithm model can not only accurately detect targets in images but also generate pixel-level segmentation masks for each detected target. It is also able to discriminate object boundaries in complex environments, improving the accuracy of detection segmentation. In addition, the Mask R-CNN algorithm model is able to perform detection and segmentation directly on the input image without additional pre-processing steps, providing a high recognition rate and high accuracy [40].

Since He et al. [27] proposed Mask R-CNN, the algorithm has been widely used in various fields. For example, Sui et al. [41] improved the Mask R-CNN model by introducing the CBAM attention module to achieve façade damage detection of complex building faces after an earthquake. In addition, Liu et al. [42] used Mask R-CNN as a base model and introduced an attention mechanism to establish an automatic landslide identification model for InSAR observation and achieved better results. In addition, Jiang et al. [43] used the Sichuan–Tibet transportation corridor as the study area and combined Mask R-CNN and migration learning methods to successfully detect old and new landslides, indicating that the migration-learning-improved Mask R-CNN can be effectively used for landslide and ice avalanche detection. Yang et al. [44] proposed a background enhancement method by adding landslide triggering factors to the data as auxiliary information and compared the applicability of Mask R-CNN, U-Net and PSP-Net methods. These research results not only provide new ideas and methods for landslide hazard monitoring, early warning and prevention but also promote the development and application of deep learning techniques.

This paper aims to perform remote sensing interpretation through combining the masked R-CNN algorithm model with the constraints of spectral, structural, morphological and physical features of landslides in remote sensing imagery. The method makes use of the stacking of data layers to effectively extract features from the data, which improves the precision and accuracy of landslide identification and demonstrates a good ability to identify small, shallow landslides in southern China. First, a landslide recognition sample dataset is established based on known landslide information, and then various deep learning models based on sample libraries and constraints, including Mask R-CNN, Faster R-CNN, U-Net and YOLOv3, are trained and tested to compare the recognition accuracy of each algorithmic model. This is followed by a comparison of the effect of different rotation angles and different combinations of constraints on the recognition of the masked R-CNN model. Finally, the Mask R-CNN model is used to automatically extract landslides from all the image data and to exclude erroneously identified blocks of bare ground based on the physical characteristics of landslides. The results of this study can be used in the investigation of shallow rainfall-type landslide disasters in mountainous areas to provide technical methods, data support and decision guidance.

2. Study Area and Data

2.1. Study Area

From 21 May to 21 June 2022, heavy rainfall occurred in Guangdong Province, China, under the influence of the "Longzhoushui" phenomenon, with a cumulative rainfall of nearly 847.2 mm; in particular, from 16 to 21 June, the average rainfall reached 294 mm, nearly seven times the normal rainfall for the period. The continuous exceptionally heavy rainfall provided rich hydroclimatic conditions for landslide hazards, resulting in a large number of landslides of varying sizes in the study area, with landslides mainly developing in granite residual soil layers, and a small proportion of landslides in alluvial ditches evolving into debris flows. Among them, 110 townships in nine counties of Shaoguan and Qingyuan cities in the Beijiang basin were severely affected, with direct economic losses amounting to USD 268.2 million, with Qingyuan being one of the most affected areas during the round of heavy rainfall, posing a serious threat to local transportation and personal property and safety [45].

The Beijiang River Basin, located between 23°10′–25°31′N and 115°55′–114°50′E, has a total area of 46,700 km^2, of which 92%, i.e., 42,900 km^2, is a fan-shaped area in Guangdong Province. The first-order tributaries are Zhenjiang, Jinjiang, Wujiang, Nanshui, Yanjiang, Lianjiang, Jiujiang, Binjiang and Suijiang, whose geographical locations are shown in Figure 1. The Beijiang basin has a typical subtropical monsoonal humid climate, with cloudy and rainy weather; it has an average annual rainfall of between 1300 and 2500 mm, with an average annual rainfall of 1800 mm in the basin, showing a decreasing trend from south to north. The annual precipitation is mainly concentrated between April and September, with the longest and most intense continuous precipitation in May. The good climatic conditions provide conditions for the growth and development of vegetation. According to statistics, it takes about six months to restore vegetation cover to the areas damaged by shallow landslides, which makes the work of remote sensing interpretation somewhat difficult. The topography is mainly mountainous and hilly. The study area is influenced by tectonic movements, with folds and fractures being more developed, while neotectonic movements are relatively complex. The special natural geographical environment, climatic conditions and geological conditions have led to the existence and development of landslide hazards in the area.

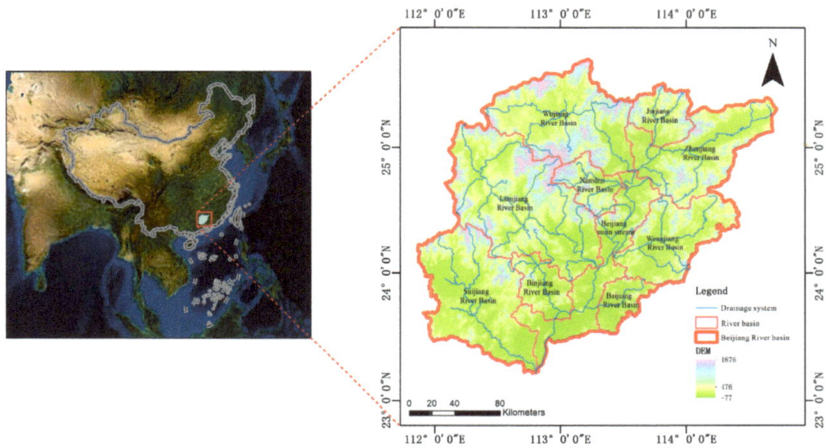

Figure 1. Location of the study area.

2.2. Data Source

In this paper, a series of datasets are used, which include remote sensing image data from the Gaofen-1 satellite, digital elevation data, lithology data, rainfall data and the boundary data of the Beijiang River Basin; the detailed sources are shown in Table 1. In addition, in this paper the slope data were extracted on the basis of the DEM.

Table 1. The sources and characteristics of the data used in the paper.

Data	Resolution	Date	Source
Panchromatic images	2 m	2022	https://www.cresda.com/ (accessed on 5 January 2023)
Multispectral images	8 m	2022	https://www.cresda.com/ (accessed on 5 January 2023)
Normalized difference Vegetation index	30 m	2022	https://www.resdc.cn/DOI (accessed on 7 January 2023)
Digital elevation model	12.5 m		https://www.resdc.cn/ (accessed on 7 January 2023)
Rainfall	1 km	2022	https://pmm.nasa.gov/precipitation-measurement-missions (accessed on 6 January 2023)
Basin boundaries			https://www.hydrosheds.org (accessed on 8 January 2023)
Stratigraphic lithology	1:500,000		http://geodata.ngac.cn (accessed on 7 January 2023)

3. Methodology

Figure 2 shows the flow chart of landslide identification based on the Mask R-CNN model and incorporating physical mechanism–optical–morphological features of landslide hazards. Firstly, data collection and processing is performed, including radiometric calibration, geometric correction, orthorectification, geometric alignment, image fusion, resampling and normalization of the image data and geographic factor data (such as landslide-inducing factor data, texture feature pattern data and geometry feature pattern data). Secondly, the sample dataset and the training model are established: the sample dataset is established by the known landslide point data, and the geographical factors applicable to landslide identification are selected and used as constraints for the establishment of the training model. Finally, automatic landslide hazard identification is carried out on remote sensing images according to the training model.

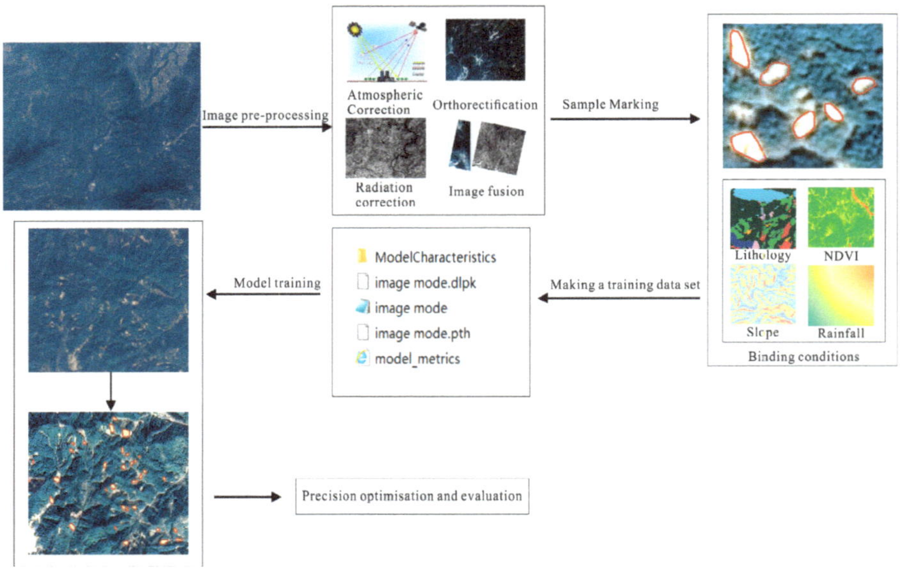

Figure 2. Flow chart of landslide identification based on Mask R-CNN model and incorporating physical mechanism–optical–morphological features of landslide hazards.

3.1. Remote Sensing Characterization of Landslides

Rainfall landslides are characterized by geomorphological features such as a wide range of distribution, small scale and multiple groups. In the landslide development area, the surface vegetation and soil are damaged by rainfall, resulting in exposed ground surface and reduced vegetation cover [46,47]. These features make the landslides exhibit distinct spectral, morphological and textural characteristics in remote sensing images and clearly distinguish them from surrounding features [48]. Therefore, in this study, landslides in a newly developed stage were selected as the object of study, and the Mask R-CNN model was used for the automatic identification of landslides based on their spectral, morphological and textural features.

3.1.1. Spectral Characteristics

In the analysis of remote sensing imagery, we were able to identify landslides of the shallow unstable, rotational and offset slide types with some accuracy. There are significant differences in tonality between the landslide area and the surrounding features. Areas where landslides occur show grey or white tones, reflecting the color characteristics of fresh soil or rock masses, while vegetation and water bodies show darker tones. Through the comparative analysis of the brightness characteristics, we are able to exclude obvious distracting features, thus improving the accuracy and precision of landslide identification. In addition to this, the vegetation around the landslide area after the occurrence of the landslide is significantly different from the vegetation in other areas, as shown in Figure 3b,c below, where the vegetation on the slope is relatively sparse, with no huge upright trees but a small number of smaller trees. This results in vegetation and landslides showing significant variability in the red light band. Therefore, the NDVI normalized vegetation index was used to enhance the delineation of vegetated and non-vegetated areas.

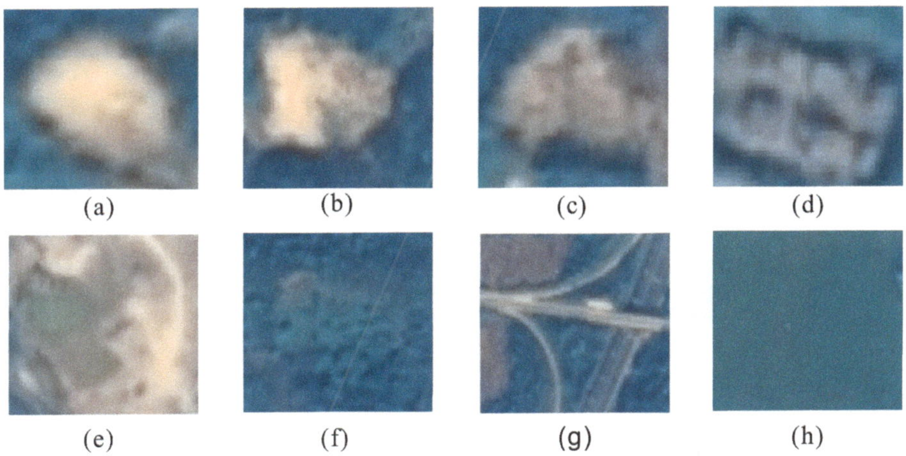

Figure 3. Spectral features of landslides on remote sensing imagery: (**a–c**) landslides; (**d**) house structures; (**e**) lakes and bare ground; (**f**) vegetation; (**g**) roads; and (**h**) water systems.

3.1.2. Textural Features

On remote sensing images, the textures of buildings, roads and other areas of human activity differ significantly from natural features such as vegetation and bare soil and can be effectively distinguished from each other (see Figure 4 below). Landslides are very similar in tone to recently tilled agricultural land, but landslide textures are coarser on the imagery, and large patchy masses can be seen on some rocky landslide imagery; thus, texture features can be used to effectively remove buildings, roads and cultivated land. In this paper, texture features are extracted in ENVI 5.6 from grey-scale co-occurrence matrix images, with the window size of the statistical image element set to 5 × 5, the shift step set to 2 and the shift direction set to 2, and the grey-scale quantization level set to 64. The texture information used for this landslide identification includes the mean value reflecting the grey scale of the image and the contrast reflecting the sharpness of the image and the depth of the grooves in the texture.

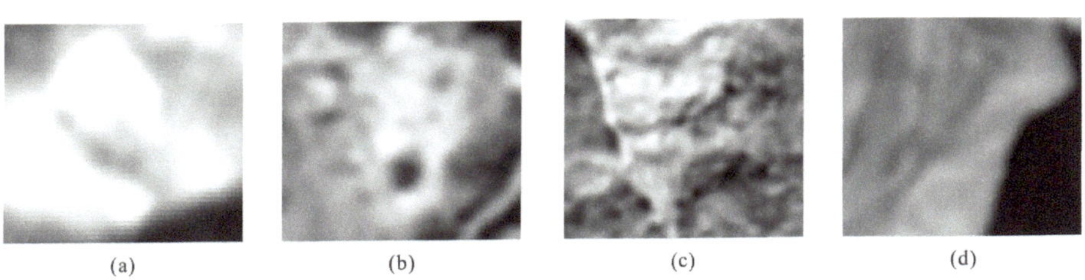

Figure 4. Map of the textural features of a landslide on remote sensing imagery: (**a**) landslide body; (**b**) bare ground; (**c**) mining area; (**d**) river floodplain.

3.1.3. Geometrical Features

In terms of geometry, landslides are mostly bumpy, horseshoe-shaped, oxbow-shaped, oval, or spatulate, without uniform morphology. As shown in Figure 5, in this study area, tongue and spoon shapes are predominant. The landslides have broken rocks, undulating topography, unevenly sunken local platforms on the slope surface, and steep and long slopes; although there are landslide platforms, their area is not large, and there is a slow dip downwards. Based on previous visual interpretations and landslide patterns in the study area, we found that the aspect ratio of shallow landslides is usually about 2.5, where the line segment with the greatest distance between any two points of a polygon patch is used as the long axis (L) of the polygon, and the short axis length is obtained by dividing the polygon area (A) by the long axis (W = A/L).

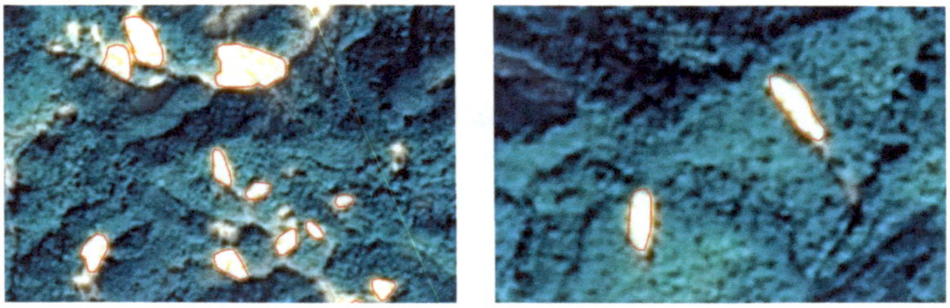

Figure 5. Geometric features of landslides on remote sensing imagery. The red circled area is the detected landslide.

3.1.4. Physical Characteristics

During the occurrence of rainfall-type landslides, the movement of landslide materials is often subject to the influence of gravity and microtopography. A certain elevation difference will be generated within the landslide area; the elevation difference of river floodplains and construction areas and roads, except for exposed rocks, is obviously small. The area (A) and elevation difference ($DEM_{max} - DEM_{min}$) of the study area, identified as a cluster of landslides, were extracted separately using Arcgis Pro software, and the statistical elevation difference was quantified using the area ($(DEM_{max} - DEM_{min})/$A) to exclude disturbing features with similar spectral, texture and shape characteristics.

3.2. Creation of a Landslide Sample Database

The landslide sample database consists of three components: stacking, data extraction and database creation [49]. As shown in Figure 6, the original one-dimensional data from each layer are stacked to form an n-dimensional dataset containing the joint data of all layers by stacking data layers representing different predictor variables. The joint dataset consists of remotely sensed imagery, landslide impact factors and landslide feature factors. Landslide samples verified in the field by drones, Google HD imagery and field surveys are then labelled at the pixel level using the LabelMe labelling tool (Figure 7), with a total of 378 landslides labelled. The labelled results are used as a 'positive sample', which includes the entire n-dimensional data of the surrounding area, and a negative sample is generated by randomly selecting areas where no landslides have been recorded. It is necessary to resample the input data to make the number of rows and columns consistent, as these data come from different sources, and the data standards are not uniform. Firstly, to facilitate training of a deep learning network, input data need to be resampled in order to have a consistent number of rows and columns. In this paper, resampling is done using linear interpolation, where image element values of four nearby points are used to perform linear interpolation, assigning different weights according to distance from interpolation point. Secondly, the input data consistency relationship should be constructed so that the data are standardized and unified, with remote sensing images as the reference, in order to facilitate the training of the deep learning network. Each factor is converted to a raster format with a spatial resolution of 2 × 2 m and normalized and classified according to the Jenks natural breakpoint method in the ArcMap program. The normalization formula is as follows:

$$x_i' = \frac{x_i - x_{min}}{x_{max} - x_{min}}$$

where x_i denotes the ith original data point, x_{min} is the minimum value in the original dataset, x_{max} is the maximum value in the original dataset and x_i' is the corresponding normalized value for the ith original data point.

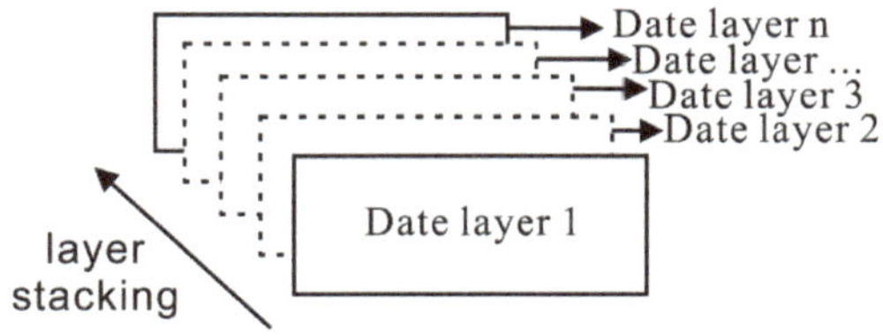

Figure 6. Laminated image. Adapted with permission from Ref. [49].

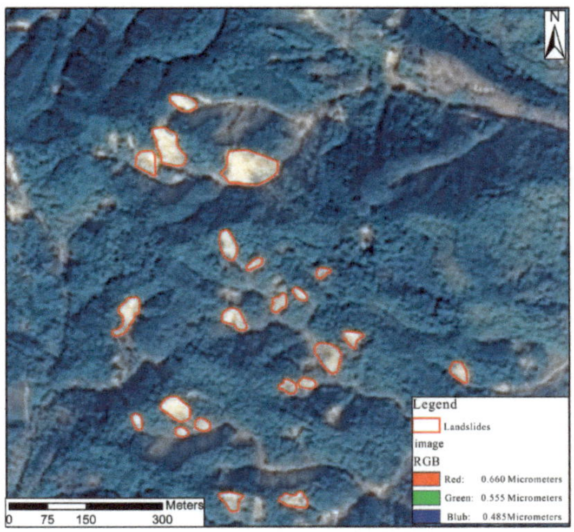

Figure 7. Landslide sample dataset tagging.

3.3. Deep Learning Models

3.3.1. Fundamentals of Mask R-CNN

Mask R-CNN is an advanced instance segmentation technique that improves on Faster R-CNN to perform both target detection and semantic segmentation. The framework of the Mask R-CNN model is shown in Figure 8. The algorithm uses a convolutional neural network as the backbone model to extract the convolutional features of an image and construct a feature pyramid. In Mask R-CNN, ResNet and FPN are used as the backbone network to process the input image and generate a feature pyramid for further feature extraction. Candidate regions of different sizes and scales are generated on the feature pyramid by RPN, and these potential targets are classified and bounding box–generated. Unlike the RoI Align used by Faster R-CNN, Mask R-CNN uses a RoI Align layer to maintain the spatial correspondence between the image and the mask, resulting in accurate pixel-level mask prediction. Finally, each RoI alignment feature mapping provides two branches: one for target classification and bounding box regression, and the other for segmentation mask prediction. Mask R-CNN uses a multi-task training approach that takes into account a combination of target classification loss, bounding box regression loss and segmentation mask loss. Together, these loss functions drive the optimization and learning process of the model to improve the performance of target detection and segmentation.

$$L = L_{cls} + L_{box} + L_{mask}$$
$$L_{mask} = Sigmoid(mask_k)$$

where L_{cls} is the target classification loss, L_{box} is the bounding box regression loss and L_{mask} is the kth mask loss. L_{mask} is defined as the average binary cross-entropy on the kth mask and is calculated from the per-pixel sigmoid on the mask.

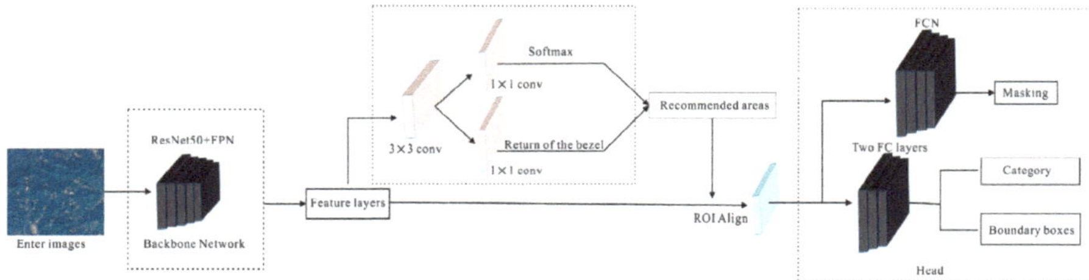

Figure 8. Overall framework of Mask R-CNN. Adapted with permission from Ref. [50].

The backbone network, as the basis of convolutional neural networks, is used to extract the depth features of an image and different convolutional neural network architectures can be selected; Mask R-CNN selects the ResNet network as the image feature extraction network, while combining with FPN to fuse the location-accurate top-level features with the semantic-rich bottom-level features [51]. The core idea of ResNet is constant mapping, i.e., replacing the original ideal mapping $f(x)$ with $f(x) - x$, which can ensure a faster subsequent feature extraction. In this paper, we use ResNet-50, a deep residual network with 50 layers, which is a subclass of convolutional networks and is the most commonly used image classification model. The advantage of FPN is that it can improve the accuracy and fast detection of small objects at multiple scales without increasing the computational effort.

RPN is based on the convolutional neural network structure to further determine the feature layers obtained by ResNet and FPN previously, filter out the locations where there may be targets, and integrate the operation of selecting candidate regions into the target detection framework. ROI Align is used to optimize the spatial location misalignment problem caused by ROI Pooling. The bilinear interpolation method is used to determine the feature value of each point in the feature map, and then pooling and other operations are performed to improve the accuracy, solving the problem of misalignment caused by extracting feature nuggets for rounding at the corresponding locations of the feature map. After ROI Align, the uniformly sized region of interest is entered into the classification regression branch and the mask branch. In the classification regression branch, two 1024-sized fully connected layers are passed, and then the specific category is determined in the classification branch. Unlike the initial category determination, the RPN only determines if it is an object. Further fine-tuning of the centroid and aspect of the position border is performed in the regression branch to precisely locate the object.

3.3.2. Faster R-CNN Model

The Faster R-CNN model was developed on the basis of the Fast R-CNN model [52]. The extraction of candidate regions in the network structure of Fast R-CNN is performed by a selective search. The selective search operation mainly uses a CPU and cannot use the high speed of a GPU; thus, the speed is still very slow. Ren et al. [26] proposed Faster R-CNN. The network is unique in that it proposes the RPN network structure; the candidate region search is similarly added to the training process, and RPN shares parameters with the convolutional layer, further improving network detection efficiency. Although it uses a convolutional neural network, it offers advantages over conventional neural networks in terms of higher detection quality and a capacity to incorporate multiple loss functions into a single training process. The model is said to be a true end-to-end object detection system because it organically combines the four modules of feature extraction, region generation suggestion, RoI merging, and classification and regression within the same network.

3.3.3. The U-Net Model

U-Net is a model based on a modified FCN proposed by Ronneberger et al. [53] in 2015 and initially applied to the task of medical image segmentation. The net structure of U-Net consists of a compressing and an expanding path, also known as the encoder–

decoder structure. The contraction path is used to obtain contextual information and perform feature extraction, while the expansion path is used to accurately locate the feature positions and to connect the feature maps in the up- and downsampling processes through branch connections to achieve the fusion of the superficial and deep semantic information to reduce the loss of edge information. The branch connections of U-Net fully incorporate the features corresponding to the superficial layers during the decoding process, so that the small samples of the data are not easy to overfit. However, U-Net's training rate is slow. The reason is that the network performs classification at each pixelpoint, and segmentation requires one patch at each pixelpoint for training. The high resemblance between neighboring pixel points causes high redundancy and thus slows down the network training.

3.3.4. YOLOv3 Model

YOLO was first proposed by Redmon et al. [54] in 2015 as a single-step target detection algorithm with a speed benefit over traditional two-step algorithms. YOLOv3 introduces a new network structure, Darknet-53, and introduces residual learning features and a high number of residual chunks to increase network depth and representativeness [55]. During training, YOLOv3 uses a number of independent logistic regression classifiers, with the classification prediction made via a binary cross-entropy loss function. This classifier design enables the model to perform multiple label classification, that is, to determine whether an object within the object boundaries belongs to the current label. In addition, YOLOv3 uses the idea of feature pyramid networks to make predictions at three different scales, 13 × 13, 26 × 26 and 52 × 52, in order to improve detection accuracy and maintain high-speed operation. These three scales allow information to be extracted, feature maps of varying sizes to be fused and recognition outputs to be produced. This design enhances detection accuracy while maintaining fast performance.

4. Experiments

4.1. Evaluation Indicators for Recognition Accuracy

In this paper, the secondary evaluation metrics precision, recall and accuracy are used as the accuracy evaluation of the recognition model [56]. Precision shows the ratio of correctly identified landslide hazards to the number of instances identified as landslide hazards; recall shows the ratio of correctly identified landslide hazards to the number of actual landslide instances; and accuracy shows the ratio of predicted landslide hazards to correct predictions.

$$Precision = \frac{TP}{TP + FP}$$

$$Recall = \frac{TP}{TP + FN}$$

$$accuracy = \frac{TP + TN}{TP + FN + TP + FN}$$

where TP is a true positive, which represents the landslide area accurately described by the applied method. FP is a false positive, which is defined as the non-landslide area detected by the applied method as the landslide area in the image. FN is a false negative, which shows the actual landslide area not detected by the applied method. TN is a true negative, which shows the non-landslide area detected by the applied method.

4.2. Experimental Design

The hardware environment of this research was a GeForce GTX 1080 graphics card, an Intel(R) Xeon(R) CPU E5-2680 v3 @ 250 GHz processor, and 40 GB of RAM. The experimental parameters were 20 epochs, 100 steps per epoch, a batch size of 4 and a 0.0001% learning rate.

A first set of comparison experiments was designed to determine the applicability and effectiveness of the proposed method. As shown in Table 2, a number of popular deep network models (MaskR-CNN, Faster-CNN, U-Net and YOLOv3) were selected for testing during the training process, and the training datasets were the original images and the image data with additional geographic factors added for overlay processing. We defined "Original satellite images" as "A" and "Original satellite images + geographical factors" as "B". In addition to analyzing the impact of this paper's approach on different deep learning models, we can also reliably assess whether the models built in this study have advantages in landslide extraction tasks.

Table 2. Comparison of first set of experiments.

No.	Deep Learning Model	Training Dataset
1	Mask R-CNN	A
2	Mask R-CNN	B
3	Faster-CNN	A
4	Faster-CNN	B
5	U-Net	A
6	U-Net	B
7	YOLOv3	A
8	YOLOv3	B

The good or bad training effect of the deep learning model is not only related to the model parameter adjustment; the size of the used dataset also has a significant impact on the training effect: the larger the dataset, the better the recognition effect of the trained model for landslide hazard recognition [57]. There are landslides with different orientations in the study area, and the features extracted from images of the same target taken at different azimuths can be transformed into each other using rotational angle processing to improve the accuracy of detecting landslide hazards with different orientations. In the second part of the work, the model will be trained again by rotating the batch size images and comparing the recognition results with different rotation angles.

In addition, a third set of comparison experiments is designed in this paper to analyze which combination of factors in the proposed method is the key to improving the recognition efficiency. Six comparison experiments are designed depending on the input training dataset, as shown in Table 3. In Experiment I, the original image, the additional landslide-inducing factor data, the texture feature sample data and the geometric feature sample data are used as the input dataset and processed with physical features. Based on Experiment I, Experiment II reduces the landslide-inducing factor data, Experiment III reduces the textural feature sampling data, Experiment IV reduces the geometry feature sampling, Experiment V reduces the physical feature processing, and Experiment VI uses only the original image as the input data.

Table 3. Comparison of third set of experiments.

No.	Deep Learning Model	Training Dataset
I	Mask R-CNN	Original satellite images + landslide-inducing factors + textural features + geometric features + physical features
II	Mask R-CNN	Original satellite images + textural features + geometric features + physical features
III	Mask R-CNN	Original satellite images + landslide-inducing factors + geometric features + physical features
IV	Mask R-CNN	Original satellite images + landslide-inducing factors + textural features + physical features
V	Mask R-CNN	Original satellite images + landslide-inducing factors + textural features + geometric features
VI	Mask R-CNN	Original satellite images

5. Results

5.1. Comparing Different Deep Learning Models

The first set of results is shown in Table 4, which compares the test results of the different models, each of which shows an improvement in all evaluation metrics. Among them, the Mask R-CNN model outperforms the Faster-CNN, U-Net and YOLOv3 models in terms of recognition performance. It improves precision and accuracy by more than 2.5% compared to the other models. Figure 9 shows the recognition results of the different models in the test set. As can be seen from the figure, Mask R-CNN not only provides more accurate detection results than the other models but can also segment each landslide shape independently and shows better landslide edge segmentation results. The mask for the recognition of large-scale landslides is more complete and fits well with the actual shape of the landslide, while also taking into account the recognition of small-scale landslides.

Table 4. Accuracy of different deep network models.

Model	Precision/%	Recall/%	Accuracy/%
1	75.51	78.69	78.13
2	87.31	86.54	91.29
3	72.84	73.65	76.37
4	81.71	84.39	85.58
5	65.72	70.46	68.94
6	78.47	81.73	82.46
7	64.29	67.05	66.78
8	73.33	76.53	77.85

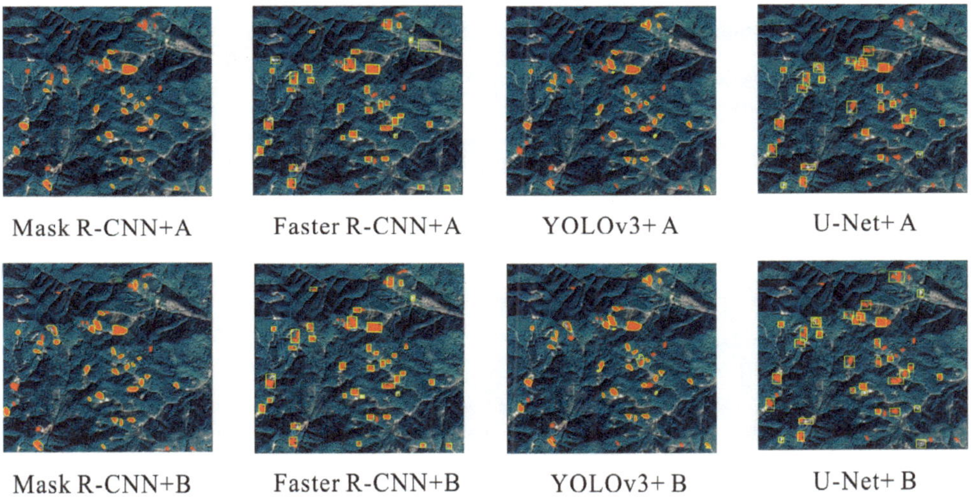

Figure 9. Results of different models. The red polygon represents the real boundary of the landslide, while the yellow wireframe represents the predicted landslide boundary.

5.2. Dataset and Sample Processing

The rotation angle can effectively improve the precision, recall and accuracy of the model. The results of the second set of experiments are shown in Table 5. In particular, when the rotation angle was set to 90°, the precision, recall and accuracy of the model achieved 87.31%, 86.54% and 91.29%, respectively. However, when the rotation angle was less than 90°, the dataset was overexpanded, the precision, recall and accuracy of the model were reduced, and the time required for model training increased significantly with the reduction of the rotation angle. Therefore, in order to avoid the redundancy of the model dataset, an appropriate rotation angle should be selected in order to improve the efficiency and accuracy of the model training.

Table 5. Results for different rotation angles.

Angle/°	Precision/%	Recall/%	Accuracy/%
0°	81.91	84.07	87.28
30°	85.17	87.34	90.55
60°	86.67	87.73	91.71
90°	87.31	86.54	91.29
120°	85.48	86.03	90.26
150°	84.84	85.27	88.53
180°	83.66	84.82	89.85

5.3. Keys Affecting Recognition Precision

From the results of the comparison experiments in Table 6, it can be seen that the landslide triggering factor data, the texture feature sample data, the geometric feature samples and the execution of the physical feature processing can effectively help the model to discriminate between the background objects and the landslides. Among them, the landslide evoking factor data and texture feature sample data improve the accuracy of landslide identification most clearly, with the precision increasing by 5.39% and 6.8%, the recall increasing by 4.59% and 5.25%, and the accuracy increasing by 5.1% and 6.24%, respectively. The increase in precision and recall is usually accompanied by an increase in the number of correctly identified landslides and decrease in the number of incorrect classifications. Therefore, based on the results obtained, it can be seen that a reasonable selection of geographical factors can effectively improve the performance of the Mask R-CNN model and reduce the number of misclassified and omitted landslides to a certain extent.

Table 6. Comparison of experimental results.

Lab No.	Precision/%	Recall/%	Accuracy/%
I	81.91	84.07	87.28
II	76.52	79.48	82.18
III	75.11	78.82	81.04
IV	78.16	82.85	85.35
V	79.84	84.01	86.16
VI	72.04	74.95	79.31

6. Discussion

6.1. Analysis of Identification Results

Based on the results of the above three sets of experiments, landslides for the June 2022 extreme rainfall events that occurred in the Beijiang River Basin were identified using a Mask R-CNN model trained on the dataset with geographical factors superimposed and rotated by 90°. The identification results are shown in Figure 10, where the number of induced landslides is 3782, and the total identified landslide area reaches 4.6×10^6 m^2. However, landslides with a landslide area of more than 10,000 m^2 account for less than 5% of the total number, and landslides with a landslide area of more than 5000 m^2 account for less than 10% of the total number, suggesting that this rainfall event mainly induced landslides of a relatively small size. Since the landslides are considered as point elements, their geographical locations can be expressed in terms of coordinates; 3782 landslide points were imported into ArcMap10.8. The results of the calculation using the average nearest neighbor tool showed that the average nearest neighbor index of landslide points in the study area was 0.398 (Figure 11a) and passed the significance test at the 0.01 confidence level. This indicates that landslide points in the study area have a strong spatial aggregation tendency. On this basis, the estimated kernel density was calculated to show the area of spatial aggregation and the degree of aggregation. The kernel density results are shown in Figure 11b. This figure shows that the spatial distribution of landslide hazards in the study area has a high spatial aggregation and spatial variability; the maximum value appeared in the Beijiang River Basin and the Lianjiang River Basin, reaching more than 1.13, which indicates that the distribution of landslides in these areas is the most concentrated. It is not difficult to see that the results have a strong correlation with the water system.

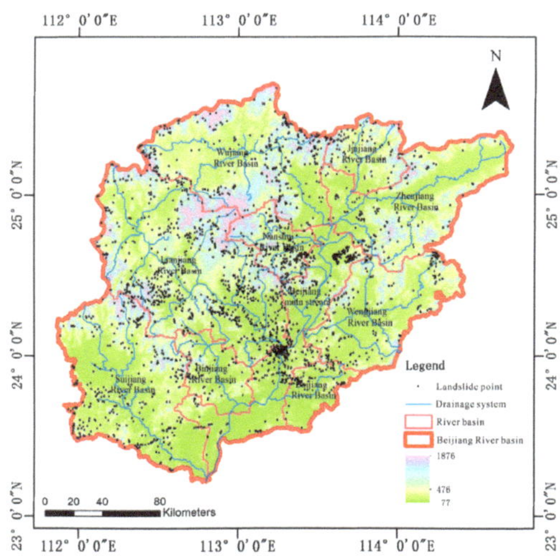

Figure 10. Landslide identification results.

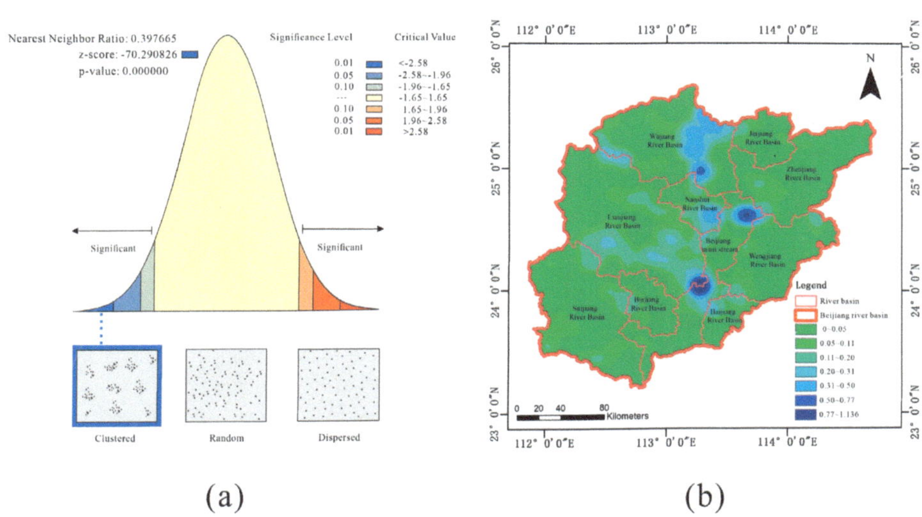

Figure 11. Landslide spatial distribution characteristics: (**a**) mean nearest neighbor index; (**b**) kernel density.

We comprehensively analyzed the training and validation sets based on the recognition accuracy evaluation index to quantitatively evaluate the landslide recognition. The experimental results showed that landslide detection precision reached 87.31%, recall reached 86.54%, and accuracy reached 91.29%. We conducted field validation in Lechang, Yangshan, Lianzhou, Luyuan and Yizhang, covering a total of 378 landslide cases, to verify the effect of the Mask R-CNN–based model with superimposed landslide recognition features on landslide recognition accuracy. After the comparative analysis, we were able to successfully assess the accuracy of 341 landslides, with a remote sensing detection accuracy of more than 90%. This further validates the feasibility and high accuracy of this research method in identifying landslides, which can be applied to identifying landslides on a large scale.

6.2. Limitations and Future Prospects

In this paper, based on datasets with different feature combinations and using physical features to highlight landslide areas, we distinguish flat bare land, mining areas, buildings, etc., to reduce the misjudgment of landslides in mountainous areas. This effectively improves the recognition efficiency and accuracy of the Mask R-CNN model for mass landslides and provides a feasible reference solution for geological disaster investigation. However, the limited sample size of the landslide hazard dataset used for deep learning in this study area resulted in some omissions and misjudgments in the recognition results. Landslides can be detected by increasing the effective sample size or obtaining the changing image information in the time period before and after the event. Moreover, the spatial resolution of remote sensing images needs to be improved to accurately identify rainfall-induced small mass landslides still. In addition, the spatial attributes are not considered; the geological–geomorphological and hydro-climatic conditions in the northern and southern parts of the Beijiang River Basin are highly variable, and the key influencing factors of landslides are closely related to the geological, hydro-meteorological and topographical features of the study area, which makes the transferability of this paper's method a challenge. This problem can be solved by increasing the sample size of complex areas or cutting the study area into smaller images and setting higher weights of network parameters for areas with complex terrain, so as to improve the problem of recognition error caused by the imbalance between samples in different areas. Previous studies have shown that the background information of remote sensing images can improve the generalization ability of the model during the recognition process, but this information also interferes with the model detection capability [58]. Therefore, it is necessary to reasonably combine effective features or add the attention mechanism in the feature pyramid to enhance the interrelationship between the features, and to select the landslide-related information from a large amount of information.

7. Conclusions

In this study, a Mask R-CNN model constructed on the basis of geographic factors affecting landslide generation and landslide remote sensing identification features is proposed to address the lack of training dataset and misdiagnosis of confusing features (e.g., bare ground, river floodplains, buildings, roads, etc.). The method was validated using data from June 2022 rainfall-induced landslides in the Beijiang River Basin. In addition, Faster-CNN, U-Net and YOLOv3 models were used to compare the applicability of different depth models with the incoming Mask R-CNN model, and the results showed that the performance of the Mask R-CNN model was better compared to the other models, with a precision rate of 81.91%, a recall rate of 84.07% and an accuracy of 87.28%. By comparing the results of model training with different rotation angles, it is found that the model has the strongest generalization ability and robustness when rotated by 90°, which can effectively improve the efficiency and accuracy of model recognition. By adding constraints, the recognition accuracy and precision can be improved. From the comparison experiments of different constraints, it can be found that superimposing landslide-evoking factor data and texture feature sample data improves the recognition accuracy of the model. This finding can bring improvement to different deep learning models, as it involves modifications at the data level. However, there are still errors in the extraction results, and how to find effective features and make reasonable combinations of them instead of simply adding various triggering factors to train the model will become a future research trend. The exceptionally heavy rainfall in June 2022 was a major factor in triggering this landslide. This study can help to understand the distribution pattern of rainfall-induced, shallow, massive landslides in the Beijiang River Basin and provide data and technology for the prevention of rainfall-induced geological hazards in the hilly areas of southeastern China.

Author Contributions: Data curation, formal analysis, methodology, writing—original draft, writing—review and editing, visualization, Z.W.; resources, writing—review and editing, S.Y. and Q.G.; formal analysis, writing—review and editing, H.L. and B.Z.; data curation, validation, J.W. All authors have read and agreed to the published version of the manuscript.

Funding: This research was funded by the National Natural Science Foundation of China (NSFC) (Grant Nos.41977413, 42101084, 42271091); Natural Science Foundation of Guangdong Province (Grant No. 2022A1515011898).

Data Availability Statement: Not applicable.

Acknowledgments: Thanks to the Guangdong Data Centre for Geographical Sciences (Guangzhou Institute of Geography) and Chen Jun for providing the relevant data.

Conflicts of Interest: The authors declare no conflict of interest.

References

1. Yu, G.; Zhang, M.; Hu, W. Analysis on the development characteristics and hydrodynamic conditions for the massive debris flow in Tianshui. *Northwest Geol.* **2014**, *47*, 185–191.
2. Liu, J. *Study on the Pattern and Mechanism of Large-Scale Landslides in Longnan Mountainous Area*; Lanzhou University: Lanzhou, China, 2020.
3. Guo, Z.; Yin, K.; Gui, L.; Liu, Q.; Huang, F.; Wang, T. Regional rainfall warning system for landslides with creep deformation in three gorges using a statistical black box model. *Sci. Rep.* **2019**, *9*, 8962. [CrossRef] [PubMed]
4. Lin, Q.; Wang, Y. Spatial and temporal analysis of a fatal landslide inventory in China from 1950 to 2016. *Landslides* **2018**, *15*, 2357–2372. [CrossRef]
5. Huang, Y.; Xu, C.; Zhang, X.; Li, L. Bibliometric analysis of landslide research based on the WOS database. *Nat. Hazards Res.* **2022**, *2*, 49–61. [CrossRef]
6. Gariano, S.L.; Guzzetti, F. Landslides in a changing climate. *Earth-Sci. Rev.* **2016**, *162*, 227–252. [CrossRef]
7. Fustos-Toribio, I.; Manque-Roa, N.; Vásquez Antipan, D.; Hermosilla Sotomayor, M.; Letelier Gonzalez, V. Rainfall-induced landslide early warning system based on corrected mesoscale numerical models: An application for the southern Andes. *Nat. Hazards Earth Syst. Sci.* **2022**, *22*, 2169–2183. [CrossRef]
8. Dai, F.; Lee, C. Frequency–volume relation and prediction of rainfall-induced landslides. *Eng. Geol.* **2001**, *59*, 253–266. [CrossRef]
9. Piciullo, L.; Calvello, M.; Cepeda, J.M. Territorial early warning systems for rainfall-induced landslides. *Earth-Sci. Rev.* **2018**, *179*, 228–247. [CrossRef]
10. Lee, M.L.; Ng, K.Y.; Huang, Y.F.; Li, W.C. Rainfall-induced landslides in Hulu Kelang area, Malaysia. *Nat. Hazards* **2014**, *70*, 353–375. [CrossRef]
11. Ray, R.L.; Lazzari, M.; Olutimehin, T. Remote sensing approaches and related techniques to map and study landslides. *Landslides-Investig. Monit.* **2020**, *2*, 1–25.
12. Ahmad, M.N.; Shao, Z.; Aslam, R.W.; Ahmad, I.; Liao, M.; Li, X.; Song, Y. Landslide hazard, susceptibility and risk assessment (HSRA) based on remote sensing and GIS data models: A case study of Muzaffarabad Pakistan. *Stoch. Environ. Res. Risk Assess.* **2022**, *36*, 4041–4056. [CrossRef]
13. Ehteram, M.; Ahmed, A.N.; Sheikh Khozani, Z.; El-Shafie, A. Convolutional Neural Network-Support Vector Machine Model-Gaussian Process Regression: A New Machine Model for Predicting Monthly and Daily Rainfall. *Water Resour. Manag.* **2023**, *37*, 3631–3655. [CrossRef]
14. Sato, H.P.; Hasegawa, H.; Fujiwara, S.; Tobita, M.; Koarai, M.; Une, H.; Iwahashi, J. Interpretation of landslide distribution triggered by the 2005 Northern Pakistan earthquake using SPOT 5 imagery. *Landslides* **2007**, *4*, 113–122. [CrossRef]
15. Mohan, A.; Singh, A.K.; Kumar, B.; Dwivedi, R. Review on remote sensing methods for landslide detection using machine and deep learning. *Trans. Emerg. Telecommun. Technol.* **2021**, *32*, e3998. [CrossRef]
16. Leshchinsky, B.A.; Olsen, M.J.; Tanyu, B.F. Contour Connection Method for automated identification and classification of landslide deposits. *Comput. Geosci.* **2015**, *74*, 27–38. [CrossRef]
17. Ju, Y.; Xu, Q.; Jin, S.; Li, W.; Su, Y.; Dong, X.; Guo, Q. Loess landslide detection using object detection algorithms in northwest China. *Remote Sens.* **2022**, *14*, 1182. [CrossRef]
18. Zhao, W.; Li, A.; Nan, X.; Zhang, Z.; Lei, G. Postearthquake landslides mapping from Landsat-8 data for the 2015 Nepal earthquake using a pixel-based change detection method. *IEEE J. Sel. Top. Appl. Earth Obs. Remote Sens.* **2017**, *10*, 1758–1768. [CrossRef]
19. Ma, Y. *An Automatic Landslide Identification Method Based on Remote Sensing Image Recognition*; Tibet University: Lhasa, China, 2022.
20. Zhang, Q.; Zhao, C. Semiautomatic object-oriented loose landslide recognition based on high resolution remote sensing images in Heifangtai, Gansu. *J. Catastrophology* **2017**, *32*, 210–215.

21. Zhang, P.; Xu, C.; Ma, S.; Shao, X.; Tian, Y.; Wen, B. Automatic extraction of seismic landslides in large areas with complex environments based on deep learning: An example of the 2018 iburi earthquake, Japan. *Remote Sens.* **2020**, *12*, 3992. [CrossRef]
22. Qigen, L.; Zhenhua, Z.; Yingqi, Z.; Ying, W. Object-oriented detection of landslides based on the spectral, spatial and morphometric properties of landslides. *Remote Sens. Technol. Appl.* **2017**, *32*, 931–937.
23. Hui, D.; Maosheng, Z.; Weihong, Z.; Tao, Z. High resolution remote sensing for the identification of loess landslides: Example from Yan'an City. *Northwestern Geol.* **2019**, *52*, 231–239.
24. Yu, B.; Chen, F.; Xu, C. Landslide detection based on contour-based deep learning framework in case of national scale of Nepal in 2015. *Comput. Geosci.* **2020**, *135*, 104388. [CrossRef]
25. Chen, L.-C.; Zhu, Y.; Papandreou, G.; Schroff, F.; Adam, H. *Proceedings of the European Conference on Computer Vision (ECCV)*; Springer: Berlin/Heidelberg, Germany, 2022; pp. 801–818.
26. Ren, S.; He, K.; Girshick, R.; Sun, J. Faster r-cnn: Towards real-time object detection with region proposal networks. *IEEE Trans. Pattern Anal. Mach. Intell.* **2017**, *39*, 1137–1149. [CrossRef] [PubMed]
27. He, K.; Gkioxari, G.; Dollár, P.; Girshick, R. Mask r-cnn. In Proceedings of the IEEE International Conference on Computer Vision, Cambridge, MA, USA, 20–23 June 1995; pp. 2961–2969.
28. Redmon, J.; Divvala, S.; Girshick, R.; Farhadi, A. You only look once: Unified, real-time object detection. In Proceedings of the IEEE Conference on Computer Vision and Pattern Recognition, San Juan, PR, USA, 17–19 June 1997; pp. 779–788.
29. Ullo, S.L.; Mohan, A.; Sebastianelli, A.; Ahamed, S.E.; Kumar, B.; Dwivedi, R.; Sinha, G.R. A new mask R-CNN-based method for improved landslide detection. *IEEE J. Sel. Top. Appl. Earth Obs. Remote Sens.* **2021**, *14*, 3799–3810. [CrossRef]
30. Qi, W.; Wei, M.; Yang, W.; Xu, C.; Ma, C. Automatic mapping of landslides by the ResU-Net. *Remote Sens.* **2020**, *12*, 2487. [CrossRef]
31. Lei, T.; Xue, D.; Lv, Z.; Li, S.; Zhang, Y.; Nandi, A.K. Unsupervised change detection using fast fuzzy clustering for landslide mapping from very high-resolution images. *Remote Sens.* **2018**, *10*, 1381. [CrossRef]
32. Fu, R.; He, J.; Liu, G.; Li, W.; Mao, J.; He, M.; Lin, Y. Fast seismic landslide detection based on improved mask R-CNN. *Remote Sens.* **2022**, *14*, 3928. [CrossRef]
33. Ding, A.; Zhang, Q.; Zhou, X.; Dai, B. Automatic recognition of landslide based on CNN and texture change detection. In Proceedings of the 31st Youth Academic Annual Conference of Chinese Association of Automation (YAC), Wuha, China, 11–13 November 2016; pp. 444–448.
34. Wang, Y.; Wang, X.; Jian, J. Remote sensing landslide recognition based on convolutional neural network. *Math. Probl. Eng.* **2019**, *2019*, 1–12. [CrossRef]
35. Huang, J.; Xin, L.; Fang, C.; Ru, C.; Huimin, L.; Bowen, D. A deep learning recognition model for landslide terrain based on multi-source data fusion. *Chin. J. Geol. Hazard Control* **2022**, *33*, 33–41.
36. Ghorbanzadeh, O.; Shahabi, H.; Crivellari, A.; Homayouni, S.; Blaschke, T.; Ghamisi, P. Landslide detection using deep learning and object-based image analysis. *Landslides* **2022**, *19*, 929–939. [CrossRef]
37. Ghorbanzadeh, O.; Blaschke, T.; Gholamnia, K.; Meena, S.R.; Tiede, D.; Aryal, J. Evaluation of different machine learning methods and deep-learning convolutional neural networks for landslide detection. *Remote Sens.* **2019**, *11*, 196. [CrossRef]
38. Gao, X.; Chen, T.; Niu, R.; Plaza, A. Recognition and mapping of landslide using a fully convolutional DenseNet and influencing factors. *IEEE J. Sel. Top. Appl. Earth Obs. Remote Sens.* **2021**, *14*, 7881–7894. [CrossRef]
39. Zhang, C.; Zhou, J.; Wang, H.; Tan, T.; Cui, M.; Huang, Z.; Wang, P.; Zhang, L. Multi-species individual tree segmentation and identification based on improved mask R-CNN and UAV imagery in mixed forests. *Remote Sens.* **2022**, *14*, 874. [CrossRef]
40. Feng, W.; Bai, H.; Lan, B.; Wu, Y.; Wu, Z.; Yan, L.; Ma, X. Spatial–temporal distribution and failure mechanism of group-occurring landslides in Mibei village, Longchuan County, Guangdong, China. *Landslides* **2022**, *19*, 1957–1970. [CrossRef]
41. Sui, H.; Huang, L.; Liu, C. Detecting building façade damage caused by Earthquake using CBAM-improved mask R-CNN. *Geomat. Inf. Sci. Wuhan Univ.* **2020**, *45*, 1660–1668.
42. Liu, Y.; Yao, X.; Gu, Z.; Zhou, Z.; Liu, X.; Chen, X.; Wei, S. Study of the automatic recognition of landslides by using InSAR images and the improved mask R-CNN model in the Eastern Tibet Plateau. *Remote Sens.* **2022**, *14*, 3362. [CrossRef]
43. Jiang, W.; Xi, J.; Li, Z.; Zang, M.; Chen, B.; Zhang, C.; Liu, Z.; Gao, S.; Zhu, W. Deep Learning for Landslide Detection and Segmentation in High-Resolution Optical Images along the Sichuan-Tibet Transportation Corridor. *Remote Sens.* **2022**, *14*, 5490. [CrossRef]
44. Yang, R.; Zhang, F.; Xia, J.; Wu, C. Landslide extraction using Mask R-CNN with background-enhancement method. *Remote Sens.* **2022**, *14*, 2206. [CrossRef]
45. Lixin, W. Lessons learnt from the defence against the "22-6" Beijiang River deluge. China Water Resources 2022. Available online: https://kns.cnki.net/kcms/detail/detail.aspx?FileName=SLZG202222005&DbName=CJFQ2022 (accessed on 17 May 2023).
46. Yan, L.; Gong, Q.; Wang, F.; Chen, L.; Li, D.; Yin, K. Integrated Methodology for Potential Landslide Identification in Highly Vegetation-Covered Areas. *Remote Sens.* **2023**, *15*, 1518. [CrossRef]
47. Bai, H.; Feng, W.; Yi, X.; Fang, H.; Wu, Y.; Deng, P.; Dai, H.; Hu, R. Group-occurring landslides and debris flows caused by the continuous heavy rainfall in June 2019 in Mibei Village, Longchuan County, Guangdong Province, China. *Nat. Hazards* **2021**, *108*, 3181–3201. [CrossRef]
48. Ji, S.; Yu, D.; Shen, C.; Li, W.; Xu, Q. Landslide detection from an open satellite imagery and digital elevation model dataset using attention boosted convolutional neural networks. *Landslides* **2020**, *17*, 1337–1352. [CrossRef]

49. Wang, H.; Zhang, L.; Yin, K.; Luo, H.; Li, J. Landslide identification using machine learning. *Geosci. Front.* **2021**, *12*, 351–364. [CrossRef]
50. Xie, B. *Research on Segmentation Algorithm for Wood Defect Detection Based on Improved Mask R-CNN*; Harbin Institute of Technology: Harbin, China, 2022.
51. Lin, T.-Y.; Dollár, P.; Girshick, R.; He, K.; Hariharan, B.; Belongie, S. Feature pyramid networks for object detection. In Proceedings of the IEEE Conference on Computer Vision and Pattern Recognition, San Juan, PR, USA, 17–19 June 1997; pp. 2117–2125.
52. Zhong, Z.; Sun, L.; Huo, Q. Improved localization accuracy by LocNet for Faster R-CNN based text detection in natural scene images. *Pattern Recognit.* **2019**, *96*, 106986. [CrossRef]
53. Ronneberger, O.; Fischer, P.; Brox, T. Medical Image Computing and Computer-Assisted Intervention–MICCAI 2015. In Proceedings of the Part III 18th International Conference, Munich, Germany, 5–9 October 2015; Springer: Berlin/Heidelberg, Germany, 2015; pp. 234–241.
54. Hinton, G.; Vinyals, O.; Dean, J. Distilling the knowledge in a neural network. *arXiv* **2015**, arXiv:1503.02531.
55. Kou, X.; Liu, S.; Cheng, K.; Qian, Y. Development of a YOLO-V3-based model for detecting defects on steel strip surface. *Measurement* **2021**, *182*, 109454. [CrossRef]
56. AlDahoul, N.; Ahmed, A.N.; Allawi, M.F.; Sherif, M.; Sefelnasr, A.; Chau, K.-w.; El-Shafie, A. A comparison of machine learning models for suspended sediment load classification. *Eng. Appl. Comput. Fluid Mech.* **2022**, *16*, 1211–1232. [CrossRef]
57. Khalifa, N.E.; Loey, M.; Mirjalili, S. A comprehensive survey of recent trends in deep learning for digital images augmentation. *Artif. Intell. Rev.* **2022**, *55*, 2351–2377. [CrossRef]
58. Hou, H.; Chen, M.; Tie, Y.; Li, W. A universal landslide detection method in optical remote sensing images based on improved YOLOX. *Remote Sens.* **2022**, *14*, 4939. [CrossRef]

remote sensing

MDPI

Article

AFRE-Net: Adaptive Feature Representation Enhancement for Arbitrary Oriented Object Detection

Tianwei Zhang [1,2,3,4], Xu Sun [2,*], Lina Zhuang [2], Xiaoyu Dong [5], Jianjun Sha [6,7], Bing Zhang [4] and Ke Zheng [8]

1. Key Laboratory of Computational Optical Imaging Technology, Aerospace Information Research Institute, Chinese Academy of Sciences, Beijing 100094, China; zhangtianwei20@mails.ucas.ac.cn
2. Key Laboratory of Digital Earth Science, Aerospace Information Research Institute, Chinese Academy of Sciences, Beijing 100094, China
3. International Research Center of Big Data for Sustainable Development Goals, Chinese Academy of Sciences, Beijing 100094, China
4. College of Resources and Environment, University of Chinese Academy of Sciences, Beijing 100049, China
5. Department of Complexity Science and Engineering, The University of Tokyo, Tokyo 277-8561, Japan
6. College of Intelligent Systems Science and Engineering, Harbin Engineering University, Harbin 150001, China; shajianjun@hrbeu.edu.cn
7. Qingdao Innovation and Development Center, Harbin Engineering University, Qingdao 266000, China
8. College of Geography and Environment, Liaocheng University, Liaocheng 252059, China

* Correspondence: sunxu@aircas.ac.cn

Abstract: Arbitrary-oriented object detection (AOOD) is a crucial task in aerial image analysis but is also faced with significant challenges. In current AOOD detectors, commonly used multi-scale feature fusion modules fall short in spatial and semantic information complement between scales. Additionally, fixed feature extraction structures are usually used following a fusion model, resulting in the inability of detectors to self-adjust. At the same time, feature fusion and extraction modules are designed in isolation and the internal synergy between them is ignored. The above problems result in feature representation deficiency, thus affecting the overall detection precision. To solve these problems, we first create a fine-grained feature pyramid network (FG-FPN) that not only provides richer spatial and semantic features, but also completes neighbor scale features in a self-learning mode. Subsequently, we propose a novel feature enhancement module (FEM) to fit FG-FPN. FEM authorizes the detection unit to automatically adjust the sensing area and adaptively suppress background interference, thereby generating stronger feature representations. Our proposed solution was tested through extensive experiments on challenging datasets, including DOTA (77.44% mAP), HRSC2016 (97.82% mAP), UCAS-AOD (91.34% mAP), as well as ICDAR2015 (86.27% F-score) and its effectiveness and high applicability are verified on all the above datasets.

Keywords: deep learning; object detection; remote sensing; feature representation enhancement

Citation: Zhang, T.; Sun, X.; Zhuang, L.; Dong, X.; Sha, J.; Zhang, B.; Zheng, K. AFRE-Net: Adaptive Feature Representation Enhancement for Arbitrary Oriented Object Detection. *Remote Sens.* **2023**, *15*, 4965. https://doi.org/10.3390/rs15204965

Academic Editors: Yanni Dong and Xiaochen Yang

Received: 15 August 2023
Revised: 30 September 2023
Accepted: 8 October 2023
Published: 14 October 2023

1. Introduction

As a fundamental task in remote sensing image understanding, arbitrary-oriented-object-detection (AOOD) is attracting the attention of researchers more and more. At the same time, with the rapid development of convolutional neural network (CNN)-based methods [1–5], many outstanding AOOD detectors stand out [6–12]. However, different from object detection in natural images, AOOD is more challenging mainly due to the following two reasons:

1. Objects in remote sensing images tend to have random orientation and larger aspect ratios, which increase the feature representation complexity of detectors.
2. Remote sensing images, due to their wide imaging range, contain complex and diverse ground objects and scenes, resulting in increased interference targets and features.

However, the existing design of AOOD detectors cannot adapt to the feature representation of remote sensing objects very well. Although AOOD detectors use the oriented bounding box (OBB) as the object's marker, which can better fit the object's spatial contour, the feature representation ability of each detection unit (DN) (i.e., feature point in multi-scale detection layers) does not change.

Take the classic anchor-based object detector as an example, as shown in Figure 1, at each position of the multi-scale detection layers, a certain number of anchors will be preset for overlap calculation with GT (ground truth). When an anchor and GT meet certain position and overlap conditions (i.e., label assignment strategy), it will be determined as positive or negative. However, no matter whether HBB (horizontal bounding box) or OBB (oriented bounding box) is used as the labeling of GT, the effective receptive field (ERF) [13] of each DN does not change; that is, no matter what shape and aspect ratio of the object appears at the current position, existing detectors use a fixed feature vector to represent it. This means that for the red high potential DN in shallow feature layers shown in Figure 1a, its feature learning area is limited and does not coincide with the space occupied by the target. This issue has been discussed by some scholars [14,15] and summarized as a feature misalignment problem; however, these researches have not conducted in-depth summary and analysis of internal causes.

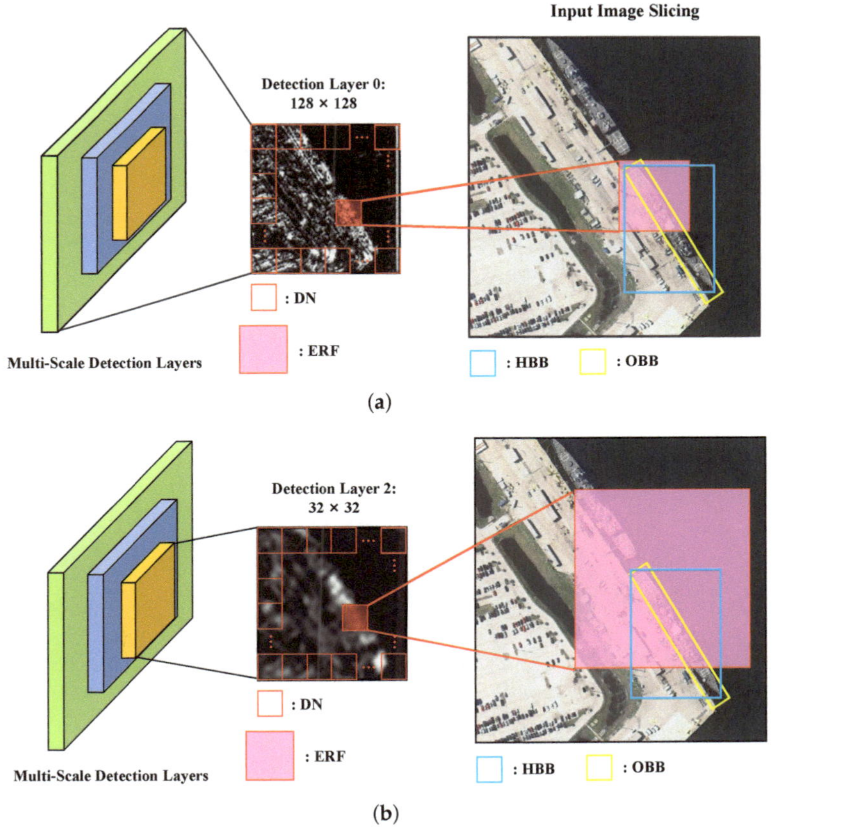

Figure 1. Illustration of the relationship between DN and GT and ERF. A part of the input image is captured inside the detector (RetinaNet-R [16]). The HBB and the OBB are the predicted box. The red box is virtual and represents only one pixel in detection layers. The ERF is calculated according to [13]. It should be noted that this is only a local scene captured from a large remote sensing image input. (**a**) Detection Layer 0; (**b**) Detection Layer 2.

For this issue, one intuitive solution is using multi-scale feature representations to compensate for the uncertainty caused by the change of image and target size. However, another problem arises, as shown in Figure 1b. With the deepening of the network and the down sampling operation, the ERF of DN has expanded. In the detection layer 2 with the size of 32 × 32, the marked DN expands its knowledge learning range but also receives more complex background information. The case in Figure 2 shows the negative impact of the disorderly expansion of ERF, which occurs in a real application scenario. Because the containers and cargo ships in the port have very similar characteristics, they are easily confused when they appear in the ERF of the same DN. Therefore, the container on the shore is also mistakenly identified as a ship with a high confidence. To deal with those problems, an ideal situation is that the field of vision focused by each DN is the whole body of the target, and does not contain additional background. However, due to the randomness of target size and input image size, it is difficult to achieve the above situation. More importantly, through the above case study, we observed that multi-scale feature fusion and feature extraction units are mutually constrained and auxiliary, because they jointly affect the ERF of each DN.

In summary, we need multi-scale fusion models to provide rich feature information to meet the size transformation of the target, and feature extraction operators to achieve the adaptive adjustment of ERF to suppress background information and highlight the key areas. However, the existing feature fusion models, such as FPN [17] and FPN-variants [18–21], cannot realize the information supplement between neighbor-scale features. The existing feature adaptive learning operators based on deformable convolution (DCN) [14,22,23] cannot achieve the synergy between the capture of key areas and background suppression, and their design is mostly separated from the fusion model, which does not form a good chemical reaction.

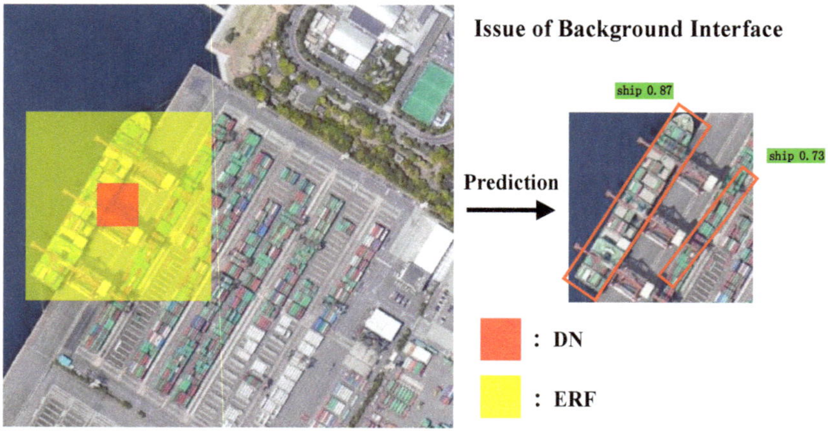

Figure 2. Example of wrong detection caused by background interference. In this case, RetinaNet-R is used. The prediction confidence threshold is 0.3. The container at the port is mistaken as a freighter because they have similar features.

Seeing the above challenges, we propose an innovative AOOD detector called AFRE-Net (adaptive feature representation enhancement network), which effectuates adaptive feature representation enhancement for DNs in multi-scale detection layers. AFRE-Net is committed to achieving feature relevance learning between adjacent scales and end-to-end ERF transformation, so as to strengthen the feature representation in the detection layer. The overall architecture of the proposed AFRE-Net is shown in Figure 3, which consists of four modules: (1) The backbone for basic feature extraction; (2) An FG-FPN (fine-grained feature pyramid network) for providing finer multi-dimensional feature maps and performing feature fusions; (3) A feature enhancement module (FEM) and (4) a rotation detection module for category prediction (CP) and bounding box regression (BBR). As opposed to the regular feature pyramid network (FPN) [17], FG-FPN is designed to make better use of the low-dimensional feature maps rich in spatial information, and it uses a more fine-grained fusion method to provide a basis of features for subsequent FEM. In FEM, we apply the ERF transformation based on DCN, and invented a background suppression and foreground enhancement algorithm named relative-Conv, to achieve automatic and adaptive object representation enhancement. Extensive experimental tests on three benchmark remote sensing datasets (DOTA, HRSC2016, UCAS-AOD), as well as a text recognition dataset (ICDAR2015) demonstrate the state-of-the-art performance of our AFRE-Net.

The contributions of our work can be concluded as follows:

1. Our systematic analysis has mined three aspects that need to be considered together to improve the detector's feature representation ability: fusion module, receptive field adjustment, and background suppression.
2. We propose a novel FG-FPN to provide finer features and fuse them in a more efficient manner. Different from FPN and its modifications, we focus on neighbor-scale information supplement to fullfil all-scale features.
3. A novel background suppression and foreground enhancement convolution module called relative conv is proposed to encourage DNs to learn the key areas adaptively.
4. We propose a new ERF transformation algorithm to make the sampling position more accurately located on the main body of the target, obtaining stronger semantic features.

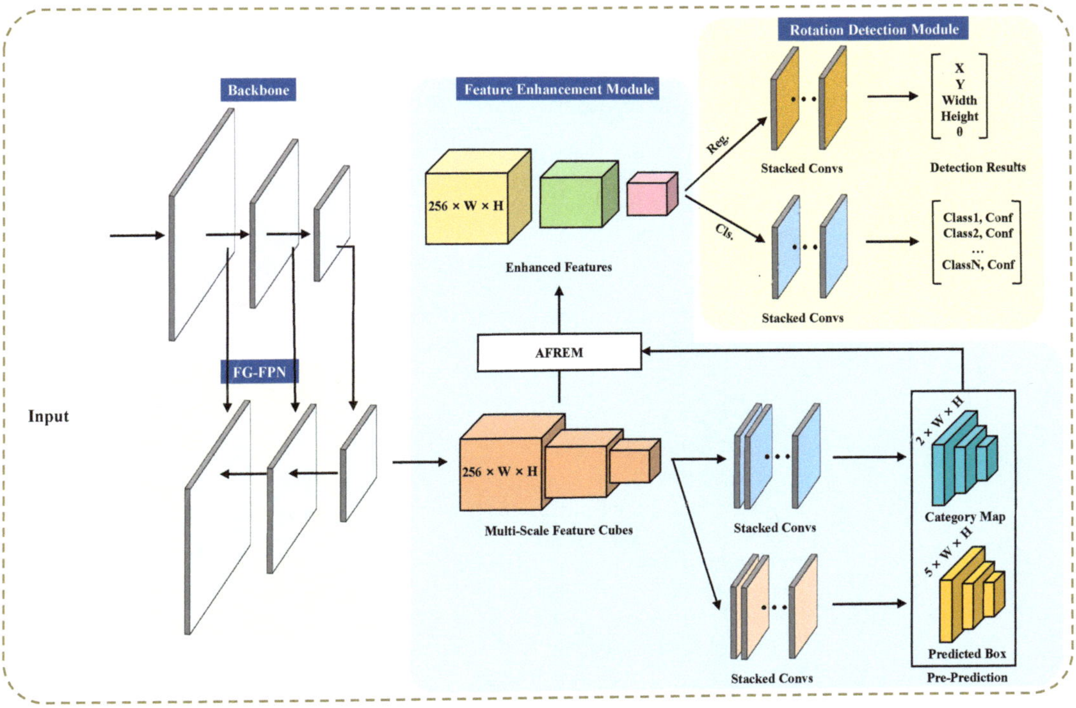

Figure 3. The overall architecture of the AFRE-Net. AFRENet is composed of a backbone, a fine-grained feature pyramid network, a feature enhancement module, and a rotation detection module. AFPEM denotes adaptive feature representation enhancement module. PB and CM denote the predicted box and classification map, respectively.

2. Related Work

2.1. Arbitrary Oriented Object Detection

AOOD is an extension of object detection tasks in natural scenes [24,25], which follows the basic natural object detector pipeline. Concretely, detectors can be divided into anchor-based methods and anchor-free methods. For anchor-based detectors, YOLO [26,27] series lead the one-stage field and have achieved remarkable effects by designing regression models that balance accuracy and speed. R-CNN series [28,29] represent two-stage detectors and use region proposal network (RPN) to filter potential DNs. The latter often achieves higher detection accuracy due to its ability to control positive and negative samples well; however, the authors of ATSS [30] point out that the sample learning strategy is the key factor. After that, plenty of intelligent positive and negative sample learning strategies have been proposed [31,32].

Additionally, numerous scholars have attempted to tackle the practical issues that arise in AOOD. For example, refs. [10–12,33] have focused on solving the discontinuous problem of angle parameter regression in the training process, and constantly refining the loss function to improve performance. Some other scholars attempt to use a target representation vector that can eliminate such boundary discontinuities to represent instances, such as polar coordinates [12], ellipse bounding box [34], and middle lines of boxes [35]. Additionally, to obtain better-refined rotated anchor boxes, RR-CNN [36], R3Det [37], and CFC-Net [38] focus on spatial alignment and anchor refinement to guide the training process. However, none of them try to explain the limitations of detection units on target feature expression in a deeper perspective.

2.2. Feature Fusion Module

Feature pyramid network (FPN) [17] is the most frequently used feature fusion structure, because it can well integrate low-level spatial information with high-level semantic information, and can well cope with the feature differences caused by the change of target size. After FPN is proposed, PANet [18] proposed a dual-path fusion mode of top-down plus bottom-up to enhance the semantic representation of low-level feature maps. BiFPN [20] refined the design of this pattern and further improved its performance. Recursive-FPN [21] adds the additional feedback connection of FPN to the bottom-up backbone layer and it convolves features with different void rates based on switchable atrous convolution. However, these methods are not designed from the perspective of scale information supplement. Considering that no feature pyramid can completely cover the full size of all targets and input images, we need to mine the feature correlation of critical scale as far as possible to make up for this deficiency.

2.3. Feature Enhancement Module

FEM in AOOD has a broader scope of reference and is not limited to using a certain class of methods or specific means to be called feature representation enhancement. For example, some scholars [14,39] focus on solving feature misalignment between classification and box regression (localization), arguing the classification score of anchor boxes cannot reflect real localization accuracy. FFN [40] enhances the model's feature expression ability for sub-optimal detection units in a creative way, but its design is too complex. Han et al. [14] were the first to attempt to alleviate this inconsistency through deformable convolutions [22] in AOOD; however, they ignore the adaptive feature representation learning. Ref. [39] proposed rotated align convolution (RAC) to improve feature representation of ship targets. However, they did not consider that the selection of sampling location should fall on the main body of the target body, and did not analyze the impact of background information interference within the detection unit on feature expression.

In addition, the performance of CNN in rotational feature extraction is known to be subpar [41]. Therefore, the research on rotation invariant feature extraction, as highlighted by works such as [42–44], plays a crucial role in improving CNN-based detectors. To achieve enhanced rotation invariant features, spatial and channel interpolation techniques are often employed [45,46]. For instance, Cheng et al. [6] were the pioneers in utilizing a fully connected layer as the rotation invariant layer, constraining it within the loss function. Furthermore, ReDet [8] takes a different approach by adaptively extracting rotation-invariant features from equivariant features based on the orientation of the Region of Interest (RoI). This adaptive extraction process contributes to the overall effectiveness of the detector. Oriented Reppoints [23] proposes an adaptive point set feature representation for AOOD tasks based on [47]. Although it can realize the adaptive transformation of ERF, its learning process is unordered and it does not realize the adaptive background suppression within a single feature extraction operator. Moreover, the above methods did not consider how to integrate the design of feature fusion with subsequent FEM as a whole. Our experiments show when the feature fusion module and ERF transformation are more fit, the feature representation ability of the detector is stronger.

3. Methodology

In this section, we will introduce the design of each independent module in AFRE-Net in detail. The overall pipeline of our detector is first introduced in Section 3.1. Then, we detail the FG-FPN in Section 3.2. Later, the adaptive feature representation enhancement module is unveiled in Section 3.4. Finally, experiment details about label assignment strategy are presented in Section 3.5.

3.1. Overall Pipeline

AFRE-Net is built based on RetinaNet-R [16] (baseline detector), which has classic object detector architecture and is easy to transfer. The overall pipeline of our AFRE-Net is shown in Figure 3. AFRE-Net consists of four main parts:

(1): A backbone network (ResNet [48] in our experiments) for basic feature extraction.

(2): A feature pyramid network for multi-scale feature fusions. We replace FPN [17] with our FG-FPN. FG-FPN is more capable of taking advantage of lower dimensional feature maps, which contain richer spatial information, can provide more fine-grained feature vectors for subsequent FEM, and enhance feature capture for small objects. FG-FPN fuses the feature maps with a top-down-top (rather than top-down only in FPN) pathway so the network constructs both rich semantic features and spacial information.

(3): A feature enhancement module (FEM). FEM is designed to reconstruct the feature vectors of DNs in the detection layer, including the automatic transformation of the FRF and adaptive background suppression.

(4): A rotation detection module (RDM). RDM converts semantic features into predicted bounding boxes and confidence of prediction categories for regions. RDM is a multi-task module and the regression targets are obtained in it. In our experiments, we adopt the five-parameter method to describe the bounding box, which is denoted as:

$$\{(x, y, w, h, \theta)\}, \tag{1}$$

where x, y, w, h are coordinates of the bounding box center, the width, and the height, respectively. Parameter $\theta \in \left[-\frac{\pi}{4}, \frac{3\pi}{4}\right]$ denotes the angle from the position direction of x to the direction of w. We have

$$\begin{aligned} t_x &= (x - x_a)/w_a, \quad t_y = (y - y_a)/h_a, \\ t_w &= \log(w/w_a), \quad t_h = \log(h/h_a), \\ t_\theta &= \tan(\theta - \theta_a). \end{aligned} \tag{2}$$

3.2. FG-FPN

3.2.1. Overall Architecture

It can be seen in Figure 4, RetinaNet-R employs the C_3, C_4, C_5 in the backbone network as the foundation for the following feature fusions. C_3, C_4, and C_5 are performed by 1×1

convolution to obtain the same channel dimension (256 in the baseline). Then, P_5 is obtained through a convolution layer of 3×3. The top-level feature P_5 is transmitted in a top-down manner and fused with low-level features layer by layer to obtain P_4 and P_3. In this way, layer P_i $(i = 3,4,5)$ is the same size as layer C_i. Based on P_5, P_6 is obtained through a 3×3 convolution with stride set to 2. P_7 is obtained through a Relu activation function and a 3×3 convolution with stride set to 2.

However, the structure of FPN has the following defects:

(1): The utilization efficiency of low-level features is insufficient. It is necessary to add lower-level features rich in spatial information to participate in the fusion process, in order to improve the feature perception ability of small objects.

(2): There are barriers between high-level features and low-level feature maps, as using only the top-down linking makes it impossible for high-dimensional feature maps to communicate directly with low-level feature maps (such as C3 and P5).

(3): Lack of mining for the correlation of features between adjacent scales.

Therefore,

(1): We re-enable the C_2 layer in ResNet (yellow layer shown in Figure 5);

(2): After performing the top-down fusion, we perform the down-top feature fusion as well, which means using top-down-top structure.

(3): We design an attention mechanism for mining inter-scale correlations, achieving the goal of simulating full-scale pyramid layers.

It should be noted that both the size and the numbers of the detection layer in FG-FPN have not changed, and C_2 is only used to generate polished P_2 (blue layer in Figure 5). All the feature channels are set to 256. Additionally, in order to better ensure the integrity of low-level spatial characteristics, depthwise convolution is used when P_2 P_3 is down-sampled and pointwise convolution is used when P_4 P_5 is down-sampled. In this way, high-level features can extract richer spatial information, and, most importantly, can greatly enhance the effectiveness of the proposed FEM in later processing stages.

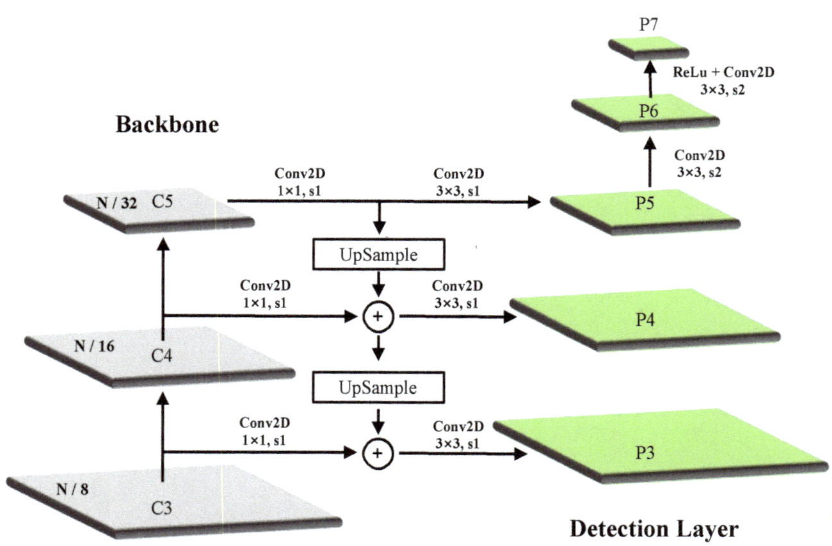

Figure 4. FPN structure of the RetinaNet. Conv2D 1×1, s1 refers to 1×1 convolution with stride set to 1. N denotes input image size. N/4 denotes the feature map resolution. $\oplus$ means addition.

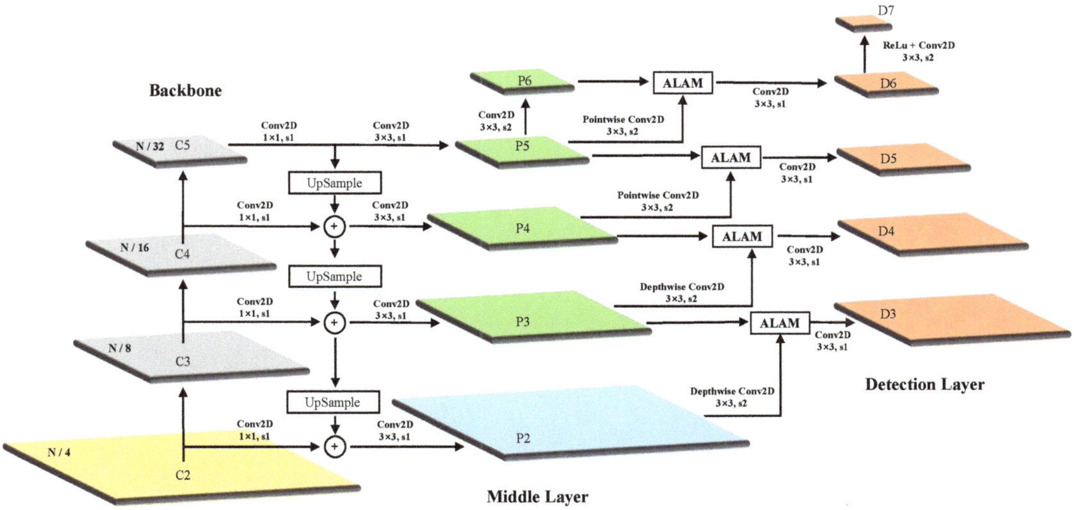

Figure 5. Structure of FG-FPN. Conv2D 1 × 1, s1 refers to 1 × 1 convolution with stride set to 1. N denotes input image size. N/4 denotes the feature map resolution. ⊕ means addition. ALAM refers to arhat layer attention module.

After strengthening the feature map of each scale, we tried to find a way to compensate for the differences between scales. Considering such an extreme situation, by building a feature pyramid with countless scales, we can cover the object scale transformation of various sizes, but this is obviously unrealistic. Therefore, we try to use the attention mechanism to generate a feature map with inter scale correlation. It is equivalent to further enhancing representation between scales.

3.2.2. ALAM

The Arhat layer attention module (ALAM) is proposed to create a strong correlation between neighbor scales. Specifically, we regard the $\mathbf{P_i}$ as the query, and $\mathbf{P_{i-1}^D}$ (downsampling output of $\mathbf{P_{i-1}}$) as the Key and Value. $\mathbf{P_{i-1}^D}, \mathbf{P_{i-1}} \in \mathbb{R}^{\mathbf{H \times W \times C}}$. Layer output AHT_i is calculated as:

$$AHT_i = softmax(\frac{\mathbf{P_i} \otimes \mathbf{P_i^D} \cdot T}{\sqrt{H \cdot W \cdot C}}) \otimes \mathbf{P_{i-1}^D}, \tag{3}$$

where, ⊗ denotes element-wise multiplication, $\mathbf{P_i^D} \cdot T$ denotes the transposition function, $i \in \{3, 4, 5\}$. $\mathbf{D_i}$ is generated from $ReLU(Conv2D(AHT_i))$. Through ALAM, $\mathbf{D_i}$ is equivalent to a correlation map between neighbor scales, which are significantly different from FPN and its variants. Because we use the attention mechanism to connect the upper and lower levels, we enrich them by building a pseudo pyramid to cover more scale information.

As shown in Figure 6, to further demonstrate the effectiveness of FG-FPN embedded with ALAM, we visualized the highest-resolution feature layers of BiFPN and FG-FPN after the same training iterations. It can be seen that FG-FPN has a more divergent heatmap distribution, but has a stronger response to the target. This is because ALAM allows for more implicated neurons in a single feature unit, which can cover a larger range of receptive fields.

3.3. FEM

FEM is a multi-stage feature enhancement module because we perform a pre-prediction before accurate detection. As shown in Figure 3, FEM takes FG-FPN output X_{LI} as the module input, where LI denotes the layer index of different feature maps, assigning multi-scale feature cubes into two separate task brunches (i.e., classification and bounding box regression). Classification and regression subnetworks are fully convolutional networks with a fixed number of stacked convolution layers (two in our experiments.) Note that here we only set one anchor in each DN and focal loss and smooth L1 loss are used for classification and bounding box regression, respectively. Softmax is used to generate category confidence and then the classification map (CM) is obtained. CM has the shape of $2 \times W \times H$. For each point (i,j) in CM, $CM(i,j)$ saves category labels and corresponding confidence, which are recorded as $CM^L(i,j)$ and $CM^C(i,j)$, respectively. For predicted box (PB), $PB(i,j)$ preserves the box position vector and has a shape of $5 \times W \times H$. Finally, FEM re-inputs X_{LI}, PB, and CM into AFREM to complete feature representation enhancement.

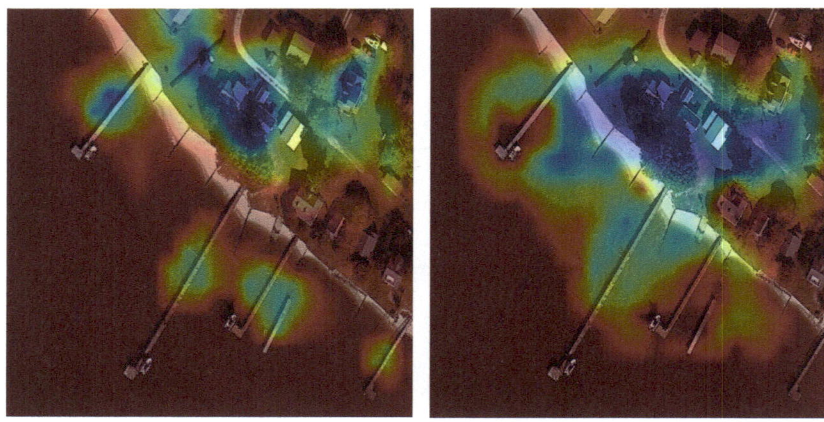

(**a**) (**b**)

Figure 6. Feature heatmap visualization of P3 in BiFPN [20] and D3 in FG-FPN. This experiment scenario is selected from baseline and AFRE-Net after 10K iterations of training on the DOTA dataset, respectively. It can be seen that FG-FPN has a stronger response to the targets' proposals, and can better distinguish the foreground and background. (**a**) Feature heatmap from BiFPN; (**b**) Feature heatmap from FG-FPN.

3.4. AFREM

The pipeline of AFREM is shown in Figure 7. To adaptively achieve ERF transformation, we design the ERF transformation algorithm based on deformable convolutional networks (DCN) [22], making it more suitable for objects in AOOD scenes. Relative convolution network (Relative Conv) is proposed to mitigate the impact of background features and magnify foreground information in a self-learning manner.

3.4.1. ERF Transformation

DCN pioneered the idea of changing the sampling position of the convolution kernel for each feature point in a self-learning manner. Outstanding works like oriented-reppoints and align-conv both achieve effective ERF transformation for remote sensing objects. However, the lack of control in the training process of the former leads to slow convergence of the network, and the lack of refined design of the latter leads to the inability of the sampling point to fall on the key area of the target of interest. Therefore, we try to use a more convenient way to make up for the two shortcomings at the same time.

Taking the convolution of 3×3 as an example, sampling offsets in both directions (18 values) are claimed at each point, making it possible to expand the ERF of DNs by providing learning capabilities in the sampling process. However, when applying DCN in AOOD, due to the larger object ratios and random orientations, intelligent ERF transformation faces challenges. As shown in Figure 8a, since PB is derived by OBB-guided regression, it basically fits the contour of the target. However, we hope that our sampling points (dark blue points in Figure 8a) can evenly fall on the target body, rather than just be limited to a certain part of the target (ERF box with mapping to original input in Figure 8a) or the boundary of the target. Alignment convolution in [14] ignores the solution to this problem. In our experiments, we try our best to place each sampling point on the main body of the object, as this ensures that the learned knowledge is focused on the object itself rather than the background features. Hence, we have designed a sample point reassignment strategy, as shown in Figure 8b. We use a shrunk PB to constrain the sampling position, making the sampling point better located inside the rotated bounding box.

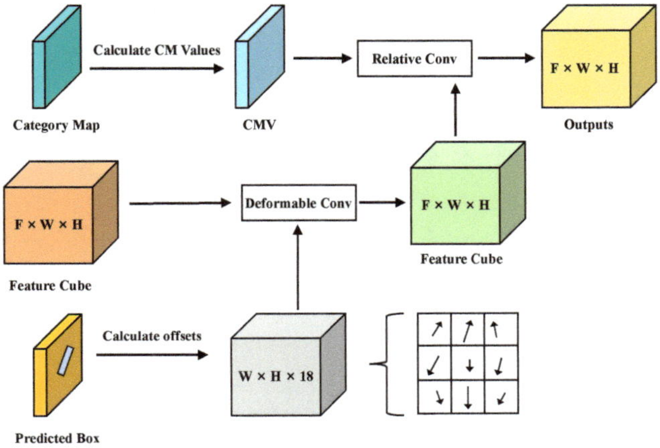

Figure 7. Pipeline of AFREM.

Given an $F \times W \times H$ (F denotes feature channels) feature cubes X_{LI}, for each position $\mathbf{P}(x_0, y_0)$, we obtain ERF transformation results FE_{LI} by

$$FE_{LI}(\mathbf{P}) = \text{Deformable}(X_{LI}, \text{offsets}(\mathbf{P})), \tag{4}$$

where offsets(**P**) is the position bias with the size of $W \times H \times 18$. The original offsets of 3×3 convolutions can be defined as $\mathcal{O}_{\mathcal{G}}$ = {(−1, -1), (−1, 0), (−1, 1), (0, −1), . . . , (1,1)},

$$\text{offsets}(\mathbf{P}) = \mathcal{O}_{\mathcal{G}} + \sigma_S, \tag{5}$$

where σ_S is the shifting vector from the original sampling box to shrunk PB. As shown in Figure 8b, PB is scaled down with a shrinkage coefficient α to obtain shrunk PB (SPB). Let (x, y, w, h, θ) represent the PB in position P, shrunk box can be defined as $(x, y, \alpha w, \alpha h, \theta)$. We set $\alpha = 0.85$ in our experiments to suitably fit the target body. Since PB is derived from horizontal anchor box $V_a(w_a, h_a)$ by

$$SPB = \alpha V_a \cdot t_{w,h} \cdot R^T(\theta), \tag{6}$$

the sampling DN S_P in position P can be calculated as

$$S_p = \frac{1}{S}(\mathbf{P} + \frac{1}{K} \cdot SPB), \tag{7}$$

where K is the kernel size and S is the down-sampling strides of current feature maps. Then, we can obtain offsets(**P**) by

$$\text{offsets}(\mathbf{P}) = S_p - \mathcal{O}_{\mathcal{G}} - \sigma_S. \tag{8}$$

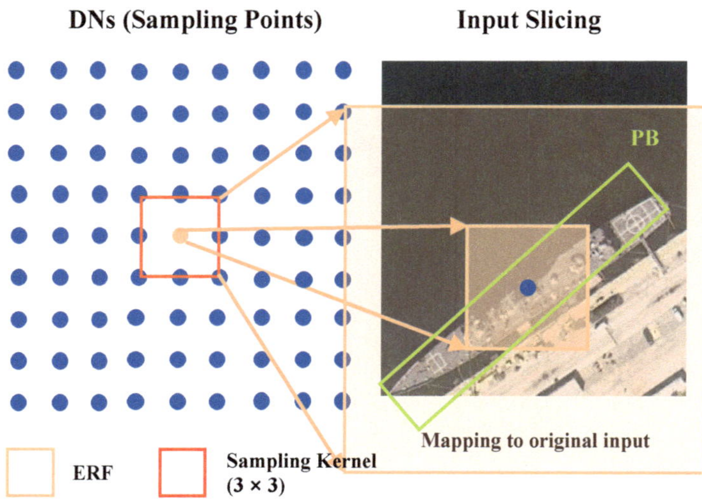

(a)

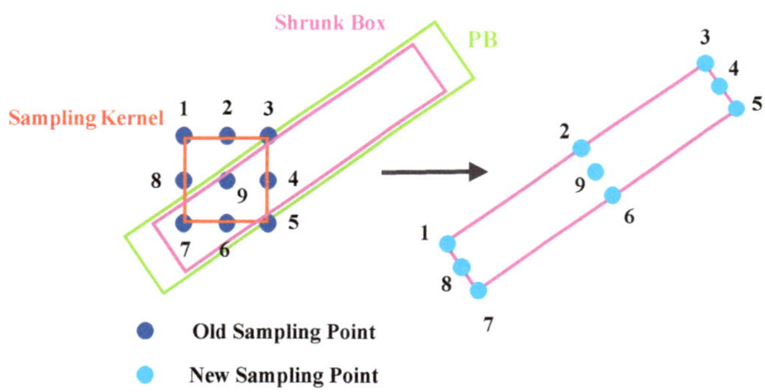

(b)

Figure 8. Illustration of ERF transformation. (**a**) Explanation diagram of the relationship between PB, sampling point, and ERF; (**b**) Identify sampling points through shrunk PB.

3.4.2. Relative ConV

For a DN, to automatically suppress background and highlight foreground targets, it is necessary to enable it to learn which regions are more important during back-propagation. In a standard 2D convolution, the output feature $\mathbf{Y}$ can be obtained by

$$\mathbf{Y}_{\mathrm{LI}}(\mathbf{P}) = \sum_{\mathbf{g} \in \mathcal{O}_{\mathcal{G}}} \mathbf{W}(\mathbf{g}) \cdot \mathbf{X}_{\mathrm{LI}}(\mathbf{P} + \mathbf{g}), \tag{9}$$

where $\mathbf{P}$ is the position of each DN, $\mathbf{P} \in \{0, 1, \ldots, W-1\} \times \{0, 1, \ldots, H-1\}$, and $\mathbf{W}(\mathbf{g})$ is the kernel weights. $\mathbf{W}(\mathbf{g})$ is the weights in an ordinary 3×3 convolution layer, which is updated and iterated with the optimization of the model. In relative convolution, we define the output $\mathbf{Y}_R$ by

$$\mathbf{Y}_R(\mathbf{P}) = \sum_{\mathbf{g} \in \mathcal{O}_{\mathcal{G}}} (\delta RJ(\mathbf{g}) + 1)\mathbf{W}(\mathbf{g}) \cdot \mathbf{X}(\mathbf{P}), \tag{10}$$

and

$$RJ(\mathbf{g}) = \begin{cases} 0, & CM^L(\mathbf{g}) \neq CM^L((0,0)) \\ 1, & CM^L(\mathbf{g}) = CM^L((0,0)) \\ & \text{and} \quad CM^L(\mathbf{g}) \neq -1 \end{cases} \tag{11}$$

CM^L is the category label obtained in FEM, δ is the accommodation coefficient and can be calculated by

$$\delta = CM^C + \eta, \tag{12}$$

where CM^C is the category prediction confidence. Symbol $\eta = -0.2$ is used to control the learning intensity. Figure 9 illustrates the operating mode of the relative convolution. By using it, detectors can obtain enhanced representations by strengthening learning about foreground targets.

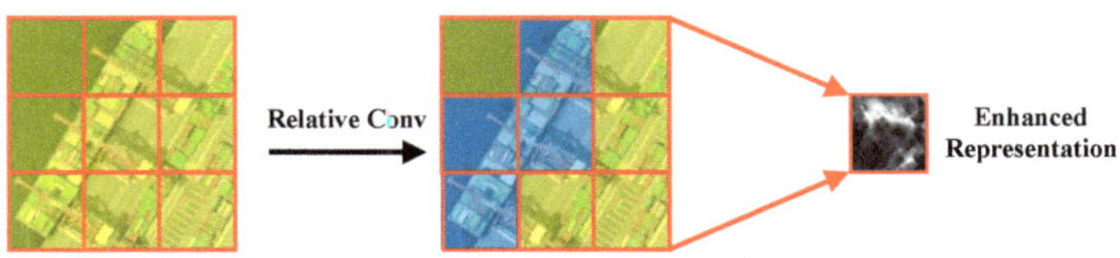

Figure 9. Illustration of Relative ConV.

3.5. Label Assignment Strategy

Label assignment strategy (LAS) has a significant impact on the overall accuracy of the model since it encourages the model to select and refine positive samples reasonably and effectively during training. To enhance the robustness of our AFRE-Net, we have decided to optimize the LAS in our detector. In the FEM, we utilize intersection over union (IoU) as the matching metric. Specifically, we set the foreground and background thresholds for determining whether an anchor is positive or negative to 0.5 and 0.4, respectively. These thresholds help us differentiate between foreground and background regions effectively. In the RDM, we employ dynamic anchor learning (DAL) [31] for intelligent anchor selection, which aims to activate more positive anchors during the refinement process. The matching degree, denoted as md, is defined as follows:

$$md = \alpha \cdot sa + (1 - \alpha) \cdot fa - u^{\gamma} \tag{13}$$

$$u = |sa - fa| \tag{14}$$

where sa denotes IoU of the anchor input, and fa represents the IoU between the GT box and the regression box. The term u is the absolute difference between the IoU of the anchor and the IoU between the GT box and the regression box. In our experiments, we set α to 0.3, and γ to 5. If the IoU of an anchor is greater than md, it will be classified as positive; otherwise, it will be classified as negative. This approach allows us to effectively determine the positive anchors based on their IoU values.

4. Experimental Results and Analysis

4.1. Datasets

Our AFRE-Net was assessed on three publicly available and challenging datasets, namely DOTA [49], HRSC2016 [50], and UCAS-AOD [51].

DOTA is an extensive dataset consisting of aerial images that capture complex scenes relevant to AOOD. It comprises a total of 2806 aerial images, with 1411 images for training, 458 images for validation, and 937 images for testing. These images contain a total of 188,281 instances belonging to 15 categories. The image size ranges from 800 × 800 to 4000 × 4000, and all instances are labeled with OBB, which exhibit variations in scales, aspect ratios, and orientations. To facilitate training, we divided the images into regular 1024 × 1024 patches with a stride of 200. The categories and corresponding IDs are as follows: Plane (PL), Baseball diamond (BD), Bridge (BR), Ground track field (GTF), Small vehicle (SV), Large vehicle (LV), Ship (SH), Tennis court (TC), Basketball court (BC), Storage tank (ST), Soccer-ball field (SBF), Roundabout (RA), Harbor (HA), Swimming pool (SP), and Helicopter (HC).

HRSC2016 is a high-resolution ship detection dataset that contains images collected from six international harbors. It consists of 1061 images, with image sizes ranging from 300 × 300 to 1500 × 900. The dataset includes 436 images for training, 541 images for validation, and 444 images for testing. All ship objects are labeled with OBB, and the substantial variation in ship sizes poses a significant challenge for detection.

UCAS-AOD is an aerial image dataset specifically designed for oriented aircraft and car detection. It comprises 1510 images, including 1000 airplane images and 510 car images. We randomly divided the dataset into training, validation, and test sets in a ratio of 5:2:3.

Additionally, to assess the scenario generalization capabilities of our AFRE-Net, we utilized the ICDAR-2015 [52] dataset as a benchmark for testing. This dataset consists of 1000 training images and 500 test images. The text boxes in this dataset are labeled with OBB and exhibit a very large aspect ratio, making them particularly challenging for detection.

4.2. Implementation Detail

For all datasets, we only set one horizontal anchor with aspect ratios of {1}, and resize all images to 1024 × 1024. Data augmentation techniques, such as random flip, rotation, and HSV color space transformation, are employed. The training optimizer used is Adam, with the initial learning rate set to 5×10^{-4}. At each decay step, the learning rate is divided by six. We utilize ResNet50 as the backbone network, which has been pre-trained on ImageNet. For DOTA, the models are trained on a single RTX 3090, and the batch size is set to two. Regarding HRSC2016, the detector undergoes a total of 12 K iterations during training, with the learning rate decaying at 8 K and 11 K, respectively. We evaluate the performance using average precision (AP) as the metric, following the same definition as the PASCAL VOC 2012 object detection challenge [53]. Unless explicitly stated, mAP refers to AP_{50}.

4.3. Ablation Studies

In this section, we conduct a series of experiments on DOTA and HRSC2016 to test our proposed AFRE-Net. We first verify the progressiveness of FG-FPN at the entire detector. Then, FEM is disassembled from the model to analyze its vital impact on overall performance. Finally, the respective capabilities of FRF expansion and RC are verified separately. Our ablation experiments demonstrate that when FG-FPN is combined with our meticulously designed FEM, our detector can achieve greater efficacy, thereby demonstrating the advantages of AFRE-Net.

To ensure fair comparisons, our baseline model adopts the same configuration as described in Section 4.2. Furthermore, we set the depth of the detection head (i.e., the rotation detection module in Figure 3) to a uniform value of 2, as it has a significant impact on the final detection result. In contrast to our AFRE-Net, which only utilizes one preset anchor with an aspect ratio of {1}, our baseline model employs three horizontal anchors with aspect ratios of {0.5, 1, 2} for matching objects. The results presented in Tables 1 and 2 demonstrate that our baseline model achieves an mAP of 68.2% on DOTA and 86.32% on HRSC2016.

4.3.1. Effectiveness of Hyper-Parameter

The parameter η in our model is used to deal with the weight imbalance caused by strengthening the learning of key areas in relative conv. As shown in Table 3, when η is around −0.2, negative compensation can achieve better performance by relative conv.

Table 1. Comparison on DOTA test dataset. R-101 represents ResNet-101 (likewise for R-50), and H-104 denotes Hourglass-104.

Methods	Backbone	PL	BD	BR	GTF	SV	LV	SH	TC	BC	ST	SBF	RA	HA	SP	HC	mAP
FR-O [49]	R-101	79.09	69.12	17.17	63.49	34.20	37.16	36.20	89.19	69.60	58.96	49.40	52.52	46.69	44.80	46.30	52.93
R-DFPN [54]	R-101	80.92	65.82	33.77	58.94	55.77	50.94	54.78	90.33	66.34	68.66	48.73	51.76	55.10	51.32	35.88	57.94
R^2CNN [36]	R-101	80.94	65.67	35.34	67.44	59.92	50.91	55.81	90.67	66.92	72.39	55.06	52.23	55.14	53.35	48.22	60.67
RRPN [55]	R-101	88.52	71.20	31.66	59.30	51.85	56.19	57.25	90.81	72.84	67.38	56.69	52.84	53.08	51.94	53.58	61.01
ICN [56]	R-101	81.36	74.30	47.70	70.32	64.89	67.82	69.98	90.76	79.06	78.20	53.64	62.90	67.02	64.17	50.23	68.16
RetinaNet-O [16]	R-50	88.67	77.62	41.81	58.17	74.58	71.64	79.11	90.29	82.13	74.32	54.75	60.60	62.57	69.67	60.64	68.43
RoI Trans. [57]	R-101	88.64	78.52	43.44	75.92	68.81	73.68	83.59	90.74	77.27	81.46	58.39	53.54	62.83	58.93	47.67	69.56
CAD-Net [58]	R-101	87.80	82.40	49.40	73.50	71.10	63.50	76.70	90.90	79.20	73.30	48.40	60.90	62.00	67.00	62.20	69.90
DRN [59]	H-104	88.91	80.22	43.52	63.35	73.48	70.69	84.94	90.14	83.85	84.11	50.12	58.41	67.62	68.60	52.50	70.70
O^2-DNet [35]	H-104	89.31	82.14	47.33	61.21	71.32	74.03	78.62	90.76	82.23	81.36	60.93	60.17	58.21	66.98	61.03	71.04
DAL [31]	R-101	88.61	79.69	46.27	70.37	65.89	76.10	78.53	90.84	79.98	78.41	58.71	62.02	69.23	71.32	60.65	71.78
SCRDet [60]	R-101	89.98	80.65	52.09	68.36	68.36	60.32	72.41	90.85	87.94	86.86	65.02	66.68	66.25	68.24	65.21	72.61
R^3Det [37]	R-152	89.49	81.17	50.53	66.10	70.92	78.66	78.21	90.81	85.26	84.23	61.81	63.77	68.16	69.83	67.17	73.74
S^2A-Net [14]	R-50	89.11	82.84	48.37	71.11	78.11	78.39	87.25	90.83	84.90	85.64	60.36	62.60	65.26	69.13	57.94	74.12
R^4Det [61]	R-152	88.96	85.42	52.91	73.84	74.86	81.52	80.29	90.79	86.95	85.25	64.05	60.93	69.00	70.55	67.76	75.84
Oriented Reppoints [23]	R-101	89.53	84.07	59.86	71.76	79.95	80.03	87.33	90.84	87.54	85.23	59.15	66.37	75.23	73.75	57.23	76.52
AFRE-Net (ours)	R-101	89.34	85.74	53.23	75.96	79.22	81.03	87.88	90.86	83.82	87.08	65.95	67.33	76.52	73.06	64.52	77.44

[1] Best results for each category are in red. Second-best results achieved by our detector are labeled in blue.

Table 2. Performance comparisons with state-of-the-art AOOD methods on the test set of HRSC2016. NA denotes the number of preset anchors of RDM. mAP (07/12): VOC2007/VOC2012 metrics.

Methods	Backbone	Size	NA	mAP (07)	mAP (12)	Params (M)	FLOPS
R^2CNN [36]	ResNet101	800 × 800	21	73.07	-	-	-
* RRD [62]	VGG16	384 × 384	13	84.30	-	27.6	176 G
RoI Trans. [57]	ResNet101	512 × 800	20	86.20	-	55.1	200 G
R^3Det [37]	ResNet101	800 × 800	21	89.26	96.01	41.9	336 G
* R-RetinaNet [16]	ResNet101	800 × 800	121	89.18	-	35.8	236 G
* DCL [9]	ResNet101	800 × 800	-	89.46	-	49.6	472 G
GWD [33]	ResNet101	800 × 800	-	89.85	97.37	47.4	456 G
DAL [31]	ResNet101	800 × 800	3	89.77	-	36.4	216 G
DRN [59]	ResNet101	768 × 768	-	92.7	-	-	-
S2A-Net [14]	ResNet50	512 × 800	1	90.17	95.01	38.6	198 G
AOPG [63]	ResNet101	800 × 800	-	90.34	96.22	-	-
ReDet [8]	-	800 × 800	-	90.46	97.63	31.6	-
O-RCNN [64]	ResNet101	800 × 800	-	90.50	97.60	41.1	199 G
* Baseline	ResNet50	800 × 800	3	86.32	91.04	31.5	199 G
AFRE-Net (ours)	ResNet50	800 × 800	1	**92.36**	97.32	42.2	323 G
AFRE-Net (ours)	ResNet101	800 × 800	1	92.18	**97.82**	51.4	469 G

[1] The instances of best detection performance are in **bold**. [2] * means that the precision and model complexity data source is from our local machine.

Table 3. Influence of hyperparameters η on HRSC2016 dataset.

η	−0.45	−0.35	−0.25	−0.20	−0.15	−0.10	NaN
AP_{50}	87.26	89.33	91.08	**92.16**	88.54	89.34	NaN
η	0.45	0.35	0.25	0.20	0.15	0.10	0.00
AP_{50}	79.32	81.26	82.77	79.98	82.32	86.02	88.06

[1] The instances of best detection performance are in **bold**. [2] Both FG-FPN and DAL are used.

4.3.2. Effectiveness of FG-FPN

Our baseline detector applies FPN as the neck to fuse multi-scale feature maps. As shown in Table 4, when replacing FPN with FG-FPN alone, the detector achieves accuracy gains of +1.01 and +0.92 on two datasets, respectively, proving that FG-FPN has a stronger feature fusion ability than FPN. In particular, it can provide low-level spatial information, which is very friendly for small targets. As can be seen in Table 1, the detection accuracy of SV has been greatly improved. However, at the same time, the replacement of FG-FPN also increased the number of parameters in the model, and we calculated that the size of the weight checkpoint file increased by 6.5 M. The detailed FG-FPN complexity is shown in Table 5. It can be seen that the introduction of FG-FPN has brought about a certain increase in model complexity, mainly caused by ALAM, as a large number of intermediate parameters are generated during the calculation process of this attention mechanism, also leading to an increase in inference time. It should be noted that this group of testing experiments did not use any feature enhancement modules, including AFREM. As shown in Table 5, when FG-FPN is used alone, it can only slightly improve the overall detection accuracy of the model. However, when FG-FPN is combined with our proposed AFREM, it can fully release the model performance, as AFREM can utilize the features rich in low-level spatial information provided by FG-FPN, obtaining more robust target feature representations.

Table 4. Ablation study of embeddings in AFRE-Net on DOTA and HRSC2016 dataset.

	Baseline	**Component Settings**				**AFRE-Net**
FPN [17]	✓			✓	✓	
FG-FPN (ours)		✓	✓			✓
FEM (ours)			✓	✓		✓
DAL [31]					✓	✓
mAP_{DOTA}	+0	+1.01	+6.92	+4.81	+1.54	+9.01
$mAP_{HRSC2016}$	+0	+0.92	+5.12	+3.79	+1.26	+6.04

[1] η is set to −2.

Table 5. Module complexity on different datasets.

Datasets	**Modules**	**Params (M)**	**FLOPS**	**Runtimes (s)**	**mAP (%)**
DOTA	B+FPN	27.6	168 G	0.42	71.27
HRSC2016	B+FPN	27.6	168 G	0.36	85.24
UCAS-AOD	B+FPN	27.6	168 G	0.35	84.72
DOTA	B+FG-FPN	34.1 (+6.5)	321 G	0.79	72.28 (+1.01)
HRSC2016	B+FG-FPN	34.1 (+6.5)	321 G	0.68	86.16 (0.92)
UCAS-AOD	B+FG-FPN	34.1 (+6.5)	321 G	0.68	86.88 (+2.16)

[1] Runtimes refer to inference time, and B denotes backbone network. [2] AFREM was not used in this set of experiments.

4.3.3. Effectiveness of FEM

FEM consists of two parts: ERF expansion and Relative ConV. We first verified how the overall detection accuracy of the detector changes when the entire FEM module is removed. As shown in the third and fourth control experiments in Table 4 (third and fourth columns), the use of different combinations of embeddings in detectors results in varying levels of detection accuracy. The combination of FG-FPN and FEM resulted in an astonishing mAP gain of +6.92 for the detector on DOTA, while FPN plus FEM achieves a +4.91 mAP improvement, which is also satisfactory. However, the former cannot be compared to the latter. Similar results also occurred on HRSC2016, where the combination of FG-FPN plus PEM achieves better performance, and improves the mAP by 5.12%.

In addition, to verify the contributions made by ERF transformation and Relative Conv in FEM, we conducted two comparative experiments, as shown in Table 6. It should be noted that both FG-FPN and DAL are used in these two comparative experiments. Inside FEM, since the size of feature cubes does not change, we only need to remove the other embedding when testing only one embedding. The experimental results show that when the two are combined, they can play a greater role. This is because by combining the two embeddings, DN not only can obtain the self-learning changes of the ERF, but also can adaptively learn the key areas in it and suppress the background information, thus obtaining better feature expression ability.

4.3.4. Effectiveness of Label Assignment Strategy

In order to eliminate the impact of LAS in our experiment, we also conducted comparative experiments to verify the universality of our proposed methods. As shown in Table 6, when using DAL in the baseline, the accuracy of the detector on DOTA increased by 1.54%

and on HRSC06 by 1.26%, indicating that the optimization of LAS is significantly helpful in improving the overall detection accuracy. However, the experimental results show that the use of LAS does not affect the improvement of the model detection accuracy brought by FG-FPN and FEM. On the contrary, when the three are combined, the maximum gain can be achieved.

Table 6. Effects of FEM structure on DOTA an HRSC2016.

	FEM Settings			AFRE-Net
ERF transformation	✓			✓
AlignConV [14]		✓		
Relative ConV		✓	✓	✓
mAP_{DOTA}	+3.77	+5.65	+2.89	+9.01
$mAP_{HRSC2016}$	+2.85	+4.26	+4.33	+6.04

[1] FG-FPN and DAL are both used in this experiments set.

4.4. Comparison with State-of-the-Art Detectors

4.4.1. Results on DOTA

We select some difficult scenarios as a demonstration of AFRE Net's detection capabilities. As shown in Figures 10 and 11, because our detector has improved the ability of feature expression, its confidence in the predicted output of the target has been greatly improved, and error detection has been effectively avoided (red circle in Figure 10a). Moreover, the detection ability for small targets has also been greatly improved.

Figure 12 also shows some tough detection scenarios in AOOD (dense, small, large aspect ratio, chaotic, and orientation random). It can be seen that AFRE-Net is able to better cope with the above challenges. Compared with other state-of-the-art AOOD detectors shown in Table 1, our model outperforms the best R^4Det [61] by an mAP of 1.6%, and achieves mAP improvement of 9.01% over the baseline detector. Compared with the anchor-free reppoints, our AFRE-Net achieves better performance on most categories.

Surprisingly, the accuracy improvement ability of AFRE-Net on specific objects is impressive. For SP, HA, RA, SBF, and ST, AFRE-Net achieves improvements of 3.39%, 13.95%, 6.73%, 11.2%, and 12.76%, respectively over the baseline. This suggests that our proposed method has a more significant and prominent effect on improving the feature expression of targets with large aspect ratio scales. The first reason is that the target with a large aspect ratio is more likely to contain more area of background information within the rectangular box of its outer contour, increasing the likelihood of interference; Secondly, the original regular feature sampling mode makes it impossible to accurately collect all the spatial scale features of the target when representing the target with a large-scale aspect ratio. AFREM achieves finer feature extractions by accurately changing the sampling points. Lastly, it can be seen that AFRE-Net has a good accuracy improvement effect on small targets, because the application of FG-FPN improves the ability of the detector to capture features in a small space range.

4.4.2. Results on HRSC2016

We evaluate the performance of our AFRE-Net on HRSC2016 with existing state-of-the-art AOOD detectors, which are divided into two categories, i.e., two-stage methods, such as R^2CNN [36], RRPN [55], R^2PN [65], RoI Trans. [57], and Gliding Vertex [66], and single-stage methods, such as DCL [9], DAL [31], DRN [59], and S2A-Net [14]. As shown in Table 2, our AFRE-Net outperforms all the detectors, especially towards two-stage methods, by a large gap up to 4.16%. Our AFRE-Net obtains an mAP of 92.36% under the condition that only ResNet50 is used, meaning that our model can achieve better feature extraction and detection results with fewer parameters of backbones. Compared with the baseline model, we improve 6.04% mAP with only one preset anchor in the FEM and RDM. In addition, Table 2 also shows the performance of our method in terms of model complexity and efficiency. The FG-FPN and AFREM has greatly improved the complexity of the model and increased the inference time of the detector, but extensive experiments have proved that our method is powerful in improving the detection performance. We have achieved 15.8FPS inference speed on a single RTX3090, proving that the model has maintained certain efficacy while improving its performance.

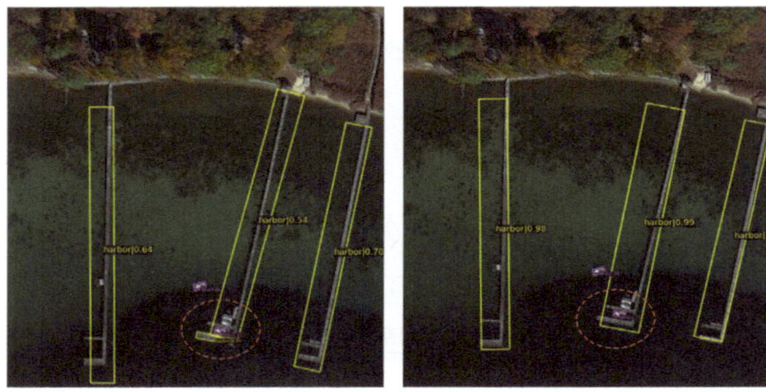

(**a**) (**b**)

Figure 10. Detection comparison between baseline detector and our AFRE-Net. AFRE-Net tends to obtain higher confidence score and more accurate predictions. (**a**) Wrong Detection by baseline; (**b**) Correct Detection by AFRE-Net.

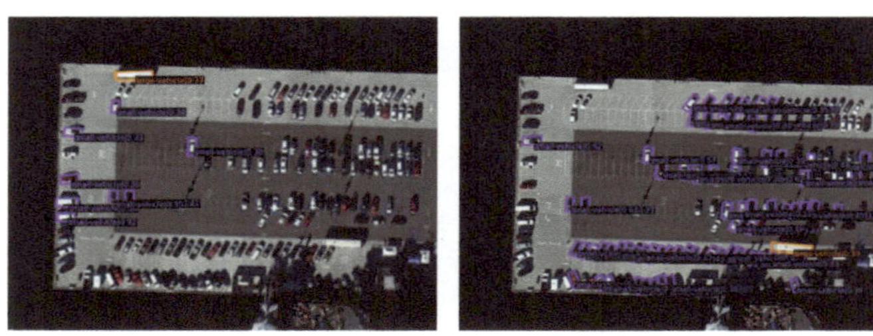

(**a**) (**b**)

Figure 11. Detection comparison between baseline detector and our AFRE-Net. AFRE-Net is more capable of generating small predictions. (**a**) Missed Detection by baseline; (**b**) Finer detection by AFRE-Net.

Figure 12. Visualization of some detection results on DOTA.

4.4.3. Results on UCAS-AOD

The distribution of vehicle targets in UCAS-AOD is relatively dense, and the spatial size is small, making detection difficult. As shown in Table 7, our baseline detector only achieved an accuracy performance of 83.22% on car detection. However, after AFRE-Net is applied, the mAP is improved to 90.62%, and the overall mAP is promoted to 91.34%, which surpasses all other comparison methods.

Table 7. Results comparison with advanced detectors on UCAS-AOD dataset.

Methods	Car	Airplane	mAP
Baseline	0.8322	0.8643	0.8472
* YOLOv3-O [26]	0.7463	0.8952	0.8208
Faster R-CNN-O [49]	0.8687	0.8986	0.8836
* DAL [31]	0.8925	0.9049	0.8987
* O-RCNN [64]	0.8874	0.9123	0.9003
* Oriented Reppoints [23]	0.8951	0.9070	0.9011
* ReDet [8]	0.9034	0.9107	0.9079
AFRE-Net (ours)	**0.9062**	**0.9143**	**0.9134**

[1] The instances of best detection performance are in **bold**. [2] * means that the precision and model complexity data source is from our local machine.

4.4.4. Results on ICDAR2015

To assess the robustness and generalization capability of our proposed AFRE-Net algorithm across various application scenarios, as well as to tackle annotation boxes with larger aspect ratio scales, we conducted training and testing on the ICDAR2015 dataset.

ICDAR2015 comprises challenging targets with significant variations in length and width, annotated in the oriented bounding box (OBB) format. As depicted in Table 8, our baseline model achieved an F-measure of 80.72 and a recall of 80.23%. Compared with other text detectors, such as EAST [67], R^2CNN [36], and R^3Det [37], AFRE-Net obtains the best recall performance at 88.82% and the best F-measure score at 86.27%, proving that our proposed solution has good migration application capabilities.

Table 8. Results comparison with text detector on ICDAR 2015.

Methods	Recall	Precision	F-Measure
RRPN [55]	82.17	73.23	77.44
EAST [67]	78.33	83.27	80.72
R^2CNN [36]	79.68	85.62	82.54
R^3Det [37]	81.64	84.97	83.27
Baseline	80.23	82.06	80.72
AFRE-Net (ours)	88.82	85.82	86.27

[1] Best results for each category are in red. Second-best results achieved by our detector are labeled in blue.

5. Conclusions

In this paper, we make several significant contributions to arbitrary-oriented object detection (AOOD). First, we identify the shortcomings and possible problems of the existing AOOD detectors in the structure design of feature extraction. Specifically, we point out that the existing models cannot automate DN learning and adjust ERF, and cannot adapt learning focus areas and suppress background information. To address these limitations, we conceive our detector AFRE-Net by designing a finer-grained feature fusion neck, and proposing ERF transformation and relative conv on this basis. These modifications enable the detector to acquire new capabilities for expressing object features. We validate the effectiveness of our algorithm on several remote sensing datasets and application scenarios. Extensive experimental results show that our method is effective and has a positive impact on the future design of feature representation enhancement strategies.

Author Contributions: Conceptualization, T.Z.; methodology, T.Z.; software, T.Z.; validation, T.Z.; formal analysis, T.Z.; investigation, T.Z.; resources, T.Z., X.S. and K.Z.; data curation, T.Z.; writing—original draft preparation, T.Z.; writing—review and editing, L.Z. and X.D.; visualization, T.Z.; supervision, J.S.; project administration, B.Z.; funding acquisition, B.Z. All authors have read and agreed to the published version of the manuscript.

Funding: This work was supported by the National Key R&D Program of China (Grant No. 2021YFB3900502).

Data Availability Statement: For all source data and code, please contact us: zhangtianwei20@mails.ucas.ac.cn.

Acknowledgments: We sincerely appreciate the constructive comments and suggestions of the anonymous reviewers, which have greatly helped to improve this paper.

Conflicts of Interest: The authors declare no conflict of interest.

References

1. He, K.; Gkioxari, G.; Dollár, P.; Girshick, R. Mask r-cnn. In Proceedings of the IEEE International Conference on Computer Vision, Venice, Italy, 22–29 October 2017; pp. 2961–2969.
2. Chen, Y.; Zhu, L.; Ghamisi, P.; Jia, X.; Li, G.; Tang, L. Hyperspectral images classification with Gabor filtering and convolutional neural network. *IEEE Geosci. Remote Sens. Lett.* **2017**, *14*, 2355–2359. [CrossRef]
3. Chen, Y.; Jiang, H.; Li, C.; Jia, X.; Ghamisi, P. Deep feature extraction and classification of hyperspectral images based on convolutional neural networks. *IEEE Trans. Geosci. Remote Sens.* **2016**, *54*, 6232–6251. [CrossRef]
4. Ma, T.Y.; Li, H.C.; Wang, R.; Du, Q.; Jia, X.; Plaza, A. Lightweight Tensorized Neural Networks for Hyperspectral Image Classification. *IEEE Trans. Geosci. Remote Sens.* **2022**, *60*, 5544816. [CrossRef]
5. Bai, L.; Liu, Q.; Li, C.; Ye, Z.; Hui, M.; Jia, X. Remote sensing image scene classification using multiscale feature fusion covariance network with octave convolution. *IEEE Trans. Geosci. Remote Sens.* **2022**, *60*, 5620214. [CrossRef]
6. Cheng, G.; Zhou, P.; Han, J. Learning rotation-invariant convolutional neural networks for object detection in VHR optical remote sensing images. *IEEE Trans. Geosci. Remote Sens.* **2016**, *54*, 7405–7415. [CrossRef]
7. Jaderberg, M.; Simonyan, K.; Zisserman, A. Spatial transformer networks. In Proceedings of the Advances in Neural Information Processing Systems, Montreal, QC, Canada, 7–12 December 2015; Volume 28.
8. Han, J.; Ding, J.; Xue, N.; Xia, G.S. Redet: A rotation-equivariant detector for aerial object detection. In Proceedings of the IEEE/CVF Conference on Computer Vision and Pattern Recognition, Nashville, TN, USA, 20–25 June 2021; pp. 2786–2795.
9. Yang, X.; Hou, L.; Zhou, Y.; Wang, W.; Yan, J. Dense Label Encoding for Boundary Discontinuity Free Rotation Detection. *arXiv* **2021**, arXiv:2011.09670.
10. Yang, X.; Yan, J. Arbitrary-oriented object detection with circular smooth label. In Proceedings of the European Conference on Computer Vision, Glasgow, UK, 23–28 August 2020; Springer: Berlin/Heidelberg, Germany, 2020; pp. 677–694.
11. Qian, W.; Yang, X.; Peng, S.; Guo, Y.; Yan, J. Learning modulated loss for rotated object detection. *arXiv* **2019**, arXiv:1911.08299.
12. Zhou, L.; Wei, H.; Li, H.; Zhao, W.; Zhang, Y.; Zhang, Y. Arbitrary-oriented object detection in remote sensing images based on polar coordinates. *IEEE Access* **2020**, *8*, 223373–223384. [CrossRef]
13. Luo, W.; Li, Y.; Urtasun, R.; Zemel, R. Understanding the effective receptive field in deep convolutional neural networks. In Proceedings of the Advances in Neural Information Processing Systems, Barcelona, Spain, 5–10 December 2016; Volume 29.
14. Han, J.; Ding, J.; Li, J.; Xia, G.S. Align Deep Features for Oriented Object Detection. *arXiv* **2020**, arXiv:2008.09397.
15. Wang, J.; Chen, K.; Yang, S.; Loy, C.C.; Lin, D. Region proposal by guided anchoring. In Proceedings of the IEEE/CVF Conference on Computer Cision and Pattern Recognition, Long Beach, CA, USA, 16–17 June 2019; pp. 2965–2974.
16. Lin, T.Y.; Goyal, P.; Girshick, R.; He, K.; Dollár, P. Focal loss for dense object detection. In Proceedings of the IEEE International Conference on Computer Vision, Venice, Italy, 22–29 October 2017; pp. 2980–2988.
17. Lin, T.Y.; Dollár, P.; Girshick, R.; He, K.; Hariharan, B.; Belongie, S. Feature pyramid networks for object detection. In Proceedings of the IEEE Conference on Computer Vision and Pattern Recognition, Honolulu, HI, USA, 21–26 July 2017; pp. 2117–2125.
18. Wang, K.; Liew, J.H.; Zou, Y.; Zhou, D.; Feng, J. Panet: Few-shot image semantic segmentation with prototype alignment. In Proceedings of the IEEE/CVF International Conference on Computer Vision, Seoul, Republic of Korea, 27 October–2 November 2019; pp. 9197–9206.
19. Ghiasi, G.; Lin, T.Y.; Le, Q.V. Nas-fpn: Learning scalable feature pyramid architecture for object detection. In Proceedings of the IEEE/CVF Conference on Computer Cision and Pattern Recognition, Long Beach, CA, USA, 16–17 June 2019; pp. 7036–7045.
20. Tan, M.; Pang, R.; Le, Q.V. Efficientdet: Scalable and efficient object detection. In Proceedings of the IEEE/CVF Conference on Computer Cision and Pattern Recognition, Seattle, WA, USA, 13–19 June 2020; pp. 10781–10790.
21. Qiao, S.; Chen, L.C.; Yuille, A. Detectors: Detecting objects with recursive feature pyramid and switchable atrous convolution. In Proceedings of the IEEE/CVF Conference on Computer Cision and Pattern Recognition, Nashville, TN, USA, 20–25 June 2021; pp. 10213–10224.

22. Dai, J.; Qi, H.; Xiong, Y.; Li, Y.; Zhang, G.; Hu, H.; Wei, Y. Deformable convolutional networks. In Proceedings of the IEEE International Conference on Computer Vision, Venice, Italy, 22–29 October 2017; pp. 764–773.
23. Li, W.; Chen, Y.; Hu, K.; Zhu, J. Oriented reppoints for aerial object detection. In Proceedings of the IEEE/CVF Conference on Computer Vision and Pattern Recognition, New Orleans, LA, USA, 18–24 June 2022; pp. 1829–1838.
24. Liu, W.; Anguelov, D.; Erhan, D.; Szegedy, C.; Reed, S.; Fu, C.Y.; Berg, A.C. Ssd: Single shot multibox detector. In Proceedings of the European Conference on Computer Vision, Amsterdam, The Netherlands, 11–14 October 2016; Springer: Berlin/Heidelberg, Germany, 2016; pp. 21–37.
25. Zhang, X.; Wan, F.; Liu, C.; Ji, R.; Ye, Q. Freeanchor: Learning to match anchors for visual object detection. In Proceedings of the Advances in Neural Information Processing Systems, Vancouver, BC, Canada, 8–14 December 2019; pp. 147–155.
26. Redmon, J.; Divvala, S.; Girshick, R.; Farhadi, A. You only look once: Unified, real-time object detection. In Proceedings of the IEEE Conference on Computer Vision and Pattern Recognition, Las Vegas, NV, USA, 27–30 June 2016; pp. 779–788.
27. Redmon, J.; Farhadi, A. YOLO9000: Better, faster, stronger. In Proceedings of the IEEE Conference on Computer Vision and Pattern Recognition, Honolulu, HI, USA, 21–26 July 2017; pp. 7263–7271.
28. Girshick, R. Fast r-cnn. In Proceedings of the IEEE International Conference on Computer Vision, Santiago, Chile, 7–13 December 2015; pp. 1440–1448.
29. Ren, S.; He, K.; Girshick, R.; Sun, J. Faster r-cnn: Towards real-time object detection with region proposal networks. In Proceedings of the Advances in Neural Information Processing Systems, Montreal, QC, Canada, 7–12 December 2015; pp. 91–99.
30. Zhang, S.; Chi, C.; Yao, Y.; Lei, Z.; Li, S.Z. Bridging the gap between anchor-based and anchor-free detection via adaptive training sample selection. In Proceedings of the IEEE/CVF Conference on Computer Vision and Pattern Recognition, Seattle, WA, USA, 13–19 June 2020; pp. 9759–9768.
31. Ming, Q.; Zhou, Z.; Miao, L.; Zhang, H.; Li, L. Dynamic anchor learning for arbitrary-oriented object detection. *arXiv* **2020**, arXiv:2012.04150.
32. Zhang, T.W.; Dong, X.Y.; Sun, X.; Gao, L.R.; Qu, Y.; Zhang, B.; Zheng, K. Performance releaser with smart anchor learning for arbitrary-oriented object detection. *Caai Trans. Intell. Technol.* **2022**. [CrossRef]
33. Yang, X.; Yan, J.; Ming, Q.; Wang, W.; Zhang, X.; Tian, Q. Rethinking rotated object detection with gaussian wasserstein distance loss. In Proceedings of the International Conference on Machine Learning, PMLR, Virtual, 18–24 July 2021; pp. 11830–11841.
34. Liu, J.; Zheng, H. EFN: Field-Based Object Detection for Aerial Images. *Remote Sens.* **2020**, *12*, 3630. [CrossRef]
35. Wei, H.; Zhou, L.; Zhang, Y.; Li, H.; Guo, R.; Wang, H. Oriented objects as pairs of middle lines. *arXiv* **2019**, arXiv:1912.10694.
36. Jiang, Y.; Zhu, X.; Wang, X.; Yang, S.; Li, W.; Wang, H.; Fu, P.; Luo, Z. R2cnn: Rotational region cnn for orientation robust scene text detection. *arXiv* **2017**, arXiv:1706.09579.
37. Yang, X.; Liu, Q.; Yan, J.; Li, A.; Zhang, Z.; Yu, G. R3det: Refined single-stage detector with feature refinement for rotating object. *arXiv* **2019**, arXiv:1908.05612.
38. Ming, Q.; Miao, L.; Zhou, Z.; Dong, Y. Cfc-net: A critical feature capturing network for arbitrary-oriented object detection in remote sensing images. *arXiv* **2021**, arXiv:2101.06849.
39. Yu, Y.; Yang, X.; Li, J.; Gao, X. A cascade rotated anchor-aided detector for ship detection in remote sensing images. *IEEE Trans. Geosci. Remote Sens.* **2020**, *60*, 5600514. [CrossRef]
40. Zhang, T.; Sun, X.; Zhuang, L.; Dong, X.; Gao, L.; Zhang, B.; Zheng, K. FFN: Fountain Fusion Net for Arbitrary-Oriented Object Detection. *IEEE Trans. Geosci. Remote Sens.* **2023**, *61*, 5609913. [CrossRef]
41. Cohen, T.; Welling, M. Group equivariant convolutional networks. In Proceedings of the International Conference on Machine Learning, PMLR, New York, NY, USA, 19–24 June 2016; pp. 2990–2999.
42. Worrall, D.E.; Garbin, S.J.; Turmukhambetov, D.; Brostow, G.J. Harmonic networks: Deep translation and rotation equivariance. In Proceedings of the IEEE Conference on Computer Vision and Pattern Recognition, Honolulu, HI, USA, 21–26 July 2017; pp. 5028–5037.
43. Weiler, M.; Hamprecht, F.A.; Storath, M. Learning steerable filters for rotation equivariant cnns. In Proceedings of the IEEE Conference on Computer Vision and Pattern Recognition, Salt Lake City, UT, USA, 18–22 June 2018; pp. 849–858.
44. Weiler, M.; Cesa, G. General e (2)-equivariant steerable cnns. In Proceedings of the Advances in Neural Information Processing Systems, Vancouver, BC, Canada, 8–14 December 2019; Volume 32.
45. Marcos, D.; Volpi, M.; Komodakis, N.; Tuia, D. Rotation equivariant vector field networks. In Proceedings of the IEEE International Conference on Computer Vision, Venice, Italy, 22–29 October 2017; pp. 5048–5057.
46. Zhou, Y.; Ye, Q.; Qiu, Q.; Jiao, J. Oriented response networks. In Proceedings of the IEEE Conference on Computer Vision and Pattern Recognition, Honolulu, HI, USA, 21–26 July 2017; pp. 519–528.
47. Yang, Z.; Liu, S.; Hu, H.; Wang, L.; Lin, S. Reppoints: Point set representation for object detection. In Proceedings of the IEEE/CVF International Conference on Computer Vision, Seoul, Republic of Korea, 27 October–2 November 2019; pp. 9657–9666.
48. He, K.; Zhang, X.; Ren, S.; Sun, J. Deep residual learning for image recognition. In Proceedings of the IEEE Conference on Computer Vision and Pattern Recognition, Las Vegas, NA, USA, 27–30 June 2016; pp. 770–778.
49. Xia, G.S.; Bai, X.; Ding, J.; Zhu, Z.; Belongie, S.; Luo, J.; Datcu, M.; Pelillo, M.; Zhang, L. DOTA: A large-scale dataset for object detection in aerial images. In Proceedings of the IEEE Conference on Computer Vision and Pattern Recognition, Salt Lake City, UT, USA, 18–22 June 2018; pp. 3974–3983.

50. Liu, Z.; Yuan, L.; Weng, L.; Yang, Y. A high resolution optical satellite image dataset for ship recognition and some new baselines. In Proceedings of the 6th International Conference on Pattern Recognition Applications and Methods (ICPRAM 2017), Porto, Portugal, 24–26 February 2017.
51. Zhu, H.; Chen, X.; Dai, W.; Fu, K.; Ye, Q.; Jiao, J. Orientation robust object detection in aerial images using deep convolutional neural network. In Proceedings of the 2015 IEEE International Conference on Image Processing (ICIP), Quebec City, QC, Canada, 27–30 September 2015; IEEE: New York, NY, USA, 2015; pp. 3735–3739.
52. Karatzas, D.; Gomez-Bigorda, L.; Nicolaou, A.; Ghosh, S.; Bagdanov, A.; Iwamura, M.; Matas, J.; Neumann, L.; Chandrasekhar, V.R.; Lu, S.; et al. ICDAR 2015 competition on robust reading. In Proceedings of the 2015 13th International Conference on Document Analysis and Recognition (ICDAR), Nancy, France, 23–26 August 2015; IEEE: New York, NY, USA, 2015; pp. 1156–1160.
53. Everingham, M.; Van Gool, L.; Williams, C.K.; Winn, J.; Zisserman, A. The pascal visual object classes (voc) challenge. *Int. J. Comput. Vis.* **2010**, *88*, 303–338. [CrossRef]
54. Yang, X.; Sun, H.; Fu, K.; Yang, J.; Sun, X.; Yan, M.; Guo, Z. Automatic ship detection in remote sensing images from google earth of complex scenes based on multiscale rotation dense feature pyramid networks. *Remote Sens.* **2018**, *10*, 132. [CrossRef]
55. Ma, J.; Shao, W.; Ye, H.; Wang, L.; Wang, H.; Zheng, Y.; Xue, X. Arbitrary-oriented scene text detection via rotation proposals. *IEEE Trans. Multimed.* **2018**, *20*, 3111–3122. [CrossRef]
56. Azimi, S.M.; Vig, E.; Bahmanyar, R.; Körner, M.; Reinartz, P. Towards multi-class object detection in unconstrained remote sensing imagery. In Proceedings of the Asian Conference on Computer Vision, Perth, Australia, 2–6 December 2018; Springer: Berlin/Heidelberg, Germany, 2018; pp. 150–165.
57. Ding, J.; Xue, N.; Long, Y.; Xia, G.S.; Lu, Q. Learning RoI transformer for oriented object detection in aerial images. In Proceedings of the IEEE Conference on Computer Vision and Pattern Recognition, Long Beach, CA, USA, 15–20 June 2019; pp. 2849–2858.
58. Zhang, G.; Lu, S.; Zhang, W. Cad-net: A context-aware detection network for objects in remote sensing imagery. *IEEE Trans. Geosci. Remote Sens.* **2019**, *57*, 10015–10024. [CrossRef]
59. Pan, X.; Ren, Y.; Sheng, K.; Dong, W.; Yuan, H.; Guo, X.; Ma, C.; Xu, C. Dynamic Refinement Network for Oriented and Densely Packed Object Detection. In Proceedings of the IEEE/CVF Conference on Computer Vision and Pattern Recognition, Seattle, WA, USA, 13–19 June 2020; pp. 11207–11216.
60. Yang, X.; Yang, J.; Yan, J.; Zhang, Y.; Zhang, T.; Guo, Z.; Sun, X.; Fu, K. Scrdet: Towards more robust detection for small, cluttered and rotated objects. In Proceedings of the IEEE International Conference on Computer Vision, Seoul, Republic of Korea, 27 October–2 November 2019; pp. 8232–8241.
61. Sun, P.; Zheng, Y.; Zhou, Z.; Xu, W.; Ren, Q. R4 Det: Refined single-stage detector with feature recursion and refinement for rotating object detection in aerial images. *Image Vis. Comput.* **2020**, *103*, 104036. [CrossRef]
62. Liao, M.; Zhu, Z.; Shi, B.; Xia, G.s.; Bai, X. Rotation-sensitive regression for oriented scene text detection. In Proceedings of the IEEE Conference on Computer Vision and Pattern Recognition, Salt Lake City, UT, USA, 18–22 June 2018; pp. 5909–5918.
63. Cheng, G.; Wang, J.; Li, K.; Xie, X.; Lang, C.; Yao, Y.; Han, J. Anchor-free oriented proposal generator for object detection. *IEEE Trans. Geosci. Remote Sens.* **2022**, *60*, 5625411. [CrossRef]
64. Xie, X.; Cheng, G.; Wang, J.; Yao, X.; Han, J. Oriented R-CNN for object detection. In Proceedings of the IEEE/CVF International Conference on Computer Vision, Montreal, BC, Canada, 11–17 October 2021; pp. 3520–3529.
65. Zhang, Z.; Guo, W.; Zhu, S.; Yu, W. Toward arbitrary-oriented ship detection with rotated region proposal and discrimination networks. *IEEE Geosci. Remote Sens. Lett.* **2018**, *15*, 1745–1749. [CrossRef]
66. Xu, Y.; Fu, M.; Wang, Q.; Wang, Y.; Chen, K.; Xia, G.S.; Bai, X. Gliding vertex on the horizontal bounding box for multi-oriented object detection. *IEEE Trans. Pattern Anal. Mach. Intell.* **2020**. [CrossRef]
67. Zhou, X.; Yao, C.; Wen, H.; Wang, Y.; Zhou, S.; He, W.; Liang, J. East: An efficient and accurate scene text detector. In Proceedings of the IEEE conference on Computer Vision and Pattern Recognition, Honolulu, HI, USA, 21–26 July 2017; pp. 5551–5560.

MDPI AG
Grosspeteranlage 5
4052 Basel
Switzerland
Tel.: +41 61 683 77 34

Remote Sensing Editorial Office
E-mail: remotesensing@mdpi.com
www.mdpi.com/journal/remotesensing

www.ingramcontent.com/pod-product-compliance
Lightning Source LLC
Chambersburg PA
CBHW042015260926
51850CB00020B/3419
* 9 7 8 3 7 2 5 8 3 2 3 1 6 *